FUEL CELLS

FUEL CELLS

CURRENT TECHNOLOGY CHALLENGES AND FUTURE RESEARCH NEEDS

More than 170 years After Its Invention, the Fuel Cell Still is Not a Viable Product: How We Can Make It Happen, and Who Will Be There First.

NORIKO HIKOSAKA BEHLING

ELSEVIER

AMSTERDAM • BOSTON • HEIDELBERG • LONDON • NEW YORK • OXFORD
PARIS • SAN DIEGO • SAN FRANCISCO • SINGAPORE • SYDNEY • TOKYO

Elsevier
Radarweg 29, PO Box 211, 1000 AE Amsterdam, The Netherlands
The Boulevard, Langford Lane, Kidlington, Oxford OX5 1GB, UK
30 Corporate Drive, Suite 400, Burlington, MA 01803, USA
525 B Street, Suite 1900, San Diego, CA 92101-4495, USA

First edition **2013**

British Library Cataloguing in Publication Data
A catalogue record for this book is available from the British Library

Library of Congress Cataloging-in-Publication Data
A catalog record for this book is available from the Library of Congress

ISBN: 978-0-444-56325-5

Printed in the United States of America

Transferred to Digital Printing, 2013

Dedication

I dedicate this book to:

My father, Riichi Hikosaka, a physicist and industrialist from the Imperial Kyoto University who died too soon with his dreams and aspirations cut short.

My mother, Chiyo Hikosaka, a teacher and a devout Catholic and Buddhist, who raised three cosmopolitan daughters on her own, Nobuko residing in Japan, Satoko in Germany, and me in the United States.

My husband, Thomas, an insightful analyst, former Deputy Under Secretary of Defense, and a patient and loving husband, who taught me how to write English from scratch, starting with a single sentence, then a paragraph, and ultimately a book.

Contents

Preface

I have worked as an analyst on science and technology policy for many years, both in and out of government.

My interest in fuel cell technology has been more recent, however. In 2000, I saw major fuel cell programs being pursued around the globe. I wrote some papers on the subjects, which were noticed by the White House and led to the FreedomCar Initiative and the Hydrogen Fuel Initiative in 2002 and 2003. These initiatives were designed to promote the development of fuel cell cars and encourage the construction of needed hydrogen fuel supply infrastructure. Since then, I have become an enthusiastic observer of fuel cell developments worldwide.

The fuel cell industry has seen many ups and downs. Many companies dedicated a lot of hard work and billions of dollars to develop fuel cell technology. Many governments also recognized the promise of fuel cells and worked in tandem with industry. The world watched these developments with great hope and expectations. But these efforts did not deliver on their promise. To date, major impediments to fuel cell commercialization remain; fuel cells still have insufficient longevity, reliability, and have unacceptably high cost.

With sadness and disappointment, I started to wonder why fuel cell technology has not become commercially successful. After all, fuel cells were invented in 1839, more than 170 years ago. Many technologies such as airplanes, computers, and automobiles, have been successfully commercialized much sooner than that. I concluded that fuel cell technology must be a more complex and challenging technology than any of these other examples and that few in the industry appreciated the difficulties they faced. I started to realize that the challenges of fuel cell technology derive from complex physical and chemical processes taking place at the atomic level within the fuel cell and its supporting components. These processes span the "seams" that define traditional scientific and engineering disciplines and therefore are much more difficult to master than problems that can be addressed within a single discipline.

Perhaps no advanced technology on the market today requires the scale, magnitude, and range of scientific, physical, and engineering knowledge as do fuel cells. Only recently have fuel cell researchers begun to recognize the true complexity of the technology. Perhaps modern science has progressed to the point that researchers are now ready to explore and discover the secrets of fuel cell, how it works, why it works, and how it can be made to work cheaply, reliably, and efficiently.

That is why I wrote this book. It took five years of research and writing to document the story of fuel cells, including the many impressive accomplishments of researchers and companies worldwide who have worked to master the technology. The book describes many unfortunate stories of companies that ultimately gave up and abandoned their fuel cell research. It evaluates the successes and failures of government policy to promote fuel cell development, and explains why government policies were unable to build the needed foundation for its commercial success. Finally, the book provides policymakers in every country a set of recommendations as to what they can do to change the current situation and help realize the ultimate promise of fuel cell technology.

I have made every effort to ensure that the material contained in this document is current, complete, and accurate. Nonetheless, errors and mistakes undoubtedly exist in the text. I apologize for any such errors and hope they will be forgiven. My intention is sincere, and I take full responsibility for the conclusions in this book. I also thank those companies which so kindly provided photographs of their automobiles, buses, and power generation systems and granted their permission for reprinting. Because of page length limitations I was unable to use them. I am very sorry for this and hope you will understand.

Lastly but not the least, I would like to thank Dr. Mark Williams, former DOE fuel cell manager and a prominent international fuel cell leader, for his kind guidance and assistance. I am particularly grateful for his fairness and understanding of my analytical judgments, which I know occasionally conflict with his own experiences.

CHAPTER

1

Introduction

OUTLINE

1.1. WILLIAM GROVE INVENTS THE FUEL CELL

The fuel cell was invented in 1839 by William Grove (1811–1890). Grove was a scientist and a lawyer in the courts of Common Law (Fig. 1.1). He was born in Swansea, Wales, and became a barrister at Lincoln's Inn and a member of the Royal Institution of Great Britain.[1] He later became a professor of experimental philosophy at the Royal Institution in London and then its vice-president. He subsequently was knighted and became a Lord Justice of the Queen's Bench.

William Grove invented two electrochemical cells, the first in 1839 and the second in 1844. In the days before electrical distribution networks became available, batteries were the only practical source of electricity, and his intent was to invent a robust, practical battery. The first invention, the "gas voltaic battery," was similar to conventional batteries in its compositions and purpose, but was based on an entirely new concept of electric generation, and indeed became the forerunner of fuel cells.

In 1839, Grove observed the production of an electric current in his experimental setup.[2] In the experiment, the two platinum electrodes were submerged halfway into a beaker of aqueous sulfuric acid, and tubes were inverted over each of the electrodes, one containing hydrogen gas and the other containing oxygen gas (Fig. 1.2). When the tubes were lowered, the gases displaced the electrolyte, leaving only a thin coating of the acid solution on the electrode. A galvanometer deflected to indicate a flow of electrons between the two electrodes, leading Grove to claim that his device "produced a voltaic current."[3] He published this process in *Philosophical Magazine*.[2] Three years later, Grove described in his paper that electric energy was generated at the three-

Fuel Cells
http://dx.doi.org/10.1016/B978-0-444-56325-5.00001-6

FIGURE 1.1 Portrait of William Robert Grove (1811–1896): Father of the fuel cell. *Source: Smithsonian Institution Libraries, US, 2012.*

phase contact where the liquid, gas, and platinum met; he sought to extend this contact by coating the electrodes with spongy platinum.

- The process was exactly the reverse of a process, electrolysis—discovered in 1800 by British scientists William Nicholson and Anthony Carlisle—that split water into hydrogen and oxygen by electricity. According to Grove, combining the gases to produce electricity and water was "a step further that any hitherto recorded."[4]
- Grove, however, could not explain the basic principles underlying the behavior of fuel cells. Grove demonstrated strong evidence that producing an electric current required both hydrogen and oxygen, but did not understand why the fuel cell produced heat and "novel gaseous and liquid products."[4]

Grove also reported on other experiments on fuel cells.[2] He experimented with stronger sulfuric acid, hydrogen, and air for oxygen. He also reported on a hydrogen–chlorine fuel cell and established that different hydrocarbon fuels, such as camphor, oils, ether, and alcohol, produced electricity. He foresaw the fuel cell as a possible source of commercial electricity, especially if hydrogen as a fuel could be replaced by coal, wood, or other combustibles. He, however, was troubled by the lack of reliability in cell performance. He also was unable to build a fuel cell that generated enough electricity to serve as a practical source of electricity. Grove, thus, never made an effort to commercialize the invention.[5]

Grove, however, made a major commercial success out of his second invention, the "Grove cell," which became a popular battery for the telegraph industry. Grove's invention in the year 1844 consisted of a zinc anode dipped in sulfuric acid and a platinum cathode dipped in nitric acid, separated by porous earthenware.[6] The Grove cell provided a high current and nearly twice the voltage of the Daniell cell and became the favorite battery of the early American telegraph (1840–1860).[5]

The first battery in history was invented in 1800 by Alessandro Volta. It was of great value for science, but was unsafe to handle because of sulfuric acid in the battery and the power of the cell diminished over time. Later batteries, starting with John Frederic Daniell's wet cell in 1836, provided more reliable currents and were adopted by industries for use in stationary devices, particularly in telegraph networks.[6]

- As telegraphic traffic increased, however, it was found that the Grove cell discharged poisonous nitric dioxide gas.[5] Platinum was also very expensive. The Grove cell was replaced by the cheaper, safer, and better performing gravity cell in the 1860s.

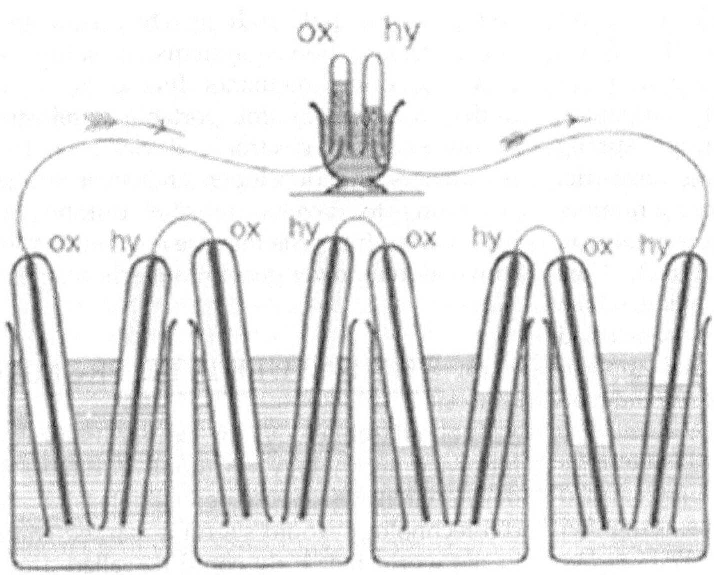

FIGURE 1.2 Sketch of William Grove's 1839 fuel cell. *Source: "1839 William Grove Fuel Cell.jpg," File, http://en.wikipedia.org/wiki/File:1839_William_Grove_Fuel_Cell.jpg, Wikipedia, no date, accessed March 2012. Originally published in the article authored by W. R. Grove, "On a Gaseous Voltaic Battery," LXXII, Third Series, the London, Edinburgh And Dublin Philosophical Magazine And Journal Of Science, Richard and John E. Taylor, London, December 1842.* (For the color version of this figure, the reader is referred to the online version of this book.)

1.2. FUEL CELLS: COMMERCIAL SUCCESS REMAINS ELUSIVE

William Grove probably never expected that his invention would keep the global science and technology community excited as well as exasperated for the next 170 years. Since Grove's invention, scientists, engineers, and businessmen have dreamed that the fuel cell could deliver low cost, clean energy to power factories, homes, cars, appliances—virtually anything that uses electric power. Despite the best of intentions, this dream remains unfulfilled.

In the first 100 years after Grove's invention, many notable European scientists followed his footsteps and attempted to develop a fuel cell with commercial viability. They explored various electrolytes and electrodes that could produce an electric current reliably and inexpensively. They also experimented with a variety of fuels, including wood, oil, coal, and coal gas. Coal was the major source of fuel in the nineteenth century, readily and easily available for public consumption at that time. Thus, the possibility of using coal in fuel cells drew particular interest among researchers and developers. Indeed, they experimented with nearly every type of fuel cell chemistry and physical design, many of which foreshadowed the fuel cells that we know of today. These researchers, however, realized that there were too many hurdles to overcome before fuel cells could be reliable and inexpensive enough to be commercialized. Over time enthusiasm for the device gradually withered away. By the end of the nineteenth century, the internal combustion engine was invented, and the fuel cell was relegated to the status of a scientific curiosity.[7]

By the mid-1900s, however, fuel cell technologies started to make significant progress, and some fuel cells began to serve practical functions. The first fuel cell used in practical application was the alkaline fuel cell (AFC) developed by British scientist Sir Francis Bacon. Bacon's

AFC technology was licensed by a US company, which developed it as the source of onboard power and drinking water for its Apollo lunar missions in 1962. A number of companies, both in the United States and Europe, sprang into action and began developing terrestrial applications of AFCs. They produced a number of prototype AFC products, but were unable to bring them to the commercial market. They found AFCs intolerant of carbon dioxide, which is omnipresent on Earth, and thus concluded AFCs to be unsuitable for terrestrial applications. Many companies abandoned AFC development by the 1970s.

In the mid-1970s, phosphoric acid fuel cells (PAFCs) came to be recognized as better suited for stationary, terrestrial applications. A number of US companies launched R&D of PAFCs. The US Government and electric and gas utilities started to assist those companies by funding their R&D and demonstrations. Japan eagerly followed the US footsteps and tried to learn the PAFC technology. By the mid-1980s, attention focused on molten carbonate fuel cells (MCFCs) and by the 1990s on solid oxide fuel cells. Also in the mid-1990s, proton exchange membrane fuel cells (PEMFCs) came to be viewed as the most exciting technology; in fact, researchers and developers were able to build an automobile powered by hydrogen-fueled PEMFCs.

Unfortunately, however, none of these fuel cells achieved commercial success on any major scale. PAFCs entered the market in the early 1990s as the "first generation" fuel cell technology to be commercialized. A few hundred power plants were installed, but sales of most of them were completed with government and public subsidies. MCFCs also hit the market as "second generation" fuel cells in the early 2000. A small number of stationary power generation systems have been sold, also with subsidies. PEMFCs have been put on the commercial market recently; they are used as a small stationary power plant and the power source for material handling vehicles and emergency backup power systems. A subtype of PEMFCs, direct methanol fuel cells, is also under development for portable applications to power small electronic devices. Fuel cell researchers and developers, however, are still struggling to develop reliable, durable, and inexpensive fuel cells that are competitive with conventional power generation technologies.

1.3. THE UNFULFILLED PROMISE

Fuel cell technology is an extraordinary, revolutionary technology. Without question, the development of a commercially feasible fuel cell technology would set off a marked change in the world that truly could be called a paradigm shift.

Fuel cells are singularly remarkable in their potential for efficiently converting the energy that is locked up in chemical bonds to electrical energy. This efficiency is achieved because fuel cells convert the chemical energy contained in a fuel into electrical energy in a single step, extracting more useful energy from the same amount of fuel than any other known device. An internal combustion engine, for example, requires several steps: converting chemical energy contained in gasoline to thermal energy via combustion, using thermal energy to expand gases within the engine's cylinders, and then converting the high pressure gas to mechanical energy via the pistons and drive train. Because conventional internal combustion engines rely on so many energy conversion steps, they are inherently inefficient and lose much energy to incomplete combustion and exhaust heat.

- A fuel cell is not only more efficient than a combustion engine, but also delivers high efficiency even while operating under partial design loads. Moreover, its high efficiency is scalable from very small units to very large units. A combustion engine, on the other hand, is most efficient under full operational loads and only at certain minimum engine sizes.

Because of these remarkable attributes and capabilities, fuel cells can end the way we generate and consume electricity today. It can deliver power that is inherently healthy for the environment and highly efficient. Because of their efficiency, fuel cells create less carbon dioxide when operating on carbon-based fuels like methane and no emissions when operating on hydrogen. In addition, fuel cells do not generate any other emissions like nitrogen oxide, which will harm the earth's atmosphere. Because of their efficiency, their widespread use would slow the depletion of the earth's fossil fuel resources.

Fuel cells most commonly use hydrogen to generate electricity, but can use various other fuels, including gasoline, natural gas, ethanol, and methanol. Hydrogen, which would have the least environmental impact, can be produced from water through renewable energy readily available in any country on earth, including sun, wind, water, geothermal energy, and biomass.

Fuel cell technology can cause a paradigm shift in quality of life on a global scale. It will give an ordinary citizen the ability to generate electricity at home instead of buying electric power from a national electrical grid system that is often delivered over great distances with high transmission losses and is commonly generated by polluting coal-fired power plants. Individuals will be able to drive cars that operate on hydrogen instead of gasoline, and help promote new industry and manufacturing sectors that can dramatically reduce carbon footprints. Thus, the fuel cell technology can spur the development of new classes of energy demand and supply systems, transportation systems, and industrial and manufacturing systems in all nations, while generating trillions of dollars in new revenue. In addition, oil importing nations can cut down on the use of oil, thereby gaining greater independence from oil-producing nations, enhancing their energy security, economic security, environmental security, and the national security. The world will then find a profoundly different paradigm for socio-economic, environmental, and geopolitical relationships that could lead to greater stability around the globe.

Many governments and corporations have recognized the promise of fuel cells and are working in tandem to develop their fuel cell technology, manufacturing capabilities, and global technological leadership. Many governments have devoted tens of billions of dollars in R&D on fuel cell technology in the past 50 years. Likewise, many corporations have invested billions of dollars in R&D, some for over half a century.

But these efforts have not delivered on the needed fuel cell technology. Despite huge investment and sacrifice, few companies have made any profit. To date, major impediments to fuel cell commercialization remain. Daunting technical challenges have frustrated every effort to create economically feasible fuel cells. The problems are wide-ranging: fuel cells have insufficient longevity, reliability, and have unacceptably high cost. For fuel cells that require special fuels (such as hydrogen), the lack of a fuel infrastructure also limits commercialization.[8]

Given the long history of fuel cell technology, it is mystery that such an important invention has not yet achieved commercial success on any significant scale. This book attempts to unravel this mystery. It explores the history of fuel cell development, evaluates the technical challenges, considers the successes and failures of government policy in promoting solutions to the technical challenges, and explains why the fuel cell industry has not been able to achieve commercial success with the technology. This book then turns to the future. It outlines a set of policy recommendations that might help realize the ultimate promise of fuel cell technology, and suggests who the potential winners might be in the race for a viable commercial product.

References

1. Grove Sir William Robert. (1811–1896), *Archives in London and the M25 Area*. US: Royal Institution of Great Britain UK; April 2001.
2. Leo J, Blomen MJ, Mugerwa Michael N. *Fuel cell systems*. US; Springer; 1993.
3. Hoogers Gregor, editor. *Fuel cell technology handbook*. UK, CRC Press LLC; July 2002.
4. *Collecting the History of Fuel Cells*. Smithsonian Institution, US; http://americanhistory.si.edu/csr/fuelcells/index.htm; 2007.
5. Sir William Grove (1811–1896), Biographies, http://www.corrosion-doctors.org/Biographies/GroveBio.htm, Kingston Technical Software, no date, accessed August 2012.
6. History of the Battery, Wikipedia, http://en.wikipedia.org/wiki/History_of_the_battery, [accessed October 2008].
7. *Fuel cells—the first 120 years*. In: *History of Fuel Cells*, vol. 24. UK, Fuel Cell Today; July 2007.
8. *Fuel cell handbook*. 7th ed., NETL, the US Department of Energy, US, November 2004.

CHAPTER

2

Fuel Cells and the Challenges Ahead

OUTLINE

Fuel Cells

http://dx.doi.org/10.1016/B978-0-444-56325-5.00002-8

After More than 170 Years Since Its Invention, Why Hasn't the Fuel Cell Become A Viable Product: How to Make It Happen, and Who Will Get There First? By Noriko Behling

2.1. WHAT IS A FUEL CELL?[i]

A fuel cell is an energy conversion device that electrochemically converts chemical energy held within two reactants, a fuel and an oxidant, into electrical energy. Fuel can be any ordinary fuel, but hydrogen is most commonly used. The oxidant is most commonly oxygen, usually in the form of ambient air. The fuel cell produces power as fuel and oxidant are supplied.

A fuel cell has physical components and characteristics that are similar to those of a typical battery. But there are two critical differences:

- Energy depletion versus continuous refueling: A battery will cease to produce electrical energy when the chemical reactants stored within the battery are consumed or discharged. (In a rechargeable battery, the reactants are regenerated by recharging from an external source.) A fuel cell, in contrast, is theoretically capable of producing electrical energy for as long as fuel and oxidant are supplied to the electrodes. Fuel cell life is typically limited by the physical degradation of the components, usually corrosion, or by malfunction of the components.
- Fixed versus dynamic electrochemistry: A battery contains a physically "fixed" set of chemicals that react with each other to produce power. A fuel cell produces energy based on an intricate, dynamic flow of chemicals and electrical properties, which can be disrupted by any number of factors.

2.1.1. The Unit Cell: A Simple But Formidable Device

Like the battery, the fuel cell is a simple device, consisting of only four basic components that support electrochemical reactions. They are as follows:

1. An electrically negative electrode, called the anode with or without a catalyst, to which a fuel is supplied;
2. An electrically positive electrode, called the cathode with or without a catalyst, to which an oxidant (oxygen) is supplied;

[i] This section draws from a range of existing documents on fuel cells, including and not limited to "Fuel Cells: Types of Fuel Cells," Hydrogen, Fuel Cells & Infrastructure Technologies Program, Energy Efficiency and Renewable Energy, US Department of Energy, http://www1.eere.energy.gov/hydrogenandfuelcells/fuelcells/fc_types.html, content last updated March 8, 2007; "Fuel Cells: Fuel Cell Basics: A Basic Overview of Fuel cell Technology," Smithsonian Institution, http://americanhistory.si.edu/fuelcells/basics.htm, accessed February 2009; and "Fuel Cell Handbook, Seventh Edition," National Energy Technology Laboratory, US Department of Energy, November 2004; "Fuel Cell Research," Princeton Applied Research, http://www.princetonappliedresearch.com/products/markets/fuel_cell.cfm, AMETEK, 2004.

3. Electrolyte (a substance that allows movement of electrically conductive ions) that separates the two electrodes;
4. An external electric circuit connecting two electrodes (Fig. 2.1).

The electrochemical reactions take place at the electrodes as fuel and oxidant are supplied, producing or consuming electrons and ions. Electrons are produced at the anode and are consumed at the cathode. Electrons move from anode to cathode via an external circuit—a process that results in the electric current that the fuel cell produces. Within the fuel cell, ions flow between anode and cathode. Ions are individual atoms that have an uneven number of protons and electrons, and thus possess a net positive charge (ions with a positive charge are called cations) or negative charge (ions with a negative charge are called anions). A hydrogen atom, which is the simplest case, has a nucleus of one proton that is balanced by a single electron in its surrounding electron shell. A hydrogen cation has no electron, and therefore carries a net positive charge. The cation flows to the cathode where it combines with electrons flowing through the external circuit. Some electrodes require electrocatalysts (commonly precious metal platinum) for the reaction to proceed at the desired rate.

In a fuel cell, ion creation is continuous and enabled by the electrolyte, which allows ions to move between the two electrodes. The electrolyte also serves as a physical barrier to prevent the fuel and oxidant gas streams from directly mixing. The process of consuming electrons and ions at the cathode also creates reaction waste products and heat, which must be transported away.

The working of a fuel cell is nothing if not complex; its various chemical and electrical

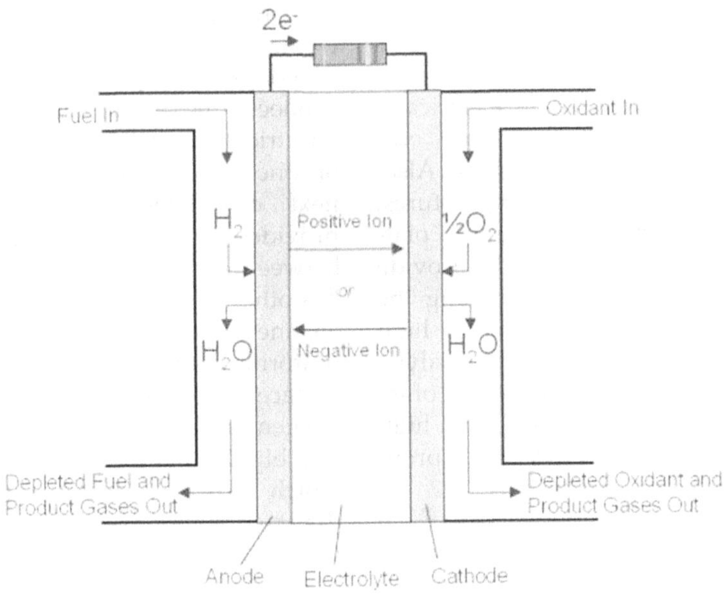

FIGURE 2.1 Schematic illustration of a single cell. *Source: J.H. Hirschenhofer, D.B. Stauffer, R.R. Engleman, and M.G. Klett, Reprinted in several editions of "Fuel Cell Handbook" produced by EG&G Technical Services, for the U.S. Department of Energy, Office of Fossil Energy, National Energy Technology Laboratory, different dates. The 7th edition, which may be the last, was published in November 2004.* (For the color version of this figure, the reader is referred to the online version of this book.)

components and processes are intricate, delicately balanced, and dynamic. The most critical part of a fuel cell is the so-called three-phase boundary or interface, which is the area where the electrodes and the electrolyte meet and the electrochemical reactions take place. For the interface area to be active, it must be exposed to the reactant (fuel or oxidant), be in electrical contact with the electrode, also be in ionic contact with the electrolyte, and contain sufficient amounts of electrocatalyst for the reaction to occur.[1]

These requirements lead to complex performance and cost trade-offs. The density or thickness of the electrolyte and the nature of the interface play a critical role in the electrochemical performance of both liquid and solid electrolyte fuel cells. In fuel cells that use liquid electrolytes, the electrolyte could wet the electrodes excessively and flood them, lowering their ability to transport reactants to the electrolyte and causing degradation of cell performance over its life.[1] In solid electrolyte fuel cells, the challenge is how to engineer the interface so that the electrodes and the electrolyte remain electrically and ionically well connected and are efficiently exposed to the reactant gases. Also, in fuel cells that operate at low temperatures, rare and precious materials—commonly platinum—are needed at the interface to provide sufficient electrocatalytic activity, making the fuel cells extraordinarily expensive. In high temperature fuel cells, the electrocatalytic activity of the bulk electrode material is often sufficient to support the reaction, but high operating temperatures can cause other problems, including longevity issues.

- Over the past 20 years, the performance of at least some fuel cells has been substantially improved based on improvements in the three-phase boundary interface. The improvements included reductions in the thickness of the electrolyte and the development of improved electrode and electrolyte materials which broaden the temperature range over which the cells can be operated.[1] These improvements, however, are incremental in nature and do not constitute a technical breakthrough.

2.1.2. Fuel Cell Stacks: Planar or Tubular Designs

A single fuel cell typically generates a small amount of electricity, but cells can be connected in series to produce greater power output. Such a collection of fuel cells in series is called a stack. There are two major types of stack technology: planar-bipolar stacking and tubular stacking.

2.1.2.1. Planar-Bipolar Stacking

For flat plate or planar fuel cells, individual cells are stacked in layers and are electrically connected with interconnects, which are called planar bipolar plates. A bipolar plate connects the entire surface of one cathode (electrically positive) and the entire surface of one anode (electrically negative) of the next cell—hence the name bipolar—and provides a series electrical connections between adjacent cells. A bipolar plate also has other functions. The bipolar plate contains channels or grooves that allow for the uniform distribution of fuel gas over the surface of the anode and the distribution of oxygen or air over the surface of the cathode. The bipolar plate also often can be hollow in which case it carries a cooling fluid through the stack with a design that keeps all the reactant gases and cooling fluids apart. The bipolar plate keeps the reactant gases contained within each cell, so the edges of the cell are designed to allow space to create a "gas-tight" seal around each cell.

- Bipolar plates, which could be made of graphite or metal, are expensive and difficult to manufacture, often coated or treated, and much attention has been focused on the materials and manufacturing processes to manufacture superior bipolar plates.

2.1.2.2. *Stacks with Tubular Cells*

Each tube consists of one cell and is generally small in diameter relative to its length. Tubular cells are stacked in arrays and connected either in series or in parallel within the stack. In one of the tubular designs, the electric current is conducted tangentially around the tube. In another tubular type, current is conducted axially along the tube, and interconnections are made between the cells at the end of each tube. The problem with tubular stacks is the long path that the electrical power has to travel through each cell and then from the anode of one cell to the cathode of the next cell and the large voids within the stack structure. Conduction of electrical current over significant distances results in a significant loss of power and, thus, low power density.

- To shorten the current path and increase power density, some tubular stack designs use flattened tubes. Tubular fuel cells, nonetheless, represent a special geometric challenge to the stack designer when it comes to achieving high power density and short current paths.
- Current collection is a major issue in all tubular designs.

2.1.3. Fuel Cell Systems

In addition to the stack, fuel cell systems require several other subsystems and components—the so-called balance of plant (BoP)—which constitute the supporting infrastructure that enables a fuel cell to operate (Fig. 2.2). Together with the stack, the BoP forms the fuel cell system. The types of BoP equipment required depend heavily on the fuel cell type, the fuel source, and the application.[1] In addition, specific operating conditions and requirements of individual cell and stack designs determine the characteristics of the BoP. Most fuel cell systems include BoP equipment that perform fuel processing or reforming, thermal management, water management, and electric power conditioning. While perhaps not the focus of most of the development effort, the BoP represents a significant fraction of the weight, volume,

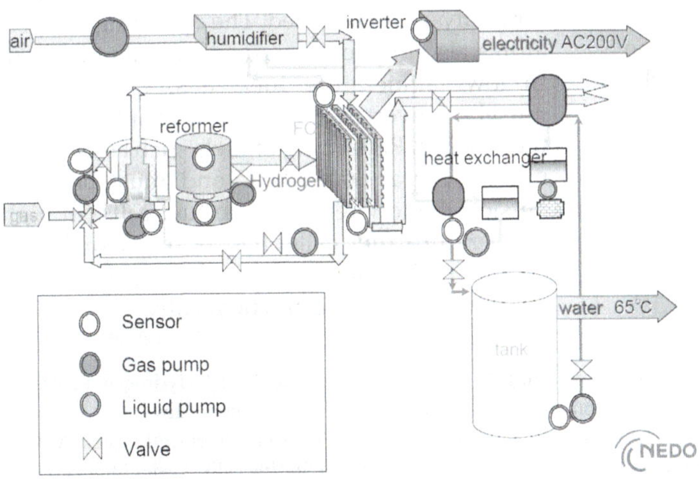

FIGURE 2.2 A fuel cell system schematic diagram. *Source: Takashi Ujiie, "Overview of Fuel Cell R&D on NEDO," New Energy and Industrial Technology Development Organization (NEDO), Japan, 27 October 2006.* (For the color version of this figure, the reader is referred to the online version of this book.)

and cost of most fuel cell systems. The BoP also plays a major role in the reliability and durability of most fuel cell systems. BoP equipment, however, often is developed and manufactured by manufacturers other than fuel cell manufacturers, which makes the integration of a total fuel cell system problematic. Still, most fuel cell systems contain the following:

- Fuel processing or reforming: Various conventional fuels need to be processed or reformed into hydrogen-rich gas which is electrochemically oxidized. Except when pure hydrogen is used, the removal of impurities from the fuel is usually required.
- Air supply: In most fuel cell systems, air compressors or blowers as well as air filters and humidifiers are used.
- Thermal management: All fuel cell systems require careful management of the fuel cell stack temperature. This is accomplished with insulation, stack cooling, air and fuel preheating, exhaust cooling, and recycle. The heat exchanger serves to cool or heat fuel cell stack, air, and fuel.
- Water management: Water is needed in some parts of the fuel cell, while water is generated in the fuel cell as a reaction product. Water management systems are required in most fuel cell systems to effectively recycle the water that was generated in the fuel cell and deliver it to where it is needed without introducing water from external sources. Water management also is required to operate heat exchangers.
- Electric power conditioning: Fuel cell stacks generate direct current electricity, which needs to be converted to alternating current that can be used by most existing facilities. This is done by some combination of inverters, converters, and/or transformers.

2.2. TYPES OF FUEL CELLS: DISTINCT TECHNOLOGIES[ii]

There are several types of fuel cells, and they are all different and are each distinct. Fuel cells are most commonly classified by the type of electrolyte used in the cell. The electrolyte can be aqueous, solid, or molten; alkaline, neutral, or acid; or polymer, chemical substance, or ceramic. The choice of the electrolyte determines the temperature range in which the fuel cell operates. The operating temperature in turn dictates the type of material used in the fuel cell. Aqueous electrolytes are limited to temperatures of about 200 °C or lower because of their high vapor pressure and the rapid degradation of these fluids that occurs at higher temperatures. Low temperature fuel cells usually have slow chemical reactions and require rare and precious metal platinum catalysts at the anode or the cathode or both to speed the reaction. Higher temperatures typically promote faster reactions, and requirements for catalysis are lessened. The operating temperatures also impact the type of fuel that is used. In low temperature fuel cells, gaseous fuel must be converted to hydrogen prior to entering the fuel cell. Reforming of fuels such as methane must occur at 600–700 °C and must be done outside the low temperature fuel cells. In high temperature fuel cells, hydrocarbon fuels can be internally converted to hydrogen or even directly oxidized electrochemically.

[ii] This section draws from existing documents on fuel cells, including "Fuel Cells: Types of Fuel Cells," Hydrogen, Fuel Cells & Infrastructure Technologies Program, EERE, US Department of Energy, http://www1.eere.energy.gov/hydrogenandfuelcells/fuelcells/fc_types.html, content last updated March 8, 2007; "Fuel Cells: Fuel Cell Basics: A Basic Overview of Fuel cell Technology," Smithsonian Institution, http://americanhistory.si.edu/fuelcells/basics.htm, accessed February 2009; and "Fuel Cell Handbook, Seventh Edition," NETL, US Department of Energy, November 2004.

- Fuel cells also sometimes are classified by the type of fuel they use. Direct methanol fuel cells (DMFCs), for example, are a subtype of polymer electrolyte membrane fuel cells (PEMFCs). DMFCs use polymers as an electrolyte and methanol as fuel.

There are six types of fuel cells currently under development including:

1) Proton exchange membrane fuel cells (PEMFCs) or polymer electrolyte fuel cells (PEFCs)
2) Direct methanol fuel cells (DMFCs)[iii]
3) Alkaline fuel cells (AFCs)
4) Phosphoric acid fuel cells (PAFCs)
5) Molten carbonate fuel cells (MCFCs)
6) Solid oxide fuel cells (SOFCs)

Table 2.1 provides a summary of the characteristics of the different types of fuel cells.

Direct methane fuel cells and direct carbon fuel cells are also being developed but at a much lower level than the other six. Methane and carbon are special fuels in which direct oxidation at the anode would be of great benefit if found feasible.

2.3. POLYMER ELECTROLYTE MEMBRANE FUEL CELLS[IV]

2.3.1. Principles of Operation and Characteristics

PEMFCs create electrochemical reactions using positive hydrogen ions as carrier ions; the direction of the flow of the ions is from anode to cathode (Fig. 2.3).

- In PEMFCs, the fuel (hydrogen, H_2) enters at the anode. There, a chemical reaction causes the hydrogen molecules to separate into positive hydrogen ions (H^+ or protons) and electrons (e^-). This reaction releases heat. The positive hydrogen ions pass through the electrolyte made of a polymer membrane and travel to the cathode. The electrons remain behind and thereby give the anode a negative charge, creating a voltage difference between the anode and the cathode. Because electrons travel from negative to positive, the electrons follow an external circuit from the anode to the cathode. At the same time, oxygen (O_2) enters the fuel cell at the cathode and combines there with the electrons, which have traveled through the external circuit, and the positive hydrogen ions, which have traveled through the electrolyte, to produce water (H_2O) at the cathode. The chemical reaction is represented here:

$$2H_2 \rightarrow 4H^+ + 4e^-$$

$$O_2 + 4e^- + 4H^+ \rightarrow 2H_2O$$

PEMFCs are regarded as best suited for fuel cell vehicles (FCVs) and small stationary applications. They have been the fuel cells most favored by researchers and developers. In 1994, when Daimler (then Daimler-Benz) demonstrated its first FCV powered by hydrogen-fueled PEMFCs, global automakers were thrilled by the commercial potential of PEMFCs as a primary power of vehicles. Major global automakers joined in the race to develop PEM-based FCVs. PEMFCs' popularity likely is partly based on their easy-to-handle characteristics, including moderate operating temperatures, use of manageable

[iii] DMFCs are a subtype of PEMFCs, in which methanol (CH_3OH) is used as a fuel without the use of a fuel processor. In addition to DMFCs, there are other PEMFC subclasses that operate on fuels other than methanol. Some operate on alcohol or fuels with a similar chemical structure, and some others operate on hydrogen stored in borohydrate or in metal storage units or use hydrogen generated from methanol by a fuel processor. Some use the term "micro fuel cells" to address DMFCs and all of the other PEMFC subsets.

[iv] PEMFC is interchangeably called PEFC.

TABLE 2.1 Comparison of Fuel Cell Technologies

Fuel cell type	Common electrolyte	Operating temperature	Typical stack size	Efficiency	Applications	Advantages	Disadvantages
Polymer electrolyte membrane (PEM)	Porfluoro sulfonic acid	50–100 °C 122–212° typically 80 °C	<1–100 kW	60% Transportation 35% Stationary	• Backup power • Portable power • Distributed generation • Transportation • Specialty vehicles	• Solid electrolyte reduces corrosion and electrolyte management problems • Low temperature • Quick start-up	• Expensive catalysts • Sensitive to fuel impurities • Low temperature waste heat
Alkaline (AFC)	Aqueous solution of potassium hydroxide soaked in a matrix	90–100 °C 194–212 °F	10–100 kW	60%	• Military • Space	• Cathode reaction faster in alkaline electrolyte, leads to high performance • Low cost components	• Sensitive to CO_2 in fuel and air • Electrolyte management
Phosphoric acid (PAFC)	Phosphoric acid soaked in a matrix	150–200 °C 302–392 °F	400 kW 100 kW module	40%	• Distributed generation	• Higher temperature enables CHP • Increased tolerance to fuel impurities	• Pf catalyst • Long start-up time • Low current and power

Molten carbonate (MCFC)	Solution of lithium, sodium, and/or potassium carbonates, soaked in a matrix	600–700 °C 1,112–1,292 °F	300 kW–3 MW 300 kW module	45–50%	• Electric utility • Distributed generation	• High efficiency • Fuel flexibility • Can use a variety of catalysts • Suitable for CHP	• High temperature corrosion and breakdown of cell components • Long start-up time • Low power density
Solid oxide (SOFC)	Yttria stabilized zirconia	700–100 °C 1,202–1,832 °F	1 KW–2 MW	60%	• Auxiliary power • Electric utility • Distributed generation	• High efficiency • Fuel flexibility • Can use a variety of catalysts • Solid electrolyte • Suitable for CHP & CHHP • Hybrid/GT cycle	• High temperature corrosion and breakdown of cell components • High temperature operation requires long start-up time and 1imits

Direct Methanol Fuel Cells (DMFC) are a subset of PEM typically used for small portable power applications with a size range of about a subwatt to 100 W and operating at 60–90 °C. *Source: "Comparison of Fuel Cell Technologies," http://www1.eere.energy.gov/hydrogenandfuelcells/fuelcells/pdfs/fc_comparison_chart.pdf, Energy Efficiency and Renewable Energy, US Department of Energy, February 2011.*

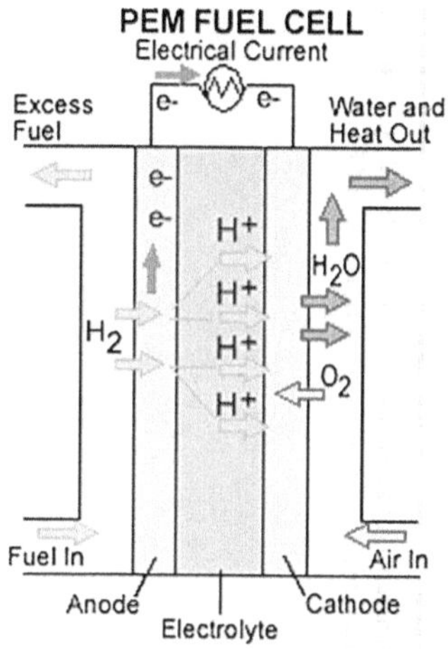

FIGURE 2.3 Polymer electrolyte membrane fuel cells (PEMFCs). *Source: Fuel Cells, Types of Fuel Cells, Energy Efficiency and Renewable Energy, U.S. Department of Energy, http://www1.eere.energy.gov/hydrogenandfuelcells/fuelcells/fc_types.html, Content Last Updated 8 March 2007.* (For the color version of this figure, the reader is referred to the online version of this book.)

polymer materials, and absence of corrosive electrolytes. Over the decade, PEMFCs for FCVs and small stationary applications received much attention and investments that likely surpass all the other types of fuel cells combined. More corporations and research institutions are involved in R&D of PEMFCs than in any other fuel cell type.

PEMFCs have many advantages. PEMFCs deliver a high power density—power generated per unit of weight, volume, or area—leading to compact size and light weight. Because of the thermal properties of the electrolyte—perfluorosulfonic acid (PFSA) polymer—PEMFCs operate at low temperatures—typically at about 60–80 °C—allowing them to warm up quickly, start up quickly, and respond to load changes rapidly. Since water is the only liquid in the PEMFCs, corrosion is minimal. In addition, PEMFCs use no corrosive fluid in the cell and, thus, are easy to seal, handle, and assemble. While PEMFCs are particularly suitable for operations using pure hydrogen as the fuel, fuel processors have been developed that will allow the use of conventional fuels such as natural gas.

- One unique design of PEMFCs allows the direct use of methanol without a fuel processor; it is the DMFC, which will be discussed in a following section.

PEMFCs, however, have disadvantages, and one of them is the need for a platinum catalyst. At low temperatures, fuel cells electrochemical reactions are slow and only a few relatively rare and precious materials, such as platinum, provide sufficient electrocatalytic activity. Both PEMFC anodes and cathodes use platinum catalysts. PEMFCs also require higher catalyst loading at both electrodes than other fuel cell types, except for DMFCs.[1] Platinum is expensive. Efforts have been made to minimize the use of platinum by extensively optimizing the platinum particle size. There also have been efforts to develop mixtures of platinum and other metals or to find substitute metals. Most of these attempts focused either on gold or on platinum alloys. So far, these efforts have not demonstrated a decisive cost advantage over pure platinum catalysts.

The effectiveness of the platinum catalyst in the PEMFC fuel cell is seriously degraded by CO even if it is present in trace amounts (as little as 10 ppm or 0.001 percent).[1] CO in the gas stream attacks the platinum catalyst at the anode, masking the catalytic activity and reducing cell performance. To minimize CO poisoning, highly purified hydrogen must be supplied to the anode. Hydrogen gas produced by reforming hydrocarbons typically contains 1 percent of CO, and extensive fuel processing is required to reduce CO content in the fuel gas. This increases overall system cost. Efforts to make the catalyst less sensitive to CO poisoning has led to the development of platinum/ruthenium catalysts with intrinsic tolerance to CO. Electrodes with platinum/ruthenium catalysts have been shown to tolerate CO up to 200 ppm or 0.02 percent, helping to lower the cost of fuel purification but this solution still does not solve the problem of CO poisoning.

2.3.2. Another Daunting Problem: Electrolyte Performance

The standard electrolyte material used in today's PEMFC systems is a PFSA polymer membrane. PFSA membranes have exceptionally high chemical and thermal stability, and remain stable against chemical attack from both strong bases and strong oxidizers and reducing acids at temperatures up to 125 °C.[1] These membranes have characteristics suited for an electrolyte for PEMFCs; they are good ionic conductors and electrical insulators as well as good fuel/oxidant separators.

The PFSA membrane, however, presents a series of daunting challenges. The membrane itself is expensive.[2] It also is low in efficiency. Nafion, for example, is available only in a limited range of thicknesses with specific ionic conductivity. Because the Nafion membrane is thick, and its internal resistance high, the ionic conductivity is low and, thus, so is the efficiency of the fuel cell. In addition, the Nafion membrane requires very precise water and thermal management, which reduces the overall efficiency of the fuel cell. A critical requirement of the membrane is that all parts must be sufficiently hydrated with water, but not flooded. If dehydration occurs, it will degrade the polymer and shorten membrane life.

Temperatures within the cell also must be kept within a narrow range in order to avoid local dehydration of the membrane, which can result in hot spots and local dead zones. Thermal management is highly challenging because of the narrow temperature range in which the membrane can operate and the relatively small temperature differences between the cell operating temperature and the ambient temperature.[1] In addition, the future supply of Nafion and other PFSA polymer is not assured. PFSA polymers contain a poisonous chemical, fluorine, a major concern when the membrane is recycled or burned. Plants that were originally built to manufacture Nafion and other PFSA polymers in the 1970s and 1970s would not get approval to be built today because of health and safety considerations.[3] Thus, future production of PFSA polymers would likely be limited to those existing plants that belong to DuPont and a few other large chemical companies.

More importantly, the PFSA-based PEMFCs have demonstrated less than satisfactory performance and durability. Total system efficiency is in range of 33–36 percent for stationary applications.[4] This is no more efficient than conventional combustion-based power plants. By recovering the waste heat from the cooling water, the overall efficiency could be raised to about 80 percent, but the resulting water temperature (about 50–70 °C) is not high enough for many combined heat and power (CHP) cogeneration applications.

PEMFC systems also suffer from lifetime performance degradation, and the nominal operating life of PEMFC systems falls far short of the lifetimes required for typical transport or stationary cogeneration applications (these requirements are at least 5000 h for transport applications and at least 40,000 h for stationary cogeneration applications—or about 10 years of daily start and stop operations).[4] According to a Japanese source as of March 2008, the operating life of PEMFC systems for transport applications is about 1000 h and that for stationary CHP cogeneration applications is about 20,000 h. Furthermore, the causes of PEMFC degradation are not fully understood, making it highly difficult to rectify the problem.

For the past several years, there has been an accelerated effort to develop PEMFC electrolytes that are better than Nafion or other similar PFSA-based polymer electrolytes. Figure 2.4 shows rapid growth in the number of patent families that refer to Nafion and fuel cells since

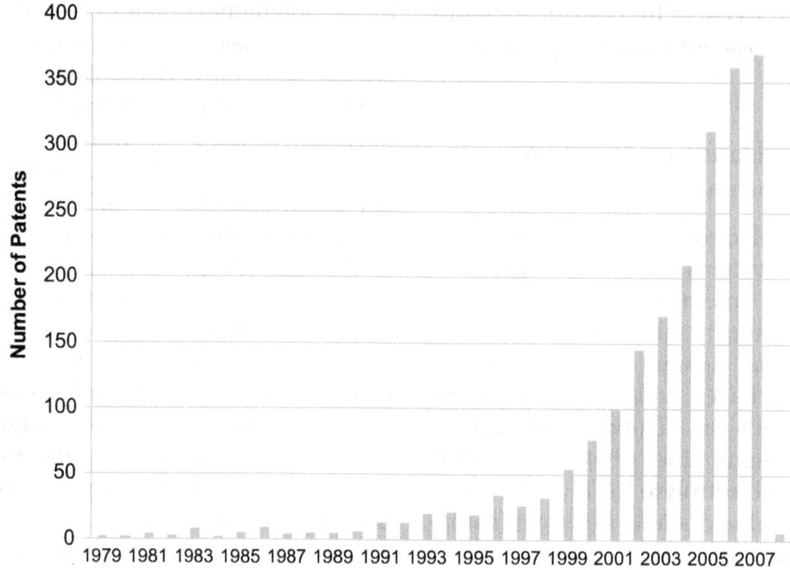

*The year denotes year of filing. Years are from January to December. The data for the full year of 2008 is not available.

FIGURE 2.4 Cumulative rates of patent filings worldwide per year for patent families referring to Nafion and fuel cells. *Source: Jonathan Butler, "2007 Patent Review Q4," Fuel Cell Today, January 2008.* (For the color version of this figure, the reader is referred to the online version of this book.)

the late 1990s.[3] The numbers are based on patents at the time of filing.[v]

From the late 1970s to the early 1990s, researchers and developers appear to have been satisfied with Nafion as a PEMFC electrolyte; Fig. 2.4 also shows there was relatively little patent activity during that period, but patent filing grew rapidly starting in the early 1990s. About ten patents were filed in 1991, around 50 in 1999, and more than 350 in 2007.[3] The rapid increase in activity undoubtedly is due to the industry's growing sentiment that Nafion and other PFSA membranes may be inadequate and are a barrier to the technical viability and commercialization of this class of fuel cell.

Many researchers and developers remain convinced that the disadvantages involving Nafion and other PFSA polymer electrolytes can ultimately be overcome. Researchers and developers have a wide range of initiatives underway. One approach focuses on increasing ionic conductivity by making the Nafion membrane thinner; the thinner the membrane, the greater the ionic conductivity and the higher the efficiency of the fuel cell.[1] Reduction in membrane thickness, however, results in weakened mechanical strength of the membrane. So researchers are introducing an internal support layer to enhance the mechanical strength of the membrane film even as they make the membranes thinner. Another approach focuses on making the Nafion and other PFSA membranes operable with low or zero hydration. They are adding some substances, such as silica and platinum, in the hope of improving this characteristic.[5] Yet another approach

[v] Since the patent applications refer to both Nafion and fuel cells, these patents likely apply to PEMFC electrolyte membranes for use in fuel cells, inclusive of modified Nafion as well as new membranes that might replace Nafion.

focuses on making Nafion and PFSA membranes operable at higher temperatures—in the 120–160 °C range. The high operating temperature is advantageous; it eliminates CO occlusion of the platinum sites and eliminates CO poisoning, increases system efficiency, and generates more heat for possible use in stationary CHP applications.[1] At such high temperatures, however, PFSA membranes would dehydrate and dry out quickly, leading to a loss of ion conductivity and severe reduction in membrane lifetimes. This has led researchers to develop hybrid materials using PFSA polymer and inorganic materials that can sustain higher operating temperatures.[5]

- All of these efforts on Nafion or other PFSA-type membranes, however, constitute a series of independent scientific activities which might be able to alter and improve some of the characteristics of the membranes, but are unlikely to be able to significantly change the fundamental shortcomings of PFSA polymer membranes.

There also are efforts to reexamine polymers based on hydrocarbons that have favorable characteristics akin to Nafion and other PFSA polymers. Hydrocarbon polymers conduct ions under properly hydrated (humid) conditions. These polymers have been known for many years and were among the first compounds to be tested as an early PEMFC electrolyte membrane, even before PFSA polymers were used in PEMFCs.[2] Hydrocarbon polymers have a number of advantages over PFSA polymers. They are far less expensive and easier to manufacture. They have greater mechanical strength and ionic conductivity at higher temperatures.[5] Hydrocarbon polymers also contain no hazardous substances.

- Hydrocarbon polymer membranes, however, have a number of major disadvantages as well. They are inferior to PFSA in chemical stability and are about equal to PFSAs in durability at low levels of humidity.[5] They also have a trade-off problem: as ionic conductivity increases, these polymers have less mechanical strength and chemical stability. Hydrocarbon polymer membranes also have physical properties that make it difficult to maintain close contact between the membrane and the catalyst.

Hydrocarbon polymers, nonetheless, have one major advantage. They possess great freedom in molecular mobility that allows for the creation of a wide variety of molecular structures. Researchers and developers are creating many varieties of hydrocarbon polymers for experimentation, including sulfonated polyimides, sulfonated aromatic polyethers, polyarylene ethers, polymerized styrene, polybenzimidazole (PBI) polymers, polymerized trifluorostyrene sulfonic acid membranes, polymerized trifluorostyrene, polymerized trifluorostyrene phosphonic acid membranes, and polyphosphazene polymers.[5]

- It is noteworthy that some hydrocarbon PEMFC electrolytes resemble other fuel cell types. PBI, for example, approximates PAFC electrolytes. PBI operates at higher operating temperatures, eliminating CO poisoning of platinum and allowing for possible use in stationary CHP.[1] Because PBIs require significantly lower water content to facilitate ion transport, an additional benefit is that water management is dramatically simplified. To achieve acceptable ionic conductivity, however, the membrane must be impregnated with phosphoric acid, which is apparently not very tightly bound to the polymer backbone. As a result, precautions must be taken to avoid liquid water and resulting corrosion—the same shortcomings inherent in PAFCs.

The most noteworthy development in the field of PEMFCs is the emergence of a systematic plan for conducting basic research. After 40 years of virtual dependence on Nafion and other PFSA membranes, researchers and developers are making a new start in the exploration of PEMFC electrolyte technology. They are experimenting with a wide range of materials, including room-temperature molten salts (or ionic liquids), inorganic materials, organic/inorganic hybrid materials, and fullerenes. They also are trying to build new materials at the molecular level.

- Room-temperature molten salts are liquid at the ambient temperature and have negligible vapor pressure and low viscosity. They are known to have high thermal and chemical stability as well as high ionic conductivity. Efforts are being made to incorporate room-temperature molten salts into organic polymers.[5]
- Researchers and developers also are exploring inorganic materials, such as glass-forming oxides. Glass-forming oxides are known to have superior ionic conductivity. To further raise conductivity, adding phosphates to glass-forming oxides is being considered.
- Researchers are also considering adding inorganic materials—such as boron, silicon, and titanium—to organic polymers to attain high durability at higher temperatures. They believe that the inorganic materials would protect the organic materials from high heat.
- They also are mixing strong acids, such as sulfonic acid, to fullerene to achieve high ionic conductivity.

The primary goal of these various efforts focuses on the creation of ion conducting electrolyte membranes that can operate at temperatures higher than 150 °C with no water. This is a temperature range that no PFSA or hydrocarbon polymer has achieved so far.[5] But no new electrolyte under exploration has demonstrated ionic conductivity and thermal or chemical stability that are equal to PFSA or hydrocarbon polymers. It is also true that some of the new candidate electrolytes, such as inorganic materials mixed with phosphoric acid, resemble PAFCs, perhaps providing an alternative to PEMFC technology.

2.3.3. Challenges with Transport Applications

PEMFCs face great challenges if they are to be used in transport applications that go well beyond the demands of stationary applications. For terrestrial transportation, PEMFCs are required to be robust enough to operate under the entire range of environmental conditions that any internal combustion engine vehicle is expected to endure on the road. FCVs will have to tolerate various contaminants, particles, dust, sand, and debris in the air. FCVs will need to be mechanically sturdy to withstand bumps on the road, vibrations, and collision impacts. FCVs will have to endure repeated start-and-stop cycles and acceleration-and-deceleration operations daily for over 10 years. FCVs also will have to withstand arid, wet, hot, cold, and freezing weather conditions as well.

If PEMFCs are to be used in transport applications, engineers will need to figure out how to refuel them. FCVs that use high density liquid fuels such as methanol, ethanol, natural gas, liquefied petroleum gas, and gasoline, will need to have onboard fuel processors to convert the fuel and purify the reformate fuel into high purity hydrogen. Considerable effort has been expended to develop liquid-fueled FCVs or to reform other fuels into hydrogen while onboard FCVs. The fuel processing operations and related balance of plant components add cost and system integration constraints due to their size and weight and maintenance requirements. They also release carbon dioxide, although less than that emitted from current gasoline engines. Most experts thus believe that onboard storage of hydrogen will be necessary for practical FCVs.[1]

A key component of the direct hydrogen FCVs is the hydrogen storage tank and the hydrogen refueling infrastructure. A wide range of hydrogen storage methods are being considered, including compressed hydrogen storage, liquid storage, storage in metal hydrides, and chemical storage.[1] Each of these options represents a compromise between energy density, weight, energy efficiency, and cost, and none of them are competitive with gasoline powered vehicles today. Special safety considerations also must be made in all cases. In addition, the volume and weight of the hydrogen storage tank needs to match the power density of internal combustion engines. Storage remains a long-term R&D challenge.

If FCVs are to become commercially practical there must be a business case which justifies establishing a widespread hydrogen refueling infrastructure. FCVs need to be refueled as conveniently as gasoline vehicles can, and the establishment of hydrogen stations that are conveniently located as gasoline stations would require a huge investment. According to a DOE data in 2005, there were 169,000 gasoline stations in the United States.[6] If a hydrogen station costs $1 million to $4 million to build, the United States would need nearly $170 billion to $680 billion for the refueling station infrastructure alone.

2.4. DIRECT METHANOL FUEL CELLS

2.4.1. Principles of Operation and Characteristics

DMFCs are a subset of PEMFCs, in which methanol (CH_3OH) is used as a fuel without the use of a fuel processor (Fig. 2.5). DMFC

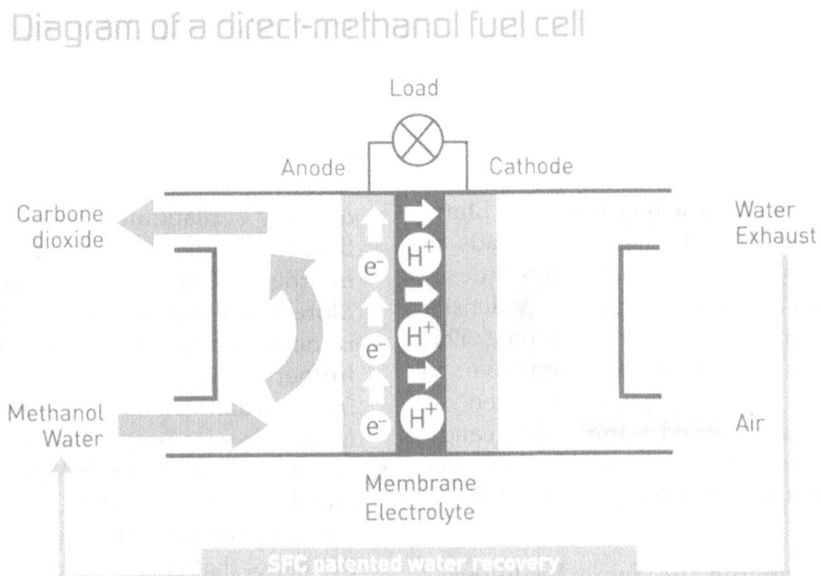

FIGURE 2.5 Direct methanol fuel cells (DMFCs). (DMFCs are a subset of PEM typically used for small portable power applications with a size range of about a subwatt to 100 W and operating at 60–90 °C.) *Source: "Technology, How Does a fuel cell Work?" www.sfc.com, SFC Energy AG, Germany, no date.* (For the color version of this figure, the reader is referred to the online version of this book.)

electrochemical reactions are similar to those of PEMFCs and PAFCs:

- In DMFCs, the carrier ion is the hydrogen ion and the direction of the flow of the ions is from the anode to the cathode—the same as for PEMFCs. When methanol enters the fuel cell at the anode, however, water is needed to enable the oxidation reaction to occur, which requires a precise metering of methanol and water at the anode. One product of this oxidation is carbon dioxide (CO_2) which must be vented from the system. The reaction also produces electrons (e^-) and hydrogen ions (H^+). The hydrogen ions pass through the electrolyte while the electrons pass through the external electrical circuit to the cathode. At the cathode, the hydrogen ions, electrons, and oxygen from the air react and produce water (H_2O), again, the same as for PEMFCs:

$$CH_3OH + H_2O \rightarrow 6H^+ + 6e^- + CO_2$$
$$1\frac{1}{2}O_2 + 6H^+ + 6e^- \rightarrow 3H_2O$$

The major advantage of the DMFC is that it uses methanol without a fuel reformer. Eliminating the need for a fuel reformer leads to a very simple system. Methanol also is a desirable fuel; it has a relatively high energy density and is a liquid under ambient conditions.[1] This makes DMFCs an attractive alternative to PEMFCs which require pure hydrogen as a fuel. DMFCs are regarded as the leading candidate for portable applications, such as cameras, notebook computers, and cell phones today.

2.4.2. Experiencing the Same Problems as PEMFCs And More

Researchers and developers of DMFCs have experienced a different set of challenges that are as daunting as those for PEMFCs. Current DMFCs are most commonly based on current state-of-the-art PEMFCs designs, which include membranes made up of Nafion or other PFSA polymers, anodes containing a platinum/ruthenium catalyst, and cathodes with platinum catalysts. Researchers and developers have come to realize that this approach has not been such a good idea. Not only have they encountered the inevitable stumbling blocks common to PEMFCs but they have also had to face a few additional problems. Indeed, they have found that for DMFCs, all of the three major fuel cell components pose great challenges:

- When methanol enters the anode, the anode needs to oxidize methanol—a compound far more complex than the typical hydrogen fuel used for PEMFCs—into hydrogen ions and electrons. The oxidation of methanol is inherently sluggish; it is a much more involved process than the oxidation of hydrogen because it takes several intermediate steps before reactions complete.[7] This process, which occurs at the anode, causes a considerable voltage loss, severely lowering efficiency. To increase the oxidation speed, a more active catalyst is required, which typically means a larger quantity of platinum, almost 10 times higher than needed in high performance direct hydrogen PEMFCs.[1] This increases cost. Moreover, the state-of-the-art platinum/ruthenium catalyst at the anode suffers further voltage loss due to self-poisoning from partially oxidized methanol—carbon monoxide—leading to lower efficiency.[8]
- The Nafion or other PFSA polymer membrane used in DMFCs also causes a major problem called methanol crossover. Methanol that enters the anode crosses through the electrolyte membrane to the cathode without fully undergoing electrochemical reaction. This happens because methanol easily mixes with water and spreads through the water in the electrolyte membrane as an aqueous methanol solution, and quickly reaches the

cathode. When feeding concentrated methanol directly, the crossover can be as high as 30–50 percent compared with the amount that is oxidized electrochemically.[1] The methanol crossover is not only a waste of fuel but also lowers the cell performance.[1] If the methanol concentration is reduced, the crossover is reduced but so is the current density.[1] Furthermore, the PFSA membrane transports a large amount of water across the membrane as part of the aqueous methanol solution, causing water loss in the anode and flooding in the cathode.[9]

- In addition, the methanol crossed over from the anode to the cathode also is oxidized by the platinum catalyst at the cathode.[8] The partially oxidized methanol poisons the platinum catalyst, leading to a voltage loss.

2.4.3. Challenges with Portable Applications

Producing commercially viable DMFCs for portable applications is a major challenge. The challenge includes the development of a more active anode catalyst that promotes methanol oxidation, more advanced electrolyte membranes that discourage methanol crossover, and a new cathode catalyst that does not oxidize methanol but has sufficient oxygen reduction performance. Much work has been done to develop suitable catalysts for the anode of DMFCs, and it is more or less standard to use a mixture of platinum and ruthenium in equal proportions as the anode catalyst.[7] The high loading of precious metals remains despite high cost because it reduces the voltage losses to reasonable levels and reduces the inefficiency caused by methanol crossover.

Some researchers are modifying Nafion membranes to reduce methanol crossover. They are treating the surface of the membrane, creating a membrane made from two or three layers, and using additives to the membrane.[7] Some are taking the approach of using different electrolyte membranes that have higher ionic conductivity but discourage methanol crossover, including experimenting with hydrocarbon polymer membranes.[5] Others continue to search for nonplatinum catalysts for the cathode that will stop the methanol from reacting at the cathode and so eliminate the voltage drop.[7] The problem of water transport must be mitigated by complex water recovery systems that detract significantly from the conceptual simplicity of the DMFCs.[1] Researchers and developers have focused on miniaturizing the balance of plant components necessary to control water balance and minimize methanol loss. Some are even developing a reformer-based portable system.

DMFCs possess high power density relative to other fuel cells that make them competitive with lithium ion (Li-ion) batteries for powering small portable electronic equipment, such as laptop computers. DMFCs have an advantage over PEMFCs in this respect. DMFCs for portable applications are not expected to be as durable or as low cost as PEMFCs for transport and stationary applications. Portable electronic equipment use Li-ion batteries almost universally today. The Li-ion battery is expensive and does not last long. As long as DMFCs can last longer or are cheaper than the Li-ion battery, they are commercially competitive. PEMFCs, however, have much tougher competition; today's gasoline engines for vehicles and electric generators are highly efficient and inexpensive. DMFCs for portable applications, however, are required to be more compact and lighter to compete with the battery counterpart.

In addition, recent rulings allowing methanol cartridges onboard airplanes would encourage the commercialization of DMFCs for portable applications. Methanol is toxic and flammable, and has been prohibited from airplanes. The International Civil Aviation Organization's Dangerous Goods Panel voted in November 2005, however, to allow passengers to carry and use methanol fuel cartridges when aboard airplanes to power laptop computers and other consumer electronic devices.[10] Similarly, the Department of Transportation issued a ruling on April 30, 2008, permitting passengers and crew to carry an approved fuel cell with an installed methanol cartridge.

2.5. ALKALINE FUEL CELLS

2.5.1. Principles of Operation and Characteristics

The AFC gets its name from the electrolyte that is used: potassium hydroxide, which is an alkaline. AFCs have electrochemical reactions similar to those of MCFCs and SOFCs, which will be discussed in the following sections. In these fuel cells, negative ions travel through the electrolyte from the cathode to the anode where they combine with hydrogen, generate water, and release electrons (Fig. 2.6):

- In AFCs, the oxygen enters the fuel cell at the cathode. Hydroxyl ions (OH^-) are produced at the cathode by a process that includes oxygen, electrons that have been separated from hydrogen at the anode and which have traveled to the cathode through the external circuit, and the water in the electrolyte. These react to form hydroxyl ions. The hydroxyl ions then migrate to the anode and react with hydrogen, produce water, and release electrons. The electrons then travel through the external circuit to the cathode and react with oxygen and water, forming new hydroxyl ions, which then repeat the same reactions continuously:

$$O_2 + 4e^- + 2H_2O \rightarrow 4OH^-$$
$$2H_2 + 4OH^- \rightarrow 4H_2O + 4e^-$$

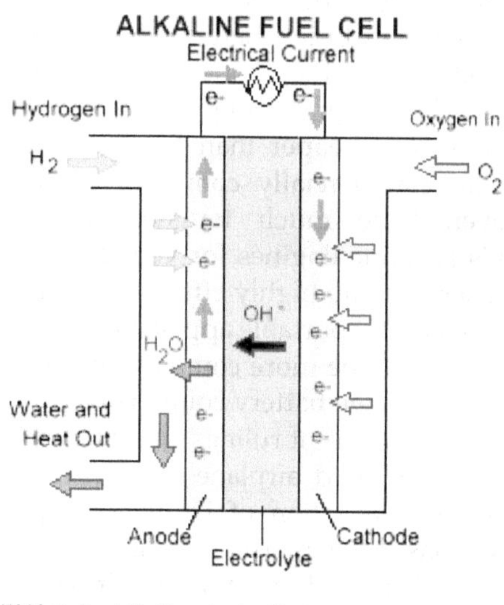

FIGURE 2.6 Alkaline fuel cells (AFCs). *Source: Fuel Cells, Types of Fuel Cells, Energy Efficiency and Renewable Energy, U.S. Department of Energy, http://www1.eere.energy.gov/hydrogenandfuelcells/fuelcells/fc_types.html, Content Last Updated 8 March 2007.* (For the color version of this figure, the reader is referred to the online version of this book.)

AFCs are one of the most efficient fuel cells for generating electricity and were the first fuel cell technology used for practical applications. Since the early 1960s, AFCs have served as onboard power generators for NASA's space missions. The high efficiency of AFCs, as high as 60 percent, is due to the kinetics of oxygen reduction at the cathode.[1] AFCs also are cheaper to manufacture compared to other fuel cells. They use inexpensive carbon-and-plastic electrodes and cheap potassium hydroxide electrolyte. They do not require expensive platinum to achieve high reactivity; for example, AFCs can use less costly catalysts, such as nickel, cobalt, and silver.[11]

2.5.2. An Early Success, Major Setbacks, Then Redemption, But...

The success of AFCs in space in the 1960s spurred global interest in developing AFCs for terrestrial applications, but a number of disadvantages of the design became apparent in the 1970s. One of the most significant was AFCs' high sensitivity to carbon dioxide and consequently the requirement for pure hydrogen and oxygen as reactants. Even a trace of carbon dioxide in the gas caused an unwanted chemical reaction that led to the formation of potassium carbonate crystals in the electrolyte. Carbonate crystals blocked pathways and pores in the electrodes, leading to a drop in performance and poor long-term stability.[12] Most methods of producing hydrogen from fossil fuels result in some carbon dioxide, and ambient air contains a significant amount of carbon dioxide, which pose an overwhelming drawback for AFCs for terrestrial applications. In the 1970s, removal of carbon dioxide required expensive devices, so AFCs were considered uneconomical for terrestrial applications.[1] In addition, the electrolyte used in AFCs is highly corrosive, which reduces cell frame life, rendering fuel cell longevity too brief for terrestrial applications. Despite much corporate investment, these problems have discouraged many researchers and developers from pursuing AFCs as a commercially feasible fuel cell:

- The leading developer of AFCs for space applications, UTC, investigated adapting the technology to terrestrial stationary power applications using air as an oxidant in the early 1970s.[1] As a result of the carbon dioxide problem, UTC abandoned its effort and focused only on spacecraft applications.
- The use of AFCs in space has been considered optimal. Carbon dioxide is less of a problem in space. A short component life is not an impediment because manned space missions do not demand a protracted service life, and cost is not a strong constraining factor in space applications. AFCs have been used by NASA starting with the Apollo manned lunar mission and continuing with the Space Shuttle mission today. NASA, however, expects to replace AFCs with PEMFCs in future space missions.

More recently, there has been a surprising development. AFC developers as well as academic researchers claim that the AFC's intolerance to carbon dioxide is no longer a stumbling block. They have devised a system for passing ambient air through an external, commercial-type soda-lime absorber that reduces the presence of CO_2 in the air stream.[1] They have also changed from an immobilized electrolyte design, where potassium hydroxide is embedded in an asbestos binder, to a circulating design, where the electrolyte moves about freely and is periodically replaced. When used in conjunction with a carbon dioxide scrubber, the buildup of CO_2 can be limited to small amounts, versus an immobilized electrolyte design where CO_2 continually builds up. AFC developers and researchers also claim that the AFC immediately returns to its original level of performance when the potassium hydroxide electrolyte is replaced.[13] These developers and researchers claim that changing the electrolyte is no more complicated or expensive than changing the oil in an internal combustion engine. Circulating electrolyte is a commonly accepted technology today for AFCs, as described in a government publication:

- The majority of AFC developers now use circulating electrolytes, which have several advantages over an immobilized electrolyte system.[1] Not only can the circulating electrolyte limit the quantity of CO_2 in the electrolyte, but CO_2 gas bubbles between electrodes and electrolyte also can be prevented because they are washed away by the circulating electrolyte. The circulating electrolyte also enables simple water and heat management; the water generated can simply mix into the electrolyte solution, and the electrolyte itself works as a cooling liquid inside the stack, requiring no dedicated heat exchanger.[1,11,13] Also, the cell will not dry out because the water content of the electrolyte remains constant inside the stack. In addition, accumulated impurities, such as carbonates, can be easily removed by changing the electrolyte.[1]

The resolution of the CO_2 intolerance issue and the advantages of the circulating electrolyte could make the development of AFCs for terrestrial applications possible, particularly for automotive applications. AFCs typically operate at temperatures between 65 and 220 °C,[1] but more recent AFCs can operate at low ambient temperatures of 23–70 °C and even at subzero temperature,[11,14] enabling the use of AFC-powered cars in colder climates. AFCs also start almost instantly.[15]

However, AFCs for terrestrial applications have suffered from low investment and research activity, leaving the state of AFC technology at a developmental level. Despite the advances that have been achieved, AFCs face many major challenges to achieve commercial viability. The markets demand adequate longevity, but today's AFC designs have a mixed record. According to data in 2000, lifetime testing, which was not completed then, was less than 1000 h for intermittent operation (a few hours per day).[16] A 2003 publication stated that AFCs could achieve longer operating life and that 15,000 h had been demonstrated.[12] In another document published in 2006, terrestrial AFCs had a rated life of 2000 h, with individual cells and electrodes tested to 5000–8000 h.[15]

2.6. PHOSPHORIC ACID FUEL CELLS

2.6.1. Principles of Operation

PAFCs use phosphoric acid as the electrolyte (Fig. 2.7). Otherwise, these cells are similar in operation to PEMFCs and DMFCs:

- Hydrogen enters at the anode where a chemical reaction separates it into positive hydrogen ions (H^+) and electrons (e^-). The hydrogen ions pass through the electrolyte and travel to the cathode. The electrons follow an external circuit to the cathode. At the same time, oxygen (O_2) enters the fuel cell at the cathode and reacts with the electrons and the hydrogen ions to produce water (H_2O):

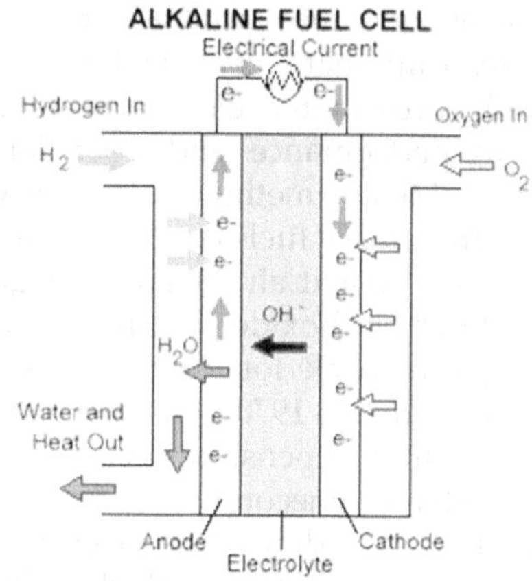

FIGURE 2.7 Phosphoric acid fuel cells (PAFCs). *Source: Fuel Cells, Types of Fuel Cells, Energy Efficiency and Renewable Energy, U.S. Department of Energy, http://www1.eere.energy.gov/hydrogenandfuelcells/fuelcells/fc_types.html, Content Last Updated 8 March 2007.* (For the color version of this figure, the reader is referred to the online version of this book.)

$$2H_2 \rightarrow 4H^+ + 4e^-$$
$$O_2 + 4e^- + 4H^+ \rightarrow 2H_2O$$

PAFCs are one of the most mature fuel cell types and the first to be placed on the commercial market. The number of units built exceeds any other fuel cell technology; hundreds of PAFC systems have been produced, sold, and used in field tests and demonstrations.[1] It is still one of the few fuel cell systems that until recently was available for purchase.

2.6.2. The Presumptive "First Generation" Commercial Fuel Cell

PAFCs were considered the first generation of modern fuel cells and were the first class of fuel cells to receive significant attention and R&D investment from national governments as well as public institutions (gas and electric utilities and related organizations). The enthusiasm that surrounded PAFCs was likely a direct result of the disappointing performance of AFCs and its intolerance for CO_2. Researchers undoubtedly felt that PAFCs finally had overcome all the shortcomings of previous fuel cell designs.

PAFCs are tolerant of CO_2 in fuel gas streams and in air and they are more tolerant of CO than PEMFCs. PAFCs can tolerate a CO concentration of about 1 percent at 200 °C, greatly simplifying the choice of fuels they can use, including natural gas, petroleum products, coal liquids, and coal gases. PAFCs are considered well suited for distributed generation (DG) because electricity can be generated very near where it is used, thereby reducing the amount of energy lost in transmitting electricity as well as reducing the construction of power grid infrastructure. Moreover, PAFCs generate high temperature waste heat that can be used for heating water, space heating, and low pressure steam in a CHP cogeneration system. The overall efficiency of PAFCs thus can rise to more than 85 percent.[17] In comparison to lower temperature fuel cells, such as AFCs and PEMFCs, PAFCs have been regarded as ideally suited for installations in urban areas (near point of use) and as an on-site cogeneration power source.[8] For these reasons, both industry and government came to conclude that PAFCs would be the first fuel cells to enter the commercial market.

2.6.3. Inferior and Expensive

PAFCs, however, possess certain limitations that have stubbornly resisted any solution. PAFCs are less efficient than any other fuel cell design, for example, in generating electricity. The PAFC electric efficiency is about 37 LHV percent (LHV refers to lower heating value), only slightly more efficient than combustion-based power plants, which are typically in the mid-1930s. UTC reports that the electric efficiency of its 200 kW power plants at the beginning of life is 40 percent, but there is a small decrease in efficiency resulting in an average total lifetime efficiency of 37 LHV percent.[1] PAFCs are also less powerful than other fuel cells, given the same weight and volume. As a result, they are typically large and heavy. PAFCs also require platinum loading in both the anode and the cathode, making it expensive. Although fuel processing requirements are less complex than for PEMFCs, PAFCs typically need a water gas shift reactor to achieve good performance, adding to costs. In addition, the highly corrosive nature of phosphoric acid requires the use of corrosion resistant materials in the stack—especially the graphite separator plates—which also are expensive to make.

In the late 1980s, a suggestion was made by A.J. Appleby and F.R. Foulkes that the key to wide commercial use of PAFCs would be to improve the cell energy conversion efficiency and reduce cost.[18] The suggested solution was to operate PAFCs at even higher temperatures and pressures. Increasing temperatures and pressures would increase efficiency, thus leading to system cost reduction. It was also suggested that the operating temperatures of about 240 °C would be high enough to permit internal reforming of methanol, eliminating fuel processing subsystem and simplifying many auxiliary subsystems, further reducing cost.

- The option for improving cell energy conversion efficiency was not pursued for the first commercial unit, the UTC 200 kW system. Higher temperature and pressurization would have required major R&D and investment.[1] The UTC 200 kW system operated at 207 °C, which is a compromise that allowed reasonable performance over an expected lifetime of 40,000 h. The economic trade-off favored operating under normal atmospheric pressure while improving the reliability of system components to achieve improved cell stability and longevity.

Since the early 1990s, UTC has installed more than two hundred and sixty 200 kW cogeneration power plants globally. The system over time has demonstrated an increasingly high level of operational viability in terms of reliability and durability. UTC's 200 kW plant installed in 2000 at St. Agnes Hospital in Bocholt, Germany, for example, has demonstrated reliability as high as 97 percent and its original fuel cell stack has operated for over 52,000 h.[19] The high power quality and reliability of this design have made it the preferred system for so-called premium power applications, such as in banks, hospitals, and computing facilities.[7]

- The UTC 200 kW PAFC stationary power plant, however, remains very expensive vis-à-vis conventional power generation technologies. The 200 kW system perhaps costs about $5,000 per installed kW, while conventional combined power and heat technologies, covering a wide capacity range from 25 kW micro-turbines to 250 MW gas turbines, cost about $500–$1000 per installed kW.[20] Indeed, most of the 200 kW PAFC systems sold were subsidized by federal, state, and local governments, utilities, and other nonprofit organizations.

Future commercial prospects for PAFCs remain uncertain. The initial excitement over PAFCs has subsided and development of PAFCs has slowed in the past 10 years, with the interest of industry and government shifting to PEMFCs and SOFCs that are thought to have better cost potential and efficiency.[1] In mid 2008, UTC placed on the market a PAFC PureCell system that produces 400 kW with an expected lifetime of 10 years or 80,000 h.[21] Orders were being accepted for delivery in 2009.[22] The latest 400 kW system was reported to incorporate less costly materials, including interconnect plates, higher power densities, and optimization of stack and system size.[23] The new model, however, appears to cost about $3,000 per kW,[21] still three times as high as conventional technologies.

2.7. MOLTEN CARBONATE FUEL CELLS

2.7.1. Principles of Operation

MCFCs operate in a manner similar to AFCs and SOFCs. In these fuel cells, negative ions travel through the electrolyte from the cathode to the anode where they combine with hydrogen to generate water, releasing electrons at the anode (Fig. 2.8).

- Unlike all the other fuel cells, carbon dioxide (CO_2) as well as oxygen needs to be supplied to the cathode. The carbon dioxide and oxygen that have entered the fuel cell at the cathode react with the electrons that have traveled from the anode to the cathode and form carbonate ions ($CO_3^{=}$). The carbonate ions then travel through the electrolyte to the anode, where hydrogen and the carbonate ions form water and carbon dioxide, releasing

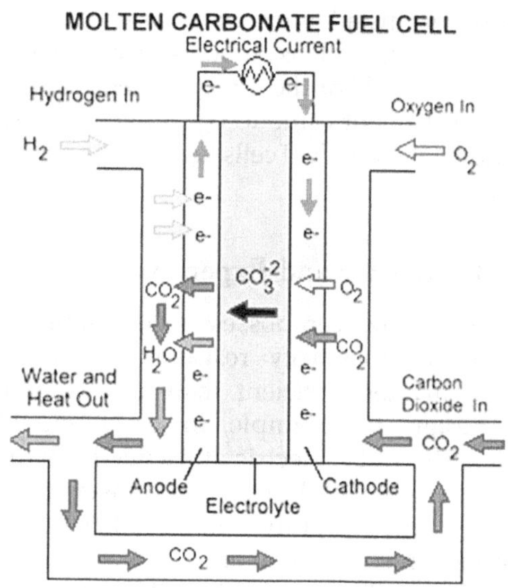

FIGURE 2.8 Molten carbonate fuel cells. *Source: Fuel Cells, Types of Fuel Cells, Energy Efficiency and Renewable Energy, U.S. Department of Energy, http://www1.eere.energy.gov/hydrogenandfuelcells/fuelcells/fc_types.html, Content Last Updated 8 March 2007.* (For the color version of this figure, the reader is referred to the online version of this book.)

electrons. In this reaction, there is a net transfer of CO_2 from the cathode to the anode. The electrons travel through the external circuit to the cathode and repeat the same reactions:

$$O_2 + 2CO_2 + 4e^- \rightarrow 2CO_3^=$$
$$H_2 + CO_3^= \rightarrow H_2O + CO_2 + 2e^-$$

2.7.2. The Presumptive "Second Generation" Commercial Fuel

MCFCs are being developed for large, natural gas or coal-based power plants for industrial, electrical utilities, and military applications. MCFCs are considered to be "second-generation" fuel cells that are expected to compete with PAFCs.[8] Like PAFCs, MCFCs also received major R&D investment and support from national governments. Their support for the development of MCFCs likely derived from their ability to operate—at least indirectly—on coal, a major advantage over PAFCs. MCFCs were likely viewed to be more robust, rugged, and of greater utility than PAFCs. Like PAFCs, MCFCs are available for purchase. About 66 units were sold worldwide as of September 2008.[24]

MCFCs operate at temperatures of about 650 °C to keep the electrolyte, a carbonate salt, in a molten state sufficient to achieve conductivity of the carbonate electrolyte.[1] A benefit associated with the high temperature is that no expensive catalysts are needed; the nickel electrodes provide sufficient activity for the electrochemical oxidation and reduction processes. A particularly strong feature of MCFCs, which distinguishes MCFCs from PAFCs, is that the high operating temperature also provides the opportunity for achieving higher overall system efficiencies and greater flexibility in the use of available fuels compared with PAFCs.[7] The high operating temperature enables internal reforming of various fuels into hydrogen without an external fuel processor, simplifying the BoP and improving electrical efficiency to about 50 LHV percent.[1] In addition, the high operating temperature allows the use of a steam turbine to further boost electric generation efficiency from the high fifties to low sixties, leading to higher power density and lower capital costs than PAFCs.[18] If the waste heat is used for air conditioning, hot water, and heating, the overall efficiency rises to 60–80 percent.[25]

- Moreover, MCFCs are not prone to carbon monoxide or carbon dioxide poisoning; they can even use carbon dioxides as fuel—making them more attractive for fueling with a variety of CO-containing fuels, such as hydrocarbons, syngas derived from biomass or coal, landfill gas, gas derived by industrial or agricultural by-products. Also, because MCFCs are more resistant to impurities than other fuel cell types, scientists believe that they could even be capable of internal reforming of coal, assuming they can be made resistant to impurities such as sulfur and particulates that result from converting coal to gas.[26]

2.7.3. Not Durable Enough and Still Expensive

MCFCs, however, have a number of serious technical challenges. In particular, MCFCs suffer from short service life and high costs. The main challenge for MCFCs stems from the very corrosive and mobile electrolyte, which requires use of nickel and high-grade stainless steel as the cell hardware. These materials are cheaper than graphite but more expensive than ferrite steels.[1] Also, the higher temperatures promote material problems, impacting mechanical stability and stack life. In addition, throughout the lifetime of MCFCs, a slow but steady loss of electrolyte occurs, which tends to cause gradual performance decay. Dissolution of the cathode also is the primary life-limiting constraint of MCFCs. The current state-of-the-art cathode, nickel oxide (NiO), slowly dissolves into the electrolyte, creating metal particles that precipitate across the electrolyte tile, a porous form of lithium aluminate filled with a molten carbonate electrolyte, eventually leading to anode–cathode shorting.[8] The mechanism by which NiO dissolves into the electrolyte is not fully understood.[5] As this occurs, the electrolyte and the cathode experience a major voltage loss,[1] leading to lower overall efficiency. Furthermore, a source of CO_2 is required at the cathode to form the carbonate ion, which requires additional BoP components.

To be commercially successful, MCFCs need to meet longevity requirements of at least 40,000 h as well as cost reduction. In data published in 2007, MCFC systems have demonstrated endurance testing for over 30,000 continuous operating hours.[27] Although current state-of-the-art MCFCs are considered to be functioning properly in full size cells at atmospheric pressure, developers are considering alternate cathode materials and electrolytes.[1] Since the mid-1970s, the materials for the electrodes and electrolyte have remained essentially unchanged. Although MCFCs require no platinum and, thus, have less capital cost than PAFCs, they still need to achieve cost reduction goals by increasing power density, decreasing operating, maintenance, stack replacement costs, and exploring less expensive manufacturing technologies.[28]

2.8. SOLID OXIDE FUEL CELLS

2.8.1. Principles of Operation and Characteristics

SOFCs have electrochemical reactions similar to AFCs and MCFCs (Fig. 2.9).

- In SOFCs, the oxygen that has entered the fuel cell at the cathode reacts with the electrons that have traveled from the anode to the cathode through the external circuit to form oxide ions ($O^=$). Oxide ions travel to the anode, react with hydrogen, and form water, while releasing electrons. The electrons travel through the external circuit to the cathode and repeat the same reaction process:

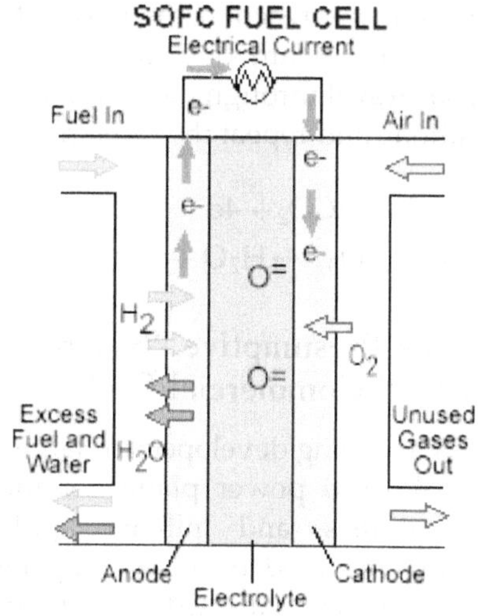

FIGURE 2.9 Solid oxide fuel cells. *Source: Fuel Cells, Types of Fuel Cells, Energy Efficiency and Renewable Energy, U.S. Department of Energy, http://www1.eere.energy.gov/hydrogenandfuelcells/fuelcells/fc_types.html, Content Last Updated 8 March 2007.* (For the color version of this figure, the reader is referred to the online version of this book.)

$$O_2 + 4e^- \rightarrow 2O^=$$
$$H_2 + O^= \rightarrow H_2O + 2e^-$$

- In addition to hydrogen, CO and other hydrocarbons such as methane (CH_4) can be used as fuels. The reactions at the anode involving CO and CH_4 produce H_2 that is easily oxidized at the anode:

$$CO + H_2O \rightarrow H_2 + CO_2$$
$$CH_4 + H_2O \rightarrow 3H_2 + CO$$

SOFCs have a number of highly attractive attributes, which have helped SOFCs enjoy the longest continuous development period of any fuel cell class—dating back to the late 1950s.[1] SOFCs are simpler in concept than all the other fuel cell types. SOFCs are solid-state devices that use a hard ceramic compound as the electrolyte and involve only two phases, gas and solid. SOFCs require no liquid electrolyte, thus eliminating material corrosion or water management problems. Also, SOFCs operate at very high operating temperatures of up to 1000 °C—the highest operating temperature of all fuel cells—enabling highly efficient conversion to power, internal reforming, and high quality by-product heat for cogeneration or for a bottoming cycle. This facilitates high electricity efficiencies ranging from around 40 to 50 LHV percent, with a theoretically expected efficiency of up to 60 LHV percent. Indeed, both simple cycle and hybrid SOFC systems have demonstrated some of the highest efficiencies of any power generation system, combined with minimal air pollutant emissions and low greenhouse gas emissions. High temperature operation also removes the need for a precious metal catalyst. SOFCs are also sulfur resistant and are not poisoned by carbon monoxide, and so can use a wide range of converted (reformed) fuels, including coal-derived gases.

- Even after many decades of development, the optimal SOFC geometric shape and design, support and temperature of operation still has not been determined. There are still many geometry and support options under development in SOFC over a range of temperatures. Some of these include: anode supported planar, electrolyte supported radial, segmented-in-series, tubular (micro-, flattened, and air electrode supported), monolithic, and metallic supported planar and tubular. This is due to both the great flexibility of the technology and the lack of fundamental understanding.

2.8.2. An Early Favorite: High Temperature Tubular Cells

The SOFC electrolyte is solid and can be cast into various geometrical shapes, including tubular, planar, segmented-in-series, or monolithic designs. The early leader among researchers and developers was the tubular design. Tubular SOFCs consist of a bundle of long thin tubes. The Siemens Westinghouse design for a high temperature tubular SOFC is the best-known and most advanced,[1] with demonstrated high reliability and durability. The Siemens Westinghouse state-of-the-art tubular design, which operates at about 1000 °C, has operated continuously for over 69,000 h with less than a 0.5 percent voltage degradation per 1000 h. A 25 kW stack has operated for over 44,000 h with less than a 0.2 percent voltage drop per 1000 h, and a 100 kW stack has operated with negligible degradation for more than 16,000 h in the Netherlands and Germany. Researchers and developers discovered, however, that the high temperature tubular SOFC systems suffered from low power density and had high production costs. The high costs were due to the specialized fabrication processes, and the low power density was a result of the long path for electrical power to travel along a long tubular cell and between cells within the stack, creating some voltage loss. High temperature tubular SOFCs were soon viewed as commercially impractical, and the excitement over them largely has subsided.

2.8.3. Brief Exploration of High Temperature Planar Cells

Researchers and developers next shifted their attention to planar SOFCs, focusing on high temperature planar SOFCs. Planar SOFCs are configured in a stack of circular, square, or rectangular plates. This design could provide superior stack performance and a much higher power density than tubular SOFCs.[7] Planar SOFCs can use bipolar electrical connections between cells, in contrast to the long current path required by tubular cells. Unfortunately, however, the high operating temperature of SOFCs places severe constraints on the materials that can be used, which results in difficult fabrication processes. SOFCs require a close match between materials that have similar thermal expansion coefficients; if there are thermal expansion mismatches among materials, stress at the interfaces between different cell and stack materials tends to cause mechanical failures.[7] Particularly challenging is the brittleness of the planar SOFCs in tension. Thermal cycling is a problem for the planar SOFC. Also, ceramic interconnects used in high temperature SOFCs are rigid and weak and do not have the flexibility to ensure good contact pressure with any of the other components.[1] Another major disadvantage of high temperature planar SOFCs is the difficulty in achieving gas-tight seals around the edges of cell components.[7] In addition, high temperature operation results in slow startup and requires significant thermal shielding to retain heat and protect personnel.

2.8.4. The Current Target: Intermediate Temperature Planar Cells, Many Problems Remain

Because of all the technical problems associated with high temperature planar SOFC designs, researchers and developers are focusing on a planar SOFC that operates at moderately high temperatures (650–800 °C). They view that the lower temperature could help reduce the cost because less expensive construction materials and low cost fabrication methods could be used.[1] Lower temperature SOFCs, for example, allow the use of metal alloys, which are less costly to fabricate than ceramic components. The lower temperature also would allow SOFCs to be used as small stationary power units (down to ~2 kW), auxiliary power units for vehicles, and mobile generators for civilian as well as military applications, and even as primary power plants for vehicles. Researchers and developers, however, are faced with the major challenge of developing low cost component materials that function properly in lower temperature planar SOFCs and that can be mass produced at low cost. Most of the major components are beset with highly challenging technical problems. There is also a serious question as to whether these SOFC cells and stacks could be scaled up to create large capacity megawatt-class power generation systems. The scale-up in cell area is believed to be practically limited by ceramic defects and the scale-up to large plants cannot occur without a large number of repeat parts if the cell area remains small. It is perhaps not unfair to say that the state of development of the intermediate planar SOFC is still at a very basic research level.

2.8.4.1. *Better Electrolytes Needed*

The current state-of-the-art electrolyte material for SOFCs is yttrium-stabilized zirconia (YSZ). YSZ has been the electrolyte of choice for high temperature tubular SOFCs since 1965. If electrolyte-supported designs and/or lower temperatures are to be used then better electrolytes are needed. At lower temperatures, however, YSZ is less conductive and produces less electrical power.[1] To increase conductivity, the electrolyte thickness would have to be reduced to films just several microns thick. Because such a thin film does not have the mechanical strength to support itself, most manufacturers have opted for supporting the electrolyte on relatively thick anodes,[7] which are often referred to as anode-supported cells. One of the problems with this type of cell is that any difference in thermal expansion between the anode material and the electrolyte becomes very significant at the high operating temperatures of SOFCs. In parallel to the effort to improve anode-supported YSZ thin film electrolytes, alternative electrolytes are being considered and are being developed.[1] Scandium-doped zirconia is more conductive than YSZ, permitting a further reduction of the operating temperature by 50–100 °C.

Gadolinium-doped ceria is even more conductive, but has disadvantages. It is partially reduced in hydrogen at temperatures above 600 °C; formation of Ce^{+3} ions generates electron holes that make ceria electronically conductive, thus short-circuiting the cell.

At the same time, efforts are being made to develop solid electrolytes that can conduct protons (H^+).[1] Current SOFC electrolytes are all oxygen transporters that pass oxide ions from the cathode to the anode, producing the reaction products—such as water—all at the anode. The reaction products dilute the fuel that is fed to the anode, thus reducing the cell performance.[vi] The efforts to develop solid proton–conducting electrolytes are still at a very early stage.

2.8.4.2. *Better Anode Needed*

Although a wide range of materials have been considered for anode materials for SOFCs, most developers today use a cermet of nickel and YSZ (Ni-YSZ).[1] Although the Ni-YSZ anode has been the state-of-the-art anode for high temperature tubular SOFCs since 1975, it has drawbacks when used as the anode for planar SOFCs. The thermal expansion coefficient of the Ni-YSZ anode is substantially higher than the electrolyte and cathode. In anode-supported designs, this can lead to mechanical and dimensional stability problems. The anode also can be easily oxidized at elevated temperatures, changing the volume of the anode and severely compromising the anode's structure and strength. In addition, the anode has poor reactivity when directly oxidizing hydrocarbons and a propensity for carbon formation when exposed to hydrocarbons. Even though these drawbacks can typically be mitigated by appropriate system design, many consider that better anodes will be needed.

2.8.4.3. *Alternate Cathode Needed...Perhaps the Greatest Long-Term Need*

The cathode is the greatest source of voltage (overpotential) loss in the SOFCs. Most cathode materials used in SOFCs today are based on doped lanthanum manganites.[1] At high temperatures, strontium-doped $LaMnO_3$ (LSM) is the cathode material of choice for SOFCs operating around 1000 °C. A serious challenge arises, however, when LSM is used as a cathode material for intermediate temperature SOFCs (700–800 °C). Intermediate temperature SOFCs often use metallic interconnects. Many of these metals contain chromium, and chromia vapors can lead to serious poisoning of the cathode, causing a significant loss of efficiency. In addition, as the temperature is lowered below 700 °C, the use of LSM as the cathode material causes a significant potential loss. It is advantageous to use mixed conducting materials that exhibit both ionic and electronic conductivity in cells operating at around 650 °C.[7] Understanding what is occurring at the cathode to create the high voltage loss is a major focus for future R&D.

2.8.4.4. *Lack of Interconnect Materials*

There are generally two types of interconnect materials for SOFCs: ceramic materials for operation at high temperature (900–1,000 °C) and metallic alloys for lower temperature operation.[1] No metal interconnect material for operating in the intermediate temperature range (650–800 °C) has yet been developed.

Ceramic materials. The ceramic interconnects exhibit high electronic conductivity and resist reduction under exposure to syngas at high temperatures. However, the problem with ceramic interconnects is that they are rigid and weak; there is no flexibility in any of the components to ensure good contact pressure. In some

[vi] The YSZ electrolyte functions as a conductor of oxide ions. Oxygen atoms are reduced into oxide ions on the porous cathode surface by electrons, and then flow through the ceramic electrolyte to the fuel rich porous anode where the oxide ions react with fuel (hydrogen), giving up electrons.

cases, a contact felt or conductive contact paste is used. However, the reliability of this component is not adequate.

Metal alloys. Lower operating temperatures open the possibility of using metallic interconnects, particularly the use of ferrite steels. Ferrite steels could reduce the materials cost; they are typically easier to process with low cost processing techniques. Developers now view that commercially available ferrite stainless steels, such as the Krupp formulation Crofer22 APU and AK Steel 441, may be viable candidates for interconnect applications.[1,29] These alloys, however, still require a number of improvements before they can function as commercially viable products in intermediate temperature SOFCs:

- Further improvement in contact resistance through contact paste development
- Further improvements in corrosion resistance
- Improved long-term (40,000 h) performance and mechanical stability of the currently used coatings
- Low cost manufacturing methods for materials, shapes, and coatings
- Improved creep strength to increase design flexibility for cells

2.8.4.5. Seal Materials Needed

A key issue with planar SOFCs is the method of sealing the ceramic components to obtain gas-tight seal. Some tubular and monolithic designs require no seals at all, but planar designs typically require multiple seals per unit.[1] Seal materials must be chemically and physically stable at operating conditions. In some applications (e.g., in on-road vehicles), the seal must also be able to withstand acceleration forces associated with road vibration and shock. Seal materials also must be low in cost and amenable to low-cost stack manufacturing methods. Seal materials in most cases are required to have the thermal expansion coefficient closely matched to all other components. To date, no publicly known satisfactory seal materials have been developed for intermediate temperature SOFCs. The technical difficulties associated with seal materials for planar SOFCs are so daunting that the benefit of a simple sealing arrangement in tubular SOFCs has started to appear more palatable despite the disadvantages of low power density and long current path.

Most planar SOFC stack designs share common requirements, material choices, and general sealing concepts. There are two different types of seals being developed for SOFCs: bonded and compressive seals.

2.8.4.5.1. BONDED SEALS

Bonded seals can be compliant or rigid. A gas-tight seal is achieved through adhesive forces between the seal material and both surfaces against which the seal is to work.[1] The seal material must have good adhesive properties, including viscous and wetting behavior. Some are designed to remain flexible over the operating range of the cell, while others are meant to be rigid.

Compliant bonded seals. If the seal is compliant, the thermal expansion coefficient matching requirements are somewhat relaxed. The bonding temperature for this type of seal should lie between the operating temperature and the stability limit for the other cell materials. There are several common subtypes of bonded seals currently under consideration for SOFC applications, but no compliant bonded seals have been identified or developed to date.

Rigid bonded seals. To use a rigid type of seal, the thermal expansion coefficient of the seal material and all other components must be closely matched. Common rigid bonded seals include glasses and glass ceramics. The composition of these materials can be tailored so that they could offer transition temperatures close to the operating temperature of the SOFC cells.[7] These materials soften as the cells are heated up and form a seal all around the cell. They are inexpensive and easy to manufacture.[1] However, these materials are brittle, leading to seal and even cell failures during cool-down. They interact with adjacent cell components, especially with the interconnects. Some of the constituents of glass, such as silica, borate, and alkali metals, migrate during operation and cause degradation in cell performance.

2.8.4.5.2. COMPRESSIVE SEALS

A compressive seal is achieved by pressing the seal material between the surfaces to be sealed. The seal material must be elastic over the operating temperature range, and sufficiently soft to fill the micro-roughness on the surfaces to be sealed.[1] There are difficulties in developing compressive seals because few materials and structures are compliant and provide a compressive seal at the SOFC operating temperatures. Recently, mica and hybrid mica seals have been developed as a viable technology. Mica seals have the ability to withstand thermal cycling, but the long-term physical and chemical stability of the seal requires additional improvement.

2.8.4.6. Stack Scale-Up is Daunting

Last but not the least is the serious question as to whether lower temperature planar SOFCs are able to scale up to a larger power generation system, ranging from several hundred kilowatts to several hundred megawatts. Currently, anode-supported planar cells for full-scale stacks are squares or rectangles about 10–20 cm, and stacks between 30 and 80 cells are the state of the art.[1] The capacities of these stacks range from 2 to 12 kW. One approach to further scale up would be to combine small stacks in a modular fashion into a larger system. Recent studies have implied, however, that this would lead to rather complex manifolding arrangements for a very large numbers of cells. For example, a 1 MW system would require at least two hundred 5 kW stacks. The complexity, cost, and pressure loss problems associated with such massive modularization undoubtedly are daunting. A number of significant issues may impede or delay progress in scaling up cells and individual single stacks:

- The larger the cells, the more severe the effects of thermal expansion mismatches among materials, creating thermal stress at the interfaces between the different cell and stack materials, causing mechanical degradation.
- Increasing the number of cells has its limits because of mechanical stability concerns.
- As the number of cells increases, minor imperfections in cell geometry will lead to uneven distribution of the contact or sealing pressure inside the stacks.
- Distributing the gas flow evenly to all cell levels will become difficult.
- Difficulties in successfully developing gas-tight seals around the edges of the cell components to prevent gas leakage or cross-leakage will delay scaling up planar design cells.

2.8.5. Are Alternative Cell Designs Feasible?

A number of alternative designs have been built and demonstrated since early 1993, aimed to reproduce the advantages of the tubular or planar SOFCs without their disadvantages. Many geometries and support types are under development. The alternative cells are of thinner, tubular form, but are either round, flattened, or delta shape. In 2000, the US Company Accumentrics built a 1000 cell stack with small, round, thin, microtubular cells. Since then, the company has built several 2–5 kW systems for computer power backup.[30] Microtubular designs are easier to supply with fuel via manifolds without leaks or contamination. They operate at 800 °C and start up rapidly. Cell interconnection and assembly issues involving microtubular cells, however, are significant, and it seems likely that microtubular systems will mainly be applicable in small systems.

- Various alternative cell designs are under development to avoid the difficulties of sealing, mechanical weakness, and thermal cyclability. All these are still in the early developmental stage.

References

1. *Fuel cell handbook.* 7th ed. US: NETL, the US Department of Energy; November 2004.
2. *Fuel cells, a handbook.* US: Office of Fossil Energy, US DOE; May 1988.
3. Jonathan Butler, 2007 Patent Review Q4, Fuel Cell Today, January 2008.
4. *NEDO fuel cell, hydrogen technology development road map 2008.* Japan: NEDO; June 2008.
5. *Fuel cells, hydrogen industrial technology trend study.* Japan: NEDO; March 2007.
6. Fact #389: September 12, 2005, More Gasoline Stations in 2005, Facts of the Week, EERE, DOE, Content Last Updated: 15 November 2005.
7. Larminie James, Dicks Andrew. *Fuel cell systems explained.* 2nd ed. West Sussex, England: Wiley; 2003.
8. Leo J, Blomen MJ, Mugerwa Michael N. *Fuel cell systems.* US: Springer; 1993.
9. Lu GQ, Liu FQ, Wang Chao-Yang. *Water transport through Nafion 112 membrane in DMFCs.* In: *Electrochemical and solid-state letters*, vol. 8(1). US: The Electrochemical Society; 10 November 2004.
10. Direct methanol fuel cell, Wikipedia, last modified on 19 September 2008.
11. Nor Jiri K. *Alkaline fuel cell-technology for the 21st century,* Canada: Astris Energi, Inc.; date unknown, www.astris.ca/IN/IN-pdf/P9A-AFC.pdf; [accessed in August 2007].
12. Cifrain M, Kordesch K. Hydrogen/oxygen (air) fuel cells with alkaline electrolytes. In: Vielstich Wolf, Lamm Arnold, Gasteiger Hubert A, editors. *Handbook of fuel cells – fundamentals, technology and applications,* vol. 1. Chichester: John Wiley & Sons, Ltd; 2003. Part 4, p. 267–280. (ISBN: 0-471-49926-9).
13. *Debunking the CO_2 myth.* UK: AFC Energy Plc.; 2007.
14. *Types of fuel cells, fuel cells, energy efficiency and renewable energy,* Department of Energy, [last updated on 8 March 2007].
15. Guelzow Erich, Nor Jiri K, Nor Peter K, Schulze Mathias. *A renaissance for alkaline fuel cells.* In: *The Fuel Cell Review,* vol 3(1) February/March 2006.
16. Cifrain M, Kordesch K, Gsellmann J, Hejze T, Aronson RR, Alkaline fuel cells with circulating electrolytes, *Poster presented at Fuel Cell Seminar 2000,* Portland, Oregon. October 30–November 2, 2000.
17. Isom J, Paul D. The PureCell fuel cell powerplant—a superior cogeneration solution from UTC power, *2006 Fuel Cell Seminar, Honolulu, Hawaii.* 13–17 November 2006.
18. Appleby AJ, Foulkes FR. *Fuel cell handbook.* New York: Van Nostrant Reinhold; 1989.
19. *"Fuel cell at German hospital tops 10 million kilowatt-hours of operation.* A Press Release, UTC Power. 4 March 2008.
20. *Review of combined heat and power technologies.* US: Office of Industrial Technologies, Office of Energy Efficiency and Renewable Energy, US Department of Energy; October 1999.
21. Samuelsen Scott. *Stationary fuel cell technology.* US: National Fuel Cell Research Center; 9 July 2008.
22. *PureCell model 400 system—overview,* UTC Power, http://www.utcpower.com/fs/com/bin/fs_com_Page/0,5672,0249,00.html; [accessed August 2008].
23. Carlson Eric, Zogg Robert, Sriramulu Suresh, Roth Kurt, Brodrick James. *Using phosphoric-acid fuel cells for distributed generation.* In: *ASHRAE Journal*, vol. 49. US: American Society of Heating, Refrigerating and Air-Conditioning Engineers, Inc.; January 2007.
24. *Delivery results of molten carbonate fuel cells,* Fuel Cells Japan (Nihon Nenryo Denchi Kabushiki Kaisha), http://www.fuelcell-japan.com/results/jisseki.php, [accessed September 2008].
25. *Merits of molten carbonate fuel cells,* Fuel Cells Japan (Nihon Nenryo Denchi Kabushiki Kaisha), http://www.fuelcell-japan.com/fuelcell/merit.html; [accessed September 2008].
26. *Types of fuel cells,* Hydrogen, Fuel Cells & Infrastructure Technologies Program, EERE, DOE, http://www1.eere.energy.gov/hydrogenandfuelcells/fuelcells/fc_types. html#phosphoric; [accessed August 2008].
27. Rolf S. The HotModule system, advancement in stationary fuel cells, A presentation at 2007 Fuel Cell Seminar & Exposition, San Antonio, Texas. 15–19 October 2007.
28. Moreno Angelo, McPhail Stephen, Bove Roberto. *International status of molten carbonate fuel cell (MCFC) technology.* France: IEA, http://www.energyagency.at/publ/pdf/fuelcell_mcfc.pdf; January 2008.
29. Rakowski JM. *Solid Oxide Fuel Cell Metallic Interconnect Systems.* Pittsburgh PA: 9th Annual SECA Workshop; 5–7 August 2008.
30. Singhal Subhash C, Kendall Kevin. *High temperature solid oxide fuel cells fundamentals, design and applications.* The Netherlands: Elsevier; 2003.

CHAPTER

3

History of Alkaline Fuel Cells

OUTLINE

Fuel Cells
http://dx.doi.org/10.1016/B978-0-444-56325-5.00003-X

3.1. OVERVIEW

The first alkaline fuel cell (AFC) was built in 1939 by British engineer Francis T. Bacon, 100 years after Grove invented the fuel cell. Bacon's research was interrupted by World War II, and it was not until 1959 that he finally built a 6 kW AFC system and demonstrated it as a power source for a forklift truck, welding machine, and a circular saw. Although Bacon did not succeed in commercialization of his invention, the US company Pratt & Whitney [part of United Technologies Corporation (UTC)], which took a license on Bacon's patents, developed an AFC to supply the electricity and water for life support, guidance, and communications for the Apollo Lunar Landing Mission.

The technological advantages of AFC technology were acknowledged around the world and a number of companies undertook AFC development in the 1950s and 1960s. Companies in the United States included UTC, Union Carbide Corporation (UCC), and Allis Chalmers. Companies in Europe included Elenco of Belgium and Siemens and Varta of Germany. Companies in Japan included Fuji Electric and Hitachi. Companies in Russia included the Kiev Research and Production Association "KVANT," S.P. Korolev Rocket and Space Corporation RSC "ENERGIA," the Ural Electrochemical Integrated Plant (UEIP), and ZAO Independent Power Technologies (IPT).

By the early 1970s, however, the companies came to recognize that AFC's intolerance of carbon dioxide—one of the major disadvantage of AFCs—and the difficulties and high cost of overcoming the problem were an insurmountable barrier for commercialization in terrestrial applications. Many left the AFC field or switched to different fuel cells, such as phosphoric acid fuel cells (PAFCs).

The oil crisis in the 1970s triggered increased public funding for fuel cell development, but most funding was channeled into the higher temperature fuel cell technology—particularly PAFCs. AFC technology was left largely on the sidelines and has been on the verge of extinction for the past 30 years.[1] There have been reports that some developers are pursuing AFC technology for transport applications using circulating electrolytes and an external, commercial type soda-lime CO_2 absorber that promises to resolve the problem of CO_2 in ambient air.[2] It is not certain if AFC researchers and developers are able to obtain sufficient resources to continue R&D so that they can commercialize the technology.

3.2. FRANCIS T. BACON BUILDS THE FIRST ALKALINE FUEL CELL

A graduate of Eton College and of Trinity College, Cambridge, Bacon became intrigued by fuel cells while reading an article about electrolysis in 1932.[3] In 1939, he built an AFC with electrolyte containing 27 percent potassium hydroxide (KOH), nickel gauze electrodes, and high-pressure equipment and seals. It operated at 3,000 psi for about 48 min. Bacon named the device the "Bacon Cell."

- It is not clear as to how Bacon decided on alkali electrolytes instead of acid electrolytes known since Grove's time. Perhaps, Bacon recognized that alkali performs as well as acid, is not as corrosive to electrodes as acid, and requires no precious metals, such as platinum.

Bacon suspended his AFC work during World War II and did not resume it until he went to Cambridge University in 1946. His research finally resulted in improved AFCs and patents in 1950 and 1955.[4] Bacon sought practical applications for his patents and built the first operational multi-kilowatt fuel cell system in 1959.[3] The AFC system had 40 cells, each of which was 10 inches in diameter, and produced

6 kW. The system was used to power a forklift, welding equipment, and a circular saw, and a demonstration was given in August 1959.

- Also in 1959, UTC obtained licenses for Bacon's patents and developed an AFC system that successfully supported the National Aeronautics and Space Administration (NASA) space mission (1963–1972).
- Even with the lengthy research work and the spectacular success of the AFCs in space, Bacon never achieved any personal commercial success from his inventions. The UK recognized Bacon's achievement, however, by awarding him Officer of the Order of the British Empire (1967). He was also elected Fellow of the Royal Society (1973), and awarded the first Grove Medal (1991).

3.3. AFC DEVELOPMENT IN THE UNITED STATES

UTC's successful deployment of AFC technology in space undoubtedly was a breathtaking event and must have encouraged many companies in the United States to get started or step up their R&D effort on AFCs. A number of major US companies produced a series of innovative prototypes powered by AFCs. In addition to UTC's efforts to develop commercial terrestrial applications, UCC and Allis Chalmers were notable AFC developers.

However, none of their R&D efforts in terrestrial applications achieved commercial success. UTC abandoned its efforts in the 1970s.[2] Both UCC and Allis Chalmers left the fuel cell work in the early 1970s. Their products became one of a kind prototypes with no commercialization.

3.3.1. United Technologies Corporation Achieves Spectacular Success with AFCs in Space

UTC was the first company in the world that succeeded in using fuel cell technology in practical applications.[i] UTC improved Bacon's AFC design by reducing the weight and making a longer lasting cell.[3] UTC dropped Bacon's design for circulating electrolyte. Instead, it adopted an immobilized electrolyte design, in which the KOH electrolyte was retained in a matrix (usually asbestos) and, thus, immobilized so that it was not susceptible to orientation.[2] In 1962, UTC won a NASA contract to power the Apollo manned spacecraft with AFCs, which successfully provided in-flight power, heat, and clean drinking water for the Apollo moon landing mission.[4]

- In the late 1950s, NASA was looking for a way to power a series of upcoming manned space flights. The first manned space flight program of NASA was the Mercury Program (1958–1963), and the capsule was battery powered.[5] Starting with the second manned program, the Gemini Program (1962–1966), NASA tested fuel cells because the flights became longer and could not be supported by batteries. The key advantage of fuel cells over batteries was evident from this first application. Gemini was designed to test man

[i] UTC is a major US technology corporation, encompassing business sectors such as aerospace, aviation, and fuel cells. Initially, Pratt and Whitney engaged in fuel cell business. The fuel cell business was taken over by a new UTC subsidiary International Fuel Cells (IFC) and its subsidiary ONSI. In this chapter, the name UTC is used, wherever possible, for the sake of simplicity. IFC was a joint venture between UTC and Toshiba of Japan (a minority stakeholder at 11.5 percent) established in 1985. IFC and Toshiba in 1990 also formed an IFC subsidiary, ONSI, which focused on fuel cell power plant production, on-site construction, and maintenance services. In 2001, IFC was renamed as UTC Fuel Cells, now part of UTC Power. The joint venture agreement between UTC and Toshiba was dissolved in 2004. Toshiba subsequently left the PAFC business.

and equipment in space flight for up to two weeks in preparation for the Apollo Lunar Landing Mission. The massive infusion of fuel cell development funding for the Gemini and Apollo missions reached a maximum of more than $20 million (in 1990 dollars) during the mid-1960s.[6]

- NASA awarded a number of research contracts to test and develop a practical working fuel cell design on the Gemini spacecraft, including AFCs and proton exchange membrane fuel cells (PEMFCs).[7] The PEMFC model cell developed by GE encountered repeated technical difficulties, and Gemini 1 through 4 flew with batteries instead.[8] GE redesigned their PEM cell, and the new model, despite malfunctions and poor performance on Gemini 5, served adequately for the remaining Gemini flights. Project Apollo mission planners, however, chose UTC's AFCs for both the command and lunar modules. AFCs demonstrated an electrical efficiency approaching 70 percent.[4]

FIGURE 3.1 Apollo fuel cell systems. *Source: "Space and Defense," http://www.utcpower.com/pressroom/downloads/space-defense, UTC Power, No Date, accessed March 2012. (Courtesy of UTC Power).* (For the color version of this figure, the reader is referred to the online version of this book.)

UTC began supplying the AFC power plant for the Apollo Command and Service Module in March 1962.[3] Three AFC stacks were used to supply the electricity for life support, guidance, and communications for the module, as well as water for the crew throughout the 2-week mission to the moon. The average power required during the mission was approximately 600 W, and each fuel cell was designed to generate 463–1,420 W for a period of at least 400 h; with this redundancy, one fuel cell module would be able to supply the electrical power to the spacecraft if two of the fuel cell modules failed (Fig. 3.1).

Since the early 1960s, UTC has continued to supply AFCs for most US space missions. By the mid-1970s, UTC had delivered 92 AFC systems, and 54 had been used in missions, including nine Apollo missions to the moon, the three Skylab space station missions (1973–1979), and the Apollo–Soyuz rendezvous and docking mission in 1975.[4] Also, UTC was the supplier of AFC systems for the Space Shuttle Program which ended in August 2011. The AFC developmental program for the shuttle mission started in January 1975, and the first development tests began in October 1975. The power plant was qualified for manned space flight in June 1979 and used in the first orbital flight in April 1981.[9] The AFC systems on the space shuttle were based on the Apollo AFC system technology and consisted of 32 cells. Each shuttle was equipped with three 12 kW stacks.[4] The AFC systems for the space shuttle, however, were substantively improved over those of the Apollo program.[2] They operated in the same pressure range as the Apollo cells, but

FIGURE 3.2 Alkaline fuel cell power plant for shuttle orbiter provided by UTC. *Source: "Space and Defense," http://www.utcpower.com/pressroom/downloads/space-defense, UTC Power, No Date, Accessed March 2012. (Courtesy of UTC Power).* (For the color version of this figure, the reader is referred to the online version of this book.)

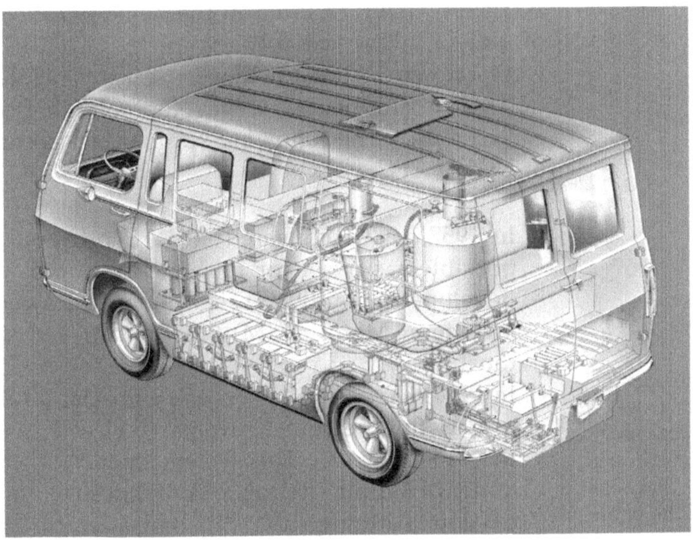

FIGURE 3.3 General Motors electrovan illustration, the first fuel cell car powered by Union Carbide alkaline fuel cells and demonstrated in 1966. *Source: "1966 The Worlds First Fuel Cell Vehicle," http://media.gm.com/media/us/en/gm/photos.html, No Date, Accessed March 2012© General Motors.* (For the color version of this figure, the reader is referred to the online version of this book.)

at lower temperatures of 85–95 °C and had a higher current density.

While enjoying its success with AFC space missions in the early 1970s, UTC investigated adapting its AFC technology to terrestrial applications, using air as an oxidant.[2] Their effort, however, was hampered by the high cost of removing carbon dioxide from the gas streams. For space missions, cost was not a major constraining factor, but it was a primary concern on terrestrial applications. Moreover, alkaline cell life—about 2600 h in space—was suitable for space missions, but too brief for terrestrial applications. As a result, UTC abandoned the idea of developing AFCs for terrestrial application.

- Now that the Space Shuttle Program has ended, UTC is likely no longer interested in AFC development.

FIGURE 3.4 Daihatsu FC show Case. *Source: Vincent Rice, "Daihatsu Kei Concepts Bet on Hydrazine as Future Fuel," Automotive, http://www.gizmag.com/daihatsu-fuel-cell-showcase-concept-car/20842/, Gizmag, April 1, 2012. (Courtesy of Daihatsu Motor).* (For the color version of this figure, the reader is referred to the online version of this book.)

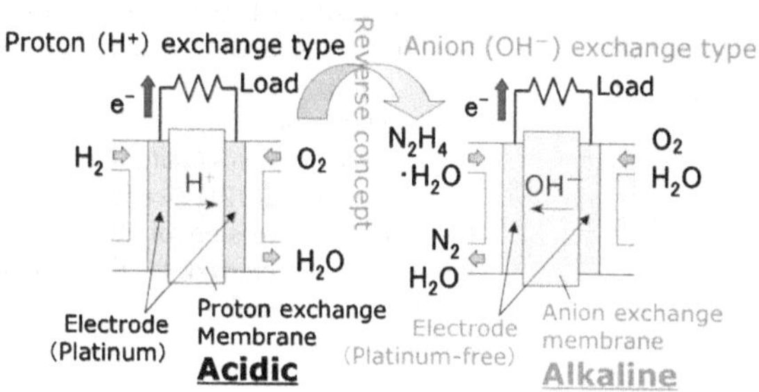

FIGURE 3.5 Daihatsu alkaline anion membrane fuel cell diagram. *Source: Vincent Rice, "Daihatsu Kei Concepts Bet on Hydrazine as Future Fuel," Automotive, http://www.gizmag.com/daihatsu-fuel-cell-showcase-concept-car/20842/, Gizmag, April 1, 2012. (Courtesy of Daihatsu Motor).* (For the color version of this figure, the reader is referred to the online version of this book.)

3.3.2. Union Carbide Corporation: Vigorous Efforts but No Successes

UCC started AFC development for terrestrial mobile applications in the late 1950s and formed the basis for circulating liquid electrolytes.[ii] Karl Kordesch and his colleagues at UCC designed AFCs with carbon gas-diffusion electrodes and demonstrated their fuel cell for military applications. The first was a fuel cell-powered mobile radar set demonstrated in 1957.[10] The new mobile radar, dubbed "Silent Sentry," was quiet, portable, and able to spot a single enemy a half mile away in total darkness, according to UCC. The US Army Signal Corps reportedly ordered the 35-pound radar sets for procurement in 1960,[11] although it is not known whether the radar sets were deployed or not. UCC also drew up plans for an undersea base that would run on fuel cells for the US Navy.[4] In 1965, Kordesch developed a thin composite carbon–metal electrode and made a number of transportation applications based on the new electrode. All of these UCC AFCs, however, resulted in the demonstration of one-of-a-kind prototypes, none of which were known to have achieved series production. UCC stopped AFC development in the early 1970s.[12] The following summarizes some of UCC's AFC work[13]:

- In 1957, UCC pioneered an AFC-powered portable radar set for the US Army and developed a 150 kW AFC system to power a General Motors (GM) "Electrovan," in the 1960s.
- In 1966, UCC supplied a 150 kW AFC unit to GM for the first hydrogen fuel cell car, the "Electrovan." The system operated on tanks of liquid hydrogen and liquid oxygen. Although the van had a driving range of 120 miles—which was respectable for 1966—and a top speed of 70 miles per hour, the system's lifetime was poor, lasting only 1000 h after the first start-up.[4] Also, the liquid hydrogen was difficult to handle. Because of safety concerns, the van was only used on the company property, where it had several mishaps along the way.[14] GM stopped working on fuel cell technology.[4]

[ii] Fluid and mobile are the circulating electrolyte and this electrolyte can circulate freely inside the electrolyte chamber.

- In 1966, Kordesch developed a motorbike powered by AFCs and traveled over 300 miles. He drove it at 25 miles an hour for 200 miles on a gallon of hydrazine.[15]
- In 1970, Kordesh built a fuel cell hybrid car powered by an AFC and a lead–acid battery which were installed in the trunk of his Austin A-40, four-door, four-passenger sedan.[13] The car had a driving range of 180 miles. Kordesch drove this car on city streets and highways for 3 years. The curb weight of the car was about 1 ton. The fuel cell extracted oxygen from the air and hydrogen from gaseous hydrogen in tanks mounted on the roof. Besides the fuel cell unit, all of the necessary accessories for automatic start-up and shut-down were installed in the rear compartment. For simplicity, the original four-gear shift transmission with clutch was retained and proved beneficial for up-hill driving.[4]

Kordesch was tireless in his efforts to commercialize AFCs and he was awarded the Fuel Cell Seminar Award in recognition for his contribution.

3.3.3. Allis Chalmers: Sharing the Same Fate as UCC

Allis Chalmers of Milwaukee, Wisconsin, also conducted major experiments on AFCs and demonstrated their fuel cells in a variety of applications. Harry Karl Ihrig of Allis Chalmers in 1959 built the first large vehicle equipped with AFCs.[4] It was a farm tractor powered by a 15 kW stack consisting of 1008 cells using compressed hydrogen and oxygen as the reactants. The tractor generated enough power to pull a weight of about 1.5 tons and was demonstrated across the United States at state fairs. It is not certain as to when Allis Chalmers started its AFC development. Subsequent to the tractor demonstration, Allis Chalmers maintained an AFC research program for some years in cooperation with the US Air Force. It developed larger, more efficient fuel cells, and built a fuel cell-powered golf cart, a submersible, and a forklift. For example, Allis Chalmers produced a 750 W fuel cell to power a one-person underwater research vessel in 1964.[16] Since the government stopped its funding, the company discontinued its AFC work in 1970.[17]

Although Allis Chalmers publicly stated it was satisfied with the efficiency and lack of pollution, only one fuel cell tractor was ever built. It is now at the Smithsonian Institution.[17]

3.4. AFC DEVELOPMENT IN EUROPE: DECADES OF WORK WITH NO SIGNIFICANT CONSEQUENCE….BUT SOME FIELD TESTS CONTINUE

A great number of European companies engaged in AFC work compared to the United States. They started AFC work in the 1950s and 1960s. Also, viewing the US success of AFCs in space, Europe envisioned the same success would be recreated in Europe. From 1984 to 1992, the European Space Agency (ESA) funded development of the AFC power source for the manned space shuttle program Hermes (1987–1993), but abandoned the program in 1993.[9,12] No companies succeeded in developing commercially viable AFCs, and all have left the AFC field, mostly by the early 1970s.

- European AFC R&D efforts were greater than for any other type of fuel cell. It is likely that the Europeans became disillusioned with fuel cell technology, however, because of the technical challenges they faced in developing applications for AFCs.

3.4.1. AFC Development in Germany

In Germany, the greatest efforts on AFCs have been carried out by Siemens and Varta.

Both Siemens and Varta started AFC development in the 1960s. Both also worked on AFCs fueled initially by hydrazine and methanol, but later came to focus on AFC work with hydrogen and oxygen. Varta abandoned AFC development in the early 1970s and Siemens left the field in the early 1990s.

3.4.1.1. *Siemens*

Siemens of Germany dedicated considerable effort to developing a fuel cell system in the 1960s and 1970s. In the late 1960s, Siemens constructed AFC-powered converter stations for television transmitters in the power range of 25–100 W.[18] During the 1960s and early 1970s, collaboration efforts on television transmitters were undertaken with Varta.[12] In the 1970s, Siemens chose to focus on AFCs using pure hydrogen and pure oxygen and developed several 6–7 kW AFC units in the second half of 1970s. The unit had a current density of 420 mA/cm^2 at a cell potential of 0.77 V. The lifetime was about 3000 h, with a performance degradation of 5 percent over 1000 h. Generators of various sizes were built based on the unit for German military (underwater) applications. By the end of the 1970s, three of the units were incorporated into a 20 kW system, designed as an emergency unit. Subsequently, a 48 kW system for an air-independent propulsion system for a diesel-powered non-nuclear submarine was built, consisting of eight units. The system used cryogenic hydrogen storage and gaseous hydrogen storage in metal hydrides. Sea trials got underway in 1988,[19] but Siemens' AFC work for submarines ended in the early 1990s.[20] Perhaps the AFC system failed to meet mission requirements.

- Siemens also cooperated with the ESA. From 1984 to 1992, the ESA funded the development of the fuel cell power source for the manned Hermes Program (1987–1993).[9,12] Following a technology selection performed by prime contractor Dornier, a German company, the low temperature, immobile electrolyte AFC technology was selected. Siemens and Elenco of Belgium each built an AFC system for the program.[4] The Hermes Program was canceled in 1993.

3.4.1.2. *Varta*

German battery manufacturer Varta started to work on AFCs in the 1960s. Varta developed a special AFC system in the early 1960s, called the Eloflux System.[4] The system had its electrolyte pressed perpendicular to the electrodes and the electrodes themselves had a complex double structure. During the 1960s and the early 1970s, collaboration efforts with Siemens were undertaken.[12] In the late 1960s, Varta, like Siemens, found applications of their 24 and 100 W AFC stacks in television repeater stations with the hydrogen and oxygen supplied from pressure cylinders. They remained operational for several years without any defects.[18] To remove waste water, the circulating electrolyte had to be replaced every 3 months, depending on the volume of the electrolyte storage vessel.[12] Varta also constructed a 3.5 kW unit, which was tested in a forklift truck. In 1973, the company decided to end its AFC work.

3.4.2. AFC Development in France

Research on AFCs started in France in the early 1960s, mainly conducted by major players in the energy sector, including Alsthom, CGE, and IFP. Alsthom developed an inexpensive AFC technology for transport and stationary applications starting in the mid-1960s. CGE and IFP developed various AFCs for military and transport applications in the late 1960s and early 1970s.[6] All the three companies experienced repeated technical failures, progressively slowed down their activities in this field, and finally abandoned the AFC field in the late 1970s and 1980s.[21]

3.4.2.1. *Alsthom*

Alsthom developed an inexpensive AFC power generation system about 1967.[6] Alsthom AFC cells had identical anodes and cathodes, low loadings of precious metal catalysts in the electrodes, and operated at atmospheric pressure conditions.[12] The fuel and oxidant were hydrazine and hydrogen peroxide, respectively.[6] Alsthom signed an agreement with Exxon to jointly develop AFC technology in 1970, which resulted in the development of a novel carbon dioxide removal technique in the 1970s. During the early 1980s, Alsthom and Oxy (Occidental Petroleum) agreed to install an AFC system in Oxy chlorine plants, using their by-product hydrogen. Alsthom's AFC system was not evaluated favorably by a review study, however. Despite the low cost, it was viewed less attractive than PAFCs, since the low operating temperature of the system would not yield enough heat to create steam for plants. The lifetime of the systems, which was expected to be 8000–10,000 h, also was viewed as insufficient for that application.[12]

3.4.2.2. *CGE*

At CGE, electrodes for low temperature AFCs were developed, using nickel and silver catalysts for the anodes and cathodes, respectively. The development led to the construction of 2 kW AFC stacks. CGE's fuel cell efforts, however, shifted toward advanced primary and rechargeable metal–air batteries.[12]

3.4.2.3. *IFP*

IFP developed AFCs, specifically, for transport applications. IFP developed nickel and silver catalysts and used hydrogen and air at atmospheric pressure.[12] The company used low catalyst loadings to achieve an acceptable cost per kilowatt. IFP performance targets in terms of lifetime were relatively modest, given the transport application, at about 3000–4000 operating hours. It appeared, however, that the company was not able to come near its targets, especially in terms of stability. IFP discontinued the AFC work in the beginning of 1980s.

3.4.3. AFC Development in Belgium: Elenco

Elenco represents a major AFC effort in Europe. In the early 1960s, a consortium of Belgian companies started AFC work. This work was taken over in 1969 by SCK/CEN, the Belgian Nuclear Research Center, which concentrated on the development of a cost-effective carbon electrode design.[12] On the basis of this work, a joint program with industry was organized, leading to the formation of the Elenco company in 1976, which is a Belgian–Dutch collaboration, consisting of the Belgium Atomic Energy Company, the Belgian company Bekaert, and the Dutch State Mines.[9] Elenco aimed to develop AFCs for transport applications. It believed that its circulating electrolyte AFC (which required periodically changing a carbon dioxide filter as well as the alkaline electrolyte) was superior to PAFCs for transport applications because of its higher efficiency, higher output, and lower susceptibility to corrosion. Elenco received about $12 million (in 1989 dollars) from the Belgian and Dutch governments during the early 1970s and expected to achieve a commercially viable application for hydrogen-fueled vehicles by about 1990.[6] Elenco completed basic development work in 1983 and investigated systems in the 1.5, 15, and 50 kW range.[9] The 1.5 kW system was intended to be a substitute for small motor generators; six 1.5 kW prototype systems were built in 1987 and 1988 for evaluation by several users. The 15 kW system was designed to power electric vehicles and was successfully tested in an electric van.[12] Modules lasted for at least 5000 h of operation with reproducible results.[22] The 50 kW system was used as an electric power source in a trailer for the Belgian Geological Service.[9] A new

Belgian project to build a prototype bus for Amsterdam started in January 1988.[22] Plans were also made for a similar project in Brussels, using a Belgian-made bus. All these AFC systems used circulating electrolyte, water removal by the outgoing air stream, and operated at an electrolyte temperature of about 70 °C.[12] Elenco also developed mass production manufacturing techniques to commercialize the product[12] and built an automated pilot plant with an annual capacity of 250,000 electrodes in 1989.[9,12]

Elenco played a major part in European Community projects. Elenco worked as a subcontractor until the ESA canceled the Hermes Program in 1993.[4] In 1994, Elenco participated in the EUREKA Bus Project and supplied the fuel cell system. According to the design parameters of the 80 passenger demonstration bus, power was delivered by a fuel cell battery hybrid, whereby the AFC delivered a power output of 80 kW while operating with a parallel battery. The nominal power of the hybrid system was estimated to be 180 kW. In 1994, the EUREKA Bus Project ended, and Elenco declared bankruptcy and was closed after more than 30 years of AFC work.

- **Zevco and ZeTek.** Elenco's assets and patents were purchased and restarted as Zevco and ZeTek in the United Kingdom in May 1994.[23] These companies demonstrated an AFC-powered Millennium London taxi that operated with five AFC systems in 1998.[20] They also contributed to the development of the "Hydra," a 22-passenger shallow-water boat to transport delegates during Expo 2000. It was equipped with a 5 kW AFC system fueled from a metal hydride storage tank.[24] Both companies have closed.[25]

3.4.4. AFC Development in Sweden: ASEA

Sweden had an AFC program centering on the company ASEA during the late 1960s and early 1970s. The Royal Institute of Technology in Stockholm developed a 50 W methanol–air AFC system, and its technology was instrumental in developing a circulating electrolyte AFC in 10 kW modules.[6] Electrodes were based on nickel catalysts, and the technical scope was very ambitious: develop relatively large units (a few hundred kilowatts) for submarine applications.[12] Because of technical problems and accidents, the program was stopped.

3.4.5. AFC Development in UK: AFC Energy

AFC Energy was incorporated in January 2006 to purchase and commercialize AFC technology and know-how developed by Elenco.[26] In April 2007, AFC Energy, based in Dunsfold near Guildford, UK, began developing a low-cost fuel cell system, targeting waste hydrogen applications.

In January 2012, AFC Energy announced that it started electrical power generation at AkzoNobel's Bitterfield plant in Germany.[27] AFC Energy's Beta fuel cell systems have been generating electrical power at the AkzoNobel site since late last year. The two Beta commercial-design fuel cell systems, installed since October 2011, have since been equipped with electrodes for trials, and were fueled with AkzoNobel's industrially produced hydrogen. This work followed a series of trials that the company had been carrying out using a Beta system installed at its UK facilities.

The field tests at AkzoNobel's plant in Bitterfeld, Germany, are currently underway.

3.5. AFC DEVELOPMENT IN RUSSIA: SUSTAINED EFFORT, BUT WITH LITTLE COMMERCIAL SUCCESS

Not very much is known about Russian fuel cell systems, but Russia appears to be more interested in AFCs than other fuel cell types. There have been several organizations that have participated in AFC's development and marketing, including the Kiev Research and Production Association "KVANT," S.P. Korolev Rocket and Space Corporation RSC "ENERGIA," the UEIP, and ZAO IPT. Some of these organizations appear to be engaged in AFC activity today.

3.5.1. Kiev Research and Production Association "KVANT"

In 1979, the Kiev Research and Production Association "KVANT" unveiled its KVANT-RAF (Riga Bus Plant) H_2 van that contained a 12 kW AFC and ran on hydrogen.[28] And in 1982, KVANT manufactured the world's first hydrogen-fueled AFC micro-bus.[29] In 1988, KVANT demonstrated a working RAF fuel cell minivan at the seventh World Hydrogen Energy Conference in Moscow.[30] Because of intense political upheaval at the time in Russia, however, the KVANT-RAF H2 van was not further developed.[28]

3.5.2. S.P. Korolev Rocket and Space Corporation RSC "ENERGIA"

The S.P. Korolev Rocket and Space Corporation RSC "ENERGIA" is a key institute for manned space flight vehicles and heavy payload rockets, including the Russian version of the space shuttle (the *Buran*).[31] ENERGIA also developed a hydrogen–oxygen (H_2–O_2) fuel cell (designated by ENERGIA as an "electrochemical generator" or "EG"), called the "Photon," for the Buran space shuttle. ENERGIA claims that the Photon AFC unit is similar to the fuel cell NASA uses in the U.S. Space Shuttle. ENERGIA has manufactured 100 Photon EGs since 1987. In test programs and on actual missions, these cells have accumulated a total of 80,000 h of operational experience.

- The only orbital launch of *Buran* occurred on 15 November 1988 from the Baikonur Cosmodrome Site.[32] It was lifted into orbit by the specially designed ENERGIA booster rocket. The life support system was partially installed (the flight was unmanned) and no software was installed to run the computer display screens. The shuttle orbited the Earth twice in 206 min of flight. It performed an automated landing on the shuttle runway at Baikonur Cosmodrome where, despite a lateral wind speed of 61.2 km (38.0 mi)/hour, it landed only 3 m (9.8 ft) laterally and 10 m (33 ft) longitudinally from the target.
- In 1989, it was planned that *Buran* would have an unmanned second flight in 1993, with a duration of 15–20 days.[32] During this period, the Soviet Union had crumbled and the funding dried up. The Buran space shuttle program completely disappeared from the government's budget after 1993.[33] On May 12, 2002, a hangar housing *Buran* in Kazakhstan collapsed, due to poor maintenance, damaging the orbiter.[32]

In the early 1990s, the ESA and the Russian Aviation and Space Agency agreed to cooperate on future launchers,[34] and ESA tested the Photon AFC system for potential use in the manned space program Hermes.[4] The tests, conducted at various power levels up to 23 kW, included current–voltage characteristics, transient behavior, autothermal start-up, and impedance measurements.[35] In addition, the waste water and the purge gas were analyzed. All test goals were met and no electrochemical limitations were apparent. However, no lifetime data is available. Also, the high noble metal loading limited the system to military and space applications.[4] Hermes was subsequently canceled and no shuttles were built.

ENERGIA proceeded with work to create more effective next-generation AFCs.[29] In 2003 at the Moscow motor show, they demonstrated an automobile with these AFCs.[29] ENERGIA now offers a modified version of the Photon AFC system for undersea applications. ENERGIA representatives state that they can design, develop, and manufacture prototype units within 1.5–3 years once an order has been received.[31]

3.5.3. Ural Electrochemical Integrated Plant

UEIP, the largest uranium enrichment enterprise in the world, appears to be engaged in marketing the Photon Fuel Cell Generator that it had developed for the *Buran* Space System.[36] UEIP states at its website that R&D had been completed in 1988, that the total cumulative operating time in the course of development and maintenance was 80,000 h, and that the Photon AFC mean-time-to-first-failure is 2000 h. UEIP also states that the Photon AFC system offers a guaranteed service life of 1200 h, service shelf life of 10 years (from the fabrication date), and a guaranteed service life of 5 years.

- UEIP indicates that the photon system can be used for space engineering, submarines, automobiles, hydrogen-fueled airplanes, and remotely controlled self-contained robotic systems working in hazardous environments.
- UEIP claims that modification of the Photon generator is possible with an increase of voltage up to 160 V and a change in nominal power to the level required for the buyer.

3.5.4. ZAO Independent Power Technologies

ZAO IPT, Moscow, Russia, developed an industrial prototype 6 kW AFC auxiliary/backup generator called the "Cascade-IP."[30] IPT is an outgrowth of the former Soviet Union's Research Production Enterprise KVANT and a designer and manufacturer of AFC-based electrochemical generators.[20] It also works closely with the Russian Rocket and Space Corporation RSC "ENERGIA." A key feature of the Cascade-IP system, claimed to be the world's first, is a built-in, "zero-waste" regenerative scrubber to remove carbon dioxide from the air, according to managing director Alexander Yuzefovsky. The company says it has spent $1 million on the development of the generator. IPT is now looking for business partners in Europe and the United States, according to the company website.[30]

3.5.5. No Commercial Success in the Near Term

It appears that the AFC technology in Russia has not shown significant advantages compared to other state-of-the-art AFC technology. Russia, moreover, has not demonstrated vigorous AFC R&D activity that could lead to breakthroughs. It is unlikely, therefore, that Russia will be able to develop cost-effective, reliable AFC systems for successful commercial applications in the near term.

3.6. AFC DEVELOPMENT IN JAPAN: LIMITED ACTIVITIES OF NO CONSEQUENCE….BUT A NEW EFFORT EMERGES

A number of Japanese companies engaged in AFC work on a limited scale. Between 1961 and 1988, Fuji, Hitachi, Japan Storage Battery, Sanyo, and Matsushita Electric Industrial developed AFC systems of about 1–15 kW size. Fuji, which received government R&D support, was the only company with substantive activities. The others appear to have engaged in exploratory activities, perhaps to test the water in the fuel cell business. All of them had abandoned the AFC field by the late 1980s.

Recently, however, a new hope emerged: one of the Japanese automakers, Daihatsu Motor, has started to develop an AFC vehicle.

3.6.1. Fuji Electric

Fuji started AFC work in 1961[37] and abandoned the field perhaps in 1988 after nearly 28 years of R&D. In 1972, Fuji developed a 10 kW AFC fuel cell system for submarine applications for the Japan Defense Agency.[37] The system used circulating electrolytes, was fueled by hydrogen and air, and operated for 1000 h.[38] Fuji developed a 2 kW AFC system in 1977 jointly with CRIEPI (the Central Research Institute of the Electric Power Industry) to test operations while connected to power grids. In 1978, Fuji started AFC work funded by the Japanese Ministry of Economy, Trade, and Industry (METI; formerly MITI, the Ministry of International Trade and Industry).[37] Between 1979 and 1981, Fuji also developed a 15 kW AFC stack for the Central Electric Power Council.[6] In early 1983, 10,000 h of operations at 70 °C were recorded. In 1984, Fuji under METI funding developed a 1 kW AFC system operating on by-product hydrogen.[39] This system employed an oxygen electrode using a silver catalyst, supported on a furnace black substrate and a zirconium-containing Raney nickel anode.[6] The system operated for 3000 h for 1 year at a remote location.[40] After the METI project ended in 1984, Fuji continued its work on AFCs on its own for a while longer, focusing on laboratory development, which was modest in comparison with the company's other fuel cell work.[12] In 1985, the company developed a 7.5 kW system for emergency power generation and completed an 18 month field test; in 1987 it developed a 3.6 kW system for welding machines.[38] Little further information is available, suggesting Fuji ended its AFC development around 1988.

3.6.2. Hitachi

Hitachi started basic fuel cell research in 1960 and worked on AFC technology until 1977.[6] In 1970, the company began work on hydrazine-fueled AFCs. This work led to the development of a 3 kW hydrazine–air AFC unit for the Japan Defense Agency. This power plant, which was used to power field communications and signal systems, has been manufactured by the Shin-Kobe Electric Machinery, a member of the Hitachi group. In 1978, Hitachi constructed a 1 kW AFC system that was tested for power grid operations in collaboration with the Chubu Electric Power Company.[38] Hitachi left the AFC field some time after that.

3.6.3. Japan Storage Battery

Japan Storage Battery focused on AFCs with immobilized electrolyte. In 1982, the company developed an 80 W system with Mitsubishi Heavy Industries for space applications.[38]

3.6.4. Sanyo

In 1965, Sanyo developed a 30 W AFC system fueled by methanol and air. It consisted of 10 fuel cells and operated at 70 mA/cm^2 at 0.45 V.[41]

3.6.5. Panasonic

In 1965 Panasonic (formerly Matsushita Electric Industrial) developed an AFC system of unknown output as a lighthouse power source that operated at 64 mA/cm^2 at 0.45 V. It reportedly operated for more than 5 years.[41] No further detail is available.

3.6.6. Daihatsu Motor

Remarkably, a new developer recently entered the AFC industry, raising hopes for the future of AFC technology.

In November 2011, Daihatsu Motor Ltd, Japan's oldest car manufacturer, now partly owned by Toyota, presented an AFC-powered vehicle, the "FC Show Case" at the 42nd Tokyo

Motor Show.[42] The automaker had been working on the AFC technology for the past several years. In September 2007, it announced that it had developed a fuel cell based on alkaline anion exchange membranes, in cooperation with Japan's National Institute of Advanced Industrial Science and Technology (AIST).[43]

According to Daihatsu, the FC Show Case has major advantages over PEMFC vehicles.[43] It can use inexpensive cobalt and nickel as a catalyst instead of expensive platinum. It also uses hydrazine hydrate ($N_2H_4 \cdot H_2O$), which can produce higher output than methanol and equals output obtained from a hydrogen fuel cell using platinum (about 0.5 W/cm^2). The company indicates that the fuel tank of the FC Show Case is filled with a granulized polymer embedded with a carbonyl group for capturing the hydrazine hydrate. When the hydrazine hydrate enters the tank, it reacts with the carbonyl group (in a dehydration–condensation reaction) and bonds with the polymer, becoming a solid called hydrazone, which can be safely stored. Circulating warm water through the hydrazone causes hydrolysis, which returns the hydrazone to the original carbonyl group and releases liquid hydrazine hydrate, which is then supplied to the fuel cell.

- Daihatsu undoubtedly faces a number of serious challenges before it can place the vehicle on the market. Nonetheless, the new activities show promise for the AFC industry.

References

1. Nor Jiri K. *Alkaline fuel cell-technology for the 21st Century.* Canada: Astris Energi, Inc., www.astris.ca/IN/IN-pdf/P9A-AFC.pdf, date unknown [accessed in August 2007].
2. *Fuel cell handbook.* 7th ed. NETL, US Department of Energy; US; November 2004.
3. Hoogers Gregor, editor. *Fuel cell technology handbook.* US; CRC Press LLC; July 2002.
4. Cifrain M, Kordesch K. Hydrogen/oxygen (air) fuel cells with alkaline electrolytes. In: Vielstich Wolf, Lamm Arnold, Gasteiger Hubert A, editors. *Handbook of fuel cells – fundamentals, technology and applications,* vol. 1, Part 4. Chichester: John Wiley & Sons, Ltd; 2003. p. 267–80 (ISBN: 0-471-49926-9).
5. *Project mercury, NASA.* http://science.ksc.nasa.gov/history/mercury/ercury-goals.txt. [accessed in August 2007].
6. Appleby AJ, Foulkes FR. *Fuel cell handbook.* New York: Van Nostrant Reinhold; 1989.
7. *"Fuel cells—the first 120 years," history of fuel cells.* Fuel Cell Today; UK; 24 July 2007.
8. *PEM fuel cells, collecting fuel cell history.* Smithsonian Institution; http://americanhistory.si.edu/fuelcells/pem/pemmain.htm; no date [accessed August 2010].
9. Kordesch Karl, Simader Gunter. *Fuel cells and their applications.* Germany: Weinheim; 1996.
10. *Quiet kilowatts for silent sentry, science service, National Museum of American History.* Smithsonian Institution; January 2007.
11. Porterfield Byron. *Portable radar ordered by army; 35-pound units developed by L.S. Company can 'see' 2–3 miles in darkness.* US; The New York Times; 19 September 1960.
12. Blomen Leo JMJ, Mugerwa Michael N. *Fuel cell systems.* US; Springer; 1993.
13. *The battery.* Apollo Energy Systems, http://www.electricauto.com/prod_hist.html; US; 2007.
14. *Hydrogen cars: 1966 GM electrovan.* H-Tech Inc, http://www.hydrogencarsnow.com/gm-electrovan.htm; US; 2005.
15. *Inventor Karl Kordesch of Union Carbide rides his alkali fuel cell motorcycle in 1967, image 059-006 from the Science Service Historical Image Collection, National Museum of American History.* US; Smithsonian Institution; January 2007.
16. Ishii Ronald K. *Stationary & transportation fuel cell applications,* Alternative Energy System Consulting, Inc., Presented at the First International Conference on Energy Efficiency and Conservation, Hong Kong Conrad Hotel, 15–17. http://www.aesc-inc.com/download/Ishii_Fuel_Cell_Paper.pdf. January 2003.
17. *Prototype tractor by Allis-Chalmers, first vehicle powered with fuel cells.* ASME Milwaukee Section, http://www.sections.asme.org/milwaukee/history/15-fuelcelltractor.html; US; 2007.
18. *From discovery to prototypes, fuel cell basics, historical background, institute of energy research, Forschungszentrum Julich.* http://www.fz-juelich.de/ief/ief-3/fuel_cells/historical/; [accessed May 2008].
19. Hoffmann Peter, Harkin Tom. *Tomorrow's energy: hydrogen, fuel cells, and the prospects for a cleaner planet.* US; MIT Press; 2001.
20. Guelzow Erich, Nor Jiri K, Nor Peter K, Schulze Mathias. A Renaissance for alkaline fuel cells. *The fuel cell review.* vol. 3(1) February/March 2006.

21. *Country paper – France, brief summary of France priorities*. DRAFT, ILC-023-04; [accessed 18.02.04].
22. Hirschenhofer JH. *International developments in fuel cells*. allbusiness.com; [accessed 01.08.89].
23. Zevco (Elenco)/Nicholas Abson, Summary, http://zevco.front.ru/; [accessed May 2008].
24. Weaver Graham, Reidy Roisin. *World fuel cells: an industry profile with market*. The Netherlands; Elsevier; 2002.
25. Kordesch Karl and Cifrain Martin, "*Alkaline fuel cells (an overview*), 204th Meeting of the Electrochemical Society, Orlando, Fl., October 12-17, 2003.
26. *History of AFC Energy, About AFC energy*, http://www.afcenergy.com/about/history/, AFC Energy, no date [accessed April 2012].
27. *AFC energy's 'beta' fuel cell generates first industrial power*, Press Release, http://www.afcenergy.com/press-releases/afc-energy%e2%80%99s-%e2%80%98beta%e2%80%99-fuel-cell-generates-first-industrial-power/n, AFC Energy, 9 January 2012.
28. *Hydrogen Car 1807-1986*, http://www.hydrogencarsnow.com/hydrogencars1807-1986.htm, Hydrogen Car Now, [accessed November 2009].
29. Mazurenko Sergey, *International partnership for the hydrogen economy ministerial meeting*, http://www.iphe.net/docs/Meetings/USA_11-03/Russia_Statement.pdf, 1st IPHE Steering Committee Meeting, Washington, DC, 19-21 November 2003.
30. *Russia's independent power technologies*, http://www.independentpower.biz/articles/alkaline.htm; no date [accessed November 2009].
31. *Research submersibles and undersea technologies*. World Technology Evaluation Center, http://www.wtec.org/loyola/subsea/toc.htm; US; 1994.
32. *Buran (spacecraft)*, http://en.wikipedia.org/wiki/Buran_(spacecraft), Wikipedia, last modified on 2 November 2009.
33. *Energia*, http://www.astronautix.com/lvs/energia.htm, ©Mark Wade, 1997 – 2008 [accessed November 2009].
34. *Hermes (spaceplane)*, http://en.wikipedia.org/wiki/Hermes_(spaceplane), Wikipedia, last modified on 6 September 2009.
35. Schautz M, Dudley G, Baron F, Popov V, Pospelov B. Testing of a Buran flight-model fuel cell. *ESA Journal*. 1944; 18(2):129–37, http://adsabs.harvard.edu/abs/1994ESAJ.18.129S (ISSN 0379-2285).
36. *Photon fuel cell generator for the space system "buran*, http://www.ueip.ru/eng/Products/159.aspx, Russia, Ural Electrochemical Integrated Plant (UEIP), [accessed November 2009].
37. Kobayashi Takashi, Alkaline fuel cells, *Fuel cell power generation technology symposium*. 84, July 1984.
38. *AFC development status, domestic*. Japan: New Energy Data File, NEDO; 1998.
39. *Hydrogen utilization international clean energy system technology (WE-NET) subtask 7, survey and study on hydrogen utilization technology*. Japan: NEDO; March 1995.
40. *Long-term energy technology strategic survey, new sunshine project systematic study and quantitative evaluation of derivative effects*; March 2001. Part One.
41. *New energy technology development related data production survey (fuel cells)*. Japan: NEDO; March 1999.
42. Michelsen Charis, *Tokyo motor show: daihatsu presents three super-cute electric prototypes, cars*, http://cleantechnica.com/2011/11/10/tokyo-motor-show-daihatsu-presents-three-super-cute-electric-prototypes/, CleanTechnica.com, 10 November 2011.
43. *Daihatsu develops new fuel cell technology that uses no precious metals —new technology has zero CO_2 emissions, conserves resources and significantly reduces cost—news*, http://www.daihatsu.com/news/n2007/07091401/index.html, Daihatsu motor, 14 September 2007.

CHAPTER

4

History of Phosphoric Acid Fuel Cells

OUTLINE

Fuel Cells

http://dx.doi.org/10.1016/B978-0-444-56325-5.00004-1

4.1. OVERVIEW

In the 1960s, acid electrolyte fuel cells, which had been known for 120 years ever since Grove's experiment, finally became the subject of intense interest in the United States and Japan. In July 1961, US researchers G. V. Elmore and H. A. Tanner—who were working at a research foundation in Ohio—described their successful experiments with phosphoric acid fuel cells (PAFCs) in an article, "Intermediate Temperature Fuel Cells," in the *Journal of the Electrochemical Society*.[1] Elmore and Tanner used an electrolyte that consisted of 35 percent phosphoric acid and 65 percent silica powder pasted into a carbon-coated Teflon gasket. Platinum or palladium was used for both the anode and the cathode.[2] The use of phosphoric acid as electrolyte was rather uncommon until that time. Most researchers who engaged in acid fuel cell research had been experimenting primarily with sulfuric acid.[1] Elmore and Tanner demonstrated that their PAFCs could operate for 6 months with no apparent deterioration because phosphoric acid, unlike sulfuric acid, was not reduced electrochemically.[2] Elmore and Tanner's PAFCs operated at atmospheric pressure (atm) at over 100 °C, allowing the water generated to escape as vapor, thus avoiding dilution of the electrolyte or flooding the pores of the electrodes.[3] The PAFC also was tolerant of carbon dioxide, offering major advantages for terrestrial applications because the cell could be fueled by hydrogen reformed from fossil fuels containing trace amounts of carbon dioxide.

In 1973, the US Department of Energy (DOE) (then the Energy Research and Development Administration) played a major role in promoting PAFC development projects. The result was the creation of a critical mass of financial and technological resources in the United States, rapidly advancing the capabilities of US corporations in PAFC technology.

The technical competence of the United States in PAFCs affected different parts of the world differently. European countries likely felt that it would be futile to compete with the United States and decided not to devote resources to launch PAFC developmental work of their own. As a result, Europe completely missed out on PAFC and MCFC development and no significant R&D programs were launched. Japan, on the other hand, decided to pursue PAFC development both by launching a national PAFC R&D program and by aggressively seeking partnership or licensing agreements with US industry.

- Major industrial players in PAFC technology in Japan at that time included Fuji Electric, Hitachi, Mitsubishi Electric (MELCO), Sanyo, and Toshiba.[4] Most of them formed partnerships with US developers. It is likely that they believed they could quickly gain PAFC expertise by relying on their US partners. Toshiba cooperated with United

Technologies Corporation (UTC)[i] and subsequently formed UTC–Toshiba joint ventures—International Fuel Cells (IFC)—which engaged in PAFC R&D and ONSI, which provided plant construction, installation, and maintenance services. Similarly, Fuji Electric teamed up with Engelhard, Sanyo teamed with ERC (Energy Research Corporation), and MELCO worked with Westinghouse.[5]

While many environmental and efficiency benefits of fuel cells were first recognized because of the excellent performance of the PAFC, in the end, PAFC systems never achieved major success either in the United States or in Japan. Although they achieved a high level of durability and reliability, PAFC systems could never meet cost-performance levels that were commercially competitive. Sales of the products were sluggish, even with subsidies that were designed to buy down the purchase price. No PAFC products were commercially deployed by the US military because the PAFC systems could not effectively adapt to the fuels that were used by the military. Many companies left the PAFC field, leaving only a few engaged in PAFC development and marketing in the world today.

4.2. PAFC DEVELOPMENT IN THE UNITED STATES: 25 YEARS OF GOVERNMENT PROGRAMS FAIL TO PRODUCE A COST-COMPETITIVE PAFC SYSTEM

Following Elmore and Tanner's successful experiments in the early 1960s, an unprecedented number of PAFC R&D and field test projects got started in the United States and largely continued throughout the 1980s. In the mid-1960s, the US military started to sponsor PAFC R&D programs that aimed to develop small-sized PAFC power sources fueled by military fuels. Also, in the mid-1960s, public research organizations affiliated with utilities—such as the Gas Research Institute (GRI, now the Gas Technology Institute (GTI) and EPRI (the Electric Power Research Institute)—started funding PAFC R&D programs.[6] In the mid-1970s, significant government funding started to bolster PAFC R&D and field test projects. The energy crisis in 1973 generated a sense of urgency for energy independence and in 1976 the Department of Energy (DOE) started to actively fund PAFC R&D, encouraging public and business-sector investment in PAFC R&D. The US focus on PAFCs was primarily to develop PAFC technology for co-generation and distributed power-station applications. The development of PAFCs for transport applications received only low-level funding.

- The DOE provided the largest share of the PAFC R&D project cost, followed by GRI and EPRI. From 1976 until it ended its R&D funding for PAFCs in 1992, DOE provided a total of about $300 million, according to DOE historical data (not adjusted for inflation).[7] Although there is no comprehensive data on the level of funding provided for PAFC development by GRI and EPRI, they probably provided close to $200 million between 1967 and 1992, based on data which indicate that GRI provided about

[i] United Technologies Corporation (UTC) is a major US technology corporation, encompassing business sectors such as aerospace, aviation, and fuel cells. Initially, Pratt and Whitney engaged in the fuel cell business, but this was taken over by a new UTC subsidiary International Fuel Cells (IFC) and its subsidiary ONSI. In this chapter, the name UTC is used wherever possible for simplicity. IFC was a joint venture between UTC and Toshiba of Japan (a minority stakeholder at 11.5 percent) established in 1985. IFC and Toshiba in 1990 also formed an IFC subsidiary, ONSI, which focuses on fuel cell power plant production, on-site construction, and services. In 2001, IFC was renamed UTC Fuel Cells, now part of UTC Power. The joint venture agreement between UTC and Toshiba was dissolved in 2004. Toshiba subsequently left the PAFC business.

a total of $13 million and EPRI about a total of $17 million for 4 years from 1981 to 1985.[5] Also, the expenses for field tests were shared by all organizations that participated, including individual utility companies in the gas and electric industry. These utility companies formed fuel cell users' groups—such as the Onsite Fuel Cell Users Group comprising primarily of gas utilities and the Electric Utility Fuel Cell Users Group comprising mostly of electric utilities.[8]

- The cost shared by UTC could have been as low as $75 million to as high as $300 million during the same period. It is the general rule that a corporate participant in a government project shares a third to half of the project cost. The corporation generally shares a third when the project entails R&D that is more oriented toward basic research and shares about half when the R&D is closer to product development. On the other hand, the utility groups' cost share might have reduced UTC's share to as low as 20 percent.

There were a number of US companies that participated in these government and public-sector funded PAFC R&D projects. They included UTC, Engelhard, ERC, and Westinghouse. The prospect that large-scale government and public funding would be available over a long period of time undoubtedly convinced these companies to commit themselves to aggressive PAFC development, creating a critical mass of financial and technical affluence. PAFC funding efforts likely inspired investment in all fuel cell technologies as well.

- UTC made the greatest progress of all. The company started PAFC development in the mid-1960s[9] and rapidly became the global leader in PAFC technology. UTC's history of PAFC development virtually parallels the US history of PAFC development. UTC also likely enjoyed the lion's share of government and public R&D funding.

The government and the public-sector funding in the United States for PAFC R&D ended in 1992. The DOE decided that PAFC technology had been sufficiently developed and was ready for commercialization and no longer needed government R&D support. Other fuel cell technologies, such as MCFC and SOFC, were also demanding DOE funding. UTC embarked upon a global commercialization effort for its most advanced prototype 200 kW power plant that year. The price of the initial plant was $500,000[5] or $2,500/kW, which was broadly acknowledged as too high to be commercially competitive. In 1993, the US Department of Defense (DoD) launched a series of programs that aimed to support the sales of the UTC systems, which continued for more than a decade.

- Except for UTC, no other US corporate players placed products on the market. Rather, just prior to or coinciding with the termination of their DOE R&D funding, Engelhard, ERC, and Westinghouse all abandoned their PAFC work. Engelhard quit fuel cell work altogether. ERC and Westinghouse moved on to another fuel cell technology, which they had likely expected to be more promising and would have a greater possibility of major government R&D funding. ERC chose to focus on MCFCs, and Westinghouse on SOFCs.

4.2.1. US Army's PAFC Programs

In the mid-1960s, the US Army started exploring the use of PAFCs powered by military fuels (diesel and JP-4). Over the next 25 years, the Army hoped to develop a silent lightweight electric power generator to support field operations and funded ERC to build developmental prototypes of a PAFC system (3 kW and 5 kW) and UTC to build preproduction prototypes

(1.5 kW).[6] However, sulfur contained in the fuels reduced the effectiveness of PAFC platinum catalyst, requiring the Army to use methanol (a nonstandard fuel) as the fuel for PAFCs. After building and testing many experimental PAFC systems, the Army recognized that for any major application of fuel cells for military use, the fuel cells must use military fuels because the Army logistics could not handle a large quantity of specialty fuel such as methanol. The Army terminated the PAFC program in around 1984.

4.2.2. US Air Force PAFC Programs

The US Air Force funded Westinghouse and ERC to develop PAFC systems to support its activities. The Air Force supported Westinghouse to investigate the feasibility of applying PAFC technology to several ground electric power and heating requirements. The Air Force also had a program with ERC to develop a 5 kW stationary fuel cell power plant for use at remote locations. Prototypes were delivered in 1984. The power plant operated on steam-reformed methanol, and was essentially a modification of ERC's 3 kW and 5 kW fuel cells originally developed for the US Army. The Air Force also reached the same conclusion as the Army that methanol was not an appropriate fuel[10] and subsequently abandoned the PAFC program.

4.2.3. PAFC Programs for Stationary Applications: United Technologies Corporation (UTC) Prevails

4.2.3.1. *TARGET Program*

In 1967, a large-scale R&D and demonstration program of PAFC power plants was initiated by an industrial partnership—consisting of 28 US, Japanese, and Canadian utility companies led by GRI—known as TARGET (Team to Advance Research for Gas Energy Transformation, Inc.).[5,6] The total budget of the 9-year program (1967–1976) was $58 million. The program aimed to develop a small combined heat and power (CHP) cogeneration power plant for residential use and test it in a large-scale field demonstration. The program aimed to establish the power plant's commercial viability in terms of cost, efficiency, pollution factor, longevity, and ease of maintenance. UTC was contracted to develop and manufacture sixty five 12.5 kW prototype natural gas-powered PAFC power plants. The power plants were developed and field tested at 35 different sites, including factories, apartment buildings, restaurants, and other locations, across the United States, Canada, and Japan between 1971 and 1973.

- Sponsoring the development of small residential power plants was a master business strategy for gas utilities. What GRI and gas utilities planned was promotion of the sale of pipeline gas by encouraging widespread distribution of small residential power plants that used natural gas. In the early 1960s, the US gas utility industry had become concerned by the prospect of inexpensive electricity from nuclear reactors and the trend toward all-electric homes.[8] Marketing fuel cells that would use natural gas to generate electricity for residential customers was their response to all-electric homes.

The technological developments achieved in the TARGET program were significant.[5] The program was able to achieve large-scale production of graphite as a cell construction material. Thin impervious graphite sheets, which could serve as bipolar plates in adjacent cells in cell stacks, were manufactured and cut to the desired size in a continuous operation, thereby significantly reducing production cost. Owing to graphite, UTC was subsequently able to increase the size of its power plant. UTC's plant size grew from 12.5 kW in 1967 to 4.5 MW in 1983, and then to 11 MW by 1985 or 1986. Also, the use of high-surface area graphite permitted a dramatic reduction in the use of platinum loading without sacrificing electrode performance, leading to additional significant cost reduction.[4] Other developments included a new structure for the electrolyte-containing matrix—a porous graphite plate—and a two-phase water cooling system.[5] TARGET also successfully demonstrated low emission and high performance for PAFCs. These positive results encouraged many utility companies to support further PAFC development.

Nonetheless, TARGET failed to meet some of its original goals: cost competitiveness, reliability, and PAFC longevity. The power plant could not reach the cost target due to its platinum content, despite the reduction in the amount of platinum used. The power plant also failed to demonstrate reliable operation due to short stack lifetimes.[5] A project evaluation indicated that those PAFC units were too technologically immature to yield the necessary production base and to permit a cost-effective market entry,[6] pointing out that the 9-year PAFC R&D project was a failure.

4.2.3.2. *GRI-DOE Project*

Despite the failure of the TARGET Project, a follow-on project called the GRI-DOE project (1976–1986) was quickly formulated and implemented. The project aimed to demonstrate the reliable operation and cost competitiveness of PAFC plants by developing and demonstrating 48 sets of 40 kW cogeneration power plants. Under this project, UTC was to scale up plant size to 40 kW to reduce per-kilowatt cost.[5] The cost of the entire project is not known; some sources report that project costs from 1981 to 1985 were about $60 million.[8] The period from 1984 to 1986 was devoted to field tests, which were funded at $35 million via support from DOE ($12 million) and GRI ($23 million).[5] For field tests, 46 plants were installed at 42 sites in the United States, including four plants at military sites under DoD sponsorship.[11] The remaining two were installed in Japan.[5] Sites selected for the field tests were customers of gas utilities, including hospitals, stores, hotels, and restaurants. Those customers served as suitable cogeneration sites because they could use both the electric power and the byproduct heat. In 1986, the field tests ended with all of the participants completing the 13-month test period.[11]

The project was considered satisfactory. The power plants showed sufficient technical performance, in both the grid-dependent mode and the independent mode.[8] Heat generated was used for domestic hot water, space heating, and pool heating. The average plant operated for about 6500 h. Fourteen power plants exceeded 8000 h of operation, with the longest running unit operating for over 15,000 h.[5]

From the commercial production point of view, however, there were problems. The 40 kW system still suffered from high capital costs.[5] A total of 46 of these plants had been produced at an average manufacturing cost of $12,500/kW.[12] This was five times higher than what would be viewed as acceptable for early commercial niche markets in 1985. UTC was on record stating that it could not produce a cost-competitive 40 kW PAFC power plant at the annual production rate the early market would likely support.[11] There were also minor problems related to the balance of plant (BoP) equipment—not the fuel cell stack itself—such as heat exchangers, boilers, and pipes.[12] The design of a system consisting of these relatively conventional components for such a new technology as the fuel cell turned out to be far more difficult than expected.

Following the demonstration of the 40 kW power plant, GRI and DOE sponsored UTC to engage in the development of a 200 kW PAFC power plant. The 200 kW unit was expected to cost less per kW than a 40 kW unit because it was supposed to take advantage of the economies of scale. UTC solicited the gas industry to join in commercialization through the purchase of 200 kW units. Potential customers were to purchase the 200 kW system with a semi-commercial price of \$500,000 for the initial plants,[5] or for \$2,500/kW. This price was expected to drop to \$300,000 with manufacturing experience. Commitments were made worldwide for over 50 units; the initial units were expected to be delivered and installed in late 1991 or early 1992. But in 1992, the DOE ended its 17-year, \$300 million funding effort for PAFC development because it deemed PAFC technology had entered a viable commercialization stage.

4.2.3.3. FCG-1 (Fuel Cell Generator 1) Project

The FCG-1 (Fuel Cell Generator 1) project was started in 1971, transitioned to the DOE Multi-Megawatt Utility Systems Program starting in 1976, and likely ended in 1988. The FCG-1 project started several years after gas utilities launched the TARGET program and ran in parallel with the TARGET and the GRI-DOE projects. In contrast to the first two projects, FCG-1 was launched by the electric power industry, the major rival of the gas utilities. In this project, nine electric utility companies (including the Edison Electric Institute) and UTC got together to develop a large-scale distributed power plant for electric grid applications.[13] Subsequently, the program enjoyed additional participants; EPRI joined the program in 1973 and the DOE in 1976. The primary factor driving the FCG-1 project was the electric utilities' increasing concern about the increasing emphasis on and expense for the control of emissions from coal-based plants, along with the rise in cost, complexity of siting requirements, and extensive procedural delays in nuclear installations.[6] Fuel cells operating on reformed natural gas or naphtha were viewed by electric utilities as having the potential for future electric power generation. The project had the goal of building a multi-megawatt PAFC power plant for commercialization by 1980.[13]

The project was nothing if not disastrous. During the life of the project, the FCG-1 project constantly changed its goals, including the size of the power plant, the program budget, and the target commercialization date, which might have been due to the participation of additional sponsors in the project after the project had launched. The initial, 1-year, \$3 million study led to the launch of a \$49 million effort to develop a 27 MW fuel cell power plant—designated the FCG-1—that was launched in 1973.[14] The 27 MW generation capacity was deemed suitable for the needs of 80 percent of municipal and rural electricity producers in the United States.[15] From 1975 through 1977, the project focused on the development and testing of a 1 MW fuel cell plant, which was expected to serve as the basic building block that would be assembled into the 27 MW plant. The 1 MW power plant was completed in late 1976 and successfully tested by early 1977.[14] In 1976, coinciding with the DOE participation in the project, the initial goal of building a 27 MW power plant was dropped and a new effort to build a 4.5 MW power plant began with commercialization targeted for 1981.[12,ii] The expected cost for the new effort was \$70 million (1980 dollars) with 48 percent of it provided by the DOE and the rest by EPRI (25 percent), UTC (20 percent), and the initial group of electric utility companies (the remaining 7 percent). The site selected for test-operation of the 4.5 MW system was Consolidated Edison Company (Con Ed), a New York electric utility located in Manhattan, New York.

- The effort to test-operate the 4.5 MW power plant started in 1978 with Con Ed's application to seek permission to build the demonstration plant in New York. The project

[ii] Some sources cite the FCG-1 goal was to build a 4.8 MW power plant. A 4.5 MW alternating current (AC) is equivalent to 4.8 MW of direct current (DC) after accounting for parasitic losses and internal power plant consumption by the DC to AC inverters.

ultimately was aborted in 1985.[16] The fuel cell never ran. The cancelation of test-operations in New York was due to the regulatory procedures of the New York Fire Department and the objections by local residents.[6] In New York, a conventional power plant would not be permitted because of siting and environmental concerns.[17] The local residents were also concerned about the possible fire hazards from the on-site fuel processing and storage systems and other untested aspects of the new technology. Authorization for the plant construction to proceed was obtained only after protracted hearings. In all, 3 years of reviews, special tests, and design studies were undertaken before the New York City Fire Department fully approved the installation. Demonstrator start-up tests began in November 1981 and were completed in mid-1983. Con Ed finally attempted to produce power at the site in April 1984. The delays were so lengthy, however, that by the time it was cleared for operation, the fuel cell stacks had deteriorated and never produced power.[18]

The technological viability of the UTC 4.5 MW power plant was fortuitously validated, however, by another UTC 4.5 MW unit operated outside the United States. In 1980, Tokyo Electric Power Company (TEPCO) purchased a 4.5 MW PAFC power plant from UTC with its own funds.[12] The TEPCO unit was constructed by Toshiba, then a minority partner in the joint venture IFC, UTC's PAFC business unit, using PAFC stacks manufactured by UTC. Construction at TEPCO's Goi Power Plant on Tokyo Bay was complete in early 1982, less than 2 years after groundbreaking. The construction cost was $25 million.[16] Checkout tests were completed the same year and the fuel cell stacks began to produce power in April 1983, the first PAFC operation at a utility scale.

- The TEPCO unit had the benefit of the Con Ed's engineering experiences, which enabled TEPCO to move its unit quickly through its critical stages. The unit also was able to take advantage of the advanced fuel cell stacks that UTC had subsequently developed. The cell stacks contained more electrolyte than the New York stacks, and the unit also had improved burner, water treatment, and water cooling systems. The 4.5 MW demonstrator at TEPCO operated satisfactorily between 25 percent and 100 percent of rated capacity, producing utility-grade power with minimal emissions.[12] It produced 5430 MW over a period of 2423 load hours between 1983 and 1985, when the test ended. The electrical efficiency was 36.7 percent based on the higher heating value (HHV) of hydrogen and the electrical output after the DC to AC inverters.[16]

Following the successful demonstration of the 4.5 MW PAFC AC power plant at TEPCO, UTC, DOE, and EPRI clearly were encouraged enough to enter the initial market acceptance phase. DOE and EPRI sponsored R&D of the cell and stack, the reformer, and the inverter that would support an 11 MW power plant.[11] UTC and Toshiba jointly developed an improved 11 MW PAFC plant for electric utility applications between 1985 and 1986.[5] Under this project, UTC hoped to build twenty three 11 MW units and solicited the electric industry to purchase them for commercial purposes. The timetable called for delivery of the first commercial 11 MW power plant by 1988,[12] 8 years later than the initially planned date of commercialization. Electric utilities gave a commitment to participate, as consortia, in the field testing of the 11 MW power plants, and the test was scheduled for the summer of 1987.[11] EPRI also made a commitment to provide substantial subsidies to the utilities to support UTC's first phase of PAFC commercialization. US electric utilities, however, did not purchase any at the end of the test. They were discouraged by the high price tag. The power plant was priced at \$4360/kW and in most cases extra installation costs added an additional \$2000/kW.[12] Furthermore, the lack of UTC guarantees that the fuel cells would operate according to UTC announcements did not contribute to a willingness to pay. In March 1988, the Electric Utility Fuel Cell Users Group, the group of electric utility companies that participated in the field tests, hoping to include PAFC power plants in their future business, disbanded.

Yet, one unit was sold to TEPCO, perhaps owing to Toshiba's sales efforts. TEPCO initiated the demonstration of the 11 MW fuel cell power plant again at Goi, Japan. TEPCO's goal was to examine the suitability of fuel cells using natural gas as a distributed or dispersed power source and also to encourage commercialization.[16] The 11 MW plant was constructed for TEPCO by Toshiba using 18 670 kW fuel cell stacks manufactured by UTC. Toshiba was responsible for the design and fabrication of the BoP equipment. Construction began in January 1989 and the designed power level of 11 MW was attained in April 1991, with the net electrical efficiency of 41.1 percent. TEPCO operated the unit for more than 23,000 h between 1991 and 1997, at which time the demonstration test ended.[19] According to TEPCO, the system supplied enough power for about 4000 households and waste heat from the unit met the air conditioning demands of 300 households or the heating requirements of 1300 households.[18] It was reported, however, that the 11 MW plant had experienced a malfunction in the stack after attaining the desired power output of 11 MW in April 1991 and had been operating at a reduced output mode until TEPCO ended the test operation in 1997.[19] TEPCO never deployed PAFC systems for its power grid applications, marking the unfortunate end of the FCG-1 project both in the United States and in Japan, more than 25 years after the commencement of the project.

4.2.4. Other PAFC Programs in the United States

The DOE provided PAFC R&D funding three other industry players, in addition to UTC. They were Engelhard, ERC, and Westinghouse.[8] All the three players left PAFC development, just prior to or coinciding with the completion of DOE R&D funding. Engelhard left the fuel cell field altogether after more than 20 years of fuel cell work; ERC and Westinghouse switched to other fuel cell technologies that were considered to be more promising.

4.2.4.1. *Engelhard*

Engelhard Corporation was engaged in the development of PAFC technology for more than 20 years,[5] starting its R&D by the mid- to late 1960s. Engelhard was engaged in the basic development work, primarily for small-capacity PAFC plants suitable for on-site cogeneration. In 1978, the DOE funded a project to improve the performance and the cost of Engelhard's PAFC technology that might lead to the development of a cost-effective, on-site power generation system for residential and commercial buildings.[10] The system was to operate on reformed methanol; Engelhard had accumulated experience with methanol in its military fuel cell development program. In 1981, Engelhard constructed a breadboard 50 kW system for commercial buildings and built a series of multi-cell stacks as well as two advanced 5 kW stacks. Economic assessments indicated that the connection of the system to the electric grid was generally favorable and, providing that the fuel cell system cost goals were met, on-site building applications could represent a sizable commercial market for fuel cells. The assessments suggested that the system had not met its cost goals, however. DOE also funded Engelhard to develop a full-size 100 kW system made up of four 25 kW fuel cell stacks, two 50 kW fuel conditioners, and two 50 kW power processors to provide adequate reliability and redundancy.[8] The project was expected to be completed in 1988 with a 25 kW PAFC on-site power plant system test.[11] In 1988, however, when Engelhard management made a strategic decision not to pursue fuel cells, the activities were discontinued and the technology was sold to Fuji Electric in Japan.[12]

4.2.4.2. *Energy Research Corporation (ERC)*

ERC has been developing air-cooled PAFC stack technology since 1969.[5] From 1979 until 1985, with funding from the US Army, ERC focused on developing small PAFC units ranging from 0.5 to 30 kW and constructed 3 kW and 5 kW portable PAFC power plants.[12] ERC also developed portable methanol reformers to fuel these plants.[5] In the late 1985, the US Army canceled its methanol fuel cell program.[12] In the early 1990s, ERC abandoned PAFC as uneconomic.[20] The basic design configuration of ERC's PAFC structure had been licensed to Westinghouse.[8] ERC's PAFC stack technology was also introduced to Sanyo Electric Company in Japan.[5] ERC changed its name to FuelCell Energy in 1999.[21]

4.2.4.3. *Westinghouse*

Westinghouse Electric Corporation started development activities of PAFC in 1978.[5] The primary goal of Westinghouse was to develop a 3–13 MW, air-cooled PAFC power plant for electric utility applications. Westinghouse was funded by DOE and EPRI in the 1980s for the development of two 7.5 MW PAFC power plants.[8,12] Westinghouse was working on this program with ERC;[5] the company believed that an air-cooled system was simpler and adopted ERC's air cooling system stack technology, but air cooling may have reduced system efficiency below desirable levels. Westinghouse hoped to demonstrate the plant in 1987 and 1988 and to have the power plant for sale before 1990.[9] The program, however, encountered many difficulties, and the demonstration schedule was significantly delayed.[5] In 1988, only one small-scale 2 kW stack had been tested. Moreover, the decay in system performance was greater than the program goals. Yet, the DOE awarded another contract for technology development in 1988.[12] Finally, in 1991, Westinghouse demonstrated a 100 kW single-module stack[6] and in 1992 developed and tested a 375–400 kW four-stack PAFC engineering module.[22] The 375–400 kW module was supposed to be a building block for the 7.5 MW plant. In 1992, the DOE ended its PAFC R&D funding, coinciding with a time of great financial stress for Westinghouse due to its large real-estate losses. Westinghouse management decided to sell the PAFC technology, including related intellectual property and the pilot manufacturing facility.[23] There were two attempts to revive the Westinghouse technology in later years by Fuel Cell America and HydroGen, but both subsequently went bankrupt in 2008.

4.2.5. PAFC Programs for Transport Applications: No Successes

There has been little government funding of possible transportation applications for PAFC technology. One noteworthy concept was the Georgetown University Fuel Cell Bus Program, which aimed to develop PAFC technology as the primary power for vehicles, specifically city buses. During the course of the project, the PAFC technology was not found well suited for transport applications and plans were dropped for a lighter, smaller, and more efficient fuel cell, the PEMFC.

In 1983, the Los Alamos National Laboratory conducted a feasibility study of the applications of fuel cells for buses.[24] The study, sponsored by the US Department of Transportation and the DOE, was completed in 1985 and concluded that urban transit buses were ideally suited for fuel cell power. This led to the implementation in 1987 of a DOE-sponsored brass board development project, in which two 25 kW PAFC systems and steam reformers for methanol were developed and tested. The participating corporations were Engelhard, which was teamed with Fuji Electric of Japan, and ERC. The brass board project was completed in 1990, and the project team recommended that the oil-cooled PAFC system be used to power a 30-foot transit bus. ERC specialized in air-cooled PAFC systems and therefore was not competitive; meanwhile, Engelhard left the PAFC field in 1988, leaving Engelhard's partner, Fuji Electric, as the contractor to be selected.

In 1991, a follow-on project to the brass board project focused on developing three hybrid fuel cell power plants for electric 30-foot buses.[25] Fuji Electric supplied 50 kW PAFC power plants for the three buses, based on the technology they had licensed from Engelhard.[26] H-Power of the United States headed the team that built the buses. The first bus was rolled out in 1994 and the other two were completed in 1995.[24]

In 1993, Georgetown University started a second follow-on project to develop two fuel cell-powered 40-foot transit buses, with the objective of demonstrating commercial feasibility. For this project, UTC was selected and in 1994 it started developing a 100 kW PAFC power plant. In 1988, one power plant was incorporated in one of the two buses for demonstration.[25] The second bus was not powered by UTC's PAFCs. Instead, Georgetown University decided to use a lighter, more compact, and more efficient PEMFC. In 1995, Georgetown awarded the contract to Ballard and the second bus debuted in 2001 using a 100 kW Ballard PEMFC as its power plant.[27]

4.2.6. US PAFC Subsidy Programs

As DOE ended its PAFC R&D program in 1992, a series of subsidy programs that aimed to promote the commercialization of PAFC power plants emerged. The first such programs were started by the US Army, an unusual party to implement programs that were largely commercial in nature. The Army supported two programs; a fuel cell power plant demonstration program and a buy-down (subsidy) program, which lasted for a total of about 10 years from 1993 to 2004. The goals of the programs were to provide a thorough evaluation of fuel cell performance and to increase production, thereby reducing the cost of PAFC power plants through economies of scale and by facilitating commercialization.[28] Both programs primarily aimed to assist UTC because other PAFC developers had left the PAFC business and other fuel cell technologies had not been sufficiently developed for commercial applications. The DOE cooperated with the DoD in administration of both programs. Following the completion of the DoD programs, the US government implemented two subsidy programs in the form of investment tax credits starting in 2005.

4.2.6.1. US Department of Defense Demonstration Program

In the early 1990s with incentives from Congress, the US Army played the role of an interested user of PAFC power plants. In 1993 and 1994, the US Congress provided the US Army about $37 million to implement a PAFC Demonstration Program. Under the program, the Army, the Navy, and the Air Force purchased and installed 30 UTC 200 kW PAFC power plants at 30 military bases spanning 17 US States (Fig. 4.1).[28] Twenty-nine power plants began operations on various dates, ranging from January 1995 to October 1997 (Table 4.1).

- Total operating time for the power plants as of January 2002 amounted to about 800,000 h—an average of about 27,600 h per plant—falling well short of the desired longevity of 40,000 h.
- The average electric efficiency was 31.8 percent, no better than the fleet efficiency of the nation's coal-based, combustion power plants.

FIGURE 4.1 ONSI PC-25 PAFC fuel cell. *Source: "UTC Fuel Cell ONSI PC25 Phosphoric Acid Fuel Cells." http://www.fuelcell.no/fuel_cell_types_pafc_eng.htm, Fuel Cell Norway ANS, no date, accessed March 2012. (Courtesy of UTC Power).* (For the color version of this figure, the reader is referred to the online version of this book.)

- The availability of the power plants was 64 percent, indicating that the plants were operating more than half the time.

The decade-long DoD fuel cell demonstration program, however, bought time for PAFC technology improvements. A closer examination of the operational performance of the power plants revealed a remarkably positive outcome. The newer power plants had made substantial technological improvements over older plants. The power plants were installed incrementally, a few at a time, over a period of 3 to 4 years. During the period, three models were manufactured (Model A, B, and C) each with incremental improvements. One Model A, 14 Model Bs, and 15 Model Cs were installed.[29] The Model A unit was shut down after 2,500 operating hours based on site-specific conditions unrelated to the performance of the power plant.[28] Model C demonstrated significantly better performance than Model B, holding out the promise that PAFC technology would evolve over time in terms of reliability, performance, and longevity.

- As of January 2002, Model B plants showed an average availability of 53.4 percent and Model C showed an availability of 73.9 percent,[4] a significant improvement. Also, average operating lifetimes totaled 25,499 for Model B and 29,216 h for Model C, making some progress toward the expected lifetime goal.
- Electrical efficiencies of Model B and Model C remained about the same; the average efficiency for Model B was about 32.2 percent, and that for Model C was 31.4 percent.

4.2.6.2. *DoD Climate Change Fuel Cell Rebate Program*

Following the demonstration program, the US Army attempted to stimulate the commercial growth and economies of scale in the PAFC industry. From 1995 to 2004, the Congress funded DoD to implement the Climate Change Fuel Cell Rebate Program. The program aimed to promote the sales of stationary fuel cell power plants by awarding partial rebates (upto \$1,000/kW or 1/3 of purchase cost, whichever was lowest) to buyers of qualified fuel cell power plants. A total of \$37.5 million was funded for the 10-year program.[30,iii]

The rebate program again helped create opportunities for PAFC technology developments in terms of both durability and performance over the decade-long program. First, the lifetime of PAFC power plants was successfully extended beyond the 40,000-hour goal. For example, the power plant installed at St. Agnes Hospital in Bocholt, Germany, by the year 2000 achieved more than 52,000 operating hours with its original fuel cell stack, according to a UTC press release on March 4, 2008.[31] Secondly, a detailed study published in 2002 based on 2

[iii] The Congress provided funding for the Climate Change Rebate Project from FY 1995 to FY 2004, but provided no funding for FY 2001.

TABLE 4.1 US DoD PAFC Demonstration Program: 1993–2002

	Through January 31, 2002						
	Service	Start date	Operating hours	MW hours output	Average kWe	Electrical efficiency (percent)	Availability (percent)
Model B units							
Naval Station Newport	Navy	1/23/95	42,375	6,387,537	150.7	30.2	76.1
US Army Soldier Systems Center	Army	1/27/95	38,608	6,379,235	165.2	31.2	61.2
US Military Academy	Army	11/17/95	28,393	4,872,371	171.6	31.5	63.0
934th Airlift Wing	Air Force	2/1/95	26,777	4,653,232	173.8	29.7	48.2
Picatinny Arsenal	Army	10/11/95	32,053	5,316,291	165.9	30.9	62.4
Naval Hospital, MCB Camp Pendleton	Marines	10/6/95	26,859	4,507,218	167.8	33.9	55.1
Naval Hospital, MCAGCC Twentynine Palms	Marines	6/20/95	21,652	3,522,419	162.7	32.3	44.6
Nellis AFB	Air Force	9/23/95	19,996	3,383,481	169.2	32.5	38.7
Watervliet Arsenal	Army	10/29/97	28,875	4,117,735	142.6	31.4	77.3
Fort Eustis	Army	9/12/95	27,705	4,256,532	157.2	31.9	50.7
Kirtland AFB	Air Force	7/20/97	16,713	2,502,970	149.8	31.0	32.5
Naval Oceanographic Office	Navy	10/7/97	19,641	3,574,854	182.0	34.6	51.4
Pine Bluff Arsenal	Army	10/21/97	9,343	1,747,040	187.0	34.9	39.6
CBC Port Hueneme	Navy	9/18/97	18,001	3,332,808	185.1	34.3	46.3
B's Total/Average			356,991	58,553,723	164.3	32.2	53.4
Model C units							
911th Airlift Wing	Air Force	12/18/96	35,234	6,037,038	171.3	31.6	79.8
Naval Hospital, NAS Jacksonville	Navy	3/18/97	33,284	6,193,403	186.1	31.6	78.4
NAS Fallon	Navy	3/30/97	31,054	4,880,720	157.2	30.8	80.1

(Continued)

TABLE 4.1 US DoD PAFC Demonstration Program: 1993–2002 (*Cont'd*)

	Through January 31, 2002						
	Service	**Start date**	**Operating hours**	**MW hours output**	**Average kWe**	**Electrical efficiency (percent)**	**Availability (percent)**
Submarine Base New London	Navy	9/30/97	32,848	5,884,840	179.2	31.8	84.8
Fort Richardson	Army	12/17/96	30,593	5,617,251	183.6	31.5	68.1
Little Rock AFB	Air Force	8/17/97	23,104	4,336,428	187.7	31.7	68.2
Westover AFB	Air Force	9/19/97	32,844	6,316,483	192.3	30.7	86.3
Barksdale AFB	Air Force	7/24/97	28,554	5,289,629	185.2	31.1	72.0
Fort Huachuca	Army	7/28/97	31,776	5,744,980	180.8	32.4	80.3
Laughlin AFB	Air Force	9/16/97	29,558	5,584,936	188.9	32.4	77.0
US Naval Academy	Navy	9/22/97	37,928	4,736,374	124.9	27.9	80.9
Edwards AFB	Air Force	7/5/97	23,866	4,603,664	192.9	32.6	59.9
Fort Bliss	Army	10/10/97	23,973	3,936,077	164.2	32.0	63.6
Davis-Monthan AFB	Air Force	10/14/97	26,462	4,513,985	170.6	32.1	71.5
NDCEE	Other	8/14/97	17,167	2,083,041	121.3	30.5	58.1
C's Total/Average			438,245	75,758,849	172.9	31.4	73.9
B+C Total/Average			795,236	134,312,570	169.0	31.8	64.0

Source: "Fuel Cell Handbook," Seventh Edition, NETL, Department of Energy, November 2004.

years of operation of the power plant at St Agnes indicated that the electrical efficiency of the system had been significantly higher than that of conventional gas engine cogeneration units; the UTC system finally achieved an electrical efficiency of almost 40 percent and a total fuel utilization of 84 percent.[32] The combined operation with the absorption chiller was a proven success in summer, enabling better heat utilization in times of low heat demand, according to the same study. The study then indicated that the UTC system showed a loss of voltage over time, resulting in performance degradation and that the lifetime of the power plant was expected to be 5–7 years. As of March 2008, however, the system had marked the eighth year of operation and operated 30 percent longer than the 40,000-hour goal.

- For another example, the UTC 200 kW PAFC power plant at the Central Park police station in New York City achieved more than 62,000 operating hours with its original fuel cell stack before a recent overhaul, according to a UTC press release on March 29, 2007.[33] The power plant was commissioned in 1999. The plant demonstrated its reliability by providing power to the police station without disruption when the Northeast Blackout of 2003 hit. The blackout was a widespread power outage that occurred throughout parts of the northeastern and midwestern United States and Ontario, Canada, on August 14.[34]

The DoD climate change fuel cell rebate program, however, was unable to realize a positive cycle of mass production, economies of scale, and cost reduction, that would lead to the growth of the PAFC industry. (Under the rebate program, a total of 274 stationary fuel cell power plants were installed, which included 112 PAFC power plants, 130 PEMFC power plants, 29 MCFC power plants, and 3 SOFC power plants) (Fig. 4.2).[30]

- In the earlier years between 1995 and 1997, all the fuel cell power plants installed under the program were PAFC power plants. There was no other commercially available fuel cell technology other than the UTC PAFC system at that time and thus, the program solely aimed to promote sales of UTC 200 kW power plants. In 1998, other types of fuel cells emerged and started to compete for the rebate. Those fuel cells came to dominate the Climate Change program by 2004, when the rebate program ended (Fig. 4.2).

4.2.6.3. *Federal and State Tax Credit Programs Implemented*

Following the end of the DoD Climate Change Fuel Cell Program, a number of tax

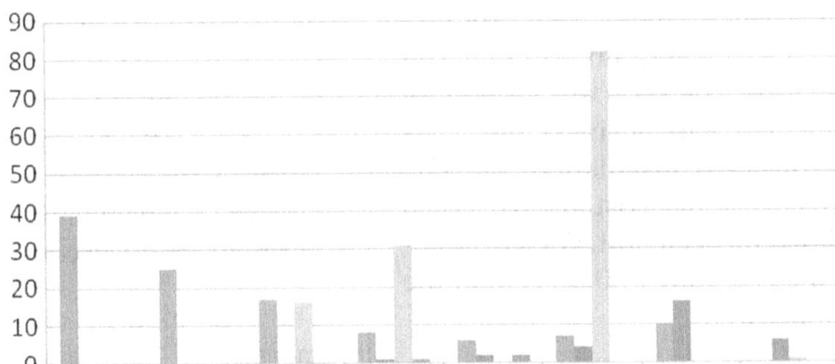

	1995	1996-97	1998	1999	2000	2002	2003	2004
PAFC	39	25	17	8	6	7	10	
MCFC				1	2	4	16	6
PEMFC			16	31		82		1
SOFC				1	2			

The figure was created based on "Climate Change Rebate Projects Overview," ERDC-CERL Fuel Cell, http://dodfuelcell.cecer.army.mil/climate/index.php, the US Department of Defense, accessed January 2009.

FIGURE 4.2 The number of fuel cell power plants installed under the DoD Climate Change Fuel Cell Rebate Program: FY1995–FY2004 (total power plants installed = 274; total PAFC power plants installed = 112). (For the color version of this figure, the reader is referred to the online version of this book.)

incentive programs were implemented. Some were initiated by the federal government and some by individual state governments.

4.2.6.3.1. FEDERAL INVESTMENT TAX CREDITS

The US government implemented two tax incentive programs to promote the commercialization of fuel cell power plants. The first program was signed into law on August 8, 2005 and took effect in January 2006. The law contained what was seen as a key step in the commercialization of fuel cell technology, the Investment Tax Credit (ITC). The legislation provided that those who acquired a qualified residential or business fuel cell power plant could write off $1000 per kW or 30 percent of the cost, whichever was less, from their tax liability. The provision entitled the taxpayer to subtract the amount of the credit, dollar for dollar, from the total federal tax liability.

The second program, which was included in the Emergency Economic Stabilization Act of 2008, was passed by the Congress and signed into law on October 3, 2008. The program extends the initial ITC program for fuel cell technology until 2016. The ITCs are differently defined for residential and business fuel cells this time. For a business installation, the ITC provides 30 percent of the cost up to $3,000 per kW of a qualified fuel cell unit. For a residential installation, the ITC allows 30 percent of the cost up to $1,000 per kW.

- It is puzzling that the amount of tax credit (up to $3000/kW for business installations) in the 8-year extension is greater than that of the earlier tax credit (up to $1000/kW). Under desirable circumstances, the tax incentives would be expected to become smaller as the technology matures. One explanation could be that the 8-year extension was drawn up with an understanding that the current fuel cell systems are more expensive than the earlier ones instead of being cheaper.

4.2.6.3.2. STATE-INITIATED CLEAN ENERGY FUND INITIATIVES

About the same time the DoD Climate Change Fuel Cell Program was entering into effect, a number of state governments became active in promoting clean, renewable energy sources, protecting the environment and stabilizing energy costs. As part of their effort, they worked to accelerate the development of energy efficiency and clean distributed generation, including renewable energy and combined heat and power systems such as fuel cell power plants. One of the tools the states have used is the clean energy fund. Clean energy funds, in most cases, are supported by small fees that are added to the electricity rates paid by customers. State governments administer the funds at their state energy offices or in cooperation with utilities, quasi-government agencies, and nonprofit organizations.

- Nationwide, about 20 states and the District of Columbia administer clean energy funds: Alaska, Arizona, California, Colorado, Connecticut, Delaware, the District of Columbia, Florida, Illinois, Maryland, Massachusetts, Minnesota, Montana, New Jersey, New Mexico, New York, Ohio, Oregon, Pennsylvania, Rhode Island, Vermont, and Wisconsin.

The size of their clean energy funds vary from state to state and range from about $1 million to more than $100 million a year.[35] Between 1998 and 2007, the state funds invested $1.5 billion in renewable energy projects and leveraged an additional $2.6 billion, bringing the total investment to over $4 billion.[36] On average, private investors provided $1.75 of capital for every $1 provided by a state fund. It appears that in general, a clean energy fund would provide a purchaser of a qualified fuel cell power plant with as much as 36 percent of the acquisition cost.

4.2.7. A Major New Hope Emerges…But Results in Little Consequence

Entering 2008, there was a significant new development that triggered a new hope that UTC might finally achieve the long-awaited commercial success. That year, UTC announced that it would produce a new 400 kW PureCell Model 400 and began taking orders. As a result, 16 PAFC power plants were sold, contracted, installed, or expected to be installed in the United States that year.[iv] Of the 16 orders, 13 were for the new 400 kW PAFC power plant.[37]

- UTC claimed that the 400 kW power plant would double the operating life and deliver twice the power and thermal energy of its predecessor 200 kW model and that the cost of the power plant would remain the same as the 200 kW system, thus cutting the cost per kW in half.
- The implication is that the new PAFC power plant would double its market competitiveness. Although the cost of the new 400 kW PAFC system is not publicly available, it might cost about $2 million for the system and installation, which was the cost of the 200 kW PAFC systems installed in July 2006 at the Los Angeles Zoo and the system installed at East Rochester High school in New York in February 2007. If this assumption is correct, the new 400 kW plant might still cost about $5000/kW.

It was hoped that the increased PAFC cost competitiveness coupled by the new tax incentives and clean energy funds might bring about commercial success for PAFC technology.

- The federal tax credit is now $3000/kW, which would reduce the estimated cost of the 400 kW power plant to $2000/kW. Combining the tax credit and the grants from clean energy funds, the cost of PAFC power plants might be lowered enough to be cost competitive with conventional power generation systems.

Contrary to these hopes, UTC did not make significant sales gains from the new system. Since 2008 through April 2012, UTC sold or contracted to sell at least 41 systems, most of which were the 400 kW systems. The level of the sales was high in 2008—reflecting customer expectations—but subsequently fell to the annual sales average seen in past years. During the past 20 years, without counting the sales made during the first 4 months in 2012, UTC sold and contracted to sell at least 174 systems, an average of 8.7 systems per year (Table 4.2).

Moreover, UTC's sales of PAFC power plants were largely dependent on federal and state subsidies. Without subsidies, UTC perhaps would have sold fewer. This suggests that the sales would be difficult if subsidies and tax credits are not available to sustain a competitive advantage, perhaps indicating that the PAFC system is not a commercially viable product today.

- A review of public data indicates that as of April 8, 2012, about 176 UTC PAFC power plants were sold or contracted to be sold in the United States (Fig. 4.3). Of the 176, 154 or about 88 percent of the plants have been or are expected to be subsidized, including 96 through the DoD Climate Change Fuel Cell Rebate Program and 58 through other forms of funding programs, including federal tax incentives or state clean energy funds or a combination of both. Regarding the remaining 22 power plants, it is not certain if they were purchased without a subsidy due to lack of information. But even if they were purchased without subsidies, this amounts to only one system a year.

[iv] To examine only the activity within the United States, the sale of twelve 400 kW systems to South Korea in October 2008 is not included.

TABLE 4.2 UTC PAFC Power Plants Sold, Contracted, or Installed in the United States: 1992–2012 (Total = 176 Systems, as of April 8, 2012)

Plant size, fuel	Installation site	Coordinator/operator	Schedule of operation	Funding/subsidies/rebates
200 kW	South Sacramento Medical Center, Sacramento, California	Sacramento Municipal Utility District, Kaiser Permanente	Perhaps in 1992	
200 kW, natural gas	Riverview Center for Jewish Seniors, Squirrel Hill, Pennsylvania	Equitable Resources	1992	
200 kW, natural gas	South Coast Air Quality Management District Office Building, Los Angeles, California	Southern California Gas	April 1992	
200 kW, natural gas	Hyatt Hotel, Irvine, California	Southern California Gas	September 1992	
200 kW, natural gas	Saint Vincent's Medical Center laundry facility, Brooklyn, New York	Brooklyn Union Gas	October 1992	
200 kW, landfill gas	Penrose Landfill power plant, Sun Valley, California	Northeast Utilities, International Fuel Cells, US Environmental Protection Agency	1993	
200 kW, natural gas	Presbyterian Nursing Home, Pittsburgh, Pennsylvania	Equitable Resources	February 1993	
200 kW, natural gas	Kraft Foods, Buena Park, California	Southern California Gas	July 1993, Was Sold to City of Mesa, Arizona in June 1996	
200 kW, natural gas	University of California, Santa Barbara, California	Southern California Gas	September 1993	
200 kW, natural gas	Riefler Concrete, Buffalo, New York	National Fuel Gas	September 1993	
200 kW, natural gas	Rochester Institute of Technology, Rochester, New York	Rochester Gas & Electric	September 1993	

200 kW, natural gas	Division of Street & Meter Repair, Chicago, Illinois	Peoples' Gas & Light	December 1993	
2 × 200 kW, natural gas	Kaiser Permanente Riverside Medical Center, Riverside, California	Kaiser Permanente Southern California Gas Co. Gas Research Institute U.S. Department of Energy	September 1994	
200 kW, natural gas	Santa Barbara jail, Santa Barbara, California	Southern California Gas Company	October 1994	
200 kW, natural gas	AT&T Research Laboratory, Morristown, New Jersey	Jersey Central Power & Light, GPU, Inc.	January 1995	
200 kW, natural gas	Kaiser Permanente Hospital Anaheim Medical Center, Anaheim, California	Toshiba Corporation, Southern California Gas, Kaiser Permanente	Perhaps in 1996	FY1995 US DoD Climate Change Fuel Cell Program
200 kW	State of Connecticut	Unknown	Perhaps in 1996	FY1995 US DoD Climate Change Fuel Cell Program
4 × 200 kW	Sun Chemical, Staten Island, New York	CLC S.r.l. Italy Ansaldo	Perhaps in 1996	FY1995 US DoD Climate Change Fuel Cell Program
200 kW	National Rural Electric Cooperative	Unknown	Perhaps in 1996	FY1995 US DoD Climate Change Fuel Cell Program
3 × 200 kW	New York Power Authority	Unknown	Perhaps in 1996	FY1995 US DoD Climate Change Fuel Cell Program
2 × 200 kW	New York Power Authority	Unknown	Perhaps in 1996	FY1995 US DoD Climate Change Fuel Cell Program
200 kW, digester gas	Flanders Road Landfill, Groton, Connecticut	Town of Groton, International Fuel Cells, US Environmental Protection Agency	June 1996	Likely funding from US EPA
2 × 200 kW	Sun Chemical, Staten Island, New York	Brooklyn Union Gas Co.	June 1996	FY1995 US DoD Climate Change Fuel Cell Program
2 × 200 kW, natural gas	Anchorage Readiness Center office-training facility, Anchorage, Alaska	State of Alaska	December 1996	FY1995 US DoD Climate Change Fuel Cell Program

(*Continued*)

TABLE 4.2 UTC PAFC Power Plants Sold, Contracted, or Installed in the United States: 1992–2012 (Total = 176 Systems, as of April 8, 2012) (*Cont'd*)

Plant size, fuel	Installation site	Coordinator/operator	Schedule of operation	Funding/subsidies/ rebates
200 kW, digester gas	Deer Island Sewage Treatment Plant, Boston, Massachusetts	Massachusetts Water Resources Authority, New England Power Company	1997	FY1995 US DoD Climate Change Fuel Cell Program
200 kW, digester gas	Westchester County wastewater treatment facility in Yonkers, New York	New York Power Authority	1997	New York Power Authority funded the fuel cell projects in partnership with the US DoD, New York State Energy Research and Development Authority, and US EPA
200 kW, natural gas	BOCES Regional Information Center, Syracuse, New York	Onondaga-Courtland-Madison Board of Cooperative Educational Services (BOCES)	January 1997	FY1995 US DoD Climate Change Fuel Cell Program
200 kW, natural gas	Double Tree Inn Hotel, Spokane, Washington	Washington Water and Power, Avista Corporation, Double Tree Inn	July 1997	FY1995 US DoD Climate Change Fuel Cell Program
2 × 200 kW, natural gas	Yankee Corporation headquarters, South Windsor, Connecticut	Yankee Gas Services	October 1997	FY1996–1997 US DoD Climate Change Fuel Cell Program
200 kW, natural gas	Hamilton Sundstrand's Data Center, Windsor Locks, Connecticut	Hamilton Sundstrand	December 1997	FY1995 US DoD Climate Change Fuel Cell Program
200 kW, natural gas	South County Hospital, Wakefield, Rhode Island	AEC South County Hospital	1999	FY1998 US DoD Climate Change Fuel Cell Program
200 kW	Navy Combat Construction Battalion Base, Gulfport, Mississippi	Reliant Energy Company, US DoD	1999	FY1996–1997 US DoD Climate Change Fuel Cell Program
200 kW, natural gas	Louisiana Gas Services Systems Operation Facility, Harvey, Louisiana	Louisiana Gas Services, Citizens Utilities	March 1999	FY1996–1997 US DoD Climate Change Fuel Cell Program

200 kW, natural gas	Central Park police headquarters, Manhattan, New York	New York Power Authority, KeySpan Energy, New York City Police Department	April 1999	The New York City Government likely paid for it from its budget
200 kW, natural gas	Cape Cod Community College library, West Barnstable, Massachusetts	Cape Cod Community College, NORESCO	April 1999	FY1995 US DoD Climate Change Fuel Cell Program
200 kW, digester gas	Columbia Blvd. Waste Water Treatment Plant, Portland, Oregon	City of Portland	July 1999	FY1996–1997 US DoD Climate Change Fuel Cell Program
200 kW, landfill gas, natural gas	Braintree Municipal Landfill, Braintree, Massachusetts	Braintree Electric Light Department	September 1999	FY1995 US DoD Climate Change Fuel Cell Program ($200,000); $100,000 grant from the Massachusetts Division of Energy Resources, Boston
4 × 200 kW, natural gas	First National Bank, Omaha, Nebraska	NORESCO, First National Bank of Omaha, Sure Power Corp.	November 1999	FY1995 US DoD Climate Change Fuel Cell Program
2 × 200 kW, methane	Las Virgenes Wastewater Treatment Plant, Calabasas, California	Energy 2000, Las Virgenes Waste Water Treatment	December 1999	FY1996–1997 US DoD Climate Change Fuel Cell Program
200 kW	Toshiba Corporation, Power Systems and Service	Toshiba Corporation, Houston, Texas	Perhaps in 2000	FY1998 US DoD Climate Change Fuel Cell Program
3 × 200 kW	Lord & Company		Perhaps in 2000	FY1998 US DoD Climate Change Fuel Cell Program
2 × 200 kW, natural gas	4th floor Conde Nast Building in Times Square, Manhattan, New York	Durst Corporation	February 2000	FY1996–1997 US DoD Climate Change Fuel Cell Program
200 kW, natural gas	Liverpool High School, Liverpool, New York	Niagara Mohawk/Plum Street Enterprises, Onondaga-	February 2000	FY1996–1997 US DoD Climate Change Fuel Cell Program
200 kW	City of Mesa Utilities Department, Mesa, Arizona	City of Mesa Utilities Department	April 2000	FY1995 US DoD Climate Change Fuel Cell Program

(Continued)

TABLE 4.2 UTC PAFC Power Plants Sold, Contracted, or Installed in the United States: 1992–2012 (Total = 176 Systems, as of April 8, 2012) (*Cont'd*)

Plant size, fuel	Installation site	Coordinator/operator	Schedule of operation	Funding/subsidies/rebates
200 kW	Texaco Energy Systems Central Plant, Bellaire, Texas	Logan Energy, Chevron Texaco	May 2000	FY1996–1997 US DoD Climate Change Fuel Cell Program
5 × 200 kW, natural gas	US Postal Service headquarters, Anchorage, Alaska	Chugach Electric Association	August 2000	FY1998 US DoD Climate Change Fuel Cell Program
200 kW, natural gas	Alcorn State University, Lorman, Mississippi	Alcorn State University, Logan Energy	October 2000	FY1995 US DoD Climate Change Fuel Cell Program
2 × 200 kW, natural gas	Ramapo College dormitory and computer center, Mahwah, New Jersey	McBride Energy, College of New Jersey	November 2000	FY1998 US DoD Climate Change Fuel Cell Program
200 kW, natural gas	North Central Bronx Hospital, Bronx, New York	New York Power Authority, KeySpan Energy, New York City Health and Hospitals Corporation	December 2000	FY1995 US DoD Climate Change Fuel Cell Program
200 kW, natural gas	Connecticut Natural Gas headquarters, Hartford, Connecticut	Connecticut Natural Gas Corp.	2001	FY1995 US DoD Climate Change Fuel Cell Program. Fuel cell provided power to Connecticut Natural Gas headquarters then was donated to the Department of Engineering at the University of Connecticut for research in 2001
6 × 200 kW, natural gas	Connecticut Juvenile Training School, Middletown, Connecticut	Connecticut Juvenile Training School	2001	FY1996–1997 US DoD Climate Change Fuel Cell Program
200 kW, natural gas	EPA's National Vehicle Emissions Laboratory, Ann Arbor, Michigan	US Environmental Protection Agency, NORESCO	March 2001, Manufactured in 1996, had numerous upgrades installed prior to its installation here	US DoD FY1995 Climate Change Fuel Cell Program

200 kW, natural gas	Lied Jungle exhibit at Henry Doorly Zoo, Omaha, Nebraska	Henry Doorly Zoo, Omaha Public Power District	August 2001	FY1999 US DoD Climate Change Fuel Cell Program
200 kW, natural gas	Ford Motor Company's North American Premier Automotive Group headquarters, Irvine, California	LOGANEnergy, Ford Motor Co.	December 2001	FY1996-1997 US DoD Climate Change Fuel Cell Program
200 kW, natural gas	New York Aquarium, Coney Island, New York	New York Power Authority, New York City Wildlife Conservation Society, the Office of the Brooklyn Borough	December 2001	Likely funded by New York Power Authority, KeySpan Energy, New York City Wildlife Conservation Society
2 × 200 kW, digester gas	Bowery Bay Waste Water Treatment Plant, New York, New York	New York Power Authority, New York State Energy Research and Development Authority	2002	Funding from New York Power Authority, New York State Energy Research and Development Authority, US EPA, US DoD
200 kW, natural gas	Chevron Data Center, San Ramon, California	Chevron	January 2002	FY2000 US DoD Climate Change Fuel Cell Program
200 kW, natural gas	Los Angeles Department of Water and Power (LADWP) Main Street Service Testing Facility, Los Angeles, California	LADWP	February 2002	FY2000 US DoD Climate Change Fuel Cell Program
200 kW	Playa Vista Project (commercial and residential development), Los Angeles, California	LADWP	March 2002	FY2000 US DoD Climate Change Fuel Cell Program
2 × 200 kW, natural gas	Mohegan Sun Casino Hotel, Uncasville, Connecticut	CTG Corporation, Mohegan Sun	March 2002	FY1996–1997 US DoD Climate Change Fuel Cell Program
200 kW, natural gas	Merck & Co. plant, Rahway, New Jersey	LOGANEnergy, Merck & Co.	June 2002	FY1996–1997 US DoD Climate Change Fuel Cell Program

(Continued)

TABLE 4.2 UTC PAFC Power Plants Sold, Contracted, or Installed in the United States: 1992–2012 (Total = 176 Systems, as of April 8, 2012) (*Cont'd*)

Plant size, fuel	Installation site	Coordinator/operator	Schedule of operation	Funding/subsidies/ rebates
200 kW, natural gas	Rebekah Baines Johnson Health Center, Austin, Texas	LOGANEnergy, Austin Energy Rebekah Baines Johnson Health Center	July 2002	FY1996-1997 DoD Climate Change Fuel Cell Program
200 kW, natural gas	South Windsor High School, South Windsor, Connecticut	Connecticut Clean Energy Fund	October 2002	Connecticut Clean Energy Fund provided $2.3 million to fund the program, including the purchase and installation of the fuel cell at the high school
200 kW, natural gas	Richard Stockton College of New Jersey, Pomona, New Jersey	Richard Stockton College of New Jersey, South Jersey Industries	May 2003	FY2002 US DoD Climate Change Fuel Cell Program
200 kW, natural gas	Oak Ridge National Laboratory, Oak Ridge, Tennessee		June 2003	FY2000 US DoD Climate Change Fuel Cell Program
2 × 200 kW, digester gas	Ward 26 Water Pollution Control Plant, Brooklyn, New York	New York Power Authority	October 2003	Funding from New York Power Authority, New York State Energy Research and Development Authority, US EPA, US DoD
200 kW, natural gas	St Francis Care hospital , Hartford, Connecticut	St. Francis Hospital	October 2003	
200 kW, natural gas	New Haven Water Pollution Control Authority (WPCA) facility, New Haven, Connecticut	WPCA New Haven, Connecticut Clean Energy Fund	November 2003	Partially funded by Connecticut Clean Energy Fund
200 kW, natural gas	Johnson & Johnson World Headquarters administrative offices, New Brunswick, New Jersey	Johnson & Johnson	December 2003	FY2003 US DoD Climate Change Fuel Cell Program

2 × 200 kW, digester gas	Red Hook Water Pollution Control Plant, Brooklyn, New York	New York Power Authority	December 2003	Funding from New York Power Authority, New York State Energy Research and Development Authority, US EPA, and US DoD
3 × 200 kW, natural gas	Guaranty Savings-owned office building housing federal offices, Fresno, California	LOGANEnergy, Guaranty Savings	May 2004	FY2003 US DoD Climate Change Fuel Cell Program
3 × 200 kW, natural gas	AB Parking Facilities, Fresno, California	AB Parking Facility, LLC,	June 2004	FY1996–1997 DoD Climate Change Fuel Cell Program
3 × 200 kW, natural gas	The College of New Jersey three building student housing facility, Ewing Township, New Jersey	The College of New Jersey	September 2004	FY2002 US DoD Climate Change Fuel Cell Program
200 kW, natural gas	East Anaheim Police Department and Community Center, Anaheim, California	Anaheim Public Utilities	February 2005	FY2002 US DoD Climate Change Fuel Cell Program
200 kW, natural gas	Camp Roberts Army National Guard Base SatCOM facility, Paso Robles, California	LOGANEnergy, Army National Guard	February 2005	FY2003 US DoD Climate Change Fuel Cell Program
2 × 200 kW, natural gas	Grand Central train station terminal, New York	New York Power Authority, Grand Central Railroad	February 2005	FY2003 US DoD Climate Change Fuel Cell Program
3 × 200 kW, digester gas	Hunts Point Water Pollution Control Plant, Bronx, New York	New York Power Authority	February 2005	Funding from New York Power Authority, New York State Energy Research and Development Authority, US EPA, US DoD
200 kW, natural gas	New York Hilton Hotel, New York, New York	Hilton New York	February 2005	FY2003 US DoD Climate Change Fuel Cell Program

(Continued)

TABLE 4.2 UTC PAFC Power Plants Sold, Contracted, or Installed in the United States: 1992–2012 (Total = 176 Systems, as of April 8, 2012) (*Cont'd*)

Plant size, fuel	Installation site	Coordinator/operator	Schedule of operation	Funding/subsidies/ rebates
200 kW, digester gas	Oakwood Beach Water Pollution Control Plant, New York, New York	New York Power Authority and New York State Energy Research and Development Authority	April 2005	Funding from New York Power Authority, New York State Energy Research and Development Authority, US EPA, US DoD
200 kW	Suffolk State Office Building, Regional Emergency Management Office, Hauppauge, New York	New York Power Authority, New York State Office of General Services	May 2005	FY2003 US DoD Climate Change Fuel Cell Program
7 × 200 kW, natural gas	Verizon call-routing center, Garden City, New York	Verizon	November 2005	FY1999 US DoD Climate Change Fuel Cell Program
200 kW, natural gas	Corona Rail Car Maintenance Facility, New York City	New York Power Authority	April 2006	FY2003 US DoD Climate Change Fuel Cell Program
200 kW	Los Angeles Zoo, Los Angeles, California	Los Angeles Department of Water and Power	July 2006	Funding included $1.4 million for purchase plus $600,000 for set up
200 kW, natural gas	East Rochester High School, East Rochester, New York	East Rochester School District	February 2007	Total funding was $2 million; $350,000 from the district Capital Reserve Fund and $833,000 from a State Energy Research and Development Authority Grant
200 kW	Fujitsu America, Sunnyvale, California	Pacific Gas & Electric	August 2007	Pacific Gas & Electric gave Fujitsu $500,000 in rebates for installing the system
4 × 200 kW	Cabela's Sporting Goods, Rentschler Field, East Hartford, Connecticut		Sales announced in September 2007; to be installed in 2008	

200 kW, natural gas	Bronx Zoo's Old Lion House, Bronx, New York	New York Power Authority	2008	FY2002 US DoD Change Fuel Cell Program
200 kW	Whole Foods Market, Glastonbury, Connecticut	Connecticut Clean Energy Fund	Sales announced in March 2008	Funding from Connecticut Clean Energy Fund
200 kW	Middletown High School and Vocational Agricultural Center, Middletown, Connecticut	Connecticut Clean Energy Fund	September 2008	Connecticut Clean Energy Fund provided $940,000 through the On-site Renewable Distributed Generation Program; $800,000 in state funding
12 × 400 kW	Freedom Tower, World Trade Center Site, New York	New York Power Authority	Sales announced in June 2008; to be delivered in January 2009	New York Power Authority provides $10.6 million to purchase the fuel cells
400 kW	St. Helena's Hospital, Napa Valley, California		Sales announced in July 2008; to be delivered in summer 2009	Partially funded by a grant from the California Self Generation Incentive Program
400 kW	Price Chopper Supermarket, Colonie, New York	New York State Energy Research Development Authority (NYSERDA)	To be commissioned in spring of 2009	Funding from NYSERDA's New Construction Program; New York offered other tax credits as well
400 kW	Star Market, West Bridgewater, Massachusetts		Slated to open in 2009	
400 kW	Whole Food Markets, Dedham, Massachusetts	Massachusetts Renewable Energy Trust Large Onsite Renewables Initiative	Sales announced in April 2009; installed in August 2009	A grant of $400,000 from the Massachusetts Renewable Energy Trust Large Onsite Renewables Initiative
2 × 400 kW	Coca Cola Enterprises' production facility in Elmsford, New York	New York State Energy Research Development Authority (NYSERDA)	Sales announced in July 2009	NYSERDA provides $2 million

(Continued)

TABLE 4.2 UTC PAFC Power Plants Sold, Contracted, or Installed in the United States: 1992–2012 (Total = 176 Systems, as of April 8, 2012) (*Cont'd*)

Plant size, fuel	Installation site	Coordinator/operator	Schedule of operation	Funding/subsidies/ rebates
200 kW	The Connecticut Science Center, Hartford, Connecticut		Commissioned in December 2009	CCEF made a grant in the amount of $613,000 to support the Science Center fuel cell
200 kW	White Plains Office Building, Westchester Country, New York	New York Power Authority (NYPA)	Installed in January 2010	A grant likely was provided by NYPA
400 kW	Whole Food Markets, South San Jose, California		Sales announced in February 2010	Unknown; but subsidy from California Self Generation Incentive Program likely
400 kW	Pitkin Plaza tower (luxury high-rise), 360 State Street. New Haven, Connecticut		Delivered in May 2010	The unit and installation cost about $4 million; CCEF paid half the cost
200 kW	Price Chopper Supermarket, Middletown, Connecticut		Shop to open in June 2010	CCEF grant likely
400 kW	Albertsons supermarket in the San Diego community of Claremont, California		Operational on September 1, 2010	Received incentive funding for clean energy alternatives
400 kW	Robert Clemente Leadership Academy and the Hill Central Music Academy, New Haven, Connecticut		Operational in September 2010	A CCEF grant likely
400 kW	Stop & Shop Supermarket, East Main Street, Torrington, Connecticut		Installed in November 2010	A CCEF grant

400 kW	Diversey, Sturtevant, Wisconsin		Project announced in November 2010	With a federal tax incentive and other grants and rebates, Diversey expects to recoup its $2.5 million investment in about 5 years
4 × 400 kW	Cox Communications, San Diego and Rancho Santa Margarita, California		Installed in February 2011	Supported by California's SGIP grant
400 kW	The Octagon, 500-unit apartment community, Roosevelt Island, New York		Installed in May 2011	Supported by $1.2 million from NYSERDA
400 kW	Hamden High School, Hamden, Connecticut		Contract agreed in May 2011	Supported by a CCEF grant
400 kW	Whole Foods Market, Fairfield, Connecticut		Sales announced in June 2011	Will be supported by a CCEF grant
2 × 400 kW	Coca Cola bottling facility, East Hartford, Connecticut		Installed in June 2011	CCEF grant likely
400 kW	CT Transit, Hartford, Connecticut	Federal Transit Administration	Sales announced in February 2012	Supported by a grant from the Transit Investments for Greenhouse Gas and Energy Reduction Program (Tigger)
400 kW	Pasadena Area Community College, Los Angeles, California		Authorized to execute a contract in March 2012	Expected to use an SGIP grant and federal tax credits

Sources: Various sources, including "Climate Change Rebate Projects Overview," ERDC-CERL Fuel Cell, http://dodfuelcell.cecer.army.mil/climate/index.php, the US Department of Defense; "Worldwide Fuel Cell Installations," Updated May 2005, Fuel Cells 2000, www.fuelcells.org; "Fuel Cell Installation Database," Fuel Cells 2000, http://www.fuelcells.org/info/data-basefront.html; and news releases from UTC Power, http://www.utcpower.com/pressroom/pressreleases, 2005–2012.

- The Climate Change Fuel Cell Rebate Program clearly encouraged power plant installation. The rebate program was started in fiscal year 1995 and it significantly increased the PAFC power plant installation in that year. The end of the program in 2004 saw significant decreases in plant installation in the following years.[v]

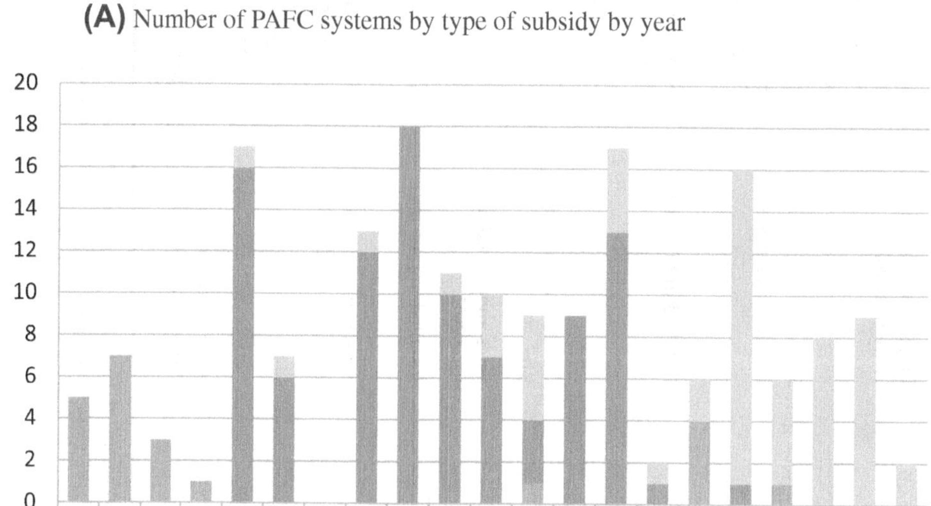

	1992	1993	1994	1995	1996	1997	1998	1999	2000	2001	2002	2003	2004	2005	2006	2007	2008	2009	2010	2011	2012
Other financial support					1	1		1		1	3	5		4	1	2	15	5	8	9	2
DoD Climate Change Fuel Cell Rebate					16	6		12	18	10	7	3	9	13	1		1				
No information on financial support	5	7	3	1								1				4		1			

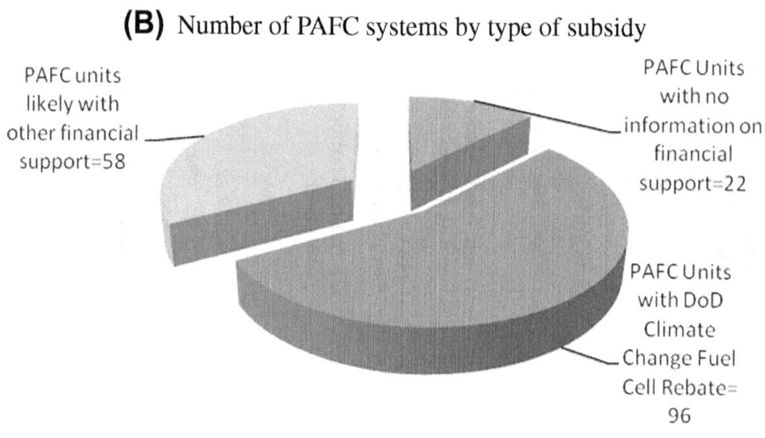

FIGURE 4.3 UTC PAFC power plant sold, contracted, or installed in the United States: 1992–2012 (total = 176, as of April 8, 2012). (For the color version of this figure, the reader is referred to the online version of this book.)

[v] The DoD Climate Change Rebate was awarded by fiscal year, but the actual plant installation was often scheduled in subsequent years.

(C) Number of PAFC systems by type of subsidy by state

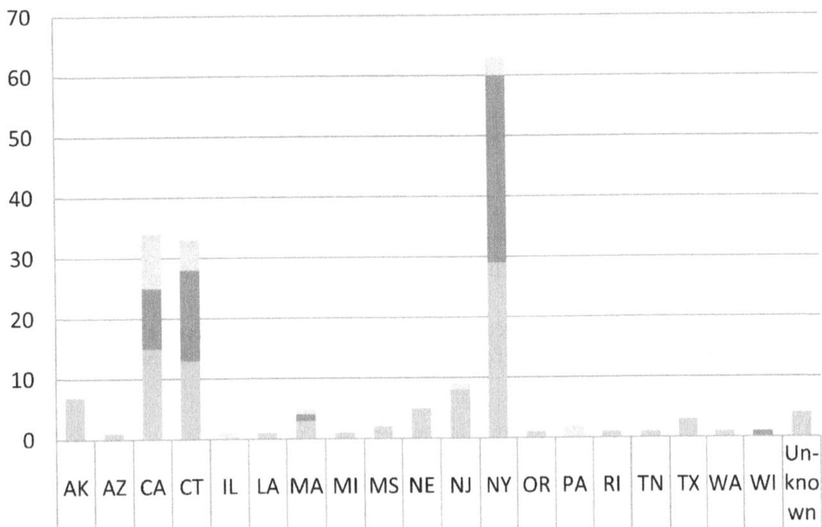

	AK	AZ	CA	CT	IL	LA	MA	MI	MS	NE	NJ	NY	OR	PA	RI	TN	TX	WA	WI	Un-known
No information on funding			9	5	1		1				1	3		2						
Other funding			10	15			1					31							1	
Climate Change Fuel Cell Rebate	7	1	15	13		1	3	1	2	5	8	29	1		1	1	3	1		4

FIGURE 4.3 (*Continued*).

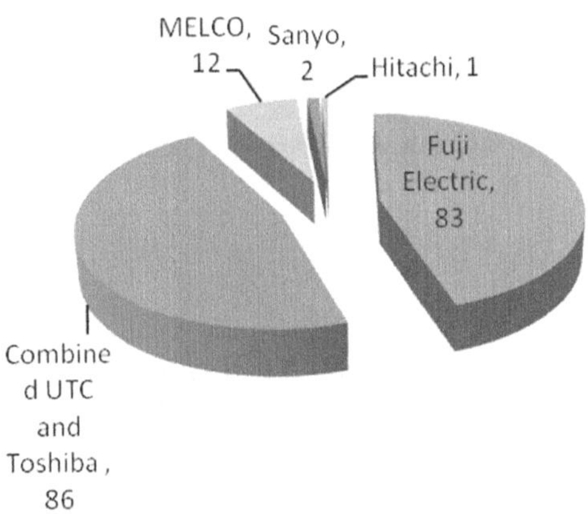

FIGURE 4.4 Japanese private-sector PAFC power plant installation by manufacturer: 1970–2012 (total PAFC power plant installed = 184, as of April 1, 2010). (For the color version of this figure, the reader is referred to the online version of this book.)

- Three states, California, Connecticut, and New York, purchased or planned to purchase a greater number of UTC PAFC systems; they are clearly more enthusiastic about fuel cell installations and offer greater financial incentives than the other states (Figs 4.3–4.6).

4.3. PAFC DEVELOPMENT IN JAPAN

4.3.1. Japanese Private-Sector PAFC Activities Begin in the 1960s

4.3.1.1. Japanese Utility Companies Engage in Field-Testing of US PAFC Power Plants

Japan began to be interested in PAFC technology in the 1960s. Other than the United

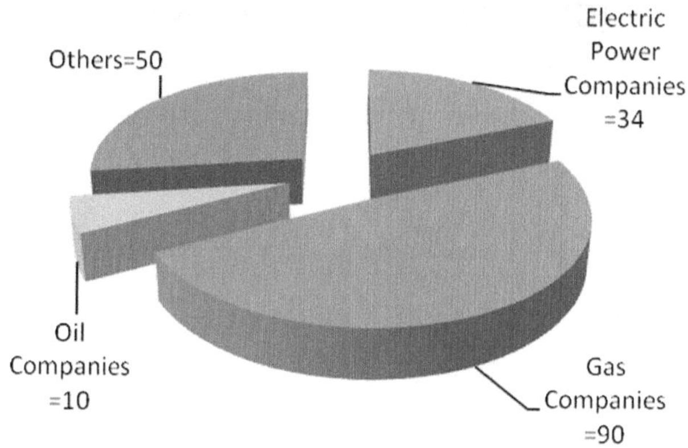

FIGURE 4.5 Japanese private-sector PAFC power plant installation by type of business: 1970–2010 (total power plants installed = 184, as of April 1, 2010). (For the color version of this figure, the reader is referred to the online version of this book.)

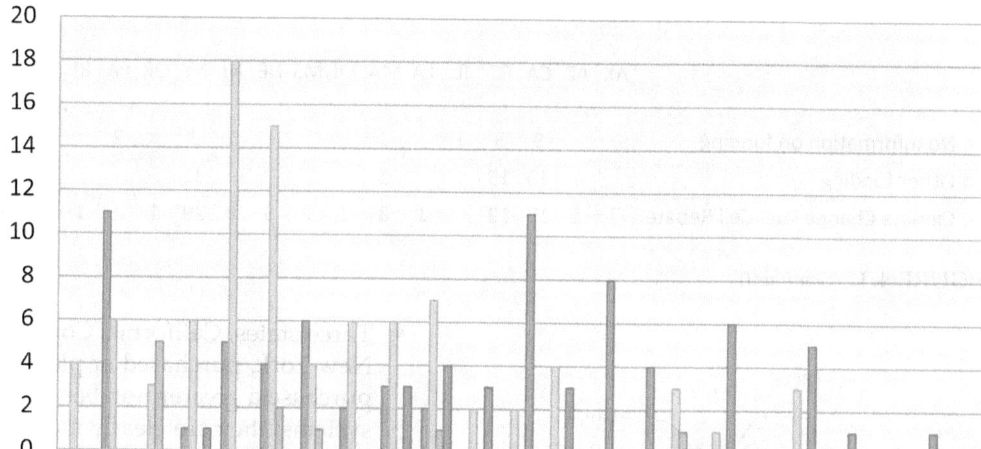

	1970-1979	1980-1989	1990	1991	1992	1993	1994	1995	1996	1997	1998	1999	2000	2001	2002	2003	2004	2005	2006	2007	2008	2009	2010
Electric Power Companies		11		1	5	4	6	2	3	2													
Gas Companies	4	6	3	6	18	15	4	6	6	7	2	2	4			3	1		3				
Oil Companies			5			2	1			1						1							
Others				1					3	4	3	11	3	8	4		6		5	1		1	

FIGURE 4.6 Japanese private-sector PAFC installation by business type by year: 1970–2010 (total PAFC power plant installed = 184, as of April 1, 2010). (For the color version of this figure, the reader is referred to the online version of this book.)

States, Japan was the only significant player in PAFC R&D. It is noteworthy that Japan was enthusiastic about PAFC technology when European countries were not and also that Japan's private sector took notice of PAFC technology decades ahead of the Japanese Government. European countries perhaps were wary of another fuel cell technology after having been disappointed in their AFC development work. The Japanese Government came to focus on PAFC technology nearly two decades after the private sector started PAFC field tests and R&D activities.

Early private-sector PAFC activity began in Japan when two major Japanese gas companies—Tokyo Gas and Osaka Gas—participated in the TARGET program (1967–1976). These two entities may have tested more fuel cells than anyone else in the world. When UTC developed the 12.5 kW PAFC power plant under the TARGET program, each of the two gas companies purchased two 12.5 kW power plants and test-operated them in Japan.

Tokyo Gas's test operation was conducted from March 1972 to August 1972 and Osaka Gas' test operation lasted from February 1973 to August 1973.[5] These were Japan's first fuel cell test operations and were widely reported, attracting major public attention.[10] The two companies also test-operated UTC's 40 kW power plants that were developed under the GRI-DOE programs (1976–1986). Tokyo Gas conducted the test operation from March 1982 to May 1987 and Osaka Gas from March 1982 to February 1985. These Japanese companies bought the power plants, installed them at their test sites, and operated them entirely with their own funds, underscoring their strong interest in PAFC technology. It was reported that the two gas companies had spent over $7 million (1982 US dollars) for the purchase of two 40 kW power plants, each from UTC. These two units operated in a grid-isolated mode close to their maximum capacity of 40 kW. Operating experience with these units generally confirmed the specified plant efficiencies and other operating characteristics. The two companies also subsequently purchased another 40 kW unit each and operated them from 1984 to 1987.[5]

Following the gas utility sector, the Japanese electric utility sector became interested in PAFC technology. A major electric power company, TEPCO, purchased a 4.5 MW power plant from UTC that was developed under the US FCG-1 Project and began plant construction at the company's Goi Thermal Power Station in Chiba Prefecture, 30 miles east of Tokyo.[5] Power plant installation started in August 1980 and was completed in early 1982. Testing of the plant was conducted from April 1982 for about a year. The 4.5 MW power plant operated successfully from April 1983 to December 1985 and demonstrated the feasible performance of a PAFC plant as a new electric power station for electric utility companies. Based on the positive experience, TEPCO contracted UTC and Toshiba to build an 11 MW power plant at the same Goi site. The construction began in January 1989 and the designed power level of 11 MW was attained in April 1991, with a net electrical efficiency of 41.1 percent,[16] an excellent performance. The 11 MW plant, however, subsequently experienced a malfunction in the stack and operated at a reduced output mode at 2 MW for 23,140 h until February 1997, at which time the demonstration test ended.[19]

4.3.1.2. *Japanese Electric Machinery Companies Launch PAFC Development*

Recognizing the enthusiasm of the utility companies and the major business opportunity that PAFC technology could present, Japanese electric machinery manufacturers—including Fuji Electric, Hitachi, MELCO, Toshiba, and Sanyo Electric—each launched PAFC R&D efforts in the 1970s. They viewed the utility companies to be their major future customers for their PAFC products and worked closely with them for R&D and field demonstration operations. Also, to quickly catch up with the state-of-the-art PAFC technology, Japanese PAFC developers looked to US companies—the global PAFC leaders—and sought partnership or licensing agreements with them. The US companies included UTC, Engelhard, ERC, and Westinghouse. Highlights of the PAFC development activities of these five Japanese electric machinery companies' are presented below.

4.3.1.2.1. FUJI ELECTRIC

Fuji started PAFC development in 1973 via a partnership with Engelhard.[11] In June 1980, Fuji launched a joint PAFC development and demonstration project with Kansai Electric Power. In October 1981, Fuji built a 30 kW air-cooled natural gas-fueled system based on Engelhard PAFC technology. After undergoing 3 months of testing at Fuji's Kawasaki factory, it was installed at Kansai Electric's Sakai-ko Thermal Electric Power Station near Osaka for on-site testing in April 1982. From December 1982 until December 1983 the plant was grid connected, and it performed satisfactorily under continuous long-term operation at various load levels. Although the initial voltage-current characteristics showed some decline with time, replenishing the phosphoric acid in the stacks after 1000 h of operation stabilized their performance. Throughout the test period, the performance of the reformer also was stable. Subsequently, Fuji developed various PAFC power plants, including 4 kW, 30 kW, 50 kW, 80 kW, 100 kW, and 500 kW,[38] and delivered more than 100 systems for experiments and field tests to gas, electric power, and oil companies.[39] For example, Fuji developed a 50 kW system using natural gas or liquefied petroleum gas (LPG) fuel supported by Tohoku Electric Power,[10] which was field tested for 2 years starting in May 1987.[5] Fuji also produced compact 4 kW methanol-fueled units in the mid- to late 1980s. In 1990, Fuji started the development of a 500 kW system with Osaka Gas and field tested it starting in 1993. In the early 1990s, Fuji was developing 50–100 kW models and subsequently focused exclusively on the 100 kW model.

4.3.1.2.2. TOSHIBA

Toshiba started PAFC R&D in the early 1960s[40] and by 1981 had built and tested 20 kW and 35 kW pressurized PAFC stacks. The company specialized in fuel reformer technology and developed an innovative catalytic reformer combustor system that burned air exhaust (with low oxygen concentration) obtained at the fuel cell cathode. This reduced the air required for compression and was claimed to increase overall plant efficiency by more than 1 percent.[10] In 1982, Toshiba fabricated a ¼ scale 50 kW burner to evaluate the best catalysts and catalyst packing methods. Based on the results of these experiments, a full-scale 50 kW exhausted air combustion reformer was built and tested in 1983 in cooperation with Mitsui Chemical. In 1985, Toshiba formed an alliance with UTC and formed joint ventures with International Fuel Cells (IFC) and its subsidiary ONSI. IFC conducted R&D and designed PAFC systems and ONSI built PAFC power plants, installed them at customer sites, and provided maintenance services. Toshiba manufactured the balance of plant that was incorporated in the UTC PAFC power plants to be marketed in Asia, including Japan. Toshiba, which never licensed UTC PAFC stack technology, developed on its own a complete PAFC power plant and built two 200 kW power plants for Chubu Electric Power for test operations in 1992 and 1994, respectively. Toshiba, however, believed that economies of scale were necessary to develop PAFC manufacturing technology that would cut the unit cost of the UTC PC25C power plant from $4,000 per kW down to $2000 per kW.[41] The company, thus, centralized all the power plant production at ONSI and undertook the bulk purchase of materials, components, and parts, as well as automated and streamlined the production line at the ONSI plant in the United States. In Japan, Toshiba developed with Mitsui & Co. Plant Systems a specialized fuel cell system—Fuel Cell Radiant Farm—a PAFC power plant using digester gases from animal wastes and launched the marketing of the system in 1998.[42] The FC Radiant Farm is a cogeneration system designed for green houses in colder regions, where excess heat could be used to warm the ground to nurture plant and bacteria growths.

4.3.1.2.3. MITSUBISHI ELECTRIC

Mitsubishi Electric (MELCO) worked with Westinghouse and started PAFC R&D in 1979, based on Westinghouse PAFC technology and built a 500 W system, including a fuel processor and inverter.[10] In late 1981, the company constructed a water-cooled 50 kW prototype pressurized PAFC system operating on reformed natural gas. In 1983, it built a 50 kW PAFC system that achieved a fuel cell module efficiency of 51 percent and an overall system efficiency of 35 percent. The company developed a 100 kW methanol-fueled power plant and demonstrated it with Hokkaido Electric Power in the late 1980s.[5] It also developed a 200 kW PAFC system and field tested it in cooperation with other electric power companies. By 1999, MELCO had field tested eleven 200 kW PAFC power plants; with four of them operating longer than 20,000 h.[43] The company also worked on developing a low-cost 200 kW prototype to be ready by 1999 for a 2-year test. It planned to start building a mass production facility in 2000, but never executed it.

4.3.1.2.4. SANYO

Sanyo started PAFC development in 1972 and developed 2 kW and 10 kW stacks in cooperation with MELCO.[10] In 1980, Sanyo licensed the US company ERC's PAFC technology to accelerate the development of air-cooled PAFC power plants. Sanyo started to cooperate with TEPCO in 1984 with plans to develop 10 MW power plants.[44] In 1986, Sanyo developed a 50 kW power plant and conducted experimental operations with TEPCO. Sanyo also developed an air-cooled 200 kW PAFC power plant in a joint venture with TEPCO;[45] the plant operated at TEPCO's Shin-Tokyo Thermal Power Station from 1987 to 1989.[5] From 1989 to 1991, Sanyo also developed 100 kW naphtha-fueled PAFC power plants and field tested them jointly with petroleum companies and the Japan Petroleum Energy Center (JPEC) at JPEC's Chita Refinery.[45] By the late 1980s, Sanyo began developing portable PAFC systems for communications. In 1991, Sanyo developed and delivered portable 5 kW and 10 kW PAFC systems to the Japan Ministry of Defense. In 1992, Sanyo developed 200 W AC and 250 W DC hydrogen-fueled, portable PAFC systems and demonstrated these at various NTT sites in Japan. In 1995, Sanyo developed a 1 kW portable system for NTT and tested it at NTT sites. In 1996, Sanyo started to develop a 3 kW household-use power plant for NEDO and delivered two of these to NTT in 1998, one fueled by natural gas and the second fueled by methanol.

4.3.1.2.5. HITACHI

Hitachi's work on PAFC technology started in 1975 and built a 100 W reformed natural gas PAFC system, which was tested by Chubu Electric Power in 1978–1979.[10] The company developed and demonstrated a 1 kW PAFC stack in 1980 and constructed and demonstrated an experimental 20 kW pressurized system in 1981. Also in 1980, Hitachi entered a cross-licensing agreement with ERC to rapidly increase its PAFC capabilities. The company built and tested a four-stack 50 kW PAFC system that operates on reformed natural gas using a steam reformer developed jointly with Toyo Engineering. The fuel cell modules, each 30 × 42.5 cm, operated at 180–190 °C and at atmospheric pressure. Hitachi developed a 100 kW power plant and field tested it with Tokyo Gas from 1990 to 1992.

4.3.2. Japanese Government Launches PAFC Program in 1981

In 1981, Japan's Ministry of Economy, Technology, and Industry (METI, formerly the Ministry of International Trade and Industry (MITI)) started a series of PAFC development programs. It implemented a 10-year PAFC R&D program under the Moonlight Project to establish basic PAFC technology, followed by programs to establish technologies for the practical use of PAFC power plants in a grid-connected environment as well as to foster the use of PAFC power plants as a distributed power system or at central power stations. Japan underwent two serious oil shocks and recognized the need to start a program to develop PAFC power plants that could generate electricity using fuels other than oil. Although the government action was considerably behind that of the private sector, the private sector undoubtedly was encouraged and reassured by the infusion of government funding for PAFC development.

4.3.2.1. *METI's Moonlight Project*

The Moonlight Project was one of the two large-scale energy R&D programs that Japan launched in reaction to the oil embargoes of the 1970s. Triggered by the oil crisis, Japan launched the Sunshine Project in 1973 and the Moonlight Project in 1978. The Sunshine Project was to address the problems caused by energy shortages or embargoes, and the Moonlight Project was to develop high-efficiency energy conversion technologies to reduce Japan's dependence on oil. These two programs were Japan's first, long-term, large-scale R&D programs, where METI would simultaneously sponsor multiple, multi-year R&D projects. In 1981, the Moonlight Project assumed responsibility for developing fuel cell technology, which was previously managed under the Sunshine Project. Prior to the shift of this responsibility, the Sunshine Project ran only a few minor AFC projects. The Sunshine Project and the Moonlight Project were consolidated into the New Sunshine Program in 1993, which continues to fund fuel cell R&D to the present day.

The 10-year PAFC R&D program under the Moonlight Project had a budget of about $116 million (¥16.3 billion[46]—at an exchange rate of $1/¥140).[vi] The program consisted of two projects: the development of two 1 MW power plants (1981–1988) and the development of two 200 kW power plants (1986–1990).[vii] METI's goal was specifically aimed at catching up to UTC, the global PAFC technology leader; METI's two projects were aimed to develop power plants closely resembling UTC power plants in terms of output capacities and operational parameters.

4.3.2.1.1. 1 MW POWER PLANT PROJECT (FISCAL YEAR 1981–1988)

The first fuel cell project implemented under the Moonlight Project was to develop two 1 MW power plants for dispersed and central power station applications, respectively. The dispersed power plant was an unmanned, automatically operated plant, and the central power plant was to be operated by one person. METI's subsequent goal following this project was to develop a 20 MW power plant operating at a hydrogen utilization rate of 80 percent, an air utilization rate of 60 percent, and an overall system efficiency of 42.3 percent (HHV).[10] The 1 MW project was initially scheduled for Japanese fiscal years 1981–1985, but was extended by 3 years. Four major electric machinery companies—Fuji Electric, MELCO, Toshiba, and Hitachi—were selected to develop the plants. Under the 1 MW project, each power plant was to be developed by a team of two companies.[47] Fuji Electric and MELCO were assigned to a team, and Hitachi and Toshiba to the second team. The project likely was an immense challenge for the Japanese companies; although they had been engaged in

[vi] Throughout the book, foreign currencies are converted into then-year US dollars to provide a ballpark figure. The exchange rates were sourced from "United States dollar," Wikipedia, http://en.wikipedia.org/wiki/U.S._Dollar, last modified on June 19, 2009, and "Historical Lookup," http://www.x-rates.com/cgi-bin/hlookup.cgi.

[vii] Japan's fiscal year starts in April and ends in March of the following year.

PAFC development for some time, they had never developed a power plant as large as a 1 MW system.[46] Nonetheless, UTC in the United States had already developed a 4.5 MW power plant and METI likely pushed the Japanese companies to catch up with UTC as quickly as possible. Illustrating Japanese eagerness to learn from US technological leadership, a system benchmark set for both 1 MW plants reflected many operating conditions for UTC's PAFC systems.[10]

- The Fuji–MELCO 1 MW plant was a low-pressure, low-temperature power generation system for dispersed plants, in which the fuel cell stacks were designed to operate at 190 °C and at 3.4 atm, which were the operating conditions in the UTC 4.5 MW demonstrator.[10]
- The second 1 MW plant to be developed by Toshiba and Hitachi was a high-pressure, high-temperature system that would operate at about 205 °C and 8.2 atm, which were the UTC 11 MW operating conditions.
- Both plants had cell areas of around 3,600 cm^2, similar to those of the UTC 4.5 MW demonstrators.[10]

The work-share arrangement for the 1 MW system development was highly unique and creative. It was designed to spread PAFC expertise widely, thereby setting up a solid PAFC industrial base in Japan quickly. In the arrangement, one partner (Partner A) in a team developed fuel cell stacks and the other partner (Partner B) the BoP equipment. The entire cost of the activities was paid by the government.[10] At the same time, however, both the partners were required to carry out an in-house program with their own funds to develop a complementary part of the total system. Thus, Partner A had to develop the BoP equipment and Partner B had to develop fuel cell stacks with their own funds. Under this arrangement, the four companies involved in the project each would develop an entire 1 MW PAFC power plant. The companies clearly were eager and invested far more than their cost share in the project; one of the four companies indicated that it invested more than the government funds and paid the researchers with in-house funds as well.[46]

- The Fuji–MELCO team: Fuji Electric worked on the BoP under METI funding; while it developed two 250 kW stacks with in-house investment. MELCO was responsible for developing two 250 kW stacks with METI support and developed the BoP with in-house funding. The outcome of the project was one 1 MW power plant, consisting of two 250 kW MELCO PAFC stacks and two 250 kW Fuji stacks.
- The Toshiba–Hitachi team: Toshiba developed the BoP under METI funding and built two 250 kW stacks with their own funds.[10] Hitachi developed its two 250 kW stacks with METI funding and the BoP with in-house funding. At the end of 1986, one 1 MW plant was built, consisting of two 250 kW Toshiba stacks and two 250 kW Hitachi stacks.

The field operations of the two 1 MW PAFC plants clearly were not successful. Some sources indicated that the principal goal for the 1 MW plants was to completely convert natural gas to grid-quality alternating current and to demonstrate stable operations and performance, and this was almost achieved in 1988.[5,47] Nonetheless, the project suffered from lack of adequate teamwork, resulting in mismatched stacks, delays in the project, and dismantling the plant without determining lifetime performance or durability:

- The Fuji–MELCO plant: The Fuji stacks and MELCO stacks did not match exactly.[10] The two companies, which were business competitors, clearly did not communicate well.[46] To remedy the situation, two separate cooling and steam systems and two parallel

DC systems had to be installed,[10] with the result that the 1 MW plant virtually had become two 500 kW power plants. The system was finally installed, tested, and certified for operation at the Sakai-ko Thermal Power Station of Kansai Electric in October 1987 and the operations started in May 1988,[48] 2 months past the scheduled end of the project. The plant operated for a total of 2045 h with 697,151 kWh of cumulative power generated. In November 1988, the plant was dismantled and was reported to have shown no discernible problems that would have impeded further operations if needed. The project was declared a success and ended in January 1989.
- The Toshiba–Hitachi plant: The plant was installed at the Chita Daini Thermal Power Station of Chubu Electric Power near Nagoya and tested throughout 1987.[48] The certification for operation was delayed until June 1988 due to problems with the cooling system of one of the stacks. The plant was finally in full operation in August and September 1988, resulting in a total of 1018 h of operation with 367,583 kWh of cumulative power generated. The plant was dismantled in October 1988 for examination and reportedly no problems were found that would have effected further operations. The project ended in December 1988.

A noteworthy fact was the project's lack of concern about the durability or lifetime of the plants and the cost competitiveness of the plants. It was simply assumed that both the plants would have operated for the expected 40,000 h based on the observed rate of cell voltage reduction.[46] Also, METI did not appear to have included cost reduction as a project goal. It is likely that the Japanese government, as well as the companies, believed that the cost could be cut once the technology became mature and mass production began.

4.3.2.1.2. 200 KW PAFC SYSTEM DEVELOPMENT (FISCAL YEAR 1986–1990)

METI's second PAFC project under the Moonlight Project was the development of two 200 kW PAFC plants for on-site cogeneration to support remote island applications. The contracts were awarded to Fuji Electric and MELCO, with each to develop one 200 kW power plant. The project started in 1986 and continued until March 1991. It was extraordinary that the development of a large power plant was followed by a smaller plant, instead of working on a simpler, smaller system first and then moving to a larger, more complex system. METI's earlier plans to scale up to a 20 MW power plant based on technology of the 1 MW power plant were never implemented, possibly because UTC by that time had decided to develop a small 200 kW power plant as its baseline commercial model.

- In this project, Fuji Electric developed a 200 kW methanol-fueled PAFC power plant. The system was a water-cooled unit operating at atmospheric pressure type[10] and was intended to be grid independent.[47] It was installed on Tokashiki Island, 20 km west of Okinawa and began operations at the beginning of 1990. The Okinawa Electric Power Company operated the power plant.
- MELCO developed a 200 kW on-site cogeneration PAFC power plant.[5] The plant was water cooled, used natural gas, and cogenerated conditioned air via an absorption chiller.[47] The plant was installed at the Hotel Plaza in Osaka at the beginning of 1990, and the on-site cogeneration program was carried out by Osaka Gas, Kansai Electric, and Kobe Steel.

The 200 kW project was deemed successful.[10] The 200 kW project benefited from the lessons learned from the 1 MW project; the power plants were compact and well designed and built with improved component technologies.[46] The two plants entered into operation in early 1990, both attaining a power output of 200 kW and achieving the efficiency goals. Both ended operations in March 1991, right on the planned schedule. The Fuji remote island plant achieved an efficiency of 39.7 percent, and the MELCO on-site cogeneration unit achieved an efficiency of 36 percent and 80.2 percent with the utilization of the waste heat. As of March 31, 1991, the 200 kW remote island plant had operated for a total of 6880 h, and the 200 kW plant for on-site cogeneration had operated for a total of 8832 h. Based on operating data, both plants were assessed to be able to operate for as long as 40,000 h, but no attempts were made to explore how long the plants actually would have operated. The project did not address cost reduction issues.

4.3.2.2. *Other METI PAFC Programs*

To complement the two PAFC R&D projects, METI launched two additional programs to foster the use of PAFC power plants in Japan. One program was designed to verify and ensure safe and stable interconnectivity of fuel cell power plants as well as other renewable power generation systems with the electric power grids. The second program was aimed at building and testing large PAFC power plants to be used either as central or as distributed power stations in urban areas. They were implemented over 10 years, from 1987 to 1997 (Table 4.7).

4.3.2.2.1. GRID INTERCONNECTION DEMONSTRATION PROJECT

From fiscal 1987 to 1993, METI implemented a project to study the feasibility of grid interconnection of a large number of dispersed advanced power generation systems, including fuel cell power plants and photovoltaic and wind power generation systems.[49] Although the study did not specifically focus on the performance of fuel cell power plants, some information on the operating hours were documented, indicating METI's new awareness that PAFC power plants needed to demonstrate durability as well as meet other performance criteria. The 1987–1993 study aimed at assessing the operational stability and viability of dispersed power sources, quality of electricity, protective coordination, security and safety, and establishing grid interconnection techniques that would be feasible for distribution lines in commercial, industrial, and residential areas. The study was deemed successful and provided a base for safe, stable, and practical grid interconnectivity that could be used once PAFC power plants were commercialized and massively installed. For this study, fourteen 50 kW PAFC power plants and one 200 kW PAFC power plants were employed, along with photovoltaic generators for 400 kW (forty 2 kW, forty 3 kW, and twenty 5 kW systems) and two 16.5 kW wind power generators, for a total capacity of 1.233 MW. These distributed power generators were incrementally installed at Rokko New Energy Test Station, which had been established by Kansai Electric near Osaka.[50] The 50 kW PAFC power plants supplied by Fuji Electric and the 200 kW PAFC power plant supplied by MELCO were installed between 1989 and 1991. The study ended in 1993, but the operation of the power plants continued until 1998 under another study, called "A Demonstration Test to Establish Next Generation Electricity Supply Systems Based on Distributed Power Sources." Available information, however, showed that the durability performance of the PAFC systems was rather poor. The Fuji 50 kW plants operated for 5–7 years, but averaged only 17,000 h each, with the longest operating 23,500 h. The MELCO 200 kW plant operated for 7 years for a total of 27,000 h.

4.3.2.2.2. URBAN ENERGY CENTER PAFC TECHNOLOGY DEVELOPMENT

In 1991, METI launched another project to establish PAFC power plants as distributed power plants in urban areas.[19] Project goals, which were planned for fiscal years

1991–1996, were to design, develop, and test operate a 5 MW pressurized PAFC power plant and a 1 MW atmospheric pressure power plant. After this project, METI never implemented another project involving megawatt-class PAFC power plants, undoubtedly marking the end of Japan's interest in large-scale PAFC systems.

- Fuji Electric, in cooperation with 10 electric power companies, developed a 5 MW power plant and installed it at Amagasaki Third Thermal Power Plant of Kansai Electric Power in 1994. The plant experienced a series of technical troubles as well as difficulties typical for pressurized plants, delaying the start of the demonstration operation for about 1 year.[19] The plant reached 41 percent electricity efficiency—lower than the expected 42 percent—in March 1995 and continued operations until March 1997, for a total of 6410 h.
- Toshiba in cooperation with four gas companies developed a 1 MW power plant and installed it at Tokyo Gas Tamachi Research Institute in 1994. The plant reached the planned power level in 1995 and continued operation until March 1997.[19] The plant operated for a total of 10,165 h, including 3247 h of continuous uninterrupted operations, and showed a power efficiency of 38 percent, higher than the expected 36 percent.

4.3.3. Government and Private Sector Join Hands in Field Test Program

As the private sector had done for many years earlier, the Japanese government started supporting field test programs in the 1990s. The government-sponsored programs likely encouraged the private sector and infused new enthusiasm in PAFC demonstration activities in Japan. PAFC demonstration activities in Japanese private sector started in the 1970s, but were not always energetic. In the early 1990s, the number of PAFC power plants field tested was as few as eight units per year. When the Japanese government field test program was started in 1992, however, the number of PAFC power plants in private-sector field tests jumped to 23 units. Although the government field demonstration and private-sector field tests were of a different scale and likely adhered to different testing qualities and standards, both came to the same conclusion in about the same time frame: both concluded in the late 1990s that Japanese PAFC power plants were ready for commercialization. The government PAFC field test program involved a total of 43 PAFC power plants.

4.3.3.1. *METI PAFC Field Test Program*

In 1992, METI launched two PAFC projects to demonstrate the reliability and durability of PAFC power plants and assess their commercial viability. They were Field Test Project I, which was called the "Fuel Cell Power Generation Field Test Project" (fiscal year 1992–1996) and Field Test Project II, which was called the "Advanced High Efficiency Energy System Field Test Project" (fiscal year 1997–2000). Field Test Project I covered a 5-year period starting in April 1992 and ending in March 1997; it tested 28 PAFC power plants of various power outputs at 24 sites.[48] Field Test Project II was implemented in the subsequent 4-year period from April 1997 to March 2001; it tested 15 PAFC power plants.[51] In Project I, METI selected mostly domestically built power plants; 20 out of the 28 units were made by Japanese manufacturers, including 17 Fuji Electric plants, two MELCO plants, and one Toshiba plant. The other eight plants—five ONSI plants and three Toshiba/ONSI plants—were partially or entirely manufactured outside Japan. It is assumed that the so-called "Toshiba plant" was made with Toshiba PAFC stacks and Toshiba BoP, the so-called "ONSI plants" were made of UTC PAFC stacks and UTC BoP, and the so-called "Toshiba/ONSI plants" were made of UTC PAFC stacks and

Toshiba BoP. In Project II, METI tested 15 PAFC power plants. Eleven of these used UTC stacks (they were Toshiba/ONSI power plants); the remaining four included three Fuji plants and one MELCO plant. METI's preference for UTC stacks likely reflected the higher performance of these plants compared to the domestic ones.

- In 1992, UTC launched the commercialization of its PAFC power plants. The shipment of commercial models started in November 1995.[19] The ONSI and Toshiba/ONSI power plants used in the METI field tests, therefore, were both precommercial prototypes and commercial models.
- A review of the test results shows that the power plants in Field Test Project I performed rather poorly but that the power plants in Field Test Project II performed considerably better. Perhaps in early 1997, METI decided that PAFC technology had matured almost to the level where the commercialization of power plants could easily be achieved.[51]

4.3.3.1.1. FIELD TEST PROJECT I (FISCAL YEAR 1992–1996)

Field Test Project I ended in March 1997 and a review of the field-test results compiled as of March 1997 showed a rather inferior overall performance (Table 4-3). The average operating time for all 28 systems was 15,038 h; the longest operating unit ran 26,741 h. The average cumulative availability was 70 percent, and the highest availability was 92 percent. The average power efficiency was about 33.6 percent and the highest was 38 percent. The average thermal efficiency was 32.6 percent and the highest was 51 percent. In summary, the durability, availability, and power efficiency were all rather poor.

A closer look at the data, however, indicates that there were considerable differences in performance among the plants, pointing to the fact that the power plants with UTC stacks demonstrated better performance than those made by the Japanese developers (Table 4.4). It is difficult to rate the overall performance of a power plant because some performance factors may be affected by a specific circumstance; for example, operating hours could be limited by late installation, and thermal efficiency could be affected by the plant users' interest in exploiting waste. In an attempt, at least, to crudely distinguish between the best and the poorest, a simplified evaluation method is employed (Table 4-4):

- This method focuses on three performance criteria—availability, overall efficiency (the sum of electric and thermal efficiencies), and the average mean time between failure (MTBF)—to rate the performance of the power plants. In each criterion, the average performance of the systems made by five manufacturers is compared. A number from 5 to 1 is assigned to assess the performance ranking among the five manufacturers in the selected performance criteria (the highest performance is 5). It is assumed that the highest total score identifies the best and the lowest the poorest.
    - Availability is selected because it is a better indicator than operating hours and because some power plants were installed earlier or later than some others, as noted earlier.
    - Overall efficiency is selected because it is a significant factor and thermal efficiency may not always be a veritable factor because of the lack of interest in exploiting waste heat in some cases.
    - Average MTBF is selected over average number of operations failures; the number of failures may be greater if power plants were installed earlier and operated longer, and the seriousness of failure and the duration of time to recover from the failure are perhaps better represented by MTBF than the number of failures.

Although none of the power plants performed favorably, the ONSI power plants appear to be the best performing among the five groups of

TABLE 4.3 METI PAFC Field Test Project I: FY1992–1996 (Through March 31,1997)

	Plant site	Plant capacity	Manufacturer	Starting date-ending date	Operating hours	Availability (%)	Efficiency (%)			Number of operations failures	MTBF
							Electric	Thermal	Overall		
1.	Tokyo Kogakuin University	50	Fuji Electric	9/22/92–2/3/96	10,338	35	33.6	37	70.6	22	470
2.	Meguro Gajoen	50	Fuji Electric	12/9/92–10/11/95	13,165	52.9	32.5	40.1	72.6	10	1,316
3.	Nippon Telegraph and Telephone Company, Yokohama Branch	100	Fuji Electric	6/5/93–3/29/96	11,328	46.7	38	39.6	77.6	16	708
4.	East Japan Railway Company, Ohi Works	100	Fuji Electric	12/17/93–3/29/96	12,498	62.6	34.6	42.9	77.5	17	735
5.	Tokyo East 21	200	Toshiba/ONSI	11/18/92–3/31/96	20,440	69.2	34.9	39.2	74.1	30	681
6.	Koshien Miyako Hotel	50	Fuji Electric	3/10/93–1/17/95	13,008	80	33.6	37.5	71.1	5	2,602
7.	Koshien Toyo Building	50	Fuji Electric	6/7/93–1/13/95	11,551	82.3	32.4	48.4	80.8	2	5,776
8.	Hankyu Corporation Head Office Building	50	Fuji Electric	3/30/93–3/31/97	23,949	68.2	30.7	40.4	71.1	9	2,661
		50	Fuji Electric	3/30/93–3/31/97	26,741	76.2	29.6	40.2	69.8	19	1,407
9.	Mitsui Garden Hotel Osaka	100	Fuji Electric	7/9/93–2/5/95	10,429	75.4	34.3	33.5	67.8	4	2,607
10.	Esteem Life Gakuenmae	100	Fuji Electric	9/20/93–1/20/95	7140	61.1	33.5	37.5	71	7	1,020
11.	Sumitomo Chemical Osaka Plant	200	Toshiba/ONSI	6/25/93–3/31/97	24,059	72.9	36	39.4	75.4	7	3,437
12.	Asia & Pacific Trading Center	500	Fuji Electric	4/11/94–3/31/97	7684	29.5	36.6	37	73.6	11	699
		500	Fuji Electric	4/11/94–3/31/97	12,919	49.6	37.2	16.2	53.4	7	1,846

13.	Port of Nagoya, Public Aquarium	100	Fuji Electric	7/23/93–3/31/97	19,244	60.1	34.7	43.7	78.4	19	1,013
14.	Toenec Corporation Education & Training	50	Fuji Electric	3/22/93–1/26/95	12,987	80.2	26.7	44.7	71.4	13	999
15.	UNEP International Environmental Technology Center	50	Fuji Electric	10/1/93–3/31/97	24,476	79.8	30.9	42.3	73.2	11	2,225
16.	Tachikawa Toshi Center	200	ONSI	9/14/94–3/31/97	16,549	74.8	30.1	0.9	31	8	2,069
		200	ONSI	9/15/94–3/31/97	19,023	86	31.1	1.5	32.6	10	1,902
17.	River Side Sumida	100	Fuji Electric	9/2/94–3/31/97	13,456	59.6	30.5	35.8	66.3	6	2,243
18.	Matsushita Electric Works	200	ONSI	1/12/94–3/31/97	23,368	82.9	35	36.1	71.1	9	2,596
19.	Osaka Red Cross Hospital	200	ONSI	10/11/94–3/31/97	18,403	85	36.6	13.9	50.5	17	1,083
20.	Tokyo Metropolitan Research Institute for Environment	200	ONSI	5/24/95–3/31/97	13,037	81.9	34.3	51	85.3	4	3,259
21.	Itabashi Ecopolis Center	50	Fuji Electric	3/1/95–3/31/97	16,191	92.2	35.9	37.7	73.6	8	2,024
22.	Kirin Brewery Kyoto Plant	200	MELCO	10/12/95–3/31/97	10,556	82	33.7	11.5	45.2	3	3,519
		200	MELCO	10/12/95–3/31/97	9772	75.9	34.8	11.9	46.7	5	1,954
23.	Nisshin Steel Sakai Plant	200	Toshiba	6/20/95–3/31/97	12,508	80.1	33.6	5.9	39.5	9	1,390
24.	Tokyo Metropolitan Misono Filtration Plant	200	Toshiba/ONSI	5/29/96–3/31/97	6260	85.1	34.6	49	83.6	3	2,087
	Total				421,079	1967.2	940	914.8	1,854.8	291	54,328
	Average of 28 power plants				15,038	70.2	33.6	32.6	66.2	10.4	1940

TABLE 4.4 METI PAFC Field Test Project I: Summary of Power Plant Performance by Manufacturers*

Manufacturer	Number of power plants	Averageoperating hours*	Average availability (%)*	Average efficiency (%)			Average number of operations failures*	Average MTBF*	Total in parentheses	Overall ranking
				Electric*	Thermal*	Overall				
Fuji Electric	17	14,536	64.2 (1)	33.3	38.5	71.8 (4)	11	1,785 (2)	7	3
ONSI	5	18,076	82 (5)	33.4	20.7	54.1 (3)	9.6	2,182 (4)	12	1
Toshiba/ ONSI	3	16,920	75.7 (2)	35.2	42.5	77.7 (5)	13.3	2,068 (3)	10	2
MELCO	2	10,164	79 (3)	34	12	46 (2)	4	2,737 (5)	10	2
Toshiba	1	12,508	80.1 (4)	33.6	5.9	39.5 (1)	9	1,390 (1)	6	4

* *Numbers in parentheses are assigned according to the order of performance among the five different manufacturers, with 5 the highest ranking and 1 the lowest in terms of performance. Numbers are assigned for selected performance areas only for simplicity.*

plants, with the Toshiba/ONSI and MELCO plants second best, the Fuji plants fourth, and the Toshiba plant perhaps the poorest. The ONSI plants demonstrated greatest availability with a relatively long MTBF, although the overall efficiency was about in the middle. The Toshiba/ONSI and MELCO plants were on a par with each other and overall were second best; the former showed the highest overall efficiency, and the latter demonstrated the longest MTBF. The Fuji plants were fourth; they demonstrated a relatively high overall efficiency but lowest availability. The Toshiba plants were fifth; they showed a relatively high availability but with lowest overall efficiency.

4.3.3.1.2. FIELD TEST PROJECT II (FISCAL YEAR 1997–2000)

For Field Test Project II, METI placed emphasis on selecting plants based on performance; it evaluated more Toshiba/ONSI power plants than domestically manufactured systems. In March 2001, Field Test Project II ended with the results demonstrating dramatic improvements over what was demonstrated during Field Test Project I (1992–1996) (Table 4.5):

- Average operating hours were 27,637 for the 15 power plants. The longest operating unit, which started operation in May 1998, had a cumulative total of 32,331 h by March 2001. The power plant with the shortest operating period (just 10,000 h) was the MELCO plant that was installed at Kirin Brewery, where the power plant was fueled with a brewery-derived digestive gas. The power plant had a defect in the fuel reformer, which could not properly remove ammonia contained in the digestive gas. The brewery also stopped its operations during holidays and weekends and did not generate sufficient amount of digestive gas for continuous power-plant operations. If the Kirin plant is excluded, the average operating time for the 14 plants increases to 28,886 h. This was a fine record and achieved more than 70 percent of the expected 40,000 hour lifetime of a stationary power plant, especially when all of the power plants continued to be in operation at the end of the field-test project period.
- The average availability of all 15 plants was 91 percent (95 percent if the Kirin plant is excluded). Not counting the Kirin plant, the overall record is an improvement of about 35 percent over the availability observed in Field Test Project I.
- The average power efficiency of the 15 plants was 38 percent. One plant had an average power efficiency of 41 percent, which was the highest power efficiency of all; the Kirin plant had the efficiency of 30 percent. The average power efficiency of the 14 plants, excluding the Kirin plant, was 38.6 percent. The average thermal efficiency at the 15 plants was 53 percent, which is highly satisfactory. One plant had a thermal efficiency of 59 percent, which was the highest. The average thermal efficiency at the 14 plants, excluding the Kirin plant, was 54.4 percent. The average overall efficiency for the 15 plants was 91 percent; excluding the Kirin plant, average overall efficiency is 93 percent, which is also an excellent record.
- There were 40 plant failures over the entire operating period, other than those for scheduled maintenance, averaging 2.7 failures per plant (or 1.8 failures per plant if the Kirin plant is excluded). Two plants had no failures, nine plants had one failure, two plants had three failures, one plant had five failures, and the Kirin plant had 20 failures. Of these, one failure was due to cell poisoning from ammonia leakage in the fuel reformer and 19 were due to an insufficient supply of digestive gas, an external factor not directly related to the power plant. The average MTBF was 6835 h for the 15 plants and 7196 h excluding Kirin. The average mean time to recovery (MTTR) was 95 h for the 15 plants and 34 h

TABLE 4.5 METI PAFC Field Test II: FY1996–2000 (Through March 31, 2001)

	Plant site	Plant capacity	Manufacturer	Start date	Operating hours	Availability (%)	Efficiency (%)			Number of operations failures	MTBF	MTTR
							Electric	Thermal	Overall			
1.	Nihon Telegraph and Telephone East Corporation Training Institute (NTT East Training Institute)	200	Toshiba/ONSI	6/29/98	30,474	93	39	59	98	1	8415	0
2.	The Energy & Earth Exploratorium	200	Toshiba/ONSI	8/4/98	29,918	93	39	55	94	3	2845	37
3.	Tachikawa Toshi Center	200	Toshiba/ONSI	7/2/98	31,017	94	39	52	91	5	1691	30
4.	TGS Akabane Building	50	Fuji Electric	7/7/98	31,185	95	36	51	87	1	8365	6
5.	Nagoya University	200	Toshiba/ONSI	5/22/98	32,331	96	37	55	92	0	8593	0
6.	Umeda Center Building	200	Toshiba/ONSI	6/29/98	30,667	93	39	57	96	1	8554	44
7.	Research Institute of Innovative Technology for the Earth (RITE)	200	Toshiba/ONSI	5/14/98	32,059	94	39	54	93	1	8514	6

8.	Senboku LNG Terminals	200	Toshiba/ONSI	4/24/98	31,847	92	39	52	91	0	8070	0
9.	Kirin Brewery Tochigi Plant	200	MELCO	5/24/99	10,146	41	30	35	65	20	1776	943
10.	Tokyo Metropolitan Institute of Technology	200	Toshiba/ONSI	3/25/99	25,116	95	39	51	90	1	8603	19
11.	Tokai Rika Factory	200	Toshiba/ONSI	3/10/99	25,972	97	39	54	93	1	8635	0
12.	Nagoya Sakae Washington Hotel Plaza	100	Fuji Electric	3/24/99	26,096	99	38	56	94	1	8550	15
13.	Mycal Bore Sanda	200	Toshiba/ONSI	3/15/99	26,003	97	38	55	93	3	2892	18
14.	Co-op Himeji Shirahama	100	Fuji Electric	3/17/99	26,398	99	41	53	94	1	8719	7
15.	Huis Ten Bosch Heating and Cooling Supply	200	Toshiba/ONSI	3/19/99	25,322	95	39	57	96	1	8296	293
	Total				414,551	1373	571	796	1367	45	102,518	1418
	Average of 15 Power Plants				27,637	91.5	38	53	91	2.7	6835	95

excluding Kirin. Of the 15 plants, four plants had a MTTR of 0 h. These data illustrate a highly reliable level of performance.

A closer look at the data indicates that Japanese PAFC technology had nearly caught up with the US if the unfortunate case of the MELCO power plant at the Kirin Brewery were to be excluded (Table 4.6). The three Fuji Electric power plants operated as well as or better than the Toshiba/ONSI power plants. The Fuji Electric plants showed better availability and MTBF than the Toshiba/ONSI plants. The Toshiba/ONSI plants demonstrated better overall efficiency than the Fuji plants. It is difficult to rank performance between the two; while the three Fuji plants demonstrated an average availability of nearly 98 percent, an impressive performance, the 11 Toshiba/ONSI plants actually demonstrated average cumulative operating time close to 30,000 h, another impressive performance.

4.3.3.2. *Private-Sector Field-Test Activities*

The Japanese private sector started its field-test activities of PAFC power plants on its own in the 1970s, but when METI launched its field-test program, the private-sector significantly accelerated its activities. As noted earlier, Japanese gas and electric utility companies started field testing UTC power plants: 12.5 kW PAFC power plants starting in 1972, 40 kW plants in 1982, a 4.5 MW plant in 1983, and an 11 MW plant in 1991. They also tested a large number of UTC 200 kW PAFC power plants while the plant was still a precommercial prototype and continued to test them when they became a commercial product in the early 1990s. Likewise, they tested a large number of Japanese PAFC power plants as the domestic PAFC demonstration models started to emerge from laboratories throughout the 1980s and the 1990s.

- Because no table is readily available, we built our own list of PAFC power plants installed in Japan (Table 4.8). Based on various data, at least 184 systems have been installed in Japan for field tests as well as commercial purposes from the 1970s until April 1, 2012 (Fig. 4.4). That suggests at least 227 PAFC power plants were installed in Japan, including the 43 units installed under the METI field test program.
- A review of the list of 184 PAFC power plants shows that at least 86 units were manufactured by either ONSI, ONSI/Toshiba, or Toshiba (which is bundled together as ONSI/Toshiba hereafter),[viii] 83 units were built by Fuji Electric, 12 by MELCO, two by Sanyo, and one by Hitachi.

TABLE 4.6 METI PAFC Field Test Project II: Summary of Average Power Plant Performance by Manufacturers

Manufacturer	Number of power plants	Operating hours	Availability (%)	Efficiency (%)			Number of operations failures	MTBF	MTTR	Total in parentheses	Overall ranking
				Electric	Thermal	Overall					
Toshiba/ONSI	11	29,157	94.5 (2)	38.7	54.6	93.3 (3)	1.5	6,828 (2)	41	7	2
Fuji Electric	3	27,893	97.7 (3)	38.3	53.3	91.6 (2)	1	8,545 (3)	9	8	1
MELCO	1	10,146	41 (1)	30	35	65 (1)	20	1,776 (1)	943	3	3

[viii] A review of sources reveals that there are major inconsistencies regarding which companies—ONSI, ONSI/Toshiba, or Toshiba—manufactured a particular power plant. Because it is difficult to determine which plant or component was manufactured by which of the three entities, all the plants that are claimed to be manufactured by any of the three entities are bundled together as ONSI/Toshiba plants hereafter.

TABLE 4.7 METI-Sponsored Major PAFC Programs

Project	Plant size, fuel, operating temperature, operating pressure, method of cooling	Manufacturers	Plant operators	Project period (FY)
Moonlight Project	1 MW, NG, 190 °C, 3.4 atmospheric pressure (atm), water	Fuji Electric, MELCO	Kansai Electric Power	1981–1990
	1 MW, NG, 205 °C, 8.2 atm, water	Toshiba, Hitachi,	Chubu Electric Power	
	200 kW, Methanol, 190 °C 1 atm, water	Fuji Electric,	Okinawa Electric Power	1986–1990
	200 kW, LNG, 190 °C, 1 atm, water	MELCO	Osaka Gas, Kansai Electric Power	
Grid Interconnection Project	14 × 50 kW	Fuji Electric	Kansai Electric Power	1987–1992
	1 × 200 kW	MELCO		
Urban Energy Center PAFC Technology Development	5 MW dispersed plant, NG, pressurized	Fuji Electric	Kansai Electric Power (jointly with CRIEPI and 8 other electric power companies)	1991–1997
	1 MW on-site plant, NG, 1 atm	Toshiba	Tokyo Gas (jointly with 3 other gas companies)	
PAFC Field Test Project	28 PAFC power plants of various types	Fuji Electric, UTC/Toshiba, MELCO	Installed at various sites	1992–1996
	15 PAFC power plants of various types	Fuji Electric, UTC/Toshiba, MELCO	Installed at various sites	1997–2000

Also, the 184 PAFC power plants were installed by four different customer groups (Fig. 4.5):

1. Gas utility companies were the most active in installing PAFC power plants. This group introduced at least 90 plants.
2. Electric utility companies were the second most active group, having installed at least 34 plants.
3. Oil companies installed at least 10 plants.
4. A fourth group included consumers, such as factories, breweries, telecommunications companies, national and local governments, and sewage treatment plants that installed power plants to test operate the plants as well as generate electricity and heat for their consumption. This group included some PAFC manufacturers that test operated systems for their own experiments. At least 50 plants were installed by this group.

- Another interesting fact is that electric utility companies lost interest in PAFC power plants around 1997 when they stopped all field-test activities (Fig. 4.6). Electric power companies likely had recognized that PAFC power plants were not suited for use in large power generation plants for use as a distributed or central power station.
- As electric power companies lost interest, a group of consumers began to be interested

TABLE 4.8 Private-Sector PAFC Power Plant Installation Activities in Japan* (as of April 1, 2010)

Plant manufacturer	Plant size, fuel, pressure, cooling method	Plant operator	Period of operation	Operation site	Relationship to US programs
Fuji Electric	50 kW, NG, liquefied petroleum gas (LPG), 2 atmospheres absolute (ata), water	Tohoku Electric Power	87/5–89/10	Tohoku Electric Niigata Thermal Power Station	
	2 × 50 kW, LPG		92–	Tohoku Electric Sendai Minami Office	
	50 kW	Tokyo Electric Power	93–	Tokyo Electric New Energy Park, Futtsu, Chiba	
	50 kW, natural gas (NG)	Chubu Electric Power	93/1–	Chubu Electric Power Technology Research Institute	
	50 kW, NG		94/4–	Chubu Electric Kita Office	
	50 kW,		97–	Chubu Electric Kamishiro Substation	
	4 kW, methanol	Hokuriku Electric Power	85–89	Tateyama-Sanroku Family Recreation & Leisure Village	
	50 kW, LPG		92–	Hokuriku Electric Research Institute	
	30 kW, NG, 1 ata, air	Kansai Electric Power	82/12–83/12	Kansai Electric Sakai-ko Thermal Power Station	
	50 kW, NG		96–	Mikuni Heights	
	4 kW, methanol	Shikoku Electric Power	87–88	Shikoku Electric Research Center	
	50 kW, LPG		92/5–95/3	Shikoku Electric Shikoku Research Institute	
	100 kW, byproduct hydrogen		96/11–	Toa Gosei Tokushima Plant	

50 kW, NG	Kyushu Electric Power	94/3–	Kyushu Electric Shin-Kokura Thermal Power Station
2 × 50 kW, liquefied natural gas (LNG)	J-Power	95/3–	J-Power Wakamatsu Research Institute
50 kW, NG	Tokyo Gas	90–91	Tokyo Gas Tamachi Test Station
50 kW, NG		91–92	
50 kW, NG		91–94/10	
50 kW, NG		92–95/12	
50 kW, NG		92–95/12	
100 kW, NG, LNG, 1 ata, water		92–95/12	
50 kW, NG		93–94/10	
100 kW, NG		93–95/12	
2 × 100 kW		00/7–	
50 kW, NG		95/1–	TGS Akabane Building
100 kW, NG		00/4–	Keio University Shonan Fujisawa Campus
100 kW		03/11–	Tokyo Gas Wonder Ship Museum
100 kW		06/3–	Tokyo Gas Science Museum
50 kW, NG	Toho Gas	92/9–	Howa Sports Land
50 kW, NG		91–95/12	Toho Gas Research Institute
100 kW		99–	
50 kW, NG		96/7–	The Scene Johoku
100 kW		97–	Toyota Motor Headquarters, Aichi

(Continued)

TABLE 4.8 Private-Sector PAFC Power Plant Installation Activities in Japan* (as of April 1, 2010) (*Cont'd*)

Plant manufacturer	Plant size, fuel, pressure, cooling method	Plant operator	Period of operation	Operation site	Relationship to US programs
	100 kW		03/7–	Kojunkai Medical Corporation Daido Hospital	
	100 kW		03/10–	Japan Red Cross Toyota Nursing College	
	100 kW		04/1–	Okazaki Credit Union Headquarters	
	100 kW		06/3–	Japan Red Cross Nagoya Daiichi Hospital	
	100 kW		06/3–	Nagoya Central Hospital	
	50 kW, NG	Osaka Gas	90/3–	Osaka Gas Nishijima Fuel Cell Center	
	50 kW, NG		91–94/11		
	50 kW, NG		91/3–		
	50 kW, NG		92–94/11		
	50 kW, NG		92–94/11		
	100 kW, NG		92–94/11		
	100 kW, NG		92/12–		
	500 kW		93–95/10		
	50 kW		97–		
	100 kW		98/8–	Initially located at Osaka Gas Nishijima Fuel Cell Center and moved to Himeji Red Cross Hospital (01/8–)	
	50 kW, NG		91–94/10	Fuji Electric Osaka Building	
	50 kW, NG		92–94/11	Daiwa Bank	
	50 kW, NG		92–94/11	Osaka Prefectural University	

50 kW, NG		93–94/12	Kinden Kyoto Research Institute
50 kW, NG		93/5–96/3	Osaka Gas Tsukaguchi Company Dormitory
50 kW, NG		93/10–	RITE
100 kW, NG		93/5–	Coop Himeji Shirahama
100 kW, NG		93/9–	NEXT21
100 kW, NG		93/12–	Mycal Bore Sanda
100 kW, NG		94–	Osaka Gas Himeji Plant
50 kW, LPG	Shikoku Gas	92/5–95/6	Shikoku Gas Shikoku Research Institute
50 kW, NG	Seibu Gas	92–95/12	Seibu Gas Kokura Office
50 kW	Idemitsu	90–95	Idemitsu Petrochemical Chiba Plant
80 kW, LPG, naphtha, kerosene, 1 ata, water	Showa Shell Sekiyu	90/3–96	Showa Shell Central Research Institute
50 kW, naphtha, 1 ata, water	Japan Petroleum Energy Center (JPEC)	90/7–96	Kanpo Waraku no Sato (Resort)
100 kW	Nippon Oil	94–95	Nippon Oil Niigata Terminal
50 kW	NTT	91–95	NTT Mitaka Research Institute
100 kW		01/3–	NTT Nishi-Nihon Motomachi Building, Hiroshima
100 kW		01/3–	NTT Nishi-Nihon Fukuromachi Building, Hiroshima

(*Continued*)

TABLE 4.8 Private-Sector PAFC Power Plant Installation Activities in Japan* (as of April 1, 2010) (*Cont'd*)

Plant manufacturer	Plant size, fuel, pressure, cooling method	Plant operator	Period of operation	Operation site	Relationship to US programs
	100 kW	Fuji Electric	97–	Fuji Electric Substation Systems Works, Chiba	
	100 kW				
	100 kW		01/12–	Fuji Electric Human Resources Development Center	
	2 × 100 kW	The City of Yamagata	02/3–	Yamagata City Sewage Treatment Center	
	100 kW	The City of Soka	04/3–	Soka City Hospital, Saitama	
	4 × 100 kW	The Prefecture of Kumamoto	06/12–	Kumamoto Hokubu Sewage Treatment Center	
	100 kW	The Minister of Land, Infrastructure, Transport and Tourism (MLIT)	07/8–	MLIT Government Building Number Eight	
	100 kW	The Ministry of Environment, Port Island, Kobe	01/7	Biogas use demonstration	
	100 kW	Unidentified office building	09/1		
UTC	4.5 MW, NG, 3.5 ata, water	Tokyo Electric Power	83/4–85/12	Tokyo Electric Goi Plant	US FCG-1 Project
	2 × 200 kW, NG, 1 ata, water		88–90, 89/3–	Tokyo Electric Shin-Tokyo Thermal Power Station	
	200 kW, NG		89–90, 90/8–1997	Initially sited at Kandenko Building and subsequently moved to Shibaura DHC (District. Heat and Cooling) plant refurbished with Toshiba fuel cell stack	

11 MW, NG, 8.4 ata, water		91/4	Tokyo Electric Goi Thermal Station	US FCG-1 Project
2 × 12.5 kW, NG, 1 ata, water	Tokyo Gas	72/3–72/8		US TARGET Program
40 kW, NG, 1 ata, water		82–87	Tsurumi Swimming Pool, Yokohama	US GRI-DOE Project
40 kW, NG, 1 ata, water		84–87	Tsukuba Science Museum	US GRI-DOE Project
200 kW		92/11–	Tokyo Gas Tamachi Test Station	
200 kW		92/12–		
200 kW		93/10–		
200 kW		96/1–		
200 kW		97/5		
200 kW		92/12–	Tokyo Gas Senju Office Technical Center	
200 kW		93/6–	Tokyo Gas Sodegaura Plant	
200 kW		96/1–		
200 kW		94/3–	Tokyo Gas Shinjuku Chirei Center	
200 kW, NG	Toho Gas	94/2–	Toho Gas Headquarters	
200 kW, NG		97/3–	Denso Nishio Plant	
2 × 12.5 kW	Osaka Gas	73/2–73/8		US TARGET Program
40 kW, NG		82/3–85/2	Royal Host Sakai Ishizu Restaurant	US GRI-DOE Project
40 kW, NG		84/11–87/10		
200 kW, NG, 1 ata, water		89/9–90/9–	Umeda Center Building	
200 kW		92/10–		
200 kW		89–91	Hotel Plaza Kobe	
200 kW, NG		92/8–	Osaka Gas Nishijima Fuel Cell Center	
200 kW, NG		92/12–		
2 × 200 kW, NG		96/3–		

(*Continued*)

TABLE 4.8 Private-Sector PAFC Power Plant Installation Activities in Japan* (as of April 1, 2010) (*Cont'd*)

Plant manufacturer	Plant size, fuel, pressure, cooling method	Plant operator	Period of operation	Operation site	Relationship to US programs
	200 kW, NG		97/4–		
	200 kW, NG		98–		
	200 kW		93/1–96/1	Kyoto Research Park	
	200 kW		97/3–		
	200 kW		93/6–	Ohtsu Tire & Rubber	
	200 kW				
	200 kW		93/2–97	NTT Kansai Network Center	
	200 kW, naphtha, 1 ata, water	Nippon Oil	90/2–91/6	Nippon Oil Yokohama Refinery	
UTC/Toshiba	170 kW	Japan Petroleum Energy Center	93–97	RITE	
	200 kW	Osaka Gas	97–	KRP Gas Building	
	200 kW		99–	Osaka National Research Institute	
	200 kW	Toho Gas	00–	Toho Gas Headquarters	
	200 kW	Toshiba	96–	Toshiba Keihin Office	FY 1995 DoD Climate Change Fuel Cell Program
	200 kW		97–		
	200 kW		96–	Toshiba Fuchu Office	FY 1995 DoD Climate Change Fuel Cell Program
	200 kW		97–	Toshiba Nuclear Power Research Institute	
	200 kW		98–	Toshiba Hama-Kawasaki Plant	
	200 kW		99–	Toshiba Nagoya Plant	

200 kW	Nippon Oil	97–	Nippon Oil Niigata Terminal
200 kW		03–	
200 kW	Asahi Brewery	98–	Asahi Brewery Shikoku Plant
200 kW	Sapporo Brewery	98–	Sapporo Brewery Chiba Plant
2 × 200 kW	Seiko Epson	99–	Seiko Epson Toyoshina Office
2 × 200 kW		99–	Seiko Epson Ina Office
200 kW	NTT	99–	NTT Musashino Research Institute
200 kW		01–	NTT Atsugi R&D Center
4 × 200 kW		04–	Aichi Expo
200 kW	The City of Kitakyushu	99–	Kitakyushu Science City
200 kW	The City of Yokohama	99–	Yokohama City Hokubu Waterworks Bureau
200 kW	The Prefecture of Osaka	00–	Osaka Prefectural Ebie Sewer Treatment Plant
200 kW	Kurita Water	00–	Kurita Water Technology Development Center
2 × 200 kW	National Center for Child Health and Development	01–	National Center for Child Health and Development
200 kW	Shinwakai Medical Corporation	01–	Shinwakai Medical Corporation Nishijima Hospital
200 kW	The City of Tokyo	02–	Tokyo Metropolitan Research Institute for Environment
200 kW	Asia Pacific Trade Center	99–	Asia Pacific Trade Center, Osaka
200 kW		02–	

(Continued)

TABLE 4.8 Private-Sector PAFC Power Plant Installation Activities in Japan* (as of April 1, 2010) (*Cont'd*)

Plant manufacturer	Plant size, fuel, pressure, cooling method	Plant operator	Period of operation	Operation site	Relationship to US programs
	200 kW, digestive gas	An identified local government	04–	Local government sewage treatment plant	
	200 kW	The City of Kawasaki	06–	Kawasaki City Tama Hospital	
Toshiba	200 kW, NG	Tokyo Electric Power	94/9–	Tokyo Electric Technology Development Center	
	200 kW, LNG	Chubu Electric Power	92–96	Chubu Electric Kawagoe Thermal Power Station	
	200 kW	Osaka Gas	94/5–	Osaka Gas Nishijima Fuel Cell Center	
	200 kW		96/5–	Osaka Teishin Hospital	
	5 × 200 kW	Tokyo Gas and four other gas companies	95–	Tokyo Gas Tamachi Test Station	
	100 kW	Japan Petroleum Energy Center (JPEC)	93–96	JPEC Oil Infrastructure Technology Institute	
	200 kW	NTT	96–	NTT Electronics Corporation	
Mitsubishi Electric (MELCO)	100 kW, methanol, 1 ata, water	Hokkaido Electric Power	87/10–89/5	Hokkaido Electric Date Thermal Power Station	
	200 kW, methanol		94/9–	Hokkaido Electric TakigawaTechnical Center	
	200 kW, NG	Tokyo Electric Power	94/2–	Kyobashi DHC (District. Heat and Cooling) plant	
	200 kW, NG	Kansai Electric Power	89–91	Hotel Plaza Kobe	
	200 kW, NG		93/6–97	Kansai Electric Research Institute	

	200 kW, NG		94/6–	Wakayama Marina City
	200 kW, LNG		96–	Kansai Electric Nanko Power Station
	200 kW, NG		97–	Nara Prefectural Medical School
	200 kW, NG	Chugoku Electric Power	93/7–	Chugoku Electric Yanai Thermal Power Station
	180 kW	The City of Yokohama	99–	Yokohama City Nanbu Sewage Sludge Treatment Plant
	200 kW	Nisshin Oillio	99–	Nisshin Oillio Yokohama Isogo Plant
	200 kW	MELCO	00–	MELCO Electricity and Industrial Systems Works, Kobe
Sanyo	220 kW, NG, 1 ata, air	Tokyo Electric Power	87–89	Tokyo Electric Shin-Tokyo Thermal Power Station
	100 kW	Japan Energy Corporation	90–91	Japan Energy Chita Refinery
Hitachi	100 kW	Tokyo Gas	90/8–92/5	Tokyo Gas Tamachi Test Station

** There are no comprehensive data available that covers three decades of Japanese PAFC installation activities; from the 1970s through the present. Most data are segmented and incomplete and have ambiguities, inconsistencies, inaccuracies, omissions, and duplications. This table is a best effort to outline the overall picture of the PAFC installation activities in Japan.*
Sources: Leo J. M. J. Blomen & Michael N. Mugerwa, "Fuel Cell Systems," Springer, 1993; "New Energy Technology Development-Related Data File Production Survey (Fuel Cells)," NEDO, March 1996, 1997, 1998, 1999, and 2000; "Proposal for Promotion of Fuel Cell Introduction," New Energy Foundation, March, 2006; "A Guide to Business-Use Fuel Cell Power Plant Systems," New Energy Foundation, 1998; "Fuel Cell Delivery Statistics Survey Report," Electrical Machinery, Japan Electrical Manufacturers' Association, November 2002, February 2004, September 2004, March 2006, March 2007, and December 2007; Kenichi Kuroda, "Operations Results and Commercialization Plan of PAFC at Fuji Electric Advanced Technology," A presentation at 2007 Fuel Cell Seminar & Exposition, San Antonio, Texas, October 15–19, 2007; Taku Fukumura, "Current Status of development of PAFC System in Fuji Electric Group," A presentation at 2008 Fuel Cell Seminar & Exposition, Phoenix, Arizona, – October 28–30, 2008.

in PAFC power plants. Many of these consumer entities were customers of gas companies and used natural gas or LPG supplied by gas utilities for their PAFC power plant operations. Some of them likely tried to use byproduct digestive gases generated from their breweries or sewage plants.

- Also, by the end of 1990s, predominantly two output classes had been installed: 100 kW power plants manufactured by Fuji Electric and 200 kW power plants by UTC/Toshiba. For Fuji Electric, this was a natural course of action; it did not choose to compete with the UTC 200 kW system and it had to focus on and a single product instead of multiple products to be competitive.

4.3.4. Japanese Fuel Cell Subsidy Programs: Funding One-Third to Two-Thirds of Acquisition Cost

In the late 1990s, the Japanese government likely felt that it needed to subsidize Japan's fledgling PAFC industry to foster commercialization. Both the government and the private sector field tests had shown that the PAFC power plants had been making considerable improvement in reliability and durability. As of February 1997, 19 UTC PAFC power plants, 14 Fuji Electric plants, and two MELCO plants had been operating for more than 20,000 h, with some approaching 30,000 h.[52] For the first time in Japan, one PAFC power plant—a UTC PC25A 200 kW power plant installed at Umeda Center Building in October 1992 and operated by Osaka Gas—had been in operation for more than 34,000 h (as of the end of February 1997) and was expected to exceed 40,000 h by the end of 1997.[19] The system had no failures, and the mean time between failures of all the operating systems proved to be longer than 3,000 h.[51] One UTC power plant that was installed at a Kandenko building in 1989, refurbished with Toshiba fuel cell stacks, and moved to a Shibaura District, Heat, and Cooling plant (DHC) in 1990 had been operating for 45,333 h as of the end of February 1997.[19]

The three Japanese PAFC developers—Fuji Electric, MELCO, and Toshiba—undoubtedly felt that their PAFC power plants were ready for commercialization and all announced in early 1997 their plans for a commercial venture starting the following year.[52] Fuji Electric announced that it would start commercial shipment in 1998 of its 100 kW model with a 40 percent power efficiency and a 5-year lifetime. Also, MELCO announced that it would focus on cost reduction and start the shipment of its 200 kW commercial prototype in 1998 and enter mass production in 2001. Toshiba, which was focused on cost reduction through bulk procurement and streamlined production lines, announced in 1997 an aggressive drive to market UTC's 200 kW PAFC power plants both in Japan and in Asia.[54]

The Japanese government, nonetheless, was aware that a PAFC power plant would cost about $8,000 per kW, excluding installation costs, in 1997.[55] This was clearly too high to compete with conventional power generation technologies. The cost had to come down to $2000 per kW to be competitive.[52] The government moved to implement measures that would help spur the PAFC market, thereby encouraging mass production, economies of scale, and eventual cost reduction. It was driven to quickly establish subsidy programs because of press reports such as that which appeared in June 1997:

- According to the Japanese media, UTC as of June 1997 had received orders for 186 PAFC power plants and that sales had reached $111 million.[52] The media also reported at that time that UTC had a total of 235 clients that had expressed interest in purchasing PAFC power plants and was faced with an urgent need to expand its production capacity to cope with the increasing interest in PAFC power plants.

In April 1997, the Japanese Diet passed a bill, Special Measures for Promotion of Utilization of New Energy (nicknamed the New Energy Policy Act), which was quickly enacted 2 months later, underlining the sense of urgency that Japanese policymakers felt. The law identified a wide range of new energy technologies, including fuel cells, which the country needed to introduce rapidly, and identified various financial measures that would promote the technologies. The financial measures included subsidies, low- or no-interest loans, loan guarantees, tax credits, tax deductions, accelerated depreciation, and government investments of capital in small businesses (Table 4.9). The law continues to be in effect today.

The financial measures are so extensive, varied, and broad that any fuel cell purchaser would likely be able to come across a subsidy measure that could be used with little difficulty:

- Some subsidies cover up to two-thirds of the acquisition cost (the cost of the fuel cell power plant plus the cost of installation) with no apparent upper monetary limit.
- Some loans are for a considerable amount extended over a long period of time at low interest rates.
- Some loans are free of interest.
- Some loan guarantees are for up to 90 percent of the total acquisition cost.
- A purchaser of a fuel cell power plant can take advantage of both a subsidy and a tax incentive.
- A subsidy typically is provided to the purchaser of a power plant, but is also provided to the builder who constructs a building that incorporates a fuel cell power plant.

4.3.5. PAFC Power Plants Are Not a Commercial Success

Japan's extensive field test programs and generous subsidy programs have not brought about the successful commercialization of PAFC power plants in the country. Indeed, the installation of PAFC power plants hardly increased when the subsidy programs were put into effect in 1997 (Fig. 4.7).

According to the industry lobbying association NEF in 2008, PAFC power plants were still too costly to be widely introduced even with available subsidies.[56] Indeed, for the past several years, the new installation of PAFC power plants has been slow. Since 2002, only four power plants on an average have been installed annually.

- The cost of a PAFC power plant had gone down considerably until 1999, but has risen since then, likely because of the costs incurred to improve performance and durability. As of 2008, a Japanese PAFC power plant costs about \$9000–\$10,000 per kW and the installation costs about \$9000–\$10,000 per kW, totaling \$18,000–\$20,000 per kW (Fig. 4.8).[57]

Moreover, the number of PAFC power plants in operations is also declining because they have been dismantled after 40,000 h of operation. According to NEF, only 43 PAFC power plants were in operation as of March 2008.[56] Most PAFC power plants installed earlier and operated for the expected lifetime of 40,000 h or 5 years have been regularly dismantled without being overhauled. According to Fuji Electric, an overhaul constitutes a replacement of main components such as the stacks and the reformer,[39] which costs about 30–40 percent of the acquisition cost of a new power plant.[56] Moreover, the annual maintenance cost is about \$90,000,[57] which is a considerable burden on the consumer. NEF stated that there were no subsidies to offset the cost for overhaul or maintenance and called for the government to subsidize the overhaul and maintenance cost as well.[56]

After more than 30 years of investment in R&D and field tests, all of the PAFC manufacturers—except one—have left the field.

TABLE 4.9 Japanese Subsidy Programs for PAFC Power Plants

Sponsoring entity	Program name (Year of implementation)	Eligible entity				Type of subsidy (Acquisition costs include the cost of the fuel cell power plant and the cost of installation)
		Local government	Private enterprise	Nonprofit organization	Private citizen	
New Energy and Industrial Technology Development Organization (NEDO)	Regional new energy, energy conservation vision planning measures	X	X	X		A set subsidy is available for studies to create regional new energy use, to explore in detail new energy technologies selected, and to investigate the feasibility of introducing the new energy technology
	Regional new energy introduction promotion measures	X		X		A subsidy up to half the acquisition cost of a fuel cell power plant (with an output capacity of 100 kW or greater) is provided to a local government or one-third for an NPO
	New energy businesses support measures		X			A subsidy up to one-third of the acquisition cost of a fuel cell power plant (with an output capacity of 100 kW or greater) is provided to businesses by METI, and a loan guarantee for up to 90 percent of the acquisition cost with a guarantee fee rate at 0.2 percent of the remaining loan is provided to businesses by NEDO
	New energy, energy conservation nonprofit activities promotion measures	X	X	X		A subsidy up to $200,000 or up to $50,000 per year for new energy educational activities

	Residential construction high efficiency energy system introduction promotion measures		X		A subsidy of up to one-third of the acquisition cost will be provided to the builder when a new building or an addition would cut the energy consumption by 10 percent or when the remodeling of an older building would cut energy consumption by 20 percent
Ministry of Economy, Trade, and Industry (METI)	Energy supply and demand structure revision investment promotion tax measures		X	X	An enterprise or a private citizen can receive special depreciation up to 30 percent or a tax credit up to 7 percent of the acquisition cost. Only a small and medium enterprise can have the tax credit benefit
Ministry of the Environment (MOE)	Local government priority project subsidy measures	X			A subsidy up to half of the acquisition cost; a subsidy up to $50,000 for a public education activities
	School ecological improvement for prevention of global warming	X			A school under local government administration can receive a subsidy up to half the acquisition cost within the limit of $1 million
	Subsidy measures to promote the establishment of facilities to achieve voluntary greenhouse gas reduction goals		X		An enterprise that volunteers to achieve its greenhouse gas reduction goal can receive a subsidy up to one-third of the acquisition cost, up to $2 million per facility (factory, office, etc.,)
	Advanced regional renewable energy introduction measures		X		An enterprise can receive a subsidy up to half the acquisition cost when a local government designates and the METI and MOE recognize that the facility would serve as a means to reduce global warming

(Continued)

TABLE 4.9 Japanese Subsidy Programs for PAFC Power Plants (*Cont'd*)

Sponsoring entity	Program name (Year of implementation)	Eligible entity				Type of subsidy (Acquisition costs include the cost of the fuel cell power plant and the cost of installation)
		Local government	Private enterprise	Nonprofit organization	Private citizen	
	Regional alternative energy, energy conservation cooperative promotion measures			X		A local cooperative can receive a subsidy up to one-third of the acquisition cost if it introduces multiple PAFC power plants to households
	Measures to promote a model regional project that facilitates a virtuous circle of a good environment and a good economy	X		X		An effort to promote a broad-based, local cooperative that would introduce new environment-friendly energy technology, such as fuel cells and would serve as a model to be emulated by other regions. large scale project can receive up to $2 million, and a small project up to $40,000
Development Bank of Japan (DBJ)	New Energy, renewable energy development	X	X			A local government and an enterprise can receive a low-interest loan up to 40 percent of the acquisition cost. The interest would be 1.85 percent for a loan over 15 years and 2.1 percent when the loan is for 20 years. Deferral of repayment is allowed for three years

	Kyoto Protocol target achievement plan promotion measures	X	X	A local government and an enterprise can receive a low-interest loan up to 40 percent of the acquisition cost. The interest would be 1.85 percent for a loan over 15 years and 2.1 percent when the loan is for 20 years. Deferral of repayment is allowed for three years
Ministry of Education, Culture, Sports, Science and Technology (MEXT), METI, MOE, and Ministry of Agriculture, Forestry, and Fisheries (MAFF)	Eco-school pilot model measures	X		A public school can receive half the acquisition cost of a new fuel cell power plant and one-third for refurbishing
Ministry of Land, Infrastructure, Transport and Tourism (MLIT)	Sewage Treatment Plant Construction Subsidy	X		A subsidy up to two-thirds of the acquisition cost of a fuelcell power plant is provided to a sanitary sewage treatment plant; a subsidy up to 55 percent of the acquisition cost is provided to a general sewage treatment plant
Japan Finance Corporation for Small and Medium Enterprise (JASME)	Environment, energy policy account		X	A small and medium enterprise can obtain a loan up to $7.2 million from JASME or up to $1.2 million from a city bank that represents JASME for a new acquisition or refurbishing of a PAFC power plant, up to a limit of no more than 80 percent of the applicable project cost. The interest is 1.05 percent for regular loans or zero percent when the project is approved under special laws or for disaster recovery. The loan period is no more than 20 years and must be approved by the prefectural governor. Deferral of payment is no more than 3 years. The loan period for a no-interest loan is 7 years

(*Continued*)

TABLE 4.9 Japanese Subsidy Programs for PAFC Power Plants (*Cont'd*)

Sponsoring entity	Program name (Year of implementation)	Eligible entity				Type of subsidy (Acquisition costs include the cost of the fuel cell power plant and the cost of installation)
		Local government	Private enterprise	Nonprofit organization	Private citizen	
	Measures based on Small Business Investment Development Corporation Law		X			Three regionally located small and medium business investment and consultation companies (SBIC) will provide capital to start or expand a business whose capital is less than \$3 million. SBICs will purchase stocks and debentures at the time of the business establishment and expansion and provide consultations. SBISs can make investments up to 50 percent of voting shares. A business with capital exceeding \$3 million can receive SBIC investment if the company conducts businesses defined in the New Energy Policy Act, including marketing fuel cell power plants, selling fuels for the power plants, and marketing the generated electricity and heat
Liquefied Petroleum Gas (LPG) Project	LPG cogeneration system introduction project		X		X	A subsidy up to half of the acquisition cost of the qualified fuel cell power plant or \$750,000, whichever is lesser, when the power output is greater than 200 kW and the generated electricity and heat is consumed by home owners or businesses

Source: "Study Guide to Introduce Commercial-Use Fuel Cell Power Plan," the FY2008 edition, New Energy Foundation, 2008; "FY2007 Report on Surveys on Outcome of PAFC Technology," Mitsubishi Research Institute, 2008.

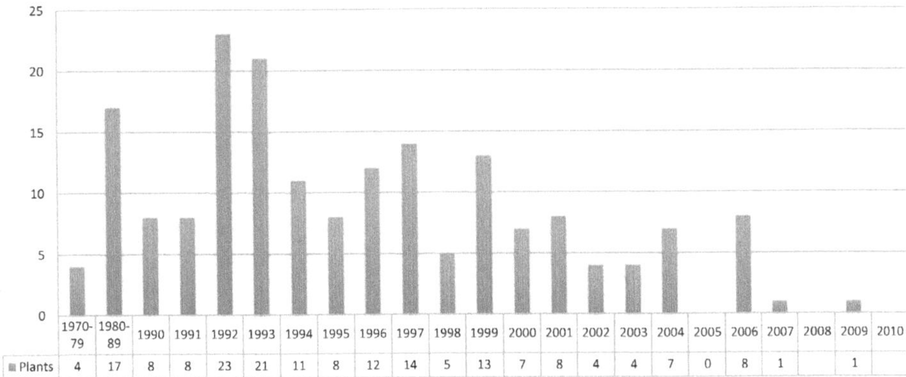

FIGURE 4.7 Trends in new installation of PAFC power plant in Japan: 1970–2010 (total PAFC power plant = 184, as of April 1, 2010). (For the color version of this figure, the reader is referred to the online version of this book.)

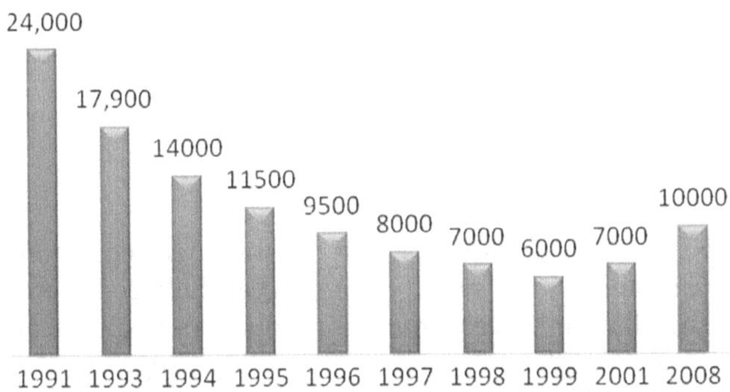

FIGURE 4.8 Changes in capital cost of Japanese PAFC power plant per kW without installation cost ($US). *Sources: The data for the period 1991-2001 is sourced from "Changes in Capital Cost of PAFC Power Plant," New Energy Data, http://www.nedo.go.jp/nedata/17fy/12/0012g002.html, NEDO, accessed November 2008. The data for 2008 is sourced from the "New Energy Guidebook," NEDO, 2008, and represents the high-end cost figure.* (For the color version of this figure, the reader is referred to the online version of this book.)

Fuji Electric is the only one still engaged in PAFC R&D in Japan today. Toshiba, which was hopeful about its PAFC business in the 1990s, is reportedly no longer actively marketing PAFC power plants today. MELCO perhaps left the PAFC field at the end of 2000. Sanyo likely left the field in the mid-1990s. Hitachi had left the PAFC field by the early 1990s.

- The sole remaining PAFC developer Fuji Electric delivered 25 commercial units between 1998 and April 1, 2010 (Fig. 4.9).[58]

Japan's subsidy programs appeared to have little effect on the number of PAFC power plants installed by Fuji Electric. In 2006, Fuji released a new commercial model that was designed to operate 60,000 h.[39] With this model, the frequency of overhaul would be reduced, thus lowering the cost. Fuji expects that the new model will run for seven and a half years, at which time the plant would be overhauled to operate for another 60,000 h. Fuji Electric installed seven of these models in 2006 and one in 2007.

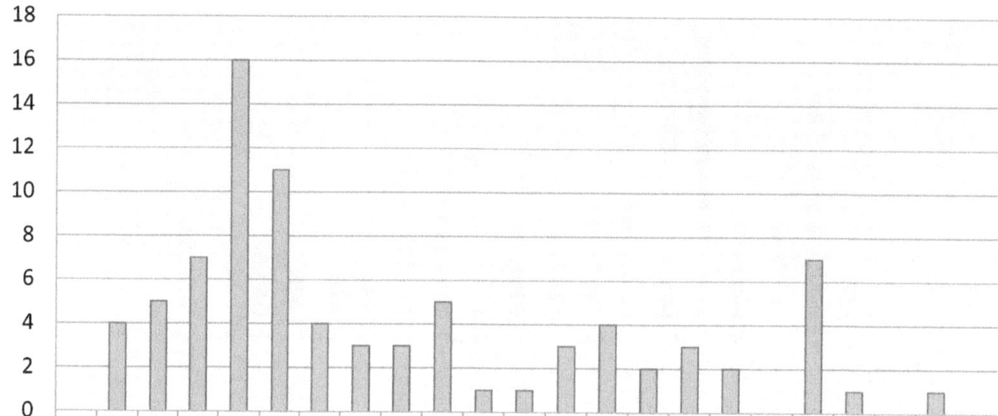

	1970 -79	1980 -89	1990	1991	1992	1993	1994	1995	1996	1997	1998	1999	2000	2001	2002	2003	2004	2005	2006	2007	2008	2009	2010
Fuji Electric	0	4	5	7	16	11	4	3	3	5	1	1	3	4	2	3	2	0	7	1		1	

*The figure was created based on Table 4-8 and from data contained in Masakazu Hasegawa and Yoshimi Horiuchi, "Development of the FP-100i Phosphoric Acid Fuel Cell," Fuji Jiho, Vol. 83, No.3, Fuji Electric, 2010.

FIGURE 4.9 PAFC power plant installation by Fuji Electric: 1970–2010* (total = 83, as of April 1, 2010). (For the color version of this figure, the reader is referred to the online version of this book.)

- In 2009, Fuji developed a more advanced FP-100i 100 kW PAFC system.[59] The new model takes half the installation space of the previous model and costs about ¥1,200,000/kW (about $14,688 at a 2012 exchange rate of $1/¥81.7) including installation, still costly for consumers. Fuji Electric has established a mass production facility in Chiba Prefecture with an annual capacity of 20 systems. In January 2009, it sold one unit for a government office building, raising Fuji's total sales to 25 units as of April 2010.
- In February 2012, Fuji Electric received an order for a 100 kW system from Daimler of Germany to be installed at its Hamburg Mercedes Benz dealership.[60] The order was valued at ¥100 million (US$1.29 million). At present, it is not certain if the new system would be a commercial success to Fuji (Fig. 4.9).
- Toshiba was confident that PAFCs would be successfully commercialized in the late 1990s. Indeed, perhaps because of Japan's extensive subsidy programs, Toshiba received an order of 10 UTC PC25C PAFC power plants in 1997 and was hoping to receive an order of 20 to 30 units in 1998.[39,ix] In reality, however, Toshiba installed 7 to 12 PAFC power plants per year from 1997 to 1999, no way near what the company was planning (Fig. 4.10). The cost of the unit system then was $4000 per kW and $80,000 per system without installation costs, likely too costly for a consumer, even with a subsidy. Toshiba's PAFC sales stagnated in the 2000s averaging one to four power plants per year during this period. Its best year was 2004 when Toshiba installed display demonstrations for the World Expo held in Aichi Prefecture in 2005. The company subsequently installed one more PAFC power system in 2006.[61]

[ix] METI's field tests do not include those units commercially sold.

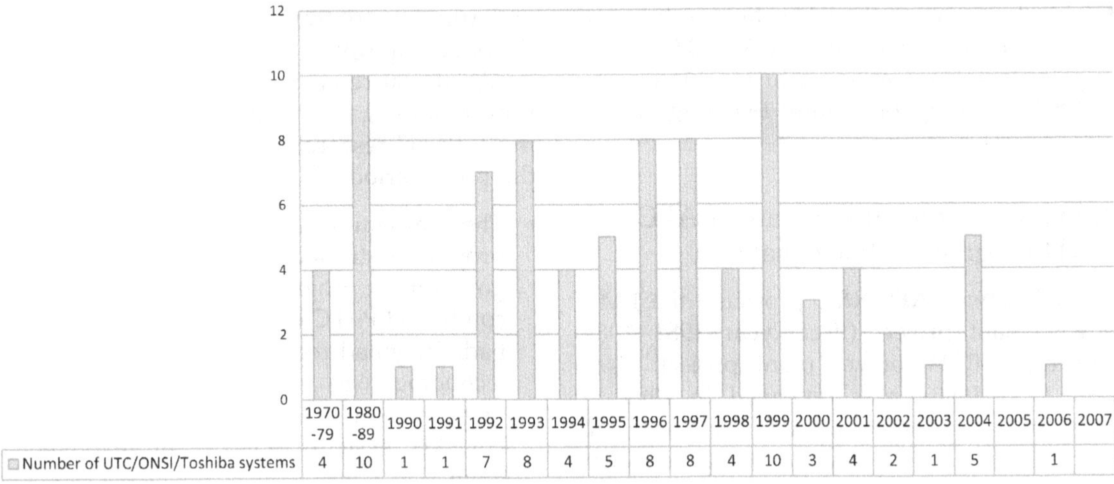

FIGURE 4.10 PAFC power plant installation by Toshiba/UTC combined: 1970–2007 (total = 86). (For the color version of this figure, the reader is referred to the online version of this book.)

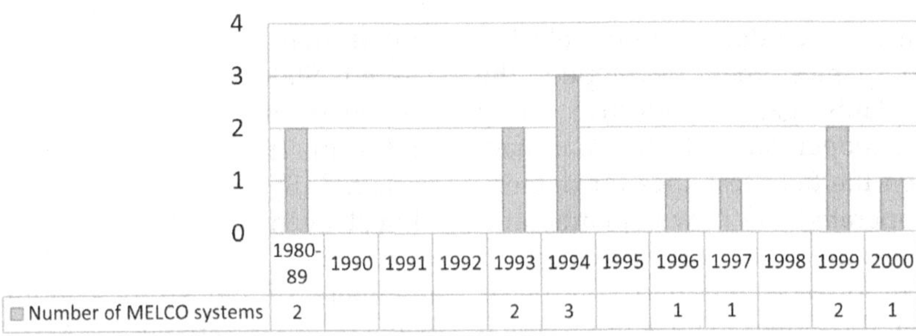

FIGURE 4.11 PAFC power plant installation by MELCO: 1980–2000 (total = 12 power plants). (For the color version of this figure, the reader is referred to the online version of this book.)

- MELCO had shipped about 12 PAFC power plants between 1989 and 2000 (Fig. 4.11). The company was planning to start mass production in 2001 to produce 200 kW PAFC power plants at a price of $4000 per kW.[54] The company, however, experienced serious problems with its PAFC units while participating in the second METI field test project (1997–2000). As noted earlier, Mitsubishi's power plant installed at the Kirin Brewery did not perform well.[51] Although most of the problems were not related to the PAFC stacks itself, it likely gave a bad impression about the company's PAFC power plants, perhaps a fatal blow to future business. The company installed two PAFC power plants in 1999 and one in 2000—probably the last Mitsubishi unit installed.
- Sanyo left the stationary PAFC field in the early 1990s to focus on portable PAFC systems. The company likely abandoned the PAFC business altogether in the mid-1990s. Sanyo installed about three PAFC power plants in the late 1980s and early 1990s.

- Hitachi likely left the PAFC field in the early 1990s. It delivered one 100 kW PAFC power plant to Tokyo Gas Tamachi Test Station in 1990, which operated until 1992 (Figs 4.10 and 4.11).

4.3.6. Government Evaluates PAFC R&D Program as Inadequate

METI's two PAFC R&D projects—the 1 MW power plant project (fiscal year 1981–1987) and the 200 kW power plant project (fiscal year 1986–1990)—was reviewed in 1999, 10 years after the end of the PAFC R&D program. A committee of 12 academics, government researchers, and industry representatives who were not involved in the PAFC R&D projects conducted the study by interviewing companies and their employees that were involved in PAFC R&D during that period. The companies included those electric machinery companies that developed the PAFC power plants, gas and electric utility companies that cooperated with the manufacturers, provided the field-test sites for demonstration, and operated the power plants, and other companies that did not participate in the government projects but had conducted PAFC R&D on their own. The study, which reflected candid views of the corporate participants, identified some positive outcomes but along with them revealed more programming flaws and failings.[46]

On the positive side, METI's PAFC program was viewed as having significantly boosted PAFC R&D activities in Japan. Those companies that participated in the government program dramatically increased their investment in PAFC R&D and employed more researchers in the PAFC field.[46] Prior to the starting of the government program in 1981, Fuji Electric, MELCO, and Toshiba had a combined total of about 50 PAFC researchers. The number increased more than eightfold to 420 during the height of the government program. The number, however, dropped rapidly in the 1990s and had fallen to about 160 researchers in 1999. The electric power and gas companies also employed more PAFC researchers and engineers during the peak period.

- The companies that participated in the program indicated that were it not for the government program they would not have conducted PAFC R&D to the extent that they had. They had recognized that PAFC technology was a difficult technology and that they would rather have imported US PAFC systems to Japan and marketed them.

The government program was also viewed as enabling Japan's industry to narrow the technological gaps between it and the United States. The companies that participated estimated that they were 10 years behind the United States in 1981. They believed that they were still several years behind the United States in 1990 when the Moonlight Project ended.[46] By 1999, they believed they were almost even with the United States in terms of reliability and durability.[46] Some academic members of the study committee, however, believed that Japanese PAFC systems were still slightly less efficient and more expensive than their US counterparts in 1999.[46] The academics indicated that Japanese systems cost about \$5,500 per kW (¥600,000 at an exchange rate of ¥110 to the dollar in 1999) while the cost of similar US systems was \$4,000 per kW.

The study, however, showed that the companies identified a number of flaws in government policy planning. Most companies viewed the 1 MW project as unsuccessful.[46] They thought that the project goals were too ambitious and the length of time allocated for the project was too short especially in the early 1980s, when the industry's PAFC technology was limited. Those companies that participated in the project stated that they had to design and build the power plants, install them at the scheduled sites, and conduct feasibility demonstration operations in just 7 years. They said that during the developmental phase, they encountered many technical problems that took much time to solve, leaving little time for demonstration operations to test reliability and durability. One company that did not participate in the project pointed out that the project had failed to show the reliability of the power plants. The company said that the plants had operated only for 1,000–2000 h and that they should have been allowed to operate for at least 15,000–20,000 h.

The members of the study committee as well as the companies that participated in the study pointed out that METI could have implemented the projects differently:[46]

- The companies believed that the 200 kW project should have been the first project under the Moonlight Project. They would have been better off by starting out with the development of a simpler, smaller power plant and moving onto developing a more complex and technically difficult power plant, such as the 1 MW project.
- The academics in the study committee acknowledged the need for PAFC basic research and stated that the government should have sponsored PAFC basic research at universities and national laboratories in parallel with the power plant development projects under the Moonlight Project.
- The companies that participated in the study also indicated that following the power plant development projects, the government should have sponsored an R&D project that aimed to reduce cost of the technology. They also stated that following the end of plant development projects, a more ambitious subsidy program should have been implemented to promote the commercialization of PAFC power plants.[46]

The study also included many criticisms of the government program management, which included the following:[46]

- According to the companies that participated in the 1 MW and 200 kW development projects, the budget was decided each fiscal year and the level of the budget fluctuated in some years. Budget downturns forced the companies to shoulder a greater financial burden to sustain the level of activities for the project.
- Also, budget requests and R&D goals were formulated half a year before the end of the fiscal year. This forced the project goals for the next fiscal year to be created on the basis of the expected outcome of the current fiscal year plan, with results that were not yet known. This created unreasonably ambitious goals each year that permitted little margin of error and no flexibility to look back and consider what went wrong.
- Moreover, the companies were tasked with onerous administrative paper work as well as the obligation to report to multiple layers of administrative authority. The administrative layers included METI's National Institute of Advanced Industrial Science and Technology (AIST), the New Energy and Industrial Technology Development Organizations (NEDO) that administered the program on behalf of METI, and the Research Cooperative that was formed by a group of companies that shared an interest in PAFC development.
- The companies also indicated that there was an atmosphere in which only positive results would be welcomed. The companies indicated that these factors made some companies averse to participation in government programs.

While they were critical about the government, the participant companies recognized that they were not blameless, either:[46]

- According to the program participants, the 1 MW project partnered two companies who were rivals, making them unwilling to exchange technical information and know-how. This resulted in major embarrassments such as the production of a mismatched pair of PAFC stacks.
- In addition, the companies indicated that they did not try to seek assistance from academic or government researchers. Instead, the companies immediately started producing fuel cells with limited knowledge about the underlying fundamentals of the technology.
- At the same time, the companies said that academia showed little interest in PAFC technology. According to them, company researchers presented their papers on PAFC technology at academic conferences but received few reactions or questions from the academic researchers. Corporate researchers felt that the academics regarded problems involving PAFC technology as engineering or manufacturing issues that were not worthy of their time. Indeed, the number of PAFC researchers in the academic sector did not increase during the 1981–1990 METI PAFC R&D project period.

4.4. PAFC DEVELOPMENT IN OTHER COUNTRIES: PRIMARILY TEST-OPERATING US AND JAPANESE PAFC POWER PLANTS

In addition to the United States and Japan, some 20 other countries showed interest in PAFC power plants starting in the late 1980s. About a dozen of them were European countries. They for the most part focused on demonstration operations of power plants manufactured by US and Japanese companies. Some governments supported PAFC R&D with modest funding. In such cases, they typically supported the development of BoP technology so that they could integrate their technology with US and Japanese fuel cell stacks to make PAFC power plants. Due to the widely recognized view that the development of PAFC stacks was close to reaching the commercial stage, these governments likely felt it would be counterproductive to start R&D on PAFC development.

4.4.1. European Countries

About a dozen European countries, including Austria, Denmark, Finland, France, Germany, Italy, the Netherlands, Norway, Spain, Sweden, Switzerland, and the UK, had conducted demonstration operations of at least 48 PAFC power plants by the end of 2008 (Table 4.10).[x] This included 37 UTC PAFC power plants, 7 Fuji PAFC power plants (25 kW, 50 kW, and 80 kW), and three others, including plants built by Ansaldo, Westinghouse, and Toshiba. Germany operated 21 power plants, more than any other European country. Italy operated at least 13, Sweden four, and Switzerland two. Another eight countries operated one each. The demonstration operations were often carried out by utility companies, with financial support from their governments and the European Community. In addition, 10 of the 37 UTC 200 kW power plants received a rebate under the US DoD Climate Change Fuel Cell Program, which amounted to $1,000 per kW.

A few countries appeared to have conducted some PAFC R&D. Germany, which was the most

[x] In 1994, Italy purchased PAFC fuel cell stacks capable of generating 1.3 MW. The purchase is counted as an equivalent of seven 200 kW PAFC power plants. Some other European countries also purchased PAFC stacks, each of which is counted as one power plant when they are incorporated into a power plant system.

TABLE 4.10 PAFC Power Plant Installations in Europe (as of May 2005)

Country	Plant size, fuel	Coordinator/project name	Plant supplier	Operator/installation site	Start of operation	DoD Climate change rebate recipient
Germany	80 kW, hydrogen	Solar Wasserstoff Bayern Hydrogen Demonstration Project	Fuji Electric (stack), Kinetics Technology International (KTI) (reformer)	Neunburg vorm Wald	90–	
	200 kW, natural gas	Ruhrgas/Stadtwerke Bochum	UTC	Testing at Ruhrgas based at Dorsten followed by a 2-year field test at Stadtwerke Bochum	92–	
	200 kW, natural gas	Thyssengas/ Stadtwerke Duren	UTC	Operated at Thyssengas facility for 1 year followed by operation for 4 years at Stadtwerke Duren	92–	
	200 kW, natural gas	Hessische Elektrizitäts-AG HEAG	UTC	Darmstadt (Eberstadt)	93–	
	200 kW, natural gas	Hamburgische Electricitäts-Werke (HEW)/Hamburger Gaswerke (HGW)	UTC	Residential buildings at Lyserstrasse, Hamburg	95–	
	200 kW, hydrogen	European Union (EU)/ HEW/HGW	UTC	Residential buildings at Lyserstrasse, Hamburg	97–	
	200 kW, natural gas	Toshiba Corporation, HEAG AG	Toshiba	Endersbach, Germany	97–	Yes
	200 kW, natural gas	CLC S.r.l. Italy Ansaldo	UTC	Housing estate, District Heat, Saarbrucken, Nachtweide	97–	Yes
	200 kW, natural gas	CLC S.r.l. Italy Ansaldo/ Gasversorgung Sachsen Anhalt/ Hamburg Gas Consult	UTC	Grovers, Halle	97–	Yes
	200 kW, natural gas	CLC S.r.l. Italy Ansaldo/ Energieversorgung Halle	UTC	Swimming pool, Halle/ Saale	97–	

(Continued)

TABLE 4.10 PAFC Power Plant Installations in Europe (as of May 2005) (*Cont'd*)

Country	Plant size, fuel	Coordinator/project name	Plant supplier	Operator/installation site	Start of operation	DoD Climate change rebate recipient
	200 kW, natural gas	CLC S.r.l. Italy Ansaldo/ BZ-BHKW Nurnberg/ Studienges Brennstoffzellen/EWAG	UTC	District Heat, Nurnberg	97–	Yes
	200 kW, natural gas	CLC S.r.l. Italy Ansaldo/ Stadtwerke Oranienburg	UTC	District Heat, Oranienburg,	97–	Yes
	200 kW, natural gas	Hamburg Gas Consult	UTC	HGW, Bargteheide	98–	
	200 kW, natural gas	PreussenElektra/ E.ON/MAINOVA AG/ Hamburg Gas Consult	UTC	Bergen-Enkheim public swimming pool, Frankfurt	98–	
	200 kW, natural gas	Erdgas Energie Systeme	UTC	School, Kaltenkirchen	99–	
	200 kW, natural gas		UTC	Malteser Hospital, Kamenz	99–	
	200 kW, digester gas	Technische Beratung Energie (TBE)	UTC	Gas, Elektrizitats und Wasserwerk, Cologne	99–	Yes
	200 kW, natural gas	Hamburg Gas Consult	UTC	Malteser Hospital, Kamenz	00–	
	200 kW, natural gas	TBE/Bildungszentrum für die Entsorgungs- und Wasserwirtschaft (BEW)	UTC	St. Agnes Hospital, Bocholt	01–	Yes
	200 kW, natural gas	Hamburg Gas Consult	UTC	Frankfurt-Hoechst industrial park, Frankfurt	01–	
	200 kW, natural gas	Rheinisch-Westfälisches Elektrizitätswerk (RWE)	UTC	RWE Fuel Cell Pavilion, Meteorit Park, Essen	04–	
Italy	1 kW & 5 kW, methanol	Italian Board for New Technologies, Energy and Environment (ENEA)	Ansaldo (stack), OTB Partecipazioni (reformer)	Ministry of Defense	90–	

	25 kW, natural gas	EU/ENEA	Fuji (stack), KTI (reformer)	ENEA Research Center	90–	
	50 kW, natural gas	Eniricerche	Fuji Electric	Eniricerche, Milan	91–	
	200 kW, natural gas	ENEA/EU	UTC	SEABO Thermorefrigera-tion plant, Bologna	93–	
	1.3 MW, LNG	Ansaldo/ENEA/ Milan Municipal Energy Company (AEM)/Milan Municipal Electric Utility	UTC (stack), Haldor Topsoe of Denmark (reformer)	Bicocca technology zone, Bicocca, Milan	94–	
	200 kW, natural gas		UTC	CLC S.r.l Piazza Carignano 2	95–	Yes
Sweden	50 kW, natural gas	Sydkraft	Fuji Electric	Sydkraft, Astrop	91–	
	200 kW, natural gas	Sydkraft	UTC	Sydkraft, Bara	92–	
	50 kW, natural gas	Vattenfall Energisystem	Fuji Electric	Vattenfall	93–	
	200 kW, natural gas	CLC S.r.l. Italy Ansaldo/ Vattenfall Energisystem	UTC	Barberg	97–	Yes
Switzerland	200 kW, natural gas	UTC/United Nations Global Environment Facility (GEF)	UTC	Service du Gaz, Geneva	93–	
	200 kW, natural gas	Asociacion Española de Banca (AEB)	UTC	AEB, Basel	00–	
Austria	200 kW, natural gas	Austria Ferngas/ Wiengas/EVN Aktiengesellschaft	UTC	Installed at Austria Ferngas in 1993 and moved to District Heating Plant of EVN in 1994 , Vienna	93–	
Denmark	200 kW, natural gas	Toftlund Fjernvarmevaer, Naturgas Syd Sonderjyllands	UTC	District Heating System, Toftlund	92–	

(Continued)

TABLE 4.10 PAFC Power Plant Installations in Europe (as of May 2005) (*Cont'd*)

Country	Plant size, fuel	Coordinator/project name	Plant supplier	Operator/installation site	Start of operation	DoD Climate change rebate recipient
Finland	200 kW, natural gas	Imatran Voima Oy/EA Technology of UK	UTC	Vanaja Power Plant, Vantaa	92–	
France	200 kW, natural gas	GDF/ Electricité de France (EDF)	UTC	Hotel Chelles, GEPPAC, Paris	99–	
The Netherlands	25 kW, natural gas	EU European Demonstration Plant	Fuji Electric (stack), KTI (reformer)	Delft University of Technology	89–	
Norway	400 kW, by-product hydrogen	Norsk Hydro	Westinghouse	Norsk Hydro's chlorine plant, Rafnes	92–	
Spain	50 kW, natural gas		Fuji Electric	ENAGAS	91–	
UK	200 kW, natural gas	Woking Borough Council	UTC	Woking Park, Woking Borough	01–	Yes

Sources: Leo J. M. J. Blomen & Michael N. Mugerwa, "Fuel Cell Systems," Springer, 1993; "New Energy Technology Development-Related Data File Production Survey, (Fuel Cells)," NEDO, March 2000; "DOD Fuel Cell ERDC/CERL Projects," http://dodfuelcell.cecer.army.mil/climate/recipients.php, Department of Defense, accessed December 2008; "Worldwide Fuel Cell Installations," Updated May 2005, Fuel Cells 2000, www.fuelcells.org; Kai Sander, "Potenziale und Perspektiven stationarer Brennstoffzellen," Institut fur Energiewirtschaft und Rationelle Energieanwendung, Universitat Stuttgart, 2004; "Brennstoffzellen-Systeme Fur Stationare Anwendungen—Workshop," BMVIT, Wirtschaftskammer Österreich, 30 Janner 2001.

active in PAFC power plant installations, for example, likely conducted no PAFC development on its own. In 1990, Kinetic Technology International (KTI) of the Netherlands constructed an 80 kW PAFC plant for the German Solar Wasserstoff Bayern Hydrogen Demonstration Project with funding from the German and Bavarian State governments. The 80 kW stack was supplied by Fuji Electric, and KTI supplied a reformer. Subsequently, Germany purchased all the PAFC power plants, primarily from UTC, for demonstration.

Still, a few individual European governments and the European Union (EU) funded some R&D efforts, but they were primarily interested in developing BoP components, centering on fuel reformers:

- In Italy, the Renewable Sources and Energy Saving Department (ENEA), the Ministry of Defense, and the EU provided some funding for PAFC R&D.[5] ENEA also managed PAFC development programs. In 1990, one 1 kW and one 5 kW PAFC power plants were developed by Ansaldo for the Ministry of Defense. Ansaldo developed the stacks, and OTB Partecipazioni developed the reformers. The goal of this effort undoubtedly was to develop PAFC power plants indigenously,[47] but no other indigenous PAFC development efforts followed. The 1.3 MW demonstration plant operated in Bicocca, Milan starting in 1994 was integrated by Ansaldo with ENEA funding.[5] UTC supplied the PAFC stacks and Haldor Topsoe of Denmark supplied the reformers. Italy appeared to have given up PAFC development by the mid-1990s.
- Two 25 kW PAFC pilot plants were constructed by KTI with funding support from the Dutch government and the EU.[5] The PAFC stacks were supplied by Fuji Electric, and the reformers were supplied by KTI. One of the plants was operated for demonstration purposes in an ENEA research center in Italy and the second plant was installed at a technical university in the Netherlands.

4.4.2. The Rest of the World

Besides the European countries, fewer than 10 other countries had operated a total of 20 PAFC power plants by the end of 2008, including Australia, Brazil, Canada, China, India, South Korea, Taiwan, Thailand, and Russia (Table 4.11). Fourteen of the 20 were UTC 200 kW PAFC systems, four were Fuji power plants, and two were Bharat Heavy Electricals Ltd. (BHEL) power plants. Five of the 14 UTC 200 kW power plants installed in these countries received a rebate under the DoD Climate Change Fuel Cell Program or a DOE financial award.

- South Korea was the most energetic of all the countries of this group. It demonstrated three small Fuji power plants in the late 1980s and early 1990s; subsequently, it switched to US power plants and conducted demonstration operations of three UTC 200 kW power plants during the 1990s.
- India was the only country that developed indigenous PAFC power plants for demonstration operations, while all the other countries purchased and operated UTC and Fuji Electric PAFC power plants. In 2000, BHEL of India developed two 25 kW PAFC power plants and tested them for 500 h, using the byproduct hydrogen from a chlor-alkali factory.[62] Earlier in 1998, BHEL operated a UTC 200 kW PAFC power plant. The power plant was purchased with funding from India's Department of Non-Conventional Energy Sources (DNES) and was operated from 1998 to 2000 using LPG.
- Russia's oil and gas pipeline engineering company Orgenergogaz, a subsidiary of Gazprom, purchased a UTC power plant in 1997, but stored it without installing it.[63] In

TABLE 4.11 PAFC Power Plant Installations in the Rest of World

Country	Plant size, fuel	Coordinator/project name	Plant manufacturer	Operator/installation site	Start of operation	DoD Climate change rebate recipient
Australia	200 kW, natural gas	Australian Technology Park, Sydney	UTC	Australian Technology Park, Sydney	98–	Yes
Brazil	200 kW, natural gas	Sieco SA/Companhia Paranaense de Energia (COPEL)	UTC	COPEL Computer System Center, Curitiba, Parana	01–	Yes
	200 kW, natural gas	Sieco SA/COPEL	UTC	Instituto de Tecnologia para o Desenvolvimento (LACTEC) research and development facility, Curitiba, Parana	02–	Yes
	200 kW, natural gas	Sieco SA/COPEL	UTC	Centro de Pesquisa e Desenvolvimento Leopoldo A Miguez de Mello (CENPES) Petrobras research and development center, Rio de Janeiro	02–	
	200 kW, natural gas	Sieco S.A/COPEL	UTC	Erasto Gaertner children's cancer hospital, Curitiba, Parana	04–	
Canada	200 kW, natural gas	Ontario Hydro	UTC	Ontario Hydro's Markham Centre	94–	
	200 kW	Northern Alberta Institute of Technology (NAIT)	UTC	NAIT's main campus Interpretive Centre, Edmonton, Canada	04–	Yes
China	200 kW, natural gas	The Institute of Energy Economics, Japan	UTC	Pig farm, Guangdong	00–	
India	200 kW, liquefied petroleum gas	Bharat Heavy Electricals Ltd. (BHEL)/Indian Department of Non-Conventional Energy Sources (DNES)/Toshiba	UTC	BHEL testing facility	98–	

	2 × 25 kW, hydrogen	BHEL	BHEL	BHEL testing facility	00–	
South Korea	5, 6 kW		Fuji Electric	Korea Institute of Energy Research (KIER)	87–	
	50 kW	Highly Advanced National Project (HAN Project)	Fuji Electric	Korea Electric Power Corporation (KEPCO)	93–	
	200 kW, natural gas	HAN Project	UTC	Korea Gas	93–	
	200 kW, natural gas		UTC	R&D Lab, Ulsan	97–	
	200 kW, natural gas		UTC	Hotel, Ulsan	99–	
	12 × 400 kW		UTC	GS Power plant, Town of Anyang, Seoul	09–	
Taiwan	200 kW, natural gas		UTC	Taiwan Power Company	95–	
Thailand	50 kW, natural gas	J-Power/ New Energy and Industrial Development Organization (NEDO)/ Japanese Government Overseas Development Aid (ODA) Project	Fuji Electric	EGAT	91–	
Russia	200 kW	Orgenergogaz (oil and gas pipeline engineering company)	UTC	Orgenergogaz facility, Russia	04–	A financial assistance award from the US Department of Energy

Sources: Leo J. M. J. Blomen & Michael N. Mugerwa, "Fuel Cell Systems," Springer, 1993; "New Energy Technology Development-Related Data File Production Survey, Fuel Cells)," NEDO, March 2000; "DOD Fuel Cell ERDC/CERL Projects," http://dodfuelcell.cecer.army.mil/climate/recipients.php, Department of Defense, accessed December 2008; "Worldwide Fuel Cell Installations," updated May 2005, Fuel Cells 2000, www.fuelcells.org.

October 2001, UTC received a financial award from the US DOE to help the Russian company with installation and start-up. After replacing some of the original parts of the power plant, which was constructed in 1996, the plant started operations in February 2004. Similar to the experience of the unit installed in New York's Central Park, Orgenergogaz never lost power at its facilities during the blackout in Russia in May 2005.[64]

4.4.3. Again, No Measurable Commercial Success

Similar to the United States and Japan, the PAFC power plants have not achieved any measurable commercial success in Europe and the rest of the world (Fig. 4.12). A review of the PAFC installations in Europe and the rest of the world as of the end of 2008 is a continuous pattern of ups and downs, followed by a hiatus in installation activity in recent years.

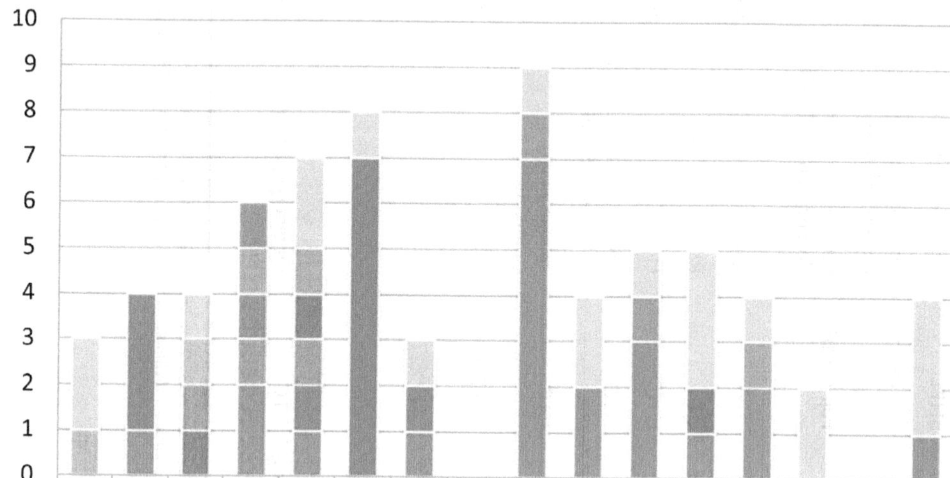

	1987-1989	1990	1991	1992	1993	1994	1995	1996	1997	1998	1999	2000	2001	2002	2003	2004
The Rest of the World	2	0	1		2	1	1		1	2	1	3	1	2		3
UK													1			
Spain			1													
Norway				1												
The Netherlands	1															
France											1					
Finland				1												
Denmark				1												
Austria					1											
Switzerland					1							1				
Sweden			1	1	1				1							
Italy		3	1		1	7	1									
Germany		1		2	1		1		7	2	3	1	2			1

FIGURE 4.12 PAFC power plant installation in Europe and the rest of the world: 1987–2004 (total = 68 units). (For the color version of this figure, the reader is referred to the online version of this book.)

The 1992 commercialization of UTC's 200 kW PAFC power plant might have caused an early spurt in installation activities, and the 1995 DoD Climate Change Rebate program might have energized the PAFC installations that occurred later in the decade. Nonetheless, Austria, Denmark, Finland, Italy, the Netherlands, Norway, and Spain all had ceased PAFC installation activities by the mid-1990s, and France, Sweden, Switzerland, and UK ceased their installation activities after 2001. Germany ended its installation activity after 2004.

More than a decade later, however, UTC announced in October 2008 that Samsung Everland, a Samsung Group affiliate, had purchased 12 UTC 400 kW PAFC power plants for its power station in the Town of Anyang, outside Seoul.[65] The PAFC power plants were operational in September 2009 and would provide power for about 5 percent of Anyang's population. It is not clear if this sale signaled the beginning of an upturn in PAFC installation activity in South Korea, but there is evidence that Samsung was interested in marketing and distributing PAFC power plants as a sole agent of a US PAFC manufacturer. Samsung, however, did not take the initiative to purchase the PAFC manufacturer when an opportunity presented itself.

- In January 2008, Samsung and the US Fuel cell maker HydroGen Corporation entered into a series of strategic cooperation agreements, under which Samsung would act as HydroGen's sole and exclusive distributor to market, distribute, and sell its PAFC power plants for hydrogen applications in Asia, the Middle East, and certain other regions.[66] Samsung also signed a nonbinding Letter of Intent for the sale of an initial 5 MW fuel cell power plant.

In October 2008, however, HydroGen LLC filed for Chapter 11 bankruptcy after failing to find a buyer or obtain additional financing, according to court filings.[67] Earlier in August, HydroGen Corp received a $2 million bridge loan from Samsung C&T Corp and the Federated Kaufmann Fund to sustain operations until other long-term financing was obtained or the company was bought.[5] But the company said in its filing that efforts to raise those additional funds through public markets or investors were unsuccessful.

References

1. *Collecting the history of fuel cells.* Smithsonian Institution, <http://americanhistory.si.edu/csr/fuelcells/index.htm>; 2007 [accessed March 2008].
2. *Phosphoric acid fuel cell, Emory University US.* <http://www.emory.edu/BUSINESS/et/552fall2003/fuelcells/phosphoric_acid_fc.htm>; [accessed in August 2012].
3. Elmore GV, Tanner HA. Intermediate temperature fuel cells, abstract, Journal of the Electrochemical Society, US.; 1961.
4. *Fuel cell handbook.* 7th ed. NETL, US Department of Energy; November 2004.
5. Blomen Leo JMJ, Mugerwa Michael N. *Fuel cell systems.* Springer; 1993.
6. O'Sullivan John B. *The status of fuel cell technology.* Canada: Presentation to the Ontario Ministry of Energy Innovations in Natural Gas Technology; 20 February 1991.
7. Enacted appropriations DOE/FE, distributed generation-fuel cells, US Department of Energy, Fossil Energy.
8. *Marine applications of fuel cell technology,* A technical memorandum, OTA-TM-O-37, Office of technical assessment, US Congress, February 1986.
9. Lens P, Westermann P, Haberbauer M, Moreno A. *Biofuels for fuel cells: renewable energy from biomass fermentation.* IWA Publishing, UK.; 2005.
10. Appleby AJ, Foulkes FR. *Fuel cell handbook.* New York: Van Nostrant Reinhold; 1989.
11. Warshay M, Status of commercial fuel cell powerplant system development, final report, prepared for the 22nd intersociety energy conversion engineering conference cosponsored by AIAA, ANS, ASME, SAE, IEEE, ACS, and AlChE Philadelphia, Pennsylvania, 10–14 August 1987.
12. *Fuel cells for the future: a contribution to technology forecasting from a technology dynamics perspective,* G.J. Schaeffer; 1998.
13. Environmental Consequences of, and Control Processes for, Energy Technologies, Pollution Technology Review, No.181, Argonne National Laboratories; 1990.

14. *4.8-MW fuel-cell demonstrator. Final report. Technical report.* United Technologies Corporation; 1 December 1981.
15. Multi-megawatt fuel cell produces electricity. *Platin Met Rev* 1983;**27**(3).
16. Hoogers Gregor, editor. *Fuel cell technology handbook.* CRC Press; 2003.
17. Fickett AP, Fuel cell power plants, Scientific America, vol/issue: 239, p.70-76, 1 December 1978.
18. Hammerli M, Fuel cell files, Natural Resources Canada, Ottawa, <http://raenergy.igc.org/fuelcl.html>; no date provided, [accessed in September 2007].
19. "Sammes N, Bove R, Stahl K, Phosphoric acid fuel cells: fundamentals and applications, Chapter 9 in Fuel Cells Compendium, Elsevier, 2005.
20. Brown RM, *A powerful opportunity,* FuelCell Energy, 23 October 2002.
21. About Us, *FuelCell Energy.* <http://www.fuelcellenergy.com/about-us.php>; [accessed August 2012].
22. *Fuel cell systems program plan, fiscal year 1993.* Fossil Energy, US Department of Energy; 1993.
23. History, HydroGen Corporation, <http://www.hydrogenllc.net/hydrov2/index.php?id=7,0,0,1,0,0–7k>; [accessed October 2007].
24. Generation I Buses, General Information, TBB Technical Details, Georgetown University. <http://fuelcellbus.georgetown.edu/gen1.cfm>; [accessed August 2012].
25. Program history, Georgetown university advance vehicle development, Georgetown University. <http://fuelcellbus.georgetown.edu/history.cfm>; [accessed August 2012].
26. Warshay M, Prokopius P, Coordinated fuel cell system programs for government and commercial applications: Are we in a new era? NASA Technical Memorandum 106699, Prepared for the 33rd Aerospace sciences meeting and exhibit sponsored by the American Institute of Aeronautics and Astronautics, Reno, Nevada, 9–12 January 1995.
27. Advanced Vehicle Development: The Fuel Cell Bus Program, Georgetown University. <http://fuelcellbus.georgetown.edu/index.cfm>; [accessed in August 2012].
28. Binder MJ, Taylor WR, Holcomb FH, Experience with the DOD fleet of 30 generators, Presented at the 2001 international gas research conference, Amsterdam, The Netherlands, 5–8 November 2001.
29. Taylor WR, Binder MJ, Holcomb FH, DOD experience from a half million operating hours of a PAFC fleet, a presentation at 2000 fuel cell seminar, Portland Oregon, 1 November 2000.
30. ERDC-CERL Fuel Cell, Department of Defense. <http://www.dodfuelcell.com/>; [accessed in August 2012].
31. *Fuel cell at German hospital tops 10 million kilowatt hours of operation, UTC power* press releases, UTC Power, 4 March 2008.
32. Stahl D-I K, Climate change fuel cell program 200 kW – PC25C fuel cell power plant for the St.-Agnes-Hospital, Bocholt, Germany final report, Reporting period: November 2, 2000 until January 2, 2002, DOE Award Number: DE-FG26–99FT40602, T.B.E. – Technische Beratung Energie GmbH, January 2002.
33. *UTC power fuel cell fleet attains 8 million hours of operation, UTC power,* press releases, UTC Power, 29 March 2007.
34. Northeast Blackout of 2003, Wikipedia, last modified on 13 April 2008, [accessed April 2008].
35. Pielli K, State clean energy funds fact sheet. State clean energy funds: An effective mechanism to encourage clean energy supply, Last updated April 12, 2007, US Environmental Protection Agency, <http://www.epa.gov/chp/state-policy/funds_fs.html>; [accessed January 2009].
36. State Clean Energy Fund Support for Renewable Energy Projects: Key Findings from the CESA National Database, Clean Energy States Alliance (CESA), January 2009.
37. *New York power authority selects UTC power to supply fuel cells for world trade center site*, press release, UTC Power, 11 June 2008.
38. New energy technology development data file production survey (fuel cells), NEDO, 1996, 1997, 1998, 1999, 2000.
39. Taku Fukumura, Current status of development of PAFC system in Fuji Electric group, A presentation at 2008 fuel cell seminar & exposition, Phoenix, Arizona, 27–30 October 2008.
40. Toshiba History Corner, Toshiba Science Museum, Toshiba Corporation. <http://kagakukan.toshiba.co.jp/history/1goki/1991plant/index.html>; [accessed December 2008].
41. Homma T. New developments in PAFC, Latest News, No. 14, FCDIC, 23 June 1997.
42. Toshiba and Mitsui Chemical launched a new business: a FC Radiant Farm, Nikkei Sangyo Newspaper, 26 February 1998.
43. *New energy technology development related data file production survey (fuel cells).* NEDO; March 2000.
44. Company Information, History, Sanyo Electric, Japan. <https://www.sanyo.co.jp/koho/doc/j/corporate/enkaku03.html>; [accessed December 2008].
45. *New energy technology development data file production survey (fuel cells).* NEDO; March 1999.
46. Large Scale Energy Conservation Technology R&D Project (Moonlight Project), Phosphoric acid fuel cell power generation R&D, 1981–1990, evaluation report Industrial Technology Council, METI, May 2000.

47. Hirschenhofer JH, International developments in fuel cells, Mechanical Engineering-CIME, 1 August 1989.
48. *Fuel cell power generation technology development, research result (revised edition)*. NEDO; February 1989.
49. *Demonstration operations to verify grid interconnection of dispersed advanced power generation systems*. NEDO; March 1994.
50. New energy technology development related data file, production and survey (fuel cells), March 1997.
51. *Advanced high efficiency energy system field test project, data collection, analysis, and evaluation*. NEDO; March 2002.
52. *New energy technology development data file production survey (fuel cells)*. NEDO; March 1998.
53. *A study to promote practical use of fuel cells*. NEDO; March 1997.
54. Homma T. *Results of PAFC demonstration tests by gas companies, latest news, No. 15*. FCDIC; 23 June 1997.
55. *Changes in the cost of PAFC power plants, FY 2005 new energy data*. NEDO, Japan. <http://www.nedo.go.jp/nedata/17fy/12/g/0012g002.html>; [accessed December 2008].
56. *Proposals on promotion of fuel cell introduction*. New Energy Foundation, Japan; March 2008.
57. *New energy guidebook*. NEDO; 2008.
58. Hasegawa M, Horiuchi Y, Phosphoric Acid Fuel Cell, the FP-100i, Fuji Jiho. vol. 83, No.3, Fuji Electric, 2010.
59. Homma T. *PAFC development and business, the latest news number 165*. Japan: Fuel Cell Development Information Center; January 2010.
60. Japan's Fuji electric to sell industrial fuel cells in European market, News. <http://www.fuelcelltoday.com/news-events/news-archive/2012/february/japan%E2%80%99s-fuji-electric-to-sell-industrial-fuel-cells-in-european-market>; Fuel Cell Today, 14 February 2012.
61. *Japan's fuel cell industry*. JETRO Japan Economic Report, JETRO; April-May 2006.
62. Cropper M, Fuel Cell Development at Bharat Heavy Electricals Ltd, India, Fuel Cell Today, 5 November 2001.
63. Trocciola JC, Pompa TN, Boyd LS, Demonstration of a PC 25 Fuel Cell in Russia, Final Report For Period: October 2001-September 2004, UTC Fuel Cells, LLC, September 2004.
64. *Fuel cell keeps power going during Russian blackout*, press release, UTC Power, 31 May 2005.
65. *Samsung Everland to install 12 UTC power fuel cell systems for GS power in South Korea*, press release, UTC Power, 29 October 2008.
66. *HydroGen corporation and samsung corporation enter into a series of strategic cooperation agreements for the sale, marketing and distribution of fuel cell power plants*, Fuel cell works, 15 January 2008.
67. HydroGen LLC files for chapter 11 bankruptcy, Reuters, 23 October 2008.

CHAPTER

5

History of Molten Carbonate Fuel Cells

OUTLINE

http://dx.doi.org/10.1016/B978-0-444-56325-5.00005-3

5.1. MCFC EFFORT STARTS IN THE NETHERLANDS IN THE 1950s

The concept of molten carbonate fuel cells (MCFCs) dates back to the 1950s when Dutch scientists G. H. J. Broers and J. A. A. Ketelaar began their fuel cell research at the University of Amsterdam under contract with TNO, the Dutch semigovernmental organization for applied research. Their initial effort was to retrace the solid oxide experiments conducted by O. K. Davtyan of Russia in the 1940s.[1] By the late 1950s, however, they concluded that near-term progress would be unlikely and abandoned their solid oxide research. They saw that solid oxide electrolyte was not really solid at operating temperatures of 650–750 °C, so they shifted their focus to molten carbonate salts, marking the beginning of MCFC technology development. In 1960, they reported that their first MCFC prototype operated continuously for 6 months.[2] The electrolyte used was an alkali metal carbonate impregnated in a disk of magnesium oxide.[3]

- Notwithstanding the promising results, Broers and Ketelaar's work was stopped by TNO in 1969 because the industry lacked interest and the technology competed poorly with conventional power and nuclear power, given their relative cost advantage at the time.

Subsequent to the end of the Dutch MCFC effort, a number of countries showed interest in MCFC technology. Starting in the mid-1960s, the US government, utility companies, and manufacturers became active in MCFC development. Likely impressed with the Dutch research, the US Army began to fund MCFC development in the mid-1960s. The Army's interest, however, waned rapidly due to the lack of positive results and it ended funding MCFC development in the late 1960s. The US Department of Energy (DOE, then ERDA) started showing interest in MCFC technology in the mid-1970s and funded the development and demonstration of MCFC power plants. US funding for MCFC development continued for nearly 25 years until 2000.

Next to the United States, Japan became the most active player in MCFC development. MCFC development in Japan was nearly exclusively conducted under the Japanese government's Moonlight Project, a large-scale undertaking nearly equal to similar projects in the United States. In 1981, Japan's METI implemented a 24-year, MCFC development program under the Moonlight Project that continued until 2004, with a total budget of about $470 million.[4,i] The ultimate goal of the program was to develop large-scale MCFC power plants that could be fueled by coal gas and replace coal-burning thermal power plants. Several companies started MCFC R&D and most of them formed close affiliations with US MCFC developers to achieve rapid technological growth.

European countries revived their interest in MCFC technology in the mid- to late 1980s. The Netherlands and Italy restarted their MCFC development efforts in 1986, and Germany in 1988. The European efforts, however, were of lesser intensity compared to those of the United States and Japan. They primarily focused on acquisition of a foreign, namely US MCFC technology, rather than developing their own technologies. The EU also played a major role in supporting MCFC development and demonstration efforts.

Elsewhere in the world, the only other major actor was South Korea, which started its MCFC development in 1993. Again, the government was more interested in acquiring global state-of-the-art technology via licensing and corporate participation with the global MCFC leader, FuelCell Energy (FCE) of the United States.

5.2. MCFC DEVELOPMENT IN THE UNITED STATES

The United States has been the most active supporter of MCFC. Key participants include DOE and the Department of Defence (DoD), utility organizations such as the Gas Research Institute (GRI), the Institute of Gas Technology (IGT), the Electric Power Research Institute (ERPI), gas and electric utilities, and electric machinery manufacturers.

5.2.1. Early Efforts

5.2.1.1. *The US Army: Early MCFC Supporter*

One of the early supporters of MCFC technology was the US Army. It supported a 15 kW Tactical Power Source Development program at Fort Belvoir in the mid-1960s.[5] Texas Instruments won the contract and developed a 1 kW prototype system. The prototype included an MCFC and a compact catalytic partial oxidation (CPOX) unit to convert lead-free gasoline to reformate. The high-temperature CPOX unit had no trouble processing gasoline. The use of a shift catalyst in the manifolds was found to enhance fuel cell performance, but the Army's standard logistics fuel, while lead-free, contained sulfur. When the system was operated with logistic fuel, unacceptable power losses occurred after several 100 hours of operation. The cause was attributed to electrolyte loss in the MCFC and poisoning from sulfur and other constituents.

[i] Dollar value of the Japanese MCFC R&D budget between 1981 and 2004 is based upon the yen/dollar exchange rate prevailing at the beginning of each year.

- Development efforts on MCFCs focused on possible solutions for operating on logistic fuels. However, low power density without the promise of significant improvements halted the development of MCFCs for military application in the late 1960s.

5.2.1.2. *The Institute of Gas Technology: Early MCFC Developer*

The Institute of Gas Technology (IGT) in Chicago has been a pioneer of MCFC technology that contributed the most to its advancement in the United States. IGT, a nonprofit R&D organization for the gas industry, started R&D on MCFC and components in 1959. Many of the basic concepts of MCFC technology have flowed from the IGT studies, including basic patents on the composition of the electrolyte, on stable anode structures, on the principle of internal reforming, and on corrosion-resistant wet seals.[6] Indeed, two of the most prominent MCFC stack concepts originated from IGT studies:

- *The externally manifolded, internally reforming configuration.*[ii] One of the MCFC pioneers, B. S. Baker, worked with IGT in the 1960s and envisioned a MCFC system that would allow fuel to be sent directly to the fuel cell without first reforming it. He subsequently turned the concept of MCFC internal reforming into a commercial reality. In 1969, he founded a company called Energy Research Corporation (ERC, now FuelCell Energy, FCE) in Danbury, Connecticut, and developed MCFC power plants integrating externally manifolded, internally reformed stacks, called the "Direct Fuel Cell (DFC)" system.
- *The internally manifolded, externally reforming configuration.*[iii] Another significant IGT contribution to MCFC development was its invention in the 1980s of what is known as the Internally Manifolded Heat Exchanger (IMHEX) separator plate concept for MCFC power plants. Internal manifolding is used to both introduce and exhaust fuel and air from the cell stack. The design calls for the fuel to be externally reformed, which virtually is the direct opposite to B.S. Baker's DFC concept. In 1987, IGT set up a dedicated MCFC developer-manufacturer M-C Power (MCP) to commercialize MCFC stacks based on the internally manifolded IMHEX stack.[7]

IGT for the most part worked on fuel cell technology as a subcontractor. It worked for United Technologies Corporation (UTC) under the TARGET program from 1967 to 1975, and later for General Electric (GE) for its DOE-sponsored studies.[6] Starting in 1977, IGT was supported by EPRI to perform research on the basic electrochemistry of MCFCs, on the effect of contaminants, particularly sulfur, on MCFC reactions, and on new materials, components, and internal reforming catalysts. IGT supported MCP's MCFC R&D activities that were conducted under DOE sponsorship as a subcontractor throughout the 1970s and 1980s.

- In 2000, IGT merged with GRI, another nonprofit research organization for the gas industry. The new organization is named the Gas Technology Institute (GTI), headquartered in Chicago.

[ii] External manifolding refers to the method of injecting the reactant fuel and oxidant into the stack from a central source and removing the exhaust gas via large inlet and outlet manifolds. Internal reforming means performing the chemical reforming process of producing hydrogen from natural gas and other fuels inside the fuel cell, near the electrochemically active sites.

[iii] Internal manifolding refers to the method of distributing gas mixtures among cells through an internal manifold; that is, a duct inside the stack itself, penetrating the separator plates. External reforming refers to the method of producing hydrogen-rich gas from natural gas and other fuels outside the stack.

5.2.2. The Department of Energy: Initiating MCFC R&D Program in 1975

DOE supported MCFC development for 30 years. DOE implemented the MCFC program in 1975, started funding in 1976, and completed the 30-year, $481-million program in 2005. The US MCFC program was administered by DOE's Fossil Energy Administration and directed by the National Energy Technology Laboratory (NETL) in Morgantown, West Virginia. The DOE MCFC program was a cost-shared, market-driven program. It was administered in close cooperation with EPRI, GRI, electric and gas utilities, and MCFC developers to fully and efficiently leverage funding for MCFC development in the United States. MCFC developers also were supported by user groups comprising utility and end-user members.

- The initial objective of the DOE MCFC program was to develop and demonstrate a multi-megawatt, coal-fired, utility MCFC plant by about 1990.[6] In reality, MCFC development took much longer than anticipated; DOE ended MCFC funding in 2005, judging that MCFC technology was near commercialization.

During the 1970s, DOE supported efforts to scale up cell components and develop MCFC stacks and awarded contracts to many companies, electric utilities, and national laboratories. Starting in 1976, DOE funded GE to conduct studies on a 675 MW coal-fired central power station, using an oxygen-blown Texaco coal gasifier and the Selexol gas purification process.[8] It also supported FCE to develop basic component materials, fabrication techniques, and cell design, with some work on performance modeling and system design (1976–1980).[6] In 1979, UTC was awarded its first major DOE MCFC contract to investigate improvements in MCFC component and system design.[8] DOE funding for MCFC development was small in those early days; its budget was about $1.5 million in 1977 and $2.2 million in 1978, although the budget gradually increased to about $12 million by 1980.

- DOE's MCFC development effort was reinforced by EPRI's effort. Starting in the 1970s, EPRI funded MCFC development independently or in conjunction with DOE. EPRI primarily supported work at three developers: (1) GE, to define utility plant parameters, (2) UTC, to establish the viability of the MCFC system, as compared with other fuel cell concepts, and to examine pressurized laboratory stack development, and (3) FCE, to investigate an alternative cell design concept.[6]
- GRI also supported the development of MCFC systems for cogeneration, complementing DOE's efforts, starting in the 1970s. GRI supported seven developers: (1) IGT, to study contaminant effects, (2) FCE, for alternative cell designs, (3) EIC Corporation and Physical Sciences Inc, to study electrode kinetics, (4) Argonne National Laboratory, to develop rigid sintered electrolyte structures, (5) Oak Ridge National Laboratory, to study ion transport mechanisms, (6) Montana Energy and MHD Research and Development Institute (MERDI), to develop fibrous ceramic materials, and (7) Pacific Gas and Electric, Southern California Edison, and others to conduct other MCFC research.[6]

5.2.2.1. MCFC Development Program in the 1980s: GE and UTC Emerge as Prime Contractors

In early 1980, DOE had high hopes for MCFC technology and envisioned the demonstration of a coal-fueled baseload MCFC power plant by the end of 1990. DOE undoubtedly was convinced that much of the knowledge about MCFC cell performance had been established via extensive R&D sponsored by DOE, EPRI, and GRI in the 1970s. The DOE MCFC program in the 1980s, however, was a tumultuous one, seriously strained by significant budget cutbacks and drastic revisions in the original MCFC development plans, as well as unexpected changes in major players in the program.

To plan the course of MCFC development, DOE in early 1980 established an ambitious 10-year roadmap to develop a power plant design and component and stack designs, to fabricate facilities for prototype stack testing, and to demonstrate two full-scale commercial, cost-efficient power plants on coal-derived gas. The 10-year plan consisted of four phases.[6]

- The time frame designed for the program envisaged the testing of a 500–700-cell prototype stack in 1983 (the first *Exploratory Development* phase), the building and testing of a several-megawatt pilot plant in 1985 (the second *Technology Development* phase), the demonstration of a small cogeneration plant or electric utility plant using volatile fuels in 1987–1989 (the third *Engineering Development* phase), and the demonstration of a 650 MW coal-fueled utility power plant incorporating a coal gasifier in about 1990 (the final *Demonstration* phase).

The 10-year program, however, was almost immediately abandoned. Budget cutbacks in the 1980s made it impossible to adhere to the schedule for the development of the MCFC power plant. The MCFC program budget, which was $12 million in 1980, was reduced to $9 million in 1981 and remained at that level throughout the 1980s. Budget cutbacks were due to the overall DOE/NETL fuel cell budget cutbacks and also because a major competing program—the PAFC program—at NETL was entering a critical field demonstration phase.[6]

Due to these budget cutbacks and because of its growing interest in natural gas MCFCs, DOE decided to divide its MCFC program into two parts. The first would involve the development of a full-size, roughly 500 kW stack operating on coal gas; the second, the development of a 10–50 MW dispersed natural gas power plant. In practice, however, DOE scaled back the size of the MCFC power plants further and removed the distinction between coal-gas-fueled and natural-gas-fueled systems. All the MCFC power plants developed were designed to use natural gas. They were subsequently expected to be modified to use coal gas.

Also, likely because of the need to economize as well as shorten the development time, DOE pushed to carry out system design in parallel with technology development to achieve relevant technology goals and a commercially attractive end product.[6] DOE also tried to use commercially available, off-the-shelf technology as much as possible, and to carry out no extensive development for components other than those for the fuel cells and stacks. It was anticipated that full-scale stack development could begin in 1986, and that start-up of the first 8 ft^2 (0.74 m^2) 20-cell short stack would begin in September 1986.

With this backdrop, DOE in 1980 awarded contracts to GE and UTC for development of a pressurized coal gas MCFC stack, with FCE receiving DOE support for materials development (in parallel to its EPRI contract to develop an alternative stack).[6] Under the 1980 DOE stack development program, GE worked to develop internally manifolded 0.1 ft^2 (93 cm^2) active area stack components and some new materials concepts, and UTC worked to develop externally manifolded MCFC stacks with the same active area.[1] The strong point of GE MCFC work was the development of a number of innovative methods and advanced processes for the production of materials. GE developed a calendaring technique that could produce a 0.5 mm-thick pliable dielectric sheet that had fewer defects.[6] GE also designed internal manifolding that presented fewer sealing and shunt current problems than did the external manifolding design used by UTC and FCE. GE also adopted crossflow manifolding to feed reactant gases to the cells in the stack, which was simpler and more effective for internal reforming than crossflow manifolding used by UTC. GE also recognized the problem of anode creep in a cell stack and resolved it by producing a nonshrinking anode consisting of nickel metal particles with a ceramic core.

- In the spring of 1984, however, GE decided to end MCFC development for business reasons. The company projected that there would not be sufficient profit either at the present time or in the near future to stay active in the MCFC market. They therefore completed their contractual obligations with DOE and closed down their Advanced Energy Systems Division, abandoning their work in areas such as sodium–sulfur batteries, fuel cells, solar energy, and wind turbines. They also sold off their interests in PEMFC technology for both fuel cells and advanced electrolyzers to the Hamilton Standard Division of UTC. GE in the early 1960s also was involved in PEMFC development for the Apollo manned lunar mission but subsequently backed out of the PEMFC effort.

With GE's departure from the DOE program, UTC and FCE were the main DOE contractors for MCFC development, with UTC receiving about twice as much support as FCE for its stack development effort.[6] UTC continued to follow DOE's emphasis on the development of utility-scale MCFC systems. The first MCFC stacks were operated by UTC in the early 1980s.[1] These stacks were 0.1 ft^2 (93 cm^2) in area, externally manifolded, and could be operated under pressure. Up to mid-1985, five subscale 20-cell pressurized stacks of conventional design (0.1 ft^2, 93 cm^2) were operated. Two of the five operated for 4000 and 5000 h, respectively.[6] UTC worked to scale up this system horizontally. In 1986, UTC operated stacks of up to 20 cells at 8 ft^2 (0.74 m^2) in active area,[iv] with a capacity of up to 25 kW, for up to 2000 h.[1] UTC further scaled it up to stacks of twenty 8 ft^2 (1.932 m^2) cells, with a design termed "thin cell," which had a capacity of 25 kW, and tested it for 2200 h. In 1990, UTC completed system studies for coal-gas and natural-gas fueled MCFC systems.[8] By 1991, UTC had successfully addressed the electrolyte migration problem and continued to address performance improvement, increased life, and component cost reduction.

- From 1987 onward, however, UTC assumed a less prominent role in MCFC development under DOE sponsorship, possibly because DOE shifted funding to ERC and MCP, which were perceived to be on a faster development track (UTC did remain involved in development of stack components).[1] Also, the 1992 DOE MCFC contracts were accompanied with an important precondition that would require a contractor to have an MCFC manufacturing facility and commercialization plans. UTC, which was planning to launch PAFC commercialization in 1992, likely did not intend to proceed with both MCFC and PAFC commercialization.

As UTC scaled back its role in the DOE MCFC program, FCE now emerged as the prime contractor for MCFC development. In 1987, FCE was awarded its first major MCFC DOE contract. By 1989, FCE successfully completed a 60-cell, 1 ft^2 (0.09 m^2) stack test that incorporated the internal reforming concept. Since then, many 2–25 kW, 4 ft^2 (0.37 m^2) internal reforming stack tests were completed using natural gas and simulated coal gas fuels. In 1990, FCE completed system studies for coal gas- and natural gas-fueled MCFC systems.[8]

- Although FCE had not played a major role in DOE's MCFC program until the late 1980s, it had done considerable MCFC development

[iv] All references to cell area are to the "active area" of the cell.

work. The company had worked on MCFC development in-house since 1969 and with external support since 1976. DOE's Office of Fossil Energy issued its first research contract to FCE in 1976, 6 years after the Danbury, Connecticut, firm originally was founded under the name of ERC.[9] In their original DOE work (1976–1980), FCE worked to develop cell component and fabrication techniques. In a contract with EPRI, FCE developed a new thin thermally cyclable cell concept and examined the possibility of using new techniques for fabricating thin layers of cell components. Since 1982–1983, FCE concentrated on developing and demonstrating the advantages of the internal reforming MCFC (what FCE called "theDirect Fuel Cell," DFC) for use with natural gas, also under EPRI contracts.[6] In addition to the use of such a fuel cell for on-site applications, FCE also examined its application to simplified atmospheric pressure-dispersed power plants in the 2 MW class. Although not chosen as a major contractor for the DOE stack development program in 1980, FCE had been receiving DOE funding in parallel with EPRI's project for cell component development.

5.2.2.2. MCFC Demonstration Program in the 1990s: Fuel Cell Energy and M-C Power as Primary Developers

In the 1990s, DOE was still hopeful that there would be a breakthrough in MCFC technology. DOE/NETL increased its fuel cell budget in FY1990 by more than 40 percent from the previous year to almost $38 million and they increased the MCFC budget by a stunning 140 percent from $7 million to $16.8 million during the same period. (The DOE/NETL fuel cell budget remained mostly between $40 million and $50 million throughout the decade). DOE also added some extra cash starting in 1993 as its support for PAFC development. Reflecting those factors, the DOE MCFC budget stayed mostly in the low to mid $30 million—a very respectable budget—in the 1990s.

With this funding, DOE implemented an energetic program characterized by three major activities: to develop and fabricate MCFC stacks, to develop and demonstrate two prototype power plants, and to improve the prototypes at lower cost to prepare for market entry.

- In 1990, DOE primarily sponsored two major MCFC companies to develop cell and stack development; FCE and MCP both signed 5-year contracts to scale up the cells to between 0.6 and 1 m^2 in active area and assemble them into 125–250 kW stacks.
- In 1992, DOE also awarded two 5-year product development test contracts. As mentioned earlier, an important precondition for the contracts was that the contractor have an MCFC manufacturing facility and commercialization plans. Such a precondition undoubtedly must have required a major commitment on the part of the contractors because MCFC technology had not yet been shown to achieve stable operations, reliability, and cost-competitiveness. Nonetheless, FCE and MCP met the preconditions[1] and DOE awarded them contracts to develop prototype MCFC power plants. Under the contracts, FCE was to showcase a 2 MW field test demonstration in Santa Clara, California, in 1996. That would be the largest MCFC demonstration in the world and the largest fuel cell power plant to operate in the United States. MCP was to demonstrate a 250 kW test plant for cogeneration in 1994. For these demonstration projects, both FCE and MCP also had coordinated arrangements with EPRI, GRI, and various gas and electric utility participants.[10]
- In 1995, DOE awarded the two companies another 5-year product design and development contract. FCE and MCP were expected to further advance cell and stack technology to improve MCFC products and

achieve the targets for cost, performance, and durability necessary for commercial power plants.[11] The companies would work to line up possible products for market entry.

In 1992, UTC concluded its last contract with DOE, in which it worked to improve MCFC stack performance.[8] UTC had plans for further developmental activities under a private sector program. This private sector effort was to be assisted through a Cooperative Research and Development Agreement with DOE, which would make certain development and testing equipment available to UTC in exchange for keeping DOE informed of program progress.[11] From this time onward, however, UTC was no longer a major player in MCFC development and eventually transferred its technology to Ansaldo with the consent of DOE (Table 5.1).

5.2.2.2.1. FUELCELL ENERGY

FCE made a major effort to qualify itself as a prime contractor for the 1992 DOE product development test contracts. In 1991, it established a wholly owned and privately financed MCFC stack manufacturing facility in Torrington, Connecticut.[8] Manufacturing capacity of the plant was 2–3 MW per year. In 1992, the manufacturing capacity increased to about 10 MW/year.[12] By 1999, it grew to 17 MW/year and reached 50 MW/year as of the end of 2008.[13] FCE also equipped itself with a 400 kW test facility in Danbury, Connecticut in 1999.[14] FCE today has the world's largest fuel cell manufacturing facility of any type in the world.

5.2.2.2.1.1. 125 KW STACK DEVELOPMENT FCE made good progress in its stack development. In 1990–1992, under the DOE sponsorship, FCE built a 2 kW stack with 4000 cm^2 components and tested it for 10,000 h. FCE then assembled a 20 kW stack with the 4000 cm^2 components. They showed good cell-to-cell uniformity.[1] The stack was tested for 400 hours at the FCE test facility, Danbury. The stack subsequently was transported to

TABLE 5.1 MCFC Development in the United States

MCFC Developer	Manifold type	Development status
FuelCell Energy	External manifolding internal reforming	20 kW stack with 0.4 m^2 cells, operated for 1000 h at Pacific Gas and Electric, San Ramon, California
		70 kW stack, 0.4 m^2 × 234 cells, operated for more than 2000 h (1992)
		125 kW stack, 0.6 m^2 × 246 cells, operated for 250 h (1993)
		2 MW system, sixteen 125 kW stacks operated for 3600 h in Santa Clara, California (1996–1997)
		375 kW stack, 0.9 m^2 × 326 cells, completed tests (1998)
		250 kW stack, 0.9 m^2 × 340 cells, tested (1999–2000)
M-C Power	Internal manifolding external reforming	20 kW stack, 1 m^2 × 19 cells, started operation (early 1992)
		250 kW stack, 1 m^2 × 250 cells, was tested (1995)
		250 kW system, operated for 2500 h, delivered 158 MW/h of direct current output and 296,500 pounds of 110 psig steam to the Miramar base
		75 kW stack tested in Miramar (1999)
UTC		25 kW stack, 6 ft^2 × 20 cells, operated for 2200 h (1986)

Pacific Gas and Electric (PG&E), San Ramon, California. The stack operated there for 1000 h and was thermally cycled several times before shutdown. FCE's MCFC development then progressed to testing of their internally reforming, externally manifolded cells in stacks of up to 70 kW using natural gas and simulated coal gas.[8] The first 70 kW stack successfully completed testing in 1992, including testing of stacks on simulated and actual coal gas. Continued testing of a 70 kW stack was conducted at PG&E's San Ramon Research Center in 1993. The 70 kW stack at the San Ramon Research Center logged more than 2000 h of cumulative running time.[15]

In April 1993, FCE passed a major milestone with a 250-hour test of a 125 kW MCFC stack, the largest tested at that time.[15] The test stack achieved a 50-percent overall efficiency level; also a record for MCFC test cells. The 125 kW prototype stack contained 246 individual flat cells assembled in 6-cell groups. Each cell package had an area of 6 ft^2 (0.6 m^2).

- The 125 kW stack was the prototype for the 2 MW natural gas–fueled power plant that was to be demonstrated in Santa Clara, California, within the next several years.

5.2.2.2.1.2. SANTA CLARA 2 MW PLANT DEMONSTRATION PROJECT The DOE objective of the 5-year demonstration project (1992–1997) was to field test a 2 MW direct carbonate fuel cell (DFC) technology at full scale to promote its viability as a commercial product.[16] FCE's externally manifolded DFC power plant was to produce power directly from hydrocarbon fuels electrochemically, without the need for external reforming or intermediate mechanical conversion steps. As a result, the DFC would have the potential to achieve very high efficiencies with very low levels of environmental emissions.[17] FCE built a 2 MW natural gas–fueled power plant with an array of 16 MCFC stacks (internally reforming) each with a capacity of 125 kW.[16] FCE entered into a joint private sector/government program to operate a 2 MW MCFC at a Santa Clara, California, municipal electric site and started plant installation in April 1994. The construction was completed in June 1995, and the plant started operations in April 1996.

The Santa Clara project was considered successful. The plant reached a number of operating goals, including peak power output of 1.93 MW AC, voltage harmonic quality, 1 ppm nitrogen oxide and no detectable sulfur oxide levels, and noise levels of less than 60 dBA at 100 feet.[16] The plant achieved an electrical efficiency of 44 percent (LHV). Because this was the first test of a large MCFC plant, two auxiliary natural gas burners, which would normally be off, were running as a precaution to ensure system stability in the event of a plant trip. Without the auxiliary burners in operation, an electrical efficiency of 49 percent would have been achieved.[18]

The test was not without problems. After 2 weeks of operation, thermal decomposition of insulating materials led to the deposition of carbon particles on electrode surfaces and connecting buses.[19] After some reconfiguration and repairs, the plant was brought back on line at 500 kW using eight stacks that had not been damaged.[16] The reduced load on the remaining eight stacks (they were collectively rated at 1 MW) allowed for lower operating temperatures, reducing the adverse impact of extreme heat on the stack. In March 1997, operation of the plant was terminated.[20] The plant operated for 4900 h of hot time (plant at normal operating temperature) with 3400 h of grid-connected operation.

The total cost of the Santa Clara demonstration project was estimated at $52 million.[17] The cost was shared among two parties: DOE and FCE (supported by a group of utilities and users). DOE invested just over $28.3 million to fabricate the first-a-kind demonstration fuel cell stacks for the Santa Clara plant. FCE via its support group of utilities and buyers—called the Fuel Cell Commercialization Group (FCCG)—provided nearly $23 million.

- The FCCG consisted of about 35 companies paying $10,000 per year to follow FCE's development, cooperate on working committees, and obtain the right to buy early commercial units.[21] A subgroup of the FCCG—consisting of five utilities, the National Rural Electric Cooperative Association, and EPRI—was sponsoring the balance of plant (BoP) design, construction, and test for the Santa Clara project. The BoP was financed by these private sponsors, as well as funding from the Santa Clara municipal utility and other prospective users.[21,22]

5.2.2.2.1.3. PRODUCT DEVELOPMENT FOR MARKET ENTRY Following the 2 MW demonstration project, DOE supported FCE with a 5-year contract that would help FCE further improve their stacks and power plants and prepare for market entry. The prototypes that were demonstrated at San Ramon and Santa Clara, including the 20 kW, 70 kW, and 125 kW plants, were technological forerunners that could be used as building blocks of FCE's expected commercial product line. FCE planned the initial target for its market entrance to be a 2 MW power plant for baseload and dispersed generation applications.[23] The company also completed the preliminary design of a 3 MW DFC power plant as a possible product. In response to market interest, FCE also thought of offering a 1 MW design. For the cogeneration market, FCE planned to develop sub-megawatt power plants. In 1998, FCE tested a 375 kW stack, consisting of 326 cells of 0.9 m^2 size.[24] In 1999–2000, two 250 kW tall stacks were built and tested. Each 250 kW stack consisted of 340 cells of 0.9 m^2 size. One of the two stacks was tested for over 2500 h with three thermocycles and no noticeable deterioration of performance.[14] The other stack operated in a grid-connected mode.[24] FCE thus envisioned their product lineup would range from 250 kW to 3 MW power plants.

- FCE subsequently upgraded the design of its 250 kW DFC power plant by raising the output to 300 kW.
- In addition, cross-licensing arrangements approved by DOE with CFC Solutions of Germany (then Motoren- und Turbinen-Union Friedrichshafen (MTU)) gave FCE access to CFC Solutions' 250 kW product "HotModule," in the United States (see Section 5.4).[14]

In 1999, the company changed its corporate name to FCE (from ERC) as part of its market entry plan. In cooperation with the American Public Power Association, FCE developed a commercialization plan that aimed to have orders for over 100 MW of early production units. The FCCG, FCE's support group of electric and gas utilities, endorsed this plan in 1999 and lobbied to secure orders for 50 of their early commercial power plants.[11]

- FCE launched MCFC commercialization in January 2003, when FCE shipped its first commercial 250 kW product off to Kirin Brewery in Japan.[25] The MCFC plant started operation in April 2003 using methane-rich digester gas generated in the process of wastewater treatment.[26]
- Today, FCE has the following products on the market: a 300 kW power plant (DFC300), a 1.4 MW power plant (DFC1500), a 2.8 MW power plant (DFC3000), and a hybrid, multi-megawatt DFC-ERG (Direct FuelCell Energy Recovery Generation) system combining a Direct FuelCell power plant with an unfired gas expansion turbine.[27]

5.2.2.2.2. M-C POWER

MCP, in Burr Ridge, Illinois, was established in 1987 by IGT to commercialize the IMHEX separator plate concept for MCFCs that IGT patented in the 1980s. This fundamental technology was transferred to MCP in 1989. MCP tested numerous sub-scale (1 ft^2) stacks containing from 3 to 70 cells, and these tests demonstrated the viability of internal manifolding, from the viewpoint of manufacturing (ease of assembly), operation (gas sealing), and performance (electrolyte management).[1] In the1990s, DOE awarded MCP three cost-shared contracts, which transitioned MCP from a supporting role to that of the primary IMHEX technology developer, a role previously held by IGT.[8] The contracts included the 1990 contract to develop a 250 kW MCFC stack, the 1992 product development test contract, and the 1994 product development and improvement contract.

5.2.2.2.2.1. 250 KW STACK DEVELOPMENT In 1990, DOE awarded MCP a contract to develop and demonstrate a 250 kW full-size stack consisting of 250 cells of 1 m^2 size (11 ft^2). The project budget was \$60.5 million with \$18.1 million from MCP.[28] In 1991, the MCP built a manufacturing facility in Burr Ridge, Illinois, which had the capacity for assembling 2–3 MW stacks per year and could make full-area 1 m^2 (11 ft^2) cells and stack components.[8] A 20 kW test facility to accommodate full-area stacks was also completed in 1991. First, sub-scale 0.09 m^2 (1 ft^2), 24-cell, and 70-cell stacks were tested in this facility in 1991, and then a 20 kW stack consisting of 1 m^2 (11 ft^2) cells was tested successfully in 1992. Subsequently, MCP assembled their first full-scale stack containing 250 cells with an electrical output rating of approximately 250 kW plus cogeneration heat.[11] This stack was installed at a test site in Brea, California, and the testing was completed in FY 1995.

- San Diego Gas and Electric and UNOCAL cooperated with MCP to design and test 250 kW units.
- IGT continued sub-scale, stack testing in support of the full-area stack testing and manufacturing program at MCP. MCP/IGT also coordinated programs with EPRI, GRI, and various gas and electric utility participants.[8]

5.2.2.2.2.2. MIRAMAR 250 KW PLANT DEMONSTRATION PROJECT In 1992, DOE awarded MCP a second contract, which included the demonstration of a MCFC plant in San Diego in 1997. The contract had a value of \$30.8 million, including \$14.2 million in cost sharing.[28] The primary goal of this product development test was to conduct a field study of a completely integrated MCFC power plant in order to identify improvement opportunities in design, performance, and cost.[16]

For this project, MCP installed a 250 kW MCFC cogeneration power plant at the Miramar Marine Corps Air Station in San Diego, which began operation on January 10, 1997. The power plant consisted of a fully integrated system including a newly designed reformer and a stack with two hundred and fifty 1 m^2 (11 ft^2) cells and operated at three atmospheres (3 atm). Two weeks later, the plant began to produce power. Problems, which were anticipated, started immediately with the inverter, turbocharger, and recycle gas blower, which caused a number of outages during the first few weeks of the operation.[16] Nonetheless, the test generated good quality data until plant operations were terminated. The unit reached 210 kW capacity, generating a total output of 158 MW hours of electricity and 346,000 pounds of steam to be used for cogeneration over 2350 h of operation.[143] This cogeneration approach raised overall efficiencies to nearly 85 percent—meaning that 85 percent of the energy value of the fuel was used.[28] The test also demonstrated excellent uniformity of voltage among the 250 cells and no undesirable electrolyte pumping in the system.[29] The stack also demonstrated good gas seal integrity and reliable pressurized operation. The fuel cell reformer also was trouble free and achieved excellent output and conversion efficiency.[16]

- Power plant operations were voluntarily terminated on May 12, 1997.[29] The plant was operated by MCP with its own funding and utility support 2½ months beyond the March 31, 1997 contract end date.[16]
- MCP felt that overall the demonstration test at Miramar was successful in showcasing many critical features of the IMHEX technology. It felt that lessons learned from this test would be very useful for improving designs and operations for future MCFC power plants.[29]

The project was funded by DOE, with the DoD also providing $2 million for the project.[31] MCP's cost-sharing obligation was shared by GRI, San Diego Gas and Electric, and Bechtel Corporation.[29] Other contributors included IGT, Stewart & Stevenson, and EPRI.[30] The Alliance to Commercialize Carbonate Technology (ACCT) also supported this effort. The group consisted of a number of utilities and end users who expected to be among the first purchasers of power systems employing the M-C Power technology.

5.2.2.2.2.3. PRODUCT DEVELOPMENT FOR MARKET ENTRY The third DOE contract, which began in 1994, was a $104 million product improvement and demonstration project, with $33 million from MCP.[28] The 5-year development effort (1994–1999) aimed to improve the design and reduce costs for the first market-entry units. For this project, MCP modified its stack design based on lessons learned during the demonstration in 1997. It constructed two 75 kW MCFC power plants to bring their technologies closer to the market.[31] One stack was used for an extended test of the BoP, while the second stack was installed in the fully integrated power plant at the same Marine Corps Air Station in Miramar, California. The objectives of these tests were to evaluate a commercial cell package and improved balance of plant components and to test plant systems' conditions and operating parameters. The 75 kW stack installed at the Miramar Marine Corp Air Station site began generating power in early July 1999.[14] The 75 kW MCFC stack was to be upgraded to 250 kW in 1999.[23] MCP subsequently was expected to install up to four commercial prototype 250 kW power plants at various locations in the second half of 2000 and early 2001. MCP planned to begin producing 250 kW and 1 MW units for commercial sale in late 2001 or early 2002.

Under DOE sponsorship, MCP became increasingly confident about their continuous process improvements and yield increases for each successive version of their MCFC stacks and expressed their intention to begin commercialization of their products. In 1999, MCP announced that it planned to commercialize MCFC power plants "during the first few years of the new millennium."[7] The company stated that it had increased the existing plant capacity to 10 MW/year in the previous year and planned to begin constructing four prototype commercial units in 2000. MCP also increased the size of its market entry power plant to 500 kW, based on relative costs of BoP equipment and supported by market research.[14]

5.2.3. Commercial Success Still Uncertain

Despite its planned commercialization and preparation, MCP closed the company in early 2000. In an announcement in March, MCP Interim President Patrick McSweeney said that for the previous several months MCP had attempted to raise capital to pay for its 20 percent cost share of a DOE contract in order to continue operations, but had failed. As a result, MCP would terminate all the power plant operations and close down the company.[32] Patrick McSweeney replaced former president Elias Camara, who resigned from the company in early January. McSweeney stressed that the reasons for the company's closure were financial and not technical problems. In an era of massive investments in alternative energy stocks, MCP had unfortunately not been able to raise the needed investment capital.

During the 2000s, FCE was the only company remaining in the MCFC field in the United States, and started its commercial activities in 2003.[v] Prior to 2003, there was one special MCFC power plant installation at the Mercedes Benz Museum and Visitor Center, Tuscaloosa, Alabama, in 2000. It was a 250 kW "HotModule" power plant manufactured by FCE's German partner CFC Solutions (then MTU CFC Solutions).[vi]

Between 2000 and 2002, FCE was preparing itself for a major commercial debut in 2003. Indeed, the company sold, installed, and/or started up operations of as many as fifteen 250 kW DFC power plants with a total output of 3.75 MW in 2003, which was a significant achievement in the first year for any fuel cell R&D company.[vii] In 2004, FCE again scored an impressive business accomplishment: it sold, installed, and/or started up twenty-seven 250 kW DFC power plants, for a total of 6.75 MW in power plants. Although FCE appeared to have had a slow year in 2005, it made a series of spectacular product introductions in the following several years accompanied by impressive sales performance.

- It introduced the 300 kW (DFC300) system in 2006, the 1.4 MW (DFC1500) in 2007, the 2.8 MW (DFC3000) in 2008, and the multi-megawatt DFC-ERG in 2009. As a result, FCE sold 5.1 MW in 2007, doubled its record in 2005. It continued with impressive sales gains in 2008 and 2009.[viii] It looked as if FCE was on a path to major success in fuel cell commercialization (Table 5.2).

In September 2009, however, FCE experienced a major setback, casting doubt on the viability of the technology and the company's long-term sustainability. The output of two 250 kW DFC300 systems installed at Santa Barbara's El Estero Wastewater Plant in 2004 dropped to nearly 5 percent of total capacity.[33] Tests showed that the chemical composition of the gas being produced by the wastewater plant contained impurities that eroded the function of the cells.

- MCFC sales during the next 2 years dropped radically; FCE sold only 13 MW worth of systems in 2010, less than half its 2009 sales, and 6 MW in 2011, barely above its record in 2007 (Fig. 5.1).

In an extraordinary development, however, FCE was rescued by its overseas partner. In May 2011, FCE announced a 2-year order for 70 MW of fuel cell kits from South Korea's POSCO Power.[34] The total value of the order

[v] Our discussion of FCE's business activities here focus on its performance in the United States in order to assess market trends domestically. FCE's business activities elsewhere in the world will be examined separately.

[vi] CFC Solutions has a licensing agreement with FCE, under which CFC Solutions integrates its BoP and FCE's proprietary MCFC stack into a complete system, called HotModule, and markets it primarily in Europe.

[vii] There have been at least two experimental projects in 2003 sponsored by DOE. One was to test the use of coal-mine methane in a MCFC power plant in Ohio, and the other to experiment with the use of coal-derived synthetic gas in Indiana. Because the projects bore more characteristics of a government test than of a commercial operation, they were not counted as a commercial sales and installation activities.

[viii] Sales performance is assessed in terms of total output instead of the number of units, because the output per unit dramatically increased.

TABLE 5.2 FuelCell Energy: MCFC Power Plant Commercialization Activities in the United States (as of July 2009)

Plant size, fuel	Project, operator, coordinator	Installation site	Date of sales, installation, operation	Funding, other notes
250 kW MTU HotModule design, natural gas	Southern Company, Alabama Municipal Electric Authority (AMEA), Mercedes Benz US, FuelCell Energy	Mercedes Benz Museum and Visitor Center, Tuscaloosa, Alabama	Began operation in June 2000	The project cost was shared by all partners
250 kW DFC, natural gas	Los Angeles Department of Water and Power (LADWP), US Army Corp of Engineers	LADWP Plant, John Ferraro Building, Los Angeles, California	Contract made in June 2000 for a pre-commercial 250 kW pre-commercial fuel cell; plant installed in August 2001. The plant replaced with a commercial 250 kW DFC in June 2003	Likely financed by LADWP entirely, except for $250,000 from DoD Climate Change Rebate Program for the commercial model that replaced the pre-commercial prototype
250 kW DFC, digester gas	LADWP, US Army Corp of Engineers	LADWP Terminal Island Wastewater Treatment Plant, Los Angeles, California	Contract signed in January 2001; plant delivered in July 2003; started operation in August 2003	$2.3 million paid by the Los Angeles Department of Water and Power (LADWP); 250,000 from DoD Climate Change Rebate Program
250 kW DFC, natural gas	LADWP, US Army Corp of Engineers	LADWP Main Street Service Center, Los Angeles, California	Contract signed in January 2001, plant delivered in July 2003, began operation in September 2003	$2.4 million paid by LADWP; $250,000 from DoD Climate Change Rebate Program
250 kW DFC, natural gas	Starwood Hotels, PPL Energy Plus	Sheraton Hotel, Parsippany, New Jersey	Delivered and installed in spring 2003; came on line in September 2003	Approximately $1.7 million from New Jersey Clean Energy Program; DoD Climate Change Rebate
250 kW DFC, natural gas	PPL EnergyPlus, US Coast Guard	US Coast Guard Air Station, Bourne, Massachusetts	Installation, startup, and commissioning in May 2003	The total project cost was $1.7 million, which was funded by federal, state, and private sources, including Massachusetts Renewable Energy Trust Fund and DoD Climate Change Rebate Program

(Continued)

TABLE 5.2 FuelCell Energy: MCFC Power Plant Commercialization Activities in the United States (as of July 2009) (*Cont'd*)

Plant size, fuel	Project, operator, coordinator	Installation site	Date of sales, installation, operation	Funding, other notes
250 kW DFC, natural gas/digester gas	Los Angeles Department of Water and Power (LADPW), Los Angeles Department of Public Works/Bureau of Sanitation, US Army Corp of Engineers	LADWP's Terminal Island Wastewater Treatment Plant located at the Port of Los Angeles, California	Installed and started operating in May 2003	$2.4 million paid by LADWP and $250,000 by DoD Climate Change Rebate Program
250 kW DFC, natural gas	Los Angeles Department of Water and Power (LADWP), US Department of Energy, US Army Corp of Engineers	LADWP Main Street Service Center, Los Angeles, California	Installed and started operation in June 2003	$2.4 million by LADWP and $250,000 by DoD Climate Change Rebate
2 × 250 kW DFC, natural gas	Zoot Enterprises, PPL EnergyPlus	Zoot Enterprises, Galactic Park Subdivision, Bozeman, Montana	Installed in June 2003	
250 kW DFC MCFC, natural gas	Generation, Inc., BP, Chugach Electric Association	BP's gas-to-liquid test facility, powering administration building and warehouse, Nikiski, Alaska	Planned to be installed in mid-2003	The total project cost was $6.5 million, of which DOE contributed $2 million and Chugach Electric Association secured $450,000 from National Rural Electric Cooperative
250 kW DFC, natural gas	Starwood Hotels, PPL Energy Plus	Sheraton Hotel, Edison, New Jersey	Came on line in August 2003	$860,000 from New Jersey Clean Energy Program; $250,000 DoD Climate Change Rebate
250 kW DFC, coal mine methane	Harrison Mining Corporation, AEP Ohio Coal, Northwest Fuel Development	AEP Ohio Coal site, Hopedale, Ohio	Operated from August to December 2003	A DOE project partnered with AEP Ohio Coal to use coal-bed methane in MCFCs
250 kW DFC, digester gas/natural gas	Quinn Power Systems Associates/Caterpillar Inc, Los Angeles County Sanitation Districts	Palmdale Water Reclamation Plant, Los Angeles, California	Sales made in August 2003; operation started in November 2004	$1,125,000 from California Public Utilities Commission (CPUC)

250 kW DFC, natural gas	Caterpillar Inc.	Caterpillar Technical Center, Peoria, Illinois	Commissioned in October 2003	
250 kW DFC, natural gas	Connecticut Clean Energy Fund (CCEF), Yale University	Yale University's archival storage facility for the Yale Peabody Museum and the Yale Environmental Science Center, New Haven, Connecticut	Commissioned in December 2003	\$1.25 million from the Connecticut Clean Energy Fund (CCEF)
250 kW DFC, natural gas	Salt River Project (SRP)	Arizona State University, Mesa, Arizona	The agreement signed in January 2004; delivery planned in late 2004 or early 2005	
250 kW DFC MCFC	Ohio Cat/Caterpillar Inc, City of Westerville Electric Division, American Municipal Power Ohio	Electric substation, Westerville, Ohio	Sales contract finalized in February 2004; operation started in November 2004	Ohio's Third Frontier Fuel Cell Initiative
250 kW DFC, natural gas	PPL EnergyPlus, Starwood Hotels & Resorts Worldwide, Inc.	Sheraton New York Hotel & Towers, Manhattan, New York	The project announced in March 2004; installed in June 2005	\$920,000 from the New York State Energy Research and Development Authority (NYSERDA)
250 kW DFC, natural gas	Grand Valley State University	Michigan Alternative and Renewable Energy Center at Grand Valley State University, Muskegon, Michigan	April 2004	\$3 million provided by the Michigan Public Service Commission and bonding from the City of Muskegon
250 kW DFC, natural gas	PPL Energy Plus, Millennium Builders (a PPL subsidiary)	Ocean County College, Toms River, New Jersey	Installed in April 2004	\$827,000 from New Jersey Clean Energy Program; \$250,000 from DoD Climate Change Rebate
4 × 250 kW DFC, natural gas/digester gas	Alliance Power, Sierra Nevada Brewing Co., PG&E	Sierra Nevada Brewing Company, Chico, California	The project announced in April 2004; planned to be delivered and operational in the first calendar half of 2005	40 percent or \$2.4 million of eligible costs from CPUC's California's Self-Generation Incentive Program (SGIP); \$1 million from DoD Climate Change Rebate

(*Continued*)

TABLE 5.2 FuelCell Energy: MCFC Power Plant Commercialization Activities in the United States (as of July 2009) (*Cont'd*)

Plant size, fuel	Project, operator, coordinator	Installation site	Date of sales, installation, operation	Funding, other notes
4 × 250 kW DFC, wastewater digester gas	King County, CH2M Hill, Brown and Caldwell, Fluor Daniel, US Environmental Protection Agency (EPA)	South Treatment Plant, Renton, Washington	The plant demonstrated for 2 years from summer 2004 to summer 2006	Of the total project cost of $22 million, $12.5 million funded by the US Environmental Protection Agency (EPA)
250 kW DFC, natural gas	US Army Construction Engineering Research Laboratory (CERL), Fuel Cell Test and Evaluation Center (FCTec), Concurrent Technology Corporation	FCTec Environmental Technologies Facility, Johnstown, Pennsylvania	Agreement signed in August 2004 for 12-month demonstration	A Cooperative Research and Development Agreement (CRADA) with the US Army Engineer Research and Development Center/ Construction Engineering Research laboratory (ERDC-CERL)
1.5 MW (6 × 250 kW DFC), natural gas	Alliance Power, Starwood Hotels and Resorts Worldwide Inc.	Sheraton San Diego Hotel & Marina, San Diego, California	Four 250 kW DFCs installed at the East Tower in September 2004 and operation started in 2005; two 250 kW DFCs added at the West Tower in July 2006	$1.25 million from California Public Utilities Commission's (CPUC) Self-Generation Incentive Program (SGIP); DoD Climate Change Rebate; $1.25 million from SGIP for the West Tower installation
2 × 250 kW DFC MCFC units	Caterpillar Inc, City of Santa Barbara, Alliance Power	El Estero Wastewater Treatment Facility, Santa Barbara, California	Started operation in September 2004	$2.25 million from California's SGIP Program; DoD Climate Change Rebate
1 MW DFC1500 MCFC (4 × 250 kW units)	Alameda County, Chevron Energy Solutions	Alameda County's Santa Rita Jail, Dublin, California	The sale announced in October 2004; installed in August 2006	Of the total project cost of $6.1 million, $1.4 million was provided from California's SGIP program and $1 million from DoD Climate Change Fuel Cell Rebate Of the remaining $3.7 million, $2.8 million was financed by the California Energy Commission's Energy Partnership Program Loan

250 kW DFC MCFC, natural gas	Chevron Energy Solutions, US Postal Service San Francisco Processing and Distribution Center (P&DC), Bonneville Power Administration	USPS Embarcadero Postal Center, San Francisco, California	The sales announced in December 2004; February 2005	Up to $625,000 from California's SGIP Program; ($250,000 from DoD Climate Change Rebate
250 kW DFC MCFC, natural gas	PPL EnergyPlus, Pepperidge Farm	Pepperidge Farm Bakery, Bloomfield, Connecticut	The project announced in March 2005; installed in January 2006	Funding from CCEF; $500,000 DoD Climate Change Fuel Cell Rebate
250 kW DFC MCFC, natural gas	State University of New York (SUNY), New York State Energy Research and Development Authority (NYSERDA), New York Power Authority (NYPA)	Walters Hall at SUNY's College Environmental Science and Forestry, Syracuse, New York	The sale announced in May 2005	$2.5 million from NYPA, $1 million grant from NYSERDA; $100,000 from Electric Power Research Institute (EPRI); $250,000 from DoD Climate Change Fuel Cell Rebate
250 kW DFC MCFC	US Department of Defense, US Department of Energy, LOGANEnergy	Marine Corp Air Ground Combat Center, training center, Twenty-nine Palms, California	The sale announced in June 2005; delivery planned in July 2006	DoD Climate Change Rebate
2 × 250 kW DFC MCFC units	South Coast Air Quality Management District (SCAQMD), California Cast Metals Association, Emergent Energy Group, Alliance Power	TST, Inc. metal foundries, Fontana, Carson, and Rancho Dominguez, California	The contract signed in July 2005; delivery planned by the first quarter of 2006	Up to $1.25 million from California's SGIP; 500,000 from South Coast Air Quality Management District (SCAQMD)
500 kW (perhaps 2 × 250 kW DFC)	Starwood Hotels & Resorts Worldwide, Inc. Alliance Star Energy	The Westin San Francisco Airport Hotel, San Mateo, California	The project announced in October 2005; installation in 2006	$1.25 million from California's SGIP; $6250 application fee refund from Pacific Gas and Electric Company
3 × 250 kW DFC, natural gas	LOGANEnergy	U.S. Marine Corps Base Camp Pendleton, Oceanside, California	The sale of two 250 kW DFC announced in October 2005; additional 250 kW ordered in June 2006; installation planned in summer 2006	Up to $1.25 million from California's SGIP Program

(*Continued*)

TABLE 5.2 FuelCell Energy: MCFC Power Plant Commercialization Activities in the United States (as of July 2009) (*Cont'd*)

Plant size, fuel	Project, operator, coordinator	Installation site	Date of sales, installation, operation	Funding, other notes
250 kW DFC with a 60 kW microturbine	DOE Vision 21 Demonstration Project, FuelCell Energy, Montana State University, Alliance Power	TST, Inc. Metal Foundries, Fontana, Carson, and Rancho Dominguez, California	2006	$1.25 million from California's SGIP and $500,000 from SCAQMD
4 × 250 kW DFC MCFC units, natural gas	Select Energy Services, Eastern Connecticut State University	Eastern Connecticut State University's central heating plant, Willimantic, Connecticut	Installed in February 2006	Received a DoD Climate Change Rebate
1 MW DFC (4 × 250 kW DFC300)	Alliance Power, California State University, Northridge	California State University, Northridge, California	The sale announced in April 2006; installed in February 2007	The total cost was $5.8 million. California's SGIP provided up to $2.25 million, and Los Angeles Department of Water and Power provided $500,000
2 × 300 kW DFC, anaerobic digester gas	Gills Onions, Southern California Gas Company	Gills Onions, Oxnard, California	The sale announced in August 2006; operation started in July 2009	$2.7 million from Southern California's SGIP Program
3 × 300 kW DFC units, anaerobic digester gas	City of Tulare, FuelCell Energy	Regional Wastewater Treatment Facility, Tulare, California	The sale announced in October 2006; planned to begin service in 2007	$4 million from California's SGIP
750 kW (3 × DFC300MA)	Alliance Power	A premier hospitality and entertainment resort, California	The sale announced in October 2006; operation planned in April 2007	$1.9 million from California's SGIP Program
600 kW DFC (perhaps 2 × 300 kW DFC) MCFC, digester gas	Alliance Power, FuelCell Energy June 2007 to install at the District's Regional Wastewater Treatment Facility	Dublin San Ramon Services District Regional Wastewater Treatment Facility, Dublin California	The sale announced in November 2006; installation in June 2007	$2.7 million from California's SGIP

600 kW (perhaps 2 × 300 kW) MCFC, biogas	San Francisco Public Utilities Commission (SFPUC), Otto H. Rosentreter Company, Alliance Power	Southeast Wastewater Treatment Plant, San Francisco, California	The project approved in January 2007; installation planned to be completed by the end of 2007	Grants from Mayor's Energy Conservation Account (MECA), San Francisco's renewable energy project special funds, San Francisco Public Utilities Commission's (SFPUC) Power Enterprise operating funds; a $2.7 million rebate from the California's SGIP
1 MW DFC1500MA	Alliance Power	Riverside Wastewater Treatment Plant, Riverside, California	The sale announced in March 2007	$4.5 million from California's SGIP
900 kW DFC (3 × 300 kW DFC), anaerobic digester gas	City of Rialto Chevron Energy Solutions FuelCell Energy	Rialto wastewater treatment facility, Rialto, California	Construction started in May 2007; start-up operation in early 2008	Of the total project cost of $15.1 million, $4.05 million was provided from California's SGIP
1.2 MW DFC power plant, Methane		Turlock Regional Water Quality Control Facility, Turlock, California	The sale announced August 2007; planned start-up in summer 2008	$4.5 million from California's SGIP
1.2 MW DFC	Pepperidge Farm, the Connecticut Clean Energy Fund, FuelCell Energy	Pepperidge Farm Bakery, Bloomfield, Connecticut	The project announced in August 2007; the plant dedicated in October 2008	$3.5 million from CCEF out of the total project cost of $5.4 million
1 × 300 kW, three 1.2 MW DFC MCFC, Methane	Linde	Four customer locations in the San Diego area which will use purified biogas (methane) from the Point Loma Wastewater Treatment Plant, California	Sales reported in November 2007; Linde would install the plants	Funding from California's SGIP
3 × 250 kW DFC anaerobic digester gas	Alliance Power	Eastern Municipal Water District (EMWD), California	The sale announced in November 2007; installation planned in first half of 2008	$3.375 from California's SGIP
2 × 300 kW DFC 300 Power Plants	Alliance Power	M&L Commodities, Inland Cold Storage, Stockton, California	The sale announced in November 2007; planned start-up in mid-2008	$1.5 million from California's SGIP

(Continued)

TABLE 5.2 FuelCell Energy: MCFC Power Plant Commercialization Activities in the United States (as of July 2009) (*Cont'd*)

Plant size, fuel	Project, operator, coordinator	Installation site	Date of sales, installation, operation	Funding, other notes
300 kW DFC300, propane	LOGANEnergy	U.S. Navy, Pacific Missile Range Facility, Island of Kauai, Hawaii	The sale announced in November 2007; installation planned by September 2009	
9 MW (3 × DFC3000) with a 1.8 MW pipeline turbo expander)	Project 150, the Connecticut Department of Public Utility Control (DPUC), Enbridge, Inc., Southern Connecticut Gas Company, Energy East	A natural gas letdown station, Milford, LLC, Milford, Connecticut	DPUC approved their financing commitment for the project in April 2008	CCEF provides $50,000
4.8 MW (2 × DFC3000)	Project 150, the Connecticut Department of Public Utility Control (DPUC)	Stamford Hospital, Stamford, Connecticut	DPUC approved their financial commitment for the projects in May 2008	CCEF provides $559,000
2.4 MW DFC3000	Project 150, the Connecticut Department of Public Utility Control	Waterbury Hospital, Waterbury, Connecticut	DPUC issued its final decision approving the project in May 2008	CCEF provides $1,015,500
300 kW DFC300		U.S. Marine Corps Air Ground Combat Center, Twenty-nine Palms, California	The sale announced in February 2009	
300 kW DFC300	Concurrent Technologies Corporation (CTC), United States Air Force Advanced Power Technology Office	Barksdale Air Force Base, Louisiana	The sale announced in April 2009; planned to be operational in early 2010	
3.2 MW DFC-ERG	Project 150, the Connecticut Department of Public Utility Control	Glastonbury, Connecticut	The final approval announced in April 2009	CCEF provides funding
3.2 MW Direct FuelCell/Turbine (DFC/T)	Project 150, the Connecticut Department of Public Utility Control	A substation, Danbury, Connecticut	The final approval announced in April 2009	CCEF provides funding
3.2 MW DFC-ERG	Project 150, the Connecticut Department of Public Utility Control	Trumbull, Connecticut	The final approval announced in April 2009	CCEF provides funding

14.3 MW DFC-ERG (5 × DFC3000)	Project 150, the Connecticut Department of Public Utility Control	Bridgeport Fuel Cell Park, Bridgeport, Connecticut	The final approval announced in April 2009	CCEF provides funding
3.4 MW DFC-ERG power plant		Natural Gas Letdown Station, Bloomfield, Connecticut	The final approval announced in April 2009	CCEF provides funding
1.4 MW (DFC1500)	AirCon Energy, Pacific Gas and Electric Company (PG&E)	A county jail and county office buildings in Santa Rosa, Sonoma County, California	The sale announced in June 2009	\$3 million grant is to be provided under California's SGIP program
300 kW DFC300	City of Tulare, FuelCell Energy	The City of Tulare wastewater treatment facility, California	The sale announced in December 2009	The project will likely receive an SGIP subsidy
300 kW DFC300	LOGANEnergy	Carla's Pasta in South Windsor, Connecticut	Sale announced in July 2010; planned to be operational by mid-2011	A \$750,000 grant is provided by CCEF
2 × 300 kW DFC300	LOGANEnergy	US Naval Submarine Base New London, Groton, Connecticut	Sale announced in August 2010	
1.4 MW DFC1500	G3 Power Systems	Olivera Egg Ranch, LLC, French Camp, California	Sale announced in August 2010; planned to be installed by mid-2011	The project will likely receive an SGIP subsidy
2 × 300 kW DFC300		Perris Valley Regional Water Reclamation Facility, EMWD, Riverside Country, California	Sale announced in August 2010; planned to be operational by mid-2011	Will receive \$2.7 million grant from the SGIP program
1.4 MW DFC1500		Ace Bowen pumping station in Temecula, California, Rancho California Water District, Southern California	Sale announced in September 2010; planned to be operational by late 2011	
4.5 MW (2.8 MW DFC3000, 1.4 MW DFC1500, and 300 kW DFC300)	BioFuels Fuel Cells, LLC	The University of California, San Diego, South Bay Water Reclamation Plant, San Diego, and Point Loma Wastewater Treatment Plant, San Diego	Sale announced in November 2010; planned to be operational by the summer of 2011	Financed by grants under the SGIP and the US Treasury Investment tax credit

(Continued)

TABLE 5.2 FuelCell Energy: MCFC Power Plant Commercialization Activities in the United States (as of July 2009) (*Cont'd*)

Plant size, fuel	Project, operator, coordinator	Installation site	Date of sales, installation, operation	Funding, other notes
2.8 MW DFC3000	UTS Bioenergy LLC	A municipal wastewater treatment plant operated by the Inland Empire Utilities Agency, Chino, California	Sale announced in November 2010; planned to be operational in early 2012	Eligible for an SGIP grant
1.4 MW DFC1500	UTS Bioenergy LLC	San Jose/Santa Clara Water Pollution Control Plant, San Jose, California	Sale announced in November 2010; planned to be operational in early 2012	Eligible for an SGIP grant
2 × 1.4 MW	Pacific Gas and Electric Company (PG&E)	California State University (CSU)-East Bay and San Francisco State University	Authorized by California Public Utilities Commission in April 2010; installed in August 2011; planned to be operational within August 2011	Likely supported by various California incentive programs; the total value of these contracts is approximately $12.6 million
1.4 MW DFC1500	Southern California Edison	CSU San Bernardino, California	Authorized in April 2010; installation ordered in February 2011; planned to be operational in early 2012	
250 kW DFC for co-production demonstration	FuelCell Energy	A wastewater treatment plant, Orange County Sanitation District, Fountain Valley, California	Commissioning event in August 2011	
1.4 MW DFC1500	Greenwood Energy	Central Connecticut State University, New Britain, Connecticut	Sale announced in October 2011; planned to be operational by December 2011	

The table was created based on various sources, including FCE press releases, http://fcel.client.shareholder.com/releases.cfm; *"Worldwide Fuel Cell Installations," updated in October 2005, Fuel Cells 2000,* www.fuelcells.org, *accessed January 2009; Ron Friesen, California Fuel Cell and Hydrogen Programs, Director, California Stationary Fuel Cell Collaborative 24 February 2006; "Proposals for Promotion of Fuel Cell Introduction," New Energy Foundation, March, 2006.*

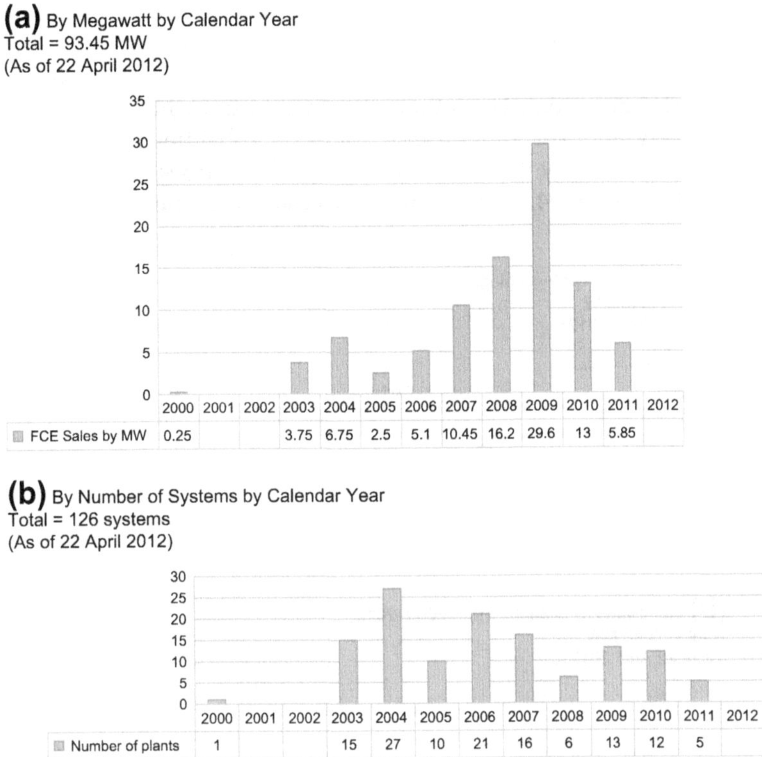

FIGURE 5.1 MCFC sales, installation, and/or start-up in the United States. (For the color version of this figure, the reader is referred to the online version of this book.)

and other commitments for services was estimated to be at least $129 million, with delivery of fuel cell kits beginning in October 2011 and occurring monthly through October 2013. Payment terms included a down payment and progress payments during the term of the contract, with approximately 40 percent of the contract value received by October 2011. As a result, FCE's 2001 financial report showed a strong revenue gain. Nonetheless, FCE's balance sheet still showed a net loss, illustrating FCE's difficulty in achieving a commercial success (Fig. 5.2).

In addition to technology viability, there are a number of concerns that cast shadows over FCE's long-term commercial prospects. One of the concerns is that FCE has a narrow domestic customer base. A review of FCE's sales and installation records reported in the press as well as in the company's press releases indicates

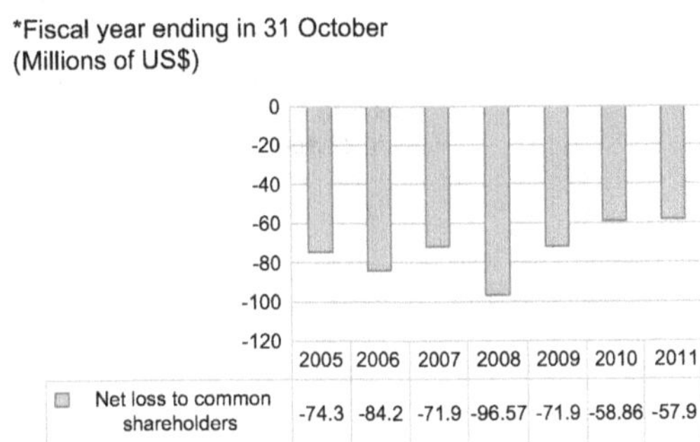

FIGURE 5.2 FCE's financial status*. (For the color version of this figure, the reader is referred to the online version of this book.)

that FCE sold its DFC systems only to 16 states (Fig. 5.3). It, moreover, shows that nearly 94 percent of the sales were made to only two states: Connecticut and California. The other 14 states made relatively modest purchases. Moreover, most of the purchases by the 14 states occurred between 2003 and 2004, when the MCFC power plants were novel and stirred major hope and interest. A few of these states have made additional purchases since then.

In addition, FCE's sales are largely dependent on subsidies for their MCFC power plant sales, which does not bode well for commercial success over the long term. FCE's sales in

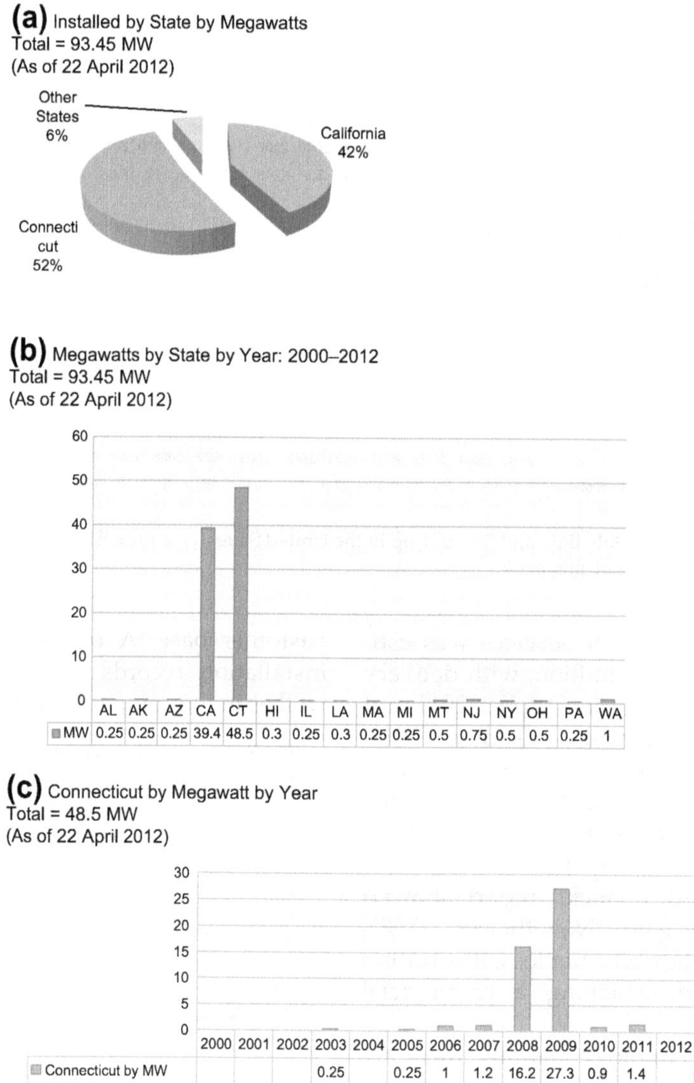

FIGURE 5.3 FCE's MCFC sales, installation, and/or start-up in the United States. (For the color version of this figure, the reader is referred to the online version of this book.)

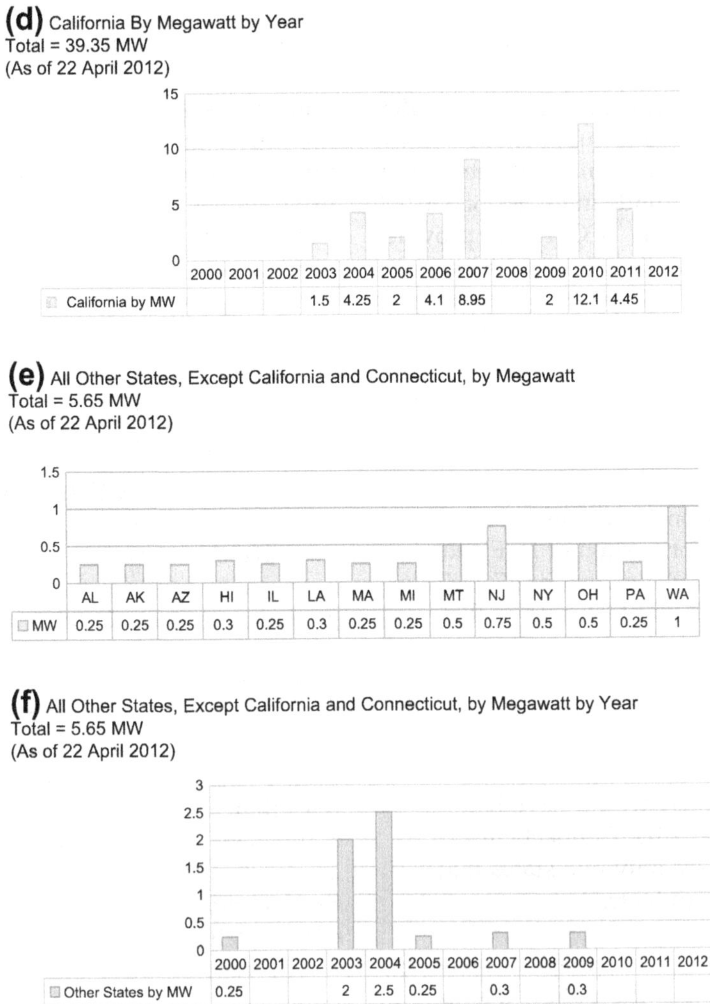

FIGURE 5.3 (*Continued*).

California and Connecticut are primarily supported by the clean energy policy and accompanying subsidies, and if there were no subsidies these sales likely would not have happened.

Connecticut. The State of Connecticut launched a clean energy policy in 2000 by creating the Connecticut Clean Energy Fund (CCEF). The CCEF promotes, develops, and invests in clean energy power sources and is funded by a surcharge on residential and commercial electric bills. In April 2009, as part of the CCEF activity, the Connecticut Department of Public Utility Control (DPUC) announced a plan to subsidize utilities that purchased MCFCs, specifically 27.3 MW worth of FCE's MCFC power plants for an estimated \$84 million.[35] This included 3.4 MW Direct Fuel Cell-Energy Recovery Generation (DFC-ERG) power plants for a natural gas letdown station in Bloomfield, 3.2 MW Direct FuelCell/Turbine (DFC/T) for a substation in Danbury, 14.3 MW DFC3000 (five 2.8 MW DFC 3000 power plants) in Bridgeport; 3.2 MW DFC-ERG in Trumbull, and 3.2 MW DFC-ERG in Glastonbury. The maximum individual project award was set at $4 million.[36] Grants of up to $50,000 per installation also were available to support site-specific technical studies and financial feasibility studies.[36] Additional funds in excess of $50,000 could also be requested.[37] CCEF also provides nonrecourse loan funding to qualified early-stage "Pre-Development" projects. Loans up to $250,000 are available for projects under 4 MW, and loans up to $500,000 are available for projects of 5 MW or greater. In addition, fuel cell projects with a capacity greater than 1 MW are eligible for an incentive up to $3.20 per watt.

- Contracts with Connecticut in 2009 constituted a major portion of FCE's total sales—more than 92 percent, or 27.3 MW of the total 29.6 MW that FCE sold in that year—indicating the importance of the state's subsidy. Moreover, in May 2008, FCE also was awarded a 16.2 MW contract from the CCFC. The 2008 contracts constituted all of the MCFC sales in that year. Without the state contracts, FCE likely would have had no sales. In 2010 and 2011, Connecticut and California were the only two states that purchased FCE products.

California. California has a major clean energy policy in place. Under the policy, the state in 2001 launched a "Self-Generation Incentive Program (SGIP)" to encourage the use of distributed generation resources, such as fuel cells.[38] The program is administered by California's utilities. The annual program budget for 2006 and 2007 was $83 million. The SGIP initially reimbursed fuel cell power plant owners $4500/kW for biogas-run units and $2500/kW for those operating on natural gas for installations of up to 1 MW. The program was revised in 2008, and fuel cell projects up to 3 MW are now eligible for the subsidy. The first 1 MW of a project is entitled to 100 percent of the incentive; the second MW of power is eligible for 50 percent of the incentive; and the third MW receives 25 percent of the incentive. In addition, the SGIP in 2008 and 2009 was supplemented with an additional $96 million of unspent SGIP funds from prior years. The program has now been extended through 2015. California is FCE's second best customer, next to Connecticut, and purchased almost 9 MW worth of power plants in 2007.

- California purchases fell, however, for 2008 and 2009, likely due to the poor economy but California represented 93 percent of FCE sales in 2010 and 76 percent in 2011.

It is clear that MCFC power plants are sold domestically only when major subsidies are available. Moreover, modest subsidies alone do not seem to generate sales. For example, the US government since October 2008 has provided a federal tax credit for fuel cells of $3000/kW, thrice the previous incentive of $1000/kW. Even the tripled government incentive does not seem big enough to attract sales. Similarly, even though most US states offer some incentives they have not led to MCFC sales either. In a nutshell, MCFC sales occur only when the aggregate of the state and federal incentives form a very large incentive, such as in the case of Connecticut and California.

The root of these subsidy-dependent sales patterns is the still the high cost of the MCFC power plant. Fuel cells still are expensive—a fact that is rarely discussed publicly. Indeed, the cost to build MCFC power plants exceeds the price at which they can be sold. In October 2011, FCE reported a product cost-to-revenue ratio of 1.11 for fiscal 2011 (Fig. 5.4). Although the ratio is declining, this still means it costs $1.11 to produce a power plant that can be sold at $1.00, resulting in a loss of $0.11 for each dollar in sales revenue. To make a profit, the cost-to-revenue ratio needs to be below 1.0, thereby yielding profits. At the same time, the cost of the MCFC power plant must be cost

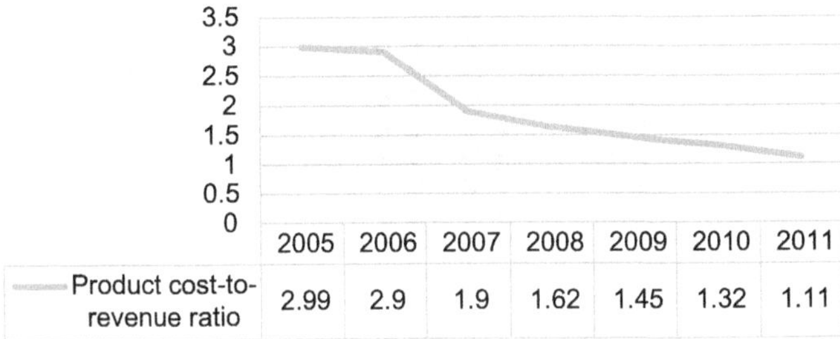

FIGURE 5.4 FCE's product cost-to-revenue ratio: fiscal 2005–2011. (For the color version of this figure, the reader is referred to the online version of this book.)

competitive compared to conventional power generation technologies. While a cost-to-revenue ratio of below 1.0 would indicate profitability, it is not necessarily commensurate with a product being cost competitive in the market place. A very expensive power plant can still be sold as long as major subsidies are available, and a profit can be made as long as the cost-to-revenue ratio is below 1.0. But as long as the cost-to-revenue ratio stays above 1.0, FCE in theory would not make money in the absence of subsidies.

- Today, FCE still suffers a net loss and it is not certain whether it will achieve commercial success in the near term.

5.3. MCFC DEVELOPMENT IN JAPAN

5.3.1. Government Plays Dominant Role—Limited Activities in Private Sector

MCFC development in Japan has been nearly exclusively conducted under the government's Moonlight Project (the current New Sunshine Project), with only modest MCFC development efforts in the private sector. MCFC development efforts under the Moonlight Project were a large-scale undertaking, nearly equal to that of the United States. In 1981, Japan's METI implemented a 24-year, MCFC development program under the Moonlight Project that continued until 2004, with a total budget of about $470 million.[4,45,ix] The ultimate goal of the program was to develop large-scale MCFC power plants that could be fueled by coal gas and replace coal-burning thermal power plants.

5.3.2. The Ministry of Economy, Trade, and Industry Starts MCFC Development Program in 1981

METI conducted the MCFC technology development plan under the New Energy and Industrial Technology Development Organization (NEDO) management. The MCFC development program proceeded in three phases (Table 5.3):

- Phase I (1981–1986) of the MCFC development program was a 6-year project to develop 10 kW MCFC stacks. The MCFC R&D budget for 1981–1986 under the Moonlight Project was about $48 million.
- Phase II (1987–1999) was a 13-year effort devoted to the development of both MCFC cells and stacks and BoP systems. The first part of the program was to develop a 200 kW internal reforming stack and a 1000 kW external reforming pilot plant. The second part was to develop BoP systems and components, stack operation and evaluation

[ix] For each year of the Japanese MCFC R&D budget between 1981 and 2004 we use the yen/dollar exchange rate that prevailed at the beginning of each year.

TABLE 5.3 MCFC Stack and Power Plant Development in Japan

Project	Project Manager/ Sponsor	MCFC stack manufacturer	Manifolding, flow configuration, pressure	Status of development	Notes
Moonlight Project Phase I, 1981–1986	New Energy and Industrial Technology Development Organization (NEDO), MCFC Research Association	Fuji Electric	External reforming	10 kW stack, twenty-four 2500 cm^2 cells in 3 blocks, giving 7.1 kW at 0.79 V per cell (1986)	Test operation and evaluation of the 10 kW stacks conducted by the Central Research Institute of Electric Power Industry (CRIEPI)
		Hitachi	External reforming	10 kW stack, thirty 3600 cm^2 cells in 3 blocks, yielding 12.4 kW at an average cell potential of 0.76 V per cell (1986)	
		IHI	External reforming	10 kW stack, twenty-nine 3600 cm^2 cells in a single block (12.5 kW at 0.8 V per cell) (1986)	
		MELCO	Internal reforming	10 kW stack, 40 2000 cm^2 cells in two blocks (9.1 kW at 0.76 V per cell) (1986)	
		Toshiba	External reforming	10 kW stack, fifty-seven 1600 cm^2 cells in one block (10.4 kW at 0.76 V per cell) (1986–1987)	
Moonlight Project Phase II, 1987–1999	NEDO	Hitachi	External reforming, crossflow configuration	100 kW stack, 1.2 $m^2 \times 88$ cells (2 modules), 5 ata, operated for 1700 h (1993)	
				Modified 25 kW stack, 1.2 $m^2 \times 22$ cells, 6 ata, operated for 5087 h (1994–1995)	
				25 kW stack, 1.2 $m^2 \times 25$ cells, 5 ata, operated for 1600 h (1996)	A preparatory stack for the 250 kW stack
				2×250 kW stacks, 1.2 $m^2 \times 300$ cells (6 modules)	Stacks used in the 1000 kW pilot plant
		IHI	External reforming, parallel flow configuration	100 kW stack, 1 $m^2 \times 102$ cells, 5 ata, operated for 5000 h (1993–1994)	
				Modified 5 kW stack, 1 $m^2 \times 5$ cells, at atmospheric pressure, operated for 16,000 h (1993–1995)	
				20 kW stack, 1 $m^2 \times 20$ cells, at atmospheric pressure, operated for 1800 h (1995–1996)	A preparatory stack for the 250 kW stack
				2×250 kW, 2×125 kW sub-stacks, each consisted of 1 $m^2 \times 140$ cells	Stacks used in the 1000 kW pilot plant

Hitachi, IHI	External reforming, pressurized operation	1000 kW pilot plant, 4 × 250 kW stacks, operated at 5 ata, generated electricity for 2669 h and operated for 4916 h (1999–2000)	Installed at Kawagoe Thermal Plant of Chubu Electric Power, used LNG
MELCO	Direct internal reforming, crossflow	1 kW stack, 0.2 m^2 × 5 cells, operated for 1000 h (1988)	
		3 kW stack, 0.5 m^2 × 5 cells, operated for 5,070 hours (1989)	
		5 kW stack, 0.1 m^2 50 cells, operated for 2650 hours (1990)	
		5 kW stack, 0.5 m^2 × 10 cells, operated for 11,500 h (1991)	
		30 kW stack, 0.5 m^2 × 57 cells, operated for 13,434 h (1992)	
	Advanced internal reforming, counterflow	10 kW stack, 0.5 m^2 × 20 cells, operated for 10,316 h (1993)	
		30 kW stack, 0.5 m^2 × 56 cells, operated for 2636 h (1994)	
		10 kW stack, 1 m^2 × 12 cells, operated for 8674 h (1996–1997)	
		30 kW stack, 1 m^2 × 26 cells, operated for 7544 h (1997–1998)	
		200 kW stack (2 × 150 kW sub-stacks, each of which consisted of 2 × 50 kW blocks with 55 cells of 1 m^2 area), operated for 5035 h (1999–2000)	
	Pressurized, advanced internal reforming, counterflow	10 kW stack, 1 m^2 × 8 cells, operated at 5 ata for 6067 h (1996)	The first pressurized, advanced, internal reforming stack
		10 kW stack, 1 m^2 × 12 cells, operated at 5 ata for 7483 h (1998–1999)	The second pressurized, advanced internal reforming stack, voltage degradation of 0.25 percent/1000 h

TABLE 5.3 MCFC Stack and Power Plant Development in Japan (*Cont'd*)

Project	Project Manager/ Sponsor	MCFC stack manufacturer	Manifolding, flow configuration, pressure	Status of development	Notes
				10 kW, 1 m^2 cell area (1999)	
Moonlight Project Phase III, 2000–2004	NEDO	IHI	Pressurized, external reforming, internal manifolding, counterflow	10 kW, 1 $m^2 \times 10$ cells, operated at 5 ata for 10,000 h (2001–2002)	
				10 kW, 1 $m^2 \times 10$ cells, operated at 12 ata for 3200 h (2002–2003)	
				150 kW, 1 $m^2 \times 125$ cells, operated at 4 ata for 4839 hours (2004)	
				170–190 kW, 1 $m^2 \times 125$ cells, operated at 12 ata (2004)	
Private-Sector Activities	IHI		External reforming, internal manifolding	10 kW, 0.36 $m^2 \times 20$ cells (1983)	
				40 kW plant, operated for 10,000 h (perhaps in the mid-1990s)	Maximum output 50 kW, efficiency at 40 percent, continuous operating hour of 1600 h
	Kansai Electric Power	MELCO	Indirect internal reforming	30 kW stack, 0.5 $m^2 \times 62$ cells, operated for 2131 h (1991)	
				100 kW stack, 0.5 $m^2 \times 192$ cells, operated for 2308 h, (1992)	
	CRIEPI, Chubu Electric Power	IHI	External reforming, parallel flow configuration	10 kW stack, 0.5 $m^2 \times 20$ cells, 5 ata, operated for 13,000 h (1998–1999)	Longevity test under pressure
	Petroleum Energy Center	Tonen, Sanyo Electric, and Toyo Engineering	Direct internal reforming	10 kW stack operated for 2700 h (1993–1994)	Used LPG
				30 kW system (1994–1995)	

The table was created based on various sources, including "Fuel Cell Power Generation Technology, Challenges for MCFC Commercialization," Denchuken Review, No. 51, CRIEPI, March 2004; "New Energy Technology Development Related Data File Production Survey (Fuel Cells)," NEDO, March 2000; "Large-Scale Fuel Cells Introduction Mode Survey (Part I)," NEDO, March 1991; Leo J. M. J. Blomen and Michael N. Mugerwa, "Fuel Cell Systems," Plenum Press, New York, 1993.

technology, and total system technology. An industry association, the MCFC Research Association, was established in February 1988 to manage the second part of the program under NEDO. The total R&D budget for Phase II activities was $353 million.

- Phase III (2000–2004) was a 5-year program focused on the development of a series of high-pressure, high-efficiency short stacks to ultimately develop a 750 kW MCFC system that would operate at 12 atmospheres absolute (ata) and achieve an efficiency of 47 percent. The total R&D budget for Phase III was $69 million.

It appears that while managing the Phase III program, NEDO also administered a subsidy program from 2000 to 2004. Although little information is available, the program was called the Demonstrative Project of Regional Power Grids with Various New Energies, under which NEDO subsidized research and development, test operation, and demonstration operations of a 300 kW-class MCFC power plant that operated at 3–4 ata.

In parallel with the Phase II program, there was a cluster of MCFC projects managed by the MCFC Research Association under the Moonlight Project. The MCFC association was established in 1988 by METI and the Agency of Industrial Science and Technology (AIST, now the National Institute of Advanced Industrial Science and Technology) to manage the cluster of MCFC projects under NEDO. The projects were designed to develop MCFC components and system technology. The association's term was set at 9 years, from April 1987 to March 1996. Participants in the MCFC development projects basically belonged to the MCFC Research Association. The association members included electric power utilities, gas utilities, electrical and industrial machinery companies, and fuel cell equipment manufacturers, and probably were divided into three groups (Table 5.4):

- The first group consisted of MCFC cells and stack developers. They were electrical and industrial machinery manufacturers, including Fuji, Hitachi, IHI, MELCO, and Toshiba. They were directly contracted by NEDO and managed by NEDO.
- The second group included the majority of the MCFC Research Association members. They developed BoP components and systems and were managed by the association. Some companies that worked to develop cells and stacks were also assigned to develop control systems and peripheral piping and measuring instruments.[39] Such companies were managed by the association for that particular part of work.
- The third group had just one member: CRIEPI. CRIEPI was commissioned by NEDO to develop MCFC operation and evaluation technology and total system technology. It was also responsible for testing and evaluation of MCFC stacks developed under the Moonlight Project. Tests were to be conducted at CRIEPI's Akagi Testing Center, Gunma Prefecture, and CRIEPI constructed the first Japanese test facility for MCFCs in 1983.[40]

5.3.2.1. Phase I MCFC Development (1981–1986): Five Companies Participate in 10 kW Stack Development

The Phase I MCFC development project started in 1981, and the goal was to develop 10 kW MCFC stacks by late 1986.[6] Two electric machinery manufacturers—Hitachi and Toshiba—participated in the project from the start. The target output of the 10 kW stacks was to achieve 150 mA/cm^2 at 0.75 V.[41] The 10 kW stacks were to be tested at CRIEPI's testing facility in Gunma for evaluation in late 1986. If the test operation was favorable, Phase II—namely the development of a 1000 kW external reforming MCFC pilot plant and a 200 kW internal reforming MCFC stack—would begin.

TABLE 5.4 MCFC Component and System Technology Development in Japan*

Type of Organization	Name of research organization or research sponsor	R&D activities	Research partner or sponsor
National or Public Research Organization	Osaka National Research Institute	Evaluation and survey of new materials, development of the basic electrochemical and materials technologies, material assessment methodology, and material testing (1981–1986)	Central Research Institute of Electric Power Industry (CRIEPI), Hitachi, MELCO, Toshiba
	National Institute of Materials and Chemical Research	Thermodynamic study of corrosion reaction, study of chemical potential	
University	Tohoku University	Electrode reaction mechanism	
		Development of electrode model, thin film model simulation	
		Metal corrosion in molten carbonate bath (study of in situ NiO generation process)	Tokyo Electric Power (1988–1989)
		Effect of coal gas impurities on electrode reaction	CRIEPI
	Yokohama National University	Measurement of cathode dissolution, measurement of metal oxide dissolution and reduction method	
	Kyoto University	Fuel cell materials	
	Kyushu Institute of Technology	Chemical and electrochemical behaviors of organic fuels in molten carbonate fuel cells	
	Tokyo University of Science	Corrosion of alloys and structure of molten carbonate bath	
	Yamanashi University	Study of electrode reaction mechanism, study of electrode-manufacturing condition optimization	

	Hosei University	Development of model to forecast fuel cell efficiency, study of heat flow dynamic	CRIEPI
Electric Power Company	Tokyo Electric Power	How to increase fuel cell longevity (study of electrolyte composition, matrix material, and electrode materials)	Partnered with Tohoku University for part of the research
		How to increase fuel cell longevity (study of metallic material corrosion control)	IHI
		Testing of efficiency of internal reforming fuel cells (1985–1987)	Hitachi
	Kansai Electric Power	Evaluation of molten carbonate corrosion (study of corrosion protection of aluminum thin film by ion plasma process)	Nisshin Electric
	J-Power	Development of gas separation (separation and collection of carbon dioxide by pressure swing adsorption process)	MCFC Research Association, CRIEPI
Other Research Organization	CRIEPI	System studies, testing of 10 kW stacks (study of gas recycle effect)	Hitachi, IHI
		Study of fuel cell efficiency determinants	MELCO, Fuji Electric, Tohoku University
		Fuel cell simulation model development (dynamic model including stack)	Partnered with Hosei University for part of the research
		MCFC power plant efficiency analysis (study of natural gas and coal gas fueled power plants and optimal power plant structure)	
		Study of commercial MCFC power plant in the future, future MCFC plant design	Partnered with NEDO and MCFC Research Association for part of the research
		Test operation of 100 kW stacks under the Moonlight Project (1991)	MCFC Research Association

(Continued)

TABLE 5.4 MCFC Component and System Technology Development in Japan* (*Cont'd*)

Type of Organization	Name of research organization or research sponsor	R&D activities	Research partner or sponsor
		Development of gas separation technology for MCFC system	Partnered with MCFC Research Association (J-Power) for part of the research
	Japan Research and Development Center for Metals	Development of metallic materials, including materials for cathode, anode, separator, and plating materials	MCFC Research Association (Kobe Steel, Nisshin Steel, NKK, Mitsubishi Materials)
	Fine Ceramic Center	Development of ceramic materials (material integration technology, ceramic cathode materials, material characteristics evaluation)	MCFC Research Association (Sumitomo Metal Mining, Matsushita Electric Industrial, Kawasaki Heavy Industries)
Manufacturers	Hitachi	Study of catalysts for internal reforming	Tokyo Electric Power (1985–1987)
		Development of MCFC component (high efficiency, high durability, corrosion-resistant, heat-resistant, pressure-resistant materials), cooling and gas seal technology	NEDO
		Development of reformer for 1,000 kW pilot plant (1991)	MCFC Research Association, Bobcock Hitachi
		Development of control system for 1,000 kW plant simulation	MCFC Research Association
		High-efficiency gas purification technology (dry gas purification technology based on titanium oxide)	MCFC Research Association
		Improvement of cell and stack durability	CRIEPI
	IHI	Development of MCFC component (high-efficiency, high-durability, corrosion-resistant materials), development of technology for high-pressure fuel cell operation	NEDO; partnered with Tokyo Electric Power for part of the research

	Development of control system for 1,000 kW pilot plant	MCFC Research Association
	Design of large-capacity plant (coal-fueled MCFC plant design)	MCFC Research Association
	Gas-recycling operation technology	CRIEPI
	2 kW mini plant development	
MELCO	Reformer control system technology (heat exchanger mode)	MCFC Research Association, Chiyoda Corporation
	Characteristics and behavior of external reforming under pressure	CRIEPI
Toshiba	Component technology development (efficiency, durability, reliability)	
	Materials development (ribbed anode)	MCFC Research Association
	Development of heat recovery system (development of heat recovery steam generator for 1,000 kW pilot plant)	MCFC Research Association
Fuji Electric, Fuji Electric Corporate Research and Development	Material development technology (development of electrolyte plates by the paper-making method)	MCFC Research Association
	Gas recycling system (differential pressure control, instrumentation technology)	MCFC Research Association
Matsushita Electric Industrial	Integration technology (Development of technology to integrate electrodes and electrolyte plate by the doctor blade method)	MCFC Research Association
Mitsubishi Materials	Development of creep resistant anode (Cu-Al-Ni dispersion and solid- solution strengthened alloys)	MCFC Research Association
NKK	Development of high-efficiency separators (Ni-Cr-Fe alloy)	MCFC Research Association

(*Continued*)

TABLE 5.4 MCFC Component and System Technology Development in Japan* (*Cont'd*)

Type of Organization	Name of research organization or research sponsor	R&D activities	Research partner or sponsor
	Kobe Steel	Development of corrosion-resistant cathodes (Fe-Cr-Ni-Al alloy)	MCFC Research Association
		Development of turbine compressor	MCFC Research Association
		Development of coal gasification technology (high temperature dry desulfurization technology)	
	Kawasaki Heavy Industries	Development of improved electrolyte plate (tungstate salt-added electrolyte plate)	MCFC Research Association
	Tonen	Development of naphtha reformer	Petroleum Energy Center, Sanyo Electric
	Toyo Engineering	Piping system design and component configuration using 3D computer-aided design	MCFC Research Association
	Bobcock Hitachi	Development of two-stage catalytic combustion reformer for 1,000 kW pilot plant (1991)	MCFC Research Association
	Ebara Corporation	Development of high temperature blower for cathode gas recycling	MCFC Research Association
	Chiyoda Corporation	Development of heat exchange type reformer for 1,000 kW pilot plant (1991)	MCFC Research Association, MELCO

Ube Industries	Development of molten carbonate sulfur scrubber, high-efficiency desulfurization technology using inorganic salt, and high-efficiency gas separation technology using inorganic salt	MCFC Research Association
Yokogawa Electric	Control system (flow-control technology for high-temperature system)	MCFC Research Association
Nisshin Steel	Plating technology for separator materials	MCFC Research Association
Sumitomo Metal Mining	Development of ceramic cathode materials	MCFC Research Association
Nisshin Electric	Assessment of corrosiveness of molten carbonate (corrosion resistance of aluminum thin film coated through ion plasma process)	Kansai Electric Power
Sumitomo Precision Products	Development of high efficiency heat exchanger (development of stainless steel plate-fin heat exchanger)	

* *Available information is ambiguous regarding the specific role of some of the organizations involved in the MCFC component and system technology development research. Also, there is little information available as to the extent to which the R&D activities were funded under the Moonlight Project or some public or industry organizations. It is highly likely that all of them were funded in some form or another under the Moonlight Project because the outcome of those R&D activities alone would most likely be of little value unless integrated into MCFC total systems that the Moonlight Project was funding to develop.*

Sources: A.J. Appleby and F.R. Foulkes, "Fuel Cell Handbook," Van Nostrand Reinhold, New York, 1989, and "Large-Scale Fuel Cell Introduction Mode Survey (Part I)," NEDO, March 1991.

- In 1984, three additional companies—Fuji, IHI, and MELCO—participated in the project to develop 10 kW MCFC stacks. The three companies, which had been developing MCFC stacks on their own, brought their technology into the government program.[41]

By the end of Phase I, five different 10 kW MCFC stacks had been developed and were ready for testing. Testing of the stacks took place over 500 h in late 1986 and early 1987.[6]

- Fuji built a stack consisting of twenty-four 0.25 m^2 cells in three blocks, which yielded 7.1 kW at 0.79 V per cell. Hitachi's stack consisted of 30 cells of 0.36 m^2 active area in three blocks, operated for 794 h, and yielded 12.4 kW at an average cell potential of 0.76 V.[42]
- IHI developed a stack with twenty-nine 0.36 m^2 cells in a single block, which operated for 600 h and generated 12.5 kW at 0.8 V per cell. MELCO used twenty 0.2 m^2 cells in two blocks, which delivered 9.1 kW at 0.76 V per cell.
- Toshiba's stack consisted of fifty-seven 0.16 m^2 cells in a single block, generating 10.4 kW at 0.76 V per cell.[6]

5.3.2.2. *Phase II MCFC Development (1987–1999): Three Companies Participate in 200 kW Internal Reforming Stack and 1000 kW Pilot Plant Development*

NEDO viewed Phase I as successful and approved the start of Phase II in 1987. Three MCFC developers out of the five were selected: MELCO, IHI, and Hitachi. MELCO was expected to develop a 200 kW internal reforming stack, and IHI and Hitachi a 1000 kW external reforming pilot plant. Fuji and Toshiba apparently dropped out of the MCFC development effort, probably because they needed to focus on the commercialization of PAFC technology. Moreover, the 10 kW stacks developed by Fuji and Toshiba did not perform as well as those developed by IHI and Hitachi. IHI and Hitachi stacks demonstrated the higher power output and cell voltage and used larger cells. MELCO was the only company that had the expertise in internally reforming MCFC stacks among the five manufactures.

The overall project proceeded in two stages. First, stacks would be scaled up from 10 kW to 30 kW for internal reforming stacks and to 100 kW for external reforming stacks. Development of the 200 kW internal reforming stack and 1000 kW external reforming pilot plant would be pursued in a second stage. To build the 200 kW stack and the 1000 kW pilot plant, cells with active cell areas of 1 m^2 were adopted. Phase II therefore was extended by 2 years to refine the 1 m^2 cell production technology.

- The initial plan called for the use of cells of 0.36 m^2 cell areas in the 200 kW and 2000 kW prototypes. By 1991, however, the use of 1 m^2 cells looked feasible because both Hitachi and IHI were successfully experimenting with stacks with 1 m^2 class cells.[42]

Also in Phase II, the MCFC Research Association was expected to develop BoP technology, including two types of fuel reformers, a high-temperature gas recycle blower, a turbine compressor, a heat recovery steam generator, and total plant control technology. The association also was expected to develop a total plant system concept for possible use of coal gas in MCFC plants.[41]

5.3.2.2.1. MELCO DEVELOPS 200 KW MCFC INTERNAL REFORMING STACK

During the first stage of Phase II (1987–1993), MELCO developed at least five stacks of various output sizes with cells ranging from 0.2 m^2 to 0.5 m^2 in area.[42] The 30 kW DIR stack with 0.5 m^2 cells, for example, was developed in 1993, operated on methane, and achieved a maximum output of 33 kW with a current density of 150 mA/cm^2 at 0.8 V. Fuel utilization was estimated at 80 percent assuming the system operated in a pressurized condition. The degradation rate was 0.6 percent per 1000 h.[41] The 30 kW stack operated for 13,434 h in 1993–1994. The project appeared to have proceeded smoothly and without major problems.

- MELCO was also developing indirect internal reforming (IIR) MCFC stacks in a private-sector joint R&D effort with Kansai Electric Power. In 1991, Mitsubishi and Kansai Electric developed a 30 kW IIR stack that consisted of sixty two 0.5 m^2 cells, which operated for 2131 h.[20] The two companies developed in 1992 a 100 kW IIR stack, which consisted of one hundred and ninety-two 0.5 m^2 cells, which reached a maximum output of 109.2 kW and operated for 2308 h.[41]

At the beginning of the second stage of Phase II, MELCO launched the development of an MCFC stack that would possess the advantages of both the DIR and IIR stacks. The DIR stack had the advantage of enabling high fuel-reforming efficiency, and the IIR stack had the advantage of facilitating uniform cell-operating temperature. MELCO developed a series of so-called "advanced internal reforming (AIR)" stacks that featured both of these characteristics using a counterflow configuration. A counterflow pattern would channel the cathode gas and the anode gas in opposite directions, enhancing uniform cell-temperature distribution.

Also, outside the METI program, MELCO worked on its own to scale up its cell area from 0.5 m^2 to 1 m^2 and developed at least four stacks ranging from 10 kW to 30 kW. The 30 kW AIR stack, which was developed in 1997, consisted of twenty-six 1 m^2 cells and operated for 7544 h between 1997 and 1998. In 1999, MELCO developed a 200 kW AIR, externally manifolded stack with a counterflow configuration.[24] The stack consisted of two 100 kW sub-stacks, each of which consisted of two 50 kW blocks. Each 50 kW block consisted of fifty-five 1 m^2 cells. In April 1999, the 200 kW stack was delivered to Kansai Electric's Amagasaki Fuel Cell Test Center for testing and evaluation. The operation of the 200 kW stack started in June 1999 at atmospheric pressure using natural gas. The stack achieved the rated output of 200 kW in July 1999. The stack showed a current density of 150 mA/cm^2 and a cell voltage goal of 0.8 V, both meeting requirements. The cell voltage degradation was 0.4 percent per 1000 h. Fuel utilization was 80 percent. Its operation ended in February 2000, after operating for 5035 h (meeting its intended goal) and generating 1134 MWh.

In addition, MELCO under the METI program also developed what was called "pressurized, advanced internal reforming (PIR) stacks." The company built at least three 10 kW stacks. The first 10 kW stack built in 1996 was made of eight 1 m^2 cells and operated for 6067 h in 1996.[20] The stack demonstrated a power density of 2.2 kW/cm^2 or a current density of 300 mV/m^2, the highest density in the world at that time.[20] The second 10 kW stack consisted of twelve 1 m^2 cells and operated for 7483 h from 1998 to 1999. This stack showed a voltage degradation of only 0.25 percent per 1000 hours. The third 10 kW stack developed in 1999 has not been much reported. NEDO also viewed the development of PIR technology as a major accomplishment.[42]

5.3.2.2.2. IHI AND HITACHI DEVELOP 1000 KW PILOT PLANT

In the first stage of Phase II (1987–1993), Hitachi and IHI were expected to scale up active cell area to 1 m^2 in a single unit or in a block and build 100 kW stacks using external reforming and internal manifolding. The 100 kW stacks could use reformed natural gas or purified coal gas. They were to operate for 5000 h under pressurized conditions or not, and they were to achieve a current density of 150 mA/cm^2 at 0.8 V and a degradation rate of less than

1 percent per 1000 hours.[41] The two companies also were assigned to develop two different cell assemblies with two different flow configurations.[40] Hitachi was expected to build a stack with eighty-eight 1.2 m^2 cells divided into four groups, each of which would be equipped with its own manifolding in a crossflow configuration. IHI was assigned to develop a stack with one hundred and two 1 m^2 cells assembled all in one unit with manifolds in a parallel flow mode.

By 1993, the two manufacturers had developed their 100 kW stacks. The stacks were tested at Akagi Testing Center in 1993. They performed rather poorly.[43] The good news was that both the stacks, which were operated in a pressurized condition (3 ata for the Hitachi stack and 5 ata for the IHI stack) with reformed natural gas, achieved a maximum output of 124 kW (the IHI stack) and 110 kW (the Hitachi stack). The bad news was that although the IHI stack operated for 5000 h, it showed a degradation rate of 1.5 percent per 1000 h. The Hitachi stack showed poorer performance. It operated only for 1700 h; the stack was taken back to a Hitachi plant, where it was modified and subsequently operated for 5000 h at a reduced temperature. Also, the Hitachi stack showed an extremely high degradation rate of 17 percent per 1000 hours.[41]

Because of these problems, a contingency plan was implemented in 1994 to decide whether to move forward to the 1000 kW pilot plant project. [42] Under the contingency plan, each company would build a smaller stack with necessary modifications and demonstrate the viability of the modified stacks. IHI assembled a 5 kW stack (five 1 m^2 cells) with a new cathode current collector plate.[43] Hitachi built a 25 kW stack (twenty-two 1.2 m^2 cells) with a new cathode current collector and a modified amount of electrolyte in a matrix of reduced thickness. Hitachi also operated the modified stack at a reduced temperature. Both modified stacks operated satisfactorily, meeting all vital goals. NEDO declared the project was a success and approved the start of the next stage of the project.[41]

- IHI's modified 5 kW stack operated at a normal atmosphere for 16,000 h from 1993 to 1994 with a degradation rate of 0.3 percent per 1000 hours.
- Hitachi's modified 25 kW stack operated at 6 ata with a fuel utilization of 60 percent, and a current density of 150 mA/cm^2 at 0.79 V. It operated for 5087 h from 1994 to 1995 with a degradation rate of 0.5 percent per 1000 hours.

With the contingency plan behind them, NEDO directed that Hitachi and IHI begin work in 1994 to build a 1000 kW pilot plant with external reforming and internal manifolding. Due to the failure to complete the first stage of Phase II on time, the second stage was extended by 2 years into 1999. During this stage, Hitachi and IHI were each expected to supply two 250 kW stacks to assemble the 1000 kW plant. Hitachi would supply two 250 kW stacks (three hundred 1.2 m^2 cells in six modules) in a crossflow configuration. IHI would supply two 250 kW stacks, each consisting of two 125 kW sub-stacks (one hundred and forty 1 m^2 cells) in a parallel flow design.[43] The 125 kW stack consisted of two groups of 70 vertically stacked cells each, which were separated by a spacer. In a preparatory exercise for building the 250 kW stacks, Hitachi and IHI each built a 25 kW stack with twenty-five 1.2 m^2 cells and a 20 kW stack of twenty 1 m^2 cells, respectively, according to the specifications established for the two different stacks. Both stacks were successfully tested from 1995 to 1996.[42]

Development of the two different 250 kW stacks progressed smoothly. The preparation of the appointed installation site at Chubu Electric' Kawagoe Thermal Power Station was complete in December 1996.[39] In 1997, all the BoP components—except for the stacks—including a fuel reformer, a gas turbine compressor, were all installed and calibrated. After 9 months of process and control tests, four 250 kW stacks were installed in 1998.[40]

The load test of the 1000 kW MCFC power plant was conducted from August 4, 1999 until January 28, 2000.[44] The plant operated on LNG at 5 ata and reached its 1000 kW load at November 5, 1999. Plant performance, however, failed to meet its goal of 5000 operating hours—it achieved only 4211 h, with one set of 500 kW stacks operating for 4916 h. The shortfall was due to an accident. After the plant operation started on August 4, 1999, adjustments were conducted at various power output levels. When the output reached 900 kW on 6 September, this level of output was maintained until 17 September, when system operations were suspended for a scheduled inspection of the cathode filter. During the inspection period, a ground fault occurred in one set of the stacks. To investigate the cause, the particular stack set was cooled down to room temperature. After one and a half months, the cause was found and the problem fixed. The stacks were reheated, and the plant resumed operation on 31 October. The plant reached a maximum output of 1000 kW on 5 November. The plant had two more scheduled outages for maintenance and continued to operate until 28 January, when the plant was officially shut down. Although the operating hours fell short of its goal, there were no signs of abnormality.[24]

- Because of the loss of operating time, all other related performance was poor. The total generating time reached 2669 h, delivering 203,120 kWh. The plant operated at 1000 kW for 1264 h, close to 50 percent of all generating hours, and produced more than 750 kW for 75 percent of its generating hours. The longest continuous period of operation at full power (1000 kW) was 528 h.

Despite this and other problems, NEDO viewed the project to be a success.[24] The plant was Japan's first 1000 kW-class pressurized power plant and the operation of a MCFC system at 5 ata for close to 5000 h set a world record.[45] The plant also was the first MCFC/gas turbine hybrid system in the world.[40] Some aspects of plant performance were satisfactory. The electrical efficiency at the generating end was 46.1 percent, which was slightly higher than the goal of 45 percent.[24] Voltage degradation was also satisfactory; the stacks developed by Hitachi showed a degradation rate of 0.65 percent per 1000 hours, and the stacks developed by IHI had a degradation rate of 0.47 percent per 1,000 hours.

- NEDO, however, recognized that the reliability of large-capacity stacks and the plant efficiency and operational performance must be increased and that cost reduction must be achieved.

5.3.2.3. *Phase III MCFC Development (2000–2004): Only One Company Remains*

METI next launched an ambitious, 5-year, Phase III MCFC development program (2000–2004). The program aimed at first developing an MCFC stack of two hundred and fifty 1 m^2 cells that would generate 300 kW, which would be the key component for two types of MCFC power plants: a small-scale MCFC power plant for near-term commercialization and another larger-scale power plant for the longer term—one to two decades in the future. The small-scale power plant in Phase III was to be a 250-cell 300 kW distributed cogeneration unit that would operate at 4 ata with an efficiency of 43 percent.[45] The stack and fuel reformer would be housed in a module, and the rest of the BoP would be combined in a compact kit, constituting a power plant that would be simpler, smaller, and less costly to build and install. The second type was a 750 kW power plant consisting of two 250-cell stacks and a reformer, contained in one module, which would be combined with a BoP kit and a gas turbine to generate a total of 750 kW. The unit would operate at high pressure (12 ata) with an efficiency of 47 percent, and would constitute a building block for larger power plants.

- In METI's vision, eight of the MCFC modules combined with gas turbines would constitute a 7 MW-class medium-size power plant and operate with a power efficiency of more than 50 percent by the 2010s.[45] In the 2020s, a greater number of 750 kW modules would be combined with gas turbines into large-scale power plants in the 90–360 MW range, with power efficiencies of 50–60 percent. Plants of these sizes could replace large thermal power plants.
- The contractor for all the ambitious, divergent projects for the Phase III was IHI. No other MCFC developers played a part in all the projects. In addition, the Phase III program was substantively shorter in time as compared to the Phase II or Phase I programs.

A review of what happened to the Phase III program indicates that it was a near disaster. Phase III consisted of five interrelated projects; the outcome of each affected the activities and goals of all the others. All the projects also were planned to proceed concurrently, with each unfortunate outcome upsetting and delaying the schedules and goals of the other projects. Indeed, a number of unfavorable outcomes cascaded through the five projects and resulted in numerous, chaotic changes in plans and goals of the entire program. There were at least four major changes in plan in the course of the 5-year program, including changes in March 2002, September 2003, March 2004, and January 2005.

The five projects in the initial METI plan were as follows:

1. Project 1 aimed at developing a tall stacking technology that could support building a 300 kW stack by vertically stacking two hundred and fifty 1 m^2 cells. This was a difficult technology to achieve for IHI. In the 1000 kW pilot plant project in Phase II, IHI stacked a total of 140 cells. The 140 cells, however, consisted of two modules of 70 cells each, which were separated by a spacer placed between the modules.
2. Project 2 was to operate a short 10 kW stack (with 1 m^2 cells) at a relatively low pressure of 5 ata, but with a longevity goal of 10,000 h and a voltage deterioration rate of 0.25 percent per 1000 hours.
3. Project 3 would operate a short 10 kW stack at a pressure of 12 ata with high current densities (200 mA/cm^2) and cell voltage greater than 0.71 V.
4. Project 4 was to operate a 250-cell, 300 kW power plant at 3 ata combined with a gas turbine for 10,000 h at a current density of 150 mA/cm^2 at an efficiency of 43 percent and a voltage deterioration rate of 0.3 percent per 1000 hours. The power plant would consist of the 300 kW stack module and a reformer, a turbocharger, and a compact kit of BoP components.[40]
5. The ultimate goal of Project 5, supported by the preceding projects, was to build a high reliability, high efficiency 750 kW MCFC pressurized module using two 300 kW stacks and a gas turbine, operating at 12 ata with an efficiency of 47 percent.

It is fair to say that there were some positive outcomes. Two of the five projects were rather successful:

- In Project 2, a 10 kW short stack (1 m^2 cell areas) was built and started operation at 5 ata in April 2001. The stack immediately experienced gas leaks, the nature of which was not clearly explained. After repairs, the stack operated with no major incidents for a total of 10,000 operating hours in June 2002.[45] The stack generated an average output of 11.5 kW with current densities of 150 mA/cm^2 and a fuel utilization rate of 60 percent. The voltage deterioration rate was

0.3 percent per 1000 hours. The postoperation dismantling analysis showed no abnormal corrosion or carbonate leakage. The analysis of the loss of electrolyte and Ni dissolution from the cathode into the electrolyte and the corrosion of the separator surface suggested that the stack could have operated for 40,000 h at 5 ata. Nonetheless, the gas leak problem caused delays in the completion of this project as well as Project 5. Due to the gas leak experience, METI/NEDO in March 2002 decided that Project 5 would use one 375 kW stack with 250 cells, instead of two stacks, in the high-pressure system operation.

- In Project 3, a 10 kW short stack was installed at the Kawagoe MCFC Fuel Cell Test Center and successfully operated at 12 ata for a total of 3100 h between February 2002 and January 2003. The stack generated about 15 kW, achieved power efficiency at the transmission end of more than 55 percent, and a current density of 200 mA/cm^2 with an initial average cell voltage of more than 0.71 V. This was reported to be the first successful high-pressure operation of a 1 m^2 stack in the world.[45]

The other three projects had disheartening results. Project 1 did not achieve its goal to deliver reliable high stacking technology. This failure became apparent in Project 4 when a 300 kW stack and reformer module was installed along with a turbo charger and other BoP equipment at the Kawagoe test center for test operations.[40] The system started operations on January 14, 2003. After 631 h, gas leaks and voltage drops occurred. The voltage drops were caused by inadequate stack conditioning prior to installation in the module and inadequate gas supply to the upper cells due to the gas leak. The cause of the gas leak was due to insufficient clamping pressure at the manifold areas. In September 2003, METI/NEDO decided that Project 4 would now use a 125-cell, 150 kW stack and the system would operate at 4 ata for 5000h. At the same time, NEDO scaled back the goals of Project 5, which would now focus on building a 180-cell, 270 kW stack and operating it at 12 ata.

- The 125-cell, 150 kW stack was built and installed at Kawagoe in early to mid 2004. The 125-cell stack, however, again had gas leaks.[45] The causes this time were traced to inadequate thickness of gaskets at the manifold, inadequate temperature control at the time of stack conditioning, and distortion and deformation in the separator.
- After repairs, the operation restarted. It generated 155 kW and met goals that were initially set for the 300 kW system.[45] It operated at 4 ata with a voltage deterioration rate of 0.3 percent per 1000 hours and a current density of 200 mA/cm^2. The efficiency at the transmission end was 43 percent. It reached 4839 operating hours—less than planned—because operations were suspended by the tripping of the cathode blower.

Despite these problems, Project 4 yielded some important achievements. Cell-stacking technology for up to 125 cells without using a spacer was believed to have been demonstrated for the first time in Project 4.[45] (There is no readily available information as to why the 250-cell stack assembled in Project 1 was not tested prior to its integration into the Project 4 power plant).

Regarding Project 5, because the stacking of 250 cells was not possible, a new plan was made in September 2003. The new plan called for a 180-cell, 270 kW stack to be operated in a high-efficiency module.[45] It was believed that stacking 180 cells would be feasible because gas leak countermeasures that were developed in Project 4 could be used for Project 5. In March 2004, however, there was a second gas leak incident in Project 4, which led to another change of plans in Project 5 to further scale down the stack to 125 cells. Project

5 was now expected to build a stack that would generate 170–190 kW and operate at 12 ata with an efficiency of 47 percent. In December 2004, the 125-cell stack was installed in a power generation system at Kawagoe, and the system started operation. The system only generated an output of about 30 kW at the onset; about 30 percent (35 cells) of the stack showed voltage degradation in varying degrees. The cause was found to be aluminum residue stuck to the separators. The annealing process for the separators was not done properly. In the annealing process, an aluminum-plated stainless steel spacer was mistakenly used and the aluminum from the spacer stuck to the separators. This problem was caught at the time of annealing, and efforts were made to remove the residue. The removal was not done well and some of the residue remained. Aluminum increases electric resistance of the cells, thus dropping the voltage. The amount of aluminum residue on the separators varied, and thus, there were varying degrees of voltage drop among the cells.

- Because there were only 3 months left before Phase III program would officially end (fiscal year 2004 ends in March 2005), NEDO decided in January 2005 that no new 125-cell stacks would be built, but a test operation of the current stack with the remaining unaffected cells (70 percent of the stack or 90 cells) would continue. The stack operated at 12 ata and generated 120 kW with an estimated efficiency of 44.4 percent.
- The initial plan for designing a 7 MW MCFC system at the conclusion of Phase III program was abandoned.

5.3.3. MCFC Commercialization in Japan Hopeless

IHI was planning to embark on MCFC commercialization in 2005 following the end of the METI-sponsored Phase III MCFC program, according to IHI's presentation at the 2005 Fuel Cell Seminar held in Palm Springs, California on 14–18 November.[46] But IHI's commercialization plans for MCFC plant installations apparently are faltering. There have been a total of 23 MCFC pre-commercial prototypes or commercial power plants installed in Japan, which is only one-third the number of PAFC installations (Table 5.5). Of the 23, six units were manufactured by IHI and 17 by FCE of the United States. FCE, the global leader in MCFC technology, exported MCFC power plants to Japan via Marubeni Corporation under agreements made in 2001. Marubeni imported complete MCFC power plants from FCE, installed them and provided maintenance services (Fig. 5.5).[47] (FCE concentrates on developing atmospheric-pressure, internal reforming MCFCs and offers three models: the 250 kW, the 1 MW, and the 2 MW model. Only 250 kW models have been sold in Japan).[48]

IHI's commercial record has been bleak. While participating in the MCFC development efforts under NEDO's Phase III program in the early 2000s, IHI started to sell pre-commercial prototype and commercial models with NEDO assistance. NEDO had a number of programs that subsidized commercialization of new energy sources, including fuel cell, solar, and wind power plants. One of the programs likely used to assist IHI MCFC power plant installation and demonstration was the "Demonstrative Project of Regional Power Grids with Various New Energies." The program provided subsidies to the purchasers of distributed clean energy sources to examine the performance of the power plants and the impact of such power sources on the grid.[49] The budget for fiscal year 2004, for example, was about \$60 million (6.36 billion yen at the January 2004 exchange rate of \$1/107 yen). Despite the subsidies, IHI sold a total of only six power plants: three 300 kW power plants in 2002 and two 300 kW units and one 10 kW unit in fiscal year 2005. Apparently, no sales

TABLE 5.5 MCFC Power Plant Installation in Japan (as of April 2012)

Country	Plant size, fuel	Project Name/ Coordinator	Plant supplier	Operator/ Installation site	Schedule of operation	Recipient of any financial support*
Japan	300 kW, MCFC, simulated coal gas	Chubu Electric Power, NEDO, MCFC Research Association	IHI	Chubu Electric's Kawagoe Thermal Power Plant	Installed in March 2002	Perhaps received a subsidy from NEDO's Demonstrative Project of Regional Power Grids with Various New Energies
	300 kW MCFC, biogas	Chubu Electric Power, NEDO, MCFC Research Association	IHI	Chubu Electric's Shin-Nagoya Thermal Power Plant, Nagoya, Japan	First Commercial prototype, installation December 2002, start-up in February 2003	Perhaps received a subsidy from NEDO's Demonstrative Project of Regional Power Grids with Various New Energies
	300 kW MCFC with a 50 kW gas turbine	Toyota, NEDO	IHI	Motomachi Environmental Center of Toyota Motor, Toyota City, Japan	Installed in October, start-up in early 2003	Perhaps received a subsidy from NEDO's Demonstrative Project of Regional Power Grids with Various New Energies
	250 kW DFC, multiple fuels, including anaerobic digester gas, propane	Marubeni, Kirin	FuelCell Energy	Kirin Brewery, Toride, Japan	January 2003	
	250 kW DFC, natural gas	Marubeni, Nippon Metal	FuelCell Energy	Nippon Metal's Sagamihara Works, Sagamihara, Japan	Start-up in fall 2003	
	250 kW DFC, anaerobic digester gas	Marubeni, City of Fukuoka, Kyushu Electric Power	FuelCell Energy	Seibu Municipal Western Water Treatment Facility, Fukuoka, Japan	January 2004 for a 2-year demonstration	

(Continued)

TABLE 5.5 MCFC Power Plant Installation in Japan (as of April 2012) (*Cont'd*)

Country	Plant size, fuel	Project Name/ Coordinator	Plant supplier	Operator/ Installation site	Schedule of operation	Recipient of any financial support*
	2 × 250 kW DFC, natural gas	Marubeni, Epson, First Energy Service Company, Seiko Epson	FuelCell Energy	Seiko Epson's Quartz Devise Division facilities, Ina, Japan	April 2004	
	250 kW DFC, natural gas	Marubeni, Japan Petroleum Exploration (JAPEX)	FuelCell Energy	JAPEX's Katakai Gas Fields, Niigata, Japan	Fourth quarter 2004	
	250 kW DFC, digester gas	Mitsubishi Heavy Industries, Bioenergy	FuelCell Energy	Food waste treatment facility, Jonan Island, Ota-ku, Tokyo, Japan	Installed in late 2004 or early 2005	
	2 × 300 kW, digester gas	Aichi Expo, IHI, Chubu Electric Power, Toyota Motor	IHI	Aichi Expo, Nagoya, Aichi, Japan	March 2005	Received a subsidy from NEDO's Demonstrative Project of Regional Power Grids with Various New Energies
	250 kW DFC, natural gas	Marubeni, Fuel Cell Japan, Kawasaki Heavy Industries, Tokyo Gas	FuelCell Energy	Akashi Works, Kawasaki Heavy Industries, Osaka, Japan	Start-up in 2005; relocated to Tokyo Gas' new R&D center in Tsurumi in the second quarter of 2006	
	250 kW DFC, anaerobic digester gas	Tokyo Eco Town Project, Tokyo Metropolitan Government (TMG), Marubeni, Bioenergy	FuelCell Energy	Tokyo Super Eco Town food recycling facility, Tokyo, Japan	2005	The Japanese government and TMG provided financial support

250 kW DFC, natural gas	Marubeni, NTT Facilities, NTT Architecture Laboratories, Tohoku Fukushi University, Sendai City	FuelCell Energy	Sendai City electric supply system site, Sendai, Japan	2005	
10 kW	CO_2 Capture Project, METI, IHI, Chubu Electric Power	IHI	Chugoku Electric Power, Misumi Plant, Shimane, Japan	October 2005	
250 kW DFC, digester gas	Kyoto Eco Energy Project, Kyoto Prefecture, NEDO, Marubeni	FuelCell Energy	Kyoto Eco Energy, Kyotango City, Kyoto, Japan	Third quarter 2005	NEDO provides the capital and installation cost
250 kW DFC	Marubeni, Fuel Cell Japan	FuelCell Energy	NTT Facilities	2006	
250 kW DFC	Marubeni, Fuel Cell Japan	FuelCell Energy	NGK, Nagoya, Aichi, Japan	2006	
4 × 250 kW DFC, natural gas	Marubeni, Fuel Cell Japan, Cenergy	FuelCell Energy	Sharp, Kameyama Daini Plant, Mie, Japan	October 2006	Japan's Ministry of Economy, Trade and Industry (METI) provides a subsidy

** Most MCFC sales in Japan were eligible for a subsidy at least up to one-third of the acquisition cost from NEDO if the plant output was greater than 100 kW.*

Sources include "Worldwide Fuel Cell Installations," updated October 2005, Fuel Cells 2000, www.fuelcells.org, accessed January 2009; "Proposals for Promotion of Fuel Cell Introduction," New Energy Foundation, March, 2006; T. HOMMA, "Market Development of MCFC," THE LATEST FUEL CELL NEWS IN JAPAN, FCDIC, FEBRUARY 2004; T. HOMMA, ""Business Deployment of MCFC," THE LATEST FUEL CELL NEWS IN JAPAN, FCDIC, OCTOBER 2004; "Japan's Fuel Cell Industry," JETRO Japan Economic Report, Japanese Economy Division, JETRO, April–May 2006; "General Trading Houses, Active Business in New Energy Area," New Energy Shimbun, http://new-energy.jp/blog/archives/2006/11/post_645.html, November 20, 2006.

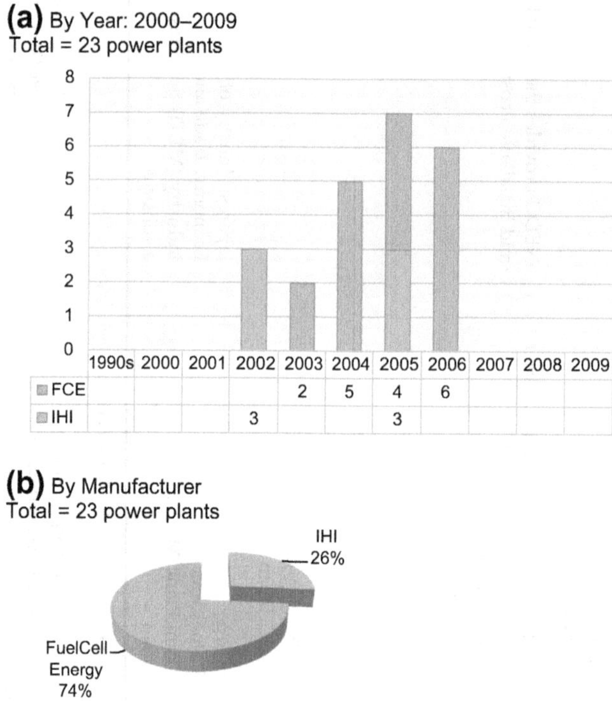

FIGURE 5.5 MCFC installation in Japan. (For the color version of this figure, the reader is referred to the online version of this book.)

have been made since then. By 2003, IHI indicated that the price of its MCFC power plant was about 500,000–600,000 yen per kW (about \$4700–\$5600 at the January 2004 exchange rate of \$1/107 yen).[50]

The six MCFC power plants IHI installed in 2002 and 2004 included the following:

- A unit at the MCFC Research Association, which purchased a 300 kW power plant for the Kawagoe MCFC Test Center of Chubu Electric Power in March 2002. Power generation began in January 2003.[50] The plant used simulated coal gas and demonstrated a lifetime of more than 10,000 h.[13]
- A second 300 kW power plant was purchased by Chubu Electric Power.[51] The plant was said to be IHI's first commercial MCFC prototype. The plant was installed in December 2002 at Chubu Electric's Shin-Nagoya Thermal Power Station in AichiPrefecture[40] and started power generation in February 2003.[52] Chubu Electric planned a 3-year field research program to test how the MCFC power plant operated with a waste-gasification device. The system demonstrated a lifetime of more than 10,000 h.[13]
- The third 300 kW MCFC plant was purchased by Toyota, installed at Toyota's Environmental Center in October 2002, and started operations in early 2003. This plant was integrated with a 50 kW micro gas turbine, which was produced by Toyota Turbine and Systems.[13]
- The fourth and fifth MCFC plants were sold to Chubu Electric Power and Toyota, respectively. They were installed at the site of the 2005 World Exposition (Aichi Expo),

Aichi Prefecture, and operated from March to September 2005. One plant was fueled by anaerobic digester gas produced from garbage using a low-temperature methane fermentation reactor,[53] and another by fermented methane gas generated from organic waste of restaurants at the Expo site.[54] The two power plants were relocated to ChubuTokoname Coastal City near Chubu International Airport, Aichi Prefecture, for a 2-year demonstration operation under the same NEDO subsidy program.[55]

- The sixth unit was a 10 kW unit installed at Chugoku Electric Power's Misumi Thermal Power Station in October 2005. The MCFC system would collect and concentrate CO_2 from the exhausted gas from Misumi's coal-burning thermal power plant.[48] The CO_2 concentration in the exhausted gas is about 13 percent, and it is increased to about 80 percent at the outlet of MCFC, a result believed to point to possible ways to effectively recover CO_2.[56]

FCE sold at least 17 MCFC power plants in Japan. FCE manufactured atmospheric-pressure, internal reforming MCFCs and offered three models: the 250 kW, the 1 MW, and the 2 MW model. The price of FCE's 250 kW MCFC power plant was 150 million yen (600,000 yen or $5600 per kW at the January 2004 exchange rate of $1/107), and the stack needed to be replaced after 3 years at a cost of 20 million yen ($187,000 at $1/107).[57] NEDO provided a subsidy for the acquisition cost. Marubeni either sold or installed two FCE 250 kW DFC power plants in 2003, five in 2004, four in 2005, and six in 2006. No additional sales apparently have been made since 2006.

Some of the noteworthy sales included the following:

- Marubeni received its first order of a 250 kW DFC in Japan in 2001 from Kirin Brewery.[48] The plant started operations in January 2003 using anaerobic digester gas produced from brewery.[58] Kirin took charge of feeding fuel to the fuel cell, Marubeni operated and maintained the fuel cell facility, while Kirin bought all the power and steam generated for use at the brewery. The price of the electric power generated by the DFC unit was the same as the price from Tokyo Electric Power, and represented about 4 percent of the total power used and 1 percent of the steam used by the brewery.
- Marubeni received the second order in 2002 from Fukuoka City Sewerage Department to install a 250 kW DFC power plant at its Western Waste water Treatment Center.[57] Marubeni, in cooperation with Kyushu Electric Power and Fukuoka City, planned to conduct a 2-year test operation to examine how well the plant would perform by using methane gas produced from the sewage treatment.[59]
- Another interesting case was the use of FCE's 250 W DFC at a food waste treatment facility. Mitsubishi Heavy Industries received an order to construct a food waste treatment facility, the largest in the country, from Bio Energy. The facility site of 4800 m^2 was procured at Jonan Island, Otaku, Tokyo, and the construction started in July 2004. The total budget for the land and facility construction was about 3.9 billion yen (about $36 million at the January 2004 exchange rate of $1/107 yen). The facility was to be completed by the end of fiscal year 2004 and would treat 110 tons of garbage a day to generate digester gas by using a fermentation reactor. The digester gas would be fed to FCE's 250 kW MCFC power plant for generating electricity for sale.[60]
- Sharp planned to install four FCE 250 kW MCFC power plants at Sharp's Third Kameyama Factory, Kameyama, Mie prefecture, with construction planned be completed in October 2006.[61] Electric

efficiency was to be 47 percent, and the steam would be supplied at 350 °C. Cenergy, a subsidiary of Chubu Electric Power, would be the owner of the fuel cells, and Fuel Cells Japan, a Marubeni subsidiary, would be in charge of maintenance. In addition to the MCFC power plants, Cenergy installed five 2800 kW natural-gas-fueled gas engines (total 14,000 kW) and five 100 kW solar cells on the roof of the factory to supply electric power and steam to the Sharp factory.

In 2004, Marubeni accelerated its efforts to establish manufacturing, sales, and maintenance operations in anticipation of growth in Japan's MCFC market.[48] This included setting up a subsidiary, Fuel Cell Japan, in August 2004 to handle MCFC sales, installations, and maintenance, as well as inviting engineers from FCE to stay in Japan on a permanent basis. On the manufacturing side, the company was working to lower costs, aiming to reduce its current system from 600,000 yen ($5600) to 300,000 yen ($2800) per kW. Marubeni expected that as sales increased, the initial costs of stack components would be lowered, in part by taking advantage of FCE's mass production facility in the United States.

- Marubeni planned to produce MCFC systems for the Japanese market with Kawasaki Heavy Industries, attempting to cut initial costs and boost the price competitiveness of its stacks and systems. Marubeni also planned to have IHI as a leading partner.[63]

Despite these initiatives, the MCFC sales and installation activity in Japan came to a halt in 2007. The large footprint and low-power density of the MCFC drove its costs higher than that of alternative fuel cell designs, even though MCFCs are made of non-precious materials. Apparently, no MCFC power plants have been sold in Japan for the past 3 years. Fewer and fewer MCFC manufacturers are engaged in the MCFC business. Moreover, it is not clear if IHI is still engaged in MCFC development or commercialization. It appears that there will be little chance for MCFC commercialization in Japan.

5.4. MCFC DEVELOPMENT IN EUROPE

Sustained MCFC development programs have been underway in Europe only since the mid-1980s. European MCFC R&D revived in 1986, when fuel cells regained popularity as an alternative to nuclear power.[6] Indeed, the Chernobyl nuclear event in April 1986 accelerated new MCFC programs in Europe, primarily in the Netherlands, Italy, and Germany. A few other countries showed some interest in MCFC development, but they served as partners and collaborators to these three countries.

Starting in 1987, the EU launched a vigorous funding drive to support MCFC R&D. The EU's approach was to directly support the projects initiated in individual nations and fund them under its Framework Program (FP). For medium- to long-term research projects, the EU FP program funds up to 50 percent of the total costs, and the participating countries and organizations supply the rest. For medium- to short-term research projects, the FP funds up to 35 percent of the total costs. It appears that for demonstration projects, about 40 percent of the total cost is funded.

European nations and the EU appear to have taken a relatively restrained approach toward MCFC technology compared to the United States and Japan. The Europeans were not interested in conducting research but rather wished to acquire US MCFC technology quickly through licensing, joint venture, or cooperative research. This approach probably was a rational one; when Europe re-engaged in MCFC development in the mid-1980s, it was lagging a decade or so behind the United States and Japan.

5.4.1. The Netherlands Revives Europe's MCFC Development Effort in 1986

In December 1985, the Dutch government approved a proposal from the Management Office for Energy Research for a 3-year MCFC development program.[63] The program was funded at €4MECU (about $4.4 million at an exchange rate of $1/1.1 ECU during this time period). The objectives were to increase knowledge of MCFC technology and create possibilities for Dutch industrial product development. To maximize cost-effectiveness, the Dutch chose to use US internal reforming technology. The Energy Research Center of the Netherlands (ECN)[x] signed an agreement in 1987 with IGT of the United States for transfer of the technology.

5.4.1.1. *European Union Framework Program Starts Funding Dutch MCFC Efforts in 1987—ECN Takes the Lead*

The EU was an energetic supporter of the Dutch effort to develop MCFC technology and started funding it in 1987 under the FP. For a dozen years until 1999, the EU FP funded at least eight MCFC projects led by or coordinated by Dutch industries or research institutions. The Dutch MCFC efforts were so strongly integrated into the EU FP that almost no efforts were undertaken outside the EU FP. Despite the seemingly strong faith in MCFC technology, the Netherlands and the EU perhaps were not entirely sure about their course of action. A review of the eight MCFC projects funded under the EU FP from 1987 to 1999 indicates that the projects lurched back and forth from pursuing applied research to product development and then back again to applied research.

Data is available only on eight of the nine EU projects implemented under the FP (Table 5.6):

1. *Development of New Cathode Materials for Use in Molten Carbonate Fuel Cells*, sponsored under the First Framework Program (FP1), which ran from 1984 to 1986.[65] The project itself ran from July 1987 to June 1989.[xi] Its aim was to reduce corrosion of the MCFC cathode and develop alternative materials and synthesis techniques. As a potential solution, the project aimed at replacing H_2O by other materials, such as doped $LiFeO_2$. Key requirements were good electrical conductivity compatibility with the electrolyte and good morphological characteristics suitable for mass production. ECN coordinated the project.
2. *Optimization of NiO Cathode Performance and Endurance in Molten Carbonate Fuel Cells*, sponsored under FP1.[65] The project, which ran from October 1987 to March 1991, studied the NiO dissolution process from the viewpoint of how to optimize the performance and endurance of cathodes. The project coordinator was the Technical University of Delft, with the participation of ECN and Centro Informazioni Studi Ed Esperienze SpA of Italy. This project was promoted to a national fuel cell program by the ECN and then funded under the EU FP.
   - EU funding under FP1 doubled the level of the 3-year Dutch national program funding to €8MECU (about $9.6 million at the 1990 exchange rate of $1/0.83 ECU).[xii,64] The Dutch national MCFC

[x] Energy Research Centre of the Netherlands (ECN) is the largest Dutch center for study into energy.

[xi] A project may run during the time period different from the time frame of the framework program. It appears that if the contract was signed within the Framework Program period, the project could run well beyond the program period.

[xii] The dollar equivalent is provided for general reference purpose only. The source of the exchange rate was "United States dollar,"Wikipedia, http://en.wikipedia.org/wiki/U.S._Dollar, last modified on June 19, 2009. In case that the exchange rate for some year was missing, the exchange rate of the closest year was used.

TABLE 5.6 EU Framework Programs: In Support of European Nations' MCFC Projects

Project title	Program	Official project period
EU Framework program in support of Dutch National MCFC projects		
Development of New Cathode Materials for Use in Molten Carbonate Fuel Cells	FP1 (1984–1986)	July 1987–June 1989
Optimization of NiO Cathode Performance and Endurance in Molten Carbonate Fuel Cells	FP1 (1984–1986)	October 1987–March 1991
A Collaborative Project to Build a 1 kW Internal Reforming Molten Carbonate Fuel Cell Stack	FP2 (1986–1990)	June 1990–August 1993
Upscaling MCFC—Technology to a 50 kW ER-MCFC Stack Demonstration	FP3 (1990–1994)	January 1993–December 1995
10 kW Demonstration of Pilot Stack Technology for Direct Internal Reforming Molten Carbonate Fuel Cell (DIR-MCFC) power plants	FP3 (1990–1994)	June 1993–December 1995
Basic Research on MCFC Corrosion and Improved Current Density	FP3 (1990–1994)	January 1994–December 1995
Selection and Development of Materials for High Endurance MCFC Bipolar Plate	FP4 (1994–1998)	January 1996–June 1998
Advanced Direct Internal Reforming (DIR) MCFC Development	FP4 (1994–1998)	February 1996–January 1999
EU Framework Program in Support of Italian-Spanish MCFC Projects		
Molten Carbonate Fuel Cell (MCFC) 100 kW Cogeneration Power Plant	Thermie 1(1990–1994)	August 1992–July 1996
100 kW MCFC-Integrated Stack Development and Testing, with an acronym "MISDAT,	FP3 (1990–1994)	January 1993–March 1998
Development of Industrially Relevant MCFC Stacks," with an acronym "DIREM"	FP4 (1994–1998)	January 1996–June 1999

MCFC "Series 500" Twinstack Powered First-of a-Kind "MCTWINS"	FP5 (1998–2002)	January 2000–April 2004
Innovative Cost-Effective Hybrid System Based on Integration of an MCFC and a Gas Turbine for High-Efficiency Dispersed CHP Generation "MOCAMI"	FP5 (1998–2002)	April 2003–April 2012
Molten Carbonate Fuel Cells for Waterborne Application "MC-WAP"	FP6 (2002–2006)	September 2005–August 2010
Biogas Integrated Concept a European Program for Sustainability "BICEPS"	FP6 (2002–2006)	January 2007–September 2011
EU Framework Program in Support of German-Danish MCFC Projects		
Development and Adaption of the Molten Carbonate Fuel Cell with Internal Reforming for the Use of CO/CO_2-Rich Gases; for example, Coal Gas	FP2 (1986–1990)	May 1990–December 1993
Design and Engineering of a 100 kW MCFC Cogeneration Plant (Natural Gas) and a 100 kW MCFC Power Plant (Coal Gas)	Thermie 1 (1990–1994)	July 1992–December 1994
MCFC Power Plants—Demonstration of a HotModule Qualification Model; Construction, Installation, and Start of a HotModule	Thermie 1 (1990–1994)	June 1994–August 1997 (The official program ended, but the project continued until February 2002)
System Development and Optimization	FP3 (1990–1994)	January 1994–June 1997
MCFC Balance of Plant Innovation—Design, Construction, and Demonstration of a DFC Module Pilot Plant for Cogeneration	FP4 (1994–1998)	August 1997–October 2001
Holistic Integration of MCFC Technology Toward a Most Effective Systems Compound Using Biogas as a Renewable Source of Energy "EFFECTIVE"	FP5 (1998–2002)	July 2000–June 2004

program immediately extended its schedule through 1992 and, with additional EU funding under FP1, increased the total budget to €22MECU ($26.5 million) with the goal of ultimately constructing a 300 kW stack.[63]

3. *A Collaborative Project to Build a 1 kW Internal Reforming Molten Carbonate Fuel Cell Stack* under FP2 (1986–1990).[66] The project ran from June 1990 to August 1993 and aimed at the development of a 1 kW internal reforming fuel cell stack. The key requirement was the development of reformer catalysts that were not poisoned by the corrosive carbonate atmosphere. The 1 kW stack was developed and demonstrated the required power density of 100 MW/cm^2. The coordinator was ECN, and other participants included the National Research Council of Italy (CNR) and British Gas of the United Kingdom. British Gas and CNR developed reformer catalysts and components. Cells up to 1000 cm^2 and stacks up to 1 kW were made by ECN. ECN tested the cells and stacks.
   - In 1990, two Dutch companies—Stork and Royal Schelde—formed a joint venture company with ECN. The objective of this new corporation—Brandstofcel Nederland (Dutch Fuel Cell Corporation—BCN)—was to commercialize ECN's MCFC technology.[67] ECN apparently believed that the Dutch MCFC technology would soon be ready for the market.[68]
   - The technology transfer from IGT helped ECN rapidly establish sufficient MCFC technology and led them to develop the first 1 kW external reforming stack in 1989,[1] a 2 kW external reforming MCFC stack in 1992,[69] a 1 kW internal reforming stack in 1993,[66] and a 10 kW external reforming stack also in 1993.[70] The stacks were made of 0.3 m^2 cells.[69]
4. *Upscaling MCFC Technology to a 50 kW ER-MCFC Stack Demonstration*, sponsored under FP3 (1990–1994).[70] The project ran from January 1993 to December 1995. ECN's state-of-the-art MCFC technology in 1993 permitted the construction of 10 kW external reforming stacks, so this project aimed to scale up the MCFC stack to 50–100 kW for commercial power generation applications. The electricity-generating area of the cells would be 1 m^2 and the 50 kW demonstration would include an operating 50-cell stack. The project also aimed to improve lifetime, performance, and start-up and operating conditions of the stacks. The project coordinator was BCN, supported by IBERDROLA and Ticnicas Reunidas of Spain. A 50 kW stack and natural-gas-fed MCFC installation was to be constructed by the Dutch partners supported by the Spanish partners.
5. *10 kW Demonstration of Pilot Stack Technology for Direct Internal Reforming Molten Carbonate Fuel Cell (DIR-MCFC) Power Plants*, sponsored under FP3 (1990–1994).[71] The project ran from June 1993 to December 1995. Because the feasibility of a 1 kW DIR-MCFC was established in a previous project (see *A Collaborative Project to Build a 1 kW Internal Reforming Molten Carbonate Fuel Cell Stack* under FP2), this project aimed to scale up to 10 kW. The project was to demonstrate the operating performance of a 10 kW stack and estimate the efficiency, performance, and lifetime of DIR-MCFC power plants in the sub-MW range. The coordinator was ECN, and other participants included CNR of Italy and British Gas.
6. *Basic Research on MCFC Corrosion and Improved Current Density*, sponsored under FP3 (1990–1994).[72] The project ran from January 1994 to December 1995. The aim was again to address and eliminate the cathode dissolution and precipitation problem by developing $LiCoO_2$ cathodes and to verify their large-area applicability. Better base materials and melt compositions were to be

studied to improve the bipolar plate corrosion resistivity. The target was to improve current density by 20 percent as compared to the state-of-the-art cells through the reduction of contact resistances and ohmic losses and the development of advanced cell concepts. ECN was the coordinator and other participants included the Royal Institute of Technology of Sweden, Technical University of Delft of the Netherlands, and Technical University of Denmark.

7. *Selection and Development of Materials for High-Endurance MCFC Bipolar Plate*, sponsored under FP4 (1994–1998).[73] The project ran from January 1996 to June 1998. The goal was to select high-chromium ferrous stainless steels required for improvements in cathode-side endurance from 25,000 to 40,000 h, selection of high-aluminum ferrous steels for 40,000 h anode-side endurance, characterization of properties and endurance behavior of coating materials, development of bond coatings and protective coatings and their application techniques for anode-side application, technical verification of selected options in small-scale MCFCs, and assessment of options that considered technical viability, endurance, and cost. The project coordinator was ECN, and other participants included Gaz de France, the Royal Institute of Technology of Sweden, and Centre National de la Recherche Scientifique of France.
8. *Advanced Direct Internal Reforming (DIR) MCFC Development*, sponsored under FP4 (1994–1998). The contract was signed in 1995, and the project ran from February 1996 to January 1999. The 3-year DIR MCFC project aimed at investigating cost reduction via integral development and innovative concepts for both system and stack. The project was a feasibility study for MCFC commercialization and consisted of a number of studies and laboratory tests, including market analysis, system design, stack verification tests, and cost analysis.[74] The project identified the most attractive initial markets for DIR-MCFC cogeneration power plants to be hospitals, hotels, leisure centers, and light process industry, with an optimum system size to be around 400 kW. A stack and system concept (SMARTER system) was designed to realize optimal reduction of system complexity and capital cost. Based on the SMARTER system requirement, a full-scale (1 m^2 cell area) stack with an expected service life of over 25,000 h was designed. Verification tests of the designed stack were carried out in 12 sub-scale (0.1 m^2) stack tests, and a 5000-hour stack endurance test was performed.[75] No actual MCFC stack or system seemed to have been built, however. The coordinator was BCN, and other participants included BG Technology of the United Kingdom, ECN, Gaz de France, Sydgas of Sweden, and two ECN subcontractors, Royal Schelde and Stork of the Netherlands.[74]

5.4.1.2. *The Netherlands Ends MCFC Development in 1999*

In 1999, the Netherlands apparently decided not to pursue MCFC development or commercialization after the completion of the last FP4.[76] It was an unfortunate turn of events, but in view of the findings of the FP4 project, it probably was a rational decision. The findings included the following:

- Plant cost estimates per kW, based on production volumes of 50 systems per year and a stack production volume of 200 MW per year, resulted in a cost of \$3000.[74] The Dutch likely considered that the high capital cost, even at high production volumes, was a show stopper.
- Technical breakthroughs, significant changes in emission regulations, or higher fuel prices were a prerequisite for further

commercialization of the technology. The Dutch likely concluded that three prerequisites would not be satisfied all at once.

In October 2005, after more than 15 years after its decision to leave the MCFC field, ECN announced that it had sold its MCFC technology to FCE of the United States.[77] According to ECN, it had accumulated an extensive technology package, covering inventions at system design level, stack development, and at the cell level. As a result of decreasing R&D budgets, the ECN management decided to concentrate on SOFC and PEMFC technology R&D.

5.4.2. Italy Starts MCFC R&D also in 1986—Ansaldo Ricerche Takes the Lead

Italy's MCFC development efforts have largely centered on Ansaldo Recerche. In December 1986, Italy implemented an ambitious 5-year national MCFC R&D effort under Project Volta, for which the Italian National Agency for New Technologies, Energy, and the Environment (ENEA) administered and coordinated all national fuel cell interests.[78,79] Project Volta had a total budget of 50 billion lire (about $40 million).[63] Fifty percent of the budget was earmarked for MCFC research to develop a 50–100 kW external reforming MCFC plant by 1992, along with advancement in materials research to reduce cost and increase reliability.[6,80] ENEA provided 80 percent of the budget, with 15 percent coming from Italian industry, and 5 percent from the EU and other sources.

- The principal contractor was Ansaldo Recerche. To meet ENEA's financing goals, Ansaldo spent €7M ECU (about $8.75 million at the 1990 exchange rate of $1/0.8 ECU) of its own funds on MCFC research in 1991.[1]
- Other participants included CISE (the Centre for Innovation and Economical Development), CNR-ITAE (the National Council of Research of Italy-Institute of Advanced Energy Technologies), and university laboratories.[79]

The 5-year program plan called for forming a joint venture with an experienced European partner and acquiring advanced technology know-how.[80] With this goal in mind, Italy had held discussions with the Netherlands,[6] but no agreement was reached. However, a technology transfer agreement was reached with UTC of the United States, and the transfer of UTC technology to Ansaldo started in early 1989.[79] The technology transfer enabled Ansaldo to construct five 20 cm^2 cells by the end of 1989. The cells delivered 0.76 V at 160 mA/cm^2. Improved cells operated at 0.8 V, 160 mA/cm^2 initially, but fell to 80 mW/cm^2 after 1800 h.[6] ENEA and CNR conducted materials research, testing, and fabrication of all-Italian cell components. In addition, CNR-ITAE obtained a complete set of state-of-the-art parts from IGT after signing a collaboration agreement. The IGT single cells operated for 30,000 h.[80]

About 1992, Italy started another ambitious, multi-year MCFC development project, called "Molten Carbonate Europe," with the acronym MOLCARE. This time, Italy found a collaborating partner in Spain and established an Italian-Spanish consortium to jointly develop MCFC technology. The consortium partners included Ansaldo, CNR, and ENEA of Italy and IBERDROLA and Babcock & Wilcox Espanola of Spain. The goal of the project was the design and development of the pressurized basic compact module to be used in MCFC plants up to 500 kW. Although the MOLCARE program was a joint venture, Italy likely took the leadership role in the program. For example, the project had a budget of 10 billion pesetas, and Italy contributed 65 percent of the budget, and Spain 35 percent.[81]

5.4.2.1. EU Framework Program Supporting the Italian MCFC Effort in 1987

The Italian-Spanish "MOLCARE" program attracted significant attention from the EU FP. The EU started to fund MOLCARE in 1987 and continued to fund it after the program achieved its goal of building a 500 kW MCFC power plant. In sum, there have been at least seven EU projects in support of the Italian and Spanish collaborative effort to develop and improve MCFC technology. Four of the projects were to focus on the development of power plants up to 500 kW, and three aimed to scale up the power plants to the MW size and demonstrate the operation of power plants in various applications prior to commercialization.

The seven projects that the EU funded benefited Ansaldo's MCFC technology, including:

1. *Molten Carbonate Fuel Cell 100 kW Cogeneration Power Plant*, which was sponsored under the program name Thermie 1.[xiii,82] The project ran from August 1992 to July 1996, and the aim was to design, manufacture, assemble, and test an innovative MCFC 100 kW power plant that would use a pressurized MCFC stack integrated with a set of ancillary devices (the Compact Unit or CU), coupled with a set of accessory BoP components.[82] The 100 kW stack used in the project was likely one of UTC's stacks; it had an external manifold, crossflow configuration and consisted of one hundred and fifty 0.74 cm^2 cells and operated under a pressure of 4 atm.[83] The stack was conditioned at San Agustin de Guadalix, Spain, and shipped back to Italy for operation on the CISE premises near Milan in 1996. Ansaldo was the primary contractor, with the participation of ENEA and IBERDROLA. The project cost was about €6 million (about $7 million at the average 1993 exchange rate of $1/€0.86) and the EU funded €2.4 million (about $2.8 million).[82]
2. *100 kW MCFC Integrated Stack Development and Testing*, or MISDAT, sponsored under FP3 (1990–1994).[84] The project, which ran from January 1993 to March 1998, aimed to develop a proof-of-concept 100 kW stack and test for use as the electrochemical module of an integrated system (the CU). The 100 kW stack was constructed at the end of 1997 in Italy at Ansaldo facilities and moved from Italy to Spain, where it was conditioned and tested during summer 1998 for about 1000 h. At the beginning of 1999, the conditioned stack was moved back to Milan, Italy, and placed in the CU Plant where experimental activities lasted for over 7 months. The project contractors included Ansaldo, Centro Informazioni Studi Ed Esperienze (CISE), and ENEA of Italy, IBERDROLA and Tecnologiay Gestion de la Innovacion (TGI) of Spain, and the British Atomic Energy Authority.
3. *Development of Industrially Relevant MCFC Stacks*, or DIREM, sponsored under FP4 (1994–1998).[85] The project, which ran from January 1996 to June 1999, was to develop MCFC stacks entirely designed and manufactured in Europe.[86] The contractors included Ansaldo, CNR-ITAE, and ENEA of Italy, and Babcock & Wilcox Espanola and IBERDROLA of Spain. Ansaldo, after manufacturing two intermediate stacks, constructed the final stack, adding significant improvements. Nicknamed STAD-3 (15 cells, active cell area 0.8 m^2), the stack had a new rectangular configuration (instead of square), simplification of the design of the wet seals, an improved manifold tightening system design, and used a thicker matrix for the porous components. The stack was assembled at Ansaldo in Genova and

[xiii] Thermie, which supported technology demonstration contracts, was run outside of, but in parallel to FP3 (1990–1994) and was incorporated in the FP, starting with FP4 (1994–1998).

shipped to San Agustin de Guadalix for conditioning and testing in the test facility. Tests started in spring 1999; the stack operated under operational loads for about 2000 h and then was completely shut down in August. A subsequent restart showed no performance decay during the following on-load operations, which lasted until November 1999. The project was extended for an additional 5 months. The 15-cell STAD-3 stack recorded a maximum output of 14.4 kW, which was more than sufficient to scale up to meet the requirements needed for a 500 kW "Series 500" power plant without stressing the power density. Stack performance corresponding to nominal operating conditions of the "Series 500" power plant would be 12.65 kW for a stack of 15 cells.[87] Scaling up the 15-cell STAD-3 stack by a factor of 10 would create a 150-cell, 126.5 kW stack, and four of these stacks would provide output of 506 kW dc, which would be adequate for the "Series 500" power plant.

4. *MCFC Series 500 Twinstack Powered First-of a-Kind* or MCTWINS, sponsored under FP5 (1998–2002).[86] The project, which ran from January 2000 to April 2004, aimed at the design, manufacture, and test of a first-of-a-kind, pre-commercial, 500 kW MCFC power plant, which Ansaldo nicknamed "Series 2 TW." The project cost was €7.7 million (about $7 million at the average 2000 exchange rate of $1/€1.1), with the EU funding €2.7 million (about $2.5 million). The heart of the Series 500, which was assembled based on the results of the previous DIREM project, were two 250 kW electrochemical modules, each comprising two 125 kW stacks (integrated with their relevant ancilliaries) linked together according to the original Twinstack configuration.[88] The plant was integrated in a skid-mounted fashion, with a pressurized vessel containing the two stacks, integrated reformers, and catalytic burners and the other BoP components installed on one skid along with the control systems and the power conditioning system. It operated at a pressure of 3.5 ata.[89] The stacks for the "First-of-a Kind" unit of the Series 2 TW were manufactured and assembled and installed at the power plant located in San Augustin de Guadalix during 2004.[90] The project started in early 2000 and officially concluded in April 2004.[91] The plant was later commissioned and started in December 2004 (with no stacks installed) and achieved complete functionality, with the stacks installed, in January 2005.

5. *Innovative Cost-Effective Hybrid System Based on Integration of an MCFC and a Gas Turbine for High-Efficiency Dispersed CHP Generation*, or MOCAMI, sponsored under FP5 (1998–2002).[92] The project was a 9-year long project, which was signed in 2001, started in April 2003, and ended in April 2012. Its aim was to develop and demonstrate a small-sized hybrid system comprising Ansaldo's 100 kW MCFC power plant and a micro gas turbine. The development of such a small hybrid system would allow lowering production costs through development of new and simpler processes with better quality control and a higher yield for different components, hopefully reducing the cost of the hybrid system by 50–70 percent. The project cost was €4.1 million ($4.65 million at the average 2003 exchange rate of $1/€0.88), about 40 percent or €1.7 million was to be funded by the EU. Contractors included Ansaldo (the primary contractor), CESI (Centro Elettrotecnico Sperimentale Italiano Giacinto Motta), and ENEL (ENEL Distribuzione) of Italy, Balcke-Durr of Germany, Bowman Power System of UK, and Tubitak-Mamara Research Center of Turkey.

6. *Molten Carbonate Fuel Cells for Waterborne Application*, or MC-WAP, sponsored under FP6 (2002–2006).[93] The project set out to

design, construct, install, and test a 500 kW auxiliary power unit (APU) prototype that would allow efficient, reliable, and safe use on board a ship. It was also to scale up to multi-MW size. Ansaldo's 500 kW MCFC power plant will be used as the base of the APU prototype. The project started in September 2005 and was to continue until August 2010. Project cost was projected to be €16.61 million (about \$20.7 million at the average 2005 exchange rate of \$1/€0.8), with the EU funding €9.9 million (about \$12.4 million). The primary contractor was CETENA SpA of Italy, which specializes in maritime research. Participants included 7 countries and 16 organizations.

- The MC-WAP project was a follow-on to a project sponsored by the Western European Armaments Organization (WEAO), the now-defunct intergovernmental defense and security organization. That project aimed to demonstrate the maritime application of a 100 kW power plant using diesel fuel. (It was demonstrated in Marmara, Turkey.)[94]

7. *Biogas Integrated Concept a European Program for Sustainability*, or BICEPS, under FP6 (2002–2006).[95] The project started in January 2007 and was planned to continue until September 2011. Its goal was to build two 1 MW Ansaldo MCFC power plants using biogas and landfill gas, respectively, generating electricity, heat, and refrigeration, thereby increasing the competitiveness of MCFC power plants for the European market. The project cost was €16.95 million (about \$23 million at the 2007 average exchange rate of \$1/€0.73), and the EU was to provide €6.98 million (about \$9.6 million). The contractors included at least 12 participants from six countries:

In addition to these seven projects, the EU funded at least five more MCFC projects that included Italian participation, but addressed European-wide R&D needs (Table 5.7). These projects focused on broader and more generic MCFC research issues that would help Europe widen its scientific and technological base and strengthen its international competiveness. Three of the five projects included Ansaldo as the project coordinator. The other two projects included an Italian or a Dutch research institution as the coordinator.

There also have been a number of projects that were not sponsored by EU but were funded at the national or regional level.

- The Italian Department of Education, Universities, and Research sponsored a project to demonstrate in Trisaia the combined operation of a biomass gasifier and a 100 kW power plant.[94] It also sponsored a project in Milan to operate a 500 kW power plant fueled by hydrogen and demonstrate the separation and management of carbon dioxide.
- The Department of Environment funded a demonstration, nicknamed "Tecnodemo," in Bosco Marengo of a 100 kW power plant that operated with natural gas.
- The local government of the Campania Region, Italy, sponsored a project demonstration in Giugliano, Naples, of a 100 kW power plant that ran on landfill gas.

5.4.2.2. *Ansaldo's MCFC Commercialization Phase Delayed*

Ansaldo's MCFC development strategy has been steadfast and unwavering for more than 20 years. It has invested as much as €100 million (about \$130 million) over the years, according to a company official.[96] It methodically moved from laboratory-scale experiments, through a number of stacks of different power levels, to the development of a 100 kW proof-of-concept power plant in 1999 with technology entirely developed in Europe.

TABLE 5.7 Other Projects Sponsored Under the Framework Program that Addressed European-Wide MCFC Research Needs*

Acronym, Project Name	Program	Project cost, EU funding	Project objective	Project period	Participants*
Experimental Study, Characterization of Catalysts and Modeling of MCFC with or Without Internal Reforming	FP1 (1984–1986)		To study the performance and characteristics of MCFC cells run with simulated fuel or with internally reformed natural gas	1986–1988	Centre National de la Recherche Scientifique (CNRS) (FR), National Research Council of Italy (CNR, the primary contractor) (IT), Polytechnic of Milan (IT)
A Collaborative Project to Build a 1 kW Internal Reforming MCFC Stack	Joule 1, FP2 (1986–1990)		To develop a 1 kW internal reforming fuel cell unit	1990–1993	CNR (IT), British Gas (UK), Energy Research Centre of the Netherlands (ECN, the primary contractor) (NL)
Investigation on MCFC Material	Joule 1, FP2 (1986–1990)		To develop materials for electrodes, electrolytes, and bipolar plates that would corrode and operate for 20,000–40,000 hours	1990–1993	Ansaldo (the primary contractor) (IT), ECN (NL), Center for Solar Energy and Hydrogen Research Baden-Wurtemberg (DE), Centro Informazioni Studi ed Esperienze (CISE) (IT), Technische Universiteit Delft (NL)

IMPROMET, Improved Technologies for Metallic Components of MCFCs	FP5 (1998–2002)	€147,000 (about $156,000), €147,000	To analyze the corrosion of MCFC anodic current collector and identify solutions and effective production technologies	2000–2004	Ansaldo (the primary contractor) (IT)
IRMATECH, Integrated Researches on Materials, Technologies, and Processes to Enhance MCFC in a Sustainable Development	FP5 (1998–2002)	€4.4 million ($4.68 million), €2.3 million ($2.4 million)	To reduce the cost of MCFC to €1,000/kW in the next 10 years, increase the lifetime to 4,000 hours, and minimize the environmental impact	2003–2005	Ansaldo (the primary contractor) (IT), A.P.I. Com S.R.L (IT), Balcke-Durr (DE), Centre National de la Recherche Scientifique (FR), Centro Elettrotecnico Sperimentale Italiano Giacinto Motta (CESI) (IT), Consejo Superior de Investigaciones Cientificas (ES), Ecole Nationale Suprieure de Chimie de Paris (FR), Enitecnologie (IT), Fraunhofer-Gesellschaft zur Foerderung der Angewandten Forshung (DE), RM Rich Mueller A/S of Denmark, Royal Institute of Technology (SE), Technip KTI (IT), University of Genova (IT)

** DE is Germany; ES is Spain; FR is France; IT is Italy; NL is the Netherlands; SE is Sweden; UK is the United Kingdom.*

With this achievement, Ansaldo likely felt confident about its technical capability and in the early 2000s drew up a road map for commercialization, with full commercialization planned for 2012. Ansaldo took several preparatory steps toward this goal. In 2003 and 2004, Ansaldo enlisted two major equity investors—EnerTAD and FINCANTIERI—to raise capital to fund major expansions. In June 2004, the company inaugurated a fuel cell manufacturing factory with an annual capacity of 3 MW in Terni, with plans to expand it to 15 MW in 2006.[97] It was a major event for Ansaldo; the Italian Minister for the Environment cut the ribbon of the new factory at the inauguration ceremony.[98] Also in 2004, Ansaldo completed the conditioning and testing of the first four full-size stacks (125 kW each), which would serve as the core of the "Series 2 TW" 500 kW power plant, a key milestone in the MCFC development under MOLCAR (Fig. 5.6).

Between 2007 and 2008, Ansaldo signed agreements that would give them major opportunities for commercialization of MCFCs. In October 2007, Ansaldo and L-3 Communications Combat Propulsion Systems (L-3 CPS) of the United States agreed to cooperate on marketing activities in the United States and on-product development in the military sector.[99] Ansaldo and its new partner were to develop a new, portable version of the Ansaldo Fuel Cell system that could supply large amounts of energy for forward-operating military bases and for emergency power.[100] In February 2008, Ansaldo and ENEL, the major Italian electric utility, agreed to build an experimental 0.5 MW plant that would supply electricity as well as heating and cooling energy by mid-2009.[101] This project represented an important product development opportunity, which could give Ansaldoa a "ready-to-market" product in the near future.[101]

Despite all the plans, however, Ansaldo's MCFC activities do not appear to have progressed very far toward its commercialization goal. As of July 2010, a 100 kW Tecnodemo pilot plant that was started in 2005 in Bosco Marengo, had accrued more than 35,000 h for balance of plant (BoP) operations and more than 12,000 h for single-stack operations.[102a,b] This is a high level of reliability, but none of the other demonstration plants had operated for much longer than 5000 h. According to a company official, Ansaldo had "no proof of evidence" that any power plant had lasted beyond that point.[96] Furthermore, Ansaldo apparently had hit a wall in terms of what it could do to reduce cost. According to an Ansaldo official, cost reduction of cell stack depended upon reducing the costs of raw materials and efficient

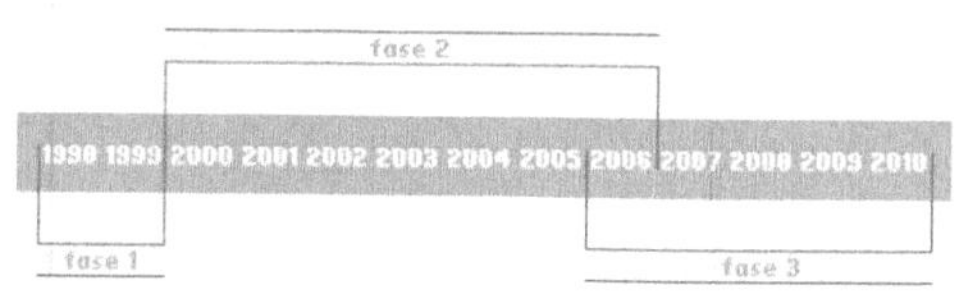

Phase 1
- Developed 100 kW "proof of concept" plant
- "Series 2TW" system defined

Phase 2
- "Series 2 TW" and "Serie 100" demo units
- New factory, refine component manufacturing processes
- Ongoing development of stacks, process and balance of plant components

Phase 3
- Move to commercialization
- Upgrade factory

FIGURE 5.6 Ansaldo's development and commercialization timeline. *Source: "Ansaldo FuelCell," Developement, http://www.ansaldofuelcells.com/cn/developement.htm, accessed July 2009.* (For the color version of this figure, the reader is referred to the online version of this book.)

manufacturing processes; reductions in BoP costs depended upon plant simplification and supplier competition and selection.[103] The official suggested that the fuel cell developers generally had borne the bigger effort in system optimization and cost reduction but did not always deliver the needed results, especially without support from the BoP developers and suppliers. With these thorny issues yet to be resolved, it is likely that Ansaldo's commercialization phase will continue to be delayed (Table 5.8).

5.4.3. Germany Starts MCFC Development in 1988—MBB (Currently CFC Solutions) Takes the Lead

Germany started MCFC development in 1988 with the goal of commercializing MCFC cogeneration power plants in the European market in the early 2000s. Noteworthy was that virtually all the German effort on MCFC development was carried out by the private sector. Messerschmitt Bolkow Blohm (MBB) of Germany,[xiv] an aerospace company and a Daimler-Benz subsidiary, took the leading role and decided that, like the Netherlands and Italy, Germany too would build on "ready-made" technology acquired from abroad rather than build it from scratch. Indeed, MBB's work is essentially based on a license and technology exchange agreement with FCE of Danbury, Connecticut.[105] In 1989, MBB, as predecessor of CFC Solutions, signed a technology transfer agreement with FCE of Danbury, Connecticut, augmented by becoming a shareholder and a strategic partner of FCE.[105]

In 1990, CFC Solutions launched the largest European program for the development and commercialization of the FCE MCFC technology, called the European Direct Fuel Cell Consortium (ARGE DFC) program.[106] The ARGE DFC Consortium *consisted of five major companies in Germany and Denmark:* CFC Solutions (then MBB); Haldor Topsoe, a Danish plant engineering company and catalyst expert; Elkraft (now DONG Energy), a Danish electric utility company; Ruhrgas, the largest German natural gas company, and RWE, the largest German electric utility. CFC Solutions served as the consortium leader and chose partners who could collaborate on the R&D effort, provide financial resources for the consortium activities, and serve as potential users of the end products.

The consortium expected that the overall program of approximately $100 million would be spent within 10 years from 1990 to 2000.[104] The consortium obviously hoped to receive funding support from the EU and their counties, but it was likely to be prepared to make its own major investment when necessary.

The consortium launched a three-phase program for the commercialization of the MCFC technology in Europe:

- The first phase of the program in the years 1990 to mid-1994 focused on basic technology R&D and resolved the fundamental materials, corrosion, and longevity problems associated with MCFC technology. In this phase, essential breakthroughs in the development of corrosion resistant long-life cell components were achieved.[107]

[xiv] Messerschmitt Bolkow Blohm GmbH (MBB), a German aerospace company, initiated the development of MCFC technology. MBB was subsequently acquired by Daimler Group, which later acquired Maybach Motorenbau, a diesel engine manufacturer, and merged its aircraft and diesel engine production under the name of Motoren und Turbinen Union (MTU) Friedrichshafen. MTU Friedrichshafen and RWE Fuel Cells formed a joint venture, MTU CFC Solutions, to join forces on MCFC development. Subsequently, MTU acquired a minority interest in RWE Fuel Cells and became the sole partner of CFC solutions. MTU CFC Solutions became CFC Solutions. The name "CFC Solutions" will be used hereafter whenever possible for simplicity.

TABLE 5.8 Demonstration Program at Ansaldo Fuel Cells*

Project	Class, Fuel	Output size	Demonstration site	Funding source (program)*	Project period	Partners	Objectives
First of a Kind	Series 2 TW, natural gas	500 kW	Guadalix, Madrid, Spain	EU FP5 (1998–2002), "MCFC Series 500 First of a Kind"	2000–2005	PEP (ES), ENEA (I), AMG (I)	Twinstack® and modular integrated reformer demonstration
Hybrid Cycle	Series 100, natural gas	125 kW	Milan, Italy	EU FP5 (1998–2002), "MOCAMI"	2002–2004	CESI (I), Balcke-Durr (D), Bowman Power System (UK), Tubitak, ENEL	Integration with microturbine
Tecnodemo	Series 100, natural gas	125 kW	Bosco Marengo, Italy	Italian Department of Environment		FN (I)	Various test with microturbine
Biomass Application	Series 100, biomass gasification	125 kW	Giugliano, Naples, Italy	Campania Region, Italy			Landfill gas application
Naval Application	Series 2 TW, diesel	500 kW	Marmara, Turkey	WEAO		MRC (TK)	Diesel application reformer, compactness
Biomass Application	Series 100, biomass gasification	125 kW	Trisaia, Italy	EU FP5 (1998–2002), "Clean Energy from Biomass"	2000 for 36 months	SUT (D), ENEA, various universities (A, F, I, UK)	Coupling of biomass gasification and fuel cell systems
Naval APU Application	Series 2 TW, diesel	500 kW	On-board operation	EU FP6 (2002–2006), "WC-WAP"	September 2005 for 60 months	CET (I), Fincantieri (I)	Multi-MW APU onboard large ships fueled by diesel oil

H_2/CO_2	Series 2 TW, hydrogen	500 kW	Milano Bicocca	Italian Ministry of Education, University and Research			CO_2 separation and management
BICEPS	MW-class, sewage gas	1 MW	Terni, Italy	EU FP6 "BICEPS"	2007–2011	ASM TERNI S.P.A (I), Balcke-Durr (D), CESPA (ES), E.On (D), Fraunhofer Institute (D), Turbec (SE), and several universities (D, RO, SE, SK)	Cost-effective purification unit for MCFC, suitable for sewage gas
	MW-class, landfill gas	1 MW	Murcia, Spain				Cost-effective purification unit for MCFC suitable for landfill gas
Bosco Marengo	FA200, natural gas	500 kW	Bosco Marengo				
PRIOLO	Series 100, syngas	125 kW					
ENEL 500	FA200, natural gas	500 kW	Livorno, Italy		2009		

* *The Framework Program (FP) period can be different from individual project periods. Projects can start within the FP period, but might end after the FP period ends.*

The table was created based on: Angelo Moreno and Stephen McPhail (ENEA—Hydrogen and Fuel Cell Project, Rome, Italy) and Roberto Bove (Joint Research Centre—Institute for Energy, Petten, The Netherlands), "International Status of Molten Carbonate. Fuel Cell (MCFC) Technology," http://www.energyagency.at/fileadmin/aea/pdf/energietechnologien/molten-carbonate.pdf, Austrian Energy Agency, January 2008.

- The second phase—from 1994 to 1997—focused on the development of a system demonstrator based on a highly innovative and integrated plant concept and incorporated cell improvements obtained in the first program phase.[107] This phase also involved the development of new production or manufacturing process in order to reduce costs.[104]
- The third phase—from 1998 to 2000—focused on the development of semicommercial field test units and product improvement.[107]

While the German private sector was energetic, the German government played a rather reserved role in MCFC development. Between 1988 and 1995, the German government allocated an annual budget of around €8 million ($9.6 million at the 1990 exchange rate of $1/€0.83) and supported the development and demonstration of three types of fuel cells: SOFCs, MCFCs, and PEMFCs.[108,123] Of the three, Germany put the least emphasis on MCFCs. The government allocated €4 million ($4.8 million) out of the €8 million for PEMFC development. It allocated the remaining €4 million ($4.8 million) to develop SOFC and MCFC technologies, of which €3 million ($3.6 million) went to SOFCs.

- In 1998, Germany started a 4-year program specifically to promote PEMFC development, indicating that the German government preferred PEMFC the most, SOFC second, and MCFC the least in the 1990s.

5.4.3.1. EU Framework Program Begins Supporting German MCFC Effort in 1990

The EU was eager to support the German ARGE DFC program and, starting in 1990, funded at least six projects to promote the effort by the ARGE DFC consortium. The consortium's progress in MCFC development was remarkably fast. They quickly built 20 kW and 100 kW stacks, designed a novel, highly integrated, and markedly simplified MCFC plant (called the HotModule) and ultimately demonstrated the operation of a 250 kW HotModule for the first time worldwide. The six projects included the following:

1. *Development and Adaption of the Molten Carbonate Fuel Cell with Internal Reforming for the Use of CO/CO_2-Rich Gases, for example, Coal Gas,* was sponsored under FP2 (1986–1990) and ran from May 1990 to December 1993.[109] The project developed the basic design and determined operating parameters for coal gas MCFC power plants, and built test facilities for a coal gas MCFC stack of up to 20 kW. Project activities included tests of an MCFC stack for CO/ CO_2-rich gases, such as coal gas from a coal gasifier, under operational conditions of power production and cogeneration, including electrical load cycles, thermal cycles, variation of fuel gas composition, and zero load periods with external heating. The project coordinator was CFC Solutions with project participants including Haldor Topsoe of Denmark.
   - The project enabled the development of a 20 kW MCFC stack in Germany that integrated FCE's Direct Fuel Cell stack and the BoP developed in Europe and demonstrated the viability of the Direct Fuel Cell system in Europe
2. *Design and Engineering of a 100 kW MCFC Cogeneration Plant (Natural Gas) and a 100 kW MCFC Power Plant (Coal Gas)* was funded under Thermie 1 (1990–1994).[110] The project ran from July 1992 to December 1994. Its aim was to design and engineer two 100 kW MCFC cogeneration plants—one designed to be fueled by natural gas and the other fed by coal gas—and demonstrate the feasibility of the two MCFC systems. The project cost was expected to be €2.6 million ($3 million at the 1993 exchange rate of $1/€0.86), €1.05 million ($1.22 million) of which was funded under Thermie 1 (1990–1994). The project coordinator was CFC Solutions.

3. *MCFC Power Plants—Demonstration of a HotModule Qualification Model: Construction, Installation, and Start of a HotModule* was also sponsored under Thermie 1 (1990–1994) and possibly was the most important project in the development of the German MCFC power plant. The project officially ran from June 1994 to August 1997.[111] The project was to install and demonstrate a full-sized MCFC HotModule qualification model (without a real stack), with the aim of demonstrating the feasibility and reliable operation of all components and subsystems necessary for operating a complete MCFC plant building unit. Next, the project would construct and install a MCFC HotModule system verifier, full size, with a real stack of almost 300 cells in the range of 270 kW.[112] The system verifier would be the first fully functional HotModule system to be fed by natural gas, producing both electricity and heat. The project cost was expected to be €11.8 million ($13.7 million at the1993 exchange rate of $1/€0.86) with EU funding €1.4 million ($1.62 million). The project coordinator was CFC Solutions (then MTU Friedrichshafen).[111]
   - The innovation of the HotModule design was seen as a radical simplification of the system, which integrated all high temperature subsystems of an internal reforming MCFC system into a common, thermally insulated vessel.[104] The HotModule design simplified and reduced the size of the system and raised thermal efficiency.[112,113]
   - It also used an MCFC stack that had been rotated 90° sitting horizontally, reducing the height and thereby enabling the entire plant to fit on the back of a lorry for easy for transport. Its single manifold was located at the bottom of the horizontal stack, sealing the manifold by the gravity.
   - The hot air exhaust was circulated to the external heat recovery steam generator (HSRG). The waste heat from the HSRG was approximately 800°F and could be used for cogeneration. The HotModule design eliminated the need for expensive high-temperature piping. All these features contributed to a lower capital cost for the system.

In 1995 and 1996, various tests on the full-scale qualification model of the HotModule confirmed the mechanical, pneumatic, and thermal viability of the integrated design.[112] In 1996, the fabrication of the MCFC HotModule system verifier started.[114] The 270 kW MCFC system verifier and stack, consisted of two hundred and ninety-two 0.6 × 1.2 m cells, was installed at Ruhrgas facilities in Dorsten, Germany, and started operation on August 5, 1997.[115] The system was connected to the grid on 16 August, and the electricity output reached 155 kW. By the end of August, however, four unscheduled shutdowns occurred; one caused by an operator error and three others by peripheral component failures.[112] The burner was found defective, which was unable to raise the operating temperature sufficiently.[115] After these outages, a high temperature was registered in the fuel entry area on one end of the cell block.[112] Insufficient temperature uniformity also was detected in the gas mixing zone, which was part of the interior design of the HotModule. Local "cool" areas also occurred in the catalytic burner. The reasons for these effects could not be adequately identified without disassembly. Thus, the system was shut down in September 1997 and sent back to MTU facilities for analysis and subsequent improvements.

- The HotModule was opened and inspected.[112] The local cool areas in the catalytic burner were caused by insufficient mixing of fresh air with anode tail gas and recycled cathode air. This led to the redesign, manufacture, and test of the gas mixing chamber with improved flow homogeneity.

The hot spot at the fuel entry side of the cell block was caused by damaged electrolyte matrices in 30 of the 292 cells, which led to immediate combustion of fuel gas within the cell block and abnormally high temperatures. The reason for the damage to the cells was traced to a weakness in the peripheral system design combined with operator error. A combination of circumstances during the first test phase led to flushing the fuel input end of the stack with large quantities of cold nitrogen instead of preheated nitrogen. While corrections in the peripheral system were easily implemented, an attempt to repair the cell block was unsuccessful.

- In March 1998, the system was returned to Dorsten and started operation.[112] The hot spot at one end of the stack persisted during the second test phase. It prevented power from exceeding levels already achieved in the first test phase, which was 155 kW. In the second test phase, nonetheless, automatic operation of the plant was achieved, and full thermal load was applied to the catalytic burner and internal gas distribution system. After 1500 hours of operation, the system was intentionally shut down for subsequent post-test analysis.
- In mid-1999, the next test unit of similar design went into operation at the University of Bielefeld, Bielefeld, Germany.[112] This test unit was a 250 kW plant, which likely used an improved stack shipped from FCE. The plant was installed at the University of Bielefeld, Germany, and operated for 16,000 h until February 2002.[116]

4. The fourth project—*System Development and Optimization*— was sponsored under FP3 (1990–1994) and ran in parallel with the third project described above, from January 1994 to June 1997.[117] The project aimed at the development of a complete set of MCFC system design tools, including simulation models of MCFC cells, MCFC stacks, and MCFC plants. The simulation models, which would allow optimization of the systems, were to be validated against experimental results. The project was to concentrate on feasibility of small cogeneration plants of less than 1 MW fed by natural gas. Field tests of different fuel cell stacks in different types of power plants and serial production and introduction to the market of fuel cell systems were also studied. The project coordinator was CFC Solutions (then MTU), working with its Danish partner Haldor Topsoe.
5. The fifth project, *MCFC Balance of Plant Innovation—Design, Construction, and Demonstration of a DFC Module Pilot Plant for Cogeneration*, also ran in parallel with the third project described above. The project was funded under FP4 (1994–1998) and implemented in two consecutive segments; the first contract segment was signed in 1997 with a project period of 21 months from August 1997 to April 1999, and the second contract segment was signed in 1998 with a project period of 30 months from May 1999 to October 2001.[87,118] The project aimed to design an innovative BoP for a DFC HotModule Pilot Plant for Cogeneration. The project called for improvements in conceptual design, detailed design and engineering, manufacturing of the components of the BoP, and the construction and installation of the BoP in a DFC HotModule Pilot Plant for Cogeneration.[119] The HotModule would be tested and demonstrated in this project for the first time worldwide. EU funding for the first segment was €1.3 million (about $1.4 million at the 1999 exchange rate of $1/€0.94) out of the total project cost of €3.3 million ($3.5 million). Funding information is not available for the second segment. The project coordinator was CFC Solutions (then MTU). The projects participants included CFC Solutions, Ruhrgas, and Haldor Topsoe.

6. The sixth project— *Holistic Integration of MCFC Technology Toward a Most Effective Systems Compound Using Biogas as a Renewable Source of Energy, using the acronym EFFECTIVE*— was funded under FP5 (1998–2002). The project contract was signed in 1999, with the project expected to run for 48 months from July 2000 to June 2004.[119] The project was coordinated by Profactor of Austria, an applied research company, with at least 10 participating companies and institutions from Denmark, Germany, Slovak Republic, and Spain. EU funding was expected to be at about €2 million (about $1.8 million at the 2000 exchange rate of $1/€1.1) out of the total project cost of €3.5 million ($3.2 million). The project aimed at the development of a gas cleaning unit that would enable an MCFC system to use biogas; a novel and innovative approach that had never before been taken into consideration.
   - The project constructed two gas cleaning units that would reduce H_2S in biogas from 50 ppm to 10 ppm. Two test beds were each comprising a 500 W MCFC provided by CFC Solutions (then MTU CFC Solutions) and a laboratory-sized gas cleaning unit.[87] One test bed was a mobile system, which was transported to and tested in Germany, Austria, and Spain.[120] The second test was a stationary system, which was tested for a longer term at University of Nitra, Slovakia Republic. Tests with different gas qualities were conducted on both test beds.
   - The mobile test bed operated for 2200 h at Industrial Research Center, Seaborne, Owschlag, Germany, and over 1000 h at Waste Water Treatment Plant in Linz, Austria. The test bed was expected to operate at Waste Treatment Plant, Urbaser, Pinto, Spain, starting in January 2004.[120] The second test bed was installed at Agricultural Biogas plant at University of Nitra, Nitra, Slovakia. The test bed operated for 2400 h in its first cycle, and over 3000 hours in the second cycle, which was ongoing as of October 2003.

5.4.3.2. German Government's MCFC Demonstration Programs Bolsters HotModule Installations

In the early 2000s, the German government began to recognize that the HotModule technology was a significant accomplishment and that it was ready for commercialization. In 2001, the government launched a 3-year (2001–2003) fuel cell demonstration program, called "Investing into the Future," abbreviated as ZIP, and funded it at €55 million ($49 million at the 2001 exchange rate of $1/€1.12).[121] The ZIP program, which subsequently was extended to 2005, aimed to jump-start the commercialization of fuel cell products by promoting the operation of pre-commercial MCFC prototype products. The German Federal Ministry of Economics and Labor (BMWA) implemented the program, rather than the more R&D-oriented ministries of Economics and Technology (MBWi) or Education and Research (BMBF). With government encouragement, at least 28 HotModule power plants were installed in Germany and elsewhere in Europe.

- Under ZIP, the German government also promoted the demonstration of other precommercial fuel cell prototypes, including SOFCs and PEMFCs, and hydrogen-energy technology products. It, however, allocated the biggest share of the ZIP budget to MCFC demonstrations. It allocated 31 percent or €17 million ($15 million) for demonstrating 10 cogeneration power plants, of which seven were MCFC projects, and three were SOFC projects. ZIP provided about €13.79 to install and operate the seven HotModule power plants.[122]
- The ZIP program financed 50 percent of the overall costs of a project. The HotModule installation projects funded under ZIP

sometimes also received additional funding from federal states, city governments, and other sources, indicating the need for alleviating the investment costs.

In addition to the seven HotModule power plant sinstalled under ZIP, at least 21 other HotModule power plants have been installed in Germany and elsewhere in Europe, including France, Norway, and Spain, since 1999. The majority of them were funded by a combination of government entities, such as the German Federal Ministry of Economics and Technology (BMWi), German federal states, city governments, district heating systems, and municipal water treatment facilities. BMWi funded those HotModule power plants that involved experimental operations, such as plants that used biofuel, methanol, and sewage gas or trying out new stack types. Federal states and city governments also funded the demonstration of HotModule power plants within their own jurisdictions.

- Federal states that provided funding included Bavaria, Baden-Wurttemberg, Hesse, North Rhine-Westphalia, Rhineland-Palatinate, and Thuringia. City governments that provided funding included Ahlen, Berlin, and Stuttgart that funded the demonstration of HotModule power plants in their federal states and cities.
- There have been at least five HotModule installations in France, Norway, and Spain. In Spain, a HotModule was installed at a shipyard in 2002 and that stack was later replaced with a new one in 2007. Also in 2007, a Spanish natural gas supplier installed a HotModule power plant. In France, an apartment complex installed a HotModule power plant in 2006. Since May 2009, a group of German and Scandinavian companies have been experimenting with a 320 kW HotModule power plant in Norway, to power a supply ship.

5.4.3.3. CFC Solutions Shuts Down its MCFC Business in December 2010

Despite all the achievements and improvements of HotModule power plants, over the last 5 years, the trends in sales and installations have been downward. A total of about 30 HotModule power plants were sold and installed in Europe in the last decade, a little less than three systems per year (Fig. 5.7). The year 2007 saw the largest number of HotModule systems sold, with seven systems ordered, delivered, installed, or beginning operations (two HotModule systems replaced older units). In that year, MTU CFC Solutions also changed its name to CFC Solutions to create an exclusive market presence as an MCFC manufacturer. The company started to fully exploit technology to operate

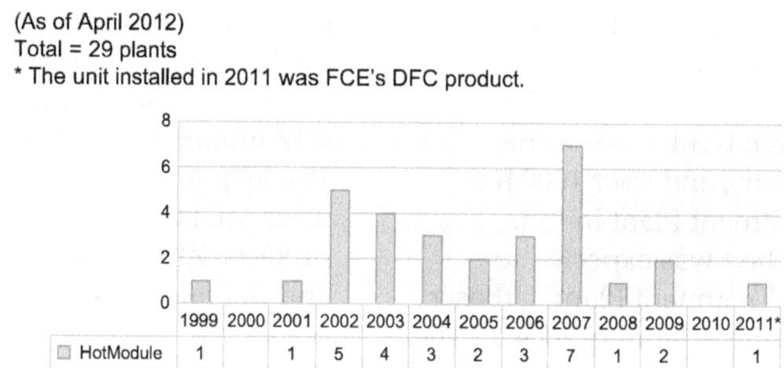

FIGURE 5.7 CFC solutions' HotModule installation in Europe*. (For the color version of this figure, the reader is referred to the online version of this book.)

HotModules with sewage gas, biogas, methanol, and biomethane, which is an innovative development in fuel cell technology to protect the environment. But these efforts did not bear fruit; there was only one sale in 2008 and two in 2009.

The small number of sales was unavoidably linked to the issue of market competitiveness. A number of sources have highlighted two basic reasons why HotModule power plants are struggling to be successful in the commercial marketplace in the near term.

- *High costs of the power plant.* While costs did come down dramatically over the years, they are still extremely high. In the early 2000s, costs ranged from €50,000/kW (about $46,300/kW at the 2000 exchange rate of $1/€1.08) to about €10,000/kW ($12,500 at the 2004 exchange rate of $1/€0.8) in 2004.[123] These costs included offsets from subsidies. In 2010, costs fell to close to half of that, but were still too high to be commercially competitive. In that year, a contract for a 345 kW HotModule power plant at Rhoen-Klinikum Giessen-Marburg University Hospital reportedly was signed for a cost of €2 million ($2.7 million at the May 2009 exchange rate of $1/€0.75). If accurate, that would put current costs for MCFC power plants at about €5800/kW ($7700/kW). Acquisition costs for the gas engine system, which is the conventional technology that competes with MCFCs, is less than €1000/kW. Although gas engines are considerably less efficient than HotModule systems, the overall high acquisition costs for these systems put it at a real disadvantage.
- *Insufficient lifetime.* The longest lifetime of a HotModule power plant with its original stack was 30,000 h, which was achieved in 2006 by a system installed at Otto-von-Guericke University Hospital, Magdeburg, Germany. While this is a notable accomplishment for a fuel cell, it still falls short of the commercially acceptable standard of 40,000 h. Since 2006, no news on advancements in the lifetime of HotModule power plants has been reported.

In December 2010, Tognum AG, the owner of CFC Solutions and a leading supplier of diesel engine and propulsion systems in Germany, announced that it had decided to discontinue its fuel cell activities for stationary power generation,[124] effectively shutting down HotModule marketing activities. After the announcement, Rolls-Royce conducted a survey on the acquisition of the fuel cell business in conjunction with its joint bid with Daimlerfor Tognum.[125] In June 2011, however, Rolls-Royce refrained from going ahead with the purchase. A search for an Asian partner also was unsuccessful. Tognum ultimately wound down its fuel cell development site in Ottobrunn, which had about 70 employees, at the end of 2011.

Nonetheless, the market for MCFCs has not disappeared. In February 2011, FuelCell Energy in the United States announced the sale of 300 kW DFC300 to be installed at the Quadrant 3 Re-development Project on Regent Street in central London, England.[126] The Quadrant 3 project is a 25,000 ft^2 mixed use retain/office/residential redevelopment project being developed by the Crown Estate. The plant, which will be installed by Logan Energy Limited, was planned to be operational by late 2011 (Table 5.9).

5.5. MCFC DEVELOPMENT IN SOUTH KOREA

5.5.1. South Korea Begins MCFC Development in 1993

South Korea's late start in MCFC development prompted the country to make major investments to make up for the delay. Its R&D efforts have been spearheaded by the Ministry of Knowledge Economy (MKE) (formerly the Ministry of Commerce, Industry, and Energy

TABLE 5.9 CFC Solutions' HotModule Installation in Europe (as of April 2012)

Plant size, fuel	Project Name, operators, partners	Installation site	Operation period	Operating hours	Funding sources
250 kW, natural gas	Field trial, CFC Solutions, Elkraft, Haldor Topsoe, Ruhrgas, RWE, University of Bielefeld, Stadtwerke Bielefeld, BEB Erdgas-Erdol	University of Bielefeld Hospital, Bielefeld, Germany	September 1999–February 2002	16,000	
250 kW, natural gas	Bavarian Hydrogen Initiative, the State of Bavaria, Ferngas Nordbayern, E.ON/Rhoen Klinikum AG	Rhoen Klinikum medical clinic, Bad Neustadt, Germany	April 2001–June 2004	21,585	Provided €1.74 million under the German Federal Ministry of Economics and Labor's (BMWA) the "Investing into the Future" ZIP Program. Also likely funded by CFC Solutions, the State of Bavaria, and Ferngas Nordbayern
250 kW, natural gas	RWE, CFC Solutions	RWE Fuel Cell Pavilion, Meteorit Park, Essen, Germany	April 2002–September 2005	26,523	Provided €2 million under the German Federal Ministry of Economics and Labor's (BMWA) ZIP program
250 kW, natural gas	DeTeImmobilien (a subsidiary of Deutsche Telekom), Deutsche Telekom, CFC Solutions	DeTeImmobilien Headquarters, Maasstrasse, Munich, Germany	October 2002–January 2006	25,477	Provided €2.56 million under BMWA's ZIP program
300 kW, natural gas			Old stack replaced with a new Eurocell stack in January 2007		Funded by the German Federal Ministry of Economics and Technology
250 kW, natural gas	Navantia (formerly IZAR; a supplier of ships and a commercial partner of CFC Solutions in Spain since 2004)	Navantia shipyard facility, Cartagena, Spain	November 2002–November 2005	24,211	A collaboration of research and development on fuel cell technology within the framework of national and European sponsorship contracts
	Polytecnic University of Cartagena	Polytecnic University of Cartagena, Spain	Old stack replaced with a new stack; Polytecnic university took over the fuel cell installation in January 2007		
250 kW, natural gas	EnBW, Michelin, Daimler, Stadtwerke Karlsruhe	Michelin tire plant, Karlsruhe, Germany	November 2002–November 2005	23,849	Funded about €1.74 million under BMWA's ZIP program. Daimler and Stadtwerke Karlsruhe likely provided some funding

250 kW, natural gas	IPF KG, CFC Solutions, Otto-von-Guericke Clinic	Otto-von-Guericke University Hospital, Magdeburg, Germany	December 2002–May 2006	30,018	Provided €1.76 million under BMWA's ZIP program
250 kW	RWE, MTA CFC Solutions	RWE Fuel Cell Pavilion, Meteorit Park	Started up in July 2003		
250 kW	RWE, VSE, CFC Solutions	A handicapped workshop, Ensdorf, Germany	2003		
250 kW, natural gas	RWE, Pfalzwerke	Gruenstadt Hospital, Gruenstadt, Germany	July 2003–July 2006	26,415	Provided €2 million Funded under BMWA's ZIP program
245 kW			The letter of intent for a follow-up test project signed in June 2007; operation planned to be started in fall 2007		Likely funded by the German Federal State Rhineland-Palatinate
250 kW, natural gas	E.ON, Rhoen Klinikum AG, CFC Solutions	Rhoen Klinikum medical clinic, Bad Berka, Germany	Operation started in October 2003; stack replacement in 2007	35,920 (as of October 21, 2008)	Likely financed by the State of Thuringia
250 kW, natural gas/ methanol	Vattenfall, Bewag, CFC Solutions	Bewag Fuel Cell Innovation Park, Berlin, Germany	September 2004–March 2007	21,900	Provided €1.99 million under BMBA's ZIP program. Also, likely funded by the City of Berlin
2 × 250 kW, natural gas	RWE, CFC Solutions, Fernwaermeversorgung Niederrhein, Stadtwerke Dinslaken	A local heating system in Krefeld-Fischeln, Germany	December 2004–June 2007	20,500 (as of May 2007)	Perhaps funded by the North Rhine-Westphalia State Economics Ministry
250 kW natural gas	RWE, CFC Solutions, Festo	Festo facility, St. Ingbert, Germany	June 2005	15,800 (as of May 2007)	
363 kW, sewage gas		Stadtwerke Ahlen municipal wastewater treatment facility, RWE, City Council Ahlen, Stadtwerke Ahlen, Ahlen, Germany	October 2005	7700 (as of May 2007)	Funded by the City of Ahlen, RWE Fuel Cells, and the North Rhine-Westphalian Ministry for the Environment commissioned the project. Daimler Group also provided financial support

(Continued)

TABLE 5.9 CFC Solutions' HotModule Installation in Europe (as of April 2012) (*Cont'd*)

Plant size, fuel	Project Name, operators, partners	Installation site	Operation period	Operating hours	Funding sources
300 kW, natural gas	Vattenfall, ARGE HafenCity	HafenCity (An inner-city development project at the docklands), Hamburg, Germany	February 2006	5200 (as of May 2007)	
250 kW, biogas	RWE, EnBW, the District of Boeblingen (Baden-Wurttemberg), Vergarungsanlagen Leonberg	The Leonberg fermentation plant, Leonberg, Germany	Trial operation started in August 2006; regular operation started in October 2006	5640 (as of October 21, 2008)	The total cost was about €3 million. As it is regarded as a demonstration project, more than 80 percent of the total cost was provided by the German Federal Government, the State of Baden-Wurttemberg, EnBW Energie Baden-Wurttemberg, RWE Fuel Cells GmbH, and Daimler AG
250 kW, natural gas	Dalkia (the energy service provider and a subsidiary of Veolia Environment and French energy supplier EDF), OPAC (Office Public d'Aménagement et de Construction)	Apartments owned by the French Housing Association OPAC, Paris, France	Delivered in autumn 2006; operation scheduled from November till March each year	550 (as of March 2007)	
250 kW, biomethane	Deutsche Telekom, PASM Power and Air Condition Solution Management GmbH & Co. KG (a subsidiary of Deutsche Telekom), Voigt & Haeffner	T-Systems Computer Center, Munich, Germany	Order received in January 2007; Installed in May 2007	5420 (as of October 21, 2008)	Funded by the German Federal Ministry of Economics and Technology and also likely funded by the Bavarian Ministry of State for People and the Environment
300 kW, sewage gas	EnEB, SES Stuttgart, the state of Baden-Wurttemberg	A water treatment plant in Moehringen near Stuttgart, Baden-Wurttemberg, Germany	Delivered in July 2007; operation started in November 2007		Funded by the German Federal Ministry of Economics and Technology and also likely funded by the City of Stuttgart, and the state of Baden-Wurttemberg

250 kW, sewage gas		A sewage plant Moosburger Klaeranlagen, Moosburg, Germany	Installed in June 2007; operation started in December 2007	6500 (as of October 21, 2008)	Funded by the German Ministry for Economics and Technology
400 kW, LNG	Det Norske Veritas (DNV), CFC Solutions, Eidesvik Offshore, Vik-Sandvik, Waertsilae Automation Norway	Supply ship Viking Energy, Norway	Order received in November 2007; an offshore test started on Viking Lady in May 2009		The DNV-led industry project FellowSHIP likely is funded by the group of Scandinavian and German companies
250 kW, natural gas	Gas Natural, Navantia (a supplier of ships and a commercial partner of CFC Solutions in Spain since 2004)	Gas Natural, Barcelona, Spain	December 2007		
250 kW, natural gas	MTU Friedrichshafen, TWF	MTU Friedrichshafen, Friedrichshafen, Germany	Operation started in May 2008		
250 kW, biogas	Erdinger Weissbrau brewery	Erdinger Weissbrau brewery, Erding, Germany	Operation started in June 2009		Likely funded by the German Federal Minister of Economics and Technology
345 kW, LNG	Rhoen-Klinikum AG	Rhoen-Klinikum, Giessen-Marburg University Hospital, Marburg, Germany	Order received in May 2009; planned to be operational in 2010		The contract is worth about €2 million Likely funded by the State of Hesse
300 kW DFC	Logan Enegy Limited, the Crown Estate	Quadrant 3 Re-Development Project, Regent Street, London, England	Sale announced in February 2011; planned to be operational by late 2011		

The table was created based on various sources, including press releases by CFC Solutions, RWE Fuel Cells, and MTU Online Energy; Michael Bode, "CUSTOM-MADE SOLUTIONS FOR HIGH-POTENTIAL FUEL CELL APPLICATIONS BASED ON MCFC," CREA Fair, Verona, October 21, 2008; "Worldwide Stationary Fuel Cell Installation Database," the Online Fuel Cell Online Resource, Fuel Cells 2000, http://www.fuelcells.org/db/, accessed March 2009; Angelo Moreno, "Hydrogen and fuel cells in CCS power plant," International Scjpp; pf Gep[jusocs—30° Course, CO_2 Capture & Storage, Erice, November 1–7, 2007; Marc Bednarz, "CUSTOM-MADE SOLUTIONS FOR HIGH POTENTIAL FUEL CELL APPLICATIONS BASED ON MCFC," 2008 Fuel Cell Seminar and Exposition, Phoenix, Arizona, October 27–30, 2008; "BRENNSTOFFZELLENSYSTEME FÜR STATIONÄRE ANWENDUNGEN Workshop," Wirtschaftskammer Österreich, 30 Jänner 2001; and Torsten Bardewyck, "Biogas—Fuel Cells: Biogas as fuel for MTU's HotModule," H_2 Expo, Hamburg, October 9–11, 2003.

(MOCIE)), which sponsored a number of MCFC development projects starting in 1993. The first development effort was a 2 kW externally manifolded, internally reforming MCFC stack developed in 1996. The stack consisted of 0.1 m^2 cells. This work led to the development of a 5 kW stack with 0.6 m^2 cells in 1998, followed by a 25 kW stack of 0.6 m^2 cells in 2000.[24] By 2004, a 100 kW demonstration plant consisted of two 50 kW stacks with 0.6 m^2 cells had been developed.[127] Participants in the MCFC development projects included Korea's major utilities and industries and research institutions, such as Korean Electric Power Company (KEPCO), the Korean Electric Power Research Institute (KEPRI, a former research unit of KEPCO), POSCO, and the Research Institute of Industrial Science and Technology (RIST).

- Since the 1990s, the Korean government has viewed fuel cell technology as one of the most important and promising technologies and provided its full support for R&D.[104] The total R&D budget for fuel cell technology invested during the period of 1990–2003 has been about US$70 million, spread out over 42 projects. R&D was conducted on MCFC, PAFC, SOFC, PEMFC, but MCFCs initially had priority and accounted for 50 percent of the funding.
- Emerging technologies, such as PEMFC, SOFC, and DMFC subsequently gained more government support.[104]

In 2004, the Korean government established the "Program by the National RD&D Organization for Hydrogen and Fuel Cell." The large-scale, ambitious program was funded by MKE in 2004 at a level of US$400 million over 8 years.[104] The program aimed to expedite the development and commercialization of hydrogen and fuel cell technology. The program launched a project to construct a 250 kW commercial prototype MCFC power plant by 2009.[127] One report indicated the budget for the 250 kW development was US$ 43.7 million.[128] The 250 kW power plant was to have an internally manifolded, externally reforming design with two 125 kW stacks, each consisting of one hundred and eighty 1 m^2 cells.[127] The contractors included KEPRI who would do the plant system integration, the Korea Institute of Science & Technology (KIST), POSCO, and RIST (responsible for stack development), Doosan, Hyosung, KEPCO, KEPRI, and Samsung (development of BoP components). In 2005, pilot-scale facilities were constructed for the fabrication of separators and cell components with a capacity of 10 MW/year. In 2007, a 75 kW stack with 0.75 m^2 cells was installed at the Boryeong test stand and operated for evaluation. The 250 kW prototype perhaps is still in a developmental stage; the prototype MCFC plant is expected to be demonstrated in 2012.[129]

MKE also launched another MCFC development project in October 2007.[130] The 3-year project was to develop a 300 kW power plant by 2010 and commercialize it by 2012.[13] The plant would be externally manifolded and internally reforming.[131] The total cost for the project was US$55.6 million. The primary contractor is Doosan Heavy Industries & Construction (DHI), a world-class steam power plant and desalination plant manufacturing company, which collaborates with KIER and Korea Midland Power. To finance the project, MKE and DHI provided $17.45 million each, and private sector investors provided another $20.7 million.[130] DHI planned to build research and production facilities necessary for cell component fabrication and stack manufacturing on the company's laboratory site in Daejeon by the first half of 2008. Little information is available regarding the project status.

5.5.2. South Korea More Interested in Rapid Acquisition of Foreign MCFC Technology for Domestic Economy and Export Growth

The Korean government's plans for MCFCs go well beyond indigenous development of

250 kW or 300 kW MCFC power plants. Under the new Lee Administration, inaugurated in December 2007, the country started to implement an aggressive green energy policy, which included wide-scale installation of MCFC power plants, rapid increase in the proportion of MCFC power generation on the national grid, and the promotion of MCFC exports.

- The goals were to increase the share of new and renewable energy (NRE) supply relative to total energy consumption, from 2.4 percent in 2008 to 11 percent by 2030.[132] It would also replace the fee-in-tariffs policy (FIT) with a Renewable Portfolio Standard (RPS) in 2012 that requires major utilities to generate power from more renewable sources. The utilities would generate 4 percent of energy from renewable sources by 2015 and 10 percent by 2022. In terms of MCFC technology, South Korea was to acquire MW-class MCFC power plants by 2010 and install these plants in large numbers in the country by 2012.[128] The 2012 target is to install 300 distributed power generation system units from 250 kW to 1 MW.[133] South Korea also aims to gain 10 percent of the global market in renewable technology by 2020,[134] be a global supplier for 20 percent of fuel cell demand by 2025, and create 560,000 jobs.[135]
- To ensure policy success, major subsidies are being provided. All renewable energy technologies will receive a 5-percent tax credit, and in 2009, import duties were halved on all components and equipment used for renewable energy power plants. The government also is providing subsidies to local governments for up to 60 percent of the installation costs for renewable facilities, as well as low-interest loans.

In 2007, POSCO Energy reached an agreement to work with FCE, an ideal partner in the United States, to enhance South Korea's MCFC business along with the government's green policy. Today, POSCO has acquired 22 percent of FCE's common shares, along with FCE's MCFC technological capabilities, and the rights to manufacture virtually all MCFC components in South Korea:

- After POSCO Energy inked a marketing and distribution partnership with FCE, it began making major purchases of FCE's DFC power plants. It purchased 12.6 MW in 2007, 25.6 MW in 2008, 30.8 MW in 2009, and 70 MW in May 2011, for a total of 139 MW as of April 2012 (Fig. 5.8). In March 2012, FCE signed a 120 MW multi-year order commitment.[136]
- POSCO will likely manufacture MCFC products in South Korea eventually and started to build its own MCFC manufacturing facility. It began the construction of a 50 MW fuel cell stack manufacturing facility in its hometown of Pohang, in the country's southeast in August 2007 and completed it in January 2008.[13] In April 2011, POSCO added another 50 MW in manufacturing capacity, bringing the annual total capacity of MFCFC stacks to 100 MW.[137] The facility was integrated with the BoP factory at the same site, which also has a 100 MW annual capacity. In addition, POSCO has a 2.4 MW fuel cell power plant, an R&D center, and

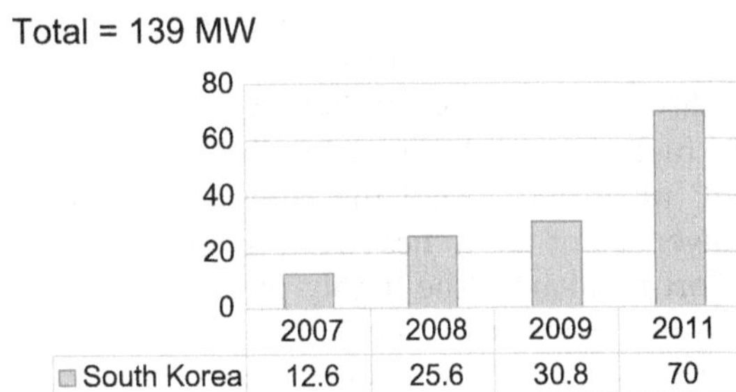

FIGURE 5.8 South Korea's purchase of FCE MCFC power plants by megawatt. *source*: (For the color version of this figure, the reader is referred to the online version of this book.)

a show room. The POSCO manufacturing facility is now larger than that of FCE's. In contrast, FCE has a 90 MW capacity fuel cell manufacturing facility.

- Meanwhile, POSCO progressively acquired rights to manufacture MCFC systems, stacks, and then cells in South Korea. In April 2008, POSCO reached an agreement for assembling MCFC systems by integrating FCE's proprietary fuel cell modules and POSCO's BoP components at the South Korean facility.[138] In June 2009, it acquired rights to assemble FCE's cell and module components into stack modules for sale in South Korea. In March 2012, FCE signed a license commitment that would provide for the manufacturing of DFC components in South Korea by October 2014.[136]
- Along with each of the major purchase orders, POSCO acquired FCE shares. In June 2009, POSCO bought 6.7 million shares of FCE common stock at $3.59 per share at $25 million.[139] In May 2012, POSCO purchased 20 million shares of FCE's common stock at a price of $1.50 per share for a total of $30 million.[140] POSCO already holds 10.8 million shares, and this investment brings POSCO Energy's common stock ownership percentage in the Company to approximately 16.6 percent.

POSCO has proceeded to promote MCFC technology exports. In September 2011, FCE announced a market expansion into Indonesia with the sale of a sub-megawatt DFC module. It will partner with POSCO for installation at a showcase location.[141] POSCO will combine the fuel cell module with a locally manufactured BoP plant and install the completed fuel cell power plant at a heavily visited waterpark resort in Jakarta. The highly public installation will demonstrate the benefits of MCFC power generation as POSCO develops a market for MW-class power plants in Southeast Asia, starting with Indonesia and with subsequent initiatives in Thailand, Malaysia, and Singapore.

5.5.3. POSCO's MCFC Strategy Still Unfolding…Too Early to Predict the Outcome

Many factors will affect how POSCO will fare with MCFC commercialization in South Korea as well as in exports in Asia, some of which are beyond POSCO's control. The factors may include the South Korean government's future green policy and subsidy program, the status of the global economy and regional environmental issues, FCE's ability to achieve breakthrough MCFC technology, and the emergence of newer, more competitive technologies. What is likely is that POSCO's future will be closely intertwined with FCE's.

References

1. Blomen L JMJ, Mugerwa MN. *Fuel cell systems*. New York: Plenum Press; 1993.
2. *Building your prototype car*. US: Instruction Manual, Office of Science, US Department of Energy; 2006.
3. Molten carbonate fuel cells (MCFCs), DoITPoMS teaching and learning packages. <http://www.doitpoms.ac.uk/tlplib/fuel-cells/mcfc_history.php>, University of Cambridge, [accessed in February 2009].
4. New sunshine project, molten carbonate fuel cells, pre-final evaluation report, industrial structure council, METI, March 1999; long-term energy technology strategy survey (A survey on new sunshine project systematic summation and quantitative assessment of after effect), Part I, NEDO, March 2001.
5. Taschek WG. History of fuel cell R&D at Fort Belvoir, Virginia, APPT - TR – 08 – 01, power technology branch, army power division. Fort Belvoir, Virginia: US Army RDECOM CERDEC C2D.
6. Appleby AJ, Foulkes FR. *Fuel cell handbook*. New York: Von Nostrand Reinhold; 1989.
7. Buss TE. *Status of M-C power's MCFC program*. US: M-C Power Corporation; 1999.
8. *Fuel cell systems program plan, fiscal year 1993*. Fossil Energy, DOE; 1993.
9. *Secretary Abraham congratulates developer, City of Los Angeles for installation of fuel cell power plant, commercial unit is culmination of 27-year DOE-industry research and development effort*. Fossil Energy Techline, U.S. Department of Energy, Office of Fossil Energy; 14 March 2003.

10. O'Sullivan JB. *The status of fuel cell technology.* Canada: Presentation to the Ontario Ministry of Energy Innovations in Natural Gas Technology; 20 February 1991.
11. *Fuel cell systems program for stationary power.* Assistant Secretary for Fossil Energy, DOE; 1996.
12. Penner SS. Report of the DOE advanced fuel-cell commercialization working group (referred to as AFC2WG or working group), work performed under U.S. DOE Grant No. DEFG03–93ER30213, U.S.: Department of Energy, March 1995.
13. Moreno A, McPhail S, Bove R. *International status of molten carbonate fuel cell (MCFC) technology.* Italy: ENEA; January 2008.
14. Williams MC, Gyorke DF. DOE/FOSSIL energy's fuel cell program: status and future plans, United States Department of Energy, Federal Energy Technology Center, 1999.
15. Kordesch K, Simader G. *Fuel cells and their applications.* Germany: VCH; February 1996.
16. Chambers A, Schnoor B, Hamilton S. *Distributed generation: a nontechnical guide.* US: PennWell Books; 2001.
17. Task 3: Investigate current uses of fuel cells in the DOD, Contract No. DAAA21-93-C-0046, Task No. N.098, CDRL Item A003, National Defense Center for Environmental Excellence (NDCEE), 2 June 1997.
18. Leo A, Skok A, O'Shea T. Santa clara direct carbon fuel cell demonstration, proceedings of the fuel cells '97 review meeting, Morgantown: West Virginia, 26–28, August 1997.
19. Bagotsky VS. *Fuel cells, problems and solutions.* US: The Electrochemical Society Series, John Wiley & Sons; 2009.
20. *New energy technology development related data file production survey (fuel cells).* NEDO; March 1997.
21. *Commercialization of fuel cells.* U.S. Department of Energy, Office of Energy Research, Office of Program Analysis; March 1995.
22. *Fuel cell dedication in California previews future use of fossil fuels.* Fossil Energy Techline, DOE; 3 June 1996.
23. Lee Willis H, Scott WG. *Distributed power generation: planning and evaluation.* UK: CRC Press; 2000.
24. *New energy technology development related data file production survey (fuel cells).* NEDO; March 2000.
25. *The wall street transcript.* FuelCell Energy, Inc. (FCEL); 16 July 2006.
26. Homma T. *Development and commercialization of MCFC.* Japan: Latest News, FCDIC; May 2003.
27. Products. <http://www.fuelcellenergy.com/products.php>, fuel cell energy, no date, [accessed May 2012].
28. Developing the Second-Generation Fuel cell—The M-C Power Project, Project Facts, PS024.0897M, Office of Fossil Energy, US DOE, 26–28 August 1997.
29. Molten carbonate fuel cell history, collecting the history of fuel cells. <http://americanhistory.si.edu/fuelcells/mc/mcfcmain.htm>. Smithsonian Institution, updated August 2008.
30. *World's first cogeneration molten carbonate fuel cell dedicated at miramar naval air station.* Fossil Energy Techline, US Department of Energy; 20 February 1997.
31. Buss TE. *Status of M-C power's MCFC program.* Morgantown: West Virginia: 1999 joint fuel cell technology review conference; 3–5 August 1999.
32. Hydrogen & Fuel Cell Letter, XV/No.3, p. 3–4, March 2000; T. Homma, MCFC, Latest News, No.43, 1 March 2000.
33. Santa Barbara to pull plug on fuel cell at treatment plant, the daily sound. <http://fuelcellsworks.com/news/2010/10/25/santa-barbara-to-pull-plug-on-fuel-cell-at-treatment-plant/>, fuel cell works, 25 October 2010.
34. Fuel cell energy announces 70 megawatt order from POSCO power and $9 million charge to second quarter 2011 earnings, press release. <http://fcel.client.shareholder.com/releasedetail.cfm?ReleaseID=581853>, fuel cell energy, 31 May 2011.
35. *Fuel cell energy reports first quarter 2009 results and latest accomplishments.* Globe News Wire; 10 March 2009.
36. Connecticut Incentives/policies for renewables & efficiency: CCEF - On-Site Renewable DG Program. <http://www.dsireusa.org/incentives/incentive.cfm?Incentive_Code=CT16F&re=1&ee=1>, database of state incentives for renewables & efficiency, State of Connecticut, last reviewed on 8 January 2009.
37. Connecticut clean energy fund: FY2009–FY2010 (July 1, 2008, through June 30, 2010) Comprehensive plan. The connecticut clean energy fund board of directors, 31 March 2008.
38. California public utilities commission raises clean energy incentive Cap to 3 megawatts, new ruling also increases funds available for larger projects by $96 million, press release. Fuel cell energy, 28 April 2008.
39. *Survey report on MCFC power generation technology development.* NEDO; September 2004.
40. Fuel cell power generation technology, challenges for MCFC commercialization. Denchuken review, No. 51, CRIEPI, March 2004.
41. *A study on practical application of fuel cells.* NEDO; March 1997.
42. *New sunshine project, fuel cell power generation technology R&D (MCFC), pre-final evaluation report.* Council of Industrial Structure, MEDI; March 1999.
43. *Long term energy technology strategy survey (New sunshine project systematic study and study of quantitative assessment of the effect) Part I.* NEDO; March 2001.

44. Ogawa T, Matsunaga A, Matsuyama T, Tooi M, Hosaka M (Ihi). Performance of the MCFC stack for 1000 kW class power plan, FCDIC fuel cell Symposium Proceedings, vol. 7. Japan: Fuel Cell Development Information Center, 2000.
45. *Molten carbonate fuel cell power generation technology development, post project evaluation report*. NEDO; September 2005.
46. Tooi M. *Progress, challenges and markets*. California: Palm Springs: 2005 Fuel Cell Seminar; 14–18 November 2005.
47. Homma T. Market development for MCFC CHP system, No. 79, FCDIC, 28, December 28, 2002.
48. Japan's fuel cell industry, industrial report, JETRO Japan Economic Report, April–May 2006.
49. Demonstrative project of regional power grids with various new energies. NEDO. <www.nedo.go.jp/english/ctivities/2_sinenergy/3/o03038e.html>, [accessed May 2009].
50. Homma T. Development and demonstration test of MCFC. The latest fuel cell news JAPAN, FCDIC, February 2003.
51. Introduction of the first molten carbonate fuel cell (MCFC) commercial unit—conducting research combining an MCFC and waste gasification system, press release, Chubu Electric Power, 18 September 2002.
52. Homma T. Development and demonstration test of MCFC. The latest fuel cell news in Japan, FCDIC, March 2003.
53. Homma T. Commercialization of MCFC. The latest fuel cell news in Japan, FCDIC, June 2004.
54. Expo 2005 Aichi, Japan E-mail newsletter, Japan association for the 2005 World Exposition. <http://www.expo2005.or.jp/ml/en/23/index.html>, [accessed May 2009].
55. Kobayashi N. Visiting new energy facilities in expo 2005 Aichi Japan. *J Heat Trans Soc Japan* July 2005;**44**(187).
56. Homma T. Development and application of MCFC. The latest fuel cell news in Japan, FCDIC, November 2005.
57. Homma T. Development and commercialization of MCFC. The latest fuel cell news in Japan, FCDIC, July 2003.
58. Homma T. Development and commercialization of MCFC. The latest fuel cell news in Japan, FCDIC, May 2003.
59. Homma T. Market development of MCFC. The latest fuel cell news in Japan, FCDIC, February 2004.
60. Homma T. Business deployment of MCFC. The latest fuel cell news in Japan, FCDIC, October 2004.
61. Homma T. Business deployment of MCFC. The latest fuel cell news in Japan, FCDIC, May 2006.
62. Homma T. Development and demonstration Test of MCFC. The latest fuel cell news in Japan, FCDIC, April 2003.
63. Hirschenhofer JH. International developments in fuel cells, <www.allbusiness.com>, August 1 1989.
64. Development of new cathode materials for use in molten carbonate fuel cells. CORDIS database. <http://cordis.europa.eu/search/index.cfm?dbname=proj>, [accessed July 2009].
65. Optimization of NiO cathode performance and endurance in molten carbonate fuel cells. CORDIS database. <http://cordis.europa.eu/search/index.cfm?dbname=proj>, [accessed July 2009].
66. A collaborative project to build a 1 kW internal reforming molten carbonate fuel cell stack. CORDIS database. <http://cordis.europa.eu/search/index.cfm?dbname=proj>, [accessed July 2009].
67. Michael PD, Maguire J. European fuel cells R&D review—final report. Contract W-31-109-Eng-38. US Department of Energy, September 1994.
68. Luiten EEM, B1ok K. Energy R&D in the Netherlands, Contract DE-AC06-76RL0 1830. U.S. Department of Energy, April 1999.
69. *Chapter 2 MCFC*. NEDO Overseas Report; July 1999.
70. Upscaling MCFC—technology to a 50kW ER-MCFC stack demonstration. CORDIS database. <http://cordis.europa.eu/search/index.cfm?dbname=proj>, [accessed July 2009].
71. 10 kW demonstration of pilot stack technology for direct internal reforming molten carbonate fuel cell (DIR-MCFC) power plants. CORDIS database. <http://cordis.europa.eu/search/index.cfm?dbname=proj>, [accessed July 2009].
72. Basic research on MCFC corrosion and improved current density. CORDIS database. <http://cordis.europa.eu/search/index.cfm?dbname=proj>, [accessed July 2009].
73. Selection and development of materials for high endurance MCFC bipolar plate. CORDIS database. <http://cordis.europa.eu/search/index.cfm?dbname=proj>, [accessed July 2009].
74. Kortbeek PJ, Ottervanger RG. Advanced DIR-MCFC development—a smarter concept for cogeneration applications, publishable final report, Contract JOE3-CT95-0033, BRANDSTOFCEL NEDERLAND BV (BCN)(Dutch Fuel Cell Corporation), 1 February 1996 to 31 January 1999.
75. European Fuel cell Projects 1995–2000, Research Directorate-General, European Commission, 2000.
76. Lequeux G. Status of the European SOFC program. In: Yokokawa H, Singhal SC, editors. Proceedings of

the seventh international symposium, Solid Oxide Fuel Cells VII (SOFC VII), Electrochemical Society. February 2001, pp. 14–27.

77. *ECN sells MCFC technology to fuel cell energy*. Petten, the Netherlands: ECN; 3 October 2005.
78. Lovering DG. The grove fuel cell symposium—pollution-free power sources attract worldwide interest for static and transport applications. *Platin Met Rev* 1989;**33**(4).
79. Zegers P. *Fuel cell research in Europe*. Long Beach, California: A Presentation at 1988 Fuel Cell Seminar; 23–26 October 1988.
80. Lovering DG. The grove fuel cell symposium—pollution-free power sources attract worldwide interest for static and transport applications. *Platin Met Rev* 1989;**33**(4):169–77.
81. *Fuel cell handbook*. Contract No. DE-AM26–99FT40575. 5th ed. U.S. Department of Energy, Office of Fossil Energy, National Energy Technology Laboratory; October 2000.
82. Molten carbonate fuel cell (MCFC) 100 kW cogeneration power plant. CORDIS database. <http://cordis.europa.eu/search/index.cfm?dbname=proj>, [accessed July 2009].
83. Dufour A, Torazza A, Regis V, Rocchini G, Hermana E. MCFC compact unit and system improvements for 100 kW—2 MW dual-fuel demonstration plant, presentato al POWER-GEN Europe, Maggio, 1995.
84. 100 kW MCFC integrated stack development and testing. CORDIS database, <http://cordis.europa.eu/search/index.cfm?dbname=proj>, [accessed July 2009].
85. Development of industrially relevant MCFC stacks. CORDIS database. <http://cordis.europa.eu/search/index.cfm?dbname=proj>, [accessed July 2009].
86. European fuel cell projects 1995–2000. EUR 19368, Research Directorate-General, European Commission, 2000.
87. Development of industrially relevant MCFC stacks. Contract JOE3-CT95-0026, Publishable Final Report, Ansaldo Ricerche, Iberdrola, Bobcock & Wilcox Espanola, ENEA, CNR-ITAE, January 2000.
88. Series 500 MCFC clean power and heat cogeneration compact plant first of a kind. CORDIS database. <http://cordis.europa.eu/search/index.cfm?dbname=proj>, [accessed June 2009].
89. Srinivasan S. *Fuel cells—from fundamentals to applications*. Heidelberg: Springer; 2006.
90. Ansaldo fuel cells SpA. <http://www.pilasde.com/ansaldo.htm, ARIEMA Energia y Medioambiente, S. L>, [accessed June 2009].
91. MCFC SERIES 500, TWINSTACK® POWERED FIRST-OF-A-KIND, MCTWINS. <http://ec.europa.eu/energy/efficiency/industry/doc/mctwins.pdf>, European Commission, no date.
92. European fuel cell and hydrogen projects 1999–2002, EUR 20718, Directorate-General for Research, European Commission, 2003; Innovative Cost-Effective Hybrid System Based on Integration of a MCFC and a Gas Turbine for High Efficiency Dispersed CHP Generation. CORDIS database. <http://cordis.europa.eu/search/index.cfm?dbname=proj>, [accessed July 2009].
93. Molten carbonate fuel cells for waterborne application, FP6 Home, CORDIS. <http://cordis.europa.eu/fp6/projects.htm>, [accessed July 2009].
94. Moreno A. *Hydrogen and fuel cells in CCS power plant*. Erice: a course material used at International School of Geophysics; 1–7 November 2007.
95. Biogas integrated concept a European program for sustainability, FP6 Home. CORDIS. <http://cordis.europa.eu/fp6/projects.htm>, [accessed July 2009].
96. Bartolomeo Marcenaro Ansaldo fuel cells, Podcast 15. <http://www.finmec.com/poddownload.php?p=Finmeccanica+Podcast+15+−Bartolomeo+Marcenaro+Ansaldo+Fuel+Cells.mp3>, Finmeccanica North America, 9 January 2008.
97. Finmeccanica. Wikipedia, last modified on 16 June 2009, [accessed July 2009].
98. Recent Achievements (October 2004). <www.ansaldofuelcells.com/files/achievements.pdf>, no date, [accessed July 2009].
99. Ansaldo fuel cells signs agreement in the energy sector with L-3 Communications Combat Propulsion Systems, press release. Finmeccanica, 9 October 2007.
100. Fuel cell systems. <http://www.finmec.com/energy/?s_id=12&p_id=55>, Finmeccanica, [accessed July 2009].
101. Ansaldo fuel cells and ENEL sign partnership agreement to build a fuel cell plant, press release. ENEL and Finmeccanica, 26 February 2008.

102a. Bedont P. (Ansaldo Fuel Cells), Ansaldo fuel cells: long lasting operating experiences on mcfc full size power plant, International symposium on the introduction of fuel cell systems to early market, WBZU-ULM, Germany, pp. 13–14 July 2010.

102b. Kastubg Ioeratubg Exoeruebces on MCFC fyekk Size Power Plnt, Ansaldo fuel cells—A Finmeccanica company, Ansaldo Energia, <http://www.ansaldoenergia.com/AnsaldoFuelCells.htm>, 6 July 2009.

103. Marcenaro B. *Development and industrialization of MCFC systems at Ansaldo fuel cells*. Phoenix, Arizona: 2008 Fuel Cell Seminar and Exposition; 27–30 October 2008.

104. *Hydrogen & fuel cells, review of national R&D programs.* International Energy Agency; 2004.
105. *The fuel cell power plant: Hot Module, Brochure for Hot Module (1).* MTU; 2003–2004.
106. Sundmacher K, Kienle A, Pesch HJ, Berndt JF, Huppmann G. *Molten carbonate fuel cells.* Wiley-VCH; 2007.
107. Rastler DM, Herman D, Goldstein R, O'Sullivan J. *State-of-the-art of fuel cell technologies for distributed power, technical and strategic assessment of products, markets, and retail competitiveness, TR-106620-R1, annual update.* US: Electric Power Research Institute; August 1997.
108. BRENNSTOFFZELLENSYSTEME FÜR STATIONÄRE ANWENDUNGEN, Austria: Workshop, BMVIT, Wirtschaftskammer Österreich, 30. Jänner 2001.
109. Development and adaption of the molten carbonate fuel cell with internal reforming for the use of CO/CO_2 rich gases, e.g. coal gas. CORDIS database. <http://cordis.europa.eu/search/index.cfm?dbname=proj>, [accessed July 2009].
110. Design and engineering of a 100 kW MCFC cogeneration plant (natural gas) and a 100 kW MCFC power plant (coal gas). CORDIS database. <http://cordis.europa.eu/search/index.cfm?dbname=proj>, [accessed July 2009].
111. MCFC power plants—demonstration of a hot module qualification model; Construction, Installation, and Start of a Hot Module. CORDIS database. <http://cordis.europa.eu/search/index.cfm?dbname=proj>, [accessed July 2009].
112. Bischoff M, Kraus P. The MTU Friedrichshafen hot module, proceedings, the 3rd international fuel cell conference, Nagoya, Japan, 30 November-3 December 1999.
113. Hot module points the way, innovation in Europe research and results, archived information, no date, The European Commission. <http://ex.europe.eu/research/success/en/ene/0265e.html>, [accessed July 2009].
114. *MCFC—overseas MCFC field test status.* NEDO. <http://www.nedo.go.jp/kankobutsu/foreigninfo/html9907/07154.html>; July 1997.
115. *MCFC—overseas MCFC field test status.* NEDO. <http://www.nedo.go.jp/kankobutsu/foreigninfo/html9907/07154.html>; July 1999.
116. Bode M. Hot module fuel cell efficient, ultra-clean, reliable. CFC Solutions, 2007.
117. System development and optimization. CORDIS database. <http://cordis.europa.eu/search/index.cfm?dbname=proj>, [accessed July 2009].
118. MCFC balance of plant innovation—design, construction, and demonstration of a DFC Module Pilot Plant for cogeneration, CORDIS database. <http://cordis.europa.eu/search/index.cfm?dbname=proj>, [accessed July 2009].
119. Holistic integration of MCFC technology towards a most effective systems compound using biogas as a renewable source of energy Effective, CORDIS database. <http://cordis.europa.eu/search/index.cfm?dbname=proj>, [accessed July 2009].
120. Trogisch S. Effective—holistic integration of MCFC technology towards a most effective systems compound using biogas as a renewable source of energy, Bio-Energy Enlarged Perspectives Budapest. 16–17 October 2003.
121. Malinowski P, Menzen G. *Fuel cells & hydrogen in the German energy research programme.* Honolulu, Hawaii: 2006 Fuel Cell Seminar; 13–17 November 2006.
122. Survey report on the development trend of fuel cells and hydrogen in Germany, FY2004 Survey Report, NEDO Paris Office, March 2005.
123. Neef H-J. IPHE Country Paper: GERMANY, ILC-021-04, 10 February 2004.
124. Shutting down fuel cells. Industry News. <http://www.dieselprogress.com/Industry-News/3098/Shutting-Down-Fuel-Cells/>, diesel progress, 30 December 2010.
125. Rolls-Royce group plc refrains from Tognum AG fuel cell business purchase-DJ. Article. <http://www.reuters.com/finance/stocks/RYCEY.PK/key-developments/article/2341209>, reuters, 10 June 2011.
126. Ultra-clean and reliable power generation drives purchase of direct fuel cell power plant in London, England, press release, <http://fcel.client.shareholder.com/releasedetail.cfm?ReleaseID=549756>, fuel cell energy, 14 February 2011.
127. Kim S-G, Choi YJ, Kuk ST, Jun J-H, Lim H-C. Status and recent progress of MCFC stack development in korea, 2008 Fuel Cell Seminar & Exposition, Phoenix, Arizona, October 2008, pp. 27–30.
128. Lee T-H. Hydrogen and fuel cell activities in Korea, 2008 Northeast Asia Energy Outlook Seminar. Washington, DC: Korea Economic Institute Policy Forum, 6 May 2008.
129. Kim TH. Status of molten carbonate fuel cell development in Korea. Seminar on IPHE, 2007.
130. *Doosan heavy tapped for Gov't fuel cell project.* Chosun Ilbo; 9 October 2007.
131. Lee T-W. Renewable energy development program in DOOSAN, Doosan Heavy Industries & Construction, 28 May 2008.
132. South Korea makes green its business. Blog. <http://www.future500.org/blog/south-korea-makes-green-its-business/>, the Fortune 500, 30 December 2011.
133. Hong S-A. Hydrogen & fuel cell activities in Korea. KIST, 29 September 2005.

134. Williamson L. South Korea's drive for renewable energy. News. <http://www.bbc.co.uk/news/business-15984399>, BBC News, 1 December 2011.
135. 15. Fuel cells and hydrogen, in 2011 update of the technology map for the SET-Plan, Browse the Newsroom, <http://setis.ec.europa.eu/newsroom-items-folder/fuel-cells-and-hydrogen>, Strategic Energy Technologies Information System, European Commission, 8 December 2011.
136. Fuel cell energy and POSCO energy announce expanded partnership, press release. <http://fcel.client.shareholder.com/releasedetail.cfm?ReleaseID=656448>, Fuel cell Energy, 12 March 2012.
137. POSCO's 100 MW fuel cell manufacturing facility now complete. Article. <http://www.fuelcelltoday.com/news-events/news-archive/2011/april/posco's-100-mw-fuel-cell-manufacturing-facility-now-complete>, fuel cell today, 20 April 2011.
138. POSCO power orders 25.6 megawatts of fuel cell energy power plant sand fuel cell modules, approximately $70 million order includes both DFC1500 and DFC3000 power plants and modules for delivery in 2009, press releases, fuel cell energy, 23 April 2008.
139. Fuel cell energy announces $24.2 million registered direct offering; 6.7 million shares to be issued at $3.59 per share, press releases, fuel cell energy, 16 June 2009.
140. Fuel cell energy announces completion of investment by South Korean partner, press release. <http://fcel.client.shareholder.com/releasedetail.cfm?ReleaseID=669562>, fuel cell energy, 2 May 2012.
141. Fuel cell energy announces Asian market expansion with sale of ultra-clean fuel cell module for installation in Indonesia, press release. <http://fcel.client.shareholder.com/releasedetail.cfm?ReleaseID=603702>, fuel cell energy, 7 September 2011.

CHAPTER

6

History of Solid Oxide Fuel Cells

OUTLINE

Fuel Cells
http://dx.doi.org/10.1016/B978-0-444-56325-5.00006-5

6.1. INTRODUCTION

The technology for solid oxide fuel cells (SOFCs) took more than a century to come to fruition. Walther Hermann Nernst was one of the earliest SOFC researchers—he was a pupil of Friedrich Wilhelm Ostwald at Leipzig's Physicochemical Institute and also a Nobel laureate. Nernst in 1899 described the first solid-state oxygen ion conductor, which had a composition of 85 percent zirconia oxide and 15 percent yttria oxide.[1] This today is known as the Nernst mass and was the prototype for the electrolyte in the present-day SOFC. Nernst discovered it by chance when he observed the evolution of oxygen at the anode with his "glower," which he developed as a commercial light source in 1897. Since those early days of discovery until the 1950s, however, there was almost no progress in the development of SOFCs. During this long hiatus, developments in MCFCs drew many researchers away from SOFC research. Many researchers seemed to believe MCFCs were less daunting and had greater near-term promise than SOFCs:

- In 1937, Swiss scientist Emil Baur and his colleague H. Preis tried to construct the first SOFC cell by using the Nernst mass as the electrolyte. That cell, with a coke anode and an iron oxide cathode, was operated at 1050 °C with hydrogen and air as reactants, and yielded a low current density of 1 mA/cm^2 at 650 mV.[1] Baur and Preis also tried several ceramics with additives, such as cerium dioxide (CeO_2) and lithium silicate but could not achieve a sufficiently high conductance. In addition, Baur recognized other difficulties. In 1939, he wrote that an unsurpassable barrier to further SOFC development might well be the difficulty of building large cell blocks from very thin and fragile elements at high temperatures. He then suggested that molten carbonate or sulfuric acid should be tried as electrolytes in a temperature range of 200–300 °C, leading him to suggest research on MCFCs.

- In the 1940s, O. K. Davtyan of Russia added monazite sand to a mix of sodium carbonate, tungsten trioxide, and soda glass "in order to increase the conductivity and mechanical strength." Davtyan's designs, however, experienced unwanted chemical reactions and short life ratings.[2] Dutch scientists G. H. J. Broers and J. A. A. Ketelaar subsequently attempted to further Davtyan's work but recognized that the electrolyte at the operating temperature of 650–750 °C was molten and not solid, concluding that an operating solid electrolyte was still beyond reach. They abandoned SOFC research and pursued MCFCs instead.

In the late 1950s, the prospect of a high temperature fuel cell that would be tolerant of carbon monoxide and which could use a stable solid electrolyte finally caught the attention of some researchers. Small-scale research on SOFCs started in the Netherlands, Consolidation Coal Company in Pennsylvania, and General Electric (GE) in New York.[3] All efforts were short lived, however, due to numerous technical hurdles. But Westinghouse Electric, which had also begun SOFC research in the late 1950s, persistently pursued small-scale research on SOFCs, focusing on zirconium compounds. By the 1960s and 1970s,[4] it had achieved major milestones in SOFC technology:

- In 1962, J. Weissbart and R. Ruka built the first modern SOFCs from a calcia-stabilized zirconia electrolyte and two porous platinum electrodes.[1] They achieved about 25 mA/cm^2 at 0.65 V.[5]
- In 1965, D. H. Archer and his colleagues demonstrated a multicell SOFC generator; they developed a 100 W solid electrolyte fuel cell employing calcia-stabilized zirconia as the electrolyte and sintered platinum as electrodes.[1] The expensive platinum electrodes used at first were subsequently replaced by a nickel zirconia cermet.
- In 1977, F. J. Rohr introduced the tubular concept of the SOFC with different anode and cathode materials which were deposited by plasma spraying onto cylindrical electrolytes.

6.2. US DEPARTMENT OF ENERGY INITIATES SOFC R&D PROGRAM IN 1977

In 1977, the Department of Energy (DOE; then ERDA) embarked on an SOFC R&D program.[3] DOE was eager to support fuel cell R&D programs that it believed could replace large-scale multi-kilowatt and multi-megawatt power-generation systems fueled by coal gas. In addition to SOFC development, DOE also launched MCFC and PAFC R&D programs.

Under the SOFC R&D program, DOE first focused on the tubular design that was under development at Westinghouse and was the most advanced of all SOFC technologies at the time. DOE funded the tubular SOFC development effort for approximately 30 years, from 1977 to 2005, with a total funding of about $264 million, mostly awarded to Westinghouse.

- DOE funding for tubular SOFC development was relatively modest, however. Throughout the 1970s and 1980s, most DOE funds went to PAFCs and MCFCs. Indeed, SOFC funding levels remained at about $5 million per year on average.
- In the late 1980s, DOE started to view PAFCs as sufficiently developed and that the private sector no longer required much support. It therefore increased support for MCFC and

SOFC development. Throughout the 1990s, however, MCFCs were always DOE's top priority. In 2005, DOE ended funding for both MCFC and SOFC development. During this period, total funding for SOFCs was about half that allocated for MCFCs.

6.2.1. DOE Taps Westinghouse to be Global Leader of SOFC Technology

In 1977, DOE launched its SOFC R&D program with a contract award to Westinghouse. Westinghouse was the natural choice for the award; it was the only SOFC developer in the United States and was demonstrating promising progress. The award allowed the company to develop a working design and by 1981 Westinghouse had developed a small SOFC system.[4] The system consisted of a set of tubular cells in a box, with the fuel electrode on the outside and the air electrode on the inside, connected to each other by nickel felts. The Westinghouse tubular design had a significant advantage over planar designs because it eliminated sealing problems.

6.2.1.1. Westinghouse Makes Major Technological Advances in 1995

Throughout the 1980s, Westinghouse made rapid progress in tubular SOFC technology and believed that commercialization would be possible by the early 1990s.[6] Westinghouse developed an electrochemical vapor deposition (EVD) process for SOFC electrolyte and electrode film preparation, reducing the thickness of the electrolyte layer to a few microns and markedly improving fuel cell performance. In the late 1980s, it began to study the development of a high power SOFC cell stack, ranging from 400 W to 3 kW output.[7] It built several such systems and starting in 1986 deployed them to customer sites for demonstration operations. Each system operated continuously for up to 5000 h, undoubtedly convincing the company to move its SOFC research from the experimental stage to commercial development:

- The first field unit was a 400 W system deployed at the Tennessee Valley Authority (TVA) in 1986.[8] The unit consisted of 24 tubular cells of 30 cm in length, which operated for 1760 h.
- In 1987, three 3 kW field test units were deployed by Japanese utilities Osaka Gas and Tokyo Gas. The units each consisted of one hundred and forty four 36 cm long tubular cells, which operated for 3012 h, 3683 h, and 4882 h, respectively.

The company continued to make progress and developed larger systems in the 20–25 kW range. The generator module used in the units was a near replica of the multikilowatt generator that was developed for DOE earlier.[9] Starting in 1992, the company built and deployed at least six field test units at customer sites. Five were deployed in Japan and one in the United States. In 1995, the company developed an air electrode supported (AES) tubular SOFC design that replaced the original tube technology in the 25 kW systems, thereby simplifying the structure of the SOFC and increasing power density threefold.[7] All the 20–25 kW systems consisted of five hundred and seventy six 50 cm cells,[8] but the 25 kW systems demonstrated higher output and longer lifetimes than the 20 kW units:[10]

- Four of the five units deployed in Japan were 20 kW units, which started operations in 1992. One of them operated for only 817 h, one for as long as 7064 h. Average operating time was 3000 h.
- The fifth unit deployed in Japan was a 25 kW unit, which started operations in 2005. The system likely incorporated the new AES cell technology and operated for as long as 13,194 h.
- The sixth unit was a 25 kW unit deployed by Southern California Edison (SCE). The unit

was initially installed at SCE's Highgrove Generating Station in the spring of 1994.[11,19] It operated for approximately 6500 h. The stack was replaced with the new AES stack in mid-1995. The new stack operated for 5000 h and was shut down in February 1996 and relocated to the National Fuel Cell Research Center (NFCRC), at the University of California, Irvine. It was restarted in January 1998 without difficulty and operated for another 8250 h through June 2002. Total lifetime of the new AES stack was 13,250 h (5000 h at SCE and 8250 h at NFCRC), virtually identical to the lifetime of the 25 kW unit deployed in Japan (13,194 h).

6.2.1.2. *Siemens Acquires Westinghouse and Launches Ambitious Commercialization Plans (1997–2002)*

Encouraged by the performance of the new air electrode tubular design, Westinghouse by 1997 had developed ambitious commercialization plans that it hoped to achieve by the early 2000s. The plans basically centered on four critical milestones that were deemed necessary to gain market acceptance:

1. Development of a 100 kW SOFC CHP (combined heat and power) cogeneration system that would operate at atmospheric pressure. The 100 kW CHP system would serve as a building block module for the future product lineup.
2. Demonstration of a pressurized and hybridized SOFC system integrated with gas turbines—the 225 kW SOFC/GT hybrid system—which the company believed would yield increased voltage and electrical efficiency, enabling lower system cost per kilowatt of electricity and achieving cost competitiveness.
3. Development and demonstration operation of a 250 kW pre-commercial SOFC CHP system, a prototype of the first product to be commercialized.[12]
4. Establishment of a large commercial SOFC manufacturing facility, capable of producing hundreds of megawatts of SOFC cells, stacks, and systems as early as in 2002.

To complement the company's ambitious commercialization plans, Westinghouse secured solid financing and technological resources. In 1997, the company obtained a favorable long-term contract award from DOE. At the same time, it became the subsidiary of a financially and technically sound parent company, Siemens AG of Germany:

- In August 1997, DOE signed a 5-year $202 million agreement with Westinghouse to continue the existing SOFC development program. Of the total, DOE funded $83 million and Westinghouse contributed about $119 million.[10] The contract provided Westinghouse with an average of a little more than $15 million a year. Compared to the 1980s, DOE funding had tripled.
- On November 14, 1997, Siemens acquired Westinghouse's fossil fuel power-generation business, including the tubular SOFC business. Westinghouse had been negotiating with a number of corporations concerning the formation of a joint venture for securing funds for DOE cost-share projects and for construction of the commercial manufacturing facility as well as augmenting existing technological capabilities.[13] The deal had clear benefits as compared to a joint venture. Siemens as a parent company would be financially more committed to Westinghouse than as a joint venture partner. Siemens also had expertise in SOFC technology to offer to Westinghouse; it had been conducting planar SOFC research since 1962. At acquisition a new company was created—Siemens Westinghouse Power Corporation (SWPC)—which would focus on tubular SOFC research, dropping the planar SOFC research in which Siemens of Germany had been engaged.

6.2.1.2.1. MILESTONE 1: 100 KW SOFC CHP COGENERATION SYSTEM FIELD TEST (1997–2008)

In 1997, SWPC initiated a $196 million[14] project to build and operate a 100 kW SOFC CHP cogeneration field test unit. The project was co-sponsored by DOE and NOVEM, an agency of the Dutch Ministry of Economic Affairs. The 100 kW CHP system represented a dramatic scale-up of what SWPC was building only 5 years earlier. The 100 kW system consisted of 1152 tubular cells 1.5 m in length—twice as many cells that were three times longer than those in the 20–25 kW units.

- The 100 kW field test unit was delivered to EDB/ELSAM, a consortium of Dutch/Danish utilities, in 1997. After the initial start-up at Westervoort in December, the system operated for 4035 h and was returned to Pittsburgh for modifications.[15] Internal voltage measurements indicated accelerating voltage decline in a segment of the stack, which necessitated repairs.[16] With a rebuilt fuel cell stack, the system was restarted at Westervoort in March 1999 and operated for 12,577 h.[15] The system generated 109 kW AC and 64 kW of thermal power with an electrical efficiency of 46 percent and thermal efficiency of 27 percent.[21] The system was shut down at the end of the contract period in November 2000. It was relocated in March 2001 to a site run by the German utility RWE in Essen, where it was restarted with the same stack and operated for another 3800 h.[15] Subsequently, the system underwent refurbishing and was moved to the facilities of Gas Turbine Technologies (GTT) in Turin, Italy, in 2005.[17] The system operated for 16,400 h there until the system was shutdown in June 2008.

The performance of the system astonished the world. The system had operated for a total of 36,812 h, approximately the 40,000 h/4-year lifetimes typically required for stationary power-generation systems. The system also consistently achieved electrical efficiencies as high as 46 percent with no performance degradation.[15] It was viewed as a remarkable feat for the first demonstration of a full-scale SOFC commercial system with 1.5 m cells. Nonetheless, a closer examination shows that the actual lifetime of the SOFC stack fell short of the required standards for stationary systems.

- One stack operated after modifications for 16,377 h (12,577 h at Westervoort and 3800 h at Essen) and another stack for 16,400 h in Turin, a little less than 2 years each.
- The system also did not operate continuously. The project took 10 years to complete, interspersed with periods of rebuilding and refurbishing.

6.2.1.2.2. MILESTONE 2: 220 KW PROOF-OF-CONCEPT SOFC/GT SYSTEM VALIDATION (2000–2002)

SWPC's second milestone was to validate a proof-of-concept pressurized 220 kW SOFC/GT hybrid system. SWPC theorized that combining a fuel cell system with a gas turbine would yield higher electrical efficiency than if the systems were operated individually. The $16-million[18] proof-of-concept demonstration was launched in 2000.

- The 220 kW SOFC/GT hybrid system was delivered to NFCRC at the University of California, Irvine, in June 2000, for a 3000-hour demonstration operation that was conducted in conjunction with Southern California Edison.[19] The system consisted of a 100 kW stack (with 1152 cells, 1.5 m in length) operating at 3 atm (atmosphere) that produced 180 kW and a microturbine generator that supplied 40 kW.[20] Testing of the system began in June 2000 and ended in April 2002, covering three separate periods of operation for a total of about 3400 h.

- During two of the three operating sessions (totaling 667 h in 2000 and 2001), the system generated a total of 181 kW AC (including turbine power of 22 kW AC) at an electrical efficiency of 52 percent.[21]

SWPC announced that the proof-of-concept demonstration validated the concept and the operating characteristics and parameters for pressurized hybrids.[19] It achieved an electrical efficiency of up to 53 percent, which was the highest known electrical efficiency achieved by any large-scale fuel cell system at that time. Still, the system did not operate long enough to establish a commercially viable benchmark for a pressurized SOFC/GT hybrid system.

6.2.1.2.3. MILESTONE 3: 250 KW PRE-COMMERCIAL PROTOTYPE CHP SYSTEM DEMONSTRATION (2003)

SWPC's third milestone was to develop and operate a 250 kW SOFC CHP pre-commercial prototype, which SWPC planned to market by 2006.[22] SWPC had earlier planned its market entry for about 2002. The prototype demonstration project was co-sponsored by the US DOE, Ontario Power Generation, Canada's intergovernment "Technology Early Action Measures (TEAM)" project, and National Resources Canada. The project cost was C$26 million[23] (about US$16.35 million at the 2002 exchange rate of US$1/C$1.59).

Under the project, Kinectrics (formerly Ontario Power Technologies) was responsible for system integration at the Kippling test facility, in Etobicoke, Toronto.[19] The system was designed by SWPC, fabricated by both SWPC and Kinectrics, and assembled in the Kinectrics' facility. SWPC supplied the SOFC module and some equipment and Kinectrics supplied the balance of the plant to SWPC specifications. The 250 kW system was the largest atmospheric pressure SOFC unit ever built, consisting of twin SOFC building block modules, each with 2304 Siemens Westinghouse prototype tubular cells.[18] The system's projected electrical efficiency was 45 percent, and its projected overall fuel conversion efficiency was as high as 75 percent, assuming heat recovery.[23] Kinectrics was to operate the system at its test facilities for 5000 h. After extensive testing at Kinectrics, the system was expected to be installed at the University of Toronto at Mississauga (UTM).[24] The unit was to produce about 8 percent of the campus' electricity needs as well as hot water.

The project encountered technical difficulties, delays, and eventual project cancelation, and was unable to establish technical and commercial viability:

- The system was assembled in October 2002, but due to numerous problems which occurred during system integration, start-up was delayed until April 2003.[23] The electrical and thermal efficiency and the output of the SOFC during early trials was less than projected. Between April and June 2003, the system operated for 1400 h over three cycles.[25] During that period, it produced only about 180 kW AC for about 700 h.[23] Even so, it was deemed that the system might meet design specifications with some fine-tuning. The system was repaired and sent to the UTM for operation for the remainder of the assessment period.

On September 12, 2004, however, a local Canadian newspaper reported that the UTM had abandoned plans to install the 250 kW SOFC system on the campus.[25] The paper implied that the cancelation was due to technical difficulties experienced at the Kinectrics facility.

6.2.1.2.4. MILESTONE 4: BUILDING COMMERCIAL SOFC MANUFACTURING FACILITY (2006)

As of 2001, SWPC was proceeding to set up a large-scale manufacturing facility to support commercial activities, which it hoped would begin in 2004.[26] In 1997, Westinghouse commissioned a pre-commercial $18-million manufacturing facility, which had an output capability of about 4 MW a year.[10] By September 2001, SWPC selected the Pittsburgh area for the new location of its SOFC business and acquired a 22-acre site for its projected 430,000 ft^2 manufacturing and office building in the Waterfront development in the borough of Munhall within Allegheny County, Pennsylvania.[27] Projected cost of the building was $122 million.[26] SWPC planned to build the new facility in phases: initially starting with 180,000 ft^2 of manufacturing and office space in 2003, growing to 430,000 ft^2 by 2006.[27] The company planned to expand the annual production capacity to 5 MW in the 2003–2004 time frame, increasing to 30 MW in 2005–2006, and bringing total capacity to 85 MW by 2007.[28] Eventually, production capacity of 100 MW/year was envisioned for the facility.

- The company had selected the site after reviewing 123 potential sites. Leaders from three states were falling over themselves offering SWPC incentives to build the new manufacturing plant in their states. SWPC was expected to employ as many as 500 workers at the facility by 2006, according to the press.[29]
- After the site was selected in Pennsylvania, the state provided SWPC a $750,000 opportunity grant to create more than 300 jobs at the new manufacturing plant.[30] Initial employment at the facility was expected to be approximately 150, expanding to 500 by 2006.[27]

6.2.1.3. *Siemens Westinghouse Hits Technical Barriers in the 2000s....Validation of Tubular SOFC Technology Fails*

SWPC[i] was so solidly in the lead as the global SOFC leader that few foresaw that its progress would be slowed or disrupted. In a startling development, however, SWPC began to show signs that it was having problems with its tubular SOFC technology. In the 2000s, SWPC was planning to demonstrate at least 10 SOFC systems of various sizes, but reportedly did not to pull off any of the major projects. A couple of the systems were installed but operated for an insignificant period of time, and the others were canceled even before they were installed. The 10 systems included three 250 kW atmospheric CHP systems, two pressurized 300 kW SOFC/GT hybrid systems, three 1 MW pressurized SOFC/GT hybrid systems, and two 125 kW CHP systems:

- *Three 250 kW atmospheric systems.*
  - In July 2003, SWPC signed a contract to install a 250 kW atmospheric CHP system for Stadtwerke Hannover and E.ON Energie in Germany.[31] Valued at around €5 million (US$4.33 million), the system was to be installed at the Herrenhausen power plant in Honnover.[31] The contract subsequently might have been modified to build a 125 kW SOFC CHP system.
  - In May 2000, SWPC and Norske Shell announced the demonstration of a 250 kW SFOC atmospheric CO_2 recovery demonstration system at Norske Shell, Norway. The system was to be installed at Norske Shell's Kollsnes gas processing plant and fueled by natural gas.[32]
  - In October 2001, SWPC signed a contract with BP to install a 250 kW CHP system at

[i] Siemens Westinghouse Power Corporation (SWPC) stopped using the name and started to use the parent company name, "Siemens Power Generation," sometime in 2005 or 2006, which it used until sometime in 2008 or 2009. It started to use Siemens Energy sometime in 2009. To avoid confusion, SWPC is used throughout the book.

BP's gas-to-liquids (GTL) test plant in Nikiski, Alaska.[33] The $6.5-million system was expected to begin operating by mid-2003 and would use natural gas as its fuel.[34] The DOE provided a grant of $2 million. The project was also awarded a DoD Climate Change Rebate for the fiscal year 2000. BP planned to use approximately 150 kW of the unit's output to power the warehouse and administration building of the GTL facility.[33] According to one source, the BP GTL plant was closed in 2004 and the projected fuel cell demonstration site was changed to Chugach Electric, Anchorage.[35] Around that time, the project apparently was modified to accommodate a 125 kW CHP system, instead of a 250 kW unit. Chugach, however, backed out of the project in the fall of 2005.

- *Two pressurized 300 kW SOFC/GT hybrid systems.*
  - In July 2000, SWPC signed a contract to build a 300 kW pressurized SOFC/GT hybrid system for Edison Spa of Milan, Italy.[36] The system was to be installed at Edison Spa's R&D facility in the town of Sinetta Marengo in northwest Italy to demonstrate cogeneration applications. Delivery of the system was expected in September 2002 with system start-up in October 2002.
  - In March 2000, SWPC signed a contract to build a 300 kW pressurized SOFC/GT system for RWE of Germany.[36] SWPC was to deliver the system by 2003 to RWE's fuel cell demonstration facility at the Meteorit Technology Park in Essen. The 300-kW system was to be combined with a microturbine generator and operate under pressure to achieve electrical efficiencies approaching 60 percent. The system reportedly operated for a short period of time at RWE's Meteorit Park facility, Essen,[37] but little else is known.
- *Three 1 MW/GT Hybrid System Demonstration Projects.*
  - In October 2000, the Secretary of DOE and the Administrator of the Environmental Protection Agency (EPA) announced that a 1 MW SOFC/GT hybrid system would be installed at Fort Meade, Maryland, in 2002.[38] DOE and EPA committed a combined $14 million for the power system. Another $10 million would be provided by the private sector partners. The price tag for the plant indicates that the capital cost for the plant amounted to about $24,000/kW, or roughly 15–20 times more than a conventional power plant. The project was canceled before the groundbreaking.
  - In December 2000, SWPC announced that it had entered into an agreement to form a consortium with four European utilities to provide a 1 MW pressurized SOFC/GT hybrid system to be demonstrated in Marbach, Germany.[36] The project was funded by the US DOE and the European Commission under the Fifth Framework Program with the participation of several major European utilities, including Energie Baden-Württemberg AG (EnBW) of Stuttgart, Germany, Electricite' de France and Gaz de France of France, and Tiroler Wasserkraftwerke Antiengesellschaft (TIWAG) of Innsbruck, Austria. The hybrid system was expected to achieve electrical efficiencies approaching 60 percent and was to be installed at an EnBW facility in Marbach, Germany.[36] The system was supposed to be operational by October 2003 and operate for a period of at least 12 months. In December 2002, the consortium decided that the 1 MW plant would not be built because a suitable microturbine generator was not available on the market.[39]

   - In April 2007, EnBW and Siemens Power Generation announced that they would again join forces to develop a 1 MW pressurized SOFC/GT hybrid demonstration plant at EnBW, Baden-Wurttemberg, Germany.[40] The goal of the project was to develop a system that could convert up to 70 percent of the fuel energy into electricity. The necessary groundwork was scheduled to be completed by 2008 and proceed in stages to a fuel cell hybrid plant with an installed capacity of 1 MW. The fuel cells would account for about 75 percent of total electricity.[41] The companies apparently envisioned that the proposed "Energy Baden-Wurttemberg" plant would eventually generate up to 2.5 MW of power, enough to power 2500 homes,[42] but clearly did not succeed.
- *Two 125 KW CHP System Demonstration Projects.*
   - In November 2005, SWPC announced an agreement with GTT to install a 125 kW CHP system at GTT in Turin, Italy. The delivery to GTT was supposed to be made in September 2007.[43]
   - In November 2005, SWPC and Meidensha agreed to collaborate in developing, manufacturing, and marketing SWPC SOFC systems in Japan.[44] In January 2008, Meidensha announced that it would demonstrate the operation of a 125 kW CHP system at Meidensha, Tokyo, Japan, in 2008.[45] Upon successful completion of the test, the two companies would establish a joint venture, and Meidensha would start marketing the systems either in 2009 or 2010.[46] Neither the SOFC system nor the joint venture materialized.

SWPC's tubular SOFC technology apparently was neither technically feasible nor commercially viable. The demonstrated lifetime of an SOFC system had reached 16,400 h,[17] less than half the requisite 40,000 h. Tubular SOFC systems were prohibitively expensive. Manufacturing the tubular SOFCs and assembly was complex and labor-intensive, not suited for low-cost mass production processes. Costs were coming down as new cell and manufacturing technologies were developed, from \$72,700/kW in 2000 (in the 220 kW proof-of-concept SOFC/GT system) to \$17,000–\$26,000 per kilowatt around 2003. But it was still substantially more expensive than conventional power-generation systems, which cost about \$1000/kW.

- Even SWPC itself concluded that the cost of the system would remain outside the acceptable range.[47]
- Moreover, tubular SOFC systems had inherent limitations in achieving high power density (HPD). The long electrical path through the tubes and the large voids within the stack caused significant ohmic losses, resulting in a large system with inferior power output (Table 6.1).

6.2.1.4. *Siemens Westinghouse Abandons Tubular SOFC Commercialization, Shuts Down Fuel Cell Business, September 30, 2010*

SWPC started to experience financial difficulty as early as in 2002. A sluggish energy market coupled with a technology that was not ready for commercialization had forced SWPC to delay opening its planned \$122 million fuel cell manufacturing plant and eventually abandon the planned construction. In March 2004, SWPC indefinitely delayed plans to complete the large-scale Munhall manufacturing and office facility. By then, SWPC had built a \$20 million, 180,000-ft^2 plant that had been sitting empty for more than a year.[29] Instead of investing an additional \$100 million to complete the Munhall facility and relocate staff to the facility, SWPC decided to invest \$2 million in 2004 to upgrade the 140,000-ft^2 office and manufacturing space it was leasing in

TABLE 6.1 List of SWPC SOFC Demonstration Projects

Year of operation or planned operation	Partners, system rating/type	Installation site	Stack rating (kW)	Operating hours or likely outcome
1986	Tennessee Valley Authority	Chattanooga, Tennessee	0.4	1,760
1987	Osaka Gas	Location unknown	3	3,012
1987	Osaka Gas	Location unknown	3	3,683
1987	Tokyo Gas	Location unknown	3	4,882
1992	Japanese Gas Utility Consortium (JGU)-1	Japan	20	817
1992	Japanese Gas and Electric Utility Group (JUG)-1	Perhaps Rokko New Energy Test Center, Kobe, Japan	20	2,601
1992	JUG-2		20	1,579
1993	JUG-3		20	7,064
1994	Southern California Edison (SCE)-1	Grand Terrace, San Bernardino, California	25	6,500
1995	SCE-2 (with a new fuel cell stack)			5,000
1998	National Fuel Cell Research Center, University of California, Irvine (NFCRC)	Irvine, California		8,250
1995	JGU-2	Japan	25	13,194
1997	EDB/ELSAM-1	Westervoort, The Netherlands	100	4,035
1999	EDB/ELSAM-2 (with a rebuilt fuel cell stack)	Westervoort, The Netherlands		12,577
2001	RWE	Essen, Germany		3,800
2005	GTT (with a refurbished fuel cell stack)	Turin, Italy		16,400

(Continued)

TABLE 6.1 List of SWPC SOFC Demonstration Projects (*Cont'd*)

Year of operation or planned operation	Partners, system rating/type	Installation site	Stack rating (kW)	Operating hours or likely outcome
2000	NFCRC/SCE, 220 kW SOFC/gas turbine (GT) hybrid	Irvine, California	180	3,400
2003	Ontario Power/ Kinectrics, 250 kW CHP	Toronto, Canada	250	1,400
2002	Edison Spa 300 kW SOFC/GT hybrid	Milan, Italy	230	Canceled
2003	RWE 300 kW SOFC/GT hybrid	Essen, Germany	230	Perhaps only for a short time
2003	BP Amoco	Nikiski, Alaska	250	Scaled down to 125 kW and changed site
	Chugach Electric	Anchorage	125	Backed out
	A US Air Force Base	Near Anchorage		Canceled
2002	Ft. Meade 1 MW hybrid	Ft. Meade, Maryland	800	Canceled
2003	EC 1MW Pressurized SOFC/GT hybrid	Marbach, Germany	800	Canceled
2003	Stadtwerke Hanover/ E.On Energie	Hanover, Germany	250	Scaled down to 125 kW
2006			125	Canceled
2003	Norske Shell 250 kW CO_2 Separating SOFC system	Kollsnes, Norway	250	Likely canceled
2007	GTT 125 kW CHP	Turin, Italy	125	Likely canceled
2008	Meidensha 125 kW CHP	Tokyo, Japan	125	Canceled
2012	EnBW 1MW-class Pressurized SOFC/GT hybrid	Germany	750	Canceled

Churchill, PA. In December 2004, the company sold the Munhall facility to US Steel at a loss of $7.2 million.[30]

It also gradually laid off its staff. The SWPC plant in Churchill had nearly 1000 employees at its peak in mid-2002.[48] By March 2009, the facility was left with a skeletal staff of 40 researchers that did bare bones research and development.[49] Only about 40 scientists and engineers remained and worked as part of a US DOE program to combine fuel cell and coal gasification technology to improve gas turbine efficiency.

In the end, the parent company, Siemens of Germany, put the SOFC business on sale. Siemens acquired the fuel cell operation from the former Westinghouse Electric in 1997 for $1.53 billion.[50] After running the fuel cell business for 10 years, Siemens decided the fuel cell business was not a core competency for Siemens' energy business.[51]

- In early 2009, following an unsuccessful 9-month effort to sell the business, the decision was made to restructure the remaining fuel cell business units.[30]
- In February 2010, Siemens agreed to reimburse $468,750 to the State of Pennsylvania for the grant it received in 2002 when the company pledged to build a fuel cell manufacturing plant in Munhall and employ more than 300 workers.

On September 30, 2010, SWPC closed its doors in Munhall, PA. Plug Power reportedly is going to get some energy storage technology from the deal, but SWPC's efforts to commercialize its tubular SOFC technology had ended.

6.2.2. DOE Launches SECA Program in 2001 in Search of New SOFC Technology

The limited progress made by SWPC in its tubular SOFC system cast doubt on the economic viability of the design and led the DOE to look for other options. In fiscal year 2000, DOE's Office of Fossil Energy launched a 10-year SOFC R&D program that focused on the planar design. The new SOFC program—called Solid State Energy Conversion Alliance (SECA)—was intended to develop an SOFC system that would cost as little as $400/kW by 2010.[52] At the time, this cost target was 10 times less than most fuel cell products.

SECA came with a set of fresh ideas to foster the program goals. It looked at planar SOFCs as having the potential to deliver higher power density at lower cost. The SOFC planar design was popular in the 1960s, but when Westinghouse's tubular design became the standard, DOE SOFC funds for planar technology dried up, leaving a small number of SOFC developers to continue planar research on their own. Those developing planar designs included the Institute of Gas Technology (IGT), Ceramatec, Ztek, Technology Management Incorporated, and Allied Signal Aerospace Corporation.[53] These developers, nonetheless, held strong patent positions on planar cell designs for low-cost manufacturing, likely providing DOE an impetus to launch a new SOFC program based on planar SOFCs.

SECA had an innovative R&D approach, which consisted of industry teams that would design and fabricate the fuel cells and a core technology team that would conduct basic research to support the industry teams. The core technology team was to direct its efforts in areas where the industry teams would most benefit, while avoiding duplication of effort. The core technology team was made up of research entities, including universities, national laboratories, small businesses, and would engage in pre-competitive applied research that addressed issues common to all industry teams.[54] Findings and inventions by the core technology team would be available to all industry teams under provisions that served to foster development.

SECA also was well funded. Between 2000 and 2010, DOE provided SECA with more than $467 million, for an average of $42.5 million per year. The annual funding level was higher than any other high temperature fuel cell technologies that DOE had ever sponsored.

6.2.2.1. *SECA Sets Off Renewal of Global Interest in SOFCs*

DOE's SECA program clearly played a major factor in encouraging SOFC activities, setting the pace and direction for SOFC R&D worldwide. The launch of the SECA program—coinciding with the rapidly increasing R&D activity in PEMFCs at that time—triggered a sense of renewed excitement in the fuel cell industry. It raised hopes that low-cost large-scale planar SOFCs would be commercialized for stationary power generation. Some countries, such as Japan, believed that fuel cells would constitute a $60 billion market by 2010 and rapidly increased government support for fuel cell development—for the first time outspending the United States (see Chapter 8).[55] The world saw the start of a global fuel cell race. Suddenly, fuel cell technology became popularly known to the general public, fuel cell conferences that were typically held once in 2 years began to be held once a year, and the race to develop economically practical fuel cells entered a critical stage.

The 10-year SECA program consisted of three phases to be completed by 2010, each phase having progressively more challenging cost and performance goals.[52]

- SECA Phase I would develop within 4 years a 3–10 kW system costing $800/kW with an efficiency of at least 35 percent and degradation rate of less than 4 percent/1000 h.
- Phase II, a 3-year effort, would trim the cost to $600/kW.
- Phase III, also a 3-year effort, would achieve the cost goal of $400/kW or less.

In each phase, fuel-to-energy efficiencies would be enhanced, ultimately reaching 60–70 percent, which would be more than twice the efficiency of most of today's fossil fuel power plants. When the goals were achieved at the end of Phase III, DOE would expect the end product to be an all solid-state fuel cell "building block" that could be mass produced, thereby dramatically lowering costs, much like advances made in solid-state electronics that had rapidly cut costs of computers and electronics in the 1980s and 1990s.

Ultimately, DOE hoped to incorporate the low-cost SOFC systems into its Vision 21 program. Started in March 2000, Vision 21 envisioned a multi-megawatt size, fossil-fuel based, ultra-clean, near-zero emission, energy plant of the future to be realized by 2015.[56] The goal of Vision 21 was to wring every possible bit of useful energy out of carbon-based feed stocks to produce energy products, while eliminating all environmental concerns regarding electricity generation, and doing so at comparatively low cost.[57] Vision 21 initially was not a line budget item but a collection of activities supported under various DOE Fossil Energy programs. By February 2003, Vision 21 had evolved into a $1-billion FutureGen initiative.[58] The initiative aimed to build a prototype of a fossil fuel power plant of the future that would combine electricity and hydrogen production with the virtual total elimination of harmful emissions, including greenhouse gases.

6.2.2.2. *SECA Soon Encounters Tough Challenges*

The SECA program had rough sailing from the outset. Problems included general delays in the development schedule, a major alteration in the program goals, changes in the industry team lineups, inadequate research within the core technology program, and drastic changes in SECA goals that left SECA programs in flux.

6.2.2.2.1. DELAYS IN PROGRAM SCHEDULE

The SECA program was to be completed in 2010, but program work did not start immediately; DOE could not award contracts with industry teams due to lack of budget at the initial phase. Much of DOE's fuel cell budget was tied up with the MCFC and tubular SOFC R&D programs; some of the funds were freed up by 2002, and all funding was freed up by 2005.[59] Eventually, DOE signed up six industry teams to start their proposed projects. According to DOE press releases, DOE granted contracts to four industry teams in August 2001[52] and two additional teams in April 2003.[60] The six industry teams were Honeywell (Torrance, California), SWPC (Pittsburgh, Pennsylvania), Delphi Automotive Systems (Flint, Michigan), and Cummins Power Generation (Minneapolis, Minnesota), FuelCell Energy (FCE) (Danbury, Connecticut), and Acumentrics (Westwood, Massachusetts).

- After Honeywell signed the SECA contract in August 2001, however, Honeywell's SOFC unit was acquired by GE as well as its right to participate in the SOFC program. GE took over Honeywell's position in SECA in December 2001.
- In a surprise move, SWPC joined SECA as one of the six participants. Under the SECA Program, SWPC started to develop the so-called high power density (HPD) SOFCs, which were either flattened tubular SOFCs or delta-shaped tubular SOFCs.

In addition to the delays in signing the SECA contracts, the industry teams took a long time to achieve program goals. GE's SOFC team was the first to complete the Phase I goals, but it took them 4.5 years.[61] All teams successfully completed Phase I by December 2007.[62]

6.2.2.2.2. CHANGE IN SECA GOALS: DEVELOPMENT OF COAL-BASED CENTRAL POWER PLANT

The SECA program was complicated by a major change in goals even before the industry teams completed Phase I tasks. In early 2005, SECA was transformed into a new "Fuel Cell Coal-Based Systems" program aimed to ultimately develop 11 MW-class SOFC fuel cell power plants that would be powered by coal gas, with the following milestones, according to news releases by DOE's National Energy Technology Laboratory (NETL):[63,94]

- By 2010, complete a 5000 hour stack test with a degradation of less than 2 percent per 1000 h of operation and costs of \$700/ or less for the system power block.
- By 2012, demonstrate a 250 kW to 1 MW fuel cell module with a 5-year lifetime.
- By 2015, build a 5 MW proof-of-concept fuel cell system to demonstrate system integration, heat recovery turbines, and power electronics.
- By 2020, build a full-scale 250–500 MW integrated gasification fuel cell (IGFC) demonstration power plant.

This dramatic change in direction was linked to the launch in March 2002 of a \$2 billion, 10-year Clean Coal Technology Initiative.[64] The Clean Coal initiative soon started to draw budget away from natural gas and oil. SECA was funded under DOE's natural gas program at the time and inevitably came under budget pressure.[59] Following a few years of uncertainty, it was incorporated into the Fossil Energy Program, but with the new coal gas goals layered on top of existing SECA cost reduction goals, making attainment of goals even more challenging. According to a DOE news release in May 2008, key system requirements to be achieved included:[65]

- A cost of \$100/kW (2002 dollars) for 40,000-hour fuel cell stack, with a 60 percent efficiency
- A cost of \$400/kW (2002 dollars) for the integrated fuel cell power block
- Achieving HPD in the large cells

Prototype MW-class SOFC power plants would be validated with testing at a FutureGen facility starting in 2012, which would demonstrate cutting-edge carbon capture and storage technology for commercial-scale combined-cycle IGFC/clean coal power plants.[66]

6.2.2.2.3. SHIFTS IN INDUSTRY TEAM LINEUP: GE LEAVES SECA PROGRAM

As the SECA program changed into the Fuel Cell Coal-Based Systems program, some of the original six industry teams in SECA Phase I found the new program less appealing. With Phase I SECA completed, two of the six industry teams—Acumentrics and Cummins—decided not to participate in the new coal-based SOFC systems program. The two had been interested in small SOFCs and perhaps did not wish to launch into MW-class SOFC systems. Surprisingly, the third industry team, GE, left the program as well. GE had previously acquired Honeywell's fuel cell unit that had been selected for the new coal-based SECA program in August 2005 and had already launched a new coal-based SECA project.

- GE closed its SOFC unit and left the SOFC field all together. Little information is available concerning the reasons for the decision. GE's move was a major shock to the fuel cell community because the SOFC unit at GE had long been a major player in planar SOFC development in the United States. GE remained in the SECA program as a nonindustry team member.

By December 2009, the new coal-based SECA program was left with four industry teams:

- SWPC, which was selected to join the new program in August 2005.[67]
- FCE in partnership with Versa Power Systems was selected in February 2006.
- Rolls Royce (RR) Fuel Cell Systems, which acquired SOFCo a year earlier, was selected in May 2008.[65]
- UTC, newly teamed up with Delphi, joined the coal-based SECA program in May 2008.

6.2.2.2.4. CORE TECHNOLOGY PROGRAM FAILS TO PROVIDE CRITICALLY NEEDED ASSISTANCE

Despite DOE's commitment to innovative approaches, the core technology program fell short of providing the critically needed assistance to the industry teams. Basic research that would seek to understand the fundamental principles and mechanisms regarding the physics of fuel cell electrochemistry was not pursued. Research projects were so small in scale and so focused on near-term solutions and product development that they could not gain knowledge and facts that would advance fuel cell technology and enable a paradigm shift in cost and longevity. As a result, the program could only generate incremental improvements and did not develop a fully functional and commercially competitive fuel cell technology. Indeed, neither the core research teams nor industry teams achieved any dramatic technological breakthroughs over the 10-year life of the SECA program.

A snapshot of the SECA core technology program shows a disappointing state of affairs. For example, the core technology program announced in May 2008 indicated that the majority of the projects were so constrained by budget that they essentially were small science projects. They generally were conducted at low levels of funding for a limited period of 2–3 years:[68]

- The core technology program announced in May 2008 provided less than $1 million per project a year on average. The nine projects announced for the core technology program in May 2008 received a total of about $6.7 million from DOE,[68] about $740,000 per project per year. One of the nine, a major corporation, received $2.8 million and two others received $600,000 and $750,000, respectively. The other six each received $300,000 to $450,000. Six projects were conducted by universities and four were conducted by corporations, including two small businesses.

- Modest funding and the short research periods were major contributing factors that determined the scale of research projects. Few researchers could be assigned to a project and little specialized infrastructure or complex tooling could be used. Short deadlines reduced the incentives to explore fundamental principles and understand new phenomena.
- Basic research was sorely needed, but not funded. The majority of the core research projects were focused on observing unfavorable phenomena—such as performance degradation—describing it, taking measurements, and trying to invent solutions that might address part of a problem and improve it incrementally or temporarily, ignoring the need to discover the underlying physics of the materials and to explain the nature of the mechanisms that rule fuel cell dynamics.

6.2.2.2.5. LATEST CHANGE WEAKENS SECA

In October 2010, DOE awarded Ameren Energy Resources Company a contract for the $1 billion FutureGen 2.0 project—the first large-scale, integrated test of an oxy-coal-fired power plant with CO_2 capture and storage.[69] Babcock & Wilcox Power Generation Group subsidiary (B&W PGG) and Air Liquide Process & Construction were included in the award. The cooperative agreement was the first step in a project to repower a 200 MW power-generation system at Ameren's Meredosia Plant near Jacksonville, Illinois, with oxy-coal combustion technology developed by B&W PGG and Air Liquide. The project was designed to produce clean energy from coal by capturing and storing approximately 1.3 million tons of CO_2 each year.

The DOE award had the effect of ending the original SECA program. It morphed the FutureGen project into an oxy-fuel project with CO_2 capture, leaving no role for attainment of SECA's broader cost reduction goals program. SECA entered a kind of limbo with its future uncertain. Indeed, its budget declined from $56.4 million in FY2009 to $50 million in FY2011, and in FY2012 the budget for SECA was zeroed out. Although the Congress restored some funds for SECA, the future of SECA would be bleak.

6.2.2.3. *Development of Commercially Viable SOFCs Under SECA Unlikely*

Fundamentally, SECA program goals rested on the flawed assumption that any technology could be developed into a commercially viable product, given sufficient investment and time to make incremental improvements. While this assumption has been valid for many technologies, such as aviation, motor transport, telecommunications, and the like, it does not apply to fuel cells. It has been more than 170 years since the first fuel cell was invented, and despite much investment in time and money, there are no truly commercially successful fuel cell products on the market. Fundamental knowledge of fuel cell mechanisms and how these mechanisms affect fuel cells performance and longevity remain elusive.

- Indeed, basic research on fuel cells has never been done. When Grove unveiled his gas battery, the scientific community was struggling to understand the basic principles of chemistry and electricity, matter and energy. A few scientists, including Christian Schönbein (1799–1868), Johann Poggendorff (1796–1877), and Friedrich Wilhelm Ostwald (1853–1932) debated the question of exactly how Grove's gas battery worked.[70] Oswald, for one, experimentally determined the interconnected roles of the various components of the fuel cell in 1893. However, none of them arrived at the theoretical understanding of how fuel cells operate. Since then, few scientists have attempted to study the basic physics of the fuel cell. History is full of stories in which a developer quickly fabricated a seemingly functional fuel cell, convinced himself of his successful mastery of the technology, and ran with it to meet an inevitable market failure.

- Without knowing the basic physics of fuel cells at the atomic level, no matter how much money and time may be spent and how many improvements may be made, ultimate success will remain beyond reach. It is as if a builder builds a castle on land with no knowledge of its geology.

Moreover, cost goals that were set by the SECA program did not factor in the very long lifetimes, reliability, and robustness that fuel cells systems would need to be commercially attractive for large-scale power generation. As a result, reported build costs for fuel cells were unrealistic and assumed heroic cost savings based on production volumes that could not be reasonably projected for the market place. Later, performance goals became confused and diluted by SECA's linkage to the Coal Gas Technology Initiative. Because the fundamental challenges of fuel cells were never resolved, meeting or achieving the specific SECA goals and milestones was utterly meaningless.

The outcomes of SECA need to be understood in this context and go some distance to explain why the achievement of SECA goals did not lead to commercialization of SOFC technology:

- In January 2006, DOE announced that the cost of GE's prototype system was estimated at $724/kW for annual productions of 50,000 units, which surpassed the SECA Phase I target of $800/kW.[61] Despite this impressive achievement, GE closed the SOFC unit and dropped out of the SECA program in 2007. We can only speculate that the developer believed that it could produce fuel cells at that price, but that the fuel cells would not be functional, efficient, or sufficiently durable.
- In February 2009, SWPC completed the first phase of the new coal-based SECA program.[71] SWPC was testing a 10 kW SOFC stack that consisted of new higher power Delta8 (HPD8) cells. The stack was operated for 2500 h as of February[63] and more than 5000 h by April 2009.[72] According to SWPC sources, the stack demonstrated significantly higher power density and no detectable degradation.[73] Nonetheless, SWPC shut down its SOFC business as of September 30, 2010. Again, we can only speculate that SWPC's SOFC HPD8 technology could not yield a successful commercial product.
- In March 2009, two fuel cell stacks developed by FCE in partnership with Versa achieved 5000 h of service in February, meeting a SECA goal, according to DOE/NETL.[74] The FCE/Versa fuel cell stacks also exhibited an overall degradation of only 1.7 percent and 2.6 percent/1000 h—much less than SECA's 2008 (interim) target of 4.0 percent/1000 h, according to DOE/NETL. These achievements were superior to those of the other SECA counterparts. Nonetheless, this performance could not guarantee that the fuel cells would operate with low enough degradation to meet the 40,000-hour operational standard.

The absence of a critical review mechanism in the SECA program was a third factor that contributed to the program's lack of meaningful achievements. While there were some review sessions, including annual peer reviews, these reviews never uncovered the fundamental issues surrounding fuel cell science. The problems might have been detected if there had been comprehensive, critical reviews by objective, independent third parties who possessed not only scientific, engineering, and physical knowledge but also perspectives, insight, and ideas, while unafraid of breaking down conventions.

- Because the US SOFC leadership and the SECA program were so admired, however, the manner in which SECA was set up became the *de facto* world standard and was adopted worldwide with little variation from country to country.

6.2.3. Meanwhile, Many US Companies Launch SOFC Development Activities

Despite all these shortcomings, DOE's SECA program did serve to dramatically revitalize SOFC R&D activities in the United States as well as globally. Some companies that had been disengaged in SOFC development were encouraged to restart R&D after DOE launched the SECA program. In contrast to SECA's goals, many of these companies developed SOFC products for small-scale, temporary, or transportable power generation.

6.2.3.1. *Acumentrics*

In 1994, Acumentrics was founded as a manufacturer of custom power products, specializing in rugged AC and DC uninterruptible power supplies (UPS) for military and industrial power conditioning and backup.[75] In 2000, the company expanded into fuel cells by acquiring intellectual property for a tubular SOFC design.

- Acumentrics employs an aggressive, versatile business style. Over the company's 10-year history, Acumentrics participated in the R&D programs of two different DOE components—the Office of Fossil Energy and the Office of Energy Efficiency and Renewable Energy (EERE)—and a Japanese METI R&D program. The two DOE components administer different types of energy sources and the participants in their programs typically affiliate with only one component, making Acumentrics one of a few companies that have access to both. The company also managed to join a highly restrictive Japanese government project. Acumentrics is focused on the military niche market now, which is more tolerant of trial products than the commercial markets.

In 2002, Acumentrics developed a 2 kW microtubular SOFC system for uninterruptible power[76] and was selected to participate in SECA in April 2003. The project was set as a 3-phase, 9-year effort for Acumentrics. Acumentrics was funded at about $38 million over the 9-year period and contributed $36.5 million in cost sharing.[60] Acumentrics' design involved a series of ceramic mini-cylinders, each roughly the size of a soda straw. Combined into a 10 kW module, the design potentially would offer exceptional ruggedness and quick start-up, making the fuel cell applicable for residential markets, military applications, broadband communication networks, and as APUs for heavy-duty trucks.

Under SECA, Acumentrics in 2006 produced a 5 kW prototype ready for testing at the NETL in Morgantown, West Virginia.[60] By mid-2006, Acumentrics' 5 kW system met and exceeded all targets in tests at NETL. The unit showed only 0.5 percent degradation per 1000 h.[77] It achieved over 52 percent total fuel-to-electrical energy efficiency when using natural gas, compared to less than a 25 percent conversion efficiency rate for standard internal combustion technology of a similar size. It also demonstrated an estimated mass production manufacturing cost below $800/kW and reduced the size and weight of fuel cell power units by more than 75 percent. The unit was shipped back and restarted at Acumentrics' Westwood, Massachusetts, facility in July 2006. In September 2008, Acumentrics reported that the system had exceeded 10,500 h of operation to become the world's longest-running small-scale SOFC system.[78]

To strengthen its technology position, Acumentrics in 2007 acquired all of the assets and employees of Fuel Cell Technologies Limited (FCT) of Ontario, Canada.[75] FCT specialized in balance of power for SOFC systems and previously was a partner of SWPC. It had developed a 5 kW SOFC CHP system under SECA Phase I.

In 2008, Acumentrics transferred out of SECA when SECA changed to a coal-based program. Acumentrics soon found a new sponsor in DOE's EERE. In April 2008, DOE EERE awarded Acumentrics a 3-year, $15.6 million (including the cost-share provided by Acumentrics) contract to advance the company's microtubular SOFC technology and to double the size of the unit to 10 kW for remote power, military uses, and micro combined heat and power (mCHP) applications.[79] The development work was funded through EERE's Hydrogen, Fuel Cells, and Infrastructure Technologies Program (HFITP).[77]

- In an HFITP peer review report in June 2010, Acumentrics reported significant progress. It reported a 24 percent increase since 2009 in cell power density to 0.421 W/cm^2, enabling 33 percent reduction in stack volume and 15 percent reduction in stack weight.[80] It also demonstrated a low degradation rate of 0.86 percent per 1000 h during 1500 h of testing[80] with 40 percent electrical efficiency.[79] In October 2010, Acumentrics reported that it had shown the ability to achieve power densities of more than 600 mW/cm^2 on a single standard production cell.[81] Owing to the increase in cell performance, the package size of a generator was reduced by 55 percent, resulting in capital cost savings and 15 percent reduction in weight for all generator types produced.
- Nonetheless, Acumentrics has not presented a commercial product that is competitive with traditional counterparts.

Acumentrics also has been interesting in business ventures, and as early as in 2003 it attempted a market entry into Japan. In October 2003, Acumentrics established a joint venture, Acumentrics Japan, with Sumitomo Corporation of Japan, a major trading house, to market Acumentrics' microtubular SOFC units in Japan.[82]

- No major success has been reported, however. Acumentrics likely has abandoned its entry into the Japanese market.

Today, Acumentrics appears to be primarily focused on a niche military market for their fuel cell products. The company produces electrical power and heat systems from 250 W to 2 kW of AC or DC output,[83] largely specializing in rugged uninterruptible power supplies (RUPS).[84] The company reported that more than sixty thousand 1–2 kW RUPS had been deployed globally at military and industrial sites.[85]

6.2.3.2. *Allied Signal Aerospace*

Allied Signal had a relatively short SOFC history, which left a sad legacy. In the early 1990s, the company made significant contributions in advancing SOFC cell manufacturing and came close to commercializing a portable power system for the military in 1999. That year, the company merged with another company, Honeywell, which then sold the SOFC division to GE in 2001. GE shut down the SOFC division in 2007, laying off all the SOFC staff.

Allied Signal's SOFC program began in 1987 in a cooperative venture between Asea-Brown Boveri (formerly Combustion Engineering), Southern California Edison, and Argonne National Laboratory.[10] The venture set out to advance Argonne's monolithic SOFC design but was repeatedly unsuccessful in fabricating multiple-cell stacks.

In 1993, Allied Signal developed a tape calendaring process,[86] a major milestone that enabled a series of rollers to roll out materials into a long, smooth tape consisting of very thin layers of electrodes and electrolytes. By cutting the tape, a large quantity of SOFC cells of equal quality could be produced, potentially fabricating low-cost planar SOFC stacks. A highly successful five-cell stack thus was built and operated without difficulty for several hundred hours, followed by efforts to build stacks with an increased number of cells.[10]

In the mid-1990s, Allied Signal secured two major SOFC contracts. GTI awarded Allied with a 4-year contract to develop and test low-temperature (600–800 °C) cells and stacks.[10] Allied was also selected by DARPA to develop a 500 W SOFC stack for mobile communications.

- In 1999, Allied tested a 26-cell stack with a cross flow configuration (10 cm × 10 cm footprint), which showed an open circuit voltage of 26.7 V at 800 °C and produced about 860 W at 18.2 V (0.7 V/cell) with a potential peak power of 1.2 kW (0.6 W/cm^2) with hydrogen fuel.[87] Based on this result, Allied planned a portable power system operating on logistic fuels for military applications. The system was to be operated at an 80 percent fuel utilization rate, produce 500 W at 28 V, weight 7 kg in a volume of 17 inches × 11 inches × 9 inches (43 cm × 28 cm × 23 cm). After the merger with Honeywell, the new company developed a 1 kW system.

6.2.3.3. *Bloom Energy*

Bloom Energy (formerly Ion America) was founded in 2001 based on work performed at the University of Arizona as part of the NASA Mars space program.[88] In early 2006, Bloom shipped its first 5 kW field trial unit to the University of Tennessee, Chattanooga. After 2 years of successful field trials in Tennessee, California, and Alaska to validate the technology, the first pre-commercial prototype product was shipped to Google in July 2008.

In February 2010, Bloom Energy announced that since the first commercial customer installation in July 2008, it had installed energy servers that generated at least 3.5 MW power output. These included 500 kW units at the Bank of America in Southern California, 500 kW units at Coca-Cola in Dinuba, California, 400 kW units at Cox Enterprises' KTVU TV station in Oakland, California, 500 kW servers at eBay in North San Jose, five 100 kW servers at an FedEx facility in Oakland, California, 400 kW units at Google's main campus, 300 kW servers at Stapes at Ontario, California, and 400 kW systems at Walmart's California retail locations.[89]

Since the announcement, Bloom has installed at least 16 more 100 kW systems in 2010. They include a 100 kW server installed at the city-run Electric Power Board (EPB), Chattanooga, Tennessee, in June 2010; three 100 kW servers installed at Cypress Semiconductor, in San Jose, California, in July;[90] twelve 100 kW servers installed at Adobe System, San Jose, California, in September.[91]

In January 2011, Bloom made another significant announcement. It started to offer a bold innovative service called "Bloom Electrons" that would allow customers to purchase the electricity at a set price for 10 years provided by the "Bloom Box" without incurring any other costs.[92] According to the press release, the immediate financial benefits of the new service had generated tremendous customer interest with more than 20 MW of power (200 systems) already secured to provide electricity to new and repeat customers. The "Bloom Electrons" customers include Coca Cola, Staples, Walmart, California Institute of Technology, Kaiser Permanente, BD (Becton, Dickinson and Company), and Southern California Gas Company.

Since the new Bloom Electrons service began until the end of April 2012, Bloom has installed or planned to install energy servers generating at least 50 MW power output. Although it is not clear how those transactions were funded, customers likely chose no cost options under the electronics service:

- The biggest deal appears to have involved Bloom Energy and the Governor of the state of Delaware, the Secretary of Delaware, Department of Natural Resources and Environmental Control, the Delaware Economic Development Office (DEDO), and Delmarva Power and Light.[93] In May 2012, a deal was approved by Delaware's Public Service Commission (PSC), paving the way to bring Bloom's fuel cell technology to Delaware. Under the plan, Bloom will build a manufacturing facility to produce its fuel cells at the former Chrysler plant in Newark. Bloom reportedly would add 30 MW of power to the regional power grid by installing fuel cells at two Delmarva Power substations by 2016. To help make those initiatives possible, the deal approved by the PSC included a renewable energy tariff added to Delmarva customers' bills. The tariff, according to a PSC consultant's report, is expected to cost Delmarva Power's residential customers an average of between \$1.30 and \$1.40 per month.

For the expected increased sales volume in the future, Bloom has also been expanding manufacturing capacity. The speed and size of expansion is also remarkable for a company that started in 2001 with only eight employees:

- In April 2011, Bloom Energy announced that it expanded its California manufacturing facility in Sunnyvale, California, to prepare for the expansion of its business.[94] Bloom increased the company's footprint by four times to over 210,000 ft^2, employing more than 1000 workers. Bloom Energy expanded its workforce by over 70 percent in 2010 alone, and has grown 525 percent over the past 4 years, company officials announced.
- Moreover, in April 2012, Bloom Energy broke ground on a new manufacturing center in Newark, Delaware, as part of the Delmarva deal.[95] Located on the 272-acer site formerly occupied by a Chrysler assembly plant, and now owned by the University of Delaware, the facility is expected to employ 900 people.[93] Once the facility in Newark is complete, Bloom would double its production capacity.

Bloom Energy has been extraordinarily successful in expanding its business, undoubtedly owing to the accomplished venture capital investor Kleiner Perkins of Silicon Valley. Kleiner is noted for helping make Google, Genentech, and Amazon.com international high-tech powerhouses.[96] Despite the strong support Bloom enjoys today, it is unclear if and when Bloom's SOFC technology will be viable, competitive, and profitable:

- Bloom has not explained the specific details of its cell and stack component technology. According to Bloom, "At the heart of every Energy Server is Bloom's patented SOFC technology. Each server consists of thousands of Bloom's fuel cells. Each cell is a flat solid ceramic square made from a common sand-like powder. Each Bloom Energy fuel cell is capable of producing about 25 W, enough to power a light bulb. For more power, the cells are sandwiched, along with metal interconnect plates into a fuel cell stack. A few stacks, together about the size of a loaf of bread, are enough to power an average home. In an energy server, multiple stacks are aggregated together into a power module, and then multiple power modules, along with a common fuel input and electrical output are assembled as a complete system."[97]

In view of the lack of information on the composition of Bloom's fuel cells, it could be assumed that Bloom is using proprietary materials for their cells. If it is using the most widely used materials today—yttria-stabilized zirconia (YSZ) for electrolyte, the perovskite LSM for the cathode, and a cermet of nickel-YSZ[98]—costs would not necessarily be a problem. Another possibility is that Bloom is using a highly expensive, rare earth element, scandium, as a replacement for yttria for electrolyte, constituting scandium-doped zirconia ceramics. Scandium is a light material with a higher melting point than aluminum, which began to be used in aerospace applications starting in the 1970s.[99] As a replacement for yttria, scandium theoretically offers an ionic conductivity increase by an order of magnitude and thus reduced layer resistance.[98] This can then be exploited to achieve higher power densities in SOFC cells as compared to an YSZ electrolyte, or could reduce operating temperatures in favor of reduced degradation and corrosion in the fuel cell device.

The actual longevity of Bloom's energy server is not fully confirmed. According to press reporting, Bloom has said that the Bloom boxes, with reasonable maintenance, would have a 10-year life span.[101] It clearly convinced one of the customers, Cypress, that the three 100 kW Bloom Energy Servers installed at Cypress were expected to pay for themselves in 5 years, at which point they would be generating nearly "free" electricity for Cypress.[90] Bloom, however, has only demonstrated that their systems operate for 2 years. The first 5 kW field test unit tested in Chattanooga was test-operated for 2 years. Little other verifiable data have been reported.

Bloom energy servers reportedly provide an electrical efficiency of about 45–55 percent.[102] In contrast, PAFC systems, which offer both heat and power, have a total efficiency as high as 90 percent, with a 10-year cell stack durability and 20-year product life.[103] Higher efficiency and durability of Bloom energy servers would lower the cost and raise the competitiveness, however.

At this time, it is difficult to confirm whether or not the Bloom electrons service will be a viable business strategy. The installation and maintenance costs of energy servers under the Bloom Electrons service could be highly costly to Bloom at least initially compared to the equipment purchase agreements. Long-term commercial success remains to be seen (Table 6.2).

6.2.3.4. *Ceramatec*

Ceramatec has a long history of SOFC development starting in 1976. It had a strong academic background and made significant technological advancements early on, which held the promise of commercial success. Ultimately, however, success did not materialize.

Ceramatec was spun off from the University of Utah and incorporated in 1976 as an R&D company, specializing in ceramic solid-state ionic technologies.[104] Between 1976 and 1990, Ceramatec developed innovative technologies with federal and corporate funding in an effort to facilitate commercialization of products.[105]

In 1990, the company became a subsidiary of Elkem, a major international materials company,[104] and started to develop SOFC technology with the sponsorship of a Norwegian based research consortium (Norcell), GTI, EPRI, and DOE.[10] Ceramatec developed a proprietary planar SOFC design, CPn, which consisted of stacks and a fuel processor and placed some cells in a series, rather than in parallel, to obtain greater efficiency.[106]

- Using CPn technology, Ceramatec attained a power density of 0.18 W/cm^2 and a current density of 250 mA/cm^2. Ceramatec tested individual planar cells for over 20,000 h with little voltage degradation.[10] A multistack configuration allowed multistaged oxidation of the fuel. Partially oxidized fuel from an upstream stack was further oxidized in downstream stacks—providing high conversion efficiencies and more effective thermal management of the fuel cell stacks with the fuel reformer and air preheater. Ceramatec

TABLE 6.2 Bloom Energy's SOFC Power Plant Installation Activities (As of April 30, 2012)

Plant size, fuel	Installation site	Schedule of operation
Installation Activities between July 2008 and February 2010		
11 MW	Customers including Bank of America, Coca-Cola, Cox Enterprises, eBay, FedEx, Google, Staples, and Walmart	Installed between July 2008 and February 2010
Installation Activities between February 2010 and December 2010		
100 kW	Electric Power Board (EPB), Chattanooga, Tennessee	Installation announced in June 2010
3 × 100 kW	Cypress Semiconductor, San Jose, California	Installation announced in July 2010
12 × 100 kW	Adobe Systems, San Jose, California	Installation announced in September 2010
"Bloom Electrons" announced in January 2011; would allow customers to purchase electricity provided by the Bloom Box without incurring any other costs; more than 20 MW of power (200 systems) are already secured to provide electricity to repeat customers, such as Coca Cola, Staples, Walmart, and new customers, such as California Institute of Technology, Kaiser Permanente, and BD (Becton, Dickinson and Company).		
5 × 100 kW	San Jose Arena Management, San Jose, California	Installation announced in July 2010
6 × 100 kW	Fireman's Fund Insurance, Navato, California	Installed in June 2011
7.5 MW	AT&T's 11 California sites (Corona, Fontana, Hyward, Pasadena, Redwood City, Rialto, San Bernardino, San Diego, San Jose, and San Ramon)	Plan announced in July 2011; to be operational by mid-2012
5 × 100 kW	NTT America Lundy Data Center, San Jose, California	Installed in July 2011
5 × 200 kW	Cox Enterprises, San Diego, California	Installed in February 2012
4.8 MW	Apple Data Center, Maiden, North Carolina	Plan confirmed in April 2012
	Owens Energy	
	Urban Outfitters	
30 MW	Delmarva Power	Planned to supply 30 MW power capacity over 21 years at a stable and competitive price, starting with 10 MW in 2012
	Washington Gas	

Note: The table was created based on press releases by Bloom Energy, http://www.bloomenergy.com/newsroom/press-releases/.

tested a 1.4 kW multi-stack module using pipeline natural gas.

In 1994, to commercialize the CPn technology, Ceramatec established SOFCo, a limited R&R partnership, with Babcock and Wilcox, a McDermott International subsidiary.[107] Ceramatec was clearly hopeful that it could achieve successful commercialization through SOFCo. Ceramatec had outlined a 5-year $20 million program that would culminate in the development of 50 kW planar SOFC-based cogeneration systems by 2000.[10] None of these plans materialized.

In 2000, Ceramatec separated from Elkem. In 2008, Ceramatec was acquired by CoorsTek, the largest technical ceramics manufacturer in North America.[104] It appears that Ceramatec recently has been conducting R&D on SOFC component technologies as a core technology team in SECA. Ceramatec is no longer engaged in the development of SOFC cells, stacks, and systems.

6.2.3.5. *Cummins Power Generation*

Cummins Power Generation, a major US manufacturer of engines and related technologies, has been developing a small SOFC system for stationary and mobile applications since the early 2000. Cummins, which does not develop SOFC stacks in-house, relies on its partners for SOFC stack supplies. During the course of a 10-year SOFC development history, Cummins has had four different SOFC partners, McDermott Technology, SOFCo, Versa, and Protonex. It has yet to make a commercial success out of its SOFC effort.

Cummins started the development of fuel cell technology in the 1960s and bolstered its SOFC activities through its association with the DOE SECA program in the 2000s.[108] In August 2001, Cummins and its partner McDermott Technology were selected for a project in SECA.[52] DOE was to provide $74.2 million for the project while the Cummins/McDermott would contribute $91.5 million. The team was interested in the consumer and commercial markets and was engaged in development of a modular, 10 kW system for stationary and mobile applications. Cummins was to integrate SOFC modules supplied by McDermott Technology into an electrolyte-supported planar SOFC system, which consisted of a seal-less stack, fabricated with thermally matched components. It was operated at 825 °C.[109]

- Cummins' partner McDermott Technology was replaced by SOFCo at the beginning of fiscal 2002. McDermott Technology went bankrupt, and SOFCo inherited McDermott Technology's assets and personnel involved in SOFC work.[110]
- During the fiscal year 2002, however, SOFCo's cell development in SECA did not go well and third-party cells were deployed to support Cummins' 2003 stack development.[111]

In 2007, Cummins successfully developed a 6 kW SOFC power system. Versa Power Systems, rather than SOFCo, supplied SOFC modules. The 6 kW system met all the SECA Phase I program goals, including durability, reliability, efficiency, and targeted manufacturing cost estimates. The prototype unit ran flawlessly for over 2000 h in CPG's test facility in Minneapolis, Minnesota.[112]

Despite its successful participation in SECA Phase I, Cummins did not participate in the subsequent, coal-based SECA program. Since 2005, Cummins has been developing an SOFC-based truck APU fueled by diesel under EERE sponsorship. Cummins' partner in the program is Versa.

- In 2009, according to press reports, a prototype 3 kW unit operated for over 2000 h at an efficiency of over 37 percent, significantly more efficient compared to conventional generator sets of similar size, where efficiencies are generally well below 30 percent.[113]

In May 2010, Cummins announced that it successfully demonstrated an SOFC stack in a hybrid-configured APU on a Class 7/8 truck for 10 h.[114] Highlights from the demonstration test included the unit running continuously on ultra low sulfur diesel, running the truck's air conditioning and lighting solely from the SOFC and hybridizing the fuel cell and battery for up to 3.8 kW AC. The average net power produced over the test period was 1.1 kW DC or 820 W AC.

- The APU unit incorporated SOFC modules supplied by Protonex Technology.
- The Cummins/Protonex APU system, however, fell far short of some of DOE's 2010 APU targets. The Cummins/Protonex fuel cell and battery was 197 kg in weight and 360 L in volume,[115] with the net power production of about 19 W/kg and 10 W/L. DOE's 2010 APU targets required the specific power of 100 W/kg and power density of 100 W/L.

Cummins said as of August 2009 that mobile power products using fuel cell technology would be commercially available within 2–3 years, with larger stationary units available in 7–10 years.[113]

6.2.3.6. Delphi

Delphi, a major US automotive manufacturer, started SOFC development in 1999, primarily focused on truck APUs. It built its first-generation gasoline-powered SOFC APU system in 2001,[116] using SOFC stacks made by an outside developer, Global Thermoelectric (Canada).[117]

In August 2001, Delphi participated in SECA and started to develop a 5 kW system that could be used for automotive and truck APUs, distributed generation, and niche markets in the military sector. DOE provided $74.6 million to Delphi while Delphi and its partners contributed $60.9 million.[52] Delphi's SOFC system was designed to be compact, had a rapid transient capability, and operated at 750 °C.[109]

Since 2003, Delphi stopped outsourcing SOFC stacks and started to develop its own.[118] In May 2006, Delphi developed a 5 kW system with its own SOFC stacks, which surpassed all SECA Phase I goals:

- Delphi's prototype system produced peak power of 4.24 kW on methane.[117] The system demonstrated a peak efficiency of 37 percent. Delphi also met the Phase I cost goal of $800/kW for total power, by achieving an estimated $770/kW, assuming high volume production. Delphi also met the durability goal with power degradation of 7 percent over 1500 h of operation.

In 2007, Delphi tested a 30-cell Generation 3.2 stack, which produced 1.44 kW with a power density greater than 450 mW/cm^2 at 0.8 V per cell.[119] Also in 2007, Delphi developed an SPU1 SOFC APU system fueled by diesel. The system generated a net rated power of 3.4 kW at an electrical efficiency of 38 percent. It operated for 1000 h with a degradation rate of 0.84 percent/500 h.

- In August 2008, Delphi successfully tested its proof-of-concept system in the Navy's unmanned undersea vehicles.[120] In the test, the Delphi system met the US Navy's targets for system size, power output, and efficiency. It successfully demonstrated that it could operate on logistic fuels—such as jet fuels—tolerate fuel impurities, generate high-quality heat for fuel reforming, and was highly efficient in converting fuel energy into electricity.

In addition to DOE's SECA and new coal-based program, Delphi in September 2004 participated in DOE's EERE Hydrogen and Fuel Cell Program to develop APU technology. In June 2008, Delphi successfully tested an SOFC APU system to operate the electrical system and air conditioning of a Peterbilt Model 386 truck under conditions simulating idling conditions for 10 h.[121] The device provided an alternative to running a truck's main diesel engine or using a truck's batteries to power auxiliary electrical loads during rest periods, thereby lowering emissions, reducing noise, and saving fuel. The APU provided an average of 800 W of electricity to the truck over the course of the test.

- The APU system showed a specific power of 40 W/kg, a power density of 52 W/L, and a power efficiency of 38 percent.[122]

In June 2010, Delphi reported that the Delphi SOFC APU system had been tested for 168 h over 640 miles on one vehicle.[123] The APU system demonstrated a net peak load of 1.4 kW and the system efficiency of 18 percent.[124] In May 2011, Delphi reported that its SOFC APU hardware was mounted on a Peterbilt Class 8 truck and driven more than 3000 miles on highways and secondary roads.[124] It also reported that an APU system completed more than 50 thermal cycles out of 250 planned at Rochester Institute of Technology (RIT) and that APU systems were tested on the equivalent of 17,000 highway miles on a vibration table; the stacks were tested for an equivalent of 3.5 million miles.

- According to Delphi, an APU system will be installed on national fleet trucks in 2012 and the technology and robustness of the SOFC will be tested for 1 year.

6.2.3.7. FuelCell Energy/Versa Power System

FCE, which is the sole developer and manufacturer of MCFCs in the United States, made an extraordinary effort to join the SECA program. In April 2003—about 2 years before the DOE MCFC program funding was to end—FCE proposed a \$139 million project to be sponsored under SECA.[125] If the project had run its full course of 10 years, more than \$67 million in federal funding would have gone to FCE and FCE would have invested nearly \$72 million in cost sharing. The company was selected to start SECA Phase I in July 2003 via a pre-award authorization from the DOE. It received a provisional SECA award from the DOE in February 2004, which enabled the company to increase the level of effort consistent with the program scope of work. DOE finally awarded the official SECA contract to FCE in September 2004.

- To qualify for obtaining a SECA contract, FCE made an extraordinary effort to sign up Versa as a partner. Versa, based in Littleton, Colorado, and Calgary, Alberta, is a leading developer of SOFC technology. It was founded in 2001 as a joint venture between GTI, EPRI, Materials and Systems Research, Inc. (MSRI), and University of Utah[126] and was a highly attractive company. As a first step, FCE in April 2003 invested \$2 million in Versa and acquired a 16-percent equity stake in Versa and enlisted Versa as a partner to win the \$139 million contract from the DOE's SECA Program.[127] Also in August 2003, FCE acquired Global Thermoelectric for US\$80 million.[128] Global Thermoelectric, based in Calgary, Canada, was the world leader in remote power generation systems.[129] Global also was leading the world in SOFC development starting in the late 1990s. In June 2004, FCE sold all the shares of Global Thermoelectric except Global's SOFC unit. In November 2004, FCE entered into an agreement to combine Global's SOFC unit with Versa, and in exchange, FCE received stock in

Versa, increasing its ownership position in Versa from 16 to 42 percent.[130]

- The series of investments had resulted in a substantial pool of intellectual property, including pioneering SOFC planar cell technology developed at the University of Utah and MSRI, cell manufacturing, stack and system design and experience developed at Global Thermoelectric, and stack cooling improvements developed by GTI.[47]

As a consequence, the FCE/Versa team excelled in the SECA program, outperforming most of the industry teams in the program:

- In April 2007, FCE and Versa completed SECA Phase I tasks 6 months ahead of schedule,[131] while the other five industry teams took longer than the expected 4 years. The FCE/Versa team announced that a prototype 3 kW fuel cell commissioned by DOE completed 2100 h of performance and endurance tests, exceeding the required 1500 operating hours. Over the operational period, the prototype met all DOE's targets, including power output, system efficiency, system availability, and overall system endurance. System cost calculations also met the target. Subsequently, the prototype was shipped to the DOE NETL in Morgantown, West Virginia, for another 1600 h of testing. Both the initial system performance tests and the factory cost estimate were audited and confirmed by independent third party consultants approved by DOE.
- In February 2006, DOE selected the FCE/Versa team for the new coal-based SECA program. Within 3 years, they successfully developed two 10 kW stacks. Both stacks reached more than 5000 h of operation by the end of February 2009.[74] The two stacks also exhibited an overall degradation of only 1.7 percent and 2.6 percent per 1000 h—much less than SECA's 2008 target of 4.0 percent. The FCE/Versa team was the third contractor to enter the new coal-based SECA program, but the first to achieve the program goals. The team took 3 years to reach the 5000-hour mark.
- The FCE/Versa team won a project continuation award in February 2009 from DOE that enabled them to move onto the second phase of the new coal-based SECA program.[63] In the second phase, the team was expected to build a minimum 25 kW SOFC stack that would form the basis for a 250 kW fuel cell power module and a 5 MW proof-of-concept system that would operate on coal gas. In October 2010, FCE/Versa reported that it had developed a stack that consisted of one hundred and twenty 550 cm^2 cells, which generated 20 kW and a peak power of 25 kW.[132] Testing under normal operating conditions was conducted at 715 °C at 200 A (0.364 A/cm^2) with fuel and air utilizations of 68 percent and 15 percent, respectively.

In April 2011, Versa completed Phase II of the Coal-Based SECA Program with development of a new generation of low-cost high-performance cell technology. It developed third-generation manufacturing processes for a 550 cm^2 active area cell (25 cm × 25 cm) based on high-performance thin cell technology. The thin anode substrate cells enhanced mechanical strength, enabling a 25 percent cost reduction in cell material. Versa tested a 120-cell stack for 1500 h demonstrating a peak power of 25.2 kW and achieving a steady-state average power degradation rate of 0.9 percent/1000 h. Versa fabricated more than 5000 of these cells in the pilot manufacturing facility with more than 95 percent product yield. Versa plans to enhance stack robustness and reliability and develop a 60 kW stack module in the SECA Phase III Project in 2012.

At its website, Versa stated that it had built and operated multiple 2 kW systems and accumulated over 30,000 operating hours, recently achieving a net fuel to AC power efficiency of 35 percent on natural gas.[133] Versa also had demonstrated a 20-cell stack life of 8000 h with a performance decay rate of 2.3 percent/1000 h at a design power density of 300 mW/cm^2.

In May 2011, the FCE/Versa team signed a contract with DOE for Phase III of the SECA program worth \$8.2 million.[131] In Phase III, a 60 kW SOFC module will be constructed at FCE's facility in Danbury, Connecticut, and begin operating in summer 2012. The design of the system is intended to allow the plant to be scaled to 250 kW output or higher.

- The FCE/Versa team is the most resilient and determined participant in the SECA program. However, the team may face substantial challenges because the future course of the SECA program is not at all certain.

6.2.3.8. General Electric

GE's involvement in SOFC development was relatively brief—about 6 years. GE entered in SOFC development by acquiring Honeywell's SOFC division in December 2001 and assumed Honeywell's place on the SECA industry team. DOE was expected to provide about \$74 million to GE over the course of 10 years with GE contributing \$59 million.[52]

- The SOFC division at Honeywell had a 14-year history of planar SOFC development at the time it was bought by GE. Honeywell's SOFC development effort originated at Allied Signal Aerospace in 1987. Allied Signal merged with Honeywell in December 1999, and the SOFC effort continued at Honeywell until GE acquired its SOFC unit.

In January 2006, GE achieved a major milestone in the SECA program by being the first industry team to successfully test an SOFC prototype and becoming the first SECA industry team to enter Phase II of the program.[61] The cost of GE's prototype system was estimated at \$724/kW for an annual production of 50,000 units, a theoretical cost approaching that of conventional stationary power technology. The prototype exceeded DOE targets for availability, efficiency, endurance, and estimated production cost.

- During testing, GE's SOFC system operated at an availability of 90 percent, exceeding the SECA Phase I target of 80 percent. A peak power of 5.4 kW was achieved on methane fuel, within the target range of 3–10 kW. An efficiency of 41 percent was achieved, exceeding the Phase I target of 35 percent for stationary applications. Observed steady-state power degradation was less than 3.6 percent/1000 h, below the Phase I target of 4 percent/1000 h.[61] At the stack and cell level, GE earlier had achieved 307 mW/cm^2 in a radial planar, 21-cell 800 °C stack and had already achieved over 400 mW/cm^2 in a single cell, exceeding its Phase I SECA targets for stack power density and utilization.[134] As of May 2007, GE reported that a single modular stack achieved a power density of 480 mW/cm^2 at 0.7 V with an 80 percent fuel utilization rate.[135]
- As a result of GE's Phase I success, DOE approved GE's application for the second phase of the coal-based SECA effort, making GE the first SECA industry team to enter Phase II.[61]

GE also was one of the first two industry teams selected for the new coal-based SECA in August 2005. The DOE award for GE was \$7.5 million for the first phase of 36 months for the new coal-based SECA.[67] The other industry team selected for new SECA was SWPC. The two teams would share the same goal: to develop the fuel cell technology required for central power stations to produce affordable, efficient, environmentally friendly electricity from coal.

- Despite the major success and the leadership role that GE's SOFC division had played in the SECA program, and in the global SOFC community, GE closed the SOFC division and dropped out of the SECA program in 2007. Little information has been released regarding the reasons and circumstances.
- Curiously, however, GE Global Research, perhaps a remnant of the former GE SOFC division or a newly created team, is still working under the SECA program. The SECA program is currently structured to include three Industry Teams focused on developing coal-based systems, FCE/Versa, RR, and UTC/Delphi. GE is no longer on the SECA Industry Team but serves to support R&D. It is focused on degradation mechanisms and development of cost-effective mitigation solutions.
- GE has engaged in virtually all the fuel cell technologies but ultimately abandoned every one of them. After abolishing the GE SOFC division in 2007, however, GE might have had a second thought and chose to continue SOFC R&D as a long-term investment.

6.2.3.9. *Protonex*

Protonex, headquartered in Boston, Massachusetts, was founded in 2000.[136] The company specializes in SOFC and PEMFC systems for portable, remote, and mobile applications in the 100–1000 W range. In April 2007, Protonex acquired Mesoscopic Devices, a developer of portable SOFC systems,[137] and significantly strengthened its SOFC position.[136] Protonex uses microtubular SOFC technology and develops products for military, commercial, and consumer markets as well as for OEM manufacturers who integrate Protonex products into their products.

Protonex' SOFC technology, compared to its PEMFC technology, is less mature, according to the company. Protonex stated in 2009 that Protonex' liquid fueled SOFC generator had a power density of 50 W/kg at 25 percent efficiency.[138] Nonetheless, the company has been energetically engaged in SOFC development under contracts, primarily with the US Army. During 3 years between February 2007 and February 2010, the US Army awarded a total of at least $15.7 million to Protonex to develop the M250-CX 250W SOFC battery charger and auxiliary power system for field use.

- In addition, Protonex worked with private partners. In September 2007, Protonex received a $1.2 million subcontract from Cummins to supply SOFC power systems for an APU.[139] Cummins was awarded a contract by DOE to develop a diesel-fueled SOFC system for on-highway truck auxiliary power (Table 6.3).

There have been few recent announcements regarding Protonex fuel cell activities. In May 2010, Cummins announced that it had operated a hybrid-configured SOFC APU for 10 h to power a Class 7/8 truck's "hotel loads" on commercial ultra-low sulfur diesel fuel. The Cummins APU incorporated Protonex SOFC modules and hybrid power electronics, system controls, and balance-of-plant components and generated up to 3.8 kW.[140]

Other reports suggest Protonex' is having technological or financial difficulties:

- In April 2010, Protonex announced that it would concentrate on the development of its propane-fueled P125-B 125W SOFC battery charger/APU system for recreational vehicle (RV) markets. It indicated that it would place "beta" P125-B units with customers and strategic OEM partners by November and targeted a full product launch in the first quarter of 2011. Derivative products on the P125-B (higher power, alternative configurations) were expected to follow later in 2011. There have been no reports of the

TABLE 6.3 Protonex: SOFC Product Development Contracts.

Date of contract announcement	Type of contract	Product for development	Amount of award
February 2007	US Army Research Office (ARO); in September 2007 Protonex subcontracted Raytheon	250 W methanol-fueled portable fuel cell battery charger and auxiliary power unit (APU)	\$3.5 million
September 2007	Subcontract from Cummins	SOFC APU to power a Class 7/8 truck	\$1.2 million
April 2008	US ARO; Raytheon continuing to work with Protonex	To advance the Pulse M250 military power system developed under the previous ARO contract	\$3.65 million for 12 months
June 2008	US ARO	To fund ruggedization and military testing activities	\$350,000 in additional funds on its previously announced \$3.65 million contract
September 2008	US ARO	To adapt its existing propane-fueled SOFC system to operate on butanol, gasoline, kerosene, and JP-8	\$400,000 for 12 months
March 2009	US Army	To adapt its SOFC power systems for operation on multiple liquid fuels, including butanol, biodiesel, gasoline, and kerosene	\$1.5 million for 24 months
September 2009	US Army	To advance the robustness of Protonex' SOFC power systems operating on liquid fuels	\$2 million for 12 months
December 2009	US Army through the University of South Carolina Research Foundation and WinTec Arrowmaker	To build, test, and deliver a fully integrated liquid fuel SOFC generator system to the US Army	\$1 million for 12 months

(Continued)

TABLE 6.3 Protonex: SOFC Product Development Contracts. (*Cont'd*)

Date of contract announcement	Type of contract	Product for development	Amount of award
January 2010	US Army (part of the American Recovery and Reinvestment Act of 2009); the contract includes three option phases	Phase one—To develop the existing 250 W, M250-CX platform by increasing its power output, integrating additional battery management and charging capabilities, reducing overall size and weight, and further ruggedizing the system for field use; To deliver multiple portable battery charger/APU systems to the US Army for further testing and field use	$1.85 million for 5 months
February 2010		Phase two—To deliver 30 additional M250-CX portable battery charger/APU products to the US Army for field use	$1.49 million for 8 months
Unknown		Part three—Unknown	Unknown

Note: The table was created based on various Protonex press releases, http://www.protonex.com/news/press-releases.aspx?year=2011, Protonex, no date, accessed May 2012.

product launch or derivative products, however.

- Protonex announced that it would cancel the commercial launch of its methanol-fueled M250-B 250W SOFC battery charger/APU system that had been planned for that spring.[141] Protonex would work instead to advance M250-CX 250 kW military-use battery charger/APU under the existing US Army contract.
- In May 2010, Protonex announced that the company would consolidate its Colorado SOFC operations into its Southborough, Massachusetts Headquarters, possibly to reduce headcounts and operating costs. Moreover, in May 2010, it canceled its admission to AIM on the London Stock Exchange, where it had been listed since July 2006.[142]

6.2.3.10. *Rolls Royce Fuel Cell Systems*

In 1993, RR Fuel Cell Systems, located in Canton, Ohio, and the United Kingdom, began to focus on developing fundamental aspects of SOFC technology. It developed a unique stack, which it called the integrated planar solid oxide fuel cell (IP-SOFC).[143] A typical IP-SOFC consists of a flattened ceramic tube with segmented-in-series electrochemical cells deposited on its outer surfaces.[144] This design concept represented a cross between tubular and planar fuel cell designs. Its integrated planar series arrangement is expected to result in high voltage, low current, and low production costs.[145] Oxygen is supplied to the cathodes from air flowing over the outside of the tubes and hydrogen diffuses from the internal fuel channels to the anodes through the porous tube structure. The IP-SOFC is intended for use in medium-scale 1 MW stationary power applications. These systems would utilize fuel cell stacks containing many thousands of tubes housed inside a pressure vessel.[144]

- In 2000, RR developed a 1 kW planar SOFC stack for laboratory testing.[76] The positive result made RR highly optimistic about the near-term commercial prospects. RR was convinced that the optimum product for market entry would be a pressurized system at MW scale.[146] RR participated in a large number of UK and EU Framework R&D projects to scale up its stack. By 2007, RR had successfully developed a 15 kW stack.

RR decided to advance its technology by capitalizing on US resources. In April 2007, RR acquired SOFCo in Ohio from McDermott International. In May 2008, RR participated in DOE's coal-based SECA program along with another new team, the UTC and Delphi team.[65] DOE included RR in the program for two major reasons.

- RR had been developing a low cost production process,[143] which produced SOFCs by screen printing onto inexpensive ceramic materials.
- RR also had been developing an SOFC gas turbine hybrid system fueled by natural gas for power-generation applications on the order of a megawatt.[147] RR's hybrid 1 MW power plant to be developed under the SECA program would reportedly consist of an 800 kW SOFC and a 200 kW RR Gas Turbine.[143]

RR started the SECA pre-award activities in September 2009, and Phase I work was extended through September 2011, according to RR.[148] The task at hand appeared to be the development and manufacturing of a 10 kW-class block.

As of July 2011, RR believed its cell technology was on track for initial commercialization.[148] RR had selected SOFC cells with area specific resistance of 0.28 ohm/cm^2 for block-scale metric tests. Since start of the program, power densities in these cells have improved by as much as 73 percent. The cells demonstrated average degradation of less than 0.5 percent/1000 h for test durations up to 8500 h. RR viewed anode durability as the top challenge for building efficient IGFC systems operating at high fuel utilization rates. Modifications were being made to further improve durability. RR prepared new, pressurized, thermally self-sustaining block rigs in Canton, Ohio, and modified a rig in the United Kingdom. As of September 2012, additional metric testing of a 20 kW-class block is being planned.

- Despite RR's significant progress, there still are challenges to overcome before reaching the SECA goal of developing a 1 MW IGFC system. If the SECA program were to end, RR's SOFC R&D activities in the United States might come to an end, likely another loss of a major SOFC developer in the country.

6.2.3.11. Siemens Westinghouse Power Corporation

In the early 2000, SWPC abandoned its effort to commercialize the tubular SOFC technology it had been developing since 1960s. The company had accumulated more experience and demonstrated higher reliability with its tubular technology than with any other high temperature fuel cell technology. It concluded, however, that the cost of the system would remain outside the acceptable range, even for introductory markets.[47]

Subsequently, SWPC made a strategic decision to pursue a new SOFC technology, a novel, nontubular SOFC design, called the HPD cells. The cells are no longer round cylinders but are flattened tubes or triangular (delta-shaped) tubes. In August 2001, SWPC participated in the DOE SECA program in partnership with Fuel Cell Technologies Limited (FCT) of Kingston, Ontario, Canada. Under the program, SWPC would develop a 3–10 kW SOFC stack and FCT would develop the BoP for the system. DOE would fund $47.8 million while SWPC would put in $32.8 million in cost share.[52]

- Under SECA, SWPC developed the HPD5 (flattened-tube) technology in 2003, the HPD10 in 2004, and the Delta9 (delta-shaped tube) technology in 2005.[73]

SWPC successfully completed SECA Phase I tasks in late 2006.[149] The company built a prototype 5 kW CHP system, consisting of cathode-supported flattened tube cells (HPD5) assembled into seal-less stacks.[109] In November 2006, the company announced the test results of the 5 kW SECA Phase I prototype system.[149] According to the company, the system met or exceeded all of the DOE technical and economic objectives for Phase 1 of the SECA program:

- The 5 kW system operated for 6300h with stable performance and no degradation,[150] the only SECA project system believed to have achieved no cell degradation in extended operations.[149] The DC peak power was 5.9 kW, the net DC efficiency was 38 percent, and the availability was 100 percent.[150] The HPD5 cells also demonstrated volumetric power density four times greater than traditional tubular cells, which translated into significantly reduced volume and system cost.[149] The system cost per kilowatt was reportedly $691/kW.[150]

The excellent performance of their new HPD cells likely encouraged SWPC to launch new demonstration projects outside the SECA program. There are indications that SWPC tried to develop some 125 kW CHP systems and a 1 MW hybrid system for demonstration in Germany, Italy, and Japan during the latter half of 2000s. SWPC reportedly was planning an updated 125 kW CHP system based on the HPD5 technology in 2006,[47] which was perhaps the system delivered to Stadtwerke Hannover, Germany in 2006. The HPD-5 or HPD-10 might have been used in the 125 kW systems to be demonstrated in Turin, Italy, in 2007 and Tokyo, Japan, in 2008. SWPC also indicated that ceramic cells with the Delta technology would be used for the 1 MW system to be demonstrated in 2012 at EnBW in Germany.[41]

- All of these projects appeared to have been canceled, however, suggesting that the HPD cells likely had not performed as well as hoped.

In August 2005, meanwhile, SWPC was selected to participate in DOE's new Fuel Cell Coal-Based Systems program. The DOE award to SWPC was $7.5 million for the first developmental phase of 3 years.[67] By 2010, SWPC would be required to develop a stack that can operate for 5000 h with a degradation of less than 2 percent per 1000 h and costs of $700/kW or less for the system power block.

- By February 2009, Siemens developed a 10 kW SOFC stack that consisted of 24 Delta (8 cells), 4 Bundles (618 cells each), and operated it for 2500 h.[63] The cell generated 550 W per cell at 0.7 V and showed a power density of 0.25 W/cm^2 at 80 percent fuel utilization using humidified hydrogen at 975 °C.[151] In April 2009, the stack achieved more than 5000 h of operation, showing an IGFC efficiency of more than 50 percent and no detectable degradation.[152] The company was selected by DOE to move onto Phase II of the new coal-based SECA program.
- After operating for 5300 h with maximum DC power of 9.8 kW, however, the stack was damaged due to failure of a control system hard drive.
- In March 2010, another 10 kW stack began operation and operated for 400 h. At 400 h, several cells failed during an air transient.

In October 2006, FCT, SWPC's partner in SECA, went out of business. FCT announced on October 16, 2006 that its workforce was being terminated, most of its board had resigned, and a financial advisor was retained to sell the company's assets or raise capital. It was reported that FCT's total revenue for the 3 months ending March 31, 2006 was $249,966, derived from the shipment of one 5 kW CHP system to Pennsylvania State University.[153]

- In 2007, Acumentrics acquired all of the assets and employees of FCT.[75]

SWPC also was undergoing business difficulties itself and gradually began to scale back its fuel cell unit, starting in 2002. At its peak of mid-2002, the SWPC plant in Churchill had nearly 1000 employees.[48] As of March 2009, about 40 remained (see detailed discussion in Section 6.2.1.4).

- On September 30, 2010, after 50 years of hard work and hundreds of millions of dollars of investment, SWPC shut down its SOFC activities.

6.2.3.12. SOFCo

SOFCo was formed in 1994 as a limited R&D partnership between a McDermott International subsidiary Babcock and Wilcox (B&W) and Ceramatec to develop planar SOFC technology.[110] B&W and Ceramatec had an alliance to develop planar SOFC development since 1986.[154] In the partnership, B&W was to provide BoP system development and marketing support to SOFCo, while Ceramatec was to advance SOFC fuel cell stack and module technology.[10] SOFCo was primarily interested in SOFC products targeted for small commercial retail markets and perhaps multifamily dwelling markets. By 1996, SOFCo had scored a number of major achievements. A single cell endurance test demonstrated up to 40000 h, and cells and stacks were scaled up from 4″ × 4″ to 6″ × 6″. Advancements also were made in reducing the internal resistance of SOFC cells using tape laminated thin film technology acquired from Dow Chemical. Also in 1997, SOFCo in partnership with Technology Management, Inc. (TMI) was awarded a 4-year,

$9 million contract from DARPA to demonstrate a 10 kW diesel-fueled SOFC mobile electric generator.

SOFCo clearly was confident that its SOFC commercialization would be successful and, in early 1997, announced the establishment of a Distributed Power Initiative and solicited members to support the initiative:[10]

- Phase I of the initiative was targeted at the development of a 10 kW SOFC technology demonstration unit. The 10 kW bread board unit was to be assembled and tested in 1998. The overall cost of the Phase I program was established as $5 million and was scheduled to be completed in 2 years. SOFCo sought two types of partners to participate in the initiative. Minimum partners would pay $50,000/year for 2 years and get access to 10 kW program reports and technical data. Potential strategic partners would have the opportunity to take ownership in the B&W Distributed Power Company (DPC); equity stakes in the B&W DPC were offered in $500,000 increments.
- Phase II of the initiative would begin about 1999 and would be aimed at demonstrating SOFCo technology in the field. Market entry products using current "conventional stack technology" were to be potentially offered in the year 2000–2001 time frame.

SOFCo's plans were never realized because SOFCo was not able to develop the 10 kW bread board unit in 1998. As of 2000, SOFCo in partnership with Cummins was able to succeed in laboratory testing of a 0.7 kW planar SOFC system for 1000 operating hours.[76]

In 2002, SOFCo had a major opportunity. It became Cummins' partner in SECA Phase I by replacing McDermott Technology, a McDermott International subsidiary. In December 2002, McDermott Technology and B&W went defunct, and all of the fuel cell technology from McDermott Technology and B&W as well as all of the McDermott International's SOFC work was taken over by SOFCo.[110] SOFCo also inherited McDermott Technology's personnel previously engaged in SOFC work.

- During the fiscal year 2002, however, SOFCo's SOFC cell development in SECA did not go well.[111] Cell breakage slowed stack development and third-party cells were supporting 2003 stack development while SOFCo was developing next-generation cells.[111]
- Subsequently, Cummins completed the SECA Phase I in partnership with Versa Power Systems, not with SOFCo.[112]

In April 2007, RR acquired SOFCo from McDermott International. In May 2008, RR was selected to lead a project under DOE's coal-based SECA program.[65]

6.2.3.13. Technology Management, Inc

TMI, located in Cleveland, Ohio, is a venture company founded in 1990.[155] TMI focused on the development and commercialization of planar SOFC technology[156] acquired from British Petroleum (BP) (formerly SOHIO).[10] TMI obtained the exclusive worldwide license from BP and planned to commercialize a compact, modular system which could be manufactured, operated, and maintained at low cost.[10]

TMI used an Interscience Radial Flow design in which each cell was made up of four layers of flat disks, with sealing being achieved through the use of rings that also formed the internal fuel and air manifolds.[107] Electrodes made from particulate materials did not require co-firing, which were sources of thermal expansion-matching problems and a constraint on all other known SOFC designs.[10] Porous electrodes allowed both air and fuel to be supplied from internal manifolds and dispersed radially outward through the electrodes before exiting into a chamber. No exhaust manifolds were needed. At approximately 50 mm in diameter, the TMI cell was full commercial size.

- Stacks of 200–300 cells would produce 0.5–1 kW of power and could be stacked in parallel to produce higher power levels.[157] Cells were operated at a current density of 200 mA/cm^2.
- Small stack testing from 1- to 10-cell stacks was performed.[107] Power densities around 0.08 W/cm^2 (75 W/ft^2) were attained.
- The system studies demonstrated that even if systems were scaled down to as low as 300 W, net efficiencies near 25 percent DC LHV (lower heating value) for natural gas fuel could be achieved, according to TMI.[158]

The most noteworthy feature of the TMI SOFC systems was that they were highly sulfur tolerant and fuel flexible. TME had made major advances in this area.[10] In 1991, TMI started to develop SOFC power-generation systems for military applications that could operate on a variety of gaseous and liquid fuels.[159] DARPA awarded TMI a multiyear program to build and test a 20 kW generator and another program to develop a mobile 500 W SOFC system. Both of these devices would be powered by logistic fuels, such as diesel fuel or JP8.[10] The programs were focused on sulfur tolerance as part of the effort to be able to operate on these fuels.

- Under the DARPA program, TMI was one of the first fuel cell developers to demonstrate a complete SOFC system operating without sulfur-removal or fuel pretreatment.[158] To date, TMI has the only publicly proven SOFC to operate on standard, untreated military JP-8.
- According to TMI, their systems could operate on a wide variety of other gas and liquid fuels, including natural gas, propane, kerosene, diesel, ethanol, biodiesel, vegetable oil (soy oil), digester biogas, ammonia, used cooking oil, and jatropha.[160]

In the 2000s, TMI received a number of federal and state funding awards and demonstrated kW-class systems:[155]

- In April 2004, TMI was awarded a $965,161 grant from the DOE and USDA for a project to demonstrate a 1 kW modular SOFC power-generation system operating on biodiesel and ethanol biofuels.
- In 2006, the State of Ohio awarded $1 million to TMI to develop small-scale SOFC systems as an APU for use in heavy.[161] The prototype unit would generate 1 kW and include a built-in device that would make hydrogen from diesel fuel with water as a by-product.
- In November 2007, TMI displayed its 1 kW class fuel cell at an Ohio farm exhibit.[162]
- In December 2009, the State of Ohio's Job Stimulus Plan announced $2.5 million funding to TMI to develop, in collaboration with Lockheed Martin and Stark State College of Technology, an SOFC product for military markets to replace kilowatt class diesel generator sets.[163] Under this program, TMI and its partners would engineer and manufacture 10 demonstration units for demonstration in Ohio. The State was hoping as many as 42 new jobs to be created.

Today, TMI is searching for a niche market for its 1–20 kW CHP systems.[160] TMI believes that sulfur tolerance would offer many advantages, including system simplicity and fuel flexibility, which directly relate to system size, weight, and cost.[158]

6.2.3.14. UTC Power/Delphi

In May 2008, UTC and Delphi partnered to work under DOE's new coal-based SECA program.[65] UTC was to lead the 2-year, $22 million SOFC technology development project. The project was among the largest DOE projects ever awarded to UTC, according to the company.[164] The project goal was to develop low-cost, modular, high density, megawatt-class fuel cell systems.

- UTC, with its expertise in developing and producing PAFC fuel cells for on-site power, transportation, space, and defense applications, was to provide fuel cell systems engineering infrastructure.

- The primary responsibility for developing SOFC modules rested with Delphi, which would bring in significant expertise in SOFC technology that it had been developing under the SECA program.[164]
- In addition, Topsoe Fuel Cells, a Danish company, would provide advanced materials to the partners.

The developmental schedule, however, was tight because all work was to be done in a 2-year time frame. The team was to design a subscale, proof-of-concept system, develop manufacturing capabilities, resolve issues related to cost, performance, and manufacturability of larger-scale fuel cells, and develop, fabricate, and test a scaled stack suitable for use in a full-scale system.[164]

In May 2008, Delphi started working toward the development of a 5 kW stationary power with a net efficiency of 45 percent under the new coal-based SECA program.[119] Delphi made a series of improvements on the Generation 3 stack that it had been developing under the SECA Phase I program.

- In 2008, Delphi developed a 30-cell "Gen 3" stack, consisting of 105 cm^2 active area cells.[165] The stack generated 1.8 kW (570 mW/cm^2) at 0.82 V/cell with a fuel mixture (48.5 percent H_2, 3 percent H_2O, and the rest N_2) at 750 °C.
- In 2009, Delphi developed a Gen 4 stack consisting of 403 cm^2 active area cells, which were much larger than Gen 3 cells. Delphi fabricated a 30-cell Gen 4 stack, which produced power densities of 500 mW/cm^2 at 570 mA/cm^2 at 0.87 V.[166]
- In 2010, a 40-cell Gen 4 stack produced 7452 W (450 mW/cm^2) at 0.81 V/cell with the same fuel mixture used by Gen 3.[123] Also in 2010, a 25-cell Gen 4 stack produced 5064 W (506 mW/cm^2) at 0.81 V. The results indicated that the stack also demonstrated a power density of 403 mW/cm^2, 70 percent fuel utilization, and a current density of 600 mA/cm^2.
- In 2011, Delphi developed a 40-cell Gen 4 stack, which demonstrated maximum power of 6.4 kW with a simulated coal gas blend.[167] The power density was 398 mW/cm^2, and the average cell voltage was 0.7 V. The stack completed 60 full thermal cycles with less than 5 percent voltage degradation. It was operated for 3000 h.

Delphi is making remarkable improvements in the stacks made with large area cells. In 2010, a 30-cell Gen 3.2 stack was tested for durability. It operated for 9700 h with fuel (48.5 percent H_2, 3 percent H_2O, and the rest N_2) at 333 mA/cm^2. Total degradation was 1.12 percent/500 h.[123] As of July 2011, Delphi indicated that degradation mechanisms during the first 1000h and after 5000 h were understood and technology solutions had been implemented.[107]

- Still, it is not certain if SECA will survive long enough for UTC/Delphi to reach their long-term goals.

6.2.3.15. *Ztek*[107]

Ztek was founded in 1984 to commercialize planar SOFC technology that was developed at the Massachusetts Institute of Technology's Lincoln Lab in the 1970s.[168] Development work at Ztek has been sponsored by DOE, DOD, EPRI, and Tennessee Valley Authority (TVA).[10] The company aimed at marketing an SOFC-based heating, ventilation, and cooling (HVAC) system in the 50–100 kW range and an SOFC small gas turbine system on the order of 250 kW. Ztek's SOFC/microturbine hybrid system was reported to have 50 percent efficiency.[169]

In the 1990s, Ztek was developing and demonstrating 1–25 kW SOFC stacks and systems with considerable success. At that time, Westinghouse and Ztek were the only two companies who successfully assembled a complete SOFC system.[10]

- In 1994 and 1995, the company developed a 1 kW stack and, along with TVA, successfully demonstrated the stack operating on a simulated reformed (hydrogen, carbon monoxide, and water) natural gas fuel in 1994 and 1995. The 1 kW stack operated for over 12,000 h at load with very little decay in cell performance (less than 3 mV/1000 h). The stack operated at 1000 °C.
- From 1998 to 2000, Ztek completed the development of a 25 kW SOFC Power System for TVA, which was installed at Huntsville Utilities Host Site for demonstration.[170] Ztek was clearly hopeful; the company intended the 25 kW module to be the building blocks for the commercial systems.[171] In March 2002, Ztek announced that it was selected by the Connecticut Clean Energy Fund to demonstrate its 25 kW SOFC system with an absorption chiller at Dinosaur State Park in Rocky Hill. The system was delivered to the park and was expected to be operational by mid 2005.[172] However, little reporting has been available regarding the operation of the 25 kW system at the Connecticut state park, suggesting that the project might have not been a success.

In recent years, the focus of their activities appears to have shifted from fuel cells to hydrogen production and marketing its proprietary steam reformer.[47] In February 2006, Ztek announced that it had reached agreement with US Department of Navy to collaborate on a hydrogen fueling station project at the Camp Pendleton Marine Corps Base. When operational, the station would serve both military and commercial hydrogen vehicles in the area.[173]

- It is likely that Ztek is no longer focused on SOFC development.

6.2.4. US Global SOFC Leadership Position Has Largely Eroded

The history of SOFC development in the United States is discouraging. Today, the US SOFC industry lacks focus and the number of SOFC developers in the United States is dwindling:

- SOFC capabilities developed by Allied Signal Aerospace, Honeywell, SOFCo, and McDermott have all disappeared as a result of mergers and acquisitions. SWPC dropped out of SOFC business. Ceramatec and Ztek no longer focus on SOFC development.

Remaining SOFC developers in the United States are Acumentrics, Bloom Energy, Delphi, GE Global Research, Protonex, Versa, RR, and TMI.

- Versa (in partnership with FuelCell Energy), RR, and Delphi (in partnership with UTC Power) are involved in the long-term, large-scale DOE coal-based SECA project. Whether the highly ambitious goals will be achieved depend heavily on whether DOE can maintain a consistent level of sufficient funding. But many people believe that it will be decades before a 100 MW coal fuel cell plant can be realized. Moreover, it is not certain if the coal-based SECA program will survive as an energy policy in the future. DOE requested no budget for SECA in FY2012.[174]
- Acumentrics and Protonex reportedly are successful in the military sector. Still, it is not clear if their prototype products meet the military—and particularly commercial—requirements.
- Bloom Energy reportedly is highly successful in marketing their products, but questions remain about production costs and longevity of their fuel cells.
- GE Global Research remains in the SOFC field and is developing SOFC cells currently.

TMI is also in the R&D stage. But their near-term commercial success is not certain.

It appears that the United States no longer is the global front runner in terms of SOFC power densities and efficiencies. The reduction of the SECA budget to only $25 million has reduced the stature of the SECA program. As we shall see in the next section, based on the available data, the power densities of US SOFC cells and stacks are generally lower than those demonstrated by their Japanese counterparts.

6.3. JAPAN LAUNCHES SOFC RESEARCH IN WAKE OF OIL CRISIS

The Japanese government began to promote SOFC technology in 1974.[ii] Prompted by the Arab oil embargo, METI launched the Sunshine Project for the purpose of finding long-term solutions to the energy crisis. METI provided modest funding for a few national laboratories at first, but in 1989 it launched a major national program to transition SOFC technology developed by the national laboratories into fully mature technologies. This program was just the beginning; a series of SOFC R&D programs followed, constituting steady government support of SOFC development for more than 36 years.

Another goal of METI was to help Japanese SOFC developers catch up with and surpass US technology—particularly SWPC's technology, which it saw as setting the global standard. At first, METI discouraged Japanese developers from competing directly with SWPC. It tried to steer Japanese developers away from tubular designs because it believed planar designs would offer greater competitive opportunities. Nonetheless, due to difficulties involved with planar systems, more and more Japanese developers engaged in research on tubular designs. Most Japanese developers that left the SOFC business primarily were planar SOFC developers.

- One tubular SOFC developer, Kyocera, emerged as the SOFC technology leader in Japan. Kyocera's SOFC cells and stacks are reported to demonstrate higher performance, and have more reliability and durability than others. It is noteworthy that, unlike SWPC, Kyocera never participated in a government SOFC R&D program. Its only involvement with the government was to allow METI to use their samples and prototypes in SOFC R&D programs.

In 2001, when the US DOE implemented SECA, METI was stunned. DOE had abandoned tubular designs and switched to the development of planar technology. METI by that time had chosen to promote both tubular and planar designs. DOE's new R&D initiative, nonetheless, focused world attention on SOFC technology, leading Japan to step up its SOFC effort as well. Indeed, METI more than doubled its SOFC R&D funding to some $10 million from relatively low levels of funding of about $4 million a year.

Today, a sense of excitement pervades the SOFC industry in Japan. Many tubular SOFC developers have announced near-term commercialization plans and a few have expressed their intention to begin commercialization as early as next year. This optimism can be traced to two reasons:

- SOFC developers are now confident in the feasibility of commercializing their technology. It has been reported, for example, that Kyocera SOFC cells might have doubled the lifetimes of their SOFC cells to 10 years.
- Second, the Japanese government has stepped up its policy to promote fuel cell

[ii] Japanese fiscal year starts 1 April and ends 31 March. Unless otherwise mentioned, any year mentioned refers to fiscal year.

commercialization. In 2007, METI launched a new subsidy program to promote the use of fuel cell power plants for residential use. By 2010, this demonstration program had more than doubled. The maximum amount of subsidy is ¥1.4 million (about $15,600 at the 2010 exchange rate of $1/¥90) per system,[175] about half the cost of a residential fuel cell power plant on the Japanese market today.

Anticipating a burgeoning market, many SOFC developers have formed alliances with one another to prepare for commercialization of their products. Understandably, many alliances centering on Kyocera have emerged this year.[iii]

6.3.1. METI Begins Modest Funding of Basic SOFC Research in 1974

While METI's initial support for SOFC development actually started a few years earlier than the US DOE, by 1974 US companies already had more than two decades of R&D experience in SOFCs compared to their Japanese counterparts.[176] Indeed, Japanese companies in general started their SOFC R&D in the late 1980s, and by that time, SWPC in the United States had already perfected its SOFC fabrication methods based on the EVD technique and the nonsealing type tubular design. R&D efforts in Japan therefore focused on other designs and manufacturing technologies than those SWPC had already established; for example, Japan chose to work on the development of planar SOFCs and a wet process to manufacture them.[iv]

The initial Japanese government effort was to incubate preliminary SOFC technology at a few national laboratories. Those that conducted basic research included the Electrotechnical Laboratory (ETL), the National Institute of Materials and Chemical Research (NIMC), and the National Institute for Resources and Environment (NIRE), all of which were independent entities under the National Institute of Advanced Industrial Science and Technology (AIST). They are part of AIST today.

- ETL began R&D on the structure of gas supply manifolds and development and verification of tubular SOFC cells as part of the Sunshine Project.[177] ETL focused their effort on the tubular SOFCs on metallic substrates and plasma spray fabrication techniques.[176] Using the plasma spray method for tubular SOFCs, ETL succeeded in the fabrication of 500 W class SOFCs and achieved a maximum power of 1.2 kW in 1986.[178]
- NIMC started R&D on planar SOFCs in 1987 as part of the New Sunshine Project.[179] NIMC focused on solving problems with SOFC component materials so that a wet method could be successfully used.[179] NIMC successfully provided the physicochemical basis that would enable using wet processes without degrading cell characteristics, thereby confirming the feasibility of wet processes on a laboratory scale.[176] NIMC also developed lanthanum chromite powders for the separator material and the co-firing

[iii] Starting in spring 2009 and coinciding with the implementation of the government fuel cell subsidy program, PEMFC developers placed their power plant products on the market. Currently, the only residential fuel cell product on the market is PEMFC power plants. The total cost of purchase and installation of a PEMFC power plant is reportedly about ¥3 million.

[iv] The following is an explanation of the dry process and the wet process extracted from the free patents online website (http://www.freepatentsonline.com/5480739.html): "The solid electrolyte film is generally formed by the dry process or the wet process. As typical dry processes, an EVD process and a spraying process may be recited. As the wet process, a tape casting process, a slip casting process and an extruding process may be recited (Energy General Engineering No. 13-2, 1990)."

technology.[178] NIMC also proposed small nonreforming SOFCs that used methane and the utilization of electrophoretic deposition techniques for small tubular cells.[176]

- NIRE conducted research on the heat recovery systems for SOFCs and studied evaluation methods for high temperature heatpipes, heatpipe-type heat exchangers, and liquid-layer-type heat exchangers.[178]

By the mid-1980s, the gap in practical applications of SOFC technology between the United States and Japan had widened. In 1986, the US SOFC developer SWPC was producing prototype SOFC demonstration systems and announced that a larger number of SWPC power plants of various output sizes would be demonstrated globally over the next several years. In fact, Osaka Gas and Tokyo Gas in 1987 imported three 3 kW SOFC power plants from SWPC for field tests.

In 1988, METI was getting ready to launch a major national R&D program. In that year, the New Energy and Industrial Technology Development Organization (NEDO) acquired new responsibilities for overseeing industrial technology R&D,[180] providing METI greater leverage in managing Japan's industrial R&D policy and program. NEDO had been created in 1980 under METI to develop new oil-alternative energy technologies via the Sunshine Project (launched in 1974) and the Moonlight Project (launched in 1978). With its newly acquired competency, the NEDO was responsible for supporting industry-sector R&D activities by providing administrative infrastructure and by implementing industrial technology policy. This surge in power had likely motivated METI to launch new large-scale R&D programs in 1989.

- It is worth noting that the MITI was equipped with a dual structure for management of R&D: AIST and NEDO.[180] AIST was charged with designing and planning major industrial R&D projects. AIST's national research laboratories contributed by evaluating and incubating elementary technology. They also carried out feasibility studies on how to use research results. NEDO, complemented by AIST, administered the industrial technology policy that AIST would design.
- In reality, the distinction between AIST and NEDO roles became blurred, with increasing duplication of paperwork and a more complex bureaucracy.

Nonetheless, by the late 1980s, both the Japanese government and the industry were ready and willing to start a race to catch up with the US SOFC technology.

6.3.2. METI Launches Long-Term SOFC R&D Programs in 1989

METI's 1989 SOFC R&D program was launched under the Moonlight Project. The goal was to develop planar SOFC technology using a "wet" method. The program was planned as a 13-year, 3-phase program (1989–2001). In retrospect, it stretched out to 16 years (1989–2004) in three phases: Phase I (1989–1991), Phase II (1992–1997) plus a Phase II extension (1998–2000), and Phase III (2001–2004).

In addition to the 16-year program, METI launched a 4-year SOFC system development program in 2004 and launched two additional programs in parallel in 2007. These parallel programs included a 4-year SOFC demonstration program (2007–2010) and a 5-year basic research program (2008–2012). One program was intended to establish the commercial feasibility of SOFC technology by field demonstrations and the other was to go back to basics to search for the underlying causes of the most challenging SOFC technological obstacle, which was performance degradation.

6.3.2.1. SOFC R&D Program Phase I (1989–1991)

Phase I of Japan's 1989 SOFC R&D Program focused on the development of 100 W class planar SOFC modules. Sanyo was to develop the module with metal interconnects, and Fuji Electric was to develop one with oxide interconnects. In addition, several different component technologies were to be investigated. Murata Manufacturing investigated co-firing technology, and Mitsui Engineering & Shipbuilding (MES) explored planar cells with interconnects made with magnesium spinel and lanthanum chromite.[176] Other participants in Phase I were NKK, Fujikura Cable, and Institute of Applied Energy.[178]

- The work carried out in Phase I seems to have moved along uneventfully and satisfactorily (Table 6.4).

6.3.2.2. SOFC R&D Program Phase II (1992–1997)

The 6-year Phase II Program began in 1992. The first goal was to develop within 4 years several-kilowatt planar SOFC modules (with power densities of 0.18 W/cm^2 at 70 percent fuel utilization) and then move on to development of 10 kW class modules (with densities of 0.2 W/cm^2 with 75 percent fuel utilization and a degradation rate of less than 1 percent/ 1000 h).[178] Fuji Electric and Sanyo Electric worked on the project. The second goal was to develop high performance cathodes, anodes, gas sealing materials, and the related manufacturing technology. Japan Fine Ceramic Center, Fujikura Cable, Mitsubishi Heavy Industries (MHI), Murata manufacturing, and Mitsui Engineering and Shipbuilding (MES) conducted the research. The third goal was to establish the SOFC system design and architecture and develop BOP components. Those that undertook the development included Central Research Institute of Electric Power Industry (CRIEPI), Electric Power Development Company (EPDC), MHI, and the Japan Research and Development Center for Metals.

The total budget figure for Phase II program is not available, but the budget for 1995 reportedly was ¥400 million (about $3.6 million at the 1993 exchange rate of $1/¥111).

Program participants had major difficulties in completing the first goal of Phase II. The evaluation conducted in 1995 found that the planar stacks encountered severe gas sealing problems and had inferior durability in thermal cycles.[176]

- One major surprise was that Fuji, which was to develop oxide interconnects, chose instead to develop metal interconnects. Although METI thought it had a tacit understanding with Fuji, the project statement did not spell out the type of interconnect materials to be used. Fuji had a strong incentive for developing metal interconnects. Outside the METI program, Fuji Electric was developing large SOFC cells (500–1,000 cm^2) with metallic separators, metallic cell substrates, and plasma-sprayed electrolytes.[178]
- Also Sanyo, which was developing a 1 kW planar SOFC module with metal separators in Phase II, noted a serious degradation problem. The module operated for 1870 h, and the power degradation rate at 1030 h was 4.4 percent/1000 h based on a current density of 0.3 A/cm^2.[181]

Due to problems experienced in the first goal, the fabrication of 10 kW class modules was suspended.[176] In 1997, another evaluation was conducted, which found that the power output of a module achieved the target goals. It was noted, however, that although gas sealing had been improved, the metal interconnects for both Sanyo and Fuji's planar stacks did not show sufficient reliability for thermal cycling and long-term stability.

- NEDO must have felt that Fuji's use of metal interconnects was particularly unfortunate. Within the same METI Phase II project, Murata had successfully developed oxide interconnects.[176] At the same time, outside the METI project, oxide interconnects had been successfully used by MHI and Tokyo Gas.

TABLE 6.4 Japanese SOFC R&D Program Phase I, II, and III: Goals and Participants

Phase	Year	Goals	Participants
I	1989–1991	100 W planar SOFC Module with metal interconnect	Sanyo
		100 W planar SOFC Module with oxide interconnect	Fuji Electric
		Co-firing technology	Murata Manufacturing
		Planar cells with interconnects made with magnesium spinel and lanthanum chromite	Mitsui Engineering and Shipbuilding
		Other developmental efforts	NKK
			Fujikura Cable
			Institute of Applied Energy
II	1992–1995	kW class planar SOFC module with current density of 0.18 W/cm^2 with fuel utilization rate of 70 percent	Fuji Electric, Sanyo
	1996–1997	10 kW class planar SOFC module with current density 0.2 W/cm^2, fuel utilization rate of 77 percent, and deterioration rate of less than 1 percent/1000 hours	Fuji Electric, Sanyo
	1992–1997	High performance cathodes, anodes, gas sealing materials, and manufacturing technology for them	Japan Fine Ceramic Center
			Fujikura Cable
			Mitsubishi Heavy Industries (MHI)
			Murata manufacturing
			Mitsui Engineering and Shipbuilding
	1992–1997	SOFC system design, architecture, and develop SOFC-related equipment	Central Research Institute of Electric Power Industry (CRIEPI)
			Electric Power Development Company
			MHI
			Japan Research and Development Center for Metals

II Extension	1998–2000	Development of kW class tubular SOFC module with high durability and reliability using wet process		TOTO, Kyushu Electric, Nippon Steel
		Evaluation of durability and reliability (thermal cycle durability of Mono-Block Layer-Built (MOLB) planar cells manufactured by MHI)		Chubu Electric
		Evaluation of durability and reliability (mechanical strength of interconnect materials manufactured by Murata)		Tokyo Gas
		Low-cost component materials (chemical properties)		CRIEPI
		Low-cost component materials (mechanical and thermal properties)		Japan Fine Ceramic Center
		Feasibility on optimization of small SOFC system		Nippon Steel
III	2001–2004	Tubular type cells by wet process	5–10 kW natural gas–fueled thermally self-supporting SOFC modules with initial average cell voltage of more than 0.7 V (200 mA/cm^2, fuel utilization rate of more than 75 percent) and degradation rate of less than 0.25 percent/1000 hours	TOTO
		MOLB type planar cells		MHI Kobe, Chubu Electric

Note: The table was created based on M. Nishikawa, "Status of SOFC Development in Japan," Solid Oxide Fuel Cells IV, edited by M. Dokiya, O. Yamamoto, G, Tagawa, and S. C. Singhal, the Electrochemical Society, 1995; H. Sasaki, and K. Takasu, "Status of SOFC Development in Japan," Solid Oxide Fuel Cells V, Edited by U. Stimming, S. C. Singhal, H. Tagawa, and W. Lennert, the Electrochemical Society, 1997; Takashi Ujiie, "Status of National Project for SOFC Development in Japan," Solid Oxide Fuel Cells IX (SOFC IX), Volume 1, Cells, Stacks, and Systems, Edited by S. C. Singhal and J. Mizusaki, the Electrochemical Society, 2005.

6.3.2.3. SOFC R&D Program Phase II Extension (1998–2000)

As the 1997 evaluation called for improvement in thermal cycling and long-term stability, the SOFC R&D Program Phase II was extended by 3 years until 2000, but development of the 10 kW class module was canceled. The goals during the 3-year extension, which included two old goals and one new goal, included:[182]

- Improvement in durability and reliability of kilowatt class SOFC modules
- Achieving cell voltage degradation rates of less than 1 percent/1000 h
- Development of tubular SOFC modules

In 1998, responsibility for tubular module development was given to TOTO, Kyushu Electric, and Nippon Steel.[183] Chubu Electric and Tokyo Gas would evaluate the reliability and durability of planar stacks, focusing on thermal cycle durability and mechanical stress, respectively. CRIEPI and JFCC would study low-cost component materials, with the former focusing on chemical properties and the latter stressing mechanical and thermal properties. Nippon Steel would conduct feasibility studies on optimization of small SOFC systems.

- TOTO, Kyushu Electric, and Nippon Steel used a wet process method to fabricate tubular SOFC stacks. Wet processes were regarded in Japan as the most promising for cost reduction and mass production.[176] Moreover, SWPC had not focused on the wet process, providing Japanese manufacturers with a chance to compete with SWPC using a different approach.
- Sanyo and Fuji, who had worked for 9 years in Phase I and Phase II, left the program.

All the project goals for the Phase II extension were met.[182] TOTO, Kyushu Electric, and Nippon Steel successfully fabricated a kilowatt class tubular module using a wet process. Also, MOLB type planar modules manufactured by MHI showed satisfactory thermal cycling. Tokyo Gas also studied planar stack reliability from the aspects of material properties, numerical evaluation, and experimental evaluation, and found that reduction in mechanical stress could be achieved by using improved interconnector materials and adopting a co-flow pattern.[184]

6.3.2.4. SOFC R&D Program Phase III (2001–2004)

METI's began Phase III of its SOFC Program just as DOE launched the SECA program. SECA in effect abandoned tubular designs, which must have been seen by Japan as opening up an opportunity to compete with the United States where it did not have a home field advantage. METI's sense of competition and excitement was reflected in the Phase III program. The goals of Phase III were somewhat similar to Phase II but set at a higher standard. The budget was about double that of SECA.

- The R&D goal for this 4-year phase was to develop a thermally self-supporting 5–10 kW module. The module would use SOFC exhaust heat for fuel reforming, contributing to stable power generation with high performance, reliability, and cost-effectiveness.[184] Two types of thermally self-supporting modules were to be developed: one using tubular cells fabricated by the wet process developed by TOTO and one using MOLB type planar cells developed by MHI and Chubu Electric. In addition, studies on advanced planar cells and stacks for rapid start-up as well as studies on advanced tubular type cells would be conducted by Tokyo Gas and MHI, respectively.[182]
- The goal for degradation was set at less than 0.25 percent/1000 h,[184] a much higher standard than SECA, which was less than 4 percent/1000 h.
- METI allocate a total of ¥4.4 billion (about $37 million at the $1/¥117.73 exchange rate,

an average exchange rate over 4 years from 2001 to 2004) for the 4-year Phase III R&D program.[185] It averaged $9 million per year while the SECA budget was about $4 million a year.

Phase III met its project goals. TOTO developed a kilowatt class thermally self-sustaining tubular SOFC module, which consisted of 20 stacks of 12 tubular cell bundles made via a wet process. The module generated 6.5 kW at 0.2 A/cm^2 with 48 percent LHV and a fuel utilization rate of 75 percent.[186] The same module, operating under different conditions, generated 6.4 kW at 0.19 A/cm^2 and 46 percent LHV. Voltage degradation was 0.1 percent/1000 h in both cases. The thermal sustainability of both operating conditions was confirmed. MHI Kobe and Chubu Electric also indicated that they were successful with their MOLB type planar SOFC module.[187] The two companies developed a kilowatt class MOLB type planar SOFC module, consisting of 10 stacks of ten 200 mm × 200 mm cells. The module generated 10.4 kW at 0.2 A/cm^2 at a fuel utilization rate of 75 percent.[185] It operated for 3236 h and showed no voltage degradation. Operations for start-up, heat up, power generation, cool down, and shutdown were performed with automatic controls. The operation was normal from start-up to shutdown without abnormal electric power output and temperature behavior of the stack.[187] It was also confirmed that MOLB type SOFC module could allow rapid start-up and shutdown procedures, faster than the typical planar SOFCs, and could change electric power output as rapidly as conventional thermal power plants.

- Although the Phase III project successfully established the technology to build an SOFC module that would constitute a building block for a thermally self-supporting SOFC system, there were problems to be solved before the modules could be commercialized, according to a post-project evaluation.[185] In case of the TOTO module, issues of enhanced reliability and cost reduction had to be addressed. Moreover, MHI and Chubu Electric's had been a number of technical problems with its module, including voltage degradation at the time of internal reforming, cell damage during shutdown and restart-up, temperature measurement errors, and performance variations among modules at the initial stage, including voltage degradation in some modules by as much as 5 percent.

The two studies conducted in Phase III were also completed with satisfactory results. Tokyo Gas established the reliability of SOFC cell stacks with alloy interconnects during thermal cycles. The company showed that a stack of 20 SOFC planar cells of 13 cm^2 showed no damage from thermal cycles at 750 °C.[185] It also affirmed that a stack of 5 cm^2 cells generated 0.2 A/cm^2 at 0.754 V with a fuel utilization rate of 84.7 percent and no degradation for 1500 h. It also confirmed that a 1 kW module of several 20-cell stacks of 13 cm^2 cells showed no cell damage from thermal cycles of 200 °C/hour. MHI was assigned to develop advanced, compact, low-cost, high performance tubular SOFC cells using tightly sealed and bonded ceramic and metallic components. It obtained results that fully met the project goals both in terms of the sealing and bonding strength and power density. The advanced tubular SOFC cells MHI developed generated 0.21 W/cm^2 using reformed natural gas at 900 °C at 0.4 MPa, also meeting the project goal.

- Although its cell stacks operated with no voltage degradation for more than 1000 h, Tokyo Gas found a number of troubling signs. The cells showed traces of iron oxides, glass seal derivatives, and chromium compound vapor formation that needed to be addressed for the stacks to operate without performance degradation for a longer period.

6.3.3. MITI Begins Ambitious System Technology Development Program (2004–2007) in 2004

In 2004, Japan launched a new, highly ambitious SOFC R&D program, called "Development of Solid Oxide Fuel Cell (SOFC) System Technology" (2004–2007).[188] The program actually had two segments: a system development segment to develop a number of different types of cogeneration plants, including a combined cycle power plant, along with a system performance evaluation technology; and a second segment to develop various component technologies that would improve system performance and solve technological problems (2005–2007).

METI's principal goal in the program was to establish the Japanese capability in building total SOFC systems, which METI viewed to be significantly behind the United States, while establishing component technologies that would support total system development. The total government budget for the new program, including the System Development Program and the Component Technology Development Program, was about $90 million to be spent over the 2004–2007 period, which was about $22.5 million a year. METI had been unable to match and exceed the SECA budget, which was about $56 million a year at that time.

- In 2004, Phase III (2001–2004) of METI's 16-year SOFC Technology R&D program was still continuing, with another year to go. Launching a new program before completing one was unusual for METI. This atypical decision suggested Japan believed that SOFC commercialization was just around corner. Indeed, METI expected SOFC power plants to be ready for commercialization in 2008.[184] It had also been reported that Phase III R&D was moving along smoothly,[187] convincing METI that no major hitch would be expected that would upset the new program or its commercialization plans.

6.3.3.1. System Development Program (2004–2007)

The focus of the first program segment was to build five different SOFC plants ranging from 10 to 200 kW in output:

- A 20 kW class wet process tubular SOFCs by Hitachi and TOTO
- A 10 kW class low temperature, flat disk SOFC plant by Kansai Electric and Mitsubishi Materials Corporation (MMC)
- A 10 kW class anode-supported tubular SOFC plant by Nippon Steel and Acumentrics Japan
- A 200 kW class MOLB planar SOFC plant by MHI Kobe and Chubu Electric
- One 200 kW class combined cycle tubular SOFC system, which was to be developed by MHI Nagasaki

The performance goals were ambitious. The first three plants were to achieve an electrical efficiency of more than 40 percent HHV, an overall efficiency of more than 80 percent HHV, and a durability of more than 40,000 h measured in terms of voltage degradation of less than 0.25 percent/1000 h as measured during continuous operations over 3000 h.[188] The goal for the 200 kW class plant included an electrical efficiency of more than 45 percent LHV, an overall efficiency of more than 80 percent LHV, and a voltage degradation of less than 0.25 percent/1000 h during continuous operations over 3000 h. The combined cycle system was expected to achieve an electrical efficiency of more than 50 percent HHV and a durability of more than 40,000 h as measured by voltage degradation of less than 0.25 percent/1000 h during continuous operations over 3000 h.[188]

The program started in April 2004 and ended in March 2008. The results were dismal. The following outlines what transpired in the 4-year program,[188] which undoubtedly taught METI that it had overreached in setting its performance goals.

- *The 200 kW MOLB planar cogeneration system.* The system was to be assembled by MHI Kobe and Chubu Electric but was never built. The stack module was too bulky and had poor power efficiency. NEDO decided that the module was too immature and canceled the project.
- *The 20 kW wet-process tubular SOFC system.* The system was built by Hitachi and TOTO. At start-up, the system experienced ground faults, triggering a voltage drop, which caused serious cell damage. The project came to a halt and the system never operated.[188]
- *The 10 kW flat disk intermediate temperature SOFC system.* Kansai Electric and MMC built a 10 kW test module and conducted a series of six performance tests. Subsequently, a 10 kW cogeneration system was built and installed at the Rokko Test Center of Kansai Electric. The system successfully generated 10 kW AC, with an electrical efficiency of 41 percent HHV, an overall efficiency of 82 percent, and a voltage degradation of 1.08 percent/1000 h during continuous operations over 3102 h.
- *The 10 kW anode-supported tubular SOFC system.* The system was built, consisting of 2 kW stacks based on two 1 kW stacks developed by Acumentrics. The system generated the maximum output of 8.2 kW AC, with an electrical efficiency of 37.5 percent and an overall efficiency of 76 percent. The system operated for 3000 h continuously, with constant current operations from 1400 to 2691 h. During that period, the voltage degradation rate ranged from −1.22 percent to 1.04 percent per 1000 h.
- *The 200 kW combined cycle system.* MHI Nagasaki built the 200 kW system consisting of an SOFC plant and a micro gas turbine (MGT). The system started operations and generated a maximum electric output of 211 kW AC at the transmission end, combining 176 kW from the SOFC system and 35 kW from the microturbine, with a power efficiency of 52.1 percent LHV. After about 90 minutes of operation, however, the system suffered a short in the direct current circuit. The system could not be operated further (Table 6.5).

6.3.3.2. Component Technology Development Program (2005–2007)

The second segment of the SOFC Technology Development Program, the Component Technology Development Program (2005–2007), was viewed as successful because the results deepened the understanding of the severity and enormity of the degradation problems. Nine projects were designed under the 3-year Component Technology Development Program, aiming to study SOFC performance from three viewpoints—reliability, power density, and functionality.[188] Two research institutes, eight SOFC manufacturers, and one university participated. The program was budgeted at ¥2,411 million (about $20.9 million at the average 2005–2007 exchange rate of $1/¥114.7).

6.3.3.2.1. PROJECTS TO INCREASE RELIABILITY

Four research teams, AIST, CRIEPI, TOTO, and MMC, conducted long-term operations of four different types of SOFC stacks and modules manufactured by four different manufacturers to understand degradation mechanisms and to establish technologies for predicting SOFC reliability and longevity (Table 6.6). Four types of SOFCs were studied, including a high temperature tubular horizontal stripe SOFC stack manufactured by MHI, an intermediate temperature flat tubular SOFC stack by Kyocera, a high temperature tubular lateral stripe SOFC module by TOTO, and an intermediate temperature flat disk SOFC module by MMC.[188] The stacks and modules were operated for two sessions of 5000 h each. They were disassembled after each operating session and examined via

TABLE 6.5 METI SOFC System Technology Development Program: System Development (2004–2007) with Results from 3000-hour Durability Operations Test

		System development (2004–2007)				
		Cogeneration system				Combined cycle system
Type of system		**20 kW class wet process tubular SOFC**	**200 kW class MOLB planar SOFC**	**10 kW class flat disk intermediate temperature SOFC**	**10 kW class anode-supported tubular SOFC**	**200 kW class combined cycle tubular SOFC**
System developer		Hitachi/TOTO	Mitsubishi Heavy Industries (MHI) Kobe/Chubu Electric	Mitsubishi Materials Corporation (MMC)/Kansai Electric	Nippon Steel/ Acumentrics Japan	MHI Nagasaki
System output	Goal	20 kW AC	200 kW AC	10 kW AC	10 kW AC	200 kW AC
	Results	No operation due to an electrical accident X	Project discontinued X	10 kW AC O	8.2 kW AC O	211 kW AC O
Electric efficiency (sending end)	Goal	More than 40% AC HHV	More than 45% AC LHV	More than 40% AC HHV	More than 36% AC HHV	More than 50% AC LHV
	Results	X	X	41% AC HHV O	37.5% AC HHV O	52.1% AC LHV O
Total efficiency	Goal	More than 80% HHV	More than 80% HHV	More than 80% HHV	More than 72% HHV	Not Applicable
	Results	X	X	82% HHV O	76% HHV O	
Durability (voltage degradation)	Goal	Less than 0.25%/1000 h after 3000 h of operation	Less than 0.25%/1000 h after 3000 h of operation	Less than 0.25%/1000 h after 3000 h of operation	Less than 0.25%/1000 h after 3000 h of operation	Less than 0.25%/1000 h after 3000 h of operation
	Results	X	X	1.08%/1000 h Δ	1.04%/1000 h Δ	An electrical accident disabled operation X

Note: O – Goal achieved; Δ – unsatisfactory results; X – failed.

Source: "'SOFC System Technology Development,' Post-Project Evaluation Report," *www.nedo.go.jp/iinkai/kenkyuu/houkoku/20h/jigo/15.pdf*, Research Evaluation Committee, New Energy

TABLE 6.6 METI SOFC System Technology Development Program: Component Technology Development (2005–2007)

Component technology development: Projects to increase reliability					
Researchers	**R&D subjects**	**Findings**			**Suggested future studies and countermeasures**
		Average degradation rate during the first 5000 h (A) and during the second 5000 h (B)*	Impurities and locations of degradation	Types of degradation	Improved Cr poisoning countermeasures (such as increased coating tolerance); improved Redox countermeasures
CRIEPI, AIST, TOTO, MMC	MHI Tubular horizontal stripe SOFC stack	**A.** 0.88%/1000 h **B.** 0.65%/1000 h	Cathode over voltage, increased Cr concentration	Cr poisoning of $LaMnO_3$ cathode/YSZ phase; dismutation of microstructure and composition	Improved cell connector metal durability; elucidation of degradation mechanism; establishment of accelerated tests for degradation; Redox countermeasures
	Kyocera flat tubular SOFC stack	**A.** 1.5%/1000 h **B.** 0.9%/1000 h	Increased polarization at cell connector metals; increased impurities, such as Si, at cathode and reaction suppression layer	Increased polarization at Interlayer/cathode; oxidation and degradation of cell connector metals; generation of $SrCrO_4$ type insulator; reactions with impurities, such as Al, Si, Cr. etc.	The relationship between atoms as an impurity and degradation; possible use of low cost insulators; improvement and adjustment of electrolyte and interconnect
	TOTO Tubular lateral stripe SOFC module	**A.** 1.6%/1000 h **B.** 1.67%/1000 h	Increased Si concentration at anode; decreased adhesion between anode and electrolyte; increased Ca between cathode and interconnect	Increased polarization and generation of electrical insulation at anode due to Si concentration; increased polarization between cathode and interconnect	Identification of sources of impurities; relationship between cause and rate of degradation; longer term studies on degradation; interaction between chemical and mechanical degradation
	MMC Flat disk SOFC Module	**A.** 0.54%/1000 h **B.** 0.81%/1000 h	Increased Si and P concentration at anode; re-oxidation at anode area; increased Cr and S concentration at cathode; abnormal oxidation of alloy interconnect	Decreased reactive areas due to reaction with impurity; re-oxidation at combustion zone and existence of impurities	

(*continued*)

TABLE 6.6 METI SOFC System Technology Development Program: Component Technology Development (2005–2007) (*Cont'd*)

Component technology development: Projects to increase power density				
SOFC developers	**SOFC types**		**Goals**	**Outcomes**
MHI	Tubular	Cell	0.5 W/cm^2	0.5 W/cm^2
		Cell stack	0.5 W/cm^2	0.65 W/cm^2
MMC, Kansai Electric, Kyushu University	Intermediate temperature flat disk	Cell	0.4 W/cm^2	0.55 W/cm^2
		Module	0.4 W/cm^2	0.4 W/cm^2
TOTO	Wet process tubular lateral stripe	Cell	0.43 W/cm^2	0.45 W/cm^2
Component technology development: Projects to increase functionality				
Researchers	**Types of functionality**		**Goals**	**Outcomes**
Sumitomo Precision/Toho Gas	Cell	Cell strength	Tolerate more than 1000 MPa	Tolerate more than 900 MPa
		Cell stack start-up time	Within 1 h	Take 1 h to start-up
	Hot module	Output	1 kW	1.2 kW
		Cell stack heat capacity reduction	More than 30%	18%
		Hot module heat capacity reduction	More than 30%	30%
		Circumference direction of temperature difference	Within 40 °C	Within 50 °C
		Start-up time	Within 2 h	2.5 h

Kyushu University	Explore major poisoning mechanisms from chemical reactions of SOFC cells stemming from sulfur impurities in fuel and performance degradation—research outcomes: 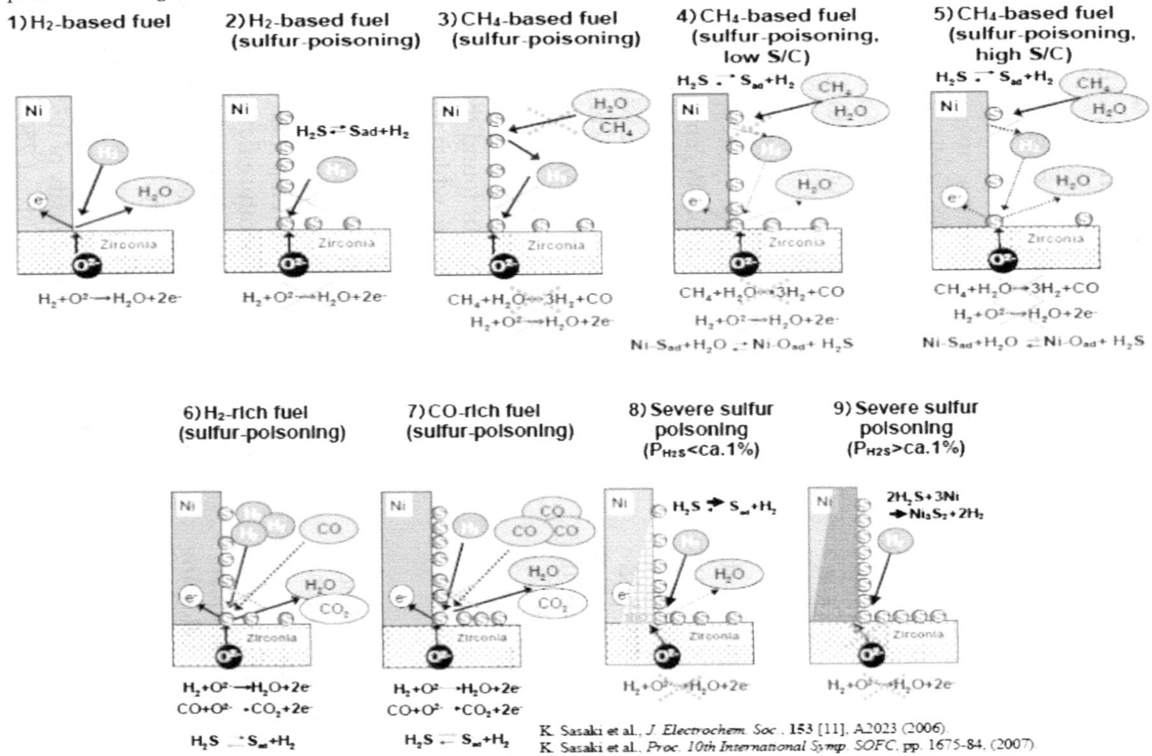

* *Two stacks and two modules of four different types of SOFCs manufactured by four different makers were tested for reliability and durability. They were operated for two sessions of 5000 h each (Session A and Session B) and disassembled after each session for degradation examination.*

Note: The table was created based on "SOFC System Technology Development,' Post-Project Evaluation Report," www.nedo.go.jp/iinkai/kenkyuu/houkoku/20h/jigo/15.pdf, Research Evaluation Committee, February 2009.

a secondary ion mass spectrometry microscope (SIMS). Impurities accumulated and the type and degree of degradation resulting from the impurities were cataloged and remedies and future studies were suggested. The findings included the following:[188]

- Performance degradation ranged between 0.5 and 1.7 percent per 1000 h. Each SOFC type for the most part had different concentrations of different impurities at different locations in cells, and suffered different types of degradation. A SIMS study identified a correlation between the voltage drop and the change in chromium (Cr) concentration in the area between a cathode and an electrolyte, suggesting future studies on the relationship between Cr poisoning mechanism and increase in polarization of electrode. All the SOFCs showed excellent durability against load changes but experienced major degradations during repeated stop/start-up operations.
- The results also showed that a high concentration of some impurities did not cause any degradation during operations. The research team, thus, suggested the need for longer term research to identify the effects of those impurities. The team also suggested further studies on the following topics:
  - Cr poisoning, cathode destabilization, and redox tolerance. The relationships between chemical and mechanical instability.
  - Impurities and their sources (within the stack and BoP). The electrochemical characteristics of impurities and their effects on durability.
  - Improvement in longevity estimate technology. Development of accelerated degradation test technology. Need for physicochemical understanding. Correlation between a specific degradation phenomenon and the electrochemical mechanism (Table 6.6).

6.3.3.2.2. PROJECTS TO INCREASE POWER DENSITY

Three research teams were assigned to develop HPD in SOFC cells and stacks by improving cell composition and structure. The teams included MHI, MMC/Kansai Electric/Kyushu University (subcontractor), and TOTO. The following are their activities:[188]

- MHI was commissioned to develop a tubular-horizontal stripe SOFC cell with a power density of 0.5 W/cm^2 in an internal reforming pressurized SOFC environment. MHI developed a cell made with a $LaMnO_3$ cathode and achieved a power density of 0.5 W/cm^2 at 0.6 MPa. MHI then assembled a cell stack with the $LaCoO_3$ cathode, which had a power density of 0.65 W/cm^2 at 0.7 MPa at 0.65 V but found that it had a durability problem, which required MHI to investigate the reliability of the $LaMnO_3$ material.
- The team of MMC, Kansai Electric, and Kyushu University (as a subcontractor) was commissioned to develop a HPD intermediate temperature flat disk SOFC stack module with a power density higher than 0.4 W/cm^2. The team developed a 120-mm diameter cell with an anode made of gadolinia-doped ceria embedded with nanometer-size ruthenium particles, which achieved a power density of 0.55 W/cm^2 with a 75 percent fuel utility at 750 °C. The team then assembled a module of 34 cells of the same composition, which achieved a maximum power density of 0.4 W/cm^2.
- TOTO was commissioned to develop a wet process tubular-lateral stripe SOFC cell with a power density of 0.43 W/cm^2. The company succeeded in improving gas permeability of the cathode interlayer, changing the anode interlayer composition,

and adjusting the anode porosity, thereby increasing the power density from 0.14 to 0.34 W/cm^2. TOTO also adopted a low temperature sintering process for the cathode interlayer microstructure and increased the power density of the cell to 0.45 W/cm^2, exceeding the goal of 0.43 W/cm^2. The company planned to achieve a power density of 0.5 W/cm^2 by lowering the internal polarization of the cathode support structure.

6.3.3.2.3. PROJECTS TO INCREASE FUNCTIONALITY

Two research teams were assigned to conduct R&D that would make SOFC system more useful and practical.[188] The team of Sumitomo Precision Industrial and Toho Gas worked to develop a planar SOFC module that could start up faster than current systems. The second team, Kyushu University, aimed to explore fuel impurities and the chemical reactions and performance degradation they cause in SOFC anodes. The project attempted to clarify the level of fuel impurities that could be tolerated by SOFC systems, thereby expanding the types of fuels that could be used and lowering the level of fuel refining needed. The outcomes included the following:

- The team of Sumitomo Precision Industrial and Toho Gas was unable to achieve all the goals but came close to establishing a rapid start-up planar SOFC hot module. The team first established a technology that could quantify the maximum thermal stress during a rapid start-up without causing cell damage. The team also developed robust cells that could tolerate unstable thermal stress and built a hot module consisting of a cell stack, a heat exchanger, and a reformer in one compact, integrated unit. The hot module was able to achieve start-up in 2.5 h, shortening the average start-up time by 1 h.
- The Kyushu University team assembled a systematic compilation of various fuel impurities and their chemical reactions and performance degradation effects on SOFC anodes. This effort was highly acclaimed as an outstanding electrochemical study and awarded the Schonbein Medal, a high academic honor presented once in 2 years. The team used thermodynamic calculations to elicit the partial pressure dependence and C–H–O ratio dependence in the reactions between Ni and various impurities. They established phase diagrams for all major elements. The team also measured the poisoning effects of various sulfur impurities, including H_2S, COS, methyl mercaptan, ammonia, TBM, DMS, chlorine, siloxane, boron, and phosphorus. The team's systematic study of the effects of sulfur—the most important impurity species—in SOFCs was the world's first. The team also conducted experiments of various impurities that could originate from external sources and observed poisoning behaviors for up to 1000 h. They explained the poisoning mechanism of each substance. The team also showed the relationship between poisoning behavior and cell structure, such as the delay in cell poisoning in anode-support cells. Finally, the team built the world's first database on various impurity types, fuel types, various operating conditions, poisoning behaviors, and equilibrium diagrams.

6.3.3.3. *The Post-Program Evaluation Report Judges the 2004–2007 SOFC R&D Program to be an Overall Failure*

In February 2009, the post-program evaluation report on the 2004–2007 SOFC System Technology Development Program was published. The reports judged that the program was a failure on the whole because it did not build a functional SOFC power-generation system or find a solution to degradation. The program made advances in testing and measurements, component technology, and SOFC performance. It also helped to better understand SOFC degradation; a number of projects in the program—including the Kyushu University team project, in particular—contributed to clarify the complex, severe, and difficult nature of degradation. However, the evaluation viewed that none of the research efforts uncovered the key to prevent, reverse, or end degradation or conceive a breakthrough that would do so. No matter how deeply one may understand the degradation mechanism, the research is moot unless the key to counter it is found or a new substance that would not degrade is created. Devoid of this vital solution, viable SOFC systems with little degradation and great longevity would never be built.

The February 2009 post-program evaluation report was a comprehensive and outstanding R&D review report that should be widely emulated. The review committee started its work in February 2005. Committee membership included seven noted academicians and one research company president, all objective third parties. The report clearly recognized and documented the limitations and deficiencies of the 2004–2007 SOFC System Technology Development Program.[189] The committee stated that many difficulties encountered in the program were due to lack of basic knowledge and facts about fuel cell electrochemistry and that it was the highest priority that Japan acquire such knowledge. To achieve the goal, the committee stressed that the researchers go back to basics. The committee stated at the same time, in addition to basic research, it was necessary that component technologies, operating experiences, and a knowledge base must be built through SOFC demonstrations in preparation for commercialization:[189]

- The evaluation report also pointed out that the program lacked leadership to lead and fine-tune the program. The committee speculated that there had not been much information sharing across the program. In a program where multiple manufacturers work toward one goal, the committee pointed out that it would be necessary to install an organization or a leader to increase the synergy and efficiency of the program, manage the program across horizontal boundaries, and promote cross-communication.
- The report also pointed out that the program lacked studies that address SOFC cost reduction, including the relationship between cost and SOFC output, reliability, and durability, and which market segments SOFC systems should first be targeted for commercialization.
- The report also mentioned that the degradation goal set by the system technology development projects—0.25 percent/1000 h (a theoretical indicator for an SOFC system to survive for 40,000 h or 5 years)—is a demanding goal but a realistic one that a system must attain. (In the US DOE SECA program, the degradation goal had been set at far less demanding level.)

The post-evaluation report took a long time, nearly 3 years, but throughout the review process, the committee frequently shared with METI administrators and program managers its review activities, findings, and preliminary recommendations, which were reflected in the formation of METI's subsequent R&D and demonstration programs. By February 2008, the entire review had been compiled with a final set of recommendations, and in February 2009 the official report was published.

6.3.4. SOFC Demonstration Research Program (2007–2010): A New Hope for Near-Term SOFC Commercialization

In 2007, METI launched a 4-year program (2007–2010) to sponsor demonstration of small SOFC power plants primarily for cogeneration for residential use, raising hopes among SOFC developers that commercialization could be achieved in the near term. The demonstration program was budgeted at ¥2.8 billion for 4 years[190] (about $27.3 million at the 2007–2010 average exchange rate of $1/¥102.7). It was a modest amount compared to the METI PEMFC system demonstration program (2005–2008), but it was the world's first, major SOFC demonstration program. NEDO and the non-profit New Energy Foundation (NEF) administered the program. The goal of the program was to accelerate the commercialization of SOFC technology. The program aimed to accumulate real-life SOFC operational experiences and knowledge—about durability, performance, reliability, and technical problems—identify and solve the problems, and advance the technology.

- NEDO also established a subsidy program to provide financial support to promote the demonstration program. A power plant that would qualify for the demonstration program was set as a 0.5—10 kW power plant fueled by natural gas and other fuels, such as LPG, with a power efficiency of more than 25 percent and a durability of more than 6 months.[191] An SOFC stack could receive under the 2007 and 2008 program a subsidy of up to ¥20 million per kilowatt (about $170,000 at the 2007 exchange rate $1/¥117.8) or up to ¥100 million per power plant system (about $850,000).[192] In 2009, the amount of the subsidy was cut, hopefully indicating that costs of SOFC stacks and systems were declining. The subsidy for an SOFC stack was ¥8.1 million per kilowatt (about $82,000 at the 2009 exchange rate of $1/¥99), and the subsidy for a power plant system was ¥52.1 million (about $526,000).[193]

6.3.4.1. *Program Helps SOFC Industry Grow*

The scale of the demonstration program grew each year. The program sponsored the installation of 29 power plants in FY2007, 36 in FY2008, and 67 in FY2009 (Fig. 6.1 and Table 6.7). In the FY2010 program, the planned number reached 101.[194] The total number of SOFC units installed over 4 years was 233. As the program grew, a greater number of industrial players participated in the program, indicating the program helped broaden the SOFC industry:

- Throughout the program, Kyocera was the dominant fuel cell stack manufacturer. Kyocera was the manufacturer of SOFC stacks for 86 percent of the SOFC units installed in FY2007 and 83 percent in FY2008.[194] In FY2009, Toyota Motor, Aisin Seiki, Kyocera, Osaka Gas formed a team to develop jointly SOFC power plants for demonstration. The team demonstrated their units integrated with Kyocera stacks, while Kyocera demonstrated some units on its own.[195] Also in FY2009, Tokyo Gas, Kyocera, Rinnai, and Gastar began demonstrating their SOFC power plants.[196] They had been jointly developing SOFC systems since 2004 by integrating a horizontal-stripe cell stack developed by Kyocera and Tokyo Gas. In FY2009, Kyocera SOFC stacks were in 69 percent of the demonstrated units. In FY2010, the Tokyo Gas/Kyocera/Rinnai/Gastar team did not participate in the demonstration program; they likely decided to wait for more technological maturity. In the final year, there were three major players, the Toyota Motor/Aisin Seiki/Kyocera/Osaka Gas team, JX Nippon Oil & Energy, and TOTO, indicating that the industry had been consolidated around stronger players.
- There were nine installers and operators, including Osaka Gas, Tokyo Gas, Hokkaido Gas, Seibu Gas, Toyo Gas, Tokyo Electric, Tohoku Electric, JX Nippon Oil & Energy, and TOTO. Osaka Gas played the leading role. It installed and operated 123 units (or more than 52 percent) of the 233 units that were installed.

(a) Number of Installed Power Plants

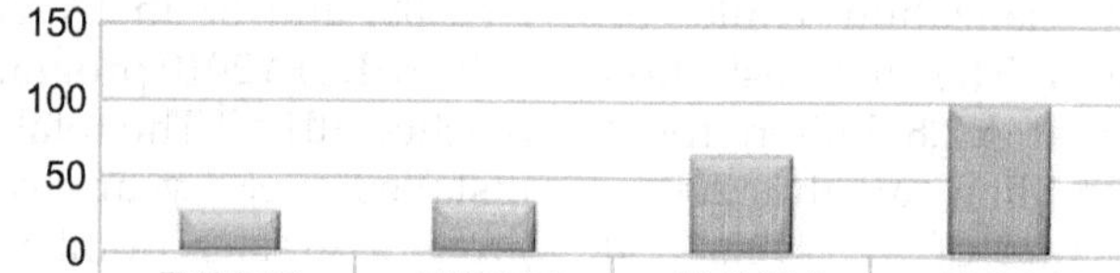

	FY2007	FY2008	FY2009	FY2010
SOFC Systems Installed	29	36	67	101

(b) Number of Plants by SOFC Manufacturer

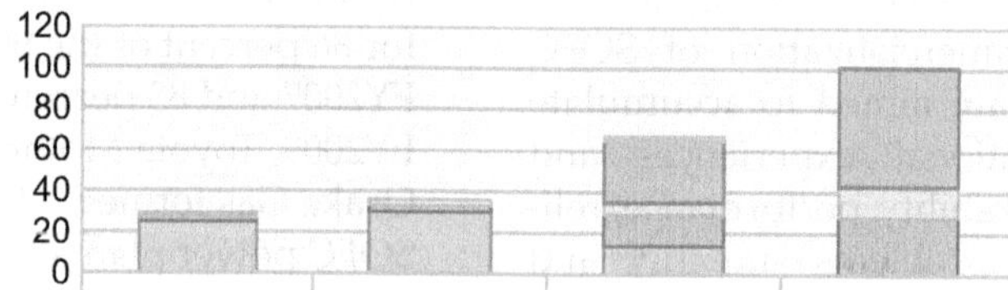

	FY2007	FY2008	FY2009	FY2010
Gastar/Rinnai	0	0	2	0
Toyota/Aisin	0	0	30	58
TOTO	2	3	6	15
Nippon Oil	2	3	15	28
Kyocera	25	30	14	0

(c) Operating Hours (as of March 2011)

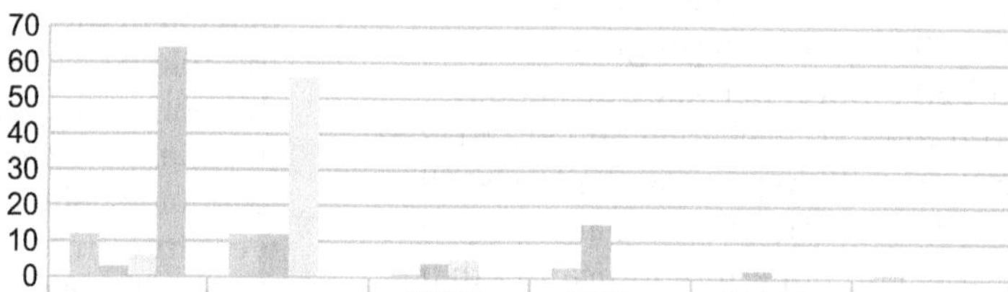

	0-5000	5000-10000	10000-15000	15000-20000	20000-25000	25000-30000
FY2007 (29 units)	12	12	1	3	0	1
FY2008 (36 units)	3	12	4	15	2	
FY2009 (67 units)	6	56	5			

FIGURE 6.1 METI SOFC Demonstration Research Program (FY2007–FY2010). The figures were created based on Makoto Okuda (New Energy Foundation), "SOFC Demonstration Research, FY2010 Result Report," in *SOFC Demonstration Research: Result Report Session*, New Energy Foundatation, 10 March 2011. For color version of this figure, the reader is referred to the online version of this book.

(d) Operating Performance (as of March 2011)

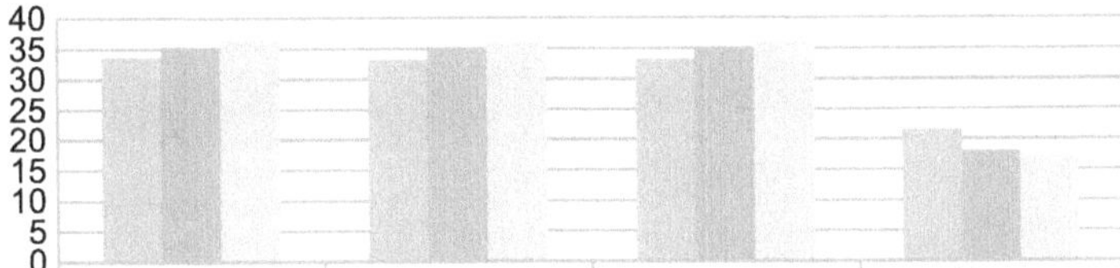

	Power Generation Efficiency (%HHV)	Electricity Use Efficiency (%HHV)	Electricity Use Efficiency (%HHV)	Heat Utilization Efficiency (%HHV)
FY2007	33.7	33.3	33.3	21.6
FY2008	35.4	35.1	35.1	18
FY2009	36.4	36.1	36.1	17.4

(e) Incidences of Technical Problems (as of March 2011)

	Stack	Fuel Processing	Air Supply	Water Supply	Heat Recovery	Inverter	Control	Others	Total
FY2007	7	5	6	11	0	0	0	7	36
FY2008	4	8	3	20	2	0	12	3	52
FY2009	3	19	6	30	8	9	29	7	111

FIGURE 6.1 *(continued)*.

6.3.4.2. *Program Results Are Mixed*

The operational results have been reported from the demonstration programs, with some positive and some negative:

Plant operating hours increased, but not long enough. NEF documents show that the number of operating hours got longer each year, demonstrating improvement in performance.[198] The unit operating longest was the unit installed in FY2007. It ran for 26,280 h (equivalent to 3 years of operation):[198]

- In FY2007, of the 29 installed units, 12 units were operated for 0–5000 h, another 12 for 5000–10,000 h, one for 10,000–15,000 h, and three for 15,000–20,000 h. The unit operating longest ran for 26,280 h, equivalent to 3 years of operation. None were operating as of December 2010.
- In the 2008 program, 36 units were installed, and 3 operated for 0–5000 h, 12 for 5000–10,000 h, 4 for 10,000–15,000 h, and 15 for 15,000–20,000 h. More units installed in FY2008 operated for more hours than those in FY2007. Of the 36, 17 were still operating as of December 2010. Several units probably exceeded 20,000 h.

TABLE 6.7 METI SOFC Demonstration Research Program (2007–2010)

Installer/operator	SOFC and system manufacturer	Type of fuel	Output size	Installation site	Number of plants installed				
					2007	2008	2009	2010	Total
Osaka Gas	Kyocera	Natural gas	0.7 kW	Residence	20	25	12		57
	Toyota/Aisin Seiki/Kyocera/ Osaka Gas						23	41	64
	TOTO							2	2
Tokyo Gas	Kyocera	Natural Gas	0.7 kW	Residence	3	2			5
	Toyota/Aisin Seiki/Kyocera/ Osaka Gas			Residence Residence			4	11	15
	Gastar/Rinnai/Kyocera/Tokyo Gas						2		2
	TOTO							1	1
Hokkaido Gas	Kyocera	Natural gas	0.7 kW	Residence	1	1			2
	Toyota/Aisin Seiki/Kyocera/ Osaka Gas						1	2	3
Seibu Gas	Kyocera	Natural gas	0.7 kW	Residence	1	1			2
	Toyota/Aisin Seiki/Kyocera/ Osaka Gas						1	2	3
Toho Gas	Toyota/Aisin Seiki/Kyocera/ Osaka Gas	Natural gas	0.7 kW	Residence			1	2	3
	TOTO							2	2
Tokyo Electric	Kyocera	Natural gas	0.7 kW	Residence		1	1		2
Tohoku Electric	Kyocera	Natural gas	0.7 kW	Residence			1		1
JX Nippon Oil & Energy	JX Nippon Oil & Energy	LPG	0.7 kW	Residence	1	2	14	27	44
	JX Nippon Oil & Energy	Kerosene	0.7 kW	Residence	1	1	1	1	4
TOTO	TOTO	Natural gas	0.7 kW	Residence		2	6	10	18
			2 kW	Small Business	2				2
			8 kW	Small Business		1			1
Total					29	36	67	101	233

Source: "SOFC Project: SOFC/Demonstration Research," Pamphlet, http://www.nef.or.jp/sofc/share/pdf/sofc_pamphlet.pdf, New Energy Foundation, February 2011.

- In the FY2009 program, 67 units were installed, 6 were operating for 0–5000 h, 56 for 5000–10,000 h, and 5 for 10,000–15,000 h, as of December 2010. Overall, 54 were still operating as of December 2010.
- In the FY2010 program, 64 units of the planned 101 units were installed as of December 2010. No further information is available.

Mediocre operational performance. The electrical efficiency and heat utilization (based on the operating data, excluding the business-use units and the kerosene unit) showed little improvement. The SOFC units operated 24/7 in response to load demand.[197] There was no Feed-In-Tariff program for excess electricity in Japan at that time, and household electricity demand significantly varied between day and night, thus forcing the units to shut down or lower operating capacity and thereby reduce electrical efficiency. The power generation efficiency as well as electricity use efficiency improved to 36.4 percent HHV and 36.1 percent HHV, respectively, for FY2009.[198] The FY2010 units were not evaluated due to the short operating period. When power output was at the rated output of 0.7 kW during the short peak demand period, efficiency was almost 40 percent HHV.[197] When power demand exceeded 0.7 kW, the electric grid filled the demand. The heat utilization ratio, however, was declining, requiring serious study to address the issue, according to NEF.

Continued technical problems. There were continued technical problems in FY2007, FY2008, and FY2009, as of December 2010.[198] The problems related to stacks declined somewhat each year, but other technical problems slightly increased.[197]

- There were a total of 36 instances in which technical problems occurred[197] or 1.24 instances per unit in FY2007, 52 instances in FY2008 (1.44 per unit), and 111 instances in FY2009 (1.65 per unit). Technical problems related to stacks declined. The most frequent problems were related to water systems, which typically originated from water pumps. The second most frequent problem was related to the control systems.
- NEF noted that some problems seemed to occur as the systems aged and indicated that longer term operating experiences and observations were required.
- NEF also indicated that some of the problems initially observed had been resolved. For example, the units installed in the FY2007 program showed a gradual degradation in electrical efficiency. Solutions were factored in the FY2008 units, significantly reducing the problem. Other initial problems were damage caused by impact shock and vibrations during transport and installation. The problems were addressed and resolved, according to NEF.

6.3.5. METI Institutes "Back-to-Basics" Research Program (2008–2012)

In 2008, METI decided to launch a program prompted by the post-program evaluation committee review of the 2004–2007 SOFC System Technology Development Program. The committee recommended that a basic research program to explore SOFC degradation be launched, and remarkably, METI accepted the idea of "going back to the basics." The 5-year program, called the "Development of System and Elemental Technology on Solid Oxide Fuel Cell Program" (2008–2012), was to support thermodynamic, chemical, and mechanical research to seek and clarify degradation mechanisms (Table 6.8).[199]

This was the first time in the world that a government had supported basic scientific research in support of the fuel cell industry. Indeed, many research institutes and companies had invested billions of dollars in intellectual

TABLE 6.8 Development of System and Elemental Technology for Solid Oxide Fuel Cell (2008–2012): Projects and Participants

	Goals	Participants
Basic research for improving durability and reliability	Identify degradation mechanisms from a micro perspective through thermodynamic, chemical, and mechanical analyses.	AIST
	Seek to clarify the degradation mechanisms, formulate countermeasures, and verify results.	CRIEPI
		TOTO
	Establish accelerated testing methods.	Mitsubishi Materials Corporation
	Establish the correlation between the degradation phenomenon and changes in the microstructure in the tri-phase interface.	Kansai Electric
		Mitsubishi Heavy Industries
	Establish durability evaluation techniques that can evaluate the degradation mechanisms from a macro perspective and that will enable users to easily assess the remaining life of the cell	Kyushu University
	• Find ways to achieve durability of 40,000 hours (defined as a voltage drop of 0.25 percent/1,000 hours).	Gifu University/Nagoya University
	• Find ways to achieve 250 start-stop operations.	Tohoku University
		Kyoto University/University of Tokyo
Technological development for improving practical utility	Technological development aimed at reducing the cost of raw materials and other materials, and at creating a low-cost cell stack module	Hitachi Metals Tokyo Institute of Technology Mitsubishi Materials Corporation
		TOTO
		Mitsubishi Heavy Industries
		Mitsubishi Materials Corporation
		AGC Seimi Chemicals
		KCM Corporation
	Stop-start technology for improving operability	TOTO
		Mitsubishi Materials Corporation
		Kansai Electric DAIHEN Corporation
	High-pressure operating technology	Mitsubishi Heavy Industries Toyota Motor

Source: FY2009 Policy Implementation Plan, Energy Innovation Program, SOFC /system Element Technology Development," P08004, NEDO, no date; "Fuel Cells & Hydrogen 2008–2009," Pamphlet, NEDO, October 2008.

property, prototype products, manufacturing equipment, and infrastructure. But going back to basics was tantamount to throwing away these efforts and starting from scratch. Such a radical shift in strategy would seem to be a major challenge to entrenched interests. But remarkably, the program launch was welcomed by the public with little resistance. Another noteworthy fact was that the new program had named a program leader—a globally renown Japanese SOFC researcher—who had the authority to control and supervise the research activities in the program. To install a leader to substantively manage an R&D program is a rather extraordinary event in Japan, where anonymity is an accepted practice, indicating METI's determination to succeed or the level of its desperation not to fail.

- In March 2008, the 5-year basic research program was announced and immediately got the attention of the fuel cell community globally. It clearly fascinated scientists and engineers with its boldness and convinced many that it was exactly what was needed to achieve a breakthrough.

6.3.5.1. *Program is Subject to Serious Constraints*

Since the program was established, however, there has been little information on the progress or status of the basic research, filling the fuel cell community with hopeful curiosity. Of course, the program is still under way, and the challenges are enormous, likely demanding more time. Nonetheless, there are some systemic problems that might seriously obstruct the program's potential:

- First, despite the program's claim of pursuing basic research, the program setup was more characteristic of applied research and product development. The Development of System and Elemental Technology SOFC Program (2008–2012) was accompanied by the SOFC demonstration program (2007–2010) discussed above.[199] The Elemental Technology SOFC Program was expected to explore the nature of SOFC degradation gathered from the demonstration operations and feed its research findings and results back to the demonstration program. NEDO established a coordination committee to link the demonstration program with the Elemental Technology Program to facilitate the feed-back mechanism and communication.[200] Three coordinating mechanisms also were established within the Elemental Technology Program—the basic research consortium, the durability technology working group, and the committee for accelerated data sharing—all aimed at better communications and data sharing between the demonstration program and the basic research program. This setup seriously constrains the basic research goals, however, and could obstruct the program's potential.
- The stated goals of the Elemental Technology Program also are more characteristic of applied research and product development. The program has a number of concrete goals to achieve, including finding specific causes of degradation and finding solutions to these problems, ensuring the effectiveness of the solutions, and establishing a technology to make SOFC products cheaper, better, and more functional. Furthermore, the primary interests of most of the program participants are in applied research and product development rather than basic research. They also are set up to conduct their research work independently from each other rather than collectively, missing the critical opportunity for cross communication.
- Third, the program is inherently constrained from reaching out and exploring basic research questions. The program has a term of 5 years, constrained by deadlines. Moreover, the budget for the program is too

small to support a major basic research effort. Furthermore, the budget is declining, giving the fuel cell community the impression that the government lacks commitment. The budget was ¥1.3 billion (about $12.5 million at the 2008 exchange rate of $1/¥104) for 2008, ¥1.2 billion (about $12 million at the 2009 exchange rate of $1/¥99) for 2009, and ¥800 million (about $8.8 million at the 2010 exchange rate of $1/¥90.8) for 2010.

- Fourth, the Elemental Technology Program lacks a strategic focus. The prime goal for SOFC basic research should first be to explore and understand the nature of the electrochemical activity at the most fundamental level, at the atomic level, to identify the mechanisms and the factors that rule, accelerate, inhibit, and control the fuel cell process. Only after this level of understanding is reached, will it be possible to identify and create new, low cost, electrochemical materials that would achieve or evolve toward high performance under all operating conditions of load, gas composition, output size, or age. But instead of taking this approach, the primary goal of the 2008–2012 program is narrowly focused on solving the degradation problem. The researchers will consider only existing state-of-the-art SOFC cells and stacks and investigate the degradation effects observed in these materials that are caused by various impurities in fuels, air, and components. Their attention then will focus on minimizing or eliminating the impurities or the effects of the impurities.

For basic research to succeed, many conditions must be present to support basic research goals. There must be unrestricted and free use of time and energy, ample financial resources, extensive research infrastructure, a large number of highly talented researchers from all related academic disciplines practically and virtually collocated in one place, and a strong, enlightened leader that ensures these components are integrated into a critical mass. It is too early to judge the success or failure of the Elemental Technology Program, 2008–2012. The program's setup, however, does appear to be much like the traditional government-industry applied R&D program that METI and NEDO have been long practiced.

6.3.6. Still Japanese SOFC Developers Press on with Their Commercialization Plans

A large number of Japanese companies, including manufacturers of ceramics, electronics, machinery, and metals, gas and power utilities, and oil companies, are engaged in SOFC development. No one company handles all the aspects of SOFC component and equipment development, production, integration, and operation. Japanese utility companies, which have been an early, active player in SOFC development in Japan, often are involved in BoP development, system integration, and operation, but some companies, such as Osaka Gas and Tokyo Gas, develop SOFC cells and stacks. They, for the most part, work jointly, taking advantage of each other's strength, centering on SOFC cell and stack manufacturers.

Most Japanese SOFC developers have participated in METI SOFC development programs, with the exception of Kyocera, which has only provided prototypes for use in METI sponsored research. Only one foreign SOFC developer established its presence in Japan, Acumentrics of the United States. It too participated in METI SOFC development programs.

6.3.6.1. *Acumentrics Japan*

Acumentrics of the United States established its business base in Japan in 2003, when Sumitomo Corporation, a general trading house, agreed to import Acumentrics' 2 kW SOFC modules. Acumentrics and Sumitomo planned to integrate the 2 kW units with BoP that conformed to Japanese specifications and test them at a number of electronic equipment manufacturers and utilities companies in Japan.[201] They planned to test 10 kW systems in the summer of 2003 and market them in 2005 and later expand their products to include the 20, 50, and 100 kW systems.[202] Acumentrics also planned to build a mass production facility in the summer of 2003 in Connecticut that would produce two hundred 10 kW modules to supply the market in Japan.[201]

To accomplish these goals, the two companies formed a joint venture, Acumentrics Japan.[202] In June 2004, Nippon Steel invested in Acumentrics Japan.[203] Acumentrics has 50 percent of Acumentrics Japan, Sumitomo has 40 percent, and Nippon Steel has 10 percent.[204] In the same month, Sanyo agreed to cooperate in the development of Acumentrics' tubular SOFC technology.[205] Sumitomo obtained exclusive rights for the sale of Acumentrics' 1–10 kW SOFC systems in Japan[182] and had overall responsibility for development, demonstration operation, and maintenance of BoP.[201] Nippon Steel was responsible for manufacturing power plants with Sumitomo, using SOFC stacks provided by Acumentrics.[203]

In June 2004, Acumentrics Japan and Nippon Steel were assigned to develop a 10 kW cogeneration power plant under the METI 2004–2007 SOFC System Development Program. This was a major opportunity for Acumentrics. METI had rarely contracted with a foreign company to work under its programs. For the SOFC system development program, Nippon Steel would match the NEDO grant in kind, and the group was to seek other Japanese investors to augment the Acumentrics venture.[202]

- The project was to develop a 10 kW tubular SOFC cogeneration power plant and, in the end, the project proved to be one of the two projects that achieved any degree of success. Acumentrics and Nippon Steel succeeded in developing a 10 kW class cogeneration power plant that achieved an electrical efficiency of 37.5 percent and a total efficiency of 76 percent. The system operated longer than the required 3000 h, but did not meet the required voltage degradation goal of less than 0.25 percent/1000 h.[188]

In late 2006, there were media reports that METI was implementing an SOFC demonstration program starting in April 2007. In September 2006, a large number of SOFC developers were reported to be participating in the SOFC demonstration program, including Acumentrics.[277]

In January 2007, Acumentrics announced that the Japan Gas Association had completed its safety test of the company's 3 kW tubular SOFC system. The test was a prerequisite to participate in the METI SOFC field demonstration test that would start in April 2007.[202] Acumentrics already had shipped more than twenty 1–10 kW power plants for demonstration operations in Japan on its own to prepare for the government demonstration program.

In April 2007, the 4-year METI SOFC Demonstration Research Program (2007–2010) was launched. Acumentrics was not among the participants.[277] There has been little information on Acumentrics' SOFC activities in Japan since then and it is likely that the company suspended SOFC activities in Japan.

6.3.6.2. *National Institute of Advanced Industrial Science and Technology*

AIST became active in SOFC technology development in 1974 when the Japanese government (namely METI) decided to foster SOFC technology and assigned three national laboratories to develop basic SOFC technology (see discussion at Section 6.3.1). In April 2001, the National Institute of Advanced Industrial Science and Technology (the new AIST) began operations.[203] AIST was set up as a public research organization to integrate the activities of 15 research institutes, including the 3 laboratories mentioned above. The new AIST is Japan's largest public research organization and has some 3200 employees in all.

AIST has demonstrated an outstanding prowess in developing SOFC technology. Many AIST SOFC advances were made in collaboration with industry players, and some independently.

- In January 2006, AIST developed a microtubular SOFC in the range of millimeters to sub-millimeters in diameter, which could operate between 500 and 600 °C. The cell with ceria-based electrolyte achieved a power density of 1 W/cm^2 at 570 °C, the world's highest power density among SOFCs with ceria-based electrolytes. In developing micro ceramic parts, the thermal shock resistance of the cell was dramatically enhanced, thus enabling the fabrication of compact SOFC modules that could be used in rapid start-up and shutdown operations (2006).[206]
- In October 2006, in collaboration with NGK Insulators, AIST developed honeycomb-type planar SOFCs with sub-millimeter sized channels. AIST incorporated more than 250 single cells per cm^3 into a honeycomb support structure 15 mm × 15 mm in size. The module achieved a power density of 0.23 W/cm^2 at 700 °C, which corresponds to the world's highest level in the middle temperature range for cathode-supported SOFCs[206] (see Section 6.3.6.11).
- In January 2007, in collaboration with Japan's Central Research Institute of Electric Power Industry (CRIEPI), AIST developed tubular SOFC cells that demonstrated a power density of 1.06 W/cm^2 using hydrogen fuel and operating in a low temperature range from 500 to 650 °C. The cells were cylindrical (3 mm in outer diameter and 3 cm in length of generating part), and a power density of 1.06 W/cm^2 was achieved by using hydrogen fuel. The air electrodes of SOFC developed were made of porous ceramics, and silver nanoparticles (the particle diameter 10–100 nm) were uniformly dispersed[207] (see Section 6.3.6.3).
- In March 2007, in collaboration with NGK Spark Plug and Toho Gas, AIST developed a small SOFC, the size of a size sugar cube, that consisted of fine ceramic tubes. The SOFC bundle generated the world's highest output power density (2 W/cm^3) at an operating temperature below 600 °C[208] (see Section 6.3.6.12).
- In February 2009, in collaboration with NGK Spark Plug, AIST developed a micro fuel cell module 1 cm^3 in size. The module achieved a power density greater than 2 W/cm^3 at 550 °C.[209] The microtubular cell was comprised of tubes 0.8 mm diameter and used ceria-based ceramic electrolyte, nickel ceria-based ceramic for the anode, and lanthanum cobalt ceria-based ceramic for the cathode.[210] Hydrogen at a temperature of 450–550 °C was input into the module, which produced power densities of 0.88–2.2 W/cm^3.
- In September 2009, in collaboration with NGK Spark Plug, Fine Ceramic Research Association, and Toho Gas, AIST developed a microtubular SOFC module capable of generating 50–200 W. The module consisted of 90 microtubular cells with a volumetric power density of 2 W/cm^3 and a power-generation efficiency of over 40 percent[211] (see Section 6.3.6.12).

6.3.6.3. *Central Research Institute of Electric Power Industry*

CRIEPI began developing an all-ceramics SOFC without using metal parts in June 2004.[212] The all-ceramic design was chosen to extend durability at operating temperatures of 1,000 °C and maintain an output density of 1 W/cm^2.[212] The all-ceramic planar SOFC stack was configured as follows and met target goals:[213]

- A dense electrolyte layer and an interconnector membrane, specially developed by CRIEPI, was formed on a porous anode substrate (5.0 cm × 5.0 cm) with gas channels made of a long-life anode material, followed by the introduction of a 12.0 cm^2 cathode layer above the electrolyte layer.
- The stacks were heated up to 1000 °C and were fed with hydrogen gas and air to evaluate the electric generation properties. With a unit cell voltage of 0.7 V, a power density exceeding the development target of 0.5 W/cm^2 was obtained. This level was maintained even after continuous operations for 150 h.[213]

Subsequently, CRIEPI decided to develop a lower temperature SOFC. In December 2006, CRIEPI and AIST announced that they had developed SOFCs that operated around 650 °C.[214] They developed a new technology by which nanoscale silver particles were uniformly dispersed on the ceramic electrode surface. They achieved a power density of 0.45 W/cm^2. Catalytic activity of the silver particles was high, and their electric resistance was low, so that the electric conductivity of the SOFC cells could be increased. In this treatment, no special equipment was needed.

- CRIPI further improved its cell technology and developed two types of cells, ceria electrolyte cells which operated at 500 °C and zirconia electrolyte cells which operated at 650 °C. In June 2008, CRIEPI reported that it had achieved a maximum power density of 0.82 W/cm^2 with the ceria-based cells and 1.06 W/cm^2 with the zirconia-based cells.[215] The research was conducted under NEDO's Ceramic Reactor Development Project.[215]

6.3.6.4. *Fuji Electric*

Fuji Electric was one of the early participants in the METI SOFC R&D Program. The company participated in the 3-year Phase I (1989–1991) and the following 6-year Phase II (1992–1997) programs. The company did well in Phase I but did not achieve Phase II goals. Fuji left SOFC development after the end of Phase II.

- The Phase I goal was to develop 100 W class planar SOFC modules. Fuji started developing 200 cm^2 flat disk SOFC cells with center manifolds and oxide separators.[216] The company developed a 200 cm^2 cell and tested it for 500 h. It generated 46 W at 300 mA/cm^2 fueled by hydrogen. A module of eight 200 cm^2 cells generated 252 W with a current density of 300 mA/cm^2, which was tested for 500 h.[216] In April 1992, Fuji succeeded in fabricating a module of ten 200 cm^2 cells, which achieved 400 W at 400 mA/cm^2 and a 20 percent fuel utilization rate.[217] The module was operated for 2000 h and showed a power density of 0.21 W/cm^2.
- In April 1992, METI started the 6-year Phase II (1992–1997) SOFC R&D Program. The goal of the first 4 years was to develop a kilowatt class planar SOFC module with a current density of 0.18 W/cm^2 and 70 percent fuel utilization. The goal of the last 2 years of the program was to scale up the module to 10 kW class with a current density of 0.2 W/cm^2 at 75 percent fuel utilization and with a degradation rate of less than 1 percent/1000 h.[178] Fuji was developing large-area (500–1000 cm^2) cells with metallic separators, metallic cell substrate, and plasma-sprayed electrolyte. The company was hoping to develop 2–5 kW planar SOFC with 20–50 cells by 1995 or 1997. The intermediate evaluation conducted in 1995 discovered severe gas sealing problems and inferior durability for thermal cycles.[176]
- METI's plan to develop a 10 kW class planar SOFC module during the last 2 years of Phase

II was canceled and instead it was decided that Fuji should continue to work on a kilowatt class module and focus on solving the gas sealing problems and durability problem associated with thermal cycles. In 1995, Fuji developed a five-cell stack with large 528 cm^2 cells and achieved a maximum output of 505 W at 0.35 A/cm^2.[218] In September 1997, Fuji developed a module of twenty 600 cm^2 cells and generated 3.31 kW at 300 mA/cm^2 and 50 percent fuel utilization. The module was operated for 1800 h and showed a power density of 0.18 W/cm^2.[217] It showed no degradation after 900 h of operation.
- In the same year, another intermediate evaluation was conducted. It found that the module achieved most of the target goals and that the gas sealing technology had been improved. The evaluation noted that the metal interconnects were unable to show the desired reliability in thermal cycles for long-term stability.[176]

At the end of FY1997 (March 1998), Fuji, having worked in the METI SOFC R&D program for 9 years, decided to end its participation. Given that Fuji Electric was Japan's leading PAFC developer, it may be the company decided it would rather focus on PAFCs.

6.3.6.5. Fujikura Cable

Fujikura Cable participated in the METI SOFC R&D Program Phase I (1989–1991), Phase II (1992–1997), and Phase II Extension (2001–2004). The company worked to develop electrolyte spray processing in Phase I[283] and component, system design, and BoP equipment in Phase II and Phase II Extension.[178] In 1992, Fujikura developed a 103 W tubular SOFC module as one of the tasks in Phase I. The module consisted of sixteen 105 cm^2 tubular cells, and achieved a current density of 143 mA/cm^2, a power density of 0.06 W/cm^2, and a fuel utilization rate of 18 percent.[217]

- Since the end of Phase II Extension in 2004, little information has been reported on Fujikura's SOFC activities. It is likely that the company has left the SOFC field.

6.3.6.6. Kyocera: The Leading SOFC Player in Japan Today

Kyocera—a developer and manufacturer of new ceramics for electronics devices such as integrated circuits—started SOFC development in the late 1980s and engaged in the development of 1 kW SOFC module in 2001.[219] In February 2003, the media reported that Kyocera had an SOFC development team of 20–30 employees directly under the Kyocera CEO. They were reportedly developing flattened tubular SOFC modules made with all in-house components.[220] According to the report, Kyocera was investing about ¥500 million a year (about $4.3 million at the 2003 exchange rate of $1/¥116) in SOFC development and planned to complete the development of a prototype 1 kW module by December 2003, test it for one year, and commercialize it in early 2005.

In December 2003, Kyocera announced that they achieved the world's highest power-generation efficiency of 54 percent DC LHV at an operating temperature of 780 °C in a 1 kW SOFC module.[221] The design was anode supported with proprietary interconnects.[47] The 1 kW SOFC module consisted of four units of 50 flattened tubular cells connected in series as a stack (the outside dimensions were 150 mm × 25 mm × 3 mm), a heat exchanger module integrated with a fuel reformer (outside dimensions were 450 mm × 320 mm × 320 mm), air blower, and steam generator.[221] Kyocera reported a high power density of 0.5 W/cm^2 at 750 °C.[222]

- This disclosure was the first public announcement of the technical details of Kyocera's fuel cells, according to the media.[221] Kyocera clearly had become Japan's leader in SOFC technology.

- Even before the press reports, the company had been known as a unique company. The small start-up had achieved remarkable success in the highly competitive, heavily capitalized, electronics industry in Japan. The success was primarily attributed to the company's founder and his inspirational leadership.

To prepare for SOFC commercialization, Kyocera has formed a number of teams with different partners. Together they have been making remarkably fast and significant technological advances, including a likely breakthrough in durability that appears to promise an SOFC lifetime of 10 years.

6.3.6.6.1. KYOCERA/OSAKA GAS

In April 2004, Kyocera started a joint development effort with Osaka Gas, aiming to develop a 1 kW SOFC cogeneration system for residential use.[219] Kyocera would develop cells and power-generation modules while Osaka Gas would develop the devices for desulfurizing natural gas and recovering heat waste as well as develop the control devices for residential use.[223] In November 2004, the 1 kW module was developed and achieved an electrical efficiency of 44 percent LHV AC.[224]

- Osaka Gas had been the most aggressive utility company in developing and field-testing SOFC systems in Japan. Starting in 1987 until the mid-1990s, the company, independently or jointly with other Japanese utilities imported from SWPC a total of seven tubular SOFC power plants ranging between 3 and 25 kW. Osaka Gas also had been developing tubular SOFCs on its own[225] perhaps since the early1990s, and in June 1995 developed a tubular 12 cm^2 SOFC cell, which was fueled by hydrogen and operated at 1000 °C at 1 A for 1080 h.[217] It achieved a maximum electric output of 8.8 W and a power density of 0.59 W/cm^2 with a fuel utilization of 30 percent and voltage degradation of 0.4 percent/1000 h. According to another source, Osaka Gas built a 1 kW tubular module by adopting the EVD techniques used for yttria-stabilized zirconia to deposition of lanthanum strontium chromite on the Ru based anode, which showed excellent performance.[176] Osaka Gas also was engaged in joint SOFC development with Murata Manufacturing starting in 1993, which perhaps lasted until the late-1990s. In addition, according to the press reports in March 2010, Osaka Gas and Mitsui & Company, a general trading company, formed a partnership with Ceramic Fuel Cells Limited (CFCL) of Australia and bought CFCL's 2 kW BlueGen cogeneration power plants for test and demonstration in Japan.[226]

In November 2005, Kyocera and Osaka Gas conducted a test operation of a 1 kW SOFC cogeneration power plant for the first time at a residence in Japan.[227] They installed one 1 kW unit at an experimental housing complex, NEXT 21, which Osaka Gas owned. The test aimed to identify the pattern of use of power and heat at an average Japanese household. The 1 kW cogeneration system operated for a total of 2000 h for about 90 days from late November 2005 to early March 2006 (except for holidays).[227]

- The 1 kW cogeneration power plant showed an average electrical efficiency of 44.1 percent LHV AC and an average thermal efficiency of 34 percent LHV. The cogeneration system consisted of 200 flattened tubular cells, used natural gas, and operated at about 750 °C.

In 2006, Kyocera and Osaka Gas developed a smaller SOFC cogeneration system with rated output of 0.7 kW, which they viewed as ideal for small, urban residences in Japan.[219] Kyocera developed a new cell with a thickness of 2 mm, 1 mm thinner than the one used in the 1 kW system. The number of cells in the stack was also considerably reduced, from 200 cells in the 1 kW stack to 126 in the 0.7 kW stack. The air induction mechanism that would supply air to the power generating unit (a metallic case containing the cell stack and reformer) was integrated into the chassis of the power generating unit. The configuration of the power generating unit was also simplified to reduce its size by about 50 percent as compared with the 1 kW model. The hot water supply/heating unit was developed by Osaka Gas and Chofu Seisakusho. The 0.7 kW system was 30 percent smaller than the 1 kW system.[228]

- The 0.7 kW power generating system had an electrical efficiency of 45 percent AC LHV with a heat recovery efficiency of over 30 percent.[229]
- The small power plant would provide about 70 percent of electric demand for small homes. The cost of electricity and heat at a household can be reduced by ¥60,000–¥70,000 (about $509–$595 at the 2007 exchange rate of $1/¥117.7), according to Osaka Gas and Kyocera.[230] The total capital and installation cost has not been published.
- In July 2007, Osaka Gas announced its participation in the METI SOFC Demonstration Research Program (2007–2010).[231] Osaka Gas would install twenty 0.7 kW SOFC cogeneration power plants at detached houses of Osaka Gas employees lived. The houses would be selected by lottery.

In August 2008, according to media reports, Kyocera and Osaka Gas developed SOFC cells that would have a life expectancy of more than 40,000 h in continuous generation.[232] The increased lifetime was achieved by adjusting the number of pores in air electrodes to increase oxygen permeability while retaining mechanical strength. Also, electric current collectors between cells were coated with heat resistant material to increase durability. Their SOFC cells so far had had a durability of 20,000 h. A 4000-hour test of a new cell conducted by Osaka Gas indicated that the cell voltage would decline from the initial 0.8 to 0.7 V after 40,000 h operation.

- According to the same report, the new heat exchanger improved the exhaust heat recovery efficiency to 40 percent, fully supplying water necessary for steam reforming, thus eliminating the need for an outside water supply and cutting cost.

6.3.6.6.2. KYOCERA/OSAKA GAS/TOYOTA MOTOR/AISIN SEIKI

In March 2009, Toyota Motor and Aisin Seiki joined Kyocera and Osaka Gas for joint SOFC development,[232] clearly accelerating system development. In the joint undertaking, Osaka Gas would be responsible for the hot water supply and heating unit using exhausted heat, while Kyocera, Toyota, and Aisin would be responsible for the power-generation unit, with Kyocera responsible for the SOFC cells and stacks.[233] By September 2009, the four companies developed a smaller, thinner, and lighter SOFC power plant,[195] different from the one developed by the Kyocera/Osaka Gas team. The components were thinner but with good thermal insulating characteristics, and the module containing stacks was smaller and lighter.[234] The power plant had a rated power-generation output of 0.7 kW with a power-generation efficiency of 45 percent LHV and a thermal efficiency of 40 percent LHV.[234]

- In December 2009, Toyota and Aisin announced that they planned to install a total of thirty 0.7 kW cogeneration power plants that the four companies had developed for the METI SOFC demonstration program.[235] Toyota and Aisin would provide the power plants to the five gas utilities: one to Hokkaido Gas, four to Tokyo Gas, one to Toho Gas, 23 to Osaka Gas, and one to Seibu Gas.

- From the 30 units Toyota and Aisin developed, Osaka Gas planned to demonstrate an additional twelve 0.7 kW power plants developed by themselves with Kyocera in FY2009 in the METI SOFC demonstration program.[236] Few details are available.

The four companies were aiming for commercialization by 2015.[233] They estimated that the sales price would be about ¥500,000–¥600,000 ($5,050–$6,060), and the system would have about a 10 year lifetime.[237] As of April 2009, their SOFC system cost about ¥10 million per kilowatt (about $101,000 at the 2009 exchange rate of $1/¥99), and an estimated lifetime of about 20,000 h.

In March 2012, Kyocera/Osaka Gas/Aisin Seiki/Chofu Seisakusho/Toyota announced that they had developed a residential-use SOFC cogeneration system with the world's highest power-generation efficiency of 46.5 percent. The ENE-FARM Type S would be on the market in April 2012. Chofu is a hot water supply and heating unit that had been working with Osaka Gas since 2006 and a recent participant on the SOFC development team. The cogeneration system, which generates 700 W of electricity and has an overall energy efficiency of 90 percent, will be sold at ¥2,751,000 (about $34,517 at a 2012 exchange rate of $1/¥79.7).

6.3.6.6.3. KYOCERA/TOKYO GAS/RINNAI/GASTAR

In February 2004, Kyocera, Tokyo Gas, Rinnai, and Gastar started joint development of a 5 kW class SOFC power plant to be operated under 750 °C.[238] The group used a low-temperature operable segmented-in-series type flattened tubular cell stack developed by Tokyo Gas and Kyocera.[239] Tokyo Gas designed the segmented-in-series electrode structure, and Kyocera developed the flattened tubular cells. The production process was simplified and significant cost reduction was expected upon mass production because the team used printing technology to form and sinter multiple cells on one ceramic base plate. The target performance as a commercial product was a net power-generation efficiency of 45 percent AC HHV.[240]

Tokyo Gas began SOFC development in 1989 but had not been successful in achieving practical applications until it teamed up with Kyocera in 2003. In 1995, the company developed a 3-cell 6.0 cm × 6.0 cm stack of cells, which had a power density of 0.39 W/cm^2.[241] In July 1998, Tokyo Gas developed a 1 kW internal reforming planar SOFC stack, which generated 1.68 kW at 1,000 °C.[242] In 1992, single cells as large as 20 cm × 20 cm were manufactured.[243] Maximum power densities obtained were 0.65 W/cm^2 and 0.36 W/cm^2 for 5.0 cm × 5.0 cm and 10 cm × 10 cm single cells, respectively. Ten-cell stacks of 5.0 cm × 5.0 cm single cells were also fabricated, and an output of 50 W was obtained. In October 2000, the company developed planar SOFC cells that were incorporated in a test apparatus made by Sulzer in Switzerland and generated 1.1 kW at 960 °C. But none demonstrated adequate mechanical reliability for practical use.[239] In 2001, Tokyo Gas switched to developing anode-supported planar SOFCs for reduced temperature operation and developed SOFCs with improved mechanical reliability by using a thin film for the electrolyte. In July 2002, as part of a NEDO project, Tokyo Gas developed a planar SOFC stack with the world's highest power-generation efficiency of 53.6 percent at that time.[244] Practical applications, however, were not possible due to problems with the long-term stability of cell stacks using metallic interconnectors.[239]

- In 2003, Tokyo Gas started to develop segmented-in-series cell stacks, incorporating the reduced temperature operation technology and the unique interconnect technology that Kyocera had been developing.[239] In August 2003, the company developed a module consisting of two stacks of forty seven 144 cm^2 cells and achieved a current density of 276 mA/cm^2, a power output of 1.33 kW, a power density of 0.14 W/cm^2, and fuel utilization of 50 percent.

In May 2006, Kyocera, Tokyo Gas, Rinnai, and Gastar announced in a press release that they had developed a 2.5 kW SOFC power module with an electrical efficiency of 56.1 percent LHV DC operated under normal atmospheric pressure.[245] The cell stacks in the module were of the horizontal stripe structure type; they started development in February 2004 and operated stably at 750 °C. The cells were sintered on a cylindrical base in the horizontal stripe structure and were connected in series along the axis of the cylinder. Therefore, according to the press release, a high voltage could be obtained with a relatively small number of cylinders, and the cell stack of high voltage and low current could be realized, so that the energy loss was reduced. The press release indicated that the materials were inexpensive and a process for stacking cells was unnecessary. Thus, the companies expected low production costs. The four companies planned to develop SOFC cogeneration systems of 50 percent LHV (45 percent HHV) electrical efficiency at the AC output by adding an inverter and other auxiliaries to the cell unit and put them out for practical use around 2008.

In December 2009, the Kyocera/Tokyo Gas/Rinnai/Gastar team announced that, starting that month, Tokyo Gas had commenced the first trial operation of two of their 0.7 kW SOFC cogeneration power plants as part of the METI FY2009 demonstration research program.[196] The announcement suggested that the companies were adjusting their SOFC development strategy to harmonize with the industry trend in Japan; the Japanese SOFC industry was increasingly focusing on smaller residential-use power plant and specifically, on the 0.7 kW cogeneration power plant. The more remarkable part of the news release was that the four companies said that they had improved the durability of their cell stacks to 5 years. According to the announcement, the durability had been extended by revising the detailed design, developing components, improving manufacturing technologies, and devising all-ceramic cell stack technologies, which avoided the use of alloy materials that might affect durability. The announcement also mentioned that they had made numerous other technological advancements, including materials that could control electrode degradation even after long-term power generation.

- Kyocera cell stacks so far had operated up to about 1 year in the METI demonstration program. It would be a breakthrough, indeed, if the cell stacks could operate for 5 years with high performance under real-life operating conditions, especially at low cost.
- Tokyo Gas also indicated that it planned to conduct trial operations of four additional 0.7 kW units from FY2010 in the METI SOFC demonstration program and to collect data from a total of six units.
- Tokyo Gas also has decided to commercialize the SOFC power plant around FY2014.[246] The company also hopes to extend the cell durability to 10 years by then.[196]

6.3.6.6.4. BUT KYOCERA'S COMMERCIAL SUCCESS IS NOT ASSURED

Kyocera has made remarkable technology progress and is regarded as the industry leader in Japan today. Nonetheless, Kyocera has not demonstrated the cost competitiveness or the needed lifetime durability with the systems that they are currently demonstrating under the METI program. Until it attains economic viability for its products, commercial success will not be assured. Kyocera also needs to commit to construction of a manufacturing facility if low-cost production can be ensured.

- US Siemens Westinghouse commercialization efforts were unsuccessful because of the high capital cost of their otherwise highly reliable and durable tubular SOFC power plants.

6.3.6.7. Mitsubishi Heavy Industries

MHI has been developing both planar and tubular SOFCs since 1984.[10] MHI Kobe Shipyard and Machinery Works (MHI Kobe) in collaboration with Chubu Electric has focused on planar SOFC technology, and MHI Nagasaki Shipyard and Machinery Works (MHI Nagasaki) in cooperation with Electric Power Development Company (EPDC; now J-Power) and Tokyo Electric has worked on tubular SOFC technology. Both efforts focus on large-scale power generation in hundreds of kilowatts or megawatts class.

6.3.6.7.1. MHI KOBE/CHUBU ELECTRIC

MHI Kobe began planar SOFC development using the wet process in 1984. In 1990, MHI Kobe started to collaborate with Chubu Electric in the development of a planar design called Mono Block Layer Build (MOLB).[247] An MOLB type SOFC fuel cell was constructed of a generation membrane (anode, electrolyte, and cathode) dimpled in three dimensions and manufactured onto an uneven surface with an interconnector that connects the generation membrane in series, sealing the gas on the cell end.[248] A MOLB type planar SOFC design was expected to deliver better and lower cost manufacturability and higher power density than tubular configurations.[10]

MHI Kobe and Chubu Electric developed the MOLB technology, rapidly scaling up to 15 kW size within 10 years. In 1992, they developed a MOLB type module consisting of three stacks of forty 150 mm square planar cells that generated 1.32 kW. The results from electric power-generation tests in 1993 showed that the maximum power was 1.275 kW and the power with a cell voltage of 0.7 V was 1.132 kW. The maximum power density was 0.236 W/cm^2.[249] In 1996, they developed a module of two stacks of forty 200 mm^2 planar cells, which generated the world's highest output of 5.1 kW.[248] By 1998, improvements were made in the cell structure, yielding the world's highest power density of 0.35 W/cm^2.[247]

In August 2000, the two companies developed a MOLB type module with a 15 kW output.[183] The module was a consolidated MOLB made by the cell layering method and had operated for 7500 h by 2001.[248]

- In 2001, MHI Kobe and Chubu Electric participated for the first time in the METI SOFC R&D Program to develop a thermally self-supporting 10 kW MOLB type planar SOFC module under the Phase III (2001–2004) SOFC System R&D Program. The thermally self-supporting module consisted of an SOFC furnace, start-up gas burner, air recuperator, and condenser.[187] Start-up and shutdown procedures were fully automated by a digital control system. The system was completed by March 2004, a year ahead of schedule.
- In September 2002, the two companies began developing the world's first 50 kW commercial prototype using MOLB type planar SOFC. MHI Kobe and Chubu Electric announced that they would start testing a 50 kW SOFC cogeneration system prior to commercialization.[250] MHI and Chubu Electric stated that they would place the 50 kW MOLB type SOFC cogeneration system on the market in 2004.[251] In addition, the two companies planned to develop a 200 kW class system by 2007.

Subsequently, MHI Kobe and Chubu Electric were commissioned by NEDO to develop a 200 kW MOLB type cogeneration power plant as part of 2004–2007 METI SOFC System Technology Development Program. The outcome, however, was discouraging. The stack module that MHI and Chubu Electric developed for the 200 kW system was found to be too large and too inefficient for the 200 kW system. NEDO assessed that the module could not be improved rapidly enough to fabricate a 200 kW cogeneration unit within the program period and canceled the effort in March 2006.[189]

MHI Kobe and Chubu Electric also demonstrated a 30 kW MOLB type planar SOFC cogeneration system at Aichi Expo in 2005.[248] Aichi Expo was held from March 25 to September 25, 2005, and the MOLB cogeneration plant was expected to operate for the full 6-month period.[252] Although the system operated for a cumulative operating time of 3887 h, the system demonstrated a low electrical efficiency of only 33 percent,[253] which was no better than the conventional power-generation efficiency. The original goal was that the system would achieve an electrical efficiency of 42 percent.[254]

- MHI Kobe reported that at Aichi Expo, the system achieved the rated output of 30 kW by 100 percent internal reforming.[248] The operators noticed, however, that the temperature difference between the fuel inlet and outlet began to increase and opted to continue operating at a reduced power output.[255] On August 12, 2005, lightening hit the transmission line nearby and caused momentary stoppage of system operation. The system subsequently restarted with no performance degradation. The system continued to operate until the end of the Expo for a total of 3887 h.
- Since the end of Aichi Expo in September 2005, there have been few reports on MOLB SOFC development. It is likely that MHI Kobe and Chubu Electric suspended or abandoned their MOLB SOFC development efforts and commercialization plans.

6.3.6.7.2. MHI NAGASAKI/J-POWER/TOKYO ELECTRIC

MHI Nagasaki has been developing tubular SOFCs using the wet plasma spraying method, spray coating method, and co-sintering method since 1984.[10,176] Major development partners have been J-Power (formerly EPDC) and Tokyo Electric, and together they made significant cell and module developments.

From 1986 to 1989, MHI developed fundamental technologies for tubular SOFC cells and modules with Tokyo Electric.[256] In 1987, a cell stack developed by MHI Nagasaki and Tokyo Gas operated with over 80 percent fuel utilization and a maximum efficiency of 40 percent at a current density of 200 mA/cm^2.[176] In 1991, MHI, Tokyo Electric, and J-Power developed a 1 kW SOFC module and operated it for 1000 h at Wakamatsu Coal Utilization Research Center of J-Power.[256] In 1993, MHI and J-Power developed improved 1 kW SOFC modules, which operated for 3000 h with two scheduled thermal cycles each at Nagasaki Works of MHI and at the Wakamatsu test facility of J-Power. The modules achieved an efficiency of 34 percent HHV and a deterioration rate of 2 percent per 1000 h. Their SOFC modules achieved 5000 h of operation in 1994.[178] In 1995, MHI and J-Power developed a 10 kW module, which operated at atmospheric pressure for 5000 h continuously.[257] The module did not need heating support to maintain the operating temperature; the module efficiency was 34 percent HHV and the deterioration rate was 2 percent/1000 h.[258] In 1998, MHI and J-Power developed a pressurized module, which continuously operated for 7000 h at a maximum output of 21 kW.[259]

In 2001, J-Power started to work on the development of an IGFC system, which would combine MHI's tubular SOFC stacks with coal gas and steam turbines,[260] under the 2001–2006 METI "EAGLE" (Coal Energy Application for Gas, Liquid & Electricity) project.[261] The IGFC system would be comprised of a triple set of power generators based on fuel cells and gas and steam turbines. The system would use a high temperature, high pressure coal gasification furnace to produce hydrogen and carbon monoxide gases that would be supplied to SOFCs. The high temperature pressurized gases that exit the fuel cells would be used in a heat recovery boiler to generate steam to power the gas turbine. The total project costs were on the order of ¥25 billion (about $206 million at the 2001 exchange rate of $1/¥121.6). The J-Power's EAGLE pilot test plant was commissioned at J-Power's Wakamatsu Operation & General Management Office in Kita-Kyushu City in February 2002.[262]

In parallel with this pilot test, studies were also in progress to develop the SOFC system required for the operation of the IGFC System.[262] In August 2001, MHI and J-Power developed an internal reforming 10 kW pressurized module by making use of cells manufactured by thermal treatment and achieved 755 h of continuous operation at an efficiency of 41.5 percent HHV. In 2006, they succeeded in fabricating a 25 kW SOFC submodule at MHI's Nagasaki Shipyard, which underwent various operational and safety tests in April and May and operated for 515 h.[263] An output density of 0.21 W/cm^2 was achieved at a pressure of 0.4 MPa.[259] In July 2006, J-Power announced that it would build a 150 kW atmospheric pressure SOFC IGFC system and test-operate it for 10,000 h for about 1 year starting in January 2007 at J-Power's Chigasaki R&D Center in Kanagawa.[263] In December 2007, J-Power indicated that it had operated the system for a cumulative 3300 h at 100 kW power output and confirmed its plans over the next year to achieve an electric output of 150 kW and more than 10,000 operating hours.[260]

- In May 2006, using results obtained through EAGLE, J-Power and Chugoku Electric started a joint effort to develop oxygen-blown coal gasification combined cycle technology (oxygen-blown IGCC).[264] In July 2009, J-Power and Chugoku Electric established a new company, Osaki CoolGen Corporation, to undertake a large-scale demonstration test of oxygen-blown IGCC and CO_2 separation and recovery technology.[265]

In 2004, NEDO commissioned MHI Nagasaki to develop a 200 kW SOFC micro gas turbine combined cycle system (SOFC-MGTCC) under the METI SOFC System Technology Development Program (2004–2007). In 2006, MHI Nagasaki fabricated a combined cycle test module consisting of a pressurized 40 kW module and a 60 kW MGT,[189] with a confirmed output of 75 kW at MHI's Nagasaki Shipyard & Machinery Works.[266] Based on the success of the module, a 200 kW SOFC MGTCC system, consisting of four 40 kW modules and a 60 kW MGT unit, was built and commissioned in 2007. At maximum output, a world's record of 229 kW was achieved, with net power efficiency of 52 percent LHV at 211 kW confirmed, the best level in its class. As mentioned in Section 6.3.3.1, the system experienced an electrical accident after about 90 min of operation and was shut down entirely.

- Subsequently, the 200 kW SOFC-MGT combined cycle system was re-commissioned. As of October 2009, according to MHI Nagasaki, the system had achieved 3224 operating hours.[267]

Over the longer term, both MHI Nagasaki and J-Power are interested in large-scale SOFC power generation plants using coal, according to MHI website.[268] They are looking at producing a 100 MW class, natural gas–fired or integrated coal gasification, SOFC/gas turbine/steam turbine combined cycle system as a future replacement for large utility thermal power stations.

For the near term, however, MHI and J-Power are focusing on the market for small/medium SOFC systems, according to the MHI website.[268] They plan to concentrate on a mid-size power cogeneration and a small-capacity SOFC system that could be used in combination with an existing gas turbine system to increase the efficiency of the system. MHI since 2008 has been developing with Toyota Motor a hybrid power-generation system combining SOFCs and an MGT for business and industrial use.[266] Toyota has an affiliate company, Toyota Turbine and Systems, which specializes in MGTs.[267]

- In October 2010, MHI reported that it had been working to improve reduction–oxidation robustness of a cell stack, long-term durability, and more compactness of modules in an effort to develop an SOFC-MGTCC.[269] Recently, an improved 200 kW SOFC-MGTCC system is being designed.[269] The improved system would use the 200 kW module that had operated for 3224 h, combined with a commercial MGT perhaps manufactured by the Toyota affiliate. It is expected that the footprint of the improved system would be about half that of the previous system and that it would achieve an efficiency of 70 percent or higher.

MHI Nagasaki and its partners are still engaged in R&D. The near term outlook for bringing a commercially viable SOFC-MGT product is dim.

6.3.6.8. Mitsubishi Materials Corporation/Kansai Electric

Mitsubishi Materials Corporation (MMC) and Kansai Electric were jointly developing SOFCs for about 10 years starting in 2001. Their goal was to develop intermediate temperature SOFC power plants made of flat discs with outputs of several dozen kilowatts.[270] The targeted temperature range was between 600 and 800 °C.[271] The two companies were supported by the Japan Fine Ceramics Center for anode development, Oita University for electrolyte development, and AIST for performance efficiency measurement technology.[272]

- MMC and Oita University began joint research on electrolytes and cathodes in 1997, and subsequently MMC switched its affiliation to Kyushu University, likely because the collaborating professor transferred to Kyushu University.
- Kansai Electric started SOFC development in 1995. Kansai Electric and Japan Fine Ceramics Center began joint research on anodes in 1997.
- MMC and Kansai Electric started collaboration in 2001 to develop SOFCs.[273]

In December 2001, MMC and Kansai Electric succeeded in developing a flat disk SOFC cell that demonstrated the world's highest power density of 0.9 W/cm^2 at 0.7 V at 700 °C and 1.8 W/cm^2 at 0.7 V at 800 °C.[273] Previously, the power density record had been Global Thermoelectric's 0.72 W/cm^2 (at 0.7 V at 700 °C) and 1.22 W/cm^2 (at 0.7 V at 800 °C). In May 2003, MMC and Kansai Electric succeeded in developing an SOFC module that generated 1 kW at a temperature below 800 °C fueled by methane. The module consisted of 41 flat disk cells 120 mm in diameter.[274] The module demonstrated an electrical efficiency of 40 percent LHV. In January 2004, MMC and Kansai Electric developed a 1 kW SOFC system with an electrical efficiency of 40 percent LHV. Also in March 2004, the two companies succeeded in developing a 3 kW SOFC module consisting of three 1 kW stacks.[275] It achieved an electrical efficiency of 55.3 percent LHV DC.

After the successful development of single-stack 1 kW modules and multi-stack 3 kW modules, MMC and Kansai Electric participated in the METI 2004–2007 SOFC System Technology Development Program to develop a 10 kW cogeneration system.[271] The project was one of the two relatively successful projects in the program. In January 2005, MMC and Kansai Electric operated a 1 kW test module fueled by natural gas for more than 1000 h with no voltage degradation.[276] In April 2006, the companies built and operated a natural gas–fueled, thermally self-supporting 10 kW module with an electrical efficiency of 41 percent AC HHV.[189] In February 2007, the 10 kW cogeneration system was fabricated and installed at Rokko Testing facilities of Kansai Electric.[278] The system contained a 10 kW SOFC module, a gas and water supply unit, a power conditioner, a control unit, and an exhaust heat recovery unit.[271]

- The 10 kW system achieved an electrical efficiency of 41 percent HHV and a total efficiency of 82 percent HHV. It had operated for 3102 h as of October 2007 and showed a voltage degradation of 1.08 percent/1000 h, which did not meet the target goal of 0.25 percent/1000 h.
- Following the first successful long-term performance evaluation of the 10 kW cogeneration system under the METI 2004–2007 program, a second 3000-hour, long-term stability test of the 10 kW system was performed in FY2008 with an improved stack module.[279] The improved system demonstrated an electrical conversion efficiency of 40 percent HHV AC at an output of 10 kW and a combined efficiency of 82 percent HHV when the exhaust heat was recovered as hot water at 60 °C. Moreover, no significant voltage degradation was observed during the operation.

In addition, MMC and Kansai Electric had been working to increase the electrochemically active area of cells so that higher output power per cell could be achieved and reduce the number of cells in a stack, thereby reducing cost.[279] In November 2009, the two companies announced that they had achieved scale-up of cells by doubling the active surface area. They manufactured 170 mm diameter cells with an electrolyte thickness of less than 200 μm, which were built into a single 32-cell stack.[279] The stack was successfully tested; electrical conversion efficiency of 50 percent HHV for a DC output of 3.15 kW was demonstrated. However, relatively large voltage degradation was observed during the 26 cyclic loadings that were performed.

In October 2010, MMC and Kansai Electric reported better success with their SOFC cells using improved high oxide ion-conducting electrolyte made of doped lanthanum gallate.[280] The two companies made progress in scaling up the electrochemically active area of cells for higher output power per stack with a reduced number of cells. Seal-less stacks were carefully redesigned together with the hot BoP components. From a single 32-cell stack, an output power of 3 kW DC was obtained. Electrical conversion efficiency of 50 percent HHV for a DC output of 3.15 kW was also demonstrated.

In parallel, a durability test of a 1 kW module made of improved Ni-gadolinia-doped ceria cells (120 mm in diameter and 0.2 mm in thickness) was conducted. The module operated with a high current density of up to 0.4 A/cm^2 and achieved fuel utilization of 70 percent. The stack test used 46 of the improved cells and showed an average degradation rate of 0.3 percent/1000 h during the 5000 h of continuous operation.

- The cells that were used generated an output of 23 W and a power density of 0.2 W/cm^2 with a current density of 0.27 A/cm^2.[281] The two companies were proceeding to develop an advanced cell 170 mm in diameter with an output of 104 W, which would achieve a power density of 0.46 W/cm^2 at the current density of 0.59 A/cm^2.

MMC and Kansai Electric have been making major progress in a relatively short time. Their technology, however, is still in the development stage. It is difficult to project when they will be able to develop commercially successful products.

6.3.6.9. Mitsui Engineering and Shipbuilding

It is not certain when MES started SOFC development but it is likely that the company was engaged in SOFC research for at least a dozen years. It perhaps started SOFC development in the late 1980s and ended this effort possibly in the early 2000s.

Mitsui participated in the METI SOFC R&D Program through Phase I (1989–1991), Phase II (1992–1997), and Phase II Extension Program (1998–2000). During Phase I, it developed planar SOFC cells with interconnects made with magnesium spinel and lanthanum chromite.[176] Also, during Phase II and Phase II Extension, the company was engaged in the development of high performance cathodes, anodes, gas sealing materials, and manufacturing technology for them.[178] While working under the METI Phase II program, MES achieved the following:

- In 1995, the company developed a module of six 100 cm^2 planar cells, which operated at 1000 °C for 1000 h and generated a maximum output of 83 W.[217] The stack demonstrated a power density of 0.14 W/cm^2 and 62 percent fuel utilization. The voltage degradation was 6 percent/1000 h.
- In 1998, it developed a module of five 140 cm^2 cells. It operated at 1000 °C and generated a maximum output of 109 W. The power density was 0.15 W/cm^2, and the fuel utilization was 53 percent.
- In 1999, MES developed a module that consisted of three stacks of eight 140 cm^2 planar cells. The module operated at 1000 °C and generated 0.5 kW with a power density of 0.149 W/cm^2 and a fuel utilization of 44 percent.[284]

In July 2001, there was a news report indicating that MES accelerated SOFC development, in which MES announced that they would develop a combined cycle generation technology in 2 years.[630] The combine cycle power generator would consist of a 10 kW class SOFC module and a gas turbine and supply 60 percent of the thermal efficiency. According to the press report, since the end of the METI program, MES had been conducting SOFC development with electric power companies.

- Since the July 2001 news report, little information has been reported on MES' SOFC activity. It is possible that MES suspended SOFC research sometime in the early 2000s.

6.3.6.10. *Murata Manufacturing/Osaka Gas*

It is not certain when Murata Manufacturing started SOFC development, but it began participating in the METI 1989–2007 SOFC R&D Program from the beginning and stayed with it for 9 years. In Phase I (1989–1991) of the METI program, Murata was assigned to develop co-firing technology.[176] In Phase II (1992–1997), Murata fabricated planar SOFC cells by co-firing technology.[178] Murata's participation in the METI program ended in 1997.

While working in the METI program, Murata Manufacturing developed co-firing technology and fabricated some planar SOFC modules:

- At an early, but unspecified, period in the METI program, Murata developed a stack of six 110 cm^2 cells, which generated an output of 117 W.[217]
- In 1994, the company fabricated planar multicell stacks with co-firing anode/electrolyte/cathode multilayers and Ni–Cr alloy interconnects.[282] The four multilayers were connected in a single cell plane and were sandwiched by the interconnects. The effective electrode area in each cell plane was 484 cm^2. It operated for 300 h using hydrogen as fuel. The maximum power of the module of five stacks of four 121 cm^2 cells was 601 W at a power density of 0.24 W/cm^2 and fuel utilization of 40 percent.

Outside the METI SOFC R&D program, Murata Manufacturing, in cooperation with Osaka Gas, was developing a unique SOFC cell stacking structure characterized by easy assembly and high crack resistance.[282]

- In January 1994, Murata and Osaka Gas developed a module of sixty five 120 cm^2 cells, which operated for 1028 h and generated an output of 872 W.[217] The stack, however, showed poor performance due to cracks in the electrolyte.[10] The module showed a current density of 213 mA/cm^2, a power density of 0.11 W/cm^2, and fuel utilization of 12 percent. The voltage degradation was high, at 6 percent/1000 h (50 cells).[217]
- In 1996, Murata and Osaka Gas developed a module of five stacks of four 120 cm^2 cells per stack which generated an output of 400 W.[283]
- In 1997, Murata and Osaka Gas developed a planar SOFC which was suitable for stacking. They fabricated a module of four stacks of twenty 113 cm^2 cells each.[284] The module generated an output of 1.31 kW, showed a current density of 0.22 A/cm^2, a power density of 0.14 W/cm^2, and fuel utilization of 70 percent. The module used a mixture of H_2, H_2O, and N_2. It operated for 500 h and demonstrated 4 percent/1000 hour voltage degradation.

While keeping a low profile, Murata Manufacturing has been continuing SOFC development. Murata researchers are clearly focused on the development of single-step co-fired SOFCs.[285] As recently as December 2010, a paper on the same topic was presented by one of the researchers at a hydrogen and fuel cell conference in Japan.[286]

- The company is no longer cooperating with Osaka Gas, which has been cooperating with Kyocera since 2004.

6.3.6.11. NGK Insulators/J-Energy/ Sumitomo Precision Products

NGK Insulators (Nihon Gaishi Kaisha) was established in Nagoya, Japan, in 1919 as a spin-off from Noritake. It also is an affiliate of NGK Spark Plugs. It is not clear when the company started SOFC development, but the company is making a remarkably rapid progress in its SOFC technology. The company is one of the Japanese SOFC leaders today.

NGK Insulators—Development of SOFCs of World's Highest Power Density (July 2003). By July 2003, it succeeded in the development of SOFC cells with the world's highest power density.[287] The cells demonstrated a power density of 0.6 W/cm^2 at 650 °C and 1.6 W/cm^2 at 750 °C. The power output could be increased by layering cells. The cells operated at a low temperature, enabling lower component costs. The high power density would also enabled a highly compact power-generation system.

NGK Insulators/AIST—Development of SOFCs of a honeycomb structure with the World's Highest Level Power Density (November 2006). In November 2006, NGK Insulators and AIST developed honeycomb-type micro SOFCs, which realized a high power output even at low temperatures around 600 °C.[206] The module of one honeycomb structure contained 256 submillimeter cells within a 15 mm × 15 mm cross section, which achieved 230 mW/cm^2 at 700 °C, the world's highest level power density in the mid temperature range for cathode-supported SOFCs. The honeycomb SOFC was highly tolerant of rapid heating and cooling.[288] It would begin operating within a few minutes starting at room temperature and withstand rapid shutdowns as well. It could be used as APU for cars, residential power-generation systems, and portable power sources.

NGK Insulators—Development of Flat-Tubular SOFCs with the World's Highest Power Efficiency (June 2009). In June 2009, NGK Insulators announced that the company had developed flat-tubular SOFC cells, which achieved the world's highest power-generation efficiency of 63 percent LHV at a fuel utilization of 90 percent, and power output of 700 W class at 800 °C.[289] The 5 μm thick zirconia electrolyte covered the fuel electrode entirely, which was the cell support, so that resistance was decreased. The air electrodes were formed on both sides of the fuel electrodes to increase generating electrode area for increased output. Space was formed in the cell for supplying fuel gas and was designed to distribute fuel gas uniformly. The cell had a card-like shape, 10 cm long, 5.5 cm wide, and 1.5 mm thick. The stack was assembled with several dozen cells. The device was offered to major domestic oil companies for performance tests.

NGK Insulators/J-Energy/Sumitomo Precision Products—Development of 3 kW Flat-Tubular SOFC Power Plant Fueled by Kerosene (November 2009). In November 2009, NGK Insulators, J-Energy, and Sumitomo Precision Products jointly developed a 3 kW flat-tubular SOFC power plant fueled by kerosene for small businesses.[290] NGK Insulators developed the SOFC stacks, J-Energy developed a de-sulfurizer/reformer, and Sumitomo Precision developed a thermal management system. The system was operated at 750 °C. The companies planned a field test demonstration from March 2010 to April 2011.

6.3.6.12. NGK Spark Plugs/AIST/Fine Ceramic Research Association/Toho Gas

NGK Spark Plugs (Nihon Tokushu Togyo Kaisha), Nagoya, is one of Japan's leading SOFC developers and is responsible for many innovative designs. NGK Spark Plugs was established in 1936 when it was spun off from NGK Insulators (Nihon Gaishi Kaisha). NGK Spark Plugs has been developing both tubular and planar SOFC technologies since 1999, with focus on microtubular SOFC technology. It has worked regularly with AIST and occasionally with other partners, such as Fine Ceramic Research Association and Toho Gas.

NGK Spark Plugs/AIST/Fine Ceramic Research Association/Toho Gas—Development of Microtubular SOFC Module of Size of Sugar Cube (March 2007). In March 2007, NGK Spark Plugs, AIST, and Fine Ceramic Research Association developed SOFC stacks, the size of sugar cubes.[208] Toho Gas, which examined the small SOFC cubes, confirmed that they produced high output power of over 2 W/cm^3 at 550 °C, the world's highest output power density at an operating temperature of below 600 °C. Microtubular SOFC cells 0.8–2 mm in diameter were stacked to form the 1 cm^3 cube. The use of porous materials decreased electric resistance and was a key to achieving the high performance.

- This research was part of the METI Advanced Ceramic Reactor project (2005–2009) administrated by NEDO. The research team indicated that future research would improve the performance of the cube, develop high precision interface construction for supplying fuel gas to each cube and collecting the current, and build stacks through integration of the cubes. In the long run, the team hoped to develop the technology for producing small SOFC systems with output power ranging from several tens of watts (several tens of cubic centimeters) to several kilowatts (several thousand cubic centimeters).

NGK Spark Plugs—Development of a 1 kW Planar SOFC Cogeneration System (October 2008). In October 2008, NGK Spark Plugs announced that it had developed a 1 kW planar type SOFC cogeneration system of one 1 kW stack with 16 planar cells that had an active area of 100 cm^2 per cell.[291] The system was packed in a thermally insulated vessel and achieved a power density exceeding 0.8 W/cm^2 (700 °C at 0.7 V) and a power efficiency of 60 percent LHV using natural gas. The company indicated that just 13 cells generated 1 kW DC power, but 16 cells were used in the stack to make up for losses converting from DC to AC. The stack was 80 mm thick. At the bottom of the stack, there was an integrated combustion layer and steam reforming layer to achieve heat balance at the top end. The 1 kW cogeneration system consisted of an electronic control unit, power conditioner, heat exchanger, and BoP contained in a compact thermally insulated vessel. The size of the system was 530 mm width × 330 mm depth × 950 mm height, which the company believed was one of the most compact 1 kW CHP units in the world. The company stated that it achieved a stable 90 percent utilization rate for fuel and air by optimizing seal materials, applying the correct torque to fastening bolts, preheating the stack, and designing gas flow channels, chambers, and current collectors to achieve effective gas distribution to active electrodes.

- In 2009, NGK Spark Plugs developed an improved, larger, and thinner anode-supported planer cell and fabricated a 1 kW module with18 cells. The module achieved a power density 1.6 W/cm^2 at 0.7 V, electrical efficiency DC LHV of 60 percent, and fuel utilization of 80 percent.[293]

NGK Spark Plugs/AIST—Development of A Microtubular Stack of 0.6 cm^3 (February 2009). In February 2009, NGK Spark Plugs and AIST developed another micro fuel cell module 1 cm^3 less than in size, consisting of microtubular SOFCs that achieved a power density of more than 2 W/cm^3 at 550 °C.[210] The cells were made of tubes 0.8 mm in diameter and used ceria-based ceramic for electrolyte, nickel ceria-based ceramic for the anode, and lanthanum cobalt ceria-based ceramic for the cathode. To form a micro bundle, the microtubular cells were integrated in a porous structure made of lanthanum cobalt ceramic. A stack (0.6 cm^3) consisted of three 10 mm × 10 mm × 2 mm micro SOFC bundles (made of five integrated microtubular cells) connected in series. A micro SOFC module was obtained by attaching a fuel manifold to the stack structure.

- Hydrogen at a temperature of 450–550 °C was injected into the module and a power density of 0.88–2.2 W/cm^3 (450–550 °C) was confirmed. The stack weighed about 2 g and provided excellent power output for its weight. With the development of the micro SOFC module, portable power sources for small-size electronic devices could be within reach, assuming that a system with tens of watts of power could be obtained by stacking the modules together.

NGK Spark Plugs/AIST/Fine Ceramics Research Center/Toho Gas—Development of 50–200 W Microtubular SOFC Modules (September 2009). In September 2009, NGK Spark Plugs, AIST, Fine Ceramics Research Center, and Toho Gas developed 50–200 W microtubular SOFC modules that worked at 650 °C.[293] The 50 W module was made with 90 cells and had a power-generation efficiency of over 40 percent and a power density of 2 W/cm^3.[294] The team also succeeded in developing modules over 200 W output.[293]

- They are planning to improve endurance and reliability of modules and develop them as auxiliary power sources for cars and small-scale cogeneration in the near term.

NGK Spark Plugs/Toho Gas—Development of 700 W Planar Cogeneration System (October 2010). In 2010, NGK Spark Plugs and Toho Gas held a field test of a 700 W SOFC CHP system for residential use.[295] A compact SOFC stack with 19 cells delivered AC 700 W with the net power efficiency of 44 percent LHV and total efficiency of 76 percent LHV. With the compact-sized stack and efficient arrangement of BoPs, this was one of the smallest CHP system units (530 mm width, 330 mm depth, and 955 mm height) to be fabricated. This CHP system unit was installed at a Toho Gas experimental house and will run according to a simulated pattern of power demand corresponding to the average resident. The system required 3 h from start-up to reach thermally self-sustaining operations and deliver the rated power output. Water vapor in the exhaust gas was condensed and used again for steam reforming for self-sustaining water utilization. The major factors for energy loss were the power consumption by BoPs and heat loss from the thermal insulated box. NGK Spark Plugs indicated that CHP systems in Japan must have a good load following capability and a high efficiency under partial load operations. In Japan, reverse current flow to the power grid is not permitted and so the CHP system frequently would operate under partial load conditions.

- NGK Spark Plugs and Toho Gas have not announced their plans for commercialization of their CHP system.

6.3.6.13. *JX Nippon Oil & Energy/Kyocera*

In 1990, JX Nippon Oil & Energy participated in the 5-year program sponsored by the Japan Petroleum Energy Center (JPEC) to develop a 10 kW SOFC module fueled by light petroleum. Fellow participants included Idemitsu Kosan and Tonen. In April 1995, JX Nippon Oil & Energy developed a planar SOFC module of five 62 cm^2 cells. The module demonstrated a current density of 470 mA/cm^2 and a power density of 0.24 W/cm^2 with a fuel utilization rate of 20 percent. In 2002, JX Nippon Oil & Energy demonstrated a 5 kW system powered by gasoline (naphtha).[296] In 2003, it conducted a demonstration test of 1 kW LPG fueled system in collaboration with Nippon Petroleum Gas.

JX Nippon Oil & Energy boosted its SOFC development activity in 2005.[297] In November 2007, JX Nippon Oil & Energy and Sanyo Electric agreed that JX Nippon Oil & Energy would acquire the fuel cell division of Sanyo and the two companies jointly establish a new company.[298] In April 2008, ENEOS Celltech was established with the acquired Sanyo fuel cell division as its core.[299] JX Nippon Oil & Energy owned 81 percent of ENEOS Celltech, Sanyo 19 percent.

JX Nippon Oil & Energy/Kyocera—Demonstration of 0.7 kW SOFC Cogeneration Power Plants (January 2008). In January 2008, JX Nippon Oil & Energy announced that the company participated in the METI 2007–2010 SOFC demonstration program.[300] In the FY2007 program, the company demonstrated one 0.7 kW cogeneration power plant fueled by LPG and one 0.7 kW unit fueled by kerosene. The systems consisted of SOFC stacks made by Kyocera and fuel reformer and system control equipment made by JX Nippon Oil & Energy. The systems were installed at company facilities in Kanagawa Prefecture and started operations in January 2008. The LPG unit operated for 5744 h and the kerosene unit for 4565 h. The LPG unit had a problem with the DC sensor that required repair.[301] The kerosene unit had six outages, all in the initial stage; they required the replacement of the SOFC module, the reformer, reformer water pipes, and the adjustment of the control parameter and the flow meter.

- In 2008, two LPG units and one kerosene unit were demonstrated with no published results. In 2009, 14 LPG units and 1 kerosene unit were to be demonstrated, but results are not known.

JX Nippon Oil & Energy/Kyocera—Plans to Develop 1 kW Class Residential Use SOFC with Kyocera (September 2009). According to the press reports in September 2009, JX Nippon Oil & Energy and Kyocera planned to develop a new kilowatt class residential-use SOFC power plant.[302] The power plant would use Kyocera SOFC cells, operate at higher than 700 °C, have an electrical efficiency of 45 percent, and take up about half the space of current SOFC power plants. The companies aimed to lower the cost of the plant to ¥1.2 million and extend the plant lifetime to 10 years by March 2012. The two companies also planned to develop several-kilowatt class power plants for small stores and restaurants.

JX Nippon Oil & Energy—Plans to Commercialize SOFC Power Plants as early as in FY2011. In December 2009, JX Nippon Oil & Energy had set its plans to launch SOFC commercialization as early as in FY2011 (starting April 2011).[303] It indicated that it would target convenience stores and small businesses that have continuous power needs. In 2009, JX Nippon Oil & Energy planned to step up its effort to assess the durability of their power plants. At the same time, the company would explore cost cutting measures, including mass production of cells and stacks and development of BoPs that can be commonly used by SOFC and PEMFC power plants. The company recently announced that it would start field tests as early as summer of 2010.[304]

6.3.6.14. Nippon Telegraph and Telephone (NTT)

NTT began SOFC development perhaps in the early 1990s. In 1995, NTT reported that it fabricated a cell that achieved a maximum power density of 0.63 W/cm^2 at 800 °C and 1.04 W/cm^2 at 850 °C.[305] By 2000, NTT researchers were studying SOFC cell materials, including lanthanum nickel oxide air polar and scandia-alumina-stabilized zirconia (SASZ) electrolyte.[306] In 2002, they established the world's leading technology for high-output co-sintered flat disk SOFC cells, which were 1 cm in diameter. Cell diameter steadily increased, first to 6 cm in 2003–2005, and then to 12 cm in 2007. NTT achieved a cell lifetime of 7000 h by about 2005.[306]

NTT was developing anode-supported SOFC cells with a metallic separator, unique sealing methods, and cell separator connecting methods.[307] In 2007, NTT developed one-cell and three-cell stacks with cells that were 60 mm in diameter. The stacks achieved a power density of up to 0.6 W/cm^2 and an electrical efficiency of 53 percent LHV DC. Demonstrated durability was over 1000 h.

NTT—Development of 1 kW Flat Disk SOFC Module (April 2008). In October 2008, NTT researchers announced that they had developed a 1 kW class SOFC stack composed of 50 anode-supported planar 120 mm diameter SOFCs.[308] Intermediate plates, which exhibited negligible deformation under operating conditions, were placed in the stack to cancel out the cumulative error related to the position and angle of the stack parts, according to the publication. The stack provided an electrical efficiency of 54 percent LHV fueled by methane and an output of 1.12 kW with a fuel utilization rate of 67 percent, current density of 0.28 A/cm^2, and operating temperature of 800 °C. The stack operated for almost 700 h.

NTT—Development of 1.5 kW Flat Disk SOFC Stack (January 2009). In January 2009, NTT reported that the company had developed a 1.5 kW class SOFC stack.[309] Anode supported cells (ASCs) 120 mm in diameter were used in the stack. The stack consisted of 40 cells and provided an electrical efficiency of 63.9 percent. A single cell was operated for 5000 h at a fuel utilization rate of 80 percent and showed an electrical efficiency of 60 percent LHV with a voltage degradation of 0.55 percent/1000 h.

NTT/Toho Gas/Sumitomo Precision Products—Development of 3 kW Planar SOFC Cogeneration System (April 2009). In April 2009, NTT, Toho Gas, and Sumitomo Precision Products developed a 3 kW natural gas–fueled SOFC module.[237]

- Since 2006, the three companies had been developing a module consisting of anode-supported flat disk SOFC cells with a diameter of 120 mm that consisted of a lanthanum nickel ferrite cathode, an SASZ electrolyte, and a nickel/SASZ anode.[310]
- The 3 kW module consisted of two 1.5 kW stacks, each comprised of 40 cells. The module was equipped with a steam generator, a reformer, a post combustor, heat exchangers, and a start-up burner,

surrounded by thermal insulators. The system was expected to achieve an electrical efficiency of 46 percent AC and an overall efficiency of 80 percent.[311]

After the start-up burner was stopped, current output of the stack reached 50 A.[311] Voltage was constant for about 200 h. At this time, the electrical conversion efficiency reached 56 percent LHV DC. Thermally self-sustainable operations were confirmed even when the electric power was reduced to 50 percent. Moreover, the maximum electrical conversion efficiency for a short time reached the world's highest level at 59 percent DC LHV. Total operating time was 750 h.

The three companies planned jointly to integrate this module with a balance of plant to fabricate several-kilowatt class SOFC systems[310] and hoped to market the products in 2 or 3 years to small businesses.[237] NTT has made a significant progress in making large-diameter cells and increasing their lifetimes. The company and its partners, however, have not had much success in developing systems. It is too early to say when they will be able to achieve commercially viable systems.

6.3.6.15. Sanyo Electric

Sanyo Electric has been in planar SOFC development for as long as 20 years. The company perhaps started its SOFC research in the 1980s and participated in the large-scale METI SOFC R&D Program when it was launched in 1989. Sanyo basically left the SOFC field in November 2007, when the company sold its fuel cell division to JX Nippon Oil & Energy.

In 1989, Sanyo Electric participated in Phase I (1989–1991) of the METI 1989–2007 SOFC R&D Program. Sanyo was assigned to develop a 100 W class planar SOFC module with metal interconnects. Sanyo developed a 20-cell planar SOFC module with internal manifolding and metal bipolar plates.[312] The module operated for 1000 h and produced 415 W. The module met the program goals.

In 1992, METI launched the 6-year Phase II program (1992–1997) and Sanyo was tasked to develop a kilowatt class planar SOFC module during the first 4 years. The target current density was 0.18 W/cm^2 at 70 percent fuel utilization.[178]

- In 1994, Sanyo developed a kilowatt class planar SOFC module with metal separators. The module consisted of a stack of thirty 160 cm^2 cells and achieved a maximum output power of 1.060 kW.[313] The module operated for 1871 h. The module exhibited severe gas leakage and voltage degradation, however.[176] The degradation reached 4.4 percent/1000 h for current densities of 300 mA/cm^2.[181]

During the remaining 2 years (1995–1997), efforts were focused on solving the gas leak and cell degradation problems.[176] The initial plan to develop a 10 kW class module with a current density of 0.2 W/cm^2 and a deterioration rate of less than 1 percent/1000 h had been suspended. In January 1996, Sanyo developed a 2 kW class module.[314] It consisted of 17 stacks of four 182 cm^2 cells and generated 2.47 kW at a current density of 300 mA/cm^2 and a 15 percent fuel utilization rate. It achieved a power density of 0.2 W/cm^2. The gas sealing technology improved.[176] It was noted, however, that the metal interconnects used in Sanyo's planar stacks were unable to demonstrate the reliability needed for thermal cycles and long-term stability. The 2 kW class module, thus, did not meet project goals.

- In 1997, Sanyo left the METI SOFC R&D Program. The company had worked in the program for 9 years (1989–1997).

In June 2004, there were press reports indicating that Sanyo agreed to cooperate in the development of Acumentrics' tubular SOFC technology.[203] In October 2003, Acumentrics of the United States established a Japanese corporation in Japan, and for the next 4 years, Acumentrics Japan and Nippon Steel, with Sanyo's cooperation, developed a 10 kW microtubular SOFC cogeneration power plant.[315] The power plant successfully met all target goals, except for the degradation rate.

In November 2007, Sanyo Electric agreed that JX Nippon Oil & Energy would acquire its fuel cell division and that the two companies jointly would establish a new company.[298] In April 2008, a new joint company, ENEOS Celltech, was established.[299]

6.3.6.16. Toho Gas/Sumitomo Precision Products/Nippon Shokubai/Daiichi Kigenso

Toho Gas has been developing scandia-stabilized zirconia (ScSZ) electrolyte since the 1990s.[316] Toho Gas devoted about 10 years of research on the properties of electrolytes, the sintering technology that controls the chemical reactions that occur when firing different kinds of materials, and improvements in mechanical strength and fracture resistance.[317] In 1997, the electrochemical performance of planar type single cells with an ScSZ electrolyte demonstrated a power density of 2 W/cm^2. In the early 2000s, Toho Gas developed scandia-stabilized tetragonal zirconia polycrystals (Sc-TZP) electrolyte, which showed great fracture resistance, and a 12 cm diameter planar electrolyte-supported cell (ESC) using the ScSZ electrolyte.

Toho Gas/Sumitomo Precision Products—Development of 1 kW Planar SOFC System (September 2003). In September 2003, Toho Gas and Sumitomo Precision Products jointly developed the first 1 kW class SOFC system (the alpha model) in Japan.[311,316,318] The system incorporated Sc-TZP. Stack structure in the unit was compact, and the volume was 0.5 kW/L, which was superior to competing products in volumetric efficiency according to Toho Gas.[318]

- The system, however, showed difficulty with thermal cycles and operations on reformed natural gas.[316]

In September 2004, an improved 1 kW system (the beta model) was developed.[316] The volume of the improved model was 40 percent less than the previous model, mainly due to the rearrangement of the BoP system. The beta model also had an oil-based heat recovery unit that was developed for use in a cogeneration system, in contrast to the alpha model that did not have this purpose. Also by controlling the adhesion strength of the gas sealing material, the beta model achieved less voltage degradation during thermal cycles. The efficiency of cell stacks in the beta model increased to 45 percent DC LHV by making the temperature distribution of cell stack uniform, while the alpha model had a 24 percent efficiency. The beta system operated on natural gas reformed by catalytic partial oxidation, and the efficiency of the system was 26 percent LHV. Two beta units were built; one was installed at the Toho Gas for further improvement and the second was installed at the 2005 World Exposition site in Aichi, Japan.

- At the Aichi Expo 2005, the 1 kW SOFC cogeneration plant commenced operation in March 2005, corresponding to the start of EXPO 2005.[319] An internal manifold 80-cell stack produced an output of 955 W using natural gas under thermally self-sustaining conditions. Maximum electrical efficiency and thermal efficiency were both 35 percent LHV. The system operated as a cogeneration system for 4459 h (96.4 percent up-time) from May 16 to September 25, 2005.
- In 2006, the 1 kW cogeneration system installed at Toho Gas was further improved and achieved an electrical efficiency of 55.7 percent DC LHV.[311]

Toho Gas/Nippon Shokubai/Daiichi Kigenso—Marketing SOFC Cells and Materials (2004 & 2005). By the mid-2000s, Toho Gas was convinced of the excellent performance of SOFC cells made of the Sc-TZP and decided to market them as commercial products:

- *SOFC Cells*. In May 2004, Toho Gas started to produce SOFC single cells jointly with Nippon Shokubai within 2004.[320] The two companies had started the joint research 3 years earlier and developed the manufacturing technology for single cells by combining the technologies for scandia-stabilized zirconia owned by Toho Gas and mass production technology owned by Nippon Shokubai.[321] In February 2005, Nippon Shokubai started to build a commercial plant to manufacture ceramic sheets for SOFCs at a rate of one million sheets a year.[322] The company at that time was operating a research-level plant that made the sheets at a rate of 100,000 a year, which were supplied to the Swiss SOFC maker Sulzer Hexis as well as to Toho Gas. The cost of the plant was reportedly ¥1 billion (approximately US$9 million at the 2005 exchange rate of $1/¥110). They planned to start mass production and commercialization of their SOFC cells in 2008.[321]
- *SOFC Powders*. Toho Gas and Daiichi Kigenso Kagaku Kogyo (DKKK—the largest zirconia supplier in Japan) agreed to pursue a joint commercial venture to market SOFC powders. In June 2005, DKKK built a pilot plant to test commercial production of scandia-stabilized zirconia ScSZ and Sc-TZP.[323] Toho Gas's electrolyte technology was transferred to DKKK.[319] DKKK and Toho Gas would co-develop the commercial products.[323]

Toho Gas—Development of SOFC Cells for Lower Temperature Operation and Long Endurance (November 2006). In November 2006, Toho Gas developed SOFC cells for lower temperature operation and long endurance and planned to market them as commercial products.[324] The amount of scandium in the electrolyte was increased by 2.5 times, so that the ion conductivity would be increased. The operating temperature was reduced to 750 °C from the conventional 800 °C, while maintaining the power-generation efficiency. Lower operating temperatures would widen the range of material choices, contributing to cost reduction. In the intermediate layer between the electrolyte and the electrodes, cerium oxide was sprayed and then sintered. By this treatment, aging was suppressed, and the aging rate was improved from the conventional rate of 3 percent/1000 h to 0.25 percent/1000 h.

Toho Gas/AIST—Development of Microtubular SOFC Module (January 2007). In January 2007, Toho Gas developed a prototype microtubular SOFC module operating at 500 °C that demonstrated a power density over 0.5 W/cm^2.[311] The micro tubes used in the module were those developed by AIST. The electrolyte was ceria-based ion conducting ceramics, and the module was made of a total of 36 micro tubes (each 2 mm in diameter and about 5 cm in length).[325] The power output per module was 10–20 W. By combining multiple modules, Toho Gas hoped to combine multiple modules to generate several hundred watts by 2009 and complete the development of home-use and business-use stationary SOFC systems in 2011–2012.

Toho Gas/NTT/Sumitomo Precision Products—Development of 3 kW Planar SOFC Cogeneration System (April 2009). In April 2009, NTT, Toho Gas, and Sumitomo Precision Products developed a 3 kW natural gas–fueled planar SOFC (see Section 6.3.6.14).[237] The three companies planned jointly to integrate this module with a balance of plant to fabricate several-kilowatt class SOFC systems[310] and hoped to market the products in 2 or 3 years targeting small businesses.[237]

Toho Gas—Participation in the METI SOFC Demonstration Research Program in FY 2009. Starting in FY2009, Toho Gas decided to participate in the METI SOFC demonstration research program and test-operate a 0.7 kW cogeneration power plant developed by Kyocera/Osaka Gas/Toyota/Aisin. In its 2009–2013 near-term management plan, Toho Gas decided to start developing residential-use SOFC power plants, while it had in the past only focused on business-use SOFC systems.[237] Also in 2010, Toho Gas decided to demonstrate field operations of a 700 kW CHP system using a NGK Spark Plugs SOFC module at a Toho Gas site.[295]

- Toho Gas has been exploring various business approaches in SOFC development but clearly is still in search of the optimum strategy. The company clearly excels in powder and cell development. When it comes to system development, however, the company is more reliant on other companies' stacks and modules. It is not certain if and when it will develop its own commercially viable SOFC systems.

6.3.6.17. *Tonen*

Tonen Corporation, a major oil refiner, started SOFC development in 1986[326] and apparently ended it in the late 1990s. Tonen aimed to develop planar SOFCs that can use reformed LNG, LPG, diesel, and other petroleum-derived fuels.[176] The company worked closely with the JPEC, a nonprofit research organization for the petroleum industry, and with METI's Agency for Natural Resources and Energy (ANRE).

In 1990, JPEC started a 5-year program for developing high temperature fuel cells. The program, likely subsidized by ANRE, aimed to develop SOFC modules using light petroleum as fuel. Tonen, JX Nippon Oil & Energy, and Idemitsu Kosan participated in the program.[327] It aimed to develop a 25 kW tubular SOFC module and a 10 kW planar SOFC module.

- In 1991, Tonen developed a planar SOFC module that achieved an output of 1.3 kW.[328] The 1.3 kW module consisted of thirty 225 cm^2 cells and had a fuel utilization rate of 50 percent and oxygen utilization of 30 percent.[329]
- Subsequently, Tonen worked to raise the electrical efficiency and reliability of its SOFC cells using hydrogen and propane as fuel and improved the cell components and cell stacking methods.[327] In 1995, Tonen developed a 5.1 kW module.[283] The 5.1 kW module consisted of sixty five 729 cm^2cells. It had a current density of 166 mA/cm^2, a power density of 0.11 W/cm^2, and 35 percent fuel utilization. The module still used hydrogen as fuel; their SOFCs were not ready for hydrocarbon fuel.

Little information has been found since then regarding Tonen's SOFC development or commercialization. The company perhaps left the SOFC fields in the late 1990s.

6.3.6.18. *TOTO/Hitachi/Kyushu Electric/ Nippon Steel*

TOTO, a leading ceramics manufacturer, began SOFC development in 1989. The company specialized in tubular SOFC cells made by a wet method.[186] The wet method enables the application of a highly homogeneous thin film deposition on a substrate more easily and at lower cost than a dry method, such as chemical vapor decomposition (CVD) or electrochemical vapor deposition (EVD).[330] TOTO's cathode was a perovskite oxide self-supporting tube molded by extrusion.[186] Zirconia electrolyte and nickel/zirconia cermet anodes were coated onto the tube by a slurry process. A vertical interconnector was coated along the tube in a stripe.

TOTO/Kyushu Electric: Development of kW Class Tubular SOFC Module Made by Wet Method. In 1991, TOTO began joint development with Kyushu Electric. In 1996, they developed a 1 kW module.[331] In 1995, TOTO and Kyushu Electric and Kyushu University successfully operated their SOFC with an output of 0.4 W/cm^2 using air as an oxidizer.[178] In 1998, TOTO and Kyushu Electric teamed with Nippon Steel. Together, they were contracted to develop a kilowatt class SOFC module made by the wet process in the Phase II Extension Period (1998–2000) of the METI SOFC R&D Program.[332] In 1999, they developed a 2.5 kW module. In 2000, they developed a 3 kW module.[333] The 3 kW module demonstrated a power density of 0.18 W/cm^2, a fuel utilization of 70 percent, and voltage degradation of 1 percent/1000 h.[334]

TOTO—Development of Thermally Self-Supporting kW Class SOFC Tubular Module. In 2001, NEDO selected TOTO to develop a thermally self-supporting SOFC tubular module by the wet method.[332] In 2004, TOTO successfully developed a 6.5 kW module, which consisted of 20 stacks of 12 tubular cell bundles and generated 6.5 kW at 0.2 A/cm^2 at 48 percent LHV with a fuel utilization of 75 percent.[186] The same module was operated under different conditions and generated 6.4 kW at 0.19 A/cm^2 and 46 percent LHV. The voltage degradation rate was 0.1 percent/1000 h in both cases. The thermal self-sustainability of both operating conditions was confirmed.

TOTO—Started to Prepare for Commercial Ventures. In 2004, TOTO was highly confident about its SOFC commercialization in the near term. It announced that it achieved the world's highest level power efficiency (55 percent) with its SOFC stack and operated it for 3000 h.[335] The company had come to believe that it established the basic SOFC stack technology and hoped to commercialize it as early as in 2007.[336] It had started the small-scale commercial production of SOFCs and trial delivery.[186] The SOFCs were supplied as a stack to SOFC system developers for the development of stationary power-generation systems.

TOTO/Hitachi: Commissioned to Develop a 20 kW SOFC Cogeneration Power plant (April 2004). TOTO in partnership with Hitachi was commissioned to develop a 20 kW SOFC cogeneration power plant under the METI 2004–2007 SOFC System Development Program.[335] The project did not work out successfully. The company had hoped to achieve a total efficiency of 80 percent with the cogeneration power plant by 2007.[336] TOTO also had hoped to increase the unit's longevity to 40,000 h at the earliest opportunity. Within the program, TOTO was responsible for developing tubular cell bundles and Hitachi for developing cogeneration systems.[337]

- First, a 10 kW cogeneration test module was successfully built and tested.[189] It generated 10 kW at 890 °C at 0.2 A/cm^2 with 80 percent fuel utilization. The electrical efficiency was 46.6 percent HHV DC.
- Following the successful 10 kW test module, a 20 kW power plant was built. Unfortunately, the 20 kW power plant experienced a ground fault while the temperature was being raised to the operational level. The fault triggered a voltage drop and cell damage, bringing the project to a full stop.[189]

TOTO/Fuel Cell Technology Limited (FCT)—Planned Development of 2–3 kW SOFC System as Commercial Product (January 2005). In 2005, TOTO and FCT of Canada signed a memorandum of understanding for development of a 2–3 kW SOFC system.[338] Under the terms of the MOU, FCT would design and build a small SOFC system for residential markets using tubular cell stacks supplied by TOTO. The MOU between the companies allowed for further expansion of FCT products into new markets. It also provided for the supply to FCT of small-size SOFC tubes well suited to power products in the 2 kW size range, which was a step in FCT's SOFC product diversification strategy. FCT had been collaborating with SWPC in the development of 5 kW cogeneration systems. FCT was acquired by Acumentrics in 2007.

TOTO—Development of Microtubular SOFC Stack Operable at 500 °C (October 2005).[339] TOTO developed a microtubular SOFC stack that operated at 500 °C and planned to commercialize it in 2008. The cell was 50 mm long and 5 mm in diameter, and several dozen cells constituted a palm-sized stack,[338] which generated 28 W at 500 °C and 37 W at 600 °C.[339] In addition to a lower operating temperature, the cell also had a short warm-up time of only about 5 min.[338] The stack would be suitable as a 50 W to 1 kW power source.[339] Potential applications would include electric wheelchairs, electric bicycles, and portable electronic devices. Liquefied propane could be used as fuel.[338]

- TOTO developed this stack as part of the METI 2005–2007 SOFC Component Technology Development.[339]
- In April 2008, TOTO planned to start the delivery of 75 W and 200 W SOFC modules to US power system manufacturers.[205] The modules would serve as a power source for portable electric devices.

TOTO—Participation in the METI 2007–2010 SOFC Demonstration Research Program (2007). TOTO was selected to participate in the METI SOFC demonstration program starting in 2007.[340] The company demonstrated two 2 kW mono-generation units in 2007 and two 0.7 kW cogeneration units and one 8 kW mono-generation unit in 2008. In 2009, it planned to demonstrate six 0.7 kW cogeneration systems. In March 2008, TOTO claimed that it had developed a residential use fuel cell module that would cost about ¥1 million per kilowatt (about US$9,670 at the 2008 exchange rate of $1/¥103.4), less than a third the cost of current residential-use SOFC cogeneration models, according to press sources.[205] The company employed its own ceramic mass-production technology to manufacture SOFCs. TOTO planned to demonstrate the performance of the module and commercialize a cogeneration power plant in 2011.

TOTO/Hitachi: Development of 10 kW SOFC Module with 360 Tubular Cells (January 2010). In January 2010, TOTO and Hitachi developed a 10 kW SOFC cogeneration power plant and operated it for more than 3000 h, according to press report.[341] The power efficiency was about 40 percent. The system could generate 60 °C hot water, raising total efficiency to more than 80 percent. The two companies hoped to commercialize the product targeting health care facilities, such as nursing homes.

- *Hitachi— Improvements to SOFCs to Raise Electrical Efficiency (January 2010).* In January 2010, the press reported that Hitachi reduced SOFC cell resistance by about 18 percent by installing an auxiliary electrode to transfer electricity.[341] The company also shortened the electricity pathway by modifying the cell structure and increased the electrical efficiency as much as 1.3 times to 0.39 W/cm^2. The auxiliary electrode and cell modification did not affect the quality of the cell.

6.3.7. Japan's Initiatives Approach Critical Mass

Japan has a large number of companies and organizations engaged in SOFC development today. They include SOFC stack and module developers, component manufacturers, system integrators, and gas and electric utility companies. They each have SOFC development programs of their own, and high levels of expertise in cell, stack, or module development.

- Their focus and interest is highly focused on small residential use systems, typically 700 W to 1 kW class.

Japanese developers started to gain modest SOFC capabilities in the early 1990s. Starting in 2005, with help from METI's large-scale SOFC demonstration program, they rapidly advanced their SOFC capabilities, demonstrating high performance at stack, module, and cell levels. Today their power densities are superior to those demonstrated by US SOFC developers, stack operating temperatures are lower, stack size is smaller, and the diversity and robustness of their designs are much greater. Japan's SOFC developers are forming partnerships and joint ventures at an accelerating rate and rapidly building and testing new designs and applications, which, taken together with their technical performance, increases their chances to reach markets before their competitors.

6.4. EUROPE RESTARTS SOFC DEVELOPMENT IN 1986

European SOFC development efforts have been modest and less resilient than those of the United States and Japan. Europe once started SOFC technology in the 1960s[342]and abruptly backed out in 1967. European SOFC developers probably had difficulty in raising the efficiency and lowering the cost of tubular SOFC cell designs and thought that near-term commercialization was not possible. Then, a renewed SOFC effort started in 1986, triggered by the Chernobyl incident in Russia in April of that year, this time focusing on a new design—planar SOFC technology. Planar SOFCs promised higher efficiency and lower production costs. SOFC development activities in Europe, however, have not been as robust as those in the United States and Japan.

- Many European SOFC developers have again dropped out from their effort since then. They were likely convinced that it was too difficult to scale up the cells, stacks, and systems, increase the reliability and durability, and lowering cost.

One of the unique characteristics of European R&D is the major role played by the European Union (EU), which provides the single largest support to European R&D. The EU, which focuses on research collaboration, runs its Framework Program (FP) in close coordination and cooperation with industrial and national SOFC development programs. Typically, industry and national research institutes develop a preliminary capability with support from their nation's research institutions. They then proceed to get their project funded under the EU FP. Because EU FP projects are typically chosen from national projects, most EU FP and national projects are analogous and complementary. Also due to this interactive relationship, it is perhaps not surprising that the levels of SOFC funding from the EU FP, individual national program funding, and industrial investment are virtually equal.

Also, there are some other characteristics that are distinct in Europe:

- While economic integration continues to progress in the European Community, national energy profiles and trends among the European Community member countries remain diverse, resulting in different fuel cell policy priorities and different levels of interest in fuel cell development. Thus, some countries are interested in SOFCs, some are interested in other fuel cell technologies, such as MCFC, PEMFC, or DMFC, and some are not much interested in fuel cells. Those interested in SOFCs include Denmark, Finland, Germany, the Netherlands, Norway, Switzerland, and the United Kingdom.
- Industrially and nationally funded SOFC projects in any individual European country may or may not include participants from other countries while EU FP projects typically include participants from multiple European countries as stipulated by EU FP regulations. Industrial competition and individual national objectives, however, often hinder efficient international collaboration.[343] Participation by many countries in an FP project helps diffuse the basic level of

knowledge throughout Europe, but it perhaps dilutes the effect of program funds and makes project coordination difficult.

- Research institutes in Europe appear to play a significantly greater role in SOFC commercialization than some of their counterparts elsewhere. Those that take an active leadership role not only in research but also commercialization include Risø National Laboratory in Denmark, Forschungszentrum Jülich (a federally and state-funded research institute) in Germany, and the Energy Research Center of the Netherlands (ECN). They conduct research as well as work with industries to bring technology to the market or go so far as to establish a company to commercialize the technology.

6.4.1. Denmark

Denmark has had a small fuel cell R&D program since 1987 with focus on SOFC technology. During the 1990s, Denmark's national program spent about €2.7 million ($2.5 million at the 2000 exchange rate of $1/€1.08) per year.[344] There are two major players in SOFC development in Denmark, Risø National Laboratory for Sustainable Energy at the Technical University of Denmark (DTU) and Haldor Topsoe, a leading catalyst company. The two have worked as close partners since 1989 and established a consortium agreement to collaborate in 2001. Since then, they have been virtually two sides of the same coin, with well defined, complementary divisions of labor. Risø conducts R&D for Haldor Topsoe, and Haldor Topsoe takes the responsibility for manufacturing SOFC stacks. Haldor Topsoe provides stacks to system developers for integration and distribution.[345]

6.4.1.1. Risø and Haldor Topsoe: Developmental Work in Partnership

Risø started SOFC R&D in 1987 following fundamental research work on electrochemical materials in the 1980s. In 1989, the Danish government decided to focus its efforts on SOFCs. With the consensus of the Danish Energy Authority, the Danish power utilities, industry, and Risø, it launched the Danish Solid Oxide Fuel Cell Program Series DK-SOFC. The participants to the program included Risø as project leader, Haldor Topsoe, IRD Fuel Cell, the University of Southern Denmark, and the DTU. Under the program, Risø and Haldor Topsoe carried out a development program focusing on SOFC components and technology. The DK-SOFC program consisted of a series of 3–4-year projects implemented in succession to establish a Danish capability in SOFC technology.[346] A total of 200 million krone (about €27 million or about $28.7 million at the 1999 exchange rate of $1/€0.94) was invested over the period 1990–1999 in the program.

Risø and Haldor Topsoe also made the most of the EU FP to augment the national program. For example, in the mid-1990s, an SOFC economic study conducted by Haldor Topsoe, ELSAM, and Risø under the sponsorship of the Danish Energy Agency and ELSAM identified methods and design modifications to reduce SOFC stack manufacturing costs to levels competitive with those of other electric power-generation technologies.[347] As early as 1996, Risø and Haldor Topsoe managed to have the EU FP sponsor a project to test the study findings. An EU FP project, "Improving durability of SOFC stacks (IDUSOFC) (1996–1998)," was implemented with its focus on extending the lifetimes of SOFC components with the collaboration of a number of leading European SOFC developers—Risø (coordinator), Siemens, Dornier, RR, Statoil, Jülich, ECN (the Netherlands), Imperial College, Oslo University, and Haldor Topsoe.[348] Significant progress in fabrication and assembling technologies was made in the field of electrical contacts between stack components (anodes to interconnect) and in electrode fabrication.[347] A new anode design was created that enabled direct conversion of natural gas to hydrogen fuel in an SOFC.

- This project was followed by another EU FP project headed by Risø, LOCO-SOFC (Low cost fabrication of SOFCs, 1997–2000), where the focus was on materials and fabrication costs and improving overall performance. Risø and Haldor Topsoe also participated in two other EU FP projects, MF-SOFC (2000–2003) to scale up stacks and systems, and CORE-SOFC (2001–2004) to increase SOFC reliability.[346]
- The participation in the Danish national program as well as EU FP projects clearly helped advance the Danish SOFC capability significantly. Risø's first-generation planar SOFCs between the 1980s and the mid-1990s were electrolyte supported.[349] In 1995, Risø demonstrated a 500 W electrolyte-supported fuel cell stack.[350] In the second half of the 1990s, Risø as well as most European planar cell developers switched to anode-supported planar cells[349] that were initially developed by Jülich.[47] In the late 1990s, Risø demonstrated a relatively cheap ASC which had a low internal resistance at a relatively low operating temperature (Table 6.9).[350]

6.4.1.2. *Forming a Consortium in 2001 for SOFC Commercialization*

Making significant progress in Denmark's SOFC capabilities, Haldor Topsoe started preparations to pursue SOFC commercialization. In 2001, Haldor Topsoe signed a formal agreement with Risø to develop and commercialize the Danish SOFC technology. The agreement formed a 5-year consortium pact that guaranteed a substantial research effort at Risø on behalf of Haldor Topsoe.[350] It also defined the division of labor between the two: Risø would focus on research and development and manufacture of cells, while Haldor Topsoe would take care of the design and production of stacks.[345] At the same time, a large-scale project was launched with financing from the Danish Energy Authority, the electricity transmission system operators, Eltra and Elkraft System, and DK-SOFC project participants.[350] One of the results of the project was the building of Risø's pre-pilot SOFC cell production facility, inaugurated in 2002. In that year, the production of SOFC stacks commenced at Haldor Topsoe.[351]

6.4.1.3. *Forming a Topsoe Fuel Cell for SOFC Commercialization in 2004*

In 2004, Haldor Topsoe reached a cooperation agreement with Wärtsilä Corporation, a Finnish manufacturer of large diesel and gas engines, to develop SOFC products for commercial use. This was another important partnership for Topsoe. Haldor Topsoe produced stacks, and Wärtsilä would integrate them into systems, thereby completing the path from R&D (undertaken by Risø) to end-use applications.[47]

Haldor Topsoe took another definitive step toward SOFC commercialization in 2004. It founded Topsoe Fuel Cell, a wholly owned subsidiary of Haldor Topsoe. In 2006, Risø and Topsoe Fuel Cell signed a renewal of their long-term collaboration agreement,[350] which would provide the framework for continued and expanded SOFC activities for the next 5 years. The agreement also secured financial support from Topsoe to Risø, while Topsoe retained exclusive rights to use Risø's patents and know-how.

- In 2004, Risø/Haldor Topsoe's stacks used a metallic alloy for the interconnect (Crofer 22APU) and a new perovskite coating.[352] This improvement reduced the overall ASR of the stack to about 0.5 ohm/cm^2 and improved durability. A 5-cell stack had no measurable voltage degradation over 1000 h and the typical initial decrease in stack voltage also had been eliminated. Due to the very compact stack design the power density reached 2.4 kW/L of stack volume under

TABLE 6.9 EU Framework Projects with Major European Participants*

Project title	Major project participants	EU Framework Program (FP)	Total project cost; project funding (€)	Dates
Projects with Danish Participation				
Improving durability of SOFC stacks (IDUSOFC)	Risø (coordinator), Energy Research Center of the Netherlands (ECN), Imperial College, Forschungszentrum Jülich (Jülich), University of Oslo, Haldor Topsoe, Dornier, Rolls Royce, Statoil R&D Center	FP4		January 1996–December 1998
Low-cost Fabrication and Improved Performance of SOFC Stack Components (LOCO-SOFC)	Risø (coordinator), Gaz De France, Rolls Royce, IRD, Napier University, Institut National Polytechnique de Grenoble	FP4		April 1997–March 2000
Scale up of a multifunctional solid oxide fuel cell to multi-tens of kilowatt levels (MF-SOFC)	Rolls Royce (coordinator), Risø, Imperial College, Gaz De France, Advanced Ceramics,	FP5	€8,933,450; €3,499,062	February 2000–July 2003
Component reliability of solid oxide fuel cell systems for commercial operation (CORE-SOFC)	Jülich (Coordinator), ECN, Risø, Haldor Topsoe, Rolls Royce	FP5	€4,305,614; €1,997,783	January 2001–August 2004
Projects with Wärtsilä Participation				
Validation of Renewable Methanol Based Auxiliary Power System for Commercial Vessels (METHAPU)	Wärtsilä (coordinator), Lloyd's Register (UK), Wallenius Marine (Sweden), University of Genova (Italy), Det Norske Veritas (Norway)	FP6	€1,930,300; €1,000,000	November 2006–October 2010
Toward a Large SOFC Power Plant (LARGE-SOFC)	Valtion Teknillinen Tutkimuskeskus (VTT) (Finland) (coordinator), Rolls Royce, Topsoe, Jülich, University of Genova, Bosal Research (Belgium), Verteco (Finland), Inmatec Technologies (Germany), Wärtsilä, the Switch Engineering (Finland)	FP6	€11,026,326; €5,799,636	January 2007–December 2009

(Continued)

TABLE 6.9 EU Framework Projects with Major European Participants* (*Cont'd*)

Project title	Major project participants	EU Framework Program (FP)	Total project cost; project funding (€)	Dates
Development and Demonstration of Manufacturing and Operation of Clean and Efficient Power Generation Based on Solid Oxide Fuel Cells (DEMO-SOFC)	Haldor Topsoe (Coordinator), Topsoe Fuel Cell, Wärtsilä	EU LIFE Environment Program	€13,400,000; €4.8 million	October 2006–September 2010
Projects under Jülich Coordination				
Component reliability of solid oxide fuel cell systems for commercial operation (CORE-SOFC)	ECN, Risøe, Haldor Topsoe, Rolls Royce	FP6	€4,305,614; €1,997,783	January 2001–August 2004
Realizing reliable, durable, energy efficient and cost effective SOFC systems (REAL-SOFC)	26 Partners (see table below)	FP6	€18,259,430; €8,999,000	February 2004–December 2008
Generic fuel cell modeling environment (GenFC)	Graz University Of Technology, etc.	FP6	€2,666,220; €1,700,000	October 2005–September 2008
Projects under ECN Coordination				
Membrane-Based Thin-Layer SOFC Technology	University Twente (the Netherlands)	FP1		January 1989–December 1990
Manufacturing Techniques for Components of Flat Plate, SOFC Reactors	EniTecnologie (Italy), Siemens (Germany)	FP2		July 1991–July 1994
Framework for system optimization studies of integrated fuel energy systems	Jülich (Germany)	FP3		January 1994–December 1995
Demonstration of SOFC stack technology for operation at 600 °C (SOFC600)	Danmarks Tekniske Universitet (DTU) Denmark, Jülich, HTceramix (Switzerland), Imperial College (UK), Topsoe Fuel Cell (Denmark), VTT (Finland)	FP6	€11,921,374/6,769,923	March 2006–February 2010

Projects with Alstom UK Participation				
Systematic analysis of gas turbine - fuel cell combinations generation with very high efficiency	Blomenco BV (the Netherlands)	FP3		January 1994–March 1995
Decentralized power generation plants based on planar SOFC technology proof of concept PROCON	Alstom UK	FP5	€2,972,000/€1,486,000	April 2000–October 2002
Integrated modeling study of fuel cell/gas turbine hybrids (IM-SOFC-GT)	Rolls Royce	FP5	€2,892,346/€1,190,506	February 2001–July 2003
Project with Royce Fuel Cell Systems Participation				
Improving durability of SOFC stacks	Risøe National Laboratory (Denmark)	FP4		January 1996–December 1998
Low-cost Fabrication and Improved Performance of SOFC Stack Components (LOCO-SOFC)	Risøe National Laboratory (Denmark)	FP4		April 1997–March 2000
High Temperature Fuel Cells Use for the Next 10 Years	Bertin et Cie S.A. (France)	FP4		October 1998–September 2000
Scale Up of a Multi-Functional Solid Oxide Fuel Cell to Multi-Tens Of Kilowatt Levels (MF-SOFC)	Rolls Royce (UK)	FP5	€8,933,450/€3,499,062	February 2000–July 2003
Integrated Modeling Study of Fuel Cell/Gas Turbine Hybrids (IM-SOFC-GT)	Rolls Royce (UK)	FP5	€2,892,346/€1,190,506	February 2001–July 2003
Component Reliability of Solid Oxide Fuel Cell Systems for Commercial Operation (CORE-SOFC)	Jülich (Germany)	FP5	€4,305,614/€1,997,783	January 2001–August 2004
Thematic Network on Solid Oxide Fuel Cell Technology (SOFCNET)	Energy Research Center Of The Netherlands (ECN) (the Netherlands)	FP5	€2,351,689/€1,683,241	January 2003–December 2005
Pressurization of IP-SOFC Technology for Second Generation Hybrid Application (PIP-SOFC)	Rolls Royce (UK)	FP5		April 2003–April 2006

(Continued)

TABLE 6.9 *EU Framework Projects with Major European Participants** (Cont'd)

Project title	Major project participants	EU Framework Program (FP)	Total project cost; project funding (€)	Dates
Pressurization of IP-SOFC technology for second-generation hybrid application - target action A (PIP-SOFC)	Rolls Royce (UK)	FP5	€9,777,061/€2,053,661	November 2006–November 2015
Realizing Reliable, Durable, Energy Efficient and Cost Effective SOFC Systems (REAL-SOFC)	Fraunhofer	FP6	€18,259,430/€8,999,000	February 2004–December 2008
Fuel Cell Power Trains And Clustering in Heavy-Duty Transports	Fraunhofer	FP6	€12,559,523/€7,943,597	April 2005–March 2008
Engineering of Balance of Plant for Pressurized SOFC Hybrid Systems	Rolls Royce (UK)	FP6	€0/€/1,023,951	January 2007–December 2010
Toward a Large SOFC Power Plant (Large-SOFC)	VTT (Finland)	FP6	€11,026,326/€5,799,636	January 2007–December 2009

* *To portray the individual participants' characteristics in EU FP participation, some major EU FP projects recur in the table.*
Note: The table was created based on "Results," Advanced Search, http://cordis.europa.eu/search/index.cfm?fuseaction=proj.resultList&q=832797EE6DDBB923EB277234C680D081&type=adv&perpage=10&page=1, CORDIS Services, no date, accessed April 2010.

conservative operating conditions of 0.38 W/cm^2. In 2010, the standard cells had dimensions of 12 cm × 12 cm with current densities of 0.22 A/cm^2 at 750 °C. Cell stacks had been tested for more than 12,000 h, including 20 full thermal cycles with a degradation of 0.6 percent/1000 h.[353]

6.4.1.4. Topsoe Fuel Cell Achieves a Number of Milestones

In 2006 and 2007, Topsoe Fuel Cell had a number of significant technological and business successes. In 2006, it was selected to develop a compact, lightweight, 2 kW class SOFC system fueled by logistic fuel JP-8 for mobile applications for the US military. The project was led by United Technologies Research Center (UTRC) under a multiyear program funded by the Air Force Research Laboratory (AFRL) and the Army Tank Automotive Research, Development and Engineering Center (TARDEC).[354] According to a 2009 publication, a demonstration unit which Topsoe delivered produced a maximum gross power of 1.7 kW on JP-8 (2 kW on hydrogen) yielding a power density of about 50 W/kg.[354] The next generation, due to be operational later in 2009, was expected to have a power density in excess of 100 W/kg, according to the same source. The source indicated that continuing development would lead toward integration of an SOFC power system into an unmanned aerial vehicle to provide primary power, with a demonstration flight scheduled for 2010.

In September 2006, Topsoe Fuel Cell in collaboration with Wärtsilä demonstrated operating results for a 4 kW SOFC unit that were in line with DOE's SECA Phase I requirements.[355] Topsoe supplied the SOFC stacks and Wärtsilä integrated them into a complete system. The system ran on natural gas from the Finnish natural gas grid and was equipped with a fuel processing system that included a Haldor Topsoe catalyst. The SOFC system had shown an availability figure well above the SECA target of 80 percent. Degradation had been on the order of 1 percent/1000 h, well below the SECA Phase I minimum requirement of 2 percent/500 h.

- Topsoe indicated that the test had shown the advantages and reliability of Topsoe stack design and froze its first-generation stack design. Topsoe indicated that it would soon initiate mass production of stacks based on this design.

Also, in 2006 and 2007, Topsoe Fuel Cell and Wärtsilä were selected to work on two major EU projects:

- In September 2006, Topsoe, Wärtsilä, and Haldor Topsoe were provided with a €4.8 million ($6 million at the 2006 exchange rate of $1/€0.8) grant by the EU LIFE Environment program to develop a 50 kW SOFC system.[356] The program, whose total budget was €13.4 million ($16.75 million), was titled "Development and Demonstration of Manufacturing and Operation of Clean and Efficient Power Generation Based on Solid Oxide Fuel Cells" ("DEMO SOFC"). It was scheduled to run from October 2006 to September 2010.
  - A major part of the project included planning, design, construction, and running-in of a new cell and stack pilot production Topsoe facility at Lyngby, Denmark. The facility was completed in December 2008 and officially inaugurated in April 2009.[357] The new facility signified a transition from lab-based to semi-industrial manufacture, featuring all the necessary operations—ceramic powder processing, continuous tape casting, screen printing, spray painting, and sintering—to complete stack modules.
  - The design capacity was 5 MW/year, or about 150,000–200,000 cells and 2000 stacks, sufficient for both internal development and expanded test and

demonstration activities with new partners and market segments.[358]
- According to current news report, Topsoe Fuel Cell would be making a CHP unit of the size of a small refrigerator, Topsoe PowerCores, for residential use to be available in 2015 with the goal to get the price to the \$10,000–\$15,000 range.[359]

- In January 2007, Topsoe and Wärtsilä were among several SOFC developers to work on the EU FP6 project, "Toward a Large SOFC Power Plant (LARGE-SOFC)," to develop SOFC units with the potential for several hundred kilowatts to several megawatts (January 2007–December 2009).[360] In late 2009, Topsoe delivered 24 Class 3 stacks to Wärtsilä to build a 50 kW system for the EU FP6 Large-SOFC project.[361] In February 2010, Topsoe delivered 24 Class 1 stacks to Wärtsilä to build a WFC20 unit, which was a 20 kW system. It was the world's first SOFC unit that ran on landfill gas. The WFC20 unit operated for more than 1500 h, successfully concluding the first phase of its validation program.[362] Wärtsilä delivered the WFC20 CHP unit to the Vaasa Housing Fair site in Suvilahti, Helsinki, Finland, for further demonstrations.[363]

Beyond all these achievements, Topsoe had been involved in an ambitious, 6-year project to install and demonstrate a large number of mCHP units at Danish residences. In 2006, the Danish government provided a grant to the 6-year Danish mCHP program (2006–2012), which was financed by a consortium of nine participating partners, including Topsoe, IRD, as well as seven other power, energy, and fuel cell technology companies. The project was to demonstrate Topsoe's SOFC and IRD's PEMFC in 1 kW mCHP units. The government grant was €6.7 million (about \$8.3 million at the 2006 exchange rate of \$1/€0.8), 30 percent of the total, while the remaining €16 million (about \$20 million) was financed by the consortium. The large share of private-sector financing indicated that the consortium's nine partners were convinced that the fuel cell technology was practical and considered the project to be a good investment opportunity for the long term.[364]

- In the program, Topsoe's SOFC cells were integrated into a 1 kW mCHP prototype by Dantherm, a Danish company specializing in customized solutions for fuel cell technology.[365] The unit was tested at a third party laboratory in 2007.[364] The prototype produced a power output of 1.07 kW with a stack volume power density of 2.4 kW/L (0.35 W/cm^2) at an average temperature of 650 °C, using a cross flow configuration with hydrogen fuel. The stacks were tested successfully for more than 13,000 h.[366]
- This phase was deemed a success. In the second phase, two SOFC field test units were to be installed in the Danish town of Sonderburg in October 2009.

Topsoe and Dantherm continued to improve the mCHP unit and install and test a small number of units in selected private homes in 2010.[364] Between 2010 and 2012, a total of 100 SOFC and PEMFC CHP units had been installed.

6.4.1.5. *But No discernible Commercial Success*

Today, Topsoe Fuel Cell is marketing stacks ranging from 1 kW to more than 10 kW for stationary and nonstationary applications, including APUs, mCHP units, and distributed generation, according to its brochure.[367] They include natural gas and diesel based PowerCore products in the low power range (1–5 kW) and larger stack modules reaching capacities of 10 kW and above. It is not clear, however, whether the products are cost competitive relative to other technologies or how successful Topsoe has been in its marketing activities.

Topsoe is a wholly owned subsidiary of Haldor Topsoe, which is a privately owned company.[368] Topsoe receives its core funding from Haldor Topsoe as well as significant contributions from public sponsors both in Denmark and the EU.[369]

- Indeed, Topsoe has significant government financial support. According to a press report in June 2010, Topsoe and Risø DTU received a grant of €7.3 million from the Danish Energy Technology Development and Demonstration Program (EUDP) to continue to develop mCHP prototypes in 2012 for the Danish Micro Combined Heat and Power project.[370]

6.4.2. Finland

Finland started its SOFC development program fairly recently. In 2002, Finland implemented a 5-year national SOFC project with the modest goals of acquiring basic SOFC know-how in selected areas and building an infrastructure that could carry out basic SOFC R&D.[371] In the national FINSOFC project (2002–2006), the VTT Technical Research Center of Finland was charged with leading a team of industry companies to build a grid-connected 5 kW SOFC laboratory demonstration plant, incorporating Forschungszentrum Jülich 1 kW stacks.[372] The total budget for the 5-year project was about €5 million (about $4.7 million at the 2002 exchange rate of $1/€1.06) and was jointly funded by the National Technology Agency of Finland (40 percent), VTT (20 percent), and industry (40 percent).[371] According to VTT in 2005, the project was highly successful. VTT stated that starting from zero in 2002, Finland by 2005 had built its SOFC capabilities to such an extent that it was able to participate in EU FP projects with leading European SOFC developers on an equal basis.[372] As a result, Wärtsilä, a Finnish manufacturer of maritime diesel and gasoline engines, has emerged as one of Europe's leading SOFC developers.

6.4.2.1. Wärtsilä Starts SOFC Development in 2000 But Soon Chooses to Outsource SOFC Stacks

Wärtsilä, the leading global ship power supplier and a major provider of solutions for marine power generation, started developing SOFC technology in 2000.[373] In 2002, Wärtsilä joined Finland's national SOFC development program FINSOFC in 2002 to further develop its SOFC capabilities. In August 2002, Wärtsilä entered a cooperation agreement with the Danish technology company Haldor Topsoe to start a joint development program in SOFC technology.[374] Because the FINSOFC project chose not to pursue the domestic development of SOFC cells and stacks but to borrow foreign technology, Wärtsilä likely chose to reach out to its Danish neighbor for SOFC technology. In the agreement with Haldor Topsoe, therefore, Wärtsilä decided to integrate Topsoe's SOFC stacks into distributed power plants and marine propulsion systems rather than develop its own. The agreement aimed to commercialize efficient and cost competitive SOFC products with power outputs above 200 kW for distributed power generation and marine applications.

Wärtsilä also actively took advantage of the EU FP sponsorship, including METHAPU, Large-SOFC, and DEMO-SOFC projects:[375]

- The EU FP6 project "Validation of a Renewable Methanol Based Auxiliary Power System for Commercial Vessels (METHAPU)," ran from November 2006 to December 2010. In the project, Wärtsilä coordinated the international consortium research and validated marine applications of a 250 kW SOFC unit fueled by methanol.[376] Research partners include Lloyd's Register, Wallenius Marine, University of Genoa, and Det Norske Veritas. Validation of the technology was carried out with a 20 kW SOFC test unit installed on board a car carrier (the *Undine*), owned by Swedish Wallenius Lines and managed by Wallenius Marine[377]. The *Undine* sailed from the German port of Bremerhaven in May 2010. From there it headed for the United States, via Sweden and the United Kingdom. The on-board validation tests were to address performance and emissions of the whole SOFC application.[376]

- In January 2007, Wärtsilä participated in the 3-year EU FP6 Large-SOFC project. Total budget of the project was €11 million (about $15 million at the 2007 exchange rate of $1/€0.73), and EU funding was about €6 million.[378] The project was coordinated by VTT and participants included Wärtsilä, RR, Topsoe, Jülich, Università di Genova, Bosal Research, Verteco, and Inmatec Technologies. The project pursued two lines of development: a pressurized SOFC/gas turbine hybrid system and a nonpressurized CHP unit.[379] The final stage, expected to be reached at the end of the decade, was the verification of a 50 kW CHP unit constructed by Wärtsilä using a 50 kW planar SOFC stack assembly manufactured by Topsoe.[379] The long-term goal for RR was the development of a 1 MW pressurized SOFC power unit based on RR's whole ceramic stack assembly. The plant would consist of four 250 kW stack assemblies.
- Wärtsilä's role in the 2006–2010 DEMO-SOFC project was to construct, test, and operate a 50 kW demonstration unit that used the first cells and stacks manufactured at the new Topsoe pilot production unit. The project was sponsored by the EU LIFE Environment Program.[355] Haldor Topsoe, the mother company of Topsoe Fuel Cell, was the project administrator.

6.4.2.2. *Wärtsilä Optimistic about Commercialization of WFC20 and WFC50 Units*

Wärtsilä has allocated significant resources for development and operation of its WFC20 (20 kW) and WFC50 (50 kW) validation units. In the three projects discussed above, Wärtsilä was responsible for the development and demonstration of one WFC20 and two WFC50 SOFC units. In the EU FP Large-SOFC and Demo-SOFC programs, Wärtsilä developed both biogas and natural gas–fueled versions of the WFC50 power unit. (Compared to WFC20 units, the nominal stack power was increased from 1.2 kW up to 3 kW by increasing the cell area from 12 cm × 12 cm to 18 cm × 18 cm.) Wärtsilä clearly is optimistic about the commercialization potential of its WFC products. According to a recent media report, Hitachi Zosen Corporatioin and Wärtsilä have planned to jointly develop fuel cells for commercialization.[380] Hitachi Zosen aims to offer by 2014 a 50 kW SOFC unit with twice the power-generation efficiency of residential-use fuel cells.

In an interesting development, in June 2011, Wärtsilä announced that it signed a cooperative agreement with Versa to develop and integrate Versa's SOFC technology into Wärtsilä products.[381] A key target was to develop CHP products for distributed energy and marine markets. Wärtsilä had launched similar pilot projects using fuel cell technology supplied by Topsoe Fuel Cell in Denmark, and it indicated that this cooperation would continue as planned.

6.4.2.3. *But Technology Yet to be Validated*

Wärtsilä likely will need some time before their WFC units prove their commercial viability. The WFC20 system that operated with landfill gas was tested for 3338 h.[382] Peak efficiency was measured as 47 percent (LHV). As of April 2011, Wartsila had developed two natural gas (NG) and biogas (BG)-operated WFC50 units, which were in laboratory validation.[383] The company's project to develop its first pre-commercial WFC50 MkII unit for stationary and marine use was also underway.

- Moreover, Wärtsilä's WFC units rely on the viability of Topsoe and Versa's SOFC cells and stacks, which have not yet fully demonstrated commercial competitiveness.

6.4.3. Germany

Germany was active in tubular SOFC development in the 1960s, stopped around 1975,[342] and restarted SOFC efforts in 1988, focusing on planar SOFC designs. German planar SOFC development has been supported via close coordination between the EU FP and the German national R&D program. The most important players in planar SOFC R&D in Germany include the federal and state co-funded research institute Forschungszentrum Jülich (Jülich) and a number of industrial corporations, including Asea Brown Boveri AG (ABB), BMW, Dornier, and Siemens. None of the industrial corporations are engaged in SOFC development today, however.

6.4.3.1. Asea Brown Boveri (ABB): Started SOFC R&D in 1968, Ended in 1993

ABB[v] conducted SOFC development for a total of 12 years. For about 8 years between 1968 and 1976, ABB (under the name of Brown Boveri & Cie, or BBC), had an extensive program on tubular SOFCs,[384] centered on the research group at its Heidelberg research facility.[385] BBC, however, had difficulty in reducing the thickness of the cell and thus was unable to increase power density.[386] It stopped its SOFC efforts in 1976. The tubular design at that time was believed not be cost effective.[384]

In 1990, ABB resumed SOFC development, focusing on planar SOFC development. From 1990 to 1993, ABB participated in three EU FP projects:

- In 1990, ABB started to work in the "Development of An Advanced 1 kW Solid Oxide Fuel Cell Prototype (ASOF)" project (1990–1991) under EU FP2. ABB created a new planar SOFC concept, optimized by using new materials for electrodes and electrolytes and new fabrication methods.[384] ABB's goal was to develop 25–40 W planar SOFC cell based on a new cell design combining the advantages of flat plate and tubular designs. A prototype demonstration of an SOFC generator in the 500 W class was prepared. The work included the development and characterization of novel ceramic materials, the construction of SOFC modules and a test generator system, tests and design studies for methane reforming in the system, mathematical modeling of the SOFC stack, system studies, cost calculations, and market research. ABB's partners included the Foundation for Technical and Industrial Research at the Norwegian Institute of Technology (Norway), Imperial Chemical Industries (United Kingdom), British Gas, Risø (Denmark), and TNO Milieuwetenschappen en Energietechnologie (the Netherlands).
- Also in 1990, ABB started to work in the "New Manufacturing Technologies for Advanced Solid Oxide Fuel Cells" (1990–1992) under EU FP2. ABB worked to develop low cost, mass production ceramic processing technologies, such as extrusion, screen printing, and tape casting.[387] ABB successfully extruded a 100 mm × 200 mm × 17 mm support structure, which had a thermal expansion coefficient optimized to closely match that of the active cell. ABB's partners included Nederlandse Organisatie voor Toegepast Natuurwetenschappelijk Onderzoek, Risø, and Imperial Chemical Industries.
- The last EU SOFC project in which ABB participated was a follow-on to the earlier FP2 project under the same title

[v] ABB, short for ASEA Brown Boveri, is a Swiss-Swedish multinational corporation headquartered in Zürich, Switzerland, operating mainly in the power and automation technology areas. ABB, Asea Brown Boveri, was the result of a merger between Asea AB of Sweden and BBC Brown Boveri Ltd. of Baden, Switzerland in 1988. ABB has a large Germany Subsidiary in BADEN-WÜRTTEMBERG KARLSRUHE Heidelberg, Germany.

"Development of An Advanced 1 kW Solid Oxide Fuel Cell Prototype." In this project (1992–1993), ABB worked to develop a 30 W SOFC module with a power density of 1 kW/m^2,[384] which was to be a building block for a 1 kW SOFC unit. In addition, the BoP system to accommodate the ceramic modules was to be designed. ABB's partners were British Gas, Nederlandse Organisatie voor Toegepast Natuurwetenschappelijk Onderzoek, Imperial Chemical Industries, and Risø.

It is clear that none of the projects delivered enough results to satisfy ABB. ABB left the SOFC field after the EU projects ended in 1993.

6.4.3.2. BMW: *Begins SOFC R&D in Late 1990s, Ends in Late 2000s*

BMW was interested in SOFCs for APU applications for about 10 years but never engaged in SOFC development directly. Instead, it commissioned others to do the work, including Delphi, DLR (German Aerospace Center), Global Thermoelectric, and Jülich:

- In April 1999, BMW and Delphi Automotive Systems announced that they had signed an agreement to develop an APU for gasoline engines using SOFCs as well as PEMFCs.[388] They developed an SOFC unit as an APU for light duty vehicles.[389] The SOFC was made by Global Thermoelectric, and the first vehicle to use the system was demonstrated in spring 2001. It appears that BMW stopped working on SOFC APUs sometime in the late 2000s, however. Today BMW is focusing on PEMFC APU applications with Delphi.
- In 2001, BMW started a series of projects to develop SOFC APUs with Jülich, the German Aerospace Center (DLR), and a number of automobile companies. BMW and the Federal Ministry of Economics and Technology (BMWi) jointly funded the projects, which included the Zellen und Stacks (ZeuS) project (2001–2003), the ZeuS-II project (2004–2006), and the ENSA project (2005–2008).[390] In 2005, DLR reported that it developed plasma sprayed cells that demonstrated a power density of about 210 mW/cm^2 at 0.7 V at 900 °C using a simulated reformate gas (H_2/N_2) and air.[391]

BMW perhaps dropped out of SOFC APU development when these projects ended.

Nonetheless, BMW could likely reconsider the use of SOFC APUs for its passenger cars if viable SOFCs are developed.

6.4.3.3. Dornier: *Begins SOFC R&D in 1988, Ends in 1995*

Dornier started SOFC development in the late 1980s. It worked on planar SOFC development for about 8 years under three EU FP projects from 1988 to 1995. The projects were implemented in close coordination with German national R&D projects.[392]

- In the "Modeling of the Mass and Energy Balances of SOFC Modules" project under EU FP1, Dornier worked to develop theoretical-numerical tools to model and calculate mass and energy balances, especially with respect to the transport of waste heat out of the system.[393] Such tools were expected to allow the assessment of the feasibility of different designs with respect to the problems of thermal controllability and eventual scale-up.
- Dornier worked to develop an all-ceramic planar SOFC in the "Operational Tests of SOFC Modules and Use of SOFC as Chemical Reactors" project (1990–1993) under EU FP2.[394] The Dornier SOFC was a conventional planar design with a thick ribbed lanthanum chromite interconnect plate.[386] This project sought to develop small planar SOFC cells up to 100 W and investigate operational behavior with regard to internal reforming.[394] Dornier partnered with the Foundation for Research and Technology-Hellas, Greece, in this project.
- From 1993 to 1995, the EU sponsored the "SOFC Integration Techniques for Up-Scaled

SOFC Modules" project to assist Dornier to upscale SOFCs to kilowatt-size systems.[392] Dornier worked in partnership with Cookson Group plc (United Kingdom) and the Foundation for Research and Technology-Hellas (Greece) on the project. Dornier cells were planar SOFCs with ceramic bipolar plates. Developing stacks with suitable scaling between the cells was a key problem. By 1995, nevertheless, Dornier was working on a 3–5 kW stack.[386] It successfully developed stacking techniques suitable for scale-up, developed manifolding and joining methods, and manufactured test stacks. Single cells were scaled up to 10 cm × 10 cm, and ten-cell stacks were operated successfully for over 1000 h. Current densities were about 200 mA/cm^2 at 0.7 V. Dornier believed that this performance could be improved upon in the future by optimizing the cell geometry.

Dornier clearly was not successful in optimizing the cell geometry and had terminated their SOFC development by the end of the 1990s.[395]

6.4.3.4. *Forschungszentrum Jülich*

The Jülich Research Center, established in 1956, is one of the largest and most prestigious research institutions in Europe and has played a wide-reaching leadership role in SOFC development globally. Jülich started working on SOFC development in 1989. It investigated all aspects of planar SOFCs, from the fundamental research to system development, including materials, processing, coating techniques, stack and system development, assembly, testing, modeling, and simulation.[396,397] Jülich is supported by the German Federal Ministry of Economics and Technology (BMWi) and the North Rhine-Westphalia federal state government[397] and often works in close association with the DLR in Stuttgart.[488] Jülich has about 160 fuel cell researchers on staff—the largest institutional fuel cell research team in Europe—about half the fuel cell staff work on the SOFC.

Since the mid-1990s, Jülich has focused on ASCs.[397] Jülich believed ASCs will lead to a reduction in the operating temperature in SOFC. Jülich viewed that ASCs would transfer the mechanical strength of the cells from the electrolyte to the anode.[398] This would allow a reduction in electrolyte thickness, which would decrease resistance of the electrolyte, enabling a lower operating temperature and thereby lowering material and manufacturing costs. Jülich believed that the operating temperature of these cells could come down to 800 °C, 200 °C lower than SOFC operating temperatures at that time.

6.4.3.4.1. JÜLICH ESTABLISHES A SERIES OF WORLD RECORDS IN SOFC DEVELOPMENT

In the early 2000s, Jülich successfully developed several ASC stack designs and constructed and tested several kilowatt class stacks, while indeed lowering the operating temperature. Jülich set a number of world records in developing ASC stack technology, many of which became global standards that were emulated by SOFC developers worldwide:

- In January 2001, Jülich achieved a breakthrough by developing an SOFC stack made of 25 cm × 25 cm cells, which were the world's largest, that generated 1.6 kW, the world's highest power output, using hydrogen and air.[399] Jülich reduced the thickness of the electrolyte to one-tenth the original thickness and reduced operating temperatures by about 200 °C. Jülich also clearly tried to get away from the expensive special materials that were then necessary for high temperature fuel cells.
- In April 2002, Jülich developed an SOFC stack of forty 50 cm × 50 cm cells that generated 5.4 kW using methane, operating at an average temperature of 850 °C, a new

record in terms of both performance and cell size.[400]

- In June 2004, Jülich set another world record.[401] It developed an SOFC stack, which could be classified as a 15 kW-class stack. The stack consisted of 60 cells that generated 13.3 kW using hydrogen (90 percent H_2 and 10 percent H_2O) at a mean operating temperature of 760 °C. The power density was 0.61 W/cm^2.[402] Fuel utilization was 51 percent. With a methane–hydrogen mixture (6 percent H_2, 31 percent CH_4, 63 percent H_2), the stack delivered 11.9 kW. The minimum operating temperature was 598 °C and the maximum temperature 809 °C. Power density was 0.55 W/cm^2 and fuel utilization was 7 percent. The stack ran for over 1100 h in continuous operation with 3 percent performance degradation.[401]

In addition to SOFC cells and stacks, Jülich developed many innovative materials, systems, and processes. The invention of a novel metal alloy stands out among all of these, however. In October 2003, Jülich invented a high temperature ferritic stainless steel that had excellent corrosion resistance at high temperatures in anode and cathode gas, a low rate of chromium vaporization, ease of working and processing, low coefficient of thermal expansion, and good electrical conductivity at the oxide layer.[403] The alloy was well suited for interconnects for SOFCs[404] and quickly became a very popular material for this purpose. The metal alloy went on the market almost immediately after it was invented. An industry partner, ThyssenKrupp, announced in September 2003, a month earlier than Jülich's own press announcement, that it would supply the metal alloy under the trade name "Crofer 22APU."[405] The Crofer22APU also played a major role in the BMWi-funded projects to develop APUs for automobiles, in which light-SOFC stacks were developed using the metal alloy for interconnects.[406]

- Jülich also made significant progress in improving thermal barrier coatings with the aid of new diagnostic methods,[407] improving reactors and systems for reforming and desulfurization, and cleaning fuel gas of middle distillates (diesel, kerosene, and heating oil), as well as advancing direct methanol fuel cell (DMFC) technology.[408]

6.4.3.4.2. JÜLICH WIDELY SOUGHT AFTER FOR ITS SOFC EXPERTISE

Jülich has become an acknowledged global leader in SOFC technology. The EU and its member countries as well as companies around the world have sought Jülich's expertise to build their own national SOFC capabilities:

In 2004, the EU selected Jülich to manage a key EU FP6 project "Realizing Reliable, Durable, Energy Efficient and Cost Effective SOFC Systems" (Real-SOFC Project) that focused on reliability and durability of SOFCs. The project goal was to solve the degradation problem in planar SOFC cells and stacks and to develop reliable, durable SOFCs. A consortium of 26 major European SOFC players, including industrial companies, research centers, and academic institutions, were given access to Jülich's SOFC cells and stacks as well those of other major manufacturers to find solutions to reduce the performance loss.[409] The consortium partners were allowed to use their findings to invent new cells and stacks—outside the project in order to secure IPR protection—which were then fed back into the project for testing and evaluation.[410] The EU funded the 4-year program at €18 million (about \$22.5 million at the 2004 exchange rate of \$1/€0.8). The project concluded in December 2008.[411] It resulted in development of improved anode, cathode, and interconnect materials, which were demonstrated successfully in stacks of HEXIS, HTceramix, RR Fuel Cell Systems, and the Topsoe Fuel Cell.

In April 2010, Jülich announced that the SOFC stack developed in the Real-SOFC Project had been running for more than 20,000 h—almost 3 years.[412] The cells provided an average power of 0.4 W/cm^2 over the electrolyte surface, about double what is currently the case in commercial systems, according to Jülich. This stack likely was one of the two stacks that reportedly reached more than 15,000 operating hours in May 2009, then a world record in terms of longevity.[413] Jülich indicated in 2009 that the EU FP Real-SOFC Project enabled the development of more efficient SOFC cells with thin functional layers, which, with the use of Jülich's special Crofer steel and the ITM steel (developed by Plansee), achieved the extended service life.

Jülich also served as coordinator in a couple of other EU projects, in which the research center served as the suppler of the interconnects and contact layers to project partners and as the integrator of European expert knowledge on mathematical modeling to create generic fuel cell modeling tools:

- In 2001, the EU implemented a 4-year, €4 million (about \$3.6 million at the 2001 exchange rate of \$1/€1.12) "Component Reliability of Solid Oxide Fuel Cell Systems for Commercial Operation (CORE-SOFC)" Project (2001–2004) under FP5 and put Jülich in charge.[414] The project included a number of prominent European SOFC developers, including ECN, Haldor Topsoe, Risøe, and RR. The project aimed to increase reliability and durability of the SOFC, which was still too low for long-term operation. The specific goal was to improve the reliability of planar SOFCs to achieve a degradation loss of less than 0.75 percent/1000 h. The project work was conducted based on identically fabricated cells, long-term tests (1000–5000 h), and a careful microstructural analysis of the tested units. Research concentrated on the influence of operating parameters and materials on contact layers and interconnects. For example, the degradation rate for a Ni-YSZ anode-supported SOFC was found to be dependent on the test procedure as well as on the fuel gas composition and current density, and if the fuel gas composition and cell temperature could be adjusted optimally, cells could be run at current densities up to 1 A/cm^2 for more than 300 h without significant degradation.[415] In addition to the electrochemical determination of degradation, thermal cycling of cells with model stacks also was carried out.
- In 2005, the EU sponsored a 3-year, €2.7 million (about \$3.4 million at the 2005 exchange rate of \$1/€0.8) "Generic Fuel Cell Modeling Environment (GenFC)" project (2005–2008) to bring the available expert knowledge on fuel cell modeling under the Jülich's leadership. The goal was to create a modeling tool for the design and manufacturing process for fuel cell and fuel cell systems and make the tool available to all the fuel cell systems developers. The user would be able to choose a modeling tool from a set of different types of fuel cells. The project participants were Aixprocess Hufschmidt (a German process and flow engineering company), Fluent Deutschland (a German expert in simulation and flow modeling), REDHADA SL (a Spanish IT company), Swiss Federal Laboratories for Materials Testing and Research (EMPA) (a research lab with computational modeling expertise), and Technische Universitat Graz (a Greek technical university with mathematical modeling and numerical simulation expertise). The fuel cell modeling tool and modeling database have been available to the public since April 2008.[416]

Beyond these major accomplishments, Jülich provided its expertise to the following companies and institutes:

- Finland's national SOFC development project, FINSOFC, heavily relied on a Forschungszentrum Jülich 1 kW stack.[372] According to a presentation made in 2005, the FINSOFC project made rapid progress in no small way due to Jülich's technology.

- South Korea also sought Jülich's SOFC technology. According to a paper published in 2006, the Korea Institute of Energy Research (KIER) designed and constructed a pressurized 5 kW SOFC gas turbine hybrid system, incorporating Jülich's 5 kW stack.[417] Undoubtedly impressed with Jülich's SOFC stack performance, KIER sought to strengthen the cooperative relationship with Jülich. In April 2009, KIER and Jülich signed a 5-year partnership agreement to develop SOFC as well as DMFC technology into practical applications.[418]
- As early as 1997, Global Thermoelectric, a Canadian manufacturer of thermoelectric generators, signed a licensing agreement with Jülich to use its SOFC technology.[419] After a series of sales and mergers, Global's SOFC unit became part of Versa, and so Jülich's SOFC expertise and technology are now imbedded in this North American company.
- In 2006, Compagnie Saint-Gobain, a French multinational high-performance materials company, entered into a licensing agreement with Jülich to use Jülich's SOFC know-how.[420] Saint-Gobain has announced that it plans to begin developing SOFC products in conjunction with Jülich.
- Also in 2006, Jülich's coating technique for wet powder spraying for tubular SOFCs was transferred to Siemens, Erlangen.[421]
- Strategic cooperation also was established with the Dutch research institution ECN for fuel gas production for fuel cell applications.
- In 2007, CFCL of Australia signed a collaboration agreement with Jülich to conduct R&D on SOFCs, including materials development and accelerated lifetime testing and modeling (Table 6.10).[397,409]

6.4.3.4.3. BUT EXPERIENCES TECHNOLOGY BARRIERS

Jülich, despite its remarkable successes, appears to have been faced with technology barriers that many others have faced. Breakthroughs are yet to occur. Jülich thus has not been able to transition its SOFC technology into commercial applications. Two episodes illustrate the challenges Jülich faces today:

Jülich's APU Development. In 2001, Jülich was involved in the German Zellen Und Stacks project (ZeuS, referenced in the section on BMW) to develop SOFC APUs for on-board power supplies in cars.[390] Lightweight SOFC stacks were to reduce fuel consumption and pollutant emissions and were considered an ideal APU for diesel vehicles. The project was carried out between 2001 and 2003 and was financed by an industrial consortium led by BMW and supported by BMWi, and the DLR. Jülich was involved as a research partner. The project was a success and as a result, a second phase, ZeuS-II, was kicked off in February 2004.[106] In 2005, Jülich and DLR began building a lightweight SOFC stack[422] and in 2006 succeeded in constructing a 20-cell stack that was thermally cycled 100 times under fast-start conditions (less than 1 hour starting-up time).[407]

- According to a 2007 Jülich publication, a 2 kW unit (pre-reformer and stack) did not operate for long. Pressure differences between the anode and the cathode side meant that leakage was occurring at an unacceptable rate and further operation of the stack became impossible. According to Jülich's 2009 publication, although solutions for the production of commercial stacks for stationary and mobile applications in APUs were becoming increasingly apparent, performance was still limited and not yet fully in line with the expectations of today's consumers.[424] In order to improve the robustness of stacks, new concepts were necessary, stated Jülich.

TABLE 6.10 EU FP6 Real-SOFC Project: List of Project Partners

Participant number	Participant name	Country
1	Forschungszentrum Jülich (Coordinator)	Germany
3	Rolls Royce Fuel Cell Systems	United Kingdom
4	Ugine-Alz (Groupe Arcelor)	France
7	Commissariat a l'energie Atomique	France
8	University Court of the University of St Andrews	United Kingdom
9	Deutsches Zentrum fur Luft- und Raumfahrt e.V.	Germany
10	EBZ Entwicklungs- und Vertriebsgesellschaft Brennstoffzelle	Germany
11	Energy Research Center of the Netherlands	The Netherlands
12	Electricite de France	France
13	Swiss Federal Laboratories for Materials Testing and Research	Switzerland
14	ENERGOPROECT AD – Science Research and Technological Institute	Bulgaria
16	Gaz de France	France
18	H.C. Starck	Germany
19	Topsoe Fuel Cells	Denmark
20	HTceramix/SOFCPower	Switzerland
21	The Imperial College of Science, Technology and Medicine	United Kingdom
22	FOUNDATION FOR RESEARCH & TECHNOLOGY HELLAS – Institute of Chemical Engineering & High Temperature Processes	Greece
25	Plansee SE	Austria
27	Risø – DTU – National Laboratory for Sustainable Energy	Denmark
28	Stiftelsen	Norway
29	Hexis AG	Switzerland
32	University of Birmingham	United Kingdom
33	University of Chemical Technology and Metallurgy	Bulgaria
37	VTT – Technical Research Center of Finland	Finland
38	Wärtsilä Finland Oy	Finland
39	University of Genoa	Italy

Source: Rolf Rosenberg, "The Publishable Executive Summary of the 1st Periodic Activity Report: M1-M12: LARGE-SOFC Towards a Large SOFC Power Plant, Project no. 019739," VTT Technical Research Center of Finland, 8 May 2008.

Jülich's 20 kW SOFC System Development. In June 2002, Jülich announced its plans to develop a 20 kW SOFC system. At the time, Jülich had successfully set a world record in developing a stack of 40 planar cells which generated 9.2 kW with hydrogen and 5.4 kW with methane.[400] The 20 kW stack consisted of four 5 kW stack modules.[425] Jülich developed an "integrated stack module," a highly compact unit that was simple in concept and which could be used for the 20 kW system. The module comprised all hot components (operating above 400 °C) of the plant: pre-reformer, recuperator, afterburner, and stack. All the components were mounted on top of each other, comparable to a stack in design, and avoided any piping. This concept would lower heat loss, improve thermal matching, and reduce volume and weight. At the end of 2006, Jülich constructed a 20 kW facility, equipped with two dummy stacks as test units for system engineering.[421] The 5 kW stacks were installed after testing the system components in the course of spring 2007.[425]

- And yet, the 20 kW system has not yet been completed, due to some unexpected problems, including problems with gas seals[421] and electrical heaters used to bring the system to operating temperature.[390] A 2009 publication indicated Jülich was ready to operate the 20 kW demonstration plant with natural gas.[424] Nonetheless, no announcement has been made regarding its operation or the status of the initiative.

6.4.3.4.4. A GLIMPSE OF NEW HOPE: WORLD'S FIRST OPERATING LIFETIME OF 40,000 H ACHIEVED

In a surprising new development in March 2012, an SOFC stack built by Jülich surpassed an operating lifetime of 40,000 h.[631] This exceeds previous records established for commercial SOFC use in stationary applications and corresponds to a period of 5 years.

- It appears that Jülich developed at least two of these stacks under the EU FP Real-SOFC project. In May 2009, the two stacks had each reached more than 15,000 h of operating time.[413] One of the two stacks, however, was shut down after 18,000 h of operation due to failure of one cell.[632] The remaining stack has now passed the 40,000-hour mark, which is a milestone in SOFC technology.

It is reported that the Jülich SOFC stake has other advantages besides longevity. The stack has an electricity efficiency of 64 percent.[432] It is fueled by hydrogen, but other gases such as methane, could yield an even higher efficiency and allow the stake to operate with a wide range of different fuels. The Jülich stack also operates at a relatively low temperature of 700 °C, which could make it possible to use relatively inexpensive materials.

- Because of these properties, this new SOFC could be used in distributed systems, households, and industrial facilities, taking advantage of waste heat to supply electricity. It could also be used as an on-board power supply in road vehicles, ships, and aircraft.

Nonetheless, there would still be challenges ahead. The stack needs to be integrated into a larger system, which hopefully can achieve 40,000 h lifetimes under everyday conditions. The cost also will need to be optimized so that it can be commercially competitive with other conventional technologies.

6.4.3.5. H.C. Starck/Staxera/Webasto (Enerday)

The history that led to the establishment of the business alliance of H.C. Starck Ceramics, Staxera, and Enerday, is an interesting one. It involves cooperative activities in pursuit of APUs among multiple parties, culminating with the establishment of a joint venture to facilitate commercialization goals:

- In 1992, Fraunhofer Institute for Ceramic Technologies and Systems (IKTS) in Dresden commenced SOFC technology development. IKTS advanced SOFC development with various partners, by testing different project elements until complete stacks could be built.[425]
- In 2000, Webasto began working on concepts for an APU for automobiles.[425] In 2002, Webasto identified the SOFC technology as the optimum approach and started a major development program in Neubrandenburg that year. As no SOFC stack supplier was able to fulfill technical and automotive business requirements, Webasto engineers started to work on SOFC stacks from scratch.[425] Fraunhofer IKTS joined the project as a partner.
- In 1996, H.C. Starck Ceramics in Selb, Germany—a subsidiary of the H.C. Starck Group in the chemical and metallurgical industry—commenced developing specialty SOFC fuel cell powders.[426] In 2003, H.C. Starck acquired InDEC in Petten, the Netherlands, and developed itself into an SOFC component supplier.[425] (InDEC was established as a spin-off from the Energy Research Center of the Netherlands (ECN) in 1999 to market ECN's SOFC technology.)[427]
- In summer 2003, Webasto, H.C. Starck, and IKTS joined forces to develop packaged SOFC units for APU applications.[47] In 2005, Webasto demonstrated the world's first functional SOFC APU that used diesel fuel cell.[425]
- In May 2005, Webasto and H.C. Starck founded Staxera, a joint venture, to help commercialize the technology, and located it on the IKTS campus in Dresden.
- In 2007, Webasto established a wholly owned subsidiary, Enerday, based on the fuel cell business unit at Webasto.[428] In November 2009, Enerday took Staxera under its wing as a wholly owned subsidiary.[425] Enerday changed its name to New Enerday in 2010. Today, Staxera and New Enerday function as independent companies under the umbrella of Webasto.

The business alliance of H.C. Starck Ceramics, Staxera, and New Enerday comprise a virtual SOFC production line. Each company has a unique responsibility. H.C. Starck produces SOFC powders and cells, Staxera assembles SOFC stacks, and New Enerday integrates stacks into SOFC systems for APUs for automobile and mCHP applications.

H.C. Starck also supplies its SOFC components and cells to the world market; it is a nonexclusive supplier of cells and ceramic components to Staxera.[425] Staxera also supplies SOFC stacks to New Enerday[428] as well as to the world market. New Enerday integrates the stacks into fuel cell systems for the world market.[429]

6.4.3.5.1. H.C. STARCK BUILDS STATE-OF-THE-ART SOFC MANUFACTURING FACILITY IN SELB, GERMANY

With the acquisition of InDEC in 2003, H.C. Starck gained SOFC know-how as well as an extensive worldwide client base.[430] Clients included the Swiss firm Sulzer (subsequently Sulzer Hexis, and then Hexis) and Siemens.

- The demand from Sulzer, in particular, has originally led ECN to establish a pilot cell-manufacturing unit for SOFC cells in the Netherlands in 1997.[430] That unit later became part of InDEC. It manufactured SOFC components for existing clients, such as Sulzer, and also sold them nonexclusively on a commercial basis to third party stack and system developers.

H.C. Starck continued SOFC component production at the InDEC pilot manufacturing plant.[426] The products were manufactured in a batch process that required significant man power. The InDEC production area also was fairly modest, approximately 350 m^2, consisting of two batch tape casting rooms and a screen print and sintering facility. In addition, InDEC had a cutting facility and a packing/characterization room. In order to reduce the cost and prepare for future cell demands, H.C. Starck clearly needed a state-of-the-art commercial-scale production facility.

In mid-2005, H.C. Starck began preparations to set up such a facility at the H.C. Starck site in Selb, Germany. The facility would have a temperature-controlled clean room in which a continuous tape-caster, an automated screen print line, and industrial-scale sintering furnaces would be installed. This facility would have an annual production capacity of more than 100,000 cell components, including cells with an active area of 100 cm^2.

- In early 2008, H.C. Starck commissioned two production lines for manufacturing SOFCs, specifically for ASCs and for ESCs, with an annual capacity of about 700,000 cells each.[431]

In January 2008, H.C. Starck gained an additional major client that assured it significant annual sales. It signed agreements with CFCL of Australia to provide a high volume supply of fuel cell components, according to a press release by CFCL and H.C. Starck.[431] Under the agreements, CFCL and H.C. Starck were to collaborate to develop advanced SOFC components, including ASCs. CFCL was to license its relevant intellectual property in order to optimize the performance of the cell. H.C. Starck was to continually improve manufacturing processes and reduce unit costs. In addition, under the supply agreement, H.C. Starck would supply fuel cells at an agreed price until December 2011, securing long-term costs for CFCL. Supply of the fuel cells was set to commence in mid-2008, with further increases in forecast volumes, and reductions in unit prices, from 2009.

- As part of CFCL's commercialization strategy, the partnership would combine CFCL's and H.C. Starck's manufacturing capabilities, allowing for mass production of high quality cells at commercially viable unit costs.
- In February 2008, CFCL began the construction of a fuel cell stack manufacturing plant in Heinsberg, Germany, and in October 2009 officially opened the facility. CFCL continued to produce fuel cells at its Melbourne R&D and pilot manufacturing facility.

In June 2010, H.C. Starck announced that it had formed a partnership with Kerafol (Keramische Folien GmbH), a developer and manufacturer of ceramic tape, and would acquire Kerafol's specialized knowledge of scandium-doped electrolytes in exchange for Starck's expertise in electrode manufacturing.[432] According to the press release, the scandium-doped zirconia used in these fuel cells is highly conductive, enabling them to have the highest output potential of all electrolyte-supported fuel cells. The result was the first high-output, scandium-doped cell with its origins in Europe. Also, in conjunction with the appropriate interconnectors, the new cells would make it possible to manufacture SOFC stacks and systems with extraordinary performance and a long service life, according to the press release.

- Scandium is a rare earth element extracted at only a few mines in Ukraine, China, and Russia.[100] The absence of reliable, secure, and stable suppliers has limited commercial applications of the metal in most countries. These reservations also limit its utility in the manufacture of fuel cells.[98]

6.4.3.5.2. STAXERA DEMONSTRATES RAPID ADVANCES IN SOFC STACK TECHNOLOGY

Established in May 2005, Staxera was based at Fraunhofer IKTS at its campus in Dresden. Despite its modest beginning, Staxera built its business presence fast. As early as in 2006, Staxera began shipping SOFC stacks to customers.[425] Between 2005 and the end of 2007, Staxera grew from 5 employees to 18. By January 2008, Staxera finished constructing of new development and production center near IKTS in Dresden and moved to the new facility. The center included labs and offices for 35 people, a complex gas supply and safety infrastructure as well as a 500 m^2 production and test hall.

- In 2007, Staxera reported that its Staxera Mk100 Stack (30 cells) demonstrated a power density of 183 mW/cm^2 and a current density of 197 mA/cm^2, with fuel utilization of 51 percent.[433] Fuel composition was 50 percent H_2, 40 percent N_2, and 10 percent H_2O.

In 2008, Staxera started to offer the Staxera 1.1 kW integrated stack modules (ISMs) to the world market,[434] which included thermal insulation, a compression system, sensors, a robust housing, and two 30-cell MK200 stacks from H.C. Starck.[435] The ISM is rated at 1.1 kW at Staxera's reference operating condition of 75 percent gas utilization, a stack voltage of 42 V (0.7 V/cell), and a fuel gas of 40 percent hydrogen in nitrogen.[429]

The ISM allows for rapid system integration by providing clearly defined interfaces for gases, sensors, and electrical current. All critical operating parameters such as compression loading, thermal insulation, and temperature measurement are addressed within the module. Using this approach Staxera guarantees 5000 h of operation and 20 thermal cycles when the ISM is operated within predefined parameters.

Since the ISM was first released onto the market in 2008, Staxera has produced about 50 units and gained thousands of hours of operating experience under a variety of conditions.[434]

- Staxera has supplied SOFC stacks to industrial customers, such as EBZ, New Enerday, and Vaillant, as well as research institutes such as ECN, CUTEC-Institut GmbH (Clausthal-Zellerfeld, Germany), and the National Research Council (NRC) Canada.[429]

In 2009, Staxera achieved a major milestone when its Mk200 stack achieved 10,000 h of operation at full power. Total degradation over this period amounted to 4 percent, or a rate of 0.4 percent/1000 h.[436] The lifetime test also included 25 thermal cycles. The stack delivered 570 W at a nominal operating temperature of 850 °C and 75 percent fuel utilization; it used a reformate consisting of 40 percent hydrogen and 60 percent nitrogen. Cathode air was supplied using blowers, drawing air from the laboratory atmosphere. Staxera expected the end product would have a lifetime of between 20,000 and 40,000 h.[429] Also in October 2009, a Staxera Mk200 stack demonstrated 150 power and thermal cycles without evidence of degradation.[437]

In April 2010, Staxera unveiled a new higher power 1.4 kW SOFC ISM.[438] The Staxera 1.4 kW ISM contains two 30-cell Mk200 stacks using ESC4 cells from H.C. Starck.[439] The ISM is otherwise identical to the 1.1 kW ISM.

- Staxera also plans to offer 3 kW and 5 kW ISM's in 2012–2013.[429]

In November 2009, Staxera began a joint SOFC system development program with EBZ (Entwicklungs- und Vertriebsgesellschaft Brennstoffzelle mbH), focusing on SOFC systems with high electrical efficiency that used natural gas steam reforming.[440] The program initially focused on a 20 kW system with an electrical efficiency between 50 and

60 percent. The basic design was to be used for other products, such as systems which operate on biogas or landfill gas. The first phase of the program involved the construction and testing of a system demonstrator based on a peak stack power of 3 kW, using two of Staxera's ISMs. The full design power of 3 kW was planned to be demonstrated in 2010, when the 1.4 kW ISMs were expected to become available. According to an April 2010 news release, Staxera was planning to present the 3 kW system demonstrator at the Hannover Messe that month.[441]

- As of April 2010, instead of a 3 kW demonstrator, Staxera introduced a 1.4 kW ISM with a performance guarantee of 1000 h and 20 thermal cycles.[438] As of April 2011, Staxera offered for delivery an ISM with a rated power output of 1.7 kW.[442] The ISM contains two new 30-cell Mk210 stacks using ESC4 cells from H.C. Starck.[443] Operating at rated power, it has a fuel utilization of 75–85 percent and is guaranteed for 5000 h and 20 thermal cycles.
- According to Staxera's website, the company plans to release a 5 kW ISM using new MK250 technology in 2012.[444]

6.4.3.5.3. NEW ENERDAY PLANS TO INTRODUCE ITS MARKET ENTRY PRODUCT, THE 1 KW-CLASS EN-1000 LPG

New Enerday specializes in commercial SOFC systems that supply electricity and heat independent of the power grid.[445] Their fuel cell system can be powered by both conventional fuels and by alternative energy sources.

New Enerday traces its origins to the fuel cell business unit of Webasto AG, one of Germany's major automotive suppliers. Webasto builds power supplies for cars that function independently of the engine. In April 2005, it demonstrated the world's first functional model of a diesel-powered fuel cell using SOFC technology. As noted earlier, in 2007 Webasto established Enerday (now New Enerday) in order to better focus its efforts on research and development in this area.

- New Enerday is based in New Brandenburg in the state of Mecklenburg-Western Pomerania, Germany. Both the developmental headquarters and future production facilities are located there. All system components for the fuel cells are developed and tested in New Brandenburg as well. The heart of its SOFC technology, the stack, is supplied by its joint venture subsidiary, Staxera, which is based in Dresden.

New Enerday has made steady progress with its SOFC systems. In April 2009, New Enerday exhibited its LPG system EN-1000 for off-grid power applications at the Hannover Fair.[446] The system was integrated with its partner H.C. Starck and their joint venture company Staxera. In June 2009, the first completely integrated New Enerday EN-1000 LPG system demonstrated an electric output of 1 kW using liquefied petroleum gas (LPG) fuel. The functionality of the system components and the control sequences were verified in a test run.

Also in July 2009, one of New Enerday's natural gas–powered SOFC systems reached a peak value of 37 percent DC efficiency. The system operated under partial-load conditions and supplied 890 W of electrical stack power. With a fuel power input (LHV) of 2 kW, 720 W of electrical stack power still could be obtained, corresponding to an efficiency of 36 percent. The peak power of the system was 1250 W at full load. Conventional natural gas from the public grid was used as fuel. An additional water supply (for the reforming process) was not needed. New Enerday views these results to be outstanding and notes the excellent partial-load operating capability of the total system.[447]

As of June 2010, New Enerday expected that its fuel cell products would be on the market beginning in 2011[448] and hoped to commercialize scaled-up products in the future. New Enerday believes that more powerful stacks have the potential to increase the SOFC's efficiency to over 50 percent and hopes that their systems will be generating power at around 5 kW in the near term. That will allow for a wide range of new, decentralized applications, such as supplying power during periods of peak demand.[449]

- New Enerday has been conducting product development of compact, high efficiency SPFC battery charging systems for mobile applications. In June 2011, the company reported that it had tested an innovative 500 W system for more than 1500 h with a gross efficiency of 38 percent.[450] In March 2012, New Enerday field tested two EN500P propane systems.[451] The systems run automatically without intervention. Both systems met or have exceeded the goals of 500 W maximum power and efficiency of 25 percent, reportedly the highest efficiency in this power range.

6.4.3.5.4. SIGNS OF COMMERCIAL SUCCESS YET TO BE SEEN

It is too early to project the success or failure for H.C. Starck, Staxera, and New Enerday SOFC products. So far none of the three have demonstrated a measurable degree of commercial success:

- Although H.C. Starck appears to have steadily increased the production of SOFC components and cells, self-sustaining, long-term commercial success based on SOFC products still is not apparent. H.C. Starck's success as an SOFC materials supplier depends on the success of its clients, such as stack and system developers Hexis, SWPC, and CFCL. SWPC was unable to commercialize its SOFC technology and its parent Siemens abandoned the fuel cell business. Hexis and CFCL could meet the same fate.
- Staxera has been rapidly advancing its SOFC stack technology, but its ISMs, which have achieved 10,000 h of operation, will have to demonstrate that they go for more than 40,000 h in stationary applications before they gain market acceptance. Staxera will also need time to scale up ISMs to large capacity units. At the same time, it needs to prove that the ISMs are reasonably priced so that they can compete with the conventional technologies.
- New Enerday has been steadily advancing its SOFC systems for market entry. It plans to commercialize the first market entry product, a 1 kW class Enerday LPG System EN-1000. But again, it is too early to project Enerday's future business success.

6.4.3.6. Siemens Efforts: Rise and Fall in 50 years

Siemens was a world leader in SOFC development for nearly 50 years. It started tubular SOFC development in 1962, ended it in 1967, and resumed planar SOFC development in 1988. Siemens made significant contributions to planar SOFC technologies through its participation in EU FP projects over a period of 12 years:

- Siemens developed a planar SOFC cell with a metal separator plate the "Development of Ceramic Oxide Fuel Cell for Power" project (1988–1990) under EU FP1.[452]
- Siemens proceeded to develop a 1 kW stack in the "Development of a Solid Oxide Fuel Cell" project (1990–1993) under EU FP2.[453] This project was part of a larger project where

fundamental SOFC material research was funded by the then German Ministry of Research and Technology (BMFT).[vi] Development of manufacturing methods for SOFC compounds were funded there as well. Siemens evidently developed a 370 W stack in the project.

- In a follow-on EU FP3 project "Development of SOFC Flat Plate Reactor with Metal Separator Plate of Alternative Materials for Low Temperature SOFC Operation" (1993–1995), Siemens attempted to develop a 20 kW module,[454] it only developed a 1.8 kW stack.[10]
- In another project under EU FP3, the "Development of Co-Fired Ceramic Cells for Low Temperature 800 °C Solid Oxide Fuel Cell Operation," funded at €1.9 million[454] (about $2.3 million at the 1994 exchange rate of €0.83/$1), Siemens developed a unique, three-layer, co-fired laminate structure PEN (Positive/Electrolyte/Negative) SOFC[455] that provided higher power densities.[10]
- In an ambitious EU FP4 project, the "Development of 50 kW Class SOFC System and Components" (1996–2000), Siemens was expected to develop a 50 kW stack.[456] It only developed a 5.4 kW module, however.[10]

Siemens' partners in these five projects included ECN (the Netherlands), DSM Research B.V (the Netherlands), Eelectricite de France, EniTecnologie S.p.A (Italy), Gas Research & Technology Center (United Kingdom), GEC Alsthom (United Kingdom), and Imperial College (United Kingdom), and Univ Aveiro (Portugal).

In November 1997, Siemens acquired the fossil power-generation division from Westinghouse of the United States and established Siemens Westinghouse Power Corporation (SWPC), headquartered in the United States, which took responsibility for Siemens' global SOFC operations.[13] (See discussion at Section 6.2.1.2 for a detailed history of SWPC activities in North America and Europe).

Subsequently, SWPC played a global role in SOFC development. It stunned the world with the stable, efficient performance of a 100 kW CHP field test unit, which operated for close to 40,000 h with no cell degradation.[15]

Despite this spectacular success, there were many disappointments:

- SWPC signed a contract in 2003 to build a 250 kW CHP system for Stadtwerke Hannover and E.ON Energie in Germany[30] but did not appear to have delivered the system.
- SWPC signed a contract in 2000 to build an RWE 300 kW SOFC/GT hybrid in Essen, Germany by 2003.[36] The system reportedly operated for only a short time.
- In 2000, SWPC formed a consortium agreement under an EU FP5 project (2000–2003) to provide a 1 MW pressurized SOFC/GT hybrid system to be demonstrated in Germany in 2004.[36] The project was suspended in December 2002 because a suitable microturbine generator was not available on the market.[39]
- In April 2007, SWPC and EnBW planned to develop a 1 MW SOFC/GT hybrid demonstration plant.[40] The same concept had been canceled in 2002 but was revived when the DLR—in cooperation with the Institute for Aviation Drive Systems at the University of Stuttgart—offered to develop suitable gas turbines based on an Italian 100 kW MGT.[41] Nonetheless, little news has been forthcoming on this project and it probably was canceled.

Siemens shut down its fuel cell business in September 2010. It paid $1.53 billion[50] in November 1997 to purchase the Westinghouse'

[vi] German ministries frequently change their name and composition by reorganizing within ministries.

fossil power division, and after running the fuel cell business unit for more than 10 years, Siemens decided that fuel cells would not constitute a core competency for its energy business.[51] The story underscores the extreme difficulty of commercializing SOFC technology.

6.4.4. The Netherlands

SOFC R&D in the Netherlands was carried out in the late 1980s by independent research institutes, such as the Energy Research Center of the Netherlands (ECN), the Netherlands Organization for Applied Scientific Research (TNO), and the University of Twente. Initially, these research and academic institutions pursued basic research and sought EU FP sponsorship to support their activities. Subsequently, the Netherlands decided to implement a small-scale national SOFC program, which was short-lived. ECN's pursuits for SOFC development continued with EU support, evolving into a productive relationship with Siemens (Germany) and Sulzer (Switzerland). ECN in the end developed a capability to produce SOFC materials and created a spin-off to begin commercial activities.

6.4.4.1. ECN Starts SOFC Activities in 1987

In early 1987, a few independent research institutes, including ECN and TNO, and a research group at the University of Twente, started to explore the possibilities of research on SOFCs.[4] During 1988, discussions took place between interested parties in the Netherlands to prepare a proposal for an EC FP project. The EU accepted a project of the University of Twente that ran from 1988 to 1990. The project was titled "Membrane-Based Thin-Layer SOFC Technology," which was to explore new production technology (electrochemical vapor deposition or ECVD) for the fabrication of thin ceramic solid oxide electrolyte layers to decrease polarization losses, increase power output, lower operating temperature, and increase cell voltage. (Later, the EU sponsored a second project under the same name, which was proposed by ECN and the University of Twente. It ran from 1998 to 1990.).[457] After the projects were completed this research ended, likely because the technology did not lead to lower production costs.

In 1990, the Netherlands implemented a national SOFC research program.[4] The Netherlands Agency for Energy and the Environment (Novem) sponsored a program with participation from TNO, the University of Twente, and ECN. The program lasted until 1992. This effort was followed by a series of Novem-ECN research projects, the main goal of which was the study of materials that allowed SOFCs to operate at lower temperatures than the standard operating temperature of 1000 °C. The rationale behind this effort was that lower temperatures would allow the use of lower cost stack materials (e.g., stainless steel for the separator plate instead of an expensive ceramic separator plate) and would also result in less stringent requirements for the surrounding system.

At the same time, in 1990, ECN started collaboration with Siemens on three successive EU FP projects: "Development of a Solid Oxide Fuel Cell" (FP2, 1990–1993), "Manufacturing Techniques For Components of Flat Plate, Solid Oxide Fuel Cell Reactors" (FP2, 1991–1994), and "Development of SOFC Flat Plate Reactor with Metal Separator Plate of Alternative Materials for Low Temperature SOFC Operation" (FP3, 1993–1995). For these projects, ECN developed single cells with reduced polarization losses that yielded power densities of 1500 mA/cm^2 at 0.7 V (on H_2/O_2). Endurance tests showed that cells decayed by about 30 μV/h, due mainly to degradation of the anode. ECN's target was to reduce this decay to less than 10 μV/h—roughly equivalent to 15 percent/year. Several ceramic component fabrication techniques, such as tape casting and screen printing, were developed by ECN and Siemens.[4]

As SOFC stack developers increased their research activities and conducted more field tests, demand for ECN cells increased. In 1994, the Swiss firm Sulzer signed a contract with ECN to purchase ECN's cells for their SOFC stacks and systems.[4] A year later, Siemens signed a similar contract with ECN. The increased demand led ECN to create an organizational and logistical split between its SOFC cell production and its research and development activities.[430] By 1998, ECN had built its pre-pilot SOFC component manufacturing line that included tape casting and screen printing techniques to fabricate state-of-the-art ESCs up to 10 cm × 10 cm in size and 12 cm diameter. The aim was that, in time, the manufacturing work for the conventional ESCs should become a business enterprise in close collaboration with European industry. This production facility supported an ambitious program of Sulzer to commercialize small-scale SOFC systems (up to 7 kW) as CHP units for individual residences by 2010.[4] It was also expected that ECN would continue to supply cells for the Siemens.

- In 1999, the Netherlands government support for SOFC R&D stopped. There have been no more government fuel cell programs as such,[458] probably because the government judged that ECN was ready to commercialize its SOFCs.

6.4.4.2. *ECN forms InDEC, a spin-off SOFC Ceramic Component Production Company, in 1999*

As noted in Section 6.4.3.5, InDEC ("Innovative Dutch Electro Ceramics") Pilot Production Ltd. was established in September 1999 as a subsidiary of ECN.[635] ECN had full control over the commercial business at InDEC.[427] InDEC manufactured planar SOFC components using tape casting and screen-printing techniques in a batch-wise production system to produce ESCs and ASCs.[459] Its strategy was to reduce manufacturing costs by translating its production know-how into a semicontinuous production line.

6.4.4.3. *InDEC Acquired by German Company H.C. Starck*

In April 2003, H.C. Starck acquired a majority holding in InDEC, which became part of H.C. Starck's subsidiary H.C. Starck Ceramics, Selb, Germany.[460] The total investment was reported to be in the double-digit million euro range.

- H.C. Starck and ECN would continue to collaborate on SOFC research, but with the sale of InDEC to H.C. Starck, the Netherlands' effort to develop a national SOFC capability essentially disappeared.

6.4.5. Switzerland

Switzerland had only a modest national program for SOFC development. According to an EU document, Switzerland, during 2000–2003, funded about €1.2 million for fuel cell development,[344] half of which focused on SOFC[177] and the other half on PEMFC. Switzerland is home to two noteworthy SOFC developers, HTceramix and Hexis AG.

HTceramix is relatively new, established in 2000 as a spin-off of an academic institution, the Swiss Federal Institute of Technology in Lausanne (EPFL). Hexis Ag, originally was part of Sulzer Ag, a Swiss industrial engineering and manufacturing firm. Sulzer's SOFC development effort led to the establishment of Sulzer Hexis, a wholly owned subsidiary, which subsequently became an independent company, Hexis AG.

6.4.5.1. *HTceramix Starts as a University Spin-Off in 2000*

HTceramix (HTc) was originally part of the prestigious Swiss state university, the Swiss Federal Institute of Technology in Lausanne (EPFL). EPFL is funded by the Swiss Federal Government and is one of Europe's leading science and technology institutions.[461] After

10 years of work within EPFL on SOFCs, HTc was spun off as a separate company.[462] HTc's main competence is its unique approach to stacking SOFC cells, the SOFConnex.[463] The SOFConnex is a ceramic gas diffusion and current collection layer with plastic properties that compensate for cell imperfections. The SOFConnex interconnection is placed on either side of the SOFC; the cell, thus, is sandwiched between two SOFConnex layers, which are in turn sandwiched between two metallic interconnects. Fully integrated gas distribution layers in the SOFConnex stack design reduce material use and the number of elements in the stack, thereby reducing the complexity of stack assembly. According to HTc, the SOFConnex allows a compact structure, low cost materials and manufacturing, and integrated fuel manifolding, current collection, and gas sealing.

As a state university spin off, HTc maintains strong ties to the national government and research communities. HTc closely cooperates and has privileged access to the test, analysis, and simulation capacity of EPFL.[464] HTc technology development is supported by the Swiss Federal Department of Economy,[463] the Swiss Federal Office of Energy (SFOE), the Swiss Commission for Technology and Innovation (CTI),[463] and the Swiss Gas Union (FOGA). Also, in the EU FP, HTc closely cooperates with academic partners, including Jülich, ECN, and Risø.[465]

- HTc worked on a number of SOFC development projects supported by domestic Swiss sponsors. In 2001–2002, the Swiss Gas Union (FOGA) supported a project to achieve a partial oxidation reactor based on oxygen separation membranes.[466] In 2001–2002, HTc engaged in development of thin film anode-supported electrolytes sponsored by the CTI. In 2002–2003, it worked on anode-supported SOFCs with thermal cycling and redox cycling stability, supported by the SFOE. In 2003–2004, FOGA sponsored a project to establish a pilot industrial manufacturing line for SOFC stacks.[466] HTc continues to work on projects sponsored by the SFOE and CTI in close collaboration with EPFL.[467]
- HTc participated in a number of EU FP projects and collaborated with major European SOFC researchers and developers. In 2004–2008, HTc worked with Jülich in the EU FP6 Real-SOFC project, in which HTc delivered several reference stacks for evaluation.[468] In 2005–2009, HTc participated in the EU FP6 Flame-SOFC project to develop a 2 kW mCHP unit, along with EPFL and major companies such as Merloni TermoSanitari (Italy), Elco Shared Services (Germany), and Fagor Electrodomesticos (Spain). In 2006–2010, HTc worked to develop SOFC technology operating at 600 °C in the EU FP6 SOFC600 project. The project involved leading SOFC academia groups in Europe, including ECN, Jülich, and Risø among many other partners.

6.4.5.1.1. HTCERAMIX PRESENTS IMPRESSIVE PRODUCT LINEUP IN 2005

HTc energetically pressed forward to commercialize the technology that it had developed under EPFL. In 2003, just 3 years after it was spun off, HTc set up a pilot manufacturing laboratory to cover the increased needs for internal development and to meet customer demand.[469] It introduced a range of SOFC products, including cells, stacks, and integrated stack modules. It also formed a number of partnerships with systems developers and integrators to develop partners' products that incorporate HTc products.

In April 2005, HTc and Fideris, a leading Swiss testing equipment maker (now TesSol, Inc. of Texas) demonstrated a turnkey SOFC testing and validation platform at the Hannover Fair in Germany.[470] The two companies had formed a partnership to develop a SOFC test platform targeting the projected $400 million-per-year SOFC testing industry. The HTc-Fideris turnkey testing system allowed researchers to install an SOFC system in approximately 4 h, enabling easier and faster access to SOFC experimental and diagnostic data. The unit demonstrated at the Hannover Fair incorporated an HTc 5-cell SOFC fuel cell stack and a Fideris test system.

- The system was demonstrated at the Hannover Fair and was available for immediate sale. It was scalable to various testing applications. Fideris and HTc planned to jointly market this platform to universities, research labs, technical schools, and potential end users for SOFC.

Also in April 2005, HTceramix and Dantherm, a Danish heat management systems manufacturer, succeeded in integrating an HTc stack into a Dantherm thermal management system.[471] HTceramix and Dantherm had been partners since 2004 in developing an mCHP. The result was a prototype Dantherm SOFC system that produced electricity and hot water. The Dantherm system was operated under thermally autonomous conditions and used hydrogen and carbon monoxide as fuel.

- Dantherm indicated that a pre-commercial product could be demonstrated within a few years.

In November 2005, the HTc presented an impressive lineup of their own commercial products. At the Palm Springs 2005 Fuel Cell Seminar, HTc exhibited individual SOFC cells, the Lilith 025 portable SOFC power-generation system, models of HTc's "R" and "S" stacks, and the HoTbox.[472]

- The Lilith 025 was a small portable power generator that runs on liquid hydrocarbon fuels such as methanol.[473] For leisure and outdoor use, the Lilith 025 started up quickly (in less than 60 min), was compact, portable (weighs about 10 kg), and thermally self-sustaining. It produced between 150 and 250 W of power and provided better resistance to thermal and redox cycling compared to other stacks, according to HTc.

The HoTbox integrated all high temperature components of an SOFC system into one unit,[636] making the integration of SOFC systems easier for system developers, according to HTc. It included a 500 W SOFC S-stack with a heat exchanger, high temperature insulation, easy temperature control independent of air flow rate, and voltage conditioning for lead acid battery compatible output.

All of the products shown at the exhibition were available for commercial sale.[472] HTc started selling SOFC stacks to educational institutions and strategic partners starting in 2005.[472]

6.4.5.1.2. SOFCPOWER OF ITALY ACQUIRES HTCERAMIX IN 2007

In February 2007, HTc entered into a partnership with an Italian company, SOFCpower, to jointly develop SOFCs.[474] SOFCpower was established in 2006 from the SOFC activities that started in 2002 within the Eurocoating-Turbocoating Group.[475] The new partnership aimed to advance the industrial production and commercialization of the HTc HoTbox.[474] The two companies said the collaboration would be carried out by a team of 25 people at newly expanded facilities at HTc headquarters in Switzerland. Meanwhile, Italian SOFCpower was implementing a trial production line in Italy for stacks and HoTboxes, which were expected to go into operation by the end of 2007.

- The partnership had an optimistic view of its commercial venture. HTc reported strong demand for its SOFC products, with orders having increased 750 percent over 2005–2006, the majority of them from the Asia-Pacific region.

In a surprisingly fast move, in May 2007, just 3 months after the two companies formed partnership, SOFCpower purchased 100 percent of

HTc shares. SOFCpower was now in a position to industrialize, produce, and commercialize the HoTbox. SOFCpower undoubtedly saw major promise in the technical expertise of HTceramix. SOFCpower is a privately held company dedicated to developing and manufacturing coatings and special processes for gas turbine and aero-engine component manufacturers (OEM). The acquisition of HTc was a major technological gain, giving SOFCpower access to the knowledge and expertise of the top Swiss fuel cell specialist and EPFL's SOFC research, including SOFC modeling, testing, and engineering.

- In December 2008, SOFCpower stated that SOFCpower/HTceramix' main product, the 500 W HoTbox, a prototype for an SOFC-based natural gas–fueled generator for on-grid applications, had been operated in a thermally self-sustaining mode for more than 1000 h.[476]

In March 2009, SOFCpower and HTc completed the technology transfer from its laboratory production in Yverdon, Switzerland, to an industrial pilot production in Mezzolombardo, Italy.[477] The first complete HoTbox generator produced in Italy was assembled and successfully tested within specifications. The establishment of the pilot manufacturing line allowed SOFCpower and HTc to provide customers with standard products and was designed to allow a steady increase in production capacity for SOFC generators, up to 2 MW by the end of 2011.

- According to the company website, the ASC-700 SOFC, with anode-supported electrolyte and a ceria protective layer and LSCF-based cathode, demonstrated power densities of 405 mW/cm^2 at 650 °C, 560 mW/cm^2 at 700 °C, and 715 mW/cm^2 at 750 °C at 0.7 V with fuel utilization of up to 70 percent. The anode-supported electrolyte with LSM-based cathode (the ASC-800) demonstrated 510 mW/cm^2 at 750 °C, 680 mW/cm^2 at 800 °C, and 890 mW/cm^2 at 850 °C at 0.7 V.[478]

6.4.5.1.3. SOFCPOWER/HTCERAMIX MAKES GOOD PROGRESS IN RAISING STACK EFFICIENCY

HTc and SOFCpower have been rapidly raising performance in terms of stack power and efficiency. In March 2009, progress made in materials and production techniques at SOFC power's pilot manufacturing plant in Trento increased stack performance to 50 percent electrical efficiency with a hydrogen nitrogen fuel mixture.[477] This progress exceeded the internal technology progress targets and would facilitate the achievement of production cost targets, according to SOFCpower and HTc.

Also in 2009, SOFCpower and HTc demonstrated an F-design stack that achieved 1.1 kW power, 40 percent efficiency, and stable operations.[479] The stack was developed as part of the EU FP6 Flame-SOFC project that aimed to develop 2 kW mCHP units. The stack was operated for more than 300 h in stable conditions. HTc said the results represented an important demonstration of the capability of the F-stack to support SOFC modules in the range of 2.5–5 kW.

In August 2009, SOFCpower and HTc operated their new larger stack design (F-design) at 1.75 kW operating with 70 percent fuel utilization.[480] Under these conditions, the electrical efficiency reached 43 percent, and the specific power density was of 440 mW/cm^2, exceeding the company's performance objectives and conforming to cost targets for mass production (less than €1,000/kW installed), according to the two companies. The same stack also operated under partial load conditions, achieving higher than 50 percent electrical efficiency at 1.4 kW and 53 percent electrical efficiency at 750 W. This result was achieved with 20 "repeat elements" in the stack, each producing 90 W, using a nitrogen–hydrogen fuel mixture. Higher stack efficiencies in the range of 60 percent could be reached by the same stack if fed with steam reformed natural gas, according to the companies. The result was assessed by experts of the EU FP6 project Flame-SOFC project in which this stack development was being pursued.

In addition, a standard SOFCpower-HTc stack module achieved 65 percent electrical efficiency based on the lower heating value of fuel in a test at EPFL.[481] The stack was fueled with steam reformed methane (steam-to-carbon ratio of 2) and operated at 750 °C with a power density of 250 mW/cm^2 and an average cell voltage of 760 mV. With such efficiencies, the distributed generation of electricity in kilowatt size units using SOFC technology is more efficient than centralized generation in megawatt size plants using the best available combined cycle gas turbine (CCGT), according to the two companies.

6.4.5.1.4. SIGNS OF COMMERCIAL SUCCESS NOT YET PRESENT

It is uncertain whether SOFCpower and HTc have achieved commercial success or not. Still, according to their reporting in the past, SOFCpower and HTc have increased sales for some years. The two companies represent a unique combination of a developer and a supplier that markets intermediate SOFC products to system developers, who in turn build system products for the global market. Sellers of intermediate products, in theory, could continue to increase sales as long as developers make more systems without making their own components in-house.

Commercial success also depends on products being sufficiently efficient, durable, reliable, and cost competitive. It is likely that SOFCpower-HTc stacks perhaps have not yet achieved this level of performance for commercial success:

- The companies have recently reported a 65 percent electrical efficiency. The efficiency, however, was achieved on reformed methane. In field situations where reformed methane or hydrogen is not readily available, a hydrocarbon fuel might be used as an alternative, with or without a reformer. This could reduce overall efficiency, increasing the stack cost per kilowatt.
- Moreover, SOFCpower-HTc cell stacks evidently have not demonstrated durability much beyond 3000 h. In January 2008, the 500 W class HoTbox was operated for 1000 h on desulfurized natural gas from the grid.[476] In April 2008, in the EU FP6 Real-SOFC project, an HTc stack operated for more than 3000 h with a degradation rate of 1 percent/1000 h.[482] An HTc S-design stack has reached 5000 h of operation, but this unit incorporated ECN cells in the stack.[483]

At the same time, a sustainable and meaningful commercial success of any manufacturer depends on the profitability of other companies that operate in the same production and supply chain. As yet, there have been no successes reported among any of the companies that have worked with SOFCpower and HTc.

But it is possible that SOFCpower and HTc may be successful in the future. Since 2004, Dantherm Power, a Danish partner of SOFCpower-HTc's, has been developing a small mCHP using the HoTbox. As of June 2010, Dantherm planned to launch commercial sales in 2012–2015.[484] The company indicated that it had developed a compact 5 kW mCHP unit for use in private households, small apartments, and small factories, and as of November 2010 Dantherm planned to have seven units in operation by early 2011 and produce 15 more units by September 2011.[485] The company believed that a major breakthrough would happen in 2013–2015, when many Danish families in 2015 would have a SOFC micro CHP plant.

6.4.5.2. Sulzer/Sulzer Hexis/Hexis

In 1989, Sulzer, a major Swiss industrial engineering and manufacturing firm, invented a theoretical SOFC power-generation concept called Hexis (Heat Exchange Integrated Stack). Sulzer started work on development of the concept in 1991.[486] In 1997, Sulzer established a wholly owned subsidiary, Sulzer Hexis, to focus on preparations for market entry. In August 2005, however, Sulzer decided that the fuel cell program would not be commercially successful and ceased financial support for Sulzer Hexis. In January 2006, Hexis was established as an independent company with a new sponsor.

6.4.5.2.1. SULZER STARTS HEXIS DEVELOPMENT IN 1991

Sulzer started the Hexis development program in 1991, focusing on materials development, process control, and system integration.[487] The Hexis design was based on circular planar cells interspaced with heat exchanger units.[386] Individual cells each contained an electrochemical cell, air preheater and after burner. Metallic or ceramic interconnects could be used, and the design was, in principle, suitable for a range of cell materials and stack sizes. By 1994, Sulzer was working to develop a 1 kW stack. At that time, Sulzer dedicated 4 man-years to SOFC development with additional resources coming from participating universities. A total budget for this program was estimated at $1 million per year, with significant funding from the Swiss Government Department of Energy.

- It appears that Sulzer procured the cell components from Ceramatec of the United States initially[386] and from ECN (subsequently from InDEC) of the Netherlands starting in 1994.[4]

6.4.5.2.2. SULZER ESTABLISHES WHOLLY OWNED SOFC SUBSIDIARY, SULZER HEXIS, IN 1997

In 1997, Sulzer succeeded in developing a pilot 1 kW field test system.[488] That year, Sulzer founded Sulzer Hexis, based on its SOFC development unit. The newly created Sulzer Hexis successfully operated the first Hexis stack for more than 12,000 h on hydrogen from 1997 to 1998.[489] The stack had a unique configuration, characterized by a disk-like shape and current collectors which also acted as heat exchangers and as fuel distribution channels. The pilot Sulzer-Hexis system, the HXS 1000 Premiere, had a nominal power of 1 kW with an electrical efficiency of about 33 percent. Sulzer Hexis installed HXS 1000 Premiere systems at several employees' homes, thereby getting relevant data on real field applications before putting its product in the market. Sulzer Hexis also demonstrated the use of biofuel with an HXS 1000 Premiere.

In 1998, Sulzer Hexis launched a second-generation field test system.[488] Starting in late 1998, Sulzer Hexis installed four field test systems at sites of different partners. The partners were city utilities—AET Basel (Switzerland) and EWE Oldenburg (Germany)—and large gas suppliers—ThyssenGas (Germany) and Tokyo Gas (Japan). All systems ran with natural gas from the low pressure grid and were connected to the electrical grid. The maximum power output was 1–1.5 kW at an electrical efficiency of 30–35 percent. The power density was 200 mW/cm^2 at 920 °C.[490] Exhaust heat was used for heating water.[488] Additional needed electrical energy could be taken from the grid. The system also had an additional gas heater if the exhaust heat (1–2 kW) was not sufficient. Sulzer Hexis planned to run the field tests for 3 years and complete them in December 2001.

- In parallel, Sulzer Hexis started focusing on system integration, cost reduction, and compatibility with industrial production. Sulzer Hexis planned to launch the first commercial 1 kW system with its field test partners in 2001.
- In addition to HXS 1000 Premiere, Sulzer Hexis was developing a more compact 1 kW system, called Galileo.[489]

At the end of 2001, Sulzer Hexis successfully concluded the 3-year international field test.[487] In December 2001, Sulzer Hexis began the production of a preliminary series ("pre-series") of fuel cell systems. Sulzer Hexis planned to install 400 pre-series HXS 1000 Premiere units in various European countries prior to market entry, which it planned for 2004.[491] By the end of 2002, Sulzer Hexis produced and delivered around 80 pre-series HXS 1000 Premiere systems.[487] Powered by natural gas, the system covered all the thermal requirements as well as the basic electricity needs of a single-family home (operating in parallel with the grid). The HXS 1000 Premiere systems, while being tested, were further developed on end-users' premises within the framework of a pre-series phase in cooperation with utilities, in Germany, Austria, Switzerland, and the Netherlands. Sulzer Hexis acknowledged having gained valuable insights from these trials regarding cyclic oxidation/reduction at the fuel cell anode, alternative fuel processors (catalytic partial oxidation as opposed to steam reforming, which required a supply of pure water), and general customer requirements.[492] The utility and installer partners included ECN (the Netherlands), EnBW Energie Baded-Wurttemberg (Germany), Energie AG Oberosterreich and Oberosterreichische Ferngas AG (Austria) in a working group, E.ON Energie (Germany), EWE (Germany), EWR Elektrizitatswerk Rheinhessen (Germany), Thissengas (Germany), and VNG-Verbundgasnetz (Germany).[493]

- Meanwhile, Sulzer Hexis continuously increased automation of its production processes. Automation steps included robotic screen printing with integrated weight control and, in the future, infrared-based layer quality control.[492]

Although Sulzer Hexis planned to enter the commercial market in 2004, it was still working to improve its pre-series 1 kW CHP unit: the HXS 1000 Premiere.[47] By June 2004, 95 units were installed in the field, having produced 250 MWh, with an availability of about 91 percent. According to a US industry review published in 2004, Sulzer by that time had accumulated more operating experience than any other planar SOFC developer and had achieved respectable (almost adequate) stack life. Sulzer Hexis claimed to have accumulated about 500,000 h of operational experience (how many stacks were used is not known). It also had made significant progress in improving its units. Stack degradation had been dramatically reduced from about 22 percent/1000 h down to about 1.5 percent/1000 h.

- However, efficiency and cost were still far from attractive, according to the same 2004 US industry review. The electrical efficiency of the unit was only about 25–30 percent. The system's target production cost was estimated to be as high as $4000 (including the high-efficiency storage water heater). It also had a reliability problem; HXS 1000 Premiere required periodic stack replacement. Furthermore, it was doubtful that the product configuration would be viable for a wide range of utility-driven applications, the same US industry review indicated.

In 2005, Sulzer Hexis developed the next generation mCHP system, Galileo.[494] The system was smaller, more cost-efficient, and easier to operate, compared to its predecessor. Galileo measured only 55 cm × 160 cm × 55 cm (L × H × D) and weighed 170 kg, showing that substantial progress had been made. Two plates were replaced by the single-plate design, and the weight of the cell stack and the costs of the current collectors were halved.[47] The thermal insulation costs had been reduced to a tenth.[494] In April 2005, Sulzer Hexis introduced the commercial mCHP prototype, Galileo 1000 N, at the Hanover Fair 2005.[495]

- The first systems of the new generation were to be delivered in autumn 2005.[494] Sulzer Hexis hoped to manufacture 1000–10,000 Galileo units per year.[487]

6.4.5.2.3. HEXIS BECOMES AN INDEPENDENT COMPANY IN 2006

In August 2005, Sulzer, the parent company of Sulzer Hexis, decided to discontinue investments in Sulzer Hexis.[496] During the previous months, Sulzer had conducted a partner search to find a financing base to support continuing operation of Sulzer Hexis, but it was not successful. Sulzer decided that from 2006 onward the venture investments in Sulzer Hexis would be stopped. Sulzer concluded that carrying the risk of further significant investments in Sulzer Hexis was beyond its scope and focus.

- According to Sulzer, it intended to start winding down the operations of Sulzer Hexis in a controlled manner if no buyer could be found. During the following weeks, discussions with employee representatives were to be held and appropriate solutions to be developed, according to Sulzer.

In January 2006, Hexis started to operate as an independent company.[497] A Swiss foundation in Winterthur—the Foundation for Art, Culture, and History—would support Hexis' effort as a 100 percent owner of the company.[498]

According to the company website information accessed as of December 2010, Hexis' Galileo 1000 N was still under development and the work focused on the improvement of the fuel cell stack.[499] Hexis' Galileo 1000 N, the website states, demonstrated 25–30 percent electrical efficiency and 90 percent total efficiency.[500]

6.4.5.2.4. ……..TOO EARLY TO FORECAST COMMERCIAL OUTCOME

Hexis has stated that its initial plans to introduce its mCHP system, the Galileo 1000 N, to the market in 2005–2006[487] were premature. According to a media release in July 2007, Hexis stated that it had yet to achieve the reliability and lifetime needed to achieve market success.[501]

- At the start of 2007, four Galileo 1000 N fuel cell heating systems were delivered to four large energy supply companies, EWE, E.ON Energie, EnBW Energie Baden Württemberg in Germany, and GVM Gasverbund Mittelland in Switzerland, for laboratory testing. Data from the tests were fed back to further improve the durability and performance of the fuel cell.[501] Hexis was clearly determined to increase the lifetime of the fuel cells to over 40,000 h and satisfactorily prove it before putting the product on the market.

As of October 2009, field tests with the newly designed Galileo 1000 N were being conducted. At that time, 17 of the mCHP systems were installed in the field, while 13 systems were installed at Hexis for laboratory tests.[502] Hexis was expecting to achieve 13,500 h of continuous operation of one newly designed Galileo 1000 N system, with a power degradation rate of 1.6 percent/1000 h. Also in a test with improved cells, Hexis expected an electrical power output of more than 1 kW (AC, net) and an electrical efficiency of more than 40 percent (DC) and 36 percent (AC, net). Another test showed that total efficiency of more than 90 percent (LHV) was possible in fuel cell operation mode.

Hexis has completed the development of Galileo 1000 N[503] and has started to demonstrate it at various venues, starting with the Gas Industry Conference in Hamburg, Germany, on 25–26 October 2011.[504] It plans to place it on the retail market in 2013.[505]

6.4.6. United Kingdom

The UK Department of Trade and Industry (DTI) and the EU through its FP have been the major supporters of UK SOFC development.

It is not entirely clear how much funding the UK DTI and EU have provided to SOFC development in the United Kingdom, but in view of various factors it seems that the UK national program funding ratio to that of the EU FP are nearly equal, as is the case in other European countries.

There have been at least three SOFC developers in the United Kingdom. Alstom, a major power engineering company based in France, no longer is in the field. Ceres Power and RR Fuel Cell Systems remain active developers. Ceres Power is a rapidly rising, small start-up firm working to develop small residential-use cogeneration power mCHP systems. It plans to market the product in mid-2011. RR, a giant manufacturer of automobiles, airplane engines, marine propulsion systems, and power generation equipment, aims to develop large megawatt class fuel cells and gas turbine hybrid (FC/GT) power plants.

6.4.6.1. Alstom: Started SOFC R&D in Mid-1990s, Ended in Early 2000s

Alstom (formerly GEC Alsthom) started SOFC development in mid-1990s and continued perhaps until 2003. SOFC activities were mainly based in the United Kingdom. Alstom is a global leader in power generation, power transmission, and distribution. It became interested in the integration of gas turbines and fuel cells to make more efficient power-generation systems. In most projects, therefore, Alstom considered SOFC stack designs in combination with gas turbines using cells developed by other SOFC manufacturers.

- Alstom also was interested in PEMFCs and installed six Ballard 250 kW fuel cell power plants for field trials in Europe between 2000 and 2002.[506] The outcome of these field trials is not clear, and no further trials were planned.

Alstom participated in at least two UK DTI projects and three EU FP projects that focused on designing SOFC stacks and systems and integration of SOFC systems with gas turbines. They were likely carried out between 2000 and 2003:

Stack and System Design Development. In a UK DTI project, "Planar SOFC Technology: Stack Design and Development for Lower Cost and Manufacturability (Contract number: ETSU F/01/00194/REP, DTI/Pub URN 02/1350)," Alstom investigated the performance of an externally manifolded SOFC stack based on ASCs developed by Jülich and InDEC. It worked to develop a new design of an SOFC stack with the potential for low cost production, to develop sealing materials and to test its performance.[507] Project partners were Jülich and ECN. The project started in 2000 and ended in 2002. In another UK DTI project, "Systems Development for Planar SOFC Based Power Plant (ETSU F/01/00195/REP, DTI/Pub URN 02/868)," and in an EU FP5 project, "Distributed Power Generation Plant based on Planar SOFC, Proof of Concept (PRO-CON)" (2000–2002), Alstom worked to assemble a 5 kW planar SOFC stack based on ASCs.[508] The stack design, which used 20 cm square ASCs, featured an internal manifold and parallel flow arrangement. It also investigated potential issues related to the stacks and system based on a conceptual 20 kW integrated system design.[509] The system was tested for 2000 h. Project partners were Jülich and Prototech (Norway).

SOFC and Gas Turbine Configuration. Alstom and Elenco (an AFC developer), and some gas turbine manufacturers worked on the EU FP3 project, "Systematic Analysis of Gas Turbine-Fuel Cell Combinations Generation with Very High Efficiency" (1994–1995). They carried out analysis of different combinations of gas turbines with different types of fuel cells to assess the economic feasibility and future commercial prospects of the systems. In an EU FP5 project, "Integrated Modeling Study of Fuel Cell/Gas Turbine Hybrids (IM-SOFC-GT)" (2001–2003), RR coordinated a group of research and industry organizations, including Alstom, to carry out an assessment of product requirements and the development of generic modeling capabilities and preliminary system designs for the three subsystems, sub-MW distributed generation (DG) systems, 1–3 MW systems suitable for CHP schemes, and 20–30 MW high efficiency DG systems.

According to an industry directory, Alstom scaled back SOFC technology work in 2002, although it continued to participate in a number of European Union projects.[506] Indeed, Alstom participated in the EU FP5 project, "Thematic Network on Solid Oxide Fuel Cell Technology (SOFCNET)" (2003–2005), which aimed to create a better mechanism to support and accelerate European SOFC development and commercialization.[343] Subsequent to the SOFCNET project, however, little reporting is available on Alstom's SOFC activities, suggesting that Alstom's interest in SOFC technology ended in the early 2000s (Table 6.11).

6.4.6.2. *Ceres Power*

Ceres Power was established in 2001 as a spin-off from the Imperial College to commercialize metal supported, intermediate temperature planar SOFC technology.[510] The SOFC technology with a multi-fuel capability had been developed by Imperial College over the previous 10 years.[511]

- Imperial College has an extensive SOFC capability and has long been a major player in SOFC R&D projects in the EU FP. Since 1987, Imperial College has participated as a collaborator in at least 14 EU FP SOFC development projects, which were coordinated by Fraunhofer (Germany), Jülich (Germany), Risø (Denmark), RR, and Siemens. It has participated in at least one project as a coordinator.
- Ceres Power has maintained its relationship with Imperial College and has access to their SOFC expertise; some Imperial College professors are principles at Ceres Power.

Ceres Power's SOFC cells use a new generation of ceramic materials known as cerium gadolinium oxide (CGO), which enables operation at 500–600 °C.[512] This also allows use of conventional stainless steel as the cell substrate, separating the functions of mechanical support and electrochemistry. Ceres Power claims that the electrochemical layers are extremely thin and have high power densities and that the stack material costs are low. Ceres also states that the heat-to-power ratio is approximately one-to-one, while other SOFCs have a higher heat-to-power ratio, making the Ceres technology more suitable for CHP applications, where the levels of electrical output need to be maintained even when heat demand is modest.

- Up to January 2004, Ceres Power had demonstrated cell performance of 0.24 W/cm^2 at 550 °C and 0.31 W/cm^2 at 600 °C in hydrogen.[513] No degradation was reported after 2000 h of testing and less than 3 percent degradation was observed after 25 thermal cycles and 7 redox cycles.

6.4.6.2.1. CERES POWER MAKES RAPID PROGRESS BACKED BY SOLID FINANCE

Since its corporate establishment in 2001, Ceres Power has been making remarkably fast technological progress, with major support from the UK DTI, the GOC Group, and the British Gas:

- In August 2003, the UK DTI implemented a 12-month project for Ceres Power to develop its cells and stacks (August 2003–November 2004).[514] The project was successful, and Ceres Power developed single cells with improved power densities as high as 400 mW/cm^2 at 600 °C and a 160 W short stack that was tested for 3300 h with less than 5 percent performance degradation. The total cost of the project was £600,000 (about $984,000 at the 2003 exchange rate of $1/£0.61), with DTI contributing £300,000

TABLE 6.11 Alstom (United Kingdom) and Rolls Royce Participation in UK DTI SOFC Development Projects

Project title	Contract number*	Cost	Period	Project description
Alstom UK participation				
Planar SOFC Technology: Stack Design and Development for Lower Cost and Manufacturability	ETSU F/01/00194/REP, DTI/ Pub URN 02/1350		Project period is not clear; but the first project report was published in 2002.	
Systems Development for Planar SOFC Based Power Plant	ETSU F/01/00195/REP, DTI/ Pub URN 02/868		Project period is not clear; but the first project report was published in 2002.	
Rolls Royce: Participation				
Integrated Planar Solid Oxide Fuel Cell (IP-SOFC)	ETSU Report F/01/00070/ REP		Project period is not clear, perhaps it started sometime in the 1990s.	
Improving the Durability of SOFC Stacks (IDU-SOFC)	ETSU Report F/01/00090/ REP		Project period is not clear, perhaps it started sometime in the 1990s and ended before 2000.	
Development of a Supported "Thick Film" Electrolyte Membrane/Electrode Assembly with Three Interconnected Solid Oxide Fuel Cells	ETSU Report F/01/00096/ REP		Project period is not clear, but the final summary report was published in 1996.	
A Proof of Concept Multi-Cell MEA Module for the IP-SOFC	ETSU Report F/01/00125/ REP		Project period is not clear, but a project report was published in 1998.	

Development of a 1 kW IP-SOFC Stack	ETSU F/01/00144/REP		Project period is not clear, but the summary report was first published in 2001.	ETSU managed the project for the DTI.
Improving the Manufacture and Durability of Multi-Cell Modules for the IP-SOFC	ETSU F/01/00179/REP DTI/Pub URN 01/1293		Project period is not clear, but the summary report was first published in 2001.	ETSU managed the project as part of the DTI Sustainable Energy Program.
Scale-Up to Multi-Tens of kW for the IP-SOFC	Scale-Up to Multi-Tens of kW for the IP-SOFC		April 2000 to September 2003; the summary report was first published in 2004.	The project was run in conjunction with an FP5 project.
Design & Build a 5 kW Thermally Self-Sustaining IP-SOFC Stack	ETSU F/01/00199/REP DTI/Pub URN 01/1161		Project period is not clear, but the summary report was first published in 2001.	
Pressurization of IP-SOFC Technology for Second Generation Hybrid Application	F/01/00278/00/00 URN Number 05/949	The total cost is £6,013k, of which DTI contribution is £2,898k	August 2003 to December 2004	DTI contract
Pressurization of IP-SOFC Technology for Second Generation Hybrid Application PIP-SOFC	NNE5-2001-00791		April 2003 to April 2006	The project was part of the EU FP

** The ETSU project was carried out under contract as part of the New and Renewable Energy Program, managed by ETSU on behalf of the Department of Trade and Industry (DTI).*

(about $492,000) and Ceres Power the rest. In October 2004, BOC, a global industrial gases giant, now a part of the Linde Group, entered an agreement with Ceres Power to run development trials on Ceres cells fueled by LPG over the following 18 months.[515]

- In February 2005, Ceres Power won a £1.9 million (about $4.5 million at the 2005 exchange rate of $1/£0.55) contract from the UK DTI to develop cheaper and lighter commercial fuel cell systems.[516] In May 2005, the company succeeded in making its fuel cell four times more efficient.[517] Also in May 2005, it announced a successful operational test which ran for more than 3000 h.[518] In March 2006, the company developed a 1 kW fuel cell stack ("Ceres Stack"), generating sufficient power for the average home.[519] Also, in March 2006, Ceres Power entered into a £2.7 million (about $5 million at the 2006 exchange rate of $1/£0.54) contract with British Gas, in part funded by the DTI, to develop cogeneration power plants for the UK residential market.[520] In May 2006, Ceres demonstrated that a complete CHP system was sufficiently compact and lightweight to be wall-mountable.[521] In September 2006, Ceres Power secured a new facility to design, build, and test complete products.[522]
- In December 2006, the company announced that it had frozen its core stack technology after their fuel cell stacks delivered high levels of durability and performance.[523] Their fuel cell stacks had completed more than 7500 h of testing. Testing of more than fifty 1 kW stacks with repeated thermal cycling (on/off) between room temperature (20 °C) and operating temperature (600 °C) did not cause performance degradation. Fuel cell stacks were able to use a range of commercially available fuels including natural gas and meet the performance requirements of a wide range of target market applications. More than 1000 stack subassemblies were produced with a zero failure rate.
- In January 2008, Ceres Power signed a new agreement with British Gas, including a funded field test program for residential CHP systems.[524] In addition, British Gas would subscribe for new shares representing 9.999 percent of Ceres Power's enlarged issued share capital, equivalent to a cash investment of around £20 million (about $37 million at the 2008 exchange rate of $1/£0.54). In September 2008, Ceres Power developed a 1 kW wall-mountable module, meeting the performance, size, and weight requirements under the Alpha Phase of the CHP program with British Gas.[525] British Gas started testing of the 1 kW Alpha model fed by natural gas in early 2009. In March 2009, Ceres Power acquired a facility for mass production in Horsham, South East England, which included a building of 50,000 ft^2 (5000 m^2).[526] In May 2009, British Gas completed the testing of the Alpha model, which successfully met most requirements of a typical home CHP system, as well as exporting and importing power to and from the grid as required.[527] All of the product requirements, including performance, size, weight, and regulatory compliance, were also achieved. British Gas issued an acceptance certificate ensuring the £2 million (about $3 million at the 2009 exchange rate of $1/£0.66) milestone payment to be made to Ceres.
- By May 2009, Ceres Power had already commenced the design and procurement activities for the beta phase of the CHP program to produce beta CHP units for in-field trials by British Gas. The preparation of the new mass manufacturing plant in Horsham was well underway with initial operations scheduled to begin in the second half of 2009. British Gas reportedly was planning to test the beta field trial units in the second half of 2009.[528] Ceres Power and British Gas together planned to achieve

market launch of the residential CHP product in the second half of 2011.[527]

6.4.6.2.2. UK GOVERNMENT INTRODUCES FEED-IN-TARIFF TO ENCOURAGE RESIDENTIAL INTRODUCTION OF MICRO COMBINED HEAT AND POWER SYSTEMS

In February 2010, the UK Government's Department of Energy and Climate Change (DECC) announced the introduction of a Feed-in Tariff (FIT) starting in April 2010 for residential CHP units, which would include Ceres Power mCHP products.[529] Under the FIT, a household installing an mCHP product would receive, for a period of 10 years, a generation payment of 10p/kWh (15.4¢/kWh) for all electricity generated plus an additional export payment of 3p/kWh (about 4.6¢/kWh) for any electricity that was not consumed in the home and was fed into the grid.

- The introduction of the FIT undoubtedly aimed to help accelerate the sale of Ceres Power's mCHP products.

6.4.6.2.3. BUT SOME TECHNICAL QUESTIONS REMAIN UNRESOLVED

The company had hoped to launch commercialization in the second half of 2011, and it is possible that Ceres Power is on its way to successful commercialization.

- As of March 2010, there reportedly were delays in the field trials of beta units because some technical questions still remained to be resolved.[528]
- Indeed, according to a recent address made by the CERES CEO, it appears that the company identified a number of issues that needed to be addressed in the in-home field trial program in 2012.[530] After these issues were addressed, Ceres would launch its first-generation product, which integrates a boiler and fuel cell module in a single wall mounted unit. The CEO indicated that that time frame would likely be in 2014–2016.

6.4.6.3. Rolls Royce Fuel Cell Systems

RR is UK's leading SOFC developer. It began basic SOFC cell and stack development in 1993[531] and has worked to develop a unique, large-scale pressurized SOFC/gas turbine (SOFC/GT) hybrid power-generation system in the megawatt range. RR's SOFC cell is a combination of tubular and planar designs, called the IP-SOFC or segmented-in-series (integrated planar) tube.[532] The IP-SOFC cell was expected to capitalize on low cost components, short current paths to achieve low cell resistance, the high power density of the planar design, and the unconstrained thermal expansion characteristic of the tubular design.

The UK DTI and the EU saw the IP-SOFC as highly promising and were eager to provide major support to encourage development. Indeed, the two organizations provided RR with funding for two dozen SOFC development projects. These projects often had similar goals and were carried out concurrently in order to improve the overall chances of success.

From 2000 to 2003, RR was involved in two concurrent projects, one sponsored by the UK DTI and one sponsored by the EU FP5, to scale up its IP-SOFC cells to kilowatt-level stacks. Extensive market and system studies on fuel cell technology during the course of the projects concluded that the optimum product for market entry would be a pressurized system at megawatt scale.[146] In 2004, RR formed a subsidiary, RR Fuel Cell Systems, to commercialize the IP-SOFC technology at the MW scale.

In the latter half of 2000s, RR turned to US DOE resources to fund the development of its IT-SOFC technology. In April 2007, it acquired SOFCo in Ohio from McDermott International, and in May 2008, RR participated in DOE's coal-based SECA program.

6.4.6.3.1. UK DEPARTMENT OF TRADE AND INDUSTRY PROVIDES SUPPORT

RR development efforts were encouraged and supported by the UK's DTI. Since the early

1990s, DTI funded as many as 10 national projects to develop the IP-SOFC technology. Initially, these projects focused on the development of modules and technical evaluations; by the early 2000s, the projects emphasized technology scale-up; and by the mid-2000s, development work focused on pressurized SOFC and gas turbine hybrid systems at megawatt levels.

- Projects that were started in the 1990s looked at ways to manufacture multicell electrolyte membrane-electrode assemblies (MEAs) and improvements in durability of the MEA modules. The development of multicell MEAs apparently was successful.[533] RR made considerable advances in process development and control, leading to the production of high quality support modules. According to company documents, however, a specified project target in module durability was not achieved.
- Attempts to scale up the modules ran into difficulty. The project "Development of a 1 kW IP-SOFC Stack (ETSU F/01/00144/REP)," which likely started in late 1990s, probably was the first attempt at scale-up. At the outset, RR demonstrated a 300 mm active length, 40-cell proof of concept module that generated 27.3 W, had a power density of 155 mW/cm^2 at 0.62 V mean cell voltage and 43 percent fuel utilization using simulated reformed hydrocarbon fuel.[534] In August 2000, the company developed a stack of nine subassemblies each carrying three 20-cell twin MEA modules (300 mm active length), which generated a maximum power of 828 W on humidified hydrogen. Operating the subassemblies individually at their maximum performance yielded a calculated maximum power of 960 W. Nonetheless, in the end, the overall stack output power fell short of project targets, which were to achieve 1 kW output and 50–80 percent fuel utilization at 950 °C.
- The next project, "Scale-Up to Multi-Tens of kW for the IP-SOFC (ETSU Report F/01/00197/REP)," also failed to achieve the project target of a fully populated multi-kilowatt stack due to technical challenges.[535] Another project, "Design & Build a 5 kW Thermally Self-Sustaining IP-SOFC Stack (ETSU F/01/00199/REP, DTI/Pub URN 01/1161)," did not result in a 5 kW stack.[536]
- In August 2003, DTI sponsored a project entitled "Pressurization of IP-SOFC Technology for Second Generation Hybrid Application (F/01/00278/00/00, URN Number 05/949)" (August 2003–December 2004). In this project, RR aimed to improve IP-SOFC materials and construction processes and design a multi-kilowatt stack to integrate into a 1 MW hybrid fuel cell system.[537] The total cost of the project was about £6 million (about $9.84 million at the 2003 exchange rate of $1/£0.61) with DTI contributing £2.9 million (about $4.75 million) and RR contributing the balance.[537] DTI's March 2005 project summary indicated that the project was successful. Major improvements were achieved in reducing leakage and increasing volumetric power density. The project confirmed that operating at elevated pressures would improve IP-SOFC performance, and an innovative hybrid system off-gas combustor concept was validated. A related project "Pressurization of IP-SOFC Technology for Second Generation Hybrid Application (PIP-SOFC) (NNE5-2001-00791)" was a collaborative effort undertaken with the EU between April 2003 and April 2006. In 2006, the pressurized 10 kW SOFC stack objective was successfully met, with two stack blocks producing 13 kW.[538] With this success, it was concluded that the pressurized SOFC concept was fundamentally sound and that it provided RR with the foundation to proceed with a 1 MW SOFC development program.

RR has gone on to a successful test of an 80 kW pressurized system, according to the company.[538] The success of the project has led to further investment in SOFC technology and development of supply chain activities.

6.4.6.3.2. EU FRAMEWORK PROGRAM PROVIDES MAJOR SUPPORT

RR also has received major support from the EU FP. Starting in 1996, RR participated in at least 13 EU FP projects. The UK DTI facilitated RR's participation in the EU FP projects by providing funds and securing positions for RR as project coordinator in some of the projects.[177] Of 13 EU FP projects in which RR participated, RR has served as coordinator in five of them. The other eight projects were coordinated by major European research organizations, including ECN (the Netherlands), Fraunhofer (Germany), Jülich (Germany), Risø (Denmark), and VTT (Finland).

- In mid- to late 1990s, RR participated in three EU FP4 projects that aimed to develop low cost and reliable SOFC cells and modules that would be suitable for use in distributed electricity production and cogeneration applications.[539] In 1996, it participated in a project to increase the durability of SOFC stacks. In 1997, it participated in a project, nicknamed LOCO-SOFC, to reduce stack cost through the use of cheap state-of-the-art materials and cheap, environmentally friendly printing techniques.[532] In 1998, it participated in a project to explore how best to facilitate SOFC technology in decentralized, electricity production and cogeneration applications within 10 years.[539]
- In 2000, RR was awarded €3.5 million (about $3.24 million at the 2000 exchange rate of $1/€1.08) under the EU "MF-SOFC (multifunctional-SOFC)" project (2000–2003) in FP5 to scale up the IP-SOFC technology from 1 to 20 kW.[540] After 42 months of work, RR developed a preliminary design for a 10–20 kW stack and prepared draft manufacturing specifications for modules.[540] It was only able to demonstrate operation up to kilowatt scale, however.[541] RR's partners in the MF-SOFC project included Risø, Imperial College, Gaz de France, and Advanced Ceramics (United Kingdom).
- In 2001, RR participated in the "IM-SOFC-GT" project (2001–2003) in FP5 to conduct an integrated modeling study to assess the performance of fuel cell/gas turbine (FC/GT) hybrids and set performance requirements for FC/GT hybrid-based generation systems through a technology characterization.[540] A wide range of system configurations were explored to find solutions that balanced system efficiency, emissions, and cost. Project funding was €2.9 million (about $2.6 million at the 2001 exchange rate of $1/€1.12). The main project finding was that pressurized configurations (vice nonpressurized) were superior in terms of efficiency, specific power, and cost of electricity, according to an EC document.[542]
- RR research continued in another EU FP project, termed "Pressurized IP-SOFC: A Path to Successful SOFC/Hybrids (PIP-SOFC)" (or "Pressurization of IP-SOFC Technology for Second Generation Hybrid Application PIP-SOFC"), which started in April 2003[541] and ran for 36 months.[538] The project aimed to develop a pressurized 10 kW SOFC block with integral internal reforming that would operate with natural gas.[541] The EU funded this project at €2 million 2003.[543] In 2006, the pressurized 10 kW SOFC stack objective was successfully met, with two stack blocks producing 13 kW.[538]
- In November 2006, RR started to work on an unusually long project "Pressurization of IP-SOFC Technology for Second Generation Hybrid Application – Target Action A (PIP-SOFC)."[537] The project is planned to run for 108 months, until November 2015. The project goals are not defined clearly and EU

funding is rather modest. It likely was designed to give RR time and autonomy to work toward developing pressurized megawatt class IP-SOFC systems combined with gas turbines. Project partners include University of Genova (Italy), Morgan Matroc (United Kingdom), Morgan Advanced Ceramics (United Kingdom), Gas de France, Imperial College, and Risø. Project cost was about €9.8 million (about $16 million at the 2003 exchange rate of $1/€0.61), and the EU grant was about €2 million (about $3.3 million). In the second quarter of 2007, RR succeeded in operating a single block 15 kW stack with an external fuel processor.[544]

RR also participated in the FP6 project, "Toward a Large SOFC Power Plant (the Large-SOFC Project)," which aimed to develop SOFC systems in the range of hundreds of kW to 1 MW.[545] (See the discussion at Section 6.4.2.1 for details of this project).

- RR's responsibility was to identify the improvements in pressurized test rigs. The project was expected to provide a clear picture of how megawatt size power plants should be planned, developed, manufactured, and brought into use, and contributed to RR's long term goal of developing a 1 MW pressurized SOFC power unit.

6.4.6.3.3. ....PROSPECTS FOR COMMERCIAL SUCCESS UNCERTAIN IN THE NEAR TERM

RR planned to introduce a 1 MW hybrid system field demonstration unit in 2007,[47] but the plan did not materialize. In 2008, RR participated in the US DOE SECA Program, hoping to develop 1 MW hybrid SOFC system under US sponsorship. So far, RR's goal to develop a large-scale pressurized SOFC/GT hybrid power-generation system in the megawatt range continues to face challenges:

- *Scaling up the stacks.* RR developed a 1 kW block in 2000, a 10 kW stack in 2006, and a 15 kW in 2007. As of July 2009, RR reported that since the fourth quarter of 2007, it had been testing a 125 kW design in a single- and two-tier configuration. It reportedly found the 125 kW scale test much more challenging than anticipated.[544]
- *Lowering stack cost.* According to industry sources in December 2004, although previous analyses had indicated that the materials and manufacturing cost of the RR design were low, published power densities have been low as well, resulting in an overall stack cost which could be higher than desired.[47]
- *Stack lifetime.* DOE indicated that RR succeeded in achieving an average degradation rate of less than 0.5 percent per 1000 h in testing under system relevant conditions for up to 8500 h. DOE noted that RR is making modifications to further improve durability.

6.4.7. Europe Lacks Clear SOFC Strategy

Europe lacks a clear strategy to transition SOFC capabilities into commercially practical applications. Several countries in Europe are involved in the SOFC business, including Denmark, Finland, Germany, Italy, the Netherlands, Switzerland, and the United Kingdom, but few of them are engaged in vigorous SOFC research. There is no "critical mass" of researchers in different countries that are pursuing SOFC development.

Instead, we have seen a zero-sum game where one European company buys the research talent of another, rather than growing their own teams. Finland is outsourcing SOFC stacks to Denmark and is not interested in developing its own capabilities. Italy has expressed an interest in establishing an SOFC industry but did not want to develop its own technology. Instead, it chose to acquire Switzerland's HTc. German H.C. Starck also did not choose to develop its own SOFC technology but acquired InDEC of the Netherlands. This has not led to increased fuel cell research efforts in Europe as a whole.

The research funding via the EU's FP is commendable, but episodic. Research teams convene around a particular project and then disperse until the next project is started. The lack of continuity and progress in technology development is self-evident.

At the same time, some SOFC expertise is leaving Europe altogether or just disappearing.

- The United Kingdom has found that RR's SOFC competence has largely been transferred to the United States. With incentives from the US SECA program, RR built a manufacturing plant in the United States and, despite substantial investments from the United Kingdom and the EU, RR SOFC development is now based in the United States.
- Siemens chose to close down its SOFC subsidiary in the United States, after investing billions of dollars in technology development, rather than transferring its core SOFC knowledge to Europe.

There is one positive development that might signify an increase in European SOFC capabilities. Germany potentially has gained an economic advantage by attracting a foreign SOFC developer to its industry base. CFCL of Australia has established large-scale manufacturing facilities and component supply chains in Germany. Germany in turn has become the virtual owner of CFCL's SOFC capabilities.

Europe also has one important competitive edge over the United States. It has more companies working on SOFCs than does the United States and thus potentially could tap into a greater research base. But without orchestrating and coordinating this talent base, success will be elusive. Using power density as a surrogate for overall fuel cell performance, European SOFC designs either are performing at about the same level as US designs or are somewhat less capable. Clearly, European SOFC companies have challenges ahead of them, both in terms of achieving key performance goals and in gaining a competitive edge relative to other global developers.

6.5. OTHER COUNTRIES

6.5.1. Australia

Australia has a relatively small fuel cell program, based on the SOFC technology of CFCL.[546] CFCL was established in 1992 by the Commonwealth Science and Industry Research Organization (CSIRO)—Australia's principal public sector research institution—and a consortium of leading energy and industrial companies. CSIRO conducted SOFC research and testing in the 1970s and 1980s,[547] building a global reputation for its basic SOFC know-how and end-to-end expertise.

6.5.1.1. CSIRO establishes Ceramic Fuel Cells Limited in 1992

CFCL, based in Melbourne, is a developer of planar SOFC technology, focusing on 1–2 kW mCHP units.[548] In the early 2000s, CFCL selected the 1 kW mCHP system as its first commercial product.[549]

- CFCL is one of the few powder-to-system developers in the world; it manufactures its own powders, cells, and stacks, and integrates them with BoP into a complete system.[550]
- CFCL has R&D, testing and pilot manufacturing facilities at its headquarters in Melbourne and additional manufacturing facilities, including a ceramic powder manufacturing plant in Bromborough, the United Kingdom, which was established in August 2006[551] and a high-volume fuel cell stack manufacturing facility in Heinsberg, Germany, established in December 2006.[552]

In the early 2000s, CFCL made remarkably rapid progress in developing kilowatt-size stacks and systems. CFCL's first complete fuel cell system was the first technical demonstrator 1 (TD1). The system validated the early design concept for the BoP and the control systems by operating continuously for about 4000 h.[553] By 2003, CFCL had developed a stand-alone CHP unit, the combined heat and power 1 (CHP1), which integrated the BoP and reduced the system size.[554] CFCL next developed CHP2, incorporating improved BoP technology.[555] In 2004, it built five demonstration CHP2 units for field trial.

- The first of these units was delivered in June 2005 to Powerco of New Zealand for field trials.[556] Two CHP2 units were dispatched to German utility EWE in December 2005. A fourth unit was installed on the premises of Energy and Telecommunications Training Australia (ETTA) in Chadstone, Melbourne, and was subsequently moved to the Szencorp site in South Melbourne in late 2005. The fifth unit remained in CFCL's laboratory for testing and engineering development.[555]
- In early April 2006, CFCL reported that the CHP2 units installed at customers' premises in Germany, Australia, and New Zealand achieved 10,000 h of operation.[557]

6.5.1.2. CFCL Develops Pre-Commercial Demonstration Units in 2005—Creates Major Interest among Utilities

During the latter half of 2000s, CFCL developed two types of pre-commercial demonstration mCHP units. In October 2005, CFCL launched the 1 kW Net~Gen mCHP unit.[558] The unit was simple in structure and compact—about the size of a standard domestic washing machine.[559] In November 2007, CFCL launched the Net~Gen Plus mCHP unit, a more advanced pre-commercial demonstration system.[560] The system had characteristics similar to its predecessor, Net~Gen, but incorporated a more powerful, higher efficiency stack module, called Gennex:

- In August 2008, CLFC announced that it had increased the output from each of its Gennex fuel cell stacks to 2 kW at minimal additional balance of plant cost, significantly reducing the unit's cost per kilowatt.[561] A 2 kW unit would provide ample power for the average household's annual "base-load" requirements, as well as additional power for export to the grid.

In February 2009, CFCL announced it had achieved electrical efficiency of 60 percent from one of its natural gas powered mCHP units with Gennex modules.[562] The unit was installed at a home, functioned as a home appliance, and exported 1.5 kW of electricity to the grid. CFCL's units also recovered heat from the production process to produce home hot water, increasing the units' efficiency to 85 percent, according to the company. It had an average power density of 320 mW/cm^2 at 0.83 V.[563]

Undoubtedly inspired by CFCL's remarkable technological developments, many global utility and electrical equipment manufacturers started to form product development agreements with CFCL to field test the mCHP pre-commercial demonstration units. There is little media reporting, however, on the extent to which these projects achieved success:

- In December 2006, CFCL signed a product development agreement with Gaz de France (GDF) and De Dietrich Thermique, part of the Remeha Group (Europe's largest boiler maker), to develop a fully integrated mCHP unit for French residential use.[564] In July 2009, CFCL extended its product development agreement with GDF SUEZ.[565] FCL and De Dietrich Remeha have been developing a fully integrated mCHP unit, which would produce up to 2 kW of electric power and provide hot water for use in French households. The mCHP unit was comprised of a fuel cell module from CFCL with a condensing boiler from the De Dietrich Remeha Group. CFCL and De Dietrich Remeha planned to deliver the first fully integrated unit to GDF SUEZ in late November 2009.

- In March 2007, it signed a product development agreement with EWE, one of Germany's top energy suppliers, and Bruns Heiztechnik GmbH (Bruns), a specialist German boiler manufacturer.[566] The project partners would design and develop a fully integrated mCHP unit for the German residential market. In December 2007, CFCL shipped a Net~Gen Plus unit to its utility partner EWE for field trials. The unit was to be installed on the premises of Bruns Heiztechnik, CFCL's German appliance partner.[567]
- In July 2007, CFCL, Nuon, Holland's leading electricity, gas, and heat utility, and De Dietrich Remeha Group agreed to jointly develop an mCHP unit for the residential market across the Netherlands and Belgium.[568] In February 2008, CFCL announced that it received a volume order from Nuon.[561] CFCL and Nuon agreed on a set of performance targets for a commercial unit, and based on CFCL's achievement of these targets, Nuon would order 50,000 fuel cell systems over a 5-year period starting in June 2009.
- In July 2007, CFCL and E.ON UK signed an agreement to develop and deploy a prototype 1 kW mCHP unit in the United Kingdom.[569] In August 2007, Gledhill Water Storage joined the project with CFCL and E.ON UK.[570] In January 2008, CFCL shipped a Net Gen Plus unit to the United Kingdom as part of this agreement.[571] In February 2009, CFCL and E.ON extended their agreement.[572] The partners planned to develop the mCHP system from 2009 to 2012, with a market launch in the United Kingdom expected to follow thereafter. E.ON and CFCL also agreed on a potential future order profile, which would secure E.ON's exclusive distribution rights for CFCL mCHP units within Great Britain. A minimum order of 100,000 units over 6 years starting in 2012 would be required for E.ON to retain exclusivity.
- In January 2008, CFCL signed an agreement with Paloma Industries of Japan to evaluate and develop integrated mCHP products for the Japanese market.[573]

6.5.1.3. CFCL Establishes Large-Scale Manufacturing Facilities and Component Supply Chains in Europe

In anticipation of strong demand for its mCHP units from Europe, CFCL established a large-scale manufacturing base in Europe and formed collaborative ties with regional component suppliers. By the end of the 2000s, CFCL was ready for high volume mCHP production in Europe.

- *Manufacturing facilities*. In October 2007, CFCL completed construction of a 20-ton high purity ceramic powder plant in Bromborough (United Kingdom)[547] to produce YSZ.[574] In October 2009, CFCL opened a manufacturing plant in Heinsberg, Germany, for commercial production of its fuel cell systems.[561] The plant was one of the first in the world to produce SOFC stacks with state-of-the-art automated manufacturing equipment. It had an initial capacity of 10,000 stacks per year, which could be increased to 160,000 per year if required.[561]
- *Component suppliers*. In January 2008, CFCL and H.C. Starck of Germany agreed to develop advanced SOFC components, particularly ASCs.[431] H.C. Starck would supply fuel cells at an agreed price until December 2011, stabilizing long-term costs for CFCL. H.C. Starck commissioned two state-of-the-art manufacturing lines, with an annual capacity of about 700,000 cells each. Also in February 2008, CFCL signed a cooperation and supply agreement with

CeramTec AG of Germany to secure a high volume of SOFC components.[431] CFCL would license its relevant intellectual property to CeramTec to optimize the performance of the cell and provide a range of ceramic powders to CeramTec to make fuel cell components for resupply to CFCL. Once optimized, CeramTec would supply fuel cells to CFCL at pre-agreed prices until December 2011.[431]

6.5.1.4. CFCL Launches New Commercial Product—2 kW BlueGen Power Plant

As soon as the manufacturing and supply infrastructures were established, CFCL made a surprise announcement. In May 2009, it launched a new commercial product—a modular generator called BlueGen—a "mini power station" about the size of a dishwasher.[575] This opened a flood gate of major purchase orders. The CFCL BlueGen product, which incorporated a Gennex module, generated up to 2 kW of electricity with a peak electrical efficiency of up to 60 percent while exporting excess electricity to the grid.[226] BlueGen is a modular unit, which can be placed at a user site with few modifications, providing an alternative to fully integrated specific European- or Japanese-style systems.[576] In addition, CFCL is working closely with utilities and appliance partners to develop an mCHP product incorporating the Gennex fuel cell module.

6.5.1.5. Commercial Success Not Yet Guaranteed

CFCL has made remarkable progress, particularly in terms of improving power density and efficiency. Between June 2007 and July 2008, the power densities of its fuel cells increased by more than 50 percent, from 350 mW/cm^2 to 500–650 mW/cm^2, at an electrical efficiency of more than 50 percent.[577] These improvements allowed CFCL to increase power output of its commercial stack design from 1 to 2 kW while reducing the cost per kilowatt of the fuel cell module.

CFCL also has made significant improvements in cell stack degradation and lifetimes. In 2008, degradation had been reduced by 35 percent over 4 months—from 1.53 percent/1000 h reported in February 2008 to less than 1 percent/1000 h in July, when operating a 1 kW stack at 750 °C with natural gas .[577] According to CFCL's 2009 financial report, its individual fuel cells have been operating at a test station for 21,000 h (about 3 years), and complete fuel cell stacks have operated for 13,700 h (one and a half years).[578]

- CFCL is committed to making further improvements. Since April 2007, it has been working with Germany's Jülich on materials development and accelerated testing and modeling.[566] They are working to optimize the lifetime of fuel cell stacks and improve the durability and performance of complete fuel cell systems. The companies have created separate projects to continue to improve the performance and reliability of fuel cells, stacks, and system components, with each project having clear targets for fiscal 2010.[578]

It is still not clear, however, if the cost of CFCL products will be competitive with other energy sources. CFCL has not yet published costs of its products. CFCL had forecast that the BlueGen would cost around $8,000 per unit in mass production with a payback time to the customer of 7 years and product lifetime of 15 years.[579] It is not certain when the cost, payback time, and lifetimes in the forecast will become a reality.

Since the May 2009 announcement on BlueGen, nonetheless, CFCL has been increasingly successful in marketing its units and Gennex-based mCHP products. In 2009, CFCL sold only five units. In 2010, it sold 250 units, and in 2011, 531 units, doubling the previous year's sales. BlueGen and Gennex are clearly popular and were sold globally, including in Australia,

Germany, Japan, the Netherlands, Switzerland, the United Kingdom, and the United States.

- Still, CFCL's annual financial reports show continuous, large net losses. Clearly, the company's products are not yet producing profits, and so the company's success and survival is uncertain in the near term (Table 6.12 and Fig. 6.2).

6.5.2. Canada

Since 1985, Natural Resources Canada has been promoting fuel cell R&D at an annual budget of approximately C$20M[177] (about $19 million at the 2010 exchange rate of $1/C$1.06). Canada has had two notable SOFC developers: Fuel Cell Technologies Ltd. and Global Thermoelectric.

6.5.2.1. *Fuel Cell Technologies Limited*

FCT, Kingston, Ontario, commenced operations in October 1994 with nine core staff drawn from Alcan's subsidiary, Alupower Canada, a developer of aluminum fuel cells, which had earlier been closed down by Alcan.[580] FCT designed and built custom fuel cell power systems for internationally known customers for several years. In May 2000, FCT completed a reverse takeover of ThermicEdge Corporation, an Edmonton ceramics manufacturer, to start SOFC development.[580] In November 2000, it signed a joint development agreement with SWPC to develop FCT's first-generation 5 kW SOFC system using SWPC tubular SOFC stacks.[581]

In 2001, a single 5 kW prototype system was designed and built.[582] FCT was responsible for system integration and BoP while SWPC was responsible for the cell stacks, recuperator, insulation systems, and reformer. In May 2002, FCT announced the successful operation of a prototype 5 kW SOFC system in Kingston, fueled by natural gas.[580] To accelerate its development of SOFC products, FCT reached a collaboration agreement with Kinectrics, the former Ontario Power Technologies, which had more than a decade of experience in SOFC power system definition, and in subsystem development and testing.

- FCT also built a manufacturing facility adjacent to its existing building during 2002.

In 2002, FCT operated the 5 kW prototype on hydrogen, methane, and natural gas and was preparing two systems for shipment to Stockholm, for demonstration at the Hammarby Sjostad project in August.[583] Also in 2002, FCT installed one 5 kW system at Presidio Trust in San Francisco, one 5 kW unit at the National Fuel Cell Research Center at the University of California, Irvine, two 5 kW units at Yosemite National Park, and one 5 kW unit at the AQMD, California.

- FCT also had a collaboration agreement with InDEC/ECN in the Netherlands to develop systems using planar SOFCs supplied by InDEC/ECN.[560] FCT perhaps hoped to reduce its reliance on SWPC's tubular technology in the future.[492]

In 2003, the four first-generation or "Alpha" demonstration units were deployed and operated at four different locations: RWE in Essen, Germany, Ford Motor in Detroit, JFE Urban Development Corporation in Yokohama, Japan, and at the University of Alaska, Fairbanks.[582] The unit delivered to the University of Alaska was deployed at a Fairbanks Natural Gas site and demonstrated the best performance of the four. It operated for 9650 h, generated the highest average electric output of 4.36 kW, and demonstrated the highest system efficiency of 39.3 percent.[584]

- Clearly, the conventional SWPC tubular SOFC technology provided a proven platform with acceptable performance in terms of efficiency and life.[47] However, the stack power density was low and the cost was high.

TABLE 6.12 Ceramic Fuel Cells Limited: BlueGen and Other mCHP Order Book (as of January 2012)

Month	Customer and activities	Country	Number
October 2009	Leading Australian energy utility Energy Australia orders a BlueGen unit to be demonstrated in a showcase sustainability home in Sydney	Australia	1
November 2009	Paloma orders a BlueGen unit to be installed in its warehouse and sales office in Sapporo, Japan	Japan	1
December 2009	Installation of a BlueGen unit with VicUrban, the Victorian Government's sustainable urban development agency along with Origin Energy. Two additional units will be installed in the future	Australia	3
Jan 2010	First sales of BlueGen in Europe—to German utility EWE	Germany	3
February 2010	BlueGen sale in the Netherlands—to Kiwa Gas Technology	The Netherlands	1
February 2010	BlueGen sale to German utility RheinEnergie AG	Germany	1
February 2010	BlueGen sale to Germany utility E.ON Ruhrgas	Germany	1
February 2010	BlueGen sale to European utility Alliander	Germany	1
March 2010	BlueGen sale to Swiss utility Cosvegas	Switzerland	1
March 2010 (the second order in July 2010)	Mitsui & Co. ordered BlueGen to be installed at the testing facility of Osaka Gas	Japan	2
March 2010	Two BlueGen sales to the German Association of Gas and Water Industry (DVGW)	Germany	2
April 2010	First BlueGen sale in the United Kingdom—to Ideal Boilers	United Kingdom	1
April 2010	Conditional order for 30 BlueGen from the Victorian Government's Office of Housing; Order confirmed in August 2010	Australia	30

July 2010	Canberra Institute of Technology	Australia	1
August 2010	A BlueGen unit to Tokyo Gas	Japan	1
August 2010	An unnamed energy company in California	United States	1
December 2010	An order of 200 CHP units was received, dependent on EWE obtaining partial funding from the German government of €700 million between 2008 and 2018; EWE received a letter of intent from the funding body in April 2011; Funding formally approved in July 2011	Germany	200
December 2010	Adelaide's Central Market, Adelaide City	Australia	1
December 2010	Three BlueGen to E.On, United Kingdom	United Kingdom	3
April 2011	One BlueGen to De Groene Bocht (Dutch office building), Amsterdam	The Netherlands	1
April 2011	An order of 200 was received, depending on EWE obtaining partial funding from the German government of €700 million between 2008 and 2018; EWE has received a letter of intent from the funding body	Germany	200
May 2011	25 BlueGen sold to Ausgrid's "Smart Grid, Smart City" Project, Newcastle	Australia	25
July 2011	An initial order for 100 BlueGen units from Sanevo Lizenz-GmbH (currently Saneve Blue Energy); the first 100 units successfully introduced to the German market in January 2012	Germany	100
September 2011	An order of 100 BlueGen received from Zestiq B.V.	The Netherlands	100
November 2011	A new 105 unit order received from E.ON United Kingdom	United Kingdom	105

Source: Various CFCL Announcements, http://www.cfcl.com.au/Recent_announcements/, Ceramic Fuel Cells Limited, no date, accessed May 2012.

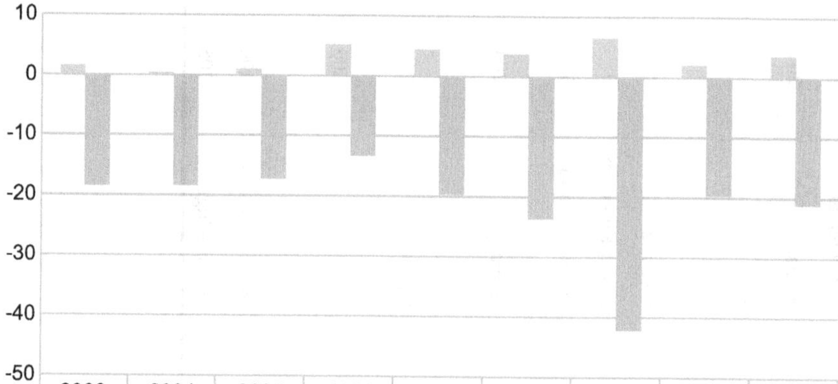

	2003	2004	2005	2006	2007	2008	2009	2010	2011
Revenues	1.570352	0.356492	0.978349	5.136239	4.426355	3.745176	6.492025	2.033	3.681
Net Loss	-18.4895	-18.3922	-17.2503	-13.3173	-19.6782	-23.6789	-42.1759	-19.652	-21.176

FIGURE 6.2 Ceramic Fuel Cell Limited: Financial Performance *,**(Millions of Australian dollars). For color version of this figure, the reader is referred to the online version of this book.

A second-generation system was developed in 2004, with plans to build and install up to 20 of the systems in 2005. Those shipped or installed include the following:[585]

- In December 2004, FCT delivered one 5 kW unit to Parker Dam State Park Fuel Cell Pavilion, Pennsylvania.
- In March 2005, FCT delivered six 5 kW systems: three systems to SWPC in Pittsburgh, one system to the Canadian Center for Housing Technology, Kingston, and one system to the Institute for Fuel Cell Innovation at the National Research Council, Vancouver, and one system to the Federal University Itajuba in Brazil, making it the first SOFC in South America.[586]
- In June 2005, FCT shipped one 5 kW unit to the University of Liège in Belgium.
- In July 2005, FCT installed four 5 kW units at the University of Toronto at Mississauga, Hydrogen Village, Canada.

While FCT seemed to be technically successful, in February 2006 it underwent restructuring.[587] Under the terms of the restructuring, FCT would sell all of its assets and business to a new subsidiary, named FCT Opco, in exchange for assumption of specified liabilities. Opco would then change its name to FCT and continue FCT's SOFC business. The restructuring was undoubtedly necessitated by FCT's mounting losses. According to available financial information, FCT's net loss grew rapidly as it shifted its business to SOFCs. FCT's financial figures swung from a net income of C$47,000 (US$46,300 at the January 1999 exchange rate of US$1/C$1.15) in 1999 to a net loss of C$1.535 million (US$1.06 million at the January 2000 exchange rate of US$1/C$1.45) in 2000 and then to a net loss of C$5.174 million (US$3.45 million at the January 2001 exchange rate of US$1/C$1.5) in 2001. Starting in 2002, FCT obtained a large number of sales contracts, and at the time of restructuring in early 2006,

* This figure was created based on CFCL annual reports. CFCL fiscal year is from July to June.

** Net loss is a negative amount by which total of costs and expenses exceeds total revenue.

FCT had a backlog of two dozen 5 kW systems to deliver, likely suggesting that the more FCT sold, the greater its losses.

- FCT, in partnership with the South Coast Air Quality Management District (SCAQMD), agreed to install, operate, and manage ten 5 kW systems in a major residential demonstration project. Two or three units were to be located at LADWP, one or two units at Sempra Utilities, and one to two units at the University of California, Irvine.[585]
- FCT also signed a letter of understanding with Border States Electric Supply (BSE) to promote and distribute FCT's 5 kW SOFCs to customers in 15 US states.[423]

In 2007, FCT was acquired by Acumentrics of the United States. FCT in Kingston is now Acumentrics Canada.

6.5.2.2. Global Thermoelectric

Global Thermoelectric, a leading supplier of thermoelectric generators in Calgary, Alberta, Canada, began SOFC development in 1997. Global's SOFC program was based on anode-supported planar SOFC technology acquired from Forschungszentrum Jülich.[492] Global's target was to develop an integrated cell manufacturing technology to produce high performance planar SOFCs.[633] By 2000, Global was able to successfully test a prototype SOFC unit capable of providing the base load power requirements of a residential home.

In July 2000, Global formed a strategic alliance with Enbridge Inc., Canada's largest natural gas distributor, to develop and distribute natural gas–fueled SOFC systems for residential use.[634] Under the terms of the alliance, Enbridge would invest C$25 million in Global to fund further technology, design, and product development required to produce a commercial product. Under the terms, Enbridge would have exclusive distribution rights for the products for all of Canada, and Global would own the technology, designs, patents, and trademarks.

- In May 2001, Global announced the delivery of a prototype SOFC system for residential power generation to Enbridge. The prototype was powered by natural gas and had an electric power rating of 2.3 kW.[492]

Over the following few years, Global developed several new stack designs.[588] Global was planning to deliver a new stack for its next generation of prototype system by mid-2003. Global also developed several 2 kW natural gas prototype systems. The latest generation system in 2003 achieved 2900 h of operation on natural gas and demonstrated 23 percent net AC efficiency. Global was expecting to achieve 30 percent net AC efficiency with its next system platform.

During the same time period, Global made significant progress in anode-supported SOFC designs. Cell power density increased dramatically. For example, the power density of a single cell at 0.7 V using H_2 as fuel was 0.15 W/cm^2 in 1998, 0.528 W/cm^2 in 1999, 0.9 W/cm^2 in April 2000, 1.093 W/cm^2 in June 2000, and 1.4 W/cm^2 in October 2002, the world's highest power density.[633] Similarly, a power density of more than 1 W/cm^2 was demonstrated at 750 °C. By 2003, Global had established a cell manufacturing process that consisted of three phases.[633] Anode substrates approximately 1 mm in thickness were produced by a single-step tape-casting process. The anode, electrolyte, and cathode functional layers were screen-printed directly on the anode substrate tape. Multilayer cells were then sintered into complete cells through a continuous co-firing process.

- In 2003, Global was able to produce cells at a rate equivalent to 5 MW/year with greater than 90 percent yield. The production pipeline was shortened from 30 days to 4 days, and cell cost per kilowatt (direct labor and materials) was reduced by 65 percent.
- In 2003, Global projected that their latest production cells would have at least a 15,000 hour service life and their latest stack design

would have a stack service life of 15,000 h of continuous operation.

Since the beginning of 2003, however, Global ran into financial difficulty, announcing a cut of 47 jobs in a bid to reduce expenditures.[589] Global, nonetheless, was clearly an attractive company to other fuel cell companies. In April 2003, Quantum Fuel Systems Technologies Worldwide Inc., an alternative energy company in Irvine, California, announced an agreement in which Quantum would acquire all of the outstanding shares of Global Thermoelectric.[590] Quantum would purchase Global in a stock-for-stock exchange valued at approximately US$75 million.[591] In a surprising move, in August 2003, FCE entered the picture and proposed that it would acquire GTI in an all-stock transaction, which was valued at approximately $80 million, or $2.72 per global common share.[592] Quantum decided not to pursue the acquisition of Global and Global paid Quantum a $2-million break-up fee.

- In August 2003, Global was acquired by FCE of the United States, which in turn transferred Global's SOFC assets to Versa Power in November 2004, in which it holds a stake.[47] Global is now Versa Power, using technology that originally was based on Jülich's anode-supported planar SOFC technology.

6.5.2.3. *Both Acquired by US Companies*

Because of financial difficulties encountered by both FCT and Global Thermoelectric, these companies ultimately were sold to US SOFC developers. As a result, Canada lost its national competence in SOFC technology.

6.5.3. South Korea

South Korean institutes and universities began fuel cell research and development in the 1980s, and SOFC research began in 1989.[593] The SOFC research has been modest and less advanced compared to other fuel cell technologies.

- In 1994, the South Korean Ministry of Commerce, Industry, and Energy launched a project to develop a 100 kW SOFC power plant.[594] It selected Ssangyong Research Institute, the research arm of Ssangyong Cement Industrial Company, to engage in development work from early 1994 to the end of 1998. The project apparently did not succeed,[595] and little information is available.

Today, South Korea is focused on small SOFC systems for residential and APU applications. Two research institutes play the primary role in carrying out SOFC development in South Korea: the Korea Electric Power Research Institute

	85	86	87	88	89	90	91	92	93	94	95	96	97	98	99	00	01	02	03	04	
PAFC	6kW Test (KIER, KEPRI)-KEPCO																				
				2kW Dev (KIER)-MOST					50kW Dev. (KIER, LG Caltex)-MOCIE												
PEMFC										1kW (KOGAS)		5kW Dev. (KIER)-MOCIE						3kW Dev. (KIER, CETI)-MOCIE			
MCFC									2kW (KEPRI, KIST)-MOCIE			25kW (KEPRI, KIST)-MOCIE						100kW (KEPRI, KIST)-MOCIE			
SOFC										100kW (Ssangyong)-MOCIE									1kW (KEPRI)-MOCIE		
																		10kW (KIER)-MOCIE			

FIGURE 6.3 History of Stationary Fuel Cell R&D in South Korea: 1988–2004. The figure was recreated based on what was presented in Sung-Chul Shin (Korea Institute of Energy Research), "Innovation In Fuel Cell Technology In Korea," OECD Conference on the Innovation of Energy Technologies, In collaboration with The National Academy of Sciences, Washington, D.C., 29-30 September 2003.

(KEPRI) and the Korea Institute of Energy Research (KIER). KEPRI focuses on planar SOFCs and KIER focuses on tubular and flat-tubular SOFCs. They are working to develop 1–5 kW CHP and APU units (Fig. 6.3).

6.5.3.1. Korea Electric Power Research Institute

KEPRI is an independent research institute that supports R&D to promote the Korea's electric power industry. In 2003, the Institute started to develop 1 kW SOFC power plants under a national project implemented by the South Korean Ministry of Commerce, Industry, and Energy (MOCIE).[594] The project, which was planned as a 3-year effort (July 2003–June 2006), aimed to develop 1 kW SOFC modules and systems for residential power applications.[593] The project apparently was delayed a year, but KEPRI had significantly improved cell performance by 2005; its cells achieved a maximum power density of 507 mW/cm^2 at 650 °C.[596] Two years later, KEPRI had achieved the project's basic goals.

- By 2007, KEPRI successfully developed a first-generation 1 kW CHP SOFC.[597] The 1 kW SOFC stack was composed of 37 anode-supported planar cells with low cost stainless steel interconnects. It had a maximum power density of 0.35 W/cm^2 at 750 °C. When the system was operated with a reformate gas, the stack delivered 1.13 kW of electrical power at 40 A and 1.3 kW at 50 A, respectively. Hot water was made via recovered heat from combustion of unreacted anode off-gas and heat captured from the fuel cell directly.
- In 2008, KEPRI developed a second-generation system.[598] The system was divided into a hot box and a cold BoP unit, along with a water reservoir. The hot box contained an SOFC stack made up of 48 improved cells and ferritic stainless steel interconnects, a fuel reformer, catalytic combustor, and heat exchangers. The thermal management and insulation system were designed for self-sustaining operations. The cold BoP unit was composed of blowers, pumps, a water trap, and system control units. When the 1 kW class SOFC system was operated at 750 °C using hydrogen, stack power reached 1.6 kW at 50 A. In self-sustaining operations, stack power was about 1.3 kW with hydrogen and 1.2 kW with city gas, respectively. The system also recovered heat equivalent to 1.1 kW to make hot water. KEPRI also manufactured a 2 kW class SOFC stack and system by scaling up the second-generation 1 kW system.

According to a 2007 publication, KEPRI was expected to develop a 5 kW class system for CHP by 2010 as part of the same national project. In this second stage, the target for system efficiency was 35 percent and the target for total efficiency including heat recovery was 75 percent.[599] In June 2010, KEPRI indicated that it had designed and manufactured a 5 kW cogeneration system which contained 2 module stacks comprised of cells 15 cm × 15 cm in area.[600] The system likely will continue to be under development for some time.

6.5.3.2. Korea Institute of Energy Research

KIER was founded in 1977 to support R&D on clean and efficient energy technology for future energy security.[601] KIER started R&D on anode-supported tubular and flat-tubular SOFC development in 1997.[602] In 2002, MOCIE sponsored a national project in which KIER would develop a 10 kW SOFC power plant. Little information is available regarding the project. It might have failed because the goal likely was too ambitious at that time.

In 2003, KIER was working to develop anode-supported flat tube cells. In November, KIER reported that they had developed the basic technology for the cell and the manufacturing processes for the anode tube, electrode, electrolyte, ceramic interconnect materials, and metallic interconnect materials.[603]

- In 2004, MOCIE implemented another national project to develop an SOFC system for APU applications.[593] KIER was the project leader of the 3-year effort (August 2004–July 2007). MOCIE likely decided to take advantage of KIER's anode-supported flat tube capabilities to develop the APUs. In 2005, KIER reported that it developed an anode-supported flat tube SOFC using an induction furnace brazing process that significantly improved power density. Performance of the flat tube cell was 210 mW/cm^2 (0.7 V, 300 mA/cm^2) at 750 °C.[604]

In an interesting development in 2006, KIER started to design and construct a pressurized 5 kW SOFC/GT hybrid system, incorporating Jülich's 5 kW stack.[417] The 5 kW hybrid system was completed in 2007. According to a paper published in 2008, the system combined an SOFC power system with an MGT.[605] At atmospheric pressure, the output of the SOFC system was 8.1 kW using H_2 and 4.7 kW using pre-reformed gas. KIER reported that it achieved an output of 5.1 kW under 3.5 atm pressure with pre-reformed gas. It considered the project a success. Little information is available on operating hours or other performance characteristics, however.

- KIER was undoubtedly impressed with Jülich's SOFC stack performance and sought to strengthen its cooperative relationship with the research institute. In April 2009, it signed a 5-year partnership agreement to develop SOFC as well as DMFC technology for practical applications.[418]

As part of the national project to develop a 1 kW APU system, KIER in 2006 reported it had developed a 500 W anode-supported flat tube solid SOFC stack which operated at intermediate temperatures (700–800 °C).[606] The anode-supported flat-tubular cell used ferritic stainless steel interconnects that were joined using an induction furnace brazing process. Current collection at the cathode was achieved by winding Ag wire, while current collection at the anode was achieved by using Ni wire and felt. A stack consisting of 12 modules was connected in series; each module was comprised of two anode-supported flat tube cells connected in parallel. The cells had an effective electrode area of 90 cm^2. Cell performance using 3 percent humidified H_2 and air at 800 °C showed a maximum power density of 0.6 W/cm^2. With this design, KIER believed that it had established a proprietary concept for the anode-supported flat tube cell stack.

In 2007, KIER reported that it had designed a complete 1 kW APU system. The system included an SOFC stack with 20 cell bundles and a reformer, heat exchanger, burner, and so on. One cell bundle was made of six tubular cells.[607]

KIER also had developed high power density flat tube SOFCs with thin interconnects. To make a stack, 37 modules were connected in series, with each module containing two anode-supported flat-tubular cells connected in parallel. The cells had an effective electrode area of 108 cm^2. Performance of the stack in 3 percent humidified H_2 and air at 800 °C showed a maximum power of 1.3 kW.

- Little additional information is available. The performance characteristic of the 1 kW APU system is unknown.

In October 2010, KIER reported that it fabricated a novel design for a flat-tubular segmented-in-series SOFC unit bundle.[608] The tubular ceramic supports were fabricated by an extrusion technique. The anode functional layer and the YSZ electrolyte were deposited onto the ceramic support using a dip-coating method. After co-sintering at 1,350 °C for 5 h, a dense, crack-free YSZ film about 10 μm thick resulted. Also, the multilayered cathode composed of LSM/YSZ composite, LSM, and LSCF, was coated onto the ceramic support by dip-coating and then sintered at 1150 °C. The porosity, mechanical property, and gas permeability of the ceramic support was measured and the cross section and surface morphology of ceramic support were analyzed by using SEM imaging.

- A single cell showed a maximum power density of 38, 67, and 102 mW/cm^2 at 700, 750, and 800 °C, respectively, using moist hydrogen (300 mL/min) as fuel and ambient air as oxidant.
- Also in 2010, KIER reported that it developed a 1.3 kW anode-supported flat-tubular SOFC stack.[609] The stack consisted of 74 cells (2 bundles of 37 cells connected in series) and successfully generated 1.3 kW in air at 800 °C. KIER is currently working to improve stack performance.

6.5.3.3. *No Plans for Near Term Commercialization*

The two research Institutes in South Korea have established the basic SOFC technology and fabricated 1–5 kW SOFC systems. Little information is available regarding the performance and the lifetime of these systems. Plans for near-term commercialization of the 1–5 kW CHP or APU systems are unknown.

6.5.4. China

The SOFC effort in China is modest compared to other fuel cell development efforts. With the launch of the 863 Program in 2001, significant funding was directed to fuel cell R&D, but most funds were directed to PEMFCs for transport applications.[610] Reportedly, 74 percent of the fuel cell funding in China goes to transportation.

While most SOFC research institutes in China still conduct basic-level research, two institutions are looking at commercial applications. They are the Ningbo Institute of Material Technology and Engineering (NIMTE) and the Shanghai Institute of Ceramics (SIC). Both are part of the Chinese Academy of Sciences. NIMTE appears to be the more advanced of the two.[611]

6.5.4.1. *Ningbo Institute of Material Technology and Engineering, Chinese Academy of Sciences*

NIMTE was established in March 2006 and in that same year it established the Fuel Cell and Energy Technology Division (FCET) to focus on SOFC technology.[612] As of October 2009, NIMTE had 45 full time scientists and engineers; they were from varied academic disciplines, including Chemistry Engineering, Applied Physics, Material Science and Engineering, Mechanical Engineering, Power Engineering, Electrical and Electronic Engineering, Control Engineering, System Science, as well as Computer Science.[612] FCET's SOFC research encompassed nanopowder preparation, single cell manufacture, cell testing, stack components and stack fabrication, fuel processing, and thermal management, as well as system integration.[612]

NIMTE's FCET has specialized in developing ultra-thin, high-strength, planar SOFC cells.[611] In August 2009, it set up China's first SOFC manufacturing line, which has a production capacity of 2,000 SOFC cells per month. Single cells ranging in size from 10 cm × 10 cm to 30 cm × 30 cm are routinely produced.[612] The cells consist of Ni-YSZ/YSZ/LSM with a thickness of 0.5 mm.[611] They have a power density ranging 0.4–0.6 W/cm^2 at 750 °C.[612] Currently, single cell products with an active area of 10 cm × 10 cm to 15 cm × 15 cm are sold to Denmark, Japan, Switzerland, Singapore, Taiwan, the United Kingdom, and the United States.[613] NIMTE believes that their planar ASCs have the highest performance/price ratio in international markets.[612] NIMTE also offers various nanopowders, including LSM, LSCF, LSC, CGO, YSZ, and SSZ.

- In 2009, NIMTE reported that it had made cells with the highest flexural strength known at that time.[613] NIMTE developed a special kind of SOFC anode support (NiO + YSZ) by doping with 0.25 weight percent Al_2O_3.The average flexural strength of the anode supports before reduction was 430 MPa, compared to 201 MPa without doping. The flexural strength of the doped anode supports after reduction was 299 MPa, which was still higher than that of nondoped anode supports at 237 MPa. Also, single cells prepared with doped anode supports (NiO + YSZ) showed the same electrical properties as single cells without doping.

NIMTE has made rapid progress in stack development. In 2009, it reported that it had developed a cell stack that showed a degradation rate of less than 1 percent per cycle and maintained 1.15 V open circuit voltage (OCV) after going through 1350 h and 46 thermal cycles at 0.1 A/cm^2 and 750 °C.[614] The OCV of a two-cell stack was above 2.2 V after 400 h and 15 thermal cycles at 0.1 A/cm^2. The degradation of a two-cell stack was also lower than 1 percent per cycle. A five-cell stack also was tested for more than 3000 h.

- In April 2009, NIMTE developed a 10-cell SOFC stack, which achieved a maximum power of 310 W (H_2, air) and a maximum power density of 0.31 W/cm^2.[615] The maximum power density and minimum power density of single cells were 0.5 and 0.25 W/cm^2, respectively. NIMTE believed the results were equal to any of the most advanced cells available.
- In October 2009, NIMTE reported that stacks of 500 W to kW power range had been built and tested.[612] It also reported that 30 cm × 30 cm cells were successfully manufactured and a multi-kilowatt single stack was under development. In addition, 2 kW power and heat cogeneration systems were under development, and the development of a 20 kW electrical generation system was also being planned in collaboration with a major Chinese electric utility.

NIMTE continued to make remarkable progresses in the past 2 years:

- In December 2010, NIMTE researchers fabricated a 30-unit standard SOFC stack module and tested it in a self-designed structure.[616] The results showed that the stack had an OCV over 33 V and a power output of 530 W (H_2, Air) at a voltage of 21 V when tested at 850 °C. With the average voltage of the unit cell at 0.7 V, the stack had a power density of 0.25 W/cm^2 and fuel utilization rate of 65.3 percent. Volume power density was 726 W/L and weight was 145 W/kg. NIMTE has started to offer the module for sale.

In April 2012, NIMT announced it had discovered that the depth of contact between cathode and interconnect greatly influences the output power density of SOFC stacks.[617] The NIMTE researchers assembled several stacks with 5–30 ASCs using a new contact method, called the Whole Soft-Contacting. The five-cell stack obtained an OCV of 5.94 V at 800 °C and a maximum power of 163.1 W with a corresponding power density of 0.52 W/cm^2. Fuel utilization was calculated to be 48 percent. For the 30-cell stack with a volume of 10 cm × 10 cm × 8.5 cm, a maximum power of 868.4 W was achieved at 800 °C and an OCV of 35.2 V. The maximum power density was 0.46 W/cm^2, with a fuel utilization of 60.4 percent.

These are significant achievements that show major promise.

6.5.4.2. *Shanghai Institute of Ceramics, Chinese Academy of Sciences*

SIC has been developing SOFC technology for more than 10 years.[618] It initially focused on planar anode-supported SOFC technology:

- In 1998, SIC was studying the properties of cathode materials for SOFCs.[619] Powder materials were synthesized, powder features and sintering behaviors were characterized, and electrical conductivity of sintered samples were measured.

- In 2002, single column SOFC stacks with internal manifolding and cross-flow structures were established and successfully operated. A 10-cell stack with cell size of 40 mm × 40 mm was fabricated and tested.[620] At 1000 °C, it showed a power output of about 10 W and a power density of about 110 mW/cm^2.

In 2006, SIC reported that its SOFC research staff had grown to 20 and that they had achieved the following:[618]

- It was able to produce planar cells with an area of 10 cm × 10 cm. SIC's ASCs with Scandia stabilized zirconia electrolyte operated at an intermediate temperature (~750 °C) with a power density close to 0.4 W/cm^2. SIC also developed sealing materials and alloy separators, and constructed, operated, and controlled SOFC stacks. It successfully fabricated and operated 60-cell stacks that produced 700 W.

SIC subsequently shifted its focus to tubular SOFC technology:

- In late 2008, SIC reported that it had successfully developed a dip-coating method to fabricate anode-supported tubular SOFCs.[621] Single cells were 10.8 cm long, 1 cm in diameter, and 0.6 mm thick. Cathode area was 15–16 cm^2 (cathode length was 4.8 cm). The cell consisted of a Ni-YSZ anode support tube, a Ni-ScSZ anode functional layer, an ScSZ electrolyte film, an LSM-ScSZ cathode functional layer, and an LSM cathode current collecting layer. A preliminary examination of the single tubular cell demonstrated acceptable performance, according to SIC. SIC reported power densities were 325, 276, 208, and 168 mW/cm^2 at 850, 800, 750, and 700 °C, respectively, when operating with humidified hydrogen.

In October 2009, SIC noted that cathode-supported tubular SOFCs (CTSOFCs) were particularly attractive compared to anode-supported SOFCs, due to the feasibility of using nickel material for current collecting.[622] SIC developed a cost-effective, facile fabrication technique to fabricate CTSOFCs and obtained good electrochemical performance. Although the performance of CTSOFC was still much lower compared to anode-supported SOFCs, the performance of CTSOFC could be improved by optimizing fabrication parameters, according to SIC. SIC also reported that by choosing 20 weight percent graphite powder as a pore former and 1260 °C as the co-sintering temperature, maximum power density could reach 530 mW/cm^2 at 850 °C.

In February 2012, SIC reported that their researchers had significantly enhanced the performance of SOFCs by using ruthenium (Ru) particles as a propane oxidation catalyst.[623] They integrated a thin film catalyst consisting of nanoscale Ru particles that were well dispersed into the ceria electrolyte matrix of the anode chamber. SOFCs that were operated on a propane–air fuel mixture yielded maximum power densities of 395, 280, and 160 W/cm^2 at 500, 450, and 400 °C, respectively. In the absence of the catalyst layer, in comparison, the output power densities dropped to 298 mW/cm^2 at 500 °C and to less than 10 mW/cm^2 at 450 °C. The substantially enhanced performance could be explained by the high catalytic activity of the Ru–CeO_2 layer for propane partial oxidation reactions, producing higher amounts of hydrogen for the fuel cell operation and thereby reducing the anode polarization resistance.

6.5.4.3. *Focused on Longer Term Commercial Success*

NIMTE and SIC have demonstrated the capability to manufacture high quality, low cost

SOFC powders, cells, and stacks and have started to market them in international markets. Their near-term activities are focused on offering low cost raw materials and components, single cells, and stacks with high power density and high stability.[624] They clearly are focused on developing power-generation systems ranging from 2 to 20 kW for international markets over the longer term.

6.5.5. India

In India, fuel cell development is primarily supported by the Ministry of Non-Conventional Energy Sources (MNES). Several universities and research organizations are involved in fuel cell research.[625] Central Glass and Ceramic Research Institute (CGCRI), Kolkata, is India's acknowledged leader in SOFC development.[626]

6.5.5.1. Central Glass and Ceramic Research Institute

CGCRI, Kolkata, was established in 1950 as one of the first national laboratories set up under the Council of Scientific and Industrial Research. CGCRI was charged with minerals research, including glass and ceramics, for product development.[627] For the past two decades or so, CGCRI has been involved in SOFC research.[626] Initially, the focus was on materials and component development. The R&D work led to the development of fine powders for different SOFC components, including Sr-substituted $LaMnO_3$ (LSM) cathodes, NiO-8 mol percent YSZ anodes, and Ca-substituted $LaCrO_3$ (LCR) interconnects. The effort culminated in the establishment of a large volume continuous powder production facility (in kilogram level). The facility is equipped with a modified spray pyrolyzer designed by CGCRI.

In December 2004, the Fuel Cell and Battery Division (FCBD) was established within CGCRI to develop SOFC stacks based on the planar anode-supported design.[628] The size of the research staff was less than 20 as of May 2010. Under a national project, FCBD made rapid progress in SOFC technology between 2004 and 2009.[626] FCBD established a world-class SOFC laboratory and demonstrated working SOFC stacks for the first time in India.[628] Some CGCRI activities are presented below[626]:

- Under the national project, ASCs 10 cm × 10 cm × 1.5 mm were fabricated using tape-casting and screen-printing techniques. A single cell demonstrated about 2 W/cm^2 at a cell voltage of 0.7 V at an operating temperature of 800 °C.
- Glass-based sealants were developed in-house. Several glass compositions with very high electrical resistivity were developed, and a bilayer glass with a compositional gradation was produced. The bilayered glass formed crack-free seals between the electrolyte and the Crofer22APU interconnects and exhibited extremely low leakage.
- Several SOFC short stacks (up to 10 cells) with a counter flow design were fabricated using 10 cm × 10 cm single cells, glass-based sealants, and ferritic steel-based metallic (Crofer22APU) interconnects and gas manifolds. A 10-cell stack was successfully operated for the first time in India. As of July 2010, CGCRI reported that a 10-cell stack has demonstrated a maximum power output of 100 W between 750 and 800 °C.[629]

In addition, CGCRI reported the following developments[626]:

- *Novel electrodes for high performance SOFCs.* CGCRI developed a novel cermet anode that contains a low volume percentage of Ni (~30 volume percent) using an electrolysis technique. CGCRI also synthesized nanocrystalline lanthanum ferrite-based (LSCF) cathode materials at a relatively low temperature by combustion synthesis. Recently, CGCRI developed nanostructured cathodes with tailor-made morphology for

improved electrical and electrochemical properties using a spray pyrolysis technique. Initial results of electrochemical experiments performed on planar ASCs fabricated with these electrodes showed a current density of more than 3 A/cm^2 at 800 °C and 0.7 V.
- *Robust glass-based sealants.* CGCRI recently developed a glass-ceramic sealant that can be sealed at a much lower temperature (~700 °C). It is currently working to develop robust sealants (with repeated thermal cycling) by adding ceramic/metallic filler materials into the glass.
- *Low temperature SOFC materials.* CGCRI prepared nanocrystalline and phase pure Gd-doped CeO_2 as well as pure $LaGaO_3$ using combustion synthesis techniques. It also synthesized nanocrystalline and phase pure CeO_2 powders with co-doping of cobalt and gadolinium. The resulting dense electrolyte demonstrated high oxygen ionic conductivity of 0.021 S/cm^2 at 60 °C.
- *Modeling and simulation.* CGCRI is engaged in modeling and simulation of the flow fields and stack designs and has been collaborating with various Indian national laboratories and academic institutes as well as foreign research institutions, including Jülich, Karlsruhe Universit, and RWTH Aachen University, Germany.

6.5.5.2. Too Early to Foretell Future Outcome

CGCRI has made remarkable rapid progress in SOFC development. The Ministry of New and Renewable Energy, Government of India, has recognized CGCRI's SOFC expertise and given CGCRI the responsibility to act as the lead institute for the development of this technology for India under a Mission-mode Project, which is an umbrella national program that encompasses wide ranging national projects aiming to advance national goals in India.[626]

6.6. JAPAN EMERGES AS THE GLOBAL SOFC LEADER; THE UNITED STATES AND EUROPE FOLLOW BEHIND

There are a number of countries discussed in this section, but they are widely different in their SOFC status. Some countries have a long history in SOFC development and have established advanced and sophisticated SOFC capabilities. Some have just started and have an elementary level of capability.

Who is ahead and who is behind in the development of SOFCs? This is a difficult question to answer. Because SOFC developers have widely varied designs and different applications, comparing and ranking individual companies can give misleading results. As an alternative, we have made an attempt to evaluate global leadership in SOFCs on a country or regional basis. The assessment is based on a broad-based general summation of available information.

Having examined available information on global SOFC development activities, we judge that Japan has emerged as the global SOFC leader today. The United States and Europe are second, followed by the rest of the world.

- This is a marked change from a decade ago. In the 1990s, the United States was the SOFC leader in the world, Japan was the second, and Europe was the third (Table 6.13).

6.6.1. Japan Promotes Close Public–Private Collaboration, Greater Breadth of Industrial Expertise and Infrastructure

Today, a large number of Japanese companies and organizations in tandem with the Japanese government are strongly committed to commercialize small, residential-use SOFC systems. Japan's METI is encouraging the industry to focus on these applications by sponsoring a massive residential demonstration program.

TABLE 6.13 A Compilation of SOFC Developers, Primary Financial Source, Corporate Size, SOFC Types, SOFC Performance, Business Outcome, and Duration of SOFC Development (As of December 2010). (For explanation of the table, see three notes provided at the end of the table on Page 394)

Country	Company (or organization)	Primarily private sector support	Major government support	Company or organization: Small	Company or organization: Large	Type of fuel cell	SOFC performance	Business outcome	Duration of SOFC development
United States	Acumentrics		X	X		Microtubular	Acumentrics' microtubular cells showed a power density of 0.072 W/cm^2 (2004), 0.099 W/cm^2 (2005), 0.261 W/cm^2 (2006), 0.339 W/cm^2 (2007), 0.421 W/cm^2 (2010), and more than 0.6 W/cm^2 (2010). Acumentrics' system had peak electrical efficiencies between 40 and 50 percent and up to 90 percent total efficiency with heat recovery (2010).	TBD	10
	Allied Signal Aerospace				X	Planar	A 26-cell planar stack of cross flow configuration (10 cm × 10 cm footprint) showed an open circuit voltage of 26.7 V at 800 °C and produced about 860 W at 18.2 V (0.7 V/cell) with a potential peak power of 1.2 kW (0.6 W/cm^2) with hydrogen fuel (1999). Allied Signal merged with Honeywell in December 1999, and the new company took the name Honeywell. GE acquired the SOFC unit from Honeywell in December 2001. GE shut down its SOFC unit in 2007.	Failed	14
	Bloom Energy	X		X		Planar	Little information is available on Bloom Energy's SOFC materials. If it is using scandium, their systems in theory have high efficiency, reliability, and durability.	TBD	10
	Ceramatec		X	X		Planar	Ceramatec attained a power density of 0.18 W/cm^2 and a current density of 250 mA/cm^2. Ceramatec single cells and multicell stacks, and individual planar cells tested for over 20,000 h, showed little voltage degradation (1997). Ceramatec likely is no longer focused on SOFC cell and stack development.	Failed	34
	Cummins Power Generation		X		X	Microtubular	The Cummins APU incorporated SOFC modules supplied by Protonex Technology Corporation, combined with Cummins' own hybrid power electronics, system controls, and balance-of-plant components (2010).	TBD	10

Delphi	X		X	Planar	In 2007, Delphi's 30-cell Generation 3.2 stack was developed under SECA phase 1 program, which produced 1.44 kW with a power density of greater than 450 mW/cm^2 at 0.8 V per cell with 48.5 percent H_2, 3 percent H_2O, rest N_2. Also Delphi developed an SPU1 SOFC APU system fueled by diesel, which generated net rated power of 3.4 kW at an electrical efficiency of 38 percent, operated for 1,000 hours with a degradation rate of 0.84 percent/500 hours (2007).	TBD	More than 10
	X				Delphi's APU system developed under SECA phase1 demonstrated a net peak load of 1.4 kW at a system efficiency of 18 percent and was tested for 168 h over 640 miles on a vehicle (2010).		
	X				Delphi's 40-cell Gen 4 stack demonstrated maximum power of 6.4 kW with simulated coal gas blend. The power density was 398 mW/cm^2, and the average cell voltage was 0.7 V (2011).		
FuelCell Energy (FCE)/Versa	X	X		Planar	Versa Power Systems built and operated multiple 2 kW systems and accumulated over 30,000 operating hours, recently achieving a net fuel to AC power efficiency of 35 percent on natural gas (2010).	TBD	7 (FCE), 10 (Versa)
	X				Versa's 20-cell stack demonstrated an operating life of 8000 h with a performance decay rate of 2.3 percent/1000 h at a power density of 300 mW/cm^2 (2010).		
General Electric (GE)	X		X	Planar	GE achieved a power density of 307 mW/cm^2 in a radial planar, 21-cell stack at 88 percent fuel utilization, and 404 mW/cm^2 in a single cell at 88 percent fuel utilization (2004).	Failed	6
	X				A system with a peak power of 5.4 kW achieved an electrical efficiency of 41 percent and a power degradation of less than 3.6 percent/1000 h (2006).		
	X				A single modular stack achieved a power density of 480 mW/cm^2 at 0.7 V and 80 percent fuel utilization (2007).		
					GE closed the fuel cell division (2007).		
Protonex	X	X		Microtubular	Protonex' liquid fueled SOFC generator had a power density of 50 W/kg at 25 percent efficiency (2009).	TBD	10

(Continued)

TABLE 6.13 A Compilation of SOFC Developers, Primary Financial Source, Corporate Size, SOFC Types, SOFC Performance, Business Outcome, and Duration of SOFC Development (As of December 2010). (For explanation of the table, see three notes provided at the end of the table on Page 394) (*Cont'd*)

Country	Company (or organization)	Primarily private sector support	Major government support	Company or organization: Small	Company or organization: Large	Type of fuel cell	SOFC performance	Business outcome	Duration of SOFC development
			X				The Cummins/Protonex APU system produced a power output of 19 W/kg and 10 W/L (2010).		
	Rolls Royce Fuel Cell Systems (RR)		X		X	Planar	RR's cells developed under SECA program demonstrated a power density of 700 mW/cm^2 at 1 A/cm^2 current density at 0.7 V. RR also demonstrated a 40 percent power improvement in stacks; a 30-cell stack showed 331 mW/cm^2 at 0.8 V while a 60-cell stack showed 470 mW/cm^2 at 0.8 V at 6 Bara at 900 °C (2009).	TBD	18
	Siemens-Westinghouse Power Corporation (SWPC)		X		X	Tubular	A 100 kW system was operated for 12,577 h at Westervoort, the Netherlands. The system generated 109 kW AC and thermal power 64 kW. The electrical efficiency was 46 percent and thermal efficiency 27 percent with no performance degradation (1999).	Failed	50
			X			Tubular	During two of the three operating sessions totaling 667 h, the 220 kW SOFC/GT system generated a total of 181 kW AC (including turbine power of 22 kW AC) at the electrical efficiency of 52 percent (2001).		
			X			Flattened tubular and Delta-shaped tubular	A 10 kW stack consisting of 24 Delta8 cells, 4 Bundles (six 18 cells each) achieved more than 5300 h of operation with maximum DC power of 9.8 kW. The cell generated 550 W per cell at 0.7 V and showed a power density of 0.25 W/cm^2 at 80 percent fuel utilization of humidified hydrogen at 975 °C. The stack showed an IGFC efficiency of more than 50 percent and no detectable degradation (2010). Siemens AG of Germany closed down SWPC (2010).		
	SOFCo		X	X		Planar	A single cell endurance test demonstrated up to 40,000 h, and cells and stacks were scaled up from 4″ × 4″ to 6″ × 6″ (1996). Rolls Royce of the United Kingdom acquired SOFCo from McDermott International (2007).	Failed	14
	Technology Management Inc. (TMI)	X		X		Planar	Small 1- to 10-cell stacks demonstrated power densities around 0.08 W/cm^2. TMI's SOFC systems are highly sulfur tolerant and fuel flexible (1997).	TBD	20

	UTC Power/ Delphi		X	X	Planar	Delphi's 30-cell Generation 3 stack was improved under the new coal-based SECA program. It consisted of 105 cm^2 active area cells and generated 1.8 kW with a power density of 570 mW/cm^2 at 0.82 V/cell with fuel (48.5 percent H_2, 3 percent H_2O, rest N_2) at 750 °C (2008).	TBD	2
			X			Delphi's Gen 4 stack consisting of 403 cm^2 active area cells, much larger than Gen 3 cells, produced a power density of 500 mW/cm^2 at 570 mA/cm^2 at 0.87 V (2009).		
			X			Delphi's 40-cell Gen 4 stack produced 7452 W with a power density of 450 mW/cm^2 and a 25-cell Gen 4 stack produced 5064 W with a power density of 506 mW/cm^2 at 0.81 V (2010).		
			X			Delphi's 30-cell Gen 3.2 stack was operated for 9700 h at 333 mA/cm^2 (2010), and the total degradation was 1.12 percent/500 h (2010).		
	Ztek	X	X		Microtubular	Ztek's SOFC with microturbine hybrid system has 50 percent efficiency (2007). It is likely that Ztek is no longer focused on SOFC development.	Failed	More than 20
Japan	Acumentrics Japan	X		X	Microtubular	Acumentrics Japan's 10 kW CHP system achieved an electrical efficiency of 37.5 percent AC HHV and the total efficiency of 76 percent HHV. It was operated for more than 3000 h and had a voltage degradation of more than 1.04 percent /1000 h (2007). Acumentrics was no longer actively engaged in business in Japan (2004).	Failed	10
	AIST		X	X		A microtubular SOFC with ceria-based electrolyte of millimeters to sub-millimeters in diameter achieved a power density of 1 W/cm^2 at 570 °C, the world's highest power density among SOFCs with ceria based electrolytes (2006).	TBD	36
	AIST/ NGK Insulators		X	X		A module of 256 cells in a honeycomb structure contained within 15 mm × 15 mm cross section achieved 230 mW/cm^2 at 700 °C, the world's highest level power density (2006).		
	AIST/CRIEPI		X	X		A tubular SOFC cell of 3 mm in outer diameter and 3 cm in length in active area demonstrated a power density of 1.06 W/cm^2 using hydrogen fuel and operating in a low temperature range from 500 to 650 °C (2007).		

(Continued)

TABLE 6.13 *A Compilation of SOFC Developers, Primary Financial Source, Corporate Size, SOFC Types, SOFC Performance, Business Outcome, and Duration of SOFC Development (As of December 2010). (For explanation of the table, see three notes provided at the end of the table on Page 394) (Cont'd)*

Country	Company (or organization)	Primarily private sector support	Major government support	Company or organization: Small	Company or organization: Large	Type of fuel cell	SOFC performance	Business outcome	Duration of SOFC development
	AIST/NGK Spark Plug/Fine Ceramic Research Association/Toho Gas		X		X		A 1 cm^3 module (sugar cube size) consisting of microtubular SOFC cells 0.8–2 mm diameter produced a high output power over 2 W/cm^3 at 550 °C, the world's highest output power density at an operating temperature below 600 °C (2007).		
	AIST/NGK Spark Plug		X		X		A microtubular module of a size of 1 cm^3 with small-size tubular SOFCs of diameters ranging from millimeter to sub-millimeter achieved power density better than 2 W/cm^3 at 550 °C. The module consisted of three 10 mm × 10 mm × 2 mm SOFC bundles (made of five integrated microtubular cells) connected in series and a fuel manifold. It weighed about 2 g. Hydrogen at a temperature of 450–550 °C was input into the module and power density of 0.88–2.2 W/cm^3 (450–550 °C) was confirmed (2009).		
	NGK Spark Plug/AIST/Fine Ceramic Research Association/Toho Gas		X		X		A module consisting of 3 cells in parallel and 15 in series (a total of 90 cells) produced an output of over 50 W with a power generation efficiency of over 40 percent and a volumetric power density of 2 W/cm^3. Modules with power output of more than 200 W were also developed (2009).		
	Fuji Electric		X		X	Planar	A five-cell stack with 528 cm^2 active areas achieved the maximum output of 505 W at 0.35 A/cm^2 (1995).	Failed	10
			X				A module of twenty 600 cm^2 cells generated 3.31 kW at 300 mA/cm^2 and 50 percent fuel utilization. The module was operated for 1800 h and showed a power density of 0.18 W/cm^2. It showed no degradation after 900 h of operation (1997). Fuji likely left SOFC development in 1998.		
	Fujikura Cable		X		X	Tubular	A module of sixteen 105 cm^2 generated 103 W. It was operated at a current density of 143 mA/cm^2, a power density of 0.06 W/cm^2, and fuel utilization of 18 percent (1992). It is likely that the company no longer engages in SOFC development in or after 2004.	Failed	15

Hosokawa Micron	X			X	Powder	A composite material for solid oxide fuel cells (SOFCs) made with the new material achieved a power density of 0.8 W/cm^2 (2005).	TBD	Unknown
Kyocera	X			X	Flat tubular	A 1 kW module achieved the world's highest power efficiency of 54 percent DC LHV at 780 °C (2003). The flat-tubular cell achieved a power density of 0.5 W/cm^2 at 750 °C (2003).	TBD	More than 20 (Kyocera), More than 20 (Osaka Gas)
	X					An AC power generator achieved power generation efficiency of 44 percent (2004).		
	X					A CHP system achieved power generation efficiency of 45 percent in an AC generator and 75 percent overall (2005).		
Osaka Gas	X			X	Tubular	A tubular cell operated at 1000 °C achieved maximum electric output of 8.8 W, power density of 0.59 W/cm^2 with fuel utilization of 30 percent and a voltage degradation of 0.4 percent/1000 h (1995).		
Kyocera/Osaka Gas	X			X	Flat tubular	A 1 kW module achieved an electrical efficiency of 44.1 percent LHV AC and an average thermal efficiency of 34 percent LHV (2006).		
	X					A 0.7 kW CHP system achieved an electrical efficiency of 45 percent AC LHV and the efficiency of the exhausted heat recovery of over 30 percent (2006).		
Kyocera/Osaka Gas/Toyota Motor/Aisin Seiki	X			X	Flat tubular	A 0.7 kW CHP system achieved an electrical efficiency of 45 percent LHV and the thermal efficiency of 40 percent LHV (2009).	TBD	More than 20 (Kyocera), More than 20 (Osaka Gas)
Kyocera/Tokyo Gas/Rinnai/Gastar	X			X	Horizontal stripe Flat tubular	A 2.5 kW SOFC power module achieved an electrical efficiency of 56.1 percent LHV DC (2004).	TBD	More than 20 (Kyocera), More than 15 (Tokyo Gas)
	X					A 0.7 kW system achieved a power generation efficiency LHV of 42 percent and heat recovery efficiency of 35 percent (2009)		
Tokyo Gas	X			X	Planar	Single planar cells of various sizes were manufactured. The maximum power densities were 0.65 W/cm^2 for 5 cm × 5 cm and 0.36 W/cm^2 for 10 cm × 10 cm, respectively (1992).		
	X					A three-cell stack of 6 cm × 6 cm cells gave a power density of 0.39 W/cm^2 (1995).		
	X					A 1 kW internal reforming planar SOFC stack generated 1.68 kW at 1000 °C (1998).		

(Continued)

TABLE 6.13 *A Compilation of SOFC Developers, Primary Financial Source, Corporate Size, SOFC Types, SOFC Performance, Business Outcome, and Duration of SOFC Development (As of December 2010). (For explanation of the table, see three notes provided at the end of the table on Page 394) (Cont'd)*

Country	Company (or organization)	Primarily private sector support	Major government support	Company or organization: Small	Company or organization: Large	Type of fuel cell	SOFC performance	Business outcome	Duration of SOFC development
			X			Segmented-in-series flat tubular	An anode-supported planar SOFC unit developed under NEDO sponsorship achieved the world's highest power generation efficiency of 53.6 percent (2002).		
	Kyocera/Tokyo Gas		X				A module of forty seven 144 cm^2 cells incorporating Kyocera's interconnect technology achieved the current density of 276 mA/cm^2, the power output of 1.33 kW, the power density of 0.14 W/cm^2, and fuel utilization of 50 percent (2003).		
	Mitsubishi Heavy Industries (MHI) Kobe/Chubu Electric	X			X	Planar MOLB	A module of three stacks of forty 140 mm^2 cells generated the maximum power output of 1.275 kW at the maximum power density of 0.236 W/cm^2 (1993).	Failed	More than 20 (MHI Kobe)
		X					A module of two stacks of forty 200 mm^2 cells generated 5.1 kW (1996).		
		X					An improved module achieved the world's highest power density of 0.35 W/cm^2 (1998).		
		X					A module generated 15 kW and was operated for 7500 h, including 2473 h of internal reforming (2001).		
			X				A thermally self-supporting 10 kW MOLB type module was developed under NEDO sponsorship (2004).		
			X				A 30 kW MOLB SOFC CHP system was operated for 3887 h at an electrical efficiency of 33 percent at Aichi Expo (2005).		
							MHI Kobe and Chubu Electric likely left SOFC development in 2005		
	Mitsubishi Heavy Industries (MHI) Nagasaki/Tokyo Gas	X			X	Tubular	An SOFC cell stack was operated with 80 percent fuel utilization and reached maximum electrical efficiency of 40 percent at the current density of 200 mA/cm^2 (1987).	TBD	26 (MHI Nagasaki)

Mitsubishi Heavy Industries (MHI) Nagasaki/J-Power/Tokyo Electric	X		X	Two 1 kW SOFC modules were operated for 1000 h each (1991).
Mitsubishi Heavy Industries (MHI) Nagasaki/J-Power	X			Improved 1 kW SOFC modules were operated for 3000 h each with two scheduled thermal cycles. The modules achieved an electrical efficiency of 34 percent HHV and a deterioration rate of 2 percent/1000 h (1993).
	X			A 1 kW SOFC module was operated for 5000 h (1994).
	X			A 10 kW SOFC module was operated for 5000 h at an electrical efficiency of 34 percent HHV with a deterioration rate of 2 percent/1000 h (1995).
	X			A pressurized module achieved a continuous operation of 7000 h at a maximum output of 21 kW (1998).
		X		An internal reforming 10 kW pressurized module achieved a continuous operation of 755 h at an electrical efficiency of 41.5 percent HHV (2001).
		X		A 25 kW SOFC sub-module was operated for power generation for 515 h. An output density of 0.21 W/cm^2 was achieved at a pressure of 0.4 MPa (2004–2007).
		X		A 100 kW system was operated for 3300 h (2007).
		X		A 200 kW SOFC/MGT combined cycle system were fabricated under a METI project and generated maximum electric output of 211 kW AC, combining 176 kW from the SOFC system and 35 kW from the micro turbine, with a power efficiency of 52.1 percent LHV. After about 90 minutes, the system suffered a short circuit and the operation was terminated (2009).
	X			• Subsequently, the 200 kW SOFC/MGT system was re-commissioned at a MHI Nagasaki site and was operated for 3224 h (2009).

(Continued)

TABLE 6.13 *A Compilation of SOFC Developers, Primary Financial Source, Corporate Size, SOFC Types, SOFC Performance, Business Outcome, and Duration of SOFC Development (As of December 2010). (For explanation of the table, see three notes provided at the end of the table on Page 394) (Cont'd)*

Country	Company (or organization)	Primarily private sector support	Major government support	Company or organization: Small	Company or organization: Large	Type of fuel cell	SOFC performance	Business outcome	Duration of SOFC development
	Mitsubishi Materials Corporation (MMC)/Kansai Electric	X			X	Flat disk	A flat disk SOFC cell demonstrated the world highest power density of 0.9 W/cm^2 at 0.7 V at 700 °C and 1.8 W/cm^2 at 0.7 V at 800 °C (2001).	TBD	13 (MMC), 13 (Kansai Electric)
		X					A module consisted of 41 flat disk cells of 120 mm in diameter generated 1 kW at a temperature below 800 °C fueled by methane with the electrical efficiency of 40 percent LHV (2003).		
		X					A 1 kW SOFC system achieved the electrical efficiency of 40 percent LHV (2004).		
		X					A 3 kW SOFC module consisting of three 1 kW stacks achieved the electrical efficiency of 55.3 percent LHV DC (2004).		
			X				A 10 kW natural gas–fueled, thermally self-supporting module developed under an METI project was operated for 3000 h and achieved the electrical efficiency of 41 percent AC HHV with a degradation rate of 1.08 percent/1000 h (2006).		
		X					•A 10 kW CHP system with an improved stack module was operated for another 3000 h and demonstrated an electrical efficiency of 40 percent HHV AC and the combined efficiency of 82 percent HHV with no significant voltage degradation (2008).		
		X					A recent MMC/Kansai Electric cell of 120 mm in diameter achieved an output of 23 W and achieved the power density of 0.2 W/cm^2 at the current density of 0.27 A/cm^2. The stack test using 46 improved cells showed an average degradation rate of 0.3 percent/1000 h during 5000 h for the constant operation (2010).		
	Mitsui Engineering and Shipbuilding (MES)		X		X	Planar	Eight 12 cm^2 cells developed under a METI project generated 300 W (year unknown).	Failed	More than 10

		X			A module of six 100 cm^2 planar cells were operated at 1000 °C for 1000 h and generated a maximum output of 83 W with a power density of 0.14 W/cm^2 at 62 percent fuel utilization rate. The voltage degradation was 6 percent/1000 h (1995).		
		X			A module of five 140 cm^2 cells was operated at 1000 °C and generated the maximum output of 109 W with the power density of 0.15 W/cm^2 and the fuel utilization at 53 percent (1998).		
		X			In 1999, MES developed a module of three stacks of eight 140 cm^2 planar cells operated at 1000 °C and generated 0.5 kW with a power density of 0.149 W/cm^2 at a fuel utilization of 44 percent (1999).		
					MES likely suspended SOFC research sometime in the early 2000s.		
Murata Manufacturing		X	X	Planar	A stack of six 110 cm^2 cells generated 117 W (year unknown).	TBD	More than 20 years (Murata)
		X			A module of five stacks of four 121 cm^2 cells generated 601 W at the power density of 0.24 W/cm^2 at fuel utilization of 40 percent (1994).		
Murata Manufacturing/ Osaka Gas	X				A module consisting of sixty five 120 cm^2 cells was operated for 1028 h, and generated 872 W. The module showed a current density of 213 mA/cm^2, power density of 0.11 W/cm^2, and fuel utilization of 12 percent. The voltage degradation was 6 percent /1000 h (1994).		
	X				A module of five stacks of four 120 cm^2 cells generated 400 W (1996).		
	X				A module of four stacks of twenty 113 cm^2 cells each generated 1.31 kW with the current density of 0.22 A/cm^2, the power density of 0.14 W/cm^2, and the fuel utilization of 70 percent. The module was operated for 500 h and demonstrated 4 percent/1000 h voltage degradation rate (1997).		
NGK Insulators	X		X	Flat tubular	A flat tubular SOFC cell demonstrated the power density of 0.6 W/cm^2 at 650 °C and 1.6 W/cm^2 at 750 °C, the world's highest power density (2003).	TBD	More than 7 (NGK Insulators)
NGK Insulators/ AIST		X		Planars	A module of 256 cells in a honeycomb structure contained within 15 mm × 15 mm cross section achieved 230 mW/cm^2 at 700 °C, the world's highest level power density (2006).		

(Continued)

TABLE 6.13 A Compilation of SOFC Developers, Primary Financial Source, Corporate Size, SOFC Types, SOFC Performance, Business Outcome, and Duration of SOFC Development (As of December 2010). (For explanation of the table, see three notes provided at the end of the table on Page 394) (Cont'd)

Country	Company (or organization)	Primarily private sector support	Major government support	Company or organization: Small	Company or organization: Large	Type of fuel cell	SOFC performance	Business outcome	Duration of SOFC development
	NGK Insulators	X				Flat tubular	A stack of several dozen flat-tubular SOFC cells generated 700 W at 800 °C. The cell was card-like shape and 10 cm long × 5.5 cm wide × 1.5 mm thick. The stack achieved the world's highest power generation efficiency of 63 percent LHV at fuel utilization of 90 percent (2009)		
	NGK Insulators/ J-Energy/ Sumitomo Precision Products	X					A 3 kW flat-tubular SOFC power plant was developed for small business use. The system was fueled by kerosene and operated at 750 °C (2009).		
	NGK Spark Plug/AIST/ Fine Ceramic Research Association/ Toho Gas		X		X	Microtubular	A 1 cm^3 cube consisted of microtubular SOFC cells of 0.8–2 mm diameter produced a high output power over 2 W/cm^3 at 550 °C, the world's highest output power density at an operating temperature of below 600 °C (2007).	TBD	More than 10 years (NGK Spark Plug)
	NGK Spark Plug	X			X	Planar	A 1 kW planar SOFC CHP system consisted of 16 cells of 100 cm^2 active area achieved the power density of 0.8 W/cm^2 (700 °C at 0.7 V) and power efficiency of 60 percent LHV, using natural gas. The system was packed in a thermally insulated vessel. The size of the system was 530 mm width × 330 mm depth × 950 mm height, which the company believed was one of the most compact 1 kW CHP units in the world. The system achieved a stable 90 percent utilization for fuel and air. It contained carefully designed gas flow channels, chambers, and current collectors for effective gas distribution to active electrodes (2008).		
		X				Planar	A 1 kW module with 18 improved, larger, and thinner anode-supported planer cells achieved a power density 1.6 W/cm^2 at 0.7 V, electrical efficiency DC LHV of 60 percent and fuel utilization of 80 percent (2009).		

NGK Spark Plug/AIST		X		Microtubular	A module in the size of 0.6 cm^3 consisted of 0.8 mm diameter micro tubes and achieved the power density of 0.88–2.2 W/cm^3 at 450–550 °C (2009). The module consisted of three 10 mm × 10 mm × 2 mm SOFC bundles (made of five integrated microtubular cells) connected in series and a fuel manifold. It weighed about 2 g (2009).		
NGK Spark Plug/AIST/ Fine Ceramic Research Association/ Toho Gas		X		Microtubular	A module consisting of 3 cells in parallel and 15 in series (a total of 90 cells) produced an output of over 50 W with a power generation efficiency of over 40 percent and a volumetric power density of 2 W/cm^3. Modules with power output of more than 200 W were also developed (2009).		
NGK Spark Plug/Toho Gas	X			Planar	A 1 kW module designed for an operation at 700 °C achieved a power density 1.6 W/cm^2 at 0.7 V (2009).		
	X				A 700 W SOFC CHP system was being demonstrated for residential use. The system consisted of 19 cells and generated 700 W AC with the net power efficiency of 44 percent LHV and total efficiency of 76 percent LHV. The CHP system was 530 mm wide, 330 mm deep, and 955 mm high, one of the smallest CHP system units (2010).		
JX Nippon Oil & Energy	X		X	Planar	A planar SOFC module of five 62 cm^2 cells demonstrated a current density of 470 mA/cm^2 and a power density of 0.24 W/cm^2 at fuel utilization of 20 percent (1995).	TBD	20 (JX Nippon Oil & Energy), More than 20 (Kyocera)
JX Nippon Oil & Energy/ Kyocera	X			Flat tubular	Two 0.7 kW CHP systems were developed with Kyocera SOFCs. One was fueled by LPG, and the second fueled by kerosene. Both demonstrated the electrical efficiency of 45 percent and exhaust heat recovery of 40 percent. The LPG unit was operated for 5744 h and the kerosene unit for 4565 h (2008).		
Nippon Telegraph and Telephone (NTT)	X		X	Flat disk	A single cell achieved the maximum power density of 0.63 W/cm^2 at 800 °C and 1.04 W/cm^2 at 850 °C (1995).	TBD	More than 10 (NTT)
	X				A one-cell and a three-cell stacks with 60 mm diameter cell achieved a power density of up to 0.6 W/cm^2 and an electrical efficiency of 53 percent LHV DC. They demonstrated the durability of over 1000 h (2007).		

(Continued)

TABLE 6.13 *A Compilation of SOFC Developers, Primary Financial Source, Corporate Size, SOFC Types, SOFC Performance, Business Outcome, and Duration of SOFC Development (As of December 2010). (For explanation of the table, see three notes provided at the end of the table on Page 394) (Cont'd)*

Country	Company (or organization)	Primarily private sector support	Major government support	Company or organization: Small	Company or organization: Large	Type of fuel cell	SOFC performance	Business outcome	Duration of SOFC development
		X					A 1 kW module consisted of 50 anode-supported planar 120 mm diameter SOFCs demonstrated the electrical efficiency of 54 percent LHV, fuel utilization of 67 percent, current density of 0.28 A/cm^2, and operating temperature of 800 °C. The stack was operated for 700 h (2008).		
		X					A 1.5 kW stack made of forty 120 mm diameter anode-supported cells achieved the electrical efficiency of 63.9 percent. A single cell was operated for 5000 h at a fuel utilization rate of 80 percent and showed an electrical efficiency of 60 percent LHV with a voltage degradation of 0.55 percent/1000 h (2009)		
	NTT/ Toho Gas/Sumitomo Precision Products	X					A 3 kW natural gas–fueled SOFC module was built, which consisted of two 1.5 kW stacks, each comprised of 40 cells. After the start-up burner was stopped, power generation was conducted under constant current conditions with the stack output finally becoming 50 A. The voltage was maintained at a constant level for about 200 hours, and the electric power of 3.1 kW was generated. At this time, the electrical efficiency reached 56 percent LHV DC (2009).		
	Sanyo Electric		X		X	Planar	A 20-cell planar SOFC module with internal manifolding and metal bipolar plates was operated for 1000 h and produced 415 W (year unknown).	Failed	More than 20
			X				A module of thirty 160 cm^2 cells generated 1.060 kW and was operated for 1871 h. The module experienced a degradation of 4.4 percent/1000 h at the current density of 3000 mA/cm^2 (1994).		
			X				A module consisting of 17 stacks of four 182 cm^2 cells generated 2.47 kW at a current density of 3000 mA/cm^2 and at 15 percent fuel utilization. It achieved a power density of 0.2 W/cm^2 (1996).		

				JX Nippon Oil & Energy acquired the majority of Sanyo's fuel cell assets and established a new company, ENEOS Celltech, in which Sanyo holds a minority stake (2008).		
Toho Gas/ Sumitomo Precision Products	X	X	Planar	A planar single SOFC cell with a ScSZ electrolyte demonstrated a power density of 2 W/cm^2 (1997).	TBD	20(Toho Gas)
	X			Toho Gas and Sumitomo Precision Products developed Japan's first 1 kW SOFC system. It had an electrical efficiency of 24 percent and volumetric efficiency of 0.5 kW/L (2003).		
	X			An improved 1 kW system had an electrical efficiency of 45 percent DC LHV, the fuel utility of 74 percent at 800 °C, and the power density of 0.28 W/cm^2 (2004).		
	X			• A 1 kW SOFC CHP system was demonstrated at the Aichi Expo 2005. The system produced an output of 955 W using natural gas and thermal self-sustaining conditions. Maximum electrical efficiency and thermal efficiency were both 35 percent LHV. The system operated as a cogeneration system for 4459 h (2005).		
	X			•An improved 1 kW CHP system achieved an electrical efficiency of 55.7 percent DC LHV (2006).		
Toho Gas	X		Microtubular	Toho Gas developed SOFC cells of low temperature operation, long endurance, and low cost. They were operated at 750 °C and had a degradation rate of 0.25 percent/1000 h (2006).		
	X			A prototype microtubular SOFC module operating at 500 °C demonstrated the power density over 0.5 W/cm^2. The micro tubes used in the module were those developed by AIST. The module made of a total of 36 micro tubes (each 2 mm in diameter and about 5 cm in length) generated 10–20 W (2007).		
Toho Gas/ NTT/Sumitomo Precision Products	X		Flat disk	A 3 kW natural gas–fueled CHP system was fabricated (2009) (see NTT's 3 kW module above).		
Tonen	X	X	Planar	A module consisting of thirty 225 cm^2 cells achieved an electrical output of 1.3 kW with the fuel utilization of 50 percent and oxygen utilization of 30 percent (1991).	Failed	10

(*Continued*)

TABLE 6.13 *A Compilation of SOFC Developers, Primary Financial Source, Corporate Size, SOFC Types, SOFC Performance, Business Outcome, and Duration of SOFC Development (As of December 2010). (For explanation of the table, see three notes provided at the end of the table on Page 394) (Cont'd)*

Country	Company (or organization)	Primarily private sector support	Major government support	Company or organization		Type of fuel cell	SOFC performance	Business outcome	Duration of SOFC development
				Small	Large				
		X					A 5.1 kW module consisting of sixty five 729 cm^2 cells demonstrated the current density of 1660 mA/cm^2, the power density of 0.11 W/cm^2, and 35 percent fuel utilization. The module still used hydrogen as fuel; not ready for hydrocarbon fuel (1995).		
							The company perhaps left SOFC fields in the late 1990s.		
	TOTO/Kyushu Electric	X				Tubular	A 1 kW module was developed (1991).	TBD	More than 20 (TOTO),
							TOTO, Kyushu Electric, and Kyushu University achieved to operate their SOFC system with an output of 0.4 W/cm^2 using air as an oxidizer (1995).		
	TOTO/ Hitachi/ Kyushu Electric/ Nippon Steel	X					A 2.5 kW module was developed (1999).		
			X				A 3 kW module demonstrated the power density of 0.18 W/cm^2, fuel utilization of 70 percent, and voltage degradation of 1 percent/ 1000 h (2000).		
			X				A 6.5 kW module consisting of 20 stacks of 12 tubular cell bundles demonstrated the electrical efficiency of 48 percent LHV at 0.2 A/cm^2 and the fuel utilization of 75 percent. The same module generated 6.4 kW at 0.19 A/cm^2 and 46 percent LHV. The voltage degradation rate was 0.1 percent/1000 h in both cases. The thermal self-sustainability of both operating conditions was confirmed (2001).		
	TOTO	X					A stack with the world's highest level power efficiency of 55 percent was developed and operated for 3000 h (2004).		

	TOTO/ Hitachi		X		Tubular	Under a METI project to build a 20 kW CHP system, a test module was built, which generated 10 kW at 890 °C at 0.2 A/cm^2 with fuel utilization of 80 percent and electrical efficiency of 46.6 percent HHV DC. Subsequently, a 20 kW power plant was built but experienced ground fault (2004–2007).		
	TOTO		X		Microtubular	A microtubular SOFC stack generated 28 W at 500 °C and 37 W at 600 °C. The cell was 50 mm long and 5 mm in diameter, and several dozen cells constituted a palm-sized stack (2005).		
			X			Two 2 kW mono-generation systems were developed and demonstrated under the METI demonstration program (2007).		
			X			Two 0.7 kW CHP units and one 8 kW mono-generation unit were demonstrated under the METI Demonstration Program (2008).		
			X			Six 0.7 kW CHP systems were planned to be demonstrated under the METI program (2009).		
	TOTO/ Hitachi	X				A 10 kW module showed an electrical power efficiency of about 40 percent and generated 60 °C hot water, raising the total efficiency to more than 80 percent (2010).		
	Hitachi	X				•Hitachi reduced SOFC cell resistance by about 18 percent by installing an auxiliary electrode to transfer electricity. The company also shortened the electricity pathway by modifying the cell structure and increased the electric efficiency as much as 1.3 times to 0.39 W/cm^2. The auxiliary electrode and cell modification did not affect the quality of the cell (2010).		
Denmark	Topsoe Fuel Cells/Risø		X	X	Planar	A new stack with a new metallic alloy for the interconnect (Crofer 22APU) and a new perovskite coating reduced the overall ASR of the stack to about 0.5 ohm/cm^2 and improved the durability. A five-cell stack had no measurable voltage degradation over 1000 h and the typical initial decrease in stack voltage was eliminated. The stack volume power density was 2.4 kW/L stack volume under a conservative safe operation condition of 0.38 W/cm^2 (2004).	TBD	23

(Continued)

TABLE 6.13 *A Compilation of SOFC Developers, Primary Financial Source, Corporate Size, SOFC Types, SOFC Performance, Business Outcome, and Duration of SOFC Development (As of December 2010). (For explanation of the table, see three notes provided at the end of the table on Page 394) (Cont'd)*

Country	Company (or organization)	Primarily private sector support	Major government support	Company or organization: Small	Company or organization: Large	Type of fuel cell	SOFC performance	Business outcome	Duration of SOFC development
			X				A stack volume power density of 2.4 kW/L stack volume has been obtained in cross flow configuration with hydrogen fuel at 0.35 W/cm^2. Standard stacks ran successfully for more than 13,000 h (2007).		
Finland	Wärtsilä		X		X		Focused on integrating Topsoe SOFC stacks into systems; Not engaged in its own SOFC development. Wärtsilä's 20 kW WFC20 system was operated with landfill gas for more than 2000 h (2010).	TBD	10
Germany	Asea Brown Boveri (ABB)		X		X		ABB worked to develop a 30 W SOFC module with a power density of 1 kW/m^2 under an EU FP project, which likely failed (1992–1993). ABB left SOFC development perhaps after the EU projects ended in 1993.	Failed	About 12
	BMW		X		X		BMW explored SOFCs made by other manufactures for APU applications, such as Delphi, DLR, Global Thermoelectric, and Jülich. BMW suspended SOFC APU development perhaps when all these projects ended in 2008.	Failed	10
	Dornier		X			Planar	Single cells were scaled up to 10 × 10 cm, and 10-cell stacks were operated successfully for over 1000 h. The performance was about 2000 mA/cm^2 at 0.7 V (1995). Dornier terminated their SOFC development by the end of the 1990s.	Failed	Less than 10
	Forschungs-zentrum Jülich		X		X	Planar	A 60-cell stack generated 13.3 kW with hydrogen at the mean operating temperature of 760 °C. The power density was 0.61 W/cm^2. The stack delivered 11.9 kW with a methane–hydrogen mixture. The minimum operating temperature was as low as 598 °C and the maximum temperature 809 °C. The power density was 0.55 W/cm^2. The stack ran at least for over 1100 h in continuous operation with 3 percent of performance degradation (2004).	TBD	21

			X				Jülich announced as of April 2010 that its SOFC stack had been running for more than 20,000 h, almost 3 years. The cells provided an average power density of 0.4 W/cm^2 of electrolyte surface, about double what is currently the case in commercial systems, according to Jülich (2010).		
			X				The SOFC stack built by Jülich surpassed an operating lifetime of 40,000 h. This exceeds previous records established for commercial SOFC use in stationary applications and corresponds to a period of 5 years (2012)		
	H.C. Starck	X			X	Planar	H.C. Starck's cells achieved the power density of 275 mW/cm^2 at 750 °C, 0.85 V, and 3250 mA/cm^2, and the fuel utilization rate was a little below 20 percent (2007).	TBD	14
	Staxera	X		X		Planar	Staxera Mk100 Stack (30 cells) demonstrated a power density of 183 mW/cm^2 at current density of 1970 mA/cm^2, with fuel utilization of 51 percent. Fuel composition was 50 percent H_2, 40 percent N_2, and 10 percent H_2O (2007).	TBD	5
	Enerday	X		X			As a system integrator, it relies on its partners for components, including H.C. Starck and Staxera.	TBD	10
	Siemens		X		X		The US subsidiary, Siemens Westinghouse Power Systems (SWPC), represented Siemens AG for the entire SOFC operations. Siemens AG closed SWPC in September 2010.	Failed	
Italy	SOFCpower				X	Planar	SOFCpower acquired HTceramix of Switzerland (2007).	TBD	3
		X					The anode-supported electrolyte with ceria protective layer and LSCF-based cathode SOFC (ASC-700) demonstrated power densities of 405 mW/cm^2 at 650 °C, 560 mW/cm^2 at 700 °C, and 715 mW/cm^2 at 750 °C at 0.7 V with a fuel utilization of up to 70 percent. The anode-supported electrolyte with LSM-based cathode (ASC-800) demonstrated 510 mW/cm^2 at 750 °C, 680 mW/cm^2 at 800 °C, and 890 mW/cm^2 at 850 °C at 0.7 V (2010).		
The Nether-lands	InDEC						H.C. Starck of Germany acquired InDEC (2003).	Failed	
Switzer-land	HTceramix		X	X		Planar	HTc's reproducible stack power density (H2, fuel, 800 °C maximum local temperature) was 0.5 W/cm^2 at 0.7 V for 67 percent fuel utilization (35 percent LHV electrical efficiency (2005).	Failed	

(Continued)

TABLE 6.13 *A Compilation of SOFC Developers, Primary Financial Source, Corporate Size, SOFC Types, SOFC Performance, Business Outcome, and Duration of SOFC Development (As of December 2010). (For explanation of the table, see three notes provided at the end of the table on Page 394) (Cont'd)*

Country	Company (or organization)	Primarily private sector support	Major government support	Company or organization: Small	Company or organization: Large	Type of fuel cell	SOFC performance	Business outcome	Duration of SOFC development
							SOFCpower of Italy acquired HTc (2007).		
	Sulzer/Sulzer Hexis/Hexis	X		X		planar	A 1 kW system demonstrated the maximum power output of 1–1.5 kW at an electrical efficiency of 30–35 percent. The power density was 200 mW/cm^2 at 920 °C (1999).	TBD	20
							Hexis' Galileo 1 kW 1000 N demonstrated 25–30 percent of electrical efficiency and 90 percent of total efficiency (2010).		
United Kingdom	Alstom (Hqs in France)		X		X	Planar	It primarily evaluated SOFC stack designs in combination with gas turbines using cells developed by other SOFC manufacturers.	Failed	More than 10
							Alstom likely lost interest in SOFC development in the early 2000s.		
	Ceres Power		X	X			Ceres Power's single cells demonstrated a power density of 400 mW/cm^2 at 600 °C and a 160 W short stack tested for 3300 h with less than 5 percent performance degradation (2005).	TBD	10
	Rolls Royce Fuel Cell Systems		X		X	Integrated planar tube	A 300 mm, 40 cell, proof of concept module generated 27.3 W and achieved 155 mW/cm^2 power density at 0.62 V mean cell voltage with a 43 percent fuel utilization from a simulated reformed hydrocarbon fuel (2001).	TBD	17
Australia; but largely Europe-based manufacturing	Ceramic Fuel Cells Ltd (CFCL)	X		X		Planar	CFCL developed an advanced anode-supported fuel cell with power densities of 500–650 mW/cm^2 at an electrical efficiency of more than 50 percent (2008).	TBD	18
		X					A CFCL 1–2 kW unit demonstrated the total efficiency of 85 percent and an average power density of 320 mW/cm^2 at 0.83 V (2009).		
Canada	Fuel Cell Technologies Ltd (FCT)	X		X		Tubular	FCT built a 5 kW system using SWPC's cells and successfully operated it with natural gas (2002).	Failed	16
		X					One 5 kW system operated for 9650 h, generated an average electrical output of 4.36 kW, and demonstrated the system efficiency of 39.3 percent (2005).		

						Acumentrics of the United States acquired FCT in 2007.		
	Global Thermoelectric	X		X	Planar	A single cell operated at 0.7 V using H_2 as fuel demonstrated a power density of 0.15 W/cm^2 in 1998, 0.528 W/cm^2 in 1999, 0.9 W/cm^2 in April 2000, and 1.093 W/cm^2 in June 2000. A power density of a stack using the same cells showed 1 W/cm^2 in 2002 (2003).	Failed	6
						FuelCell Energy of the United States acquired Global in August 2003.		
South Korea	Korea Electric Power Research Institute (KEPRI)		X	X	Planar	KEPRI's cell showed a maximum power density of 507 mW/cm^2 at 650 °C (2005).	TBD	21
			X			A 1 kW SOFC stack composed of 37 anode-supported planar cells and low cost stainless steel interconnects showed the maximum power density of 0.35 W/cm^2 at 750 °C (2007).		
	Korea Institute of Energy Research (KIER)		X	X	Tubular and flat tubular	KIER developed induction furnace brazed anode-supported flat tube cell that showed 210 mW/cm^2 (0.7 V, 3000 mA/cm^2) at 750 °C (2005).	TBD	21
			X			The performance of unit cell in 3 percent humidified H_2 and air at 800 °C showed maximum power density of 0.6 W/cm^2 (2006)		
			X		Segmented-in-series tubular	A single cell showed a maximum power density of 38, 67, and 102 mW/cm^2 at 700, 750, and 800 °C, respectively, using moist hydrogen (300 ml/minute) as fuel and ambient air as oxidant (2010).		
China	Ningbo Institute of Material Technology and Engineering (NIMTE)		X	X	Planar	An ultra-thin, high-strength, planar cell in size ranging from 10 cm × 10 cm to 30 cm × 30 cm showed the power density ranging 0.4–0.6 W/cm^2 at 750 °C (2009).	TBD	4
			X			A 10-cell SOFC stack achieved a maximum power of 310 W (H_2, Air) and a maximum power density of 0.31 W/cm^2. The maximum power density and minimum power density of single cells achieved 0.5 W/cm^2 and 0.25 W/cm^2, respectively (2009).		

(Continued)

TABLE 6.13 *A Compilation of SOFC Developers, Primary Financial Source, Corporate Size, SOFC Types, SOFC Performance, Business Outcome, and Duration of SOFC Development (As of December 2010). (For explanation of the table, see three notes provided at the end of the table on Page 394) (Cont'd)*

Country	Company (or organization)	Primarily private sector support	Major government support	Company or organization		Type of fuel cell	SOFC performance	Business outcome	Duration of SOFC development
				Small	Large				
	Shanghai Institute of Ceramics (SIC)		X		X	Planar	A 2-cell stack and a 10-cell stack with cell size of 40 mm × 40 mm were fabricated and tested. The 10-cell stack at 1000 °C showed a power output of about 10 W and a power density of about 110 mW/cm^2 (2002).	TBD	10
			X				A planar cell with an area of 10 cm × 10 cm was operated at intermediate temperature (~750 °C) with the power density close to 0.4 W/cm^2 (2006).		
			X			Tubular	A single anode-supported tubular cell demonstrated the maximum power density of 325, 276, 208, and 168 mW/cm^2 at 850, 800, 750, and 700 °C, when operating with humidified hydrogen (2008).		
			X				A cathode-supported tubular SOFC cell showed the maximum power density of 530 mW/cm^2 at 850 °C (2009).		
India	Central Glass and Ceramic Research Institute (CGCRI)		X		X	Planar	Anode-supported cells of 10 cm × 10 cm × 1.5 mm were fabricated using tape-casting and screen-printing techniques. A thin (~10 μm) layer of electrolyte (YSZ) film was adhered to the 1.5 mm NiO-YSZ porous anode support on one side and LSM-based cathode on the other side. A single cell demonstrated about 2 W/cm^2 at a cell voltage of 0.7 V and at an operating temperature of 800 °C (2009).	TBD	More than 20

Note:
1. The table attempts to categorize which country or region has a global leadership role in SOFC development today. The ranking is shown in the table by means of different background shading: dark gray background indicates the top global leader, medium gray indicates the second-tier leader, and light gray background indicates a third-tier developer. Today, Japan is ranked as the top global leader, and the United States and Europe are second-tier leaders. Companies in the rest of the world are third-tier developers but are catching up rapidly. (For the color version of this table, the reader is referred to the online version of this book. In the online version, the darkest blue indicates the top global leader, the medium blue indicates the second-tier leader, and the lightest blue indicates the third-tier developer.)
2. Whether a company (or organization) is small or large is based on commonly accepted views.
3. "TBD" indicates when commercial outcome is uncertain, yet to be determined, too early to tell, etc. "Failed" indicates the company or organization closed, sold all or the majority of SOFC assets or business, or is no longer focused on SOFC development.

The tendency of Japanese companies to team up creates a force multiplier effect and promotes rapid innovation. In addition, many of Japan's gas and electric utility companies, utility research institutions, and oil companies are engaged in SOFC development themselves. They do not fit the typical profile for small, single-purpose SOFC developers in other countries. Rather, these companies have the potential to be large-scale consumers or users of fuel cell products. The combination of government support, the industry's own tendency for collaboration, and the diversity and strength of SOFC developers' expertise and infrastructure all give Japan a global competitive edge.

6.6.2. The United States Loses Its Edge Through the Slow Erosion of Its Base

For about 45 years from the 1960s, the United States was the unquestionable leader in SOFC development worldwide. From the 1960s to 2000, Siemens Westinghouse was the predominant player. Its achievements in tubular designs were respected worldwide and were so far-reaching that few in global fuel cell industry wished to compete with them. From 2000 to 2007, GE became the global leader, demonstrating cutting-edge planar SOFC technology that led the world. But today neither of these companies is a player in the SOFC field. Moreover, many other SOFC developers in the United States have ceased their SOFC activities. Only a handful of developers remain today.

- Nonetheless, the United States still holds an edge in large stationary power-generation development.

6.6.3. Europe Is Fragmented, Uncoordinated, and Ineffective

Europe, while a progressive end-user for advanced forms of energy production, is disorganized and unable to focus development on any one fuel cell technology. Individual European countries have their own fuel cell policies, uncoordinated with each other, and are unable to create a multiplier effect or establish synergies to increase European capabilities. While the European Commission has provided substantial support for SOFC manufacturing, its engagement has been episodic and it has failed to coordinate European SOFC development as a whole. It sponsors a large number of SOFC R&D projects, but the projects have produced few worthwhile results. Projects reflect contradictory priorities, goals, and methods, include too many participants that are unable to effectively collaborate or share information, spread resources too widely and too thinly, and have time frames that are too short. Europe thus has not significantly improved its SOFC capabilities and remains in the second or third position.

A number of SOFC-developing countries in the rest of the world are bundled together and ranked in a third tier.

- Australia has a strong SOFC developer in CFCL, but the country lacks a large consumer base for SOFCs. As a result, CFCL has chosen to concentrate on business opportunities in Europe. To serve the European market, CFCL has established a large-scale manufacturing and supplier-chain base in Germany.
- Canada had two strong SOFC developers, FCT and Global Thermoelectric. But Canada lost core SOFC capabilities when these companies were acquired by the US companies.
- South Korea, China, and India possess basic levels of SOFC knowledge and capabilities. Their future prospects are unknown, but there are possibilities that one or more might achieve a breakthrough that would rapidly advance their capabilities.

6.6.4. But No Country Has a Viable SOFC Product Yet

Irrespective of the ranking order in SOFC capabilities, no region has a bright outlook as yet. No SOFC technology has demonstrated credible efficiency, lifetime durability, and bona fide cost competitiveness. And most important, no SOFC product has demonstrated net profitability.

References

1. Blomen LJMJ, Mugerwa MN. *Fuel cell systems.* Springer; 1993.
2. Collecting the history of fuel cells. Smithsonian National Museum of American History. http://americanhistory.si.edu/fuelcells/index.htm, updated August 2009.
3. Tubular solid oxide fuel cell technology. http://www.fossil.energy.gov/programs/powersystems/fuelcells/fuelcells_solidoxide.html, Fossil Energy Office of Communication, US Department of Energy, page updated 28 April 2009.
4. Schaeffer GJ. Fuel cells for the future: a contribution to technology forecasting from atechnology dynamics perspective. ISBN 90 365 12 30 1, 1998.
5. Weissbart J, Ruka R. A solid electrolyte fuel cell. *J Electrochem Soc* 1962;**109**(8):723–6.
6. Marine applications for fuel cell technology—a technical memorandum. OTA-TM-O-37, U.S. Congress, Office of Technology Assessment, February 1986.
7. Fuel cell power generation technology on the research report (d). http://www.feedmode.com/user/blogs/view/name_besblo001/id_3589/title_Fuel-cell-power-generation-technology-on-the/, Feedmode, 24 July 2009.
8. SOFC tests and demonstrations: summary. Siemens AG. http://www.powergeneration.siemens.com/products-solutions-services/products-packages/fuel-cells/demonstrations/demonstrations-summary/; [accessed September 2009].
9. Bratton RJ, Singh P. Tubular solid oxide fuel cell developments, DOE/MC/28055-95/CO496. Prepared for the 97th Annual Meeting of the American Ceramic Society, Cincinnati, OHIO 1–3 May 1995.
10. Rastler DM. State-of-the-art of fuel cell technologies for distributed power technical and strategic assessment of products: markets, and retail competitiveness. Annual Update, Electric Power Research Institute, TR-106620–R1, August 1997.
11. Siemens Westinghouse: 25 kW tubular solid oxide fuel cell first SOFC pre-commercial prototype and research platform overview. Research Archives, National Fuel Cell Research Center, University of California, Irvine. http://www.nfcrc.uci.edu/2/ACTIVITIES/RESEARCH_STUDIES/RESEARCH_ARCHIVES/westinghouse.aspx; [accessed October 2009].
12. Siemens Westinghouse to demo fuel cell for ontario power generation. Energy User News, 1 February 2001.
13. Veyo S. Westinghouse SOFC field unit status. Fuel cells '97 Review Meeting, The Federal Energy Technology Center (FETC), DIE, Morgantown, West-Virginia, 26–28 August 1997.
14. Carbasho T. Energy lab contributes to futuristic power source cells in Netherlands.Pittsburgh Business Times, 14 April 2000.
15. Siemens Westinghouse's 100 kW SOFC system completes operating period. Business Wire, 5 January 2001. http://www.allbusiness.com/energy-utilities/utilities-industry-electric-power/6026822-1.html.
16. Singhal SC. Status of Siemens Westinghouse SOFC technology and demonstration units. 1999 joint fuel cell technology review conference. Federal Energy Technology Center (FETC), Department of Energy, Chicago, Illinois; August 3–5, 1999.
17. 100 kW solid oxide fuel cell CHP system operating in Italy, <https://www.canyontrailchuckwagon.com/index.cfm/lev1/621, Turbo services network, 12 August 2005; Siemens successfully restarts SOFC CHP system in Italy, Fuel Cells Today, 17 August 2005.
18. Siemens Westinghouse and Southern California Edison announce successful testing of the world's first fuel cell/gas turbine hybrid. Business Wire, 17 April 2000.
19. Demonstrations. Siemens, AG. http://www.powereneration.siemens.com/products-solutions-services/products-packages/fuel-cells/demonstrations/, [accessed October 2009].
20. Hybrid - 220 kW hybrid solid oxide fuel cell/micro turbine generator system. Research Archives. National fuel cell research center, University of California, Irvine, CA. http://www.nfcrc.uci.edu/2/ACTIVITIES/RESEARCH_STUDIES/RESEARCH_ARCHIVES/hybrid.aspx; [accessed October 2009].
21. Singhal SC, Kendall K. *High temperature solid oxide fuel cells, fundamentals, design and applications*. Elsevier; 2003.
22. Ontario power generation and SIEMENS Westinghouse launch SOFC power plant. Power Group Online Article, PennWell Corporation, Tulsa, OK, 5 March 2003.
23. Prototype fuel-cell heat and power plant. TEAM Projects – International Initiatives, http://www.team.gc.ca/english/dbprojects/viewProject.asp?id=401&typ=,T, Technology Early Action Measures (TEAM), Government of Canada, [accessed October 2009].
24. Largest SOFC plant for U. of Toronto. Fuel Cell Technology News, 1 September 2003.

25. UTM abandons plans to install solid oxide fuel cell. Fuel Cells Today, 16 September 2004.
26. Fitzpatrick D. Siemens Trims Munhall, Pa., Plant Start-Up, Job targets. Knight Ridder/Tribune Business News, November, 2002.
27. Siemens Westinghouse chooses Pittsburgh area for expansion of its fuel cell operation. Business Wire, 26 September 2001.
28. Solid Oxide: Siemens aims for 85 MW to 100 MW by 2006. Fuel Cell Technology News, 1 March 2002.
29. Siemens won't use new plant in Munhall, Pa., but will keep fuel cell operation. Knight Ridder/Tribune Business News, January, 2004.
30. Lindeman TF. Siemens to refund $468,750 to state, http://www.post-gazette.com/pg/10054/1037809-28.stm, Pittsburgh Post-Gazette, 23 February 2010.
31. Fuel-cell power plant. Distributed Generation Bulletin, No. 06, Center for Advanced Engineering, Christchurch, New Zealand, 4 July 2003.
32. Norway to get SIEMENS SOFC Plant. Fuel Cell Technology News, 1 May 2000.
33. Siemens Tests Alaskan Climate. Fuel Cell Technology News, 1 October 2001.
34. Highlights of 6th natural gas conversion symposium, northern gas pipelines: GTL news and reference, http://www.arcticgaspipeline.com/GTL.htm, Northern Gas Pipelines, 31 March 2001.
35. Witmer D. Hydrogen and fuel cell demonstrations in Alaska. Arctic Energy Summit, University of Alaska Fairbanks, October 2007.
36. Siemens Westinghouse power corporation announces key Contract awards for solid oxide fuel cell systems. Business Wire, 6 December 2000.
37. Review of technical, socio-economic and safety findings from stationary fuel cell demonstration activities. Feed-back and feed-forward from WP 4/5 and demonstration activities—stationary, roads2hycom, the European Commission, November 2007.
38. Utilities and power generation EPA showcases fuel cell power plant. Energy Conservation News, 1 November 2000.
39. Suitable microturbine generator not available on the market, pre-COMMERCIAL 1-MW-fuel cell power plant in Marbach/Neckar will not be built, consortium continues to follow the developments of SOFC-Technology. Press release, EnBW Energie Baden-Württemberg AG, 9 December 2002.
40. EnBW, Siemens plan first ever megawatt-class fuel-cell power plant, http://www.physorg.com/news77298251.html>, Physorg, 12 September 2006.
41. Hybrid Future. Pictures of the future, Siemens AG, Spring 2007.
42. Templeton D. Siemens readies fuel cell as source of power. Pittsburgh Post-Gazette, 7 March 2007.
43. Siemens wins third solid oxide fuel cell power system order from gas turbine technologies. Press release, Siemens Power Generation, Orlando, 28 November 2005.
44. Meidensha Siemens power generation to join hands in fuel cell business in Japan. JCNN News Summaries, 10 November 2005.
45. Homma T. Meidensha to launch SOFC power generation business with siemens. SOFC business development, No. 141, Latest News, FCDIC, 31 January 2008.
46. Siemens SOFC surfaces in Japan with 125 kW CHP, http://fuelcellintel.blogspot.com/2007/11/siemens-sofc-surfaces-in-japan-with-125.html. Fuel Cell Intel. 6 November 2007.
47. Rastler D. Update on fuel cell development—review of major and stealth fuel cell players' activities: Stealth Player Reviews. 1008429, Technical Update. EPRI, December 2004.
48. Siemens turbine plant to lay off another 190 employees, http://ecmweb.com/news/electric_siemens_turbine_plant/, Industry News, EC&M, 7 March 2003.
49. Olson T. Siemens energy to lay off 75 in Churchill. Pittsburgh Tribune-Review, 18 March 2009.
50. Siemens power to put SOFC unit on block, Pittsburgh Tribune-Review, 30 July 2008.
51. Siemens looks to sell SOFC business, Pittsburgh Business Times, 30 June 2008.
52. Four industry teams begin quest for low-cost, breakthrough fuel cell, could broaden market acceptance of cutting edge technology cited in president's climate change policy. News release, NETL, DOE, 8 August 2001.
53. Williams MC. Solid oxide fuel cell commercialization in the United States. 1995 Department of Energy/Agency of Industrial Science and Technology International Fuel Cell Symposium, Yokohama, Japan, 12–16 June 1995.
54. Future fuel cells R&D, http://www.fossil.energy.gov/programs/powersystems/fuelcells/index.html, Fossil Energy, DOE, Page updated on 8 July 2009.
55. The global fuel cell race: can the United States Win? Centra Technology, Inc. Arlington, Virginia, September 2001.
56. DOE selects first vision 21 projects to design the energy plant of the future. News Release, NETL, DOE, 7 March 2000.
57. Rudins G. Fossil energy mission and the fuel cell program. Office of Fossil Energy, DOE, Solid State Energy Conversion Alliance Workshop, 1–2 June 2000.

58. Abraham announces pollution-ree power plant of the future $1 Billion living prototype to showcase cutting-edge technologies to advance president's climate change, hydrogen initiatives. News release, NETL, DOE, February 27, 2003.
59. Our budget. Office of Fossil Energy, DOE, http://www.fossil.energy.gov/aboutus/budget/index.html, update June 08, 2009.
60. DOE continues implementing president's hydrogen initiative with addition of two new fuel cell projects, goal is to move fuel cells into mainstream energy markets. News release, NETL, DOE, 23 April 2003.
61. Significant milestone achieved in SECA fuel cell development program, GE successfully tests solid oxide fuel cell prototype, becomes first SECA industry team to enter second phase of program. News release, NETL, DOE, 19 January 2006.
62. SECA fuel cell development program completes phase I, six industry teams surpass fuel cell prototype test targets. News release, NETL, DOE, 16 May 2007.
63. SECA fuel cell program moves two key projects into next phase, projects continue push for low-cost, environmentally friendly coal power. News release, NETL, DOE, 5 February 2009.
64. President's clean coal initiative underway: energy department's solicitation calls on industry to match between $300 to $400 million in federal funding. NETL, DOE, 4 March 2002.
65. SECA fuel cell program selects two projects, low-cost fuel cell systems to address energy security, climate and water challenges. News release, NETL, DOE, 29 May 2008.
66. DOE announces restructured future gen approach to demonstrate carbon capture and storage technology at multiple clean coal plants, affirms commitment to clean coal technology investments; requests $648 million for coal research, development and deployment for FY09 budget – largest coal budget in more than 25 years. News release, NETL, DOE, 30 January 2008.
67. Coal-based fuel cells: a giant leap for fuel cell technology, new program to develop multi-megawatt fuel cell systems. News release, NETL, DOE, 11 August 2005.
68. DOE selects projects to advance fuel cell technology. News release, NETL, DOE, 22 May 2008.
69. B&W technology tapped for use in DOE's Future Gen 2.0 CO2 capture, storage project. Power, http://www.pennenergy.com/index/power/display/6502839838/articles/pennenergy/power/coal/2010/10/b_w-technology_tapped.html, Penn Energy, October 7, 2010.
70. Fuel cell origins: 1840–1890. Collecting the history of fuel cells, http://americanhistory.si.edu/fuelcells/origins/origins.htm, Smithsonian Institution, no date, [accessed November 2011].
71. SECA fuel cell program moves two key projects into next phase, projects continue push for low-cost, environmentally friendly coal power. News release, NETL, DOE, 5 February 2009.
72. Pierre J. Siemens energy. 10th Annual SECA Workshop, Pittsburgh, PA, 15 July 2009.
73. Vora SD. SECA program review—siemens. The 10th Annual SECA Workshop. Pittsburgh, PA, 15 July 2009.
74. Fuel cell stacks still going strong after 5000 hours, SECA fuel cell stacks meet, surpass doe goals, News release, NETL, DOE, 24 March 2009.
75. Company history and overview. Acumentrics. http://www.acumentrics.com/about-company.htm; [accessed December 2009].
76. Fuel cell handbook, 7th ed. NETL, Department of Energy, November 2004.
77. Acumentrics receives $15.6 million doe fuel cell development contract, Targets further cost reduction, durability and increased system size. News & Event, http://www.acumentrics.com/d31585c3-997c-408c-a6cb-c7523ac83d4c/news-and-events-press-releases-2008-detail.htm, Acumentrics, 4 June 2008.
78. Acumentrics fuel cell tops 10,000 operating hours, Longest-running 5kW SOFC system for remote, auxiliary & backup power. Press release, Acumentrics, WestWood, MA, 8 September 2008.
79. Bessette N (Acumentrics). Development of a low cost 3–10 kW tubular SOFC power system, http://www.hydrogen.energy.gov/pdfs/review10/fc032_bessette_2010_o_web.pdf, 2010.Annual Merit Review Proceedings, 7–11 June 2010.
80. Farmer R. Fuel cell technologies program overview. 2010 Annual Merit Review and Peer Evaluation Meeting, EERE, DOE, http://www1.eere.energy.gov/hydrogenandfuelcells/pdfs/amrplenary_farmer.pdf; [accessed 7 June 2010].
81. Schmidt D, Bessette NF, Foster RE, Litka AF, Rawson J (Acumentrics). Progress in Tubular SOFCs at Acumentrics. 2010 Fuel cell seminar & exposition, San Antonio, Texas, 18–21 October 2010.
82. SECA member enters joint venture to market fuel cells in Japan. News release, NETL, DOE, 3 October 2003.
83. Overview. SOFC Products, http://www.acumentrics.com/sofc-products-overview.htm, Acumentrics, 2010.
84. Acumentrics history. About us, http://www.acumentrics.com/about-us-company-history.html, Acumentrics, 2010 [accessed June 2010].
85. Trusted Power Innovations. Trusted Power Innovations, http://www.acumentrics.com/about-us-company-

overview.htm>, Acumentrics, 2010 [accessed June 2010].

86. Minh N, Wentzel W, Gibson R. Tape-calendaring monolithic and flat plate SOFC's. In: Singhal SC, Iwahara H, editors. *Proceedings of the third international symposium on SOFCs*; 1993. p. 801–9.

87. Minh N, Anumakonda A, Chung B, Doshi R, Ferrall J, Guan J, (AlliedSignal Aerospace). Development of reduced-temperature solid oxide fuel cell power systems. In: Singhal SC, Dokiya M, editors. *Solid oxide fuel cells (SOFC VI). The electrochemical society*; 1999.

88. NASA technology comes to earth. Company history, http://www.bloomenergy.com/about/company-history/, Bloom Energy, 2010 [accessed June 2010].

89. Bloom energy announces industry-leading customers: bank of America, coca-cola, cox enterprises, eBay, FedEx, Google, Staples, and Walmart among first companies to deploy bloom energy server. Press release, http://c0688662.cdn.cloudfiles.rackspacecloud.com/downloads-pdf-release-bloom-foundation-customer-2-24-10_1.pdf; Bloom Energy, 24 February 2010.

90. Cypress's CEO issues declaration of energy independence, planning 100 percent self-generation by 2015, renewable electricity from new bloom energy fuel cells and existing solar panels now powers 75 percent of San Jose company's campus. Press release, Cypress Semiconductor, 2 July 2010.

91. Adobe powers San Jose headquarters with bloom energy fuel cells, largest bloom box installation to date will generate 30 percent of adobe HQ electricity needs, promote sustainability. Press release, Adobe Systems, 28 September 2010.

92. Industry leading companies choose bloom electrons^{sm} for immediate cost savings and carbon reduction benefits. Press release, http://www.bloomenergy.com/newsroom/press-release-01-20-11-b/, Bloom Energy [accessed 20 January 2011].

93. Hilgen J. Public service commission gives bloom energy deal its approval. Article, http://www.delawarefirst.org/18847-bloom-hearing-delaware, DFM News, 18 May 2012.

94. Bloom energy quadruples size of its manufacturing operations providing over 1000 California jobs. Press release, http://www.bloomenergy.com/newsroom/press-release-04-15-11/, Bloom Energy, 15 April 2011.

95. Bloom energy expands to East Coast. Press release, http://www.bloomenergy.com/newsroom/press-release-04-30-12/, Bloom Energy, 30 April 2012.

96. Delaware attracting bloom energy for high-tech manufacturing hub. News, http://governor.delaware.gov/news/2011/1106june/20110609-bloom_energy.shtml, Governor Jack Markell, 9 June 2011.

97. Energy server architecture. Products, http://www.bloomenergy.com/products/architecture/, Bloom Energy, 2010 [accessed June 2010].

98. Steinberger-Wilckens R, Lugovy M, Klein D. Development of SOFC based On Scandia Doped Zirconia Ceramics, http://perso.ensem.inpl-nancy.fr/Olivier.Lottin/FDFC08/CD/Contributions/FCHSTEI11-09_38M.pdf, Fundamentals and developments of fuel cells conference 2008 (FDFC2008), Nancy, France, 10–12 December 2008.

99. Scandia stabilized Zirconia (6 Mol. %): SCZ, http://www.americanelements.com/scz.html, American Elements, no date [accessed June 2010].

100. Scandium. Article, http://en.wikipedia.org/wiki/Scandium, Wikipedia, last modified on 19 June 2010, [accessed June 2010].

101. Woody T. Bloom energy claims a new fuel cell technology. Business, http://www.nytimes.com/2010/02/24/business/energy-environment/24bloom.html. The New York Times, 23 February 2010.

102. Than K. Bloom Box Launch Is Big Hype—Invention Nothing New? News Blogs, http://news.nationalgeographic.com/news/2010/02/100224-bloom-box-launch-bloom-energy-press-conference-update/, National Geographic News, 24 February 2010.

103. What are phosphoric acid fuel cells? Fuel cells 101, http://www.utcpower.com/knowledge-library/what-are-phosphoric-acid-fuel-cells, UTC Power, no date [accessed May 2012].

104. History of ceramatec, http://www.ceramatec.com/about/history.php, Ceramatec, [accessed December 2009].

105. Technology development, http://www.ceramatec.com/technology-development/index.php, Ceramatec, [accessed December 2009].

106. Williams MC. U.S. solid oxide fuel cell powerplant development and commercialization. In: Stimming U, Singhal SC, Tagawa H, Lennert W, editors. *Solid oxide fuel cells v.* The Electrochemical Society; 1997.

107. Williams MC. U.S. solid oxide fuel cell powerplant development and commercialization DOE/FETC/C-97/7281, Fifth international symposium on solid oxide fuel cells, Aachen, Germany, 2–4 June 1997.

108. Fuel Cell Research. Cummins, http://www.cummins.com/cmi/content.jsp?siteId=1&langId=1033&menuId=82&overviewId=4&anchorId=11&contName=285&dataId=2971&menuIndex=none&dataId=2971&feed=1&feed=1>, 23 July 2009.

109. Solid State Energy Conversion Alliance (SECA)—Industry Teams, http://www.netl.doe.gov/technologies/coalpower/fuelcells/seca/industry.html, NETL, DOE, [accessed December 2009].

110. D. Norrick, C. Vesely, T. Romine, B. Palmer, G. Rush, E. Barringer, et al. 10 kW SOFC power system commercialization. SECA Semi-Annual Report 41244R02, Cummins Power Generation, February 2003.
111. Norrick D (CPG). Cummins power generation 10 kWe SOFC power system commercialization program, http://www.netl.doe.gov/publications/proceedings/03/seca/Daniel Norrick.pdf. The Fourth Annual SECA Meeting, Seattle, WA, 15–16 April 2003.
112. Fuel cell research, fuel cell power systems, http://www.cummins.com/cmi/content.jsp?siteId=1&overviewId=4&dataId=2971&contName=285&langId=1033&menuId=82&anchorId=11&feed=1&menuIndex=none&. Cummins, 23 July 2009.
113. Pollok M. Cummins says fuel cell mobile power available in 2 to 3 years, http://www.khl.com/magazines/international-rental-news/detail/item43133/Cummins-says-fuel-cell-mobile-power-available-in-2-to-3-years/. International Rental News, 24 August 2009.
114. Cummins and Protonex Demonstrate Fuel Cell Truck APU, http://www.fuelcelltoday.com//online/news/articles/2010-05/Cummins-Protonex-Fuel-Cell-Truck, Fuel Cell Today, 11 May 2010.
115. Norrick D (Cummins Power Generation). Diesel Fueled SOFC System for Class 7/Class 8 On-Highway Truck Auxiliary Power, http://www.hydrogen.energy.gov/pdfs/review10/fc061_norrick_2010_p_web.pdf, Annual Merit Review Proceedings, 2010.
116. Mukerjee S, Haltiner K, Shaffer (Delphi) S, Meinhardt K, Chick L, Sprenkle V, (Battelle Memorial Institute-Pacific Northwest Division). Development update on Delphi's SOFC stack. In: Singhal SC, Mizusaki J, editors. *Solid oxide fuel cells IX vol. 1 (SOFC IX)Cells, stacks, and systems.* The Electrochemical Society; 2005.
117. Delphi reaches developmental milestone for fuel-flexible solid oxide fuel cell power system. Press release, Delphi, 20 June 2006.
118. Zizelman J. Development update on Delphi's solid oxide fuel cell system: from gasoline to electric power. The 4th SECA meeting, 15–16 April 2003.
119. Shaffer S (Delphi). Development update on Delphi's solid oxide fuel cell systems, http://www.netl.doe.gov/publications/proceedings/07/SECA_Workshop/pdf/Dev%20Update%20on%20Delphi%27s%20SOFC%20Systems-Steven%20Shaffer%2C%20Delphi.pdf. Eighth Annual SECA Workshop, 7–9 August 2007.
120. SECA fuel cell proves successful in Navy's proof-of-concept testing, Spinoff applications mark the road to widespread commercialization. News release, NETL, DOE, 5 August 2008.
121. Delphi and peterbilt successfully power truck cab and sleeper with solid oxide fuel cell Auxiliary power unit. Peterbilt News, http://www.peterbilt.com/newsdetails.aspx?id=226. Peterbilt, 22 July 2008.
122. Shaffer S (Delphi Corporation). Solid oxide fuel cell development for auxiliary power in heavy duty vehicle applications, http://www.hydrogen.energy.gov/pdfs/progress08/v_i_3_shaffer.pdf; 2008 Annual progress report, EERE, DOE, 2008.
123. Shaffer S. Solid oxide fuel cell development at Delphi. 2010 Fuel cell seminar & exposition, http://www.fuelcellseminar.com/media/5055/hrd33-3_shaffer.pdf. San Antonio, Texas, 18–20 October 2010.
124. Hennessy D (Delphi), Solid oxide fuel cell development for auxiliary power in heavy duty vehicle applications. http://www.hydrogen.energy.gov/pdfs/review10/fc062_blake_2010_p_web.pdf, 2010 Annual Merit Review Proceedings, 8 June 2010.
125. Fuel cell energy and department of energy finalize phase 1 award for solid state energy conversion alliance program. Business Wire, 16 September 2004.
126. Versa power systems, company overview and background, http://www.versa-power.com/company.htm. Versa Power Systems, [accessed December 2009].
127. Transaction will move VPS from R&D to prototype development stage company, News, Versa Power Systems, http://www.versa-power.com/news/Versa_Power_Systems_10-21-04_Rev1.pdf, 21 October 2004.
128. Fuel cell energy acquires global thermoelectric Inc., http://www.fuelcelltoday.com/online/industry-directory/organisations/gl/Global-Thermoelectric-Inc-. Industry Directory. Fuel Cell Today, [accessed December 2009].
129. Global Thermoelectric—History, http://www.globalte.com/index.php?pageId=1&sId=8. Global Thermoelectric, [accessed December 2009].
130. Fuel cell energy to combine its Canadian solid oxide fuel cell operations into versa power systems. Business Wire, 2 November 2004.
131. Fuel cell energy and versa power successfully complete SECA phase I solid oxide fuel cell program six months early. Green Jobs, 24 April 2007.
132. Borglum B (Versa Power Systems). The status of SOFC development at versa power systems. Fuel Cell Seminar & Exposition, San Antonio, Texas, 18–20 October, 2010.
133. Investors, http://www.versa-power.com/investors.htm. Versa, no date [accessed November 2010].
134. Williams M. Ceramic energy: advancing SOFCs for distributed generation. Feature Article, http://www.ceramicindustry.com/Articles/Feature_Article/

9dc17442bbac7010VgnVCM100000f932a8c0____. Ceramic Industry, [accessed 1 September 2005].
135. Guan J, Ramamurthi B, Ruud J, Hong J, Riley P, Weng D, et al. (GE Global Research Center). High performance flexible reversible solid oxide fuel cell, http://www.hydrogen.energy.gov/pdfs/review07/pdp_15_guan.pdf. Hydrogen program review, 15 May 2007.
136. Overview. Company, http://www.protonex.com/company/index.aspx. Protonex, 2010 [accessed June 2010].
137. Mesoscopic devices. Industry directory, Fuel Cell Today, 12 May 2010.
138. Poshusta JC, Bruinsma D, Booten CW, Martin JL, Martin CM (Protonex). Fuel Flexible Portable SOFC Generators, http://www.fuelcellseminar.com/assets/2009/HRD42-3_1130AM_Poshusta.pdf, 2009, Fuel Cell Seminar, 19 November 2009.
139. Protonex teams with Cummins to integrate diesel-fueled SOFC power systems into commercial applications. Online News. Fuel Cell Today, 24 September 2007.
140. Cummins demonstrates solid oxide fuel cell auxiliary power system. News releases, Cummins, 5 May 2010.
141. Strategic Update. Press releases, http://www.protonex.com/_assets/pressrelease/16473351-9677-4e8f-8450-6a29855871d9.pdf. Protonex, 9 April 2010.
142. Protonex to power Cummins fuel cell demo vehicles, stories, http://www.masshightech.com/stories/2007/09/24/daily2-Protonex-to-power-Cummins-fuel-cell-demo-vehicles.html. American City Business Journals, 24 September 2007.
143. Rolls-Royce fuel cell systems. Industry Directory, http://www.fuelcelltoday.com/online/industry-directory/organisations/ro/Rolls-Royce-Fuel-Cell-Systems. Fuel Cell Today, [accessed December 2009].
144. Hawkes GL, O'Brien JE, Haberman BA, Marquis AJ, Martinez Baca C, Travis RP. *Numerical prediction of the performance of integrated planar solid-oxide fuel cells, with comparisøns of results from several codes*. Denver, CO: 6th International Conference on Fuel Cell Science, Engineering, and Technology; 16–18 June 2008.
145. Goettler R, Ohrn T. Update on the rolls-royce coal-based SECA program. http://www.netl.doe.gov/publications/proceedings/10/seca/Presentations/Goettler%26Ohrn_Presentations.pdf. 11th Annual SECA Workshop, 27–29 July 2010.
146. Hart NT. Scale-up of the IP-SOFC to multi-kilowatt levels, F/01/00197/REP, URN 04/556. Rolls Royce, 2004.
147. Solid oxide fuel cell. Wikipedia, http://en.wikipedia.org/wiki/Solid_oxide_fuel_cell. Last modified on 13 December 2009.
148. Ohrn T, Liu Z (Rolls Royce). Rolls-Royce coal-based SECA program update, http://www.netl.doe.gov/publications/proceedings/11/seca/pdf/Tue%20PM/Ohrn.RRFCS%20SECA%202011%20Workshop%20Final.pdf. 12th Annual SECA Workshop, US Department of Energy, 26–28 July 2011.
149. Siemens achieves major milestone in Solid Oxide Fuel Cell program, http://www.powergeneration.siemens.com/press/press-releases/fossil-power-generation/2006/PG200611-005.htm. Press release, Siemens AG, 17 November 2006.
150. Vora S.D. Siemens: SECA program review Presented at the 8th Annual SECA Workshop san Antonio, TX. 7–9 August 2007.
151. Vora S.D (Siemens Energy). SECA program review. The 10th Annual SECA workshop, 14–16 July 2009.
152. Vora S.D. Office of fossil energy fuel cell program – solid state energy conversion alliance (SECA) clean economic energy for a carbon challenged world. http://www.netl.doe.gov/publications/proceedings/10/seca/Presentations/Vora%20_Presentation.pdf. The 11th annual SECA workshop, 27–29 July 2010.
153. Fuel cell technologies Ltd. Reports 2006 first quarter results. Green jobs, http://www.greenjobs.com/Public/IndustryNews/inews00777.htm, Kingston, Ontario, Canada, 30 May 2006.
154. Khandkar A, Elangovan S, Hartvigsen J. *Status of sofco's planar SOFC development*. Chicago, Illinois: 1999 Joint fuel cell technology review conference proceedings; 3–5 August 1999.
155. Technology Management, Inc., Company Profile. http://www.tmifuelcellsystems.com/company.asp, Technology Management, Inc., [accessed December 2009].
156. Ohio's fuel cell roadmap. Ohio Department of Development Technology Division, 1 September 2004.
157. Milliken C, Ruhl R, Hillman J. Technology Management Inc., Low cost multi-layer fabrication method for solid oxide fuel cells, Final Report, DE-AC26-00NT40707, 15 June 2002.
158. Ward S. Department of energy office of energy efficiency and renewable energy: high efficiency, low emissions, solid oxide fuel cell systems for multiple applications—major participant michael A. Petrik technology management Inc., 41009TPRG093002(1), 2004.
159. Milliken C, Ruhl R, Petrik MA, Lee BP. *TMI solid oxide fuel cell portable power generation*. Technology Management, Inc.. Chicago, IL: 1999 Joint fuel cell technology review conference proceedings http://www.netl.doe.gov/publications/proceedings/99/99fuelcell/fc6-3.pdf; 3–5 August 1999
160. Democratized energy: the social value of distributed fuel cell systems by Technology Management Inc.

http://docs.google.com/viewer?a=v&q=cache: pzsP7DEtuqUJ:ohioinnovationsummit.com/2010/presos/UCEAO_OIS_April_2010_Lee_Tues_AM.pdf+Technology+Management,+Inc+(TMI)+sofc&hl=en&gl=us&pid=bl&srcid=ADGEEShhKiHFh F9kaR0XFcm0sZp SuenBqGkBcVbAG2r GBRFN-nIpIW6VMdEO5X_2Y2NApI1iLt8w2TRjW-ea-kcOuJP7VIupGAqFsHd_OVU7CT_8Qejj TKoDwna0ODiZe65OcYgx3btr&sig=AHIEtbT0MTf7AukGItx0N2YzM1scU3S-hg, UCEAO – Ohio Innovation Summit, 20 April 2010.

161. Announcement of financial and non-monetary awards, third frontier makes $1 million award for fuel cell, national idling reduction network news, Argonne National Laboratory, April 2006.
162. Development fuel cells for use anywhere, anytime, by anyone, News. http://www.tmifuelcellsystems.com/news.asp, Technology Management, Inc., [accessed December 2009].
163. Project name: technology management/fuel cell systems, FA C T S H E E T, Ohio Bipartisan Job Stimulus Plan, http://www.ohioairquality.org/advanced_energy_program/pdf/TMI_Factsheet_12_7.pdf, [accessed December 2009].
164. Department of energy selects UTC Power-led team for $22 million technology development project, Press release, http://www.utcpower.com/fs/com/Attachments/PR_060508.pdf, UTC Power, 5 June 2008.
165. Shaffer S (Delphi). Delphi SOFC development update, http://www.netl.doe.gov/publications/proceedings/08/seca/Presentations/Shaffer.pdf, 9th annual solid state energy conversion alliance (SECA) workshop, 5–7 August 2008.
166. Kerr R. Delphi SOFC stack development update, http://www.netl.doe.gov/publications/proceedings/09/seca/presentations/Kerr_Presentation.pdf. 10th Annual Solid State Energy Conversion Alliance (SECA) Workshop, 14–16 July 2009
167. Narasimhamurthy P (UTC Power), Kerr R(Delphi). Coal-based IGFC project phase I, http://www.netl.doe.gov/publications/proceedings/11/seca/pdf/Tue%20PM/Praveen.SECA%20Workshop%207-26-11_Combined.pdf. 12th annual Solid State Energy Conversion Alliance (SECA) Workshop, US Department of Energy, 26–28 July 2011.
168. Reshaping energy production with the pure power of hydrogen. Brochure, http://www.ztekcorporation.com/PDF/Brochure2003.pdf, Ztek Corporation, 2003.
169. Fuel cells for power generation, http://www.usfcc.com/download_a_file/download_a_file/E&M-FuelCellsforPowerGeneration-07-020.pdf, US Fuel Cell Council, 2007.
170. Fuel cells, http://www.ztekcorporation.com/sofc.htm. Ztek Corporation, [accessed December 2009].
171. Hsu M.S. Ztek advanced planar solid oxide fuel cell for distributed generation 1999 Joint fuel cell technology review conference proceedings, Chicago, IL, 3–5 August 1999.
172. Ztek corporation selected to provide solid oxide fuel cell for connecticut test site project will advance commercialization of Ztek SOFC yechnology. Press release, Ztek Corporation, http://www.ztekcorporation.com/PDF/CTPressRelease.pdf, 13 March 2002.
173. Ztek corporation to collaborate with U.S. department of navy on hydrogen fueling Station, project at camp pendleton will advance commercialization in Southern California. Press release, http://www.ztekcorporation.com/press_releases/PressRelease. Ztek Corporation, 27 February 2006.
174. Department of energy FY 2012 congressional budget request: budget highlights. Office of Chief Financial Officer, US Department of Energy, February 2012.
175. Residential use fuel cells: subsidy and running cost. Fuel-Cell Share 2007, http://fuel-cell.share2007.net/; no date [accessed March 2010].
176. Yokokawa H. Recent development of solid oxide fuel cells in Japan. In: Singhal SC, Dokiya M, editors. *Solid oxide fuel cells (SOFC VI)*. The Electrochemical Society; 1999.
177. Hydrogen & fuel cells: review of national R&D programs. International Energy Agency, OECD, 2004; ETL100 Year History. http://www.etl.go.jp/jp/gen-info/history/nenshi/1974-17.html, updated 31 March 2001, [accessed March 2010].
178. Nishikawa M. Status of SOFC development in Japan, Solid Oxide Fuel Cells IV. In: Dokiya M, Yamamoto O, Tagawa H, Shinghal SC, editors; 1995.
179. Teruhisa H, Yokokawa H. Research and development of solid oxide fuel cell. http://www.nimc.go.jp/publication/news98/31-2.html; no date [accessed March 2010].
180. HarayamaY. Japanese technology policy: history and a new perspective. RIETI Discussion Paper Series 01-E-001, August 2001.
181. Miyake Y, Yasuo T, Akiyama Y, Taniguchi S, Kadowaki M, Kawamura H. Development of a planar solid oxide fuel cell module at Sanyo, Solid Oxide Fuel Cells IV. In: Dokiya M, Yamamoto O, Tagawa G, Singhal SC, editors. *The electrochemical society*; 1995.
182. Fujii H. Status of national project for SOFC development in Japan Solid Oxide Fuel Cells VIII (SOFC VIII). The Electrochemical Society, 2003.

183. Nakayama T, Suzuki M. Current status of SOFC R&D program at NEDO, Solid oxide fuel cells (SOFC VII). In: Singhal SC, Yokokawa H, editors. *The electrochemical society*; 2001.
184. Ujiie T. Status of national project for SOFC development in Japan, " In: Solid oxide fuel cells IX (SOFC IX). In: Singhal SC, Mizusaki J, editors. *Cells, stacks, and systems. The electrochemical society* 2005;vol. 1.
185. *Outline of SOFC R&D projects. The 1st "SOFC R&D" post evaluation sub working group, Material 5–2*. Japan: NEDO; 22 December 2005.
186. Saito T, Abe T, Fujinaga K, Miyao M, Kuroishi M, Hiwatashi K. Development of tubular SOFC at TOTO, Solid oxide fuel cells IX (SOFC IX). In: Singhal SC, Mizusaki J, editors. *Cells, stacks, and systems. The electrochemical society* 2005;vol. 1.
187. Nakanishi A, Hattori M, Sakaki Y, Kimura K, Ando Y, Miyamoto H. Development of MOLB Type SOFC. In: Singhal SC, Mizusaki J, editors. *Solid oxide fuel cells IX (SOFC IX). Cells, stacks, and systems*, vol. 1. The Electrochemical Society; 2005.
188. SOFC system technology development, post-project evaluation report, <www.nedo.go.jp/iinkai/kenkyuu/houkoku/20h/jigo/15.pdf. Research Evaluation Committee, February 2009.
189. SOFC system technology development, post-project evaluation report. NEDO, Japan, February 2009.
190. SOFC Demonstration Research, FY2007 Policy Implementation, P07003, http://www.nedo.go.jp/activities/portal/gaiyou/p07003/h19jisshi.pdf, NEDO, no date.
191. Okuda M. FY2007-2008 SOFC demonstration research result report Workshop. New Energy Foundation, 3 March 2009.
192. On public announcement on applications for subsidy by developers for SOFC demonstration research, Public Announcement, http://www.nedo.go.jp/informations/koubo/190322_3/190322_3.html, NEDO, 22 March 2007.
193. "On Solicitation of Installors/Operators for SOFC Demonstration Research, public notice, <https://app3.infoc.nedo.go.jp/informations/koubo/koubo/FA/nedokouboplace.2008-03-13.0143901654/nedokoubo.2009-01-28.7962306829/, NEDO, 3 February 2009.
194. SOFC Project: SOFC/Demonstration Research, Pamphlet, http://www.nef.or.jp/sofc/share/pdf/sofc_pamphlet.pdf. New Energy Foundation, February 2011.
195. Osaka gas, Kyocera, Toyota and Aisin announce agreement on co-development of solid oxide fuel cell (SOFC) cogeneration system for residential use. News release, http://global.kyocera.com/news/2009/0508_qazx.html, Osaka Gas Co., Ltd., Kyocera Corporation, Toyota Motor Corporation. Aisin Seiki Co., Ltd., 25 March 2009.
196. Tokyo Gas, Kyocera, Rinnai and Gastar Co-Develoed SOFC System Begins Residential Trial Operations, News release, Ltd.,http://global.kyocera.com/news/2009/1204_kode.html, Tokyo Gas Co., Ltd., Kyocera Corporation, Rinnai Corporation, Gastar Co., 15 December 2009.
197. SOFC demonstration research. FY2009 result reporting session, New Energy Foundation, 8 March 2010.
198. OkudaM. SOFC Demonstration Research Project, FY2010 Results, http://www.nef.or.jp/sofc/share/pdf/h22y.pdf, Fuel Cell Division, Planning Headquarters, New Energy Foundation, 10 March 2011.
199. Hosoi K, Nakabaru M. *Status of national project for SOFC development*. Vienna, Austria: 216th ECS Meeting; 4–9 October 2009.
200. Sato Y. SOFC R&D trend. FY 2008 SOFC demonstration research result report workshop, New Energy Foundation, 3 March 2009.
201. Homma T. *Sumitomo corporation. Latest news, No. 81*. Japan: FCDIC; 28 February 2003.
202. Acumentrics Ships Solid Oxide Fuel Cell System to Sumitomo of Japan for Testing. Press release, http://www.acumentrics.com/cc83cc2a-544e-44ce-87d8-bb44220f7e0e/press-release-archives-release-details.htm. Acumentrics, 31 January 2005,
203. Crawley G. Opening doors to fuel cell commercialization, solid oxide fuel cells (SOFC). Fuel Cell Today, January 2007.
204. Sumitomo and Nippon steel jointly develop sofc fuel cells for Japanese market. http://www.nikkeibp.co.jp/archives/311/311560.html. Nikkei Bp.net, 4 June 2004.
205. Homma T. *SOFC development and commercialization. Latest news, no. 98*. Japan: FCDIC; 25 June 2004.
206. Successful Development of a High Efficient Micro Tubular Fuel Cell—Toward the realization of high power micro solid-oxide fuel cells. Press release, http://www.aist.go.jp/aist_e/latest_research/2006/20060216/20060216.html. AIST, 18 January 2006.
207. Homma T. CEPRI and AIST. The latest fuel cell news in Japan, February 2007 Low-cost and small-size SOFC for home, FCDIC, February 2007.
208. Successful development of a small, high-performance micro fuel cell bundle—realizes more than two watts of output power in a unit size of a sugar cube (one cubic centimeter) below 600 °C. Press release, http://www.aist.go.jp/aist_e/latest_research/2009/20090306_2/20090306_2.html. AIST, 29 March 2007.
209. Development of a high-efficiency micro fuel cell module —a great step toward realizing high-power small-size solid oxide fuel cell systems. Press release,

http://www.aist.go.jp/aist_e/latest_research/2009/20090306_2/20090306_2.html, AIST, 12 February 2009.

210. Awano M, Suzuki T. Development of a High-efficiency Micro Fuel Cell Module- A great step toward realizing high-power small-size solid oxide fuel cell systems -, Translation of AIST press release of February 12, 2009, http://www.aist.go.jp/aist_e/latest_research/2009/20090306_2/20090306_2.html, Research Results, AIST, no date [accessed March 2010].

211. Micro-tubular solid oxide fuel cell integrated compact modules operable at low temperatures—potential applications include domestic power supplies and auxiliary power units of automobiles. Press release, http://www.aist.go.jp/aist_e/latest_research/2009/20091204/20091204.html. AIST, 10 September 2009.

212. FCTESTNET Fuel Cell Survey by region, Part 3 JAPAN, http://ie.jrc.ec.europa.eu/fctestnet/docs/Finals_WG_8/3_JAPAN.pdf, Edited by W. Winkler (leader of FCTESTNET WP 8), http://ie.jrc.ec.europa.eu/fctestnet/docs/Finals_WG_8/3_JAPAN.pdf. FCTESTNET, European Commission,March 2006.

213. R&D on all ceramics SOFC with high power density under 1000 °C operation, principal research results. http://criepi.denken.or.jp/en/publications/annual/2004/04kiban23.pdf. Central Research Institute of Electric Power Industry, 2004.

214. Homma T. *CRIEPI and AIST, the latest fuel cell news in Japan, January 2007 R&D of high-performance hydrocarbon membrane are activated*. Japan: FCDIC; January 2007.

215. Mori M. Advancing performance of intermediate temperature SOFC, report no. Q07010, http://criepi.denken.or.jp/jp/kenkikaku/report/leaflet/Q07010.pdf. Central Research Institute of Electric Power Industry, June 2008.

216. Iwata T, Shundo H. *Development of planar type SOFC with a ceramic separator and its module concept*. Proceedings of International Fuel Cell Symposium, NEDO, Makuhari Messe Conference Hall, Chiba, Japan; March 1992.

217. *New energy technology development related data file production survey (fuel cells)*. Japan: NEDO; March 1998.

218. Sasaki H, Takasu K. Status of SOFC development in Japan. In: Stimming U, Singhal SC, Tagawa H, Lennert W, editors. *Solid oxide fuel cells V (SOFC V)*. the Electrochemical Society; 1997.

219. Special feature: creating new energy source—developing fuel cells, environmental report, assuring more effective environmental protection in visible ways, Kyocera CSR report. http://global.kyocera.com/ecology/report/pdf2007/46_47.pdf, Kyocera, 2007.

220. Homma T. *SOFC development: Kyocera. Latest news, no. 81*. Japan: FCDIC; February 2003.

221. Homma T. *SOFC: Kyocera corporation. The latest news*. Japan: FCDIC; January 2004.

222. Pursue Growth by Creating New Businesses, Kyocera Group Investor Meeting, http://global.kyocera.com/ir/presentations/pdf/pp0602-2_e.pdf, Kyocera, 3 February 2006.

223. Two firms develop cogeneration system for household use, http://www.kippo.or.jp/kansaiwindowhtml/news/2004-e/20041208_NEWS.HTML. vol. 11, No.485, Kansai Window, Kansai News, Kippo News, Japan, 8 December 2004.

224. Osaka gas and Kyocera to start joint development of 1 kW SOFC cogeneration system for household use, http://www.digital-research.co.jp/fuelcell_onl_2004_11.htm. Fuel Cell Newspaper Online, November 2004.

225. Ipponmatsu M. *Current status of SOFC development of Osaka gas*. Japan: The report on international fuel cell, NEDO; March 1992.

226. Ceramic fuel cells – CFU partnership with Mitsui & Osaka gas. Interactive Investor, http://www.iii.co.uk/investment/detail/?display=news&code=cotn:CFU.L&action=article&articleid=7802985, 22 March 2010.

227. On the results of Japan's first residential operation of SOFC cogeneration system. Press release, Osaka Gas and Kyocera, 16 May 2006.

228. Osaka gas reduced 1 kW residential use SOFC cogeneration system by morethan 30 percent, http://www.digital-research.co.jp/fuelcell_onl_2007_11.html. Fuel Cell Newspaper Online, November 2007.

229. Homma T. *Osaka Gas*. The Latest Fuel Cell News in Japan, FCDIC; February 2008.

230. Homma T. *Osaka Gas and Kyocera*. The Latest Fuel Cell News in Japan, FCDIC; February 2007.

231. Homma T. *Osaka Gas*. The Latest Fuel Cell News in Japan, FCDIC; July 2007.

232. Homma T. *Osaka Gas*. Japan: Latest News, FCDIC; 18 August 2008.

233. Agreement on joint development of solid oxide fuel cell (SOFC) cogeneration system for residential use. Topics of Interest 2008, KYOCERA CSR Report 2009, Kyocera, Japan, 2009.

234. Homma T. *Osaka Gas and 3 other companies*. THE Latest News Number 161, 2009, FCDIC; 21 September 2009.

235. TMC, aisin provide household fuel cells for test program. News release, http://pressroom.toyota.com/pr/tms/document/TMC.AisinNews Release.pdf. Toyota Motor Corporation Aisin Seiki Co., Ltd., 17 December 2009.

236. FY2009 test operation sites. SOFC demonstration research installation map. http://sofc.nef.or.jp/map/. NEF, Japan, [accessed March 2010].

237. Homma T. SOFC technology development the latest fuel cell news in Japan, number 156, Fuel Cell Development Information Center (FCDIC), 17 April 2009.
238. Homma T. Development and commercialization of SOFC. The Latest Fuel Cell News in Japan, FCDIC, March 2004.
239. Ishikawa T, Yamashita S, Sobue T, Hase K. SOFC development by Tokyo Gas, Kyocera, Rinnai and Gastar, http://www.igu.org/html/wgc2006/pdf/paper/add10759.pdf, 23rd world gas conference, Amsterdam. 2006.
240. Homma T. Development and Commercialization of SOFC, The Latest Fuel Cell News in Japan, March 2004.
241. Matsuzaki Y, Ogasawara (Tokyo Gas) K. Current status of planar SOFC development at Tokyo gas. In: Dokiya M, Yamamoto O, Hagawa H, Shinghal SC, editors. *Solid oxide fuel cells (SOFC IV)*; 1995.
242. Honma T. SOFC development. Latest News, No. 28, 31 August 1998.
243. Yasuda I. Research and development of planar solid oxide fuel cells. http://irs.ub.rug.nl/dbi/4581697c93ead. The University of Groningen, the Netherlands, 16 November 1992.
244. Tokyo Gas successfully developed the cell stack, with an efficiency exceeding 50%. Effort of Tokyo Gas, http://www.tokyo-gas.co.jp/pefc_e/dev-fc_36.html,Tokyo Gas, no date [accessed February 2010].
245. Achieved the world's highest level electric efficiency from a normal atmospheric pressure SOFC power plant with low temperature, horizontal stripe cell stacks. News release, http://www.kyocera.co.jp/news/2006/0504.html. Tokyo Gas, Kyocera, Rinnai, and Gastar, May 2006.
246. Homma T. *Tokyo Gas*. Latest News, FCDIC, No.155; 11 March 2009.
247. Planar SOFC achieved the world highest electric output of 15 kW. Press release, Chubu Electric, 11 September 2000.
248. Development Status: mono-block layer built (MOLB) type SOFC, http://www.mhi.co.jp/en/power/technology/sofc_system/contents/development_situation.html. Mitsubishi Heavy Industries, Japan, no date [accessed February 2010].
249. Sasaki Y, Esaki Y, Hattori (RPDC) M, Miyamoto (MHI Takasago) H, Satake T, Nanjo F. Development of 5 kW class MOLB type SOFC. In: Stimming U, Singhal SC, Tagawa H, Lehnert W, editors. *Solid oxide fuel cells V*. The Electrochemical Society; 1987.
250. Development of an solid oxide fuel cell (SOFC): launching the development of the world's first 50-kW commercial unit as a flat-type SOFC. Press release, http://www.chuden.co.jp/corpo/publicity/press2002/0930_2.html. Chubu Electric, 30 September 2002.
251. Homma T. Mitsubishi heavy industries, Ltd. development and production of SOFC, http://www.fcdic.com/eng/news/200408.html#3. The latest fuel cell news in Japan, FCDIC, Japan, August 2004.
252. New energy power generation plant in the exposition. The 2005 World Exposition, Aichi, Japan, 25 March–25 September 2005.
253. FY2005 November Press Conference by President Kawaguchi, the Federation of Electric Power Companies of Japan: material 3: results of fuel cell demonstration tests at Aichi Expo. Press release, http://www.chuden.co.jp/corpo/publicity/other/1118_4.html. Chubu Electric, 18 November 2005.
254. FY2005 March Press Conference by Vice President Yamauchi, the Federation of Electric Power Companies of Japan: material 2 Fuel Cell Demonstration Tests at Aichi Expo. Press release, http://www.chuden.co.jp/corpo/publicity/other/0311_3.html. Chubu Electric, 11 March 2005.
255. The results of the SOFC Demonstration in the Electric Power Pavilion of EXPO 2005 in Aichi, Japan, aiming at practical application of an SOFC cogeneration system. Technology Development News, No. 118, http://www.chuden.co.jp/torikumi/study/library/news/pdf/list118/N11813.pdf. Chubu Electric, January 2006.
256. Mori H, Omura H, Hisatome N, Ikeda K, Tomida K. Pressurized 10 kW class module of SOFC. In: Singhal SC, Dokiya M, editors. *Solid oxide fuel cells (SOFC VI)*. The Electrochemical Society; 1999.
257. Iritani J, Kogami K, Komiyama N, Nagata N, Ikeda K, Tomida K. Pressurized 10 kW class module of SOFC. In: Yokokawa H, Singhal SC, editors. *Solid oxide fuel cells VII (SOFC VII)*. The Electrochemical Society; 2001.
258. Kakigami S, Kurihara T, Hisatome N, Nagata K. Development of SOFC module. In: Stimming U, Singhal SC, Tagawa H, Lennert W, editors. *Solid oxide fuel cells V*. The Electrochemical Society; 1997.
259. Gengo T, Hisatome N, Kobayashi Y, Kanehira S, Ando Y (Mitsubishi Heavy Industries). Latest state of fuel cells that open the way to an environmentally friendly society in the 21 century. technical review, vol. 42(3), http://www.mhi.co.jp/technology/review/pdf/e423/e423104.pdf, Mitsubishi Heavy Industries, Ltd., October 2005.
260. Atmospheric pressure SOFC system generated Japan's highest power output of more than 100 kW and achieved cumulative operation of 3300 hours. Press Release, J-Power, 6 December 2007.
261. J-POWER's initiative for commercialization of coal gasification technology. 2006 Annual report, http://

www.jpower.co.jp/english/ir/pdf/2006-08.pdf. J-Power, 2006.

262. Commissioning of pilot test plant for developing production technology for coal gas used for fuel cells. News release, http://www.jpower.co.jp/english/news_release/news/news146.pdf. J-Power, 22 February 2002.
263. J-Power started test-operating the world's largest atmospheric pressure 150 kW SOFC system. Fuel Cell Newspaper Online, http://www.digital-research.co.jp/fuelcell_onl_2006_07.htm. July 2007.
264. Chugoku Electric, J-POWER to cooperate on Setouchi project. News release, http://www.jpower.co.jp/english/news_release/news/news060531.pdf. Chugoku Electric and Electric Power Development Company, 31 May 2006.
265. Establishment of Osaki Cool Gen Corporation. News release, http://www.jpower.co.jp/english/news_release/news/news090729.pdf. Electric Power Development Co. and Chugoku Electric, 29 July 2009.
266. Development status: tubular type SOFC, http://www.mhi.co.jp/en/power/technology/sofc_system/contents/development_situation.html. Mitsubishi Heavy Industries, Japan, no date [accessed February 2010].
267. MHI achieves 3000-hour operation, unprecedented in Japan, Of SOFC-MGT combined-cycle power generation system, http://www.mhi.co.jp/en/news/story/0910011315.html. Mitsubishi Heavy Industries, Japan, no date [accessed February 2010].
268. Implementation initiatives, http://www.mhi.co.jp/en/power/technology/sofc_system/contents/approach_practical.html. Mitsubishi Heavy Industries, Japan, no date [accessed February 2010].
269. Tomida K, Nishiura M, Koga S, Miyamoto K, Teramoto Y, Yoshida S, et al. (MHI Nagasaki and Yokohama). Development of SOFC-GT combined cycle system with tubular type cell stack. Fuel Cell Seminar & Exposition, San Antonio, Texas, 18–21 October 2010.
270. Development of 10 kW intermediate temperature SOFC module. R&D topics, <www.kepco.co.jp/rd/topics/2006/0410.html. Kansai Electric and Mitsubishi Materials Corporation, 10 April 2006.
271. Inagaki T, Nishiwaki F, Kano J, Nishi J, Onishi Y, Yamasaki S, et al. Development of IT-SOFC based on Lanthanum Gallate electrolyte by Mitsubishi materials. 2008 Fuel Cell Seminar & Exposition, Phoenix, Arizona, 28 October–30 October 2008.
272. Development of 1 kW intermediate temperature SOFC system. Press release, <www.kepco.co.jp/presre/2004/0113-1j.html. Kansai Electric, 13 January 2004.
273. Development of high efficiency intermediate temperature SOFC cell. Press release, http://www.kepco.co.jp/pressre/2001/1217-1j.html. Kansai Electric, Mitsubishi Materials Corporation, and Japan Fine Ceramic Center, 17 December 2001.
274. Development of 1 kW low temperature SOFC module, succeeded in generation 1 kW at low temperature, achieved the world's highest level power efficiency. Press release, http://www.kepco.co.jp/pressre/2003/0507-1_2j.html, Kansai Electric, Mitsubishi Materials Corporation, 7 May 2003.
275. Development 3 kW intermediate temperature SOFC module, Press Release, http://www.kepco.co.jp/pressre/2004/0329-1j.html, Mitsubishi Materials Corporation and Kansai Electric, 29 March 2004.
276. Verifying intermediate temperature SOFC durability. Press release, http://www.kepco.co.jp/pressre/2005/0111-1j.html. Kansai Electric and Mitsubishi Materials Corporation, 13 January 2004.
277. Homma T. SOFC residential demonstration. Latest news, FCDIC, No. 125, September 22, 2006.
278. Project story: fuel cell development project. Inventing tomorrow, http://www.mmc.co.jp/recruit/inventing/project/no5.html. Mitsubishi Material, no date [accessed February 2010].
279. Eto H, Akbay T, Hirata K, Sato M, Inagaki T, Nishiwaki F. HRD25a-17 progress in development of IT-SOFC based on Lanthanum Gallate electrolyte by Mitsubishi electric and Kansai electric. 2009 Fuel Cell Seminar & Exposition, Palm Springs, CA,16–20 November 2009.
280. Yoshida H, Kawano M, Ueno D, Kinugasa A, Inagaki T, Eto H (Kansai Electric, and Mitsubishi Materials). Characterization of cyclic startup/shutdown tests of SOFC modules based on Lanthanum-Gallate electrolyte. 2010 Fuel Cell Seminar & Exposition, San Antonio, TX, 18–20 October 2010.
281. Yoshida H, Kawano M, Ueno D, Kinugasa A, Inagaki T (Kansai Electric), Akikusa J, et al. Characterization of cyclic startup/shutdown tests of SOFC modules based on Lanthanum-Gallate electrolyte. http://www.fuelcellseminar.com/media/4997/hrd32-2_yoshida.pdf, 2010 Fuel Cell Seminar & Exposition, San Antonio, TX, 18–20 October 2010.
282. Akagi K, Shiratori A, Iha M, Chikagawa (Osaka Gas) O. Development of Osaka gas type planar SOFC. In: Dokiya M, Yamamoto O, Tagawa H, Singhal SC, editors. *Proceedings of the fourth international symposium on solid oxide fuel cells, SOFC IV*. The Electrochemical Society; 1995.
283. New Energy Technology Development Related Data File Production Survey (Fuel Cells), NEDO, Japan, March 1997.
284. New Energy Technology Development Related Data File Production Survey (Fuel Cells), NEDO, Japan, March 2000.

285. Mori N, Iha M, Harada J, Takagi H (Murata Manufacturing). Development of new concept SOFCs at Murata. http://www.electrochem.org/meetings/scheduler/abstracts/216/1647.pdf. 216th ECS Meeting—Vienna,Austria, 4–9 October 2009.
286. Michiaki Iha (Murata Manufacturing), Current status of single-step co-fired SOFC development, Hydrogen and fuel cell system research meeting, The first regular session, http://www.kansai-scej.org/2010/22-1fc.pdf, The Society of Chemical Engineering, Japan, Kansai Branch, [accessed 9 December 2010].
287. NGK insulators developed SOFC cells. News, http://www.ngk.co.jp/news/2003/0701.html, NGK Insulators, 1 July 2003.
288. AIST and NGK insulators developed SOFC of a honeycomb structure. Nikkei Electronics, 1 November 2006.
289. Homma T. SOFC Development. The latest fuel cell news in Japan, Number 158, Fuel Cell Development Information Center (FCDIC), 18 June 2009.
290. J energy, sumitomo precision products, NGK insulators developed SOFC system using kerosene, http://www.j-energy.co.jp/cp/release_new/2009/20091119_1500.php. News release, J Energy, 19 November 2009.
291. Todo Y, Ishikawa H, Shibata M, Uematsu H, Gonda I, Sumi H, et al. Development of solid oxide fuel cells at NGK spark plugs Co., Ltd. 2008 Fuel Cell Seminar & Exposition, Phoenix, AZ, 28 October–30 October 2008.
292. D. Nishijima, H. Ishikawa, M. Shibata, H. Uematsu, H. Sumi, Y. Okuyama, et al. Development of SOFC stack at NGK Spark Plug Co., Ltd. http://www.electrochem.org/meetings/scheduler/abstracts/216/1212.pdf, The 216th ECS meeting, Vienna, Austria, 4–9 October 2009.
293. Homma T. National research institute for advanced industrial science and technology, NGK Plug Co., Ltd., etc. The Latest News, No. 161, 21 September 2009.
294. National Institute of Advanced Industrial Science and Technology (AIST) Develops micro-tubular solid oxide fuel cell integrated compact modules operable at low temperatures. http://fuelcellsworks.com/news/2009/12/15/ational-institute-of-advanced-industrial-science-and-technology-aist-develops-micro-tubular-solid-oxide-fuel-cell-integrated-compact-modules-operable-at-low-temperatures/, Fuel Cells Work, 15 December 2009.
295. Nishijima D, Shibata M, Gonda I, Matsuno T, Todo Y, Kobayashi N, et al. Development of SOFC for residential use at NGK Spark Plugs Co., Ltd. Co. 2010 Fuel Cell Seminar & Exposition, San Antonio, TX, 18–21 October 2010.
296. Meakin I. Fuel cell development in Japan. An outline of public & private sector activities. <www.fuelcellmarkets.com/content/images/articles/FCJapan04-2003.doc. Fuel Cell Markets, April 2003.
297. SOFC development story. http://www.eneos.co.jp/company/rd/special/story17-2.html. Nippon Oil Corporation, no date [accessed March 2010].
298. Nippon oil virtually acquired fuel cell division of Sanyo electric. Response, http://response.jp/article/2007/11/29/102563.html, IRI Commerce and Technology, [accessed 29 November 2007].
299. On establishment of the new company for stationary fuel cell. News release, Nippon Oil, 28 February 2008.
300. Nippon oil to start demonstration test operation of next generation fuel cell SOFC system. News release, Nippon Oil, 16 January 2008.
301. Nippon oil: development and installation operation status of SOFC systems fueled by oil derivatives. SOFC Demonstration Research Result Report Workshop, New Energy Foundation, Japan, 3 March 2009.
302. Homma T. Nippon oil and kyocera. Latest news, No. 162, FCDIC, 21 October 2009.
303. Homma T. Nippon oil. Latest News, FCDIC, No.164, 29 December 2009.
304. Homma T. Nippon oil. Latest News, FCDIC, No.165, 28 January 2010.
305. Ishii T, Chiba R. SOFC using cubic stabilized Zirconia in a CrO2-Sc2O3-Al2O3 system as an electrolyte. In: Dokiya M, Yamamoto O, Hagawa H, Shinghal SC, editors. *Solid oxide fuel cells (SOFC IV)*; 1995.
306. Maruno T. R&D approaches to clean energy, NTT Energy and Environment Laboratories, <https://www.ntt-review.jp/library/ntttechnical.php?contents=ti0804rdfws05e.pdf, 8 February 2008.
307. Sugita S, Arai H, Yoshita Y, Orui H, Arakawa M. Anode-supported planar-type SOFC development at NTT. In: Williams MC, Krist K, editors. *ECS transactions, the 30th fuel cell seminar*, vol. 5. The Electrochemical Society; 2007. No. 1.
308. Hayashi K, Yokoo M, Yoshida Y, Tabata Y, Orui H, Nozaki Y, et al. Development of a 1 kW solid oxide fuel cell stack with anode supported planar cells at NTT. 2008 Fuel Cell Seminar & Exposition, Phoenix, AZ, 27–30 October 2008.
309. Yoshida Y, Yokoo M, Orui H, Hayashi K, Arai H. Study on high power SOFC stack at NTT, http://ci.nii.ac.jp/naid/110007131299/en. Technical report of IEICE, Energy engineering in electronics and communications, 108(402) p. 35–39 20090116. The Institute of Electronics, Information and Communication Engineers, 16 January 2009.
310. Arai H, Hayashi K, Yoshida Y, Yokoo M, Mizutani Y, Ukai K, et al. 3 kW SOFC power generation module developed by NTT/THG/SPP, http://www.electrochem.org/meetings/scheduler/abstracts/

216/1216.pdf. Solid Oxide Fuel Cells, Eleventh International Symposium (SOFC-XI), The 216th ECS Meeting, Vienna, Austria, 4–9 October 2009.

311. Kurachi S, Mizutani Y, Hiroyama T, Katsurayama K, Okada F, Ukai K, Development of a small-scale solid oxide fuel cell (SOFC). 24th World Gas Conference, Argentina, 2009.
312. Lovering D.G. Third grove fuel cell symposium platinum metals-loaded polymer membranefuel cells progress to electric traction. Platinum Metals Review, vol. 37(4). http://www.platinummetalsreview.com/pdf/pmr-v37-i4-197-209.pdf; 1993.
313. New energy technology developed related data file compilation survey (Fuel Cells). NEDO, March 1996.
314. New energy technology developed related data file compilation survey (Fuel Cells), NEDO, March 1998.
315. Japan reaffirms commitment to fuel cell commercialization. Fuel Cell Catalyst. vol. 5(1). US Fuel Cell Council. Fall 2004.
316. Yokoyama M, Ukai K, Hisada K, Mizuani Y. Current status of 1kW class SOFC system at Toho gas and sumitomo precision products. In: Singhal SC, Mizusaki J, editors. *Solid oxide fuel cells IX (SOFC-IX). Cells, stacks, and systems*, vol. 1. The Electrochemical Society; July 2005.
317. Current status of SOFC development at Toho gas. http://www.fuelcellstore.com/products/fcj/FCJ1.pdf. vol/No.1. Fuel Cell Japan, September 2004.
318. Homma T. Toho Gas. The Latest Fuel Cell News in Japan, October 2003.
319. Mizutani Y, Hisada K, Yasuhara M, Ukai K, Yokoyama M. First demonstration program of solid oxide fuel cells in Japan at Aichi EXPO 2005. 23rd World Gas Conference, http://www.igu.org/html/wgc2006/pdf/paper/add10232.pdf. Amsterdam. 2006.
320. Homma T. Toho gas. The latest fuel cell news in Japan, FCDIC, Japan, June 2004.
321. Homma T. Toho gas and Nippon shokubai. The latest fuel cell news in Japan, FCDIC, Japan, August 2004.
322. Nippon Shokubai builds solid electrolyte plant. Advanced ceramics report, appeared in high bean research, http://www.highbeam.com/doc/1G1-128059011.html, 1 February 2005.
323. SOFC: DKKK builds pilot ScSZ plant, BNET, http://resources.bnet.com/topic/daiichi+kigenso+kagaku+kogyo+co.+ltd.html, 1 June 2005.
324. Homma T. SOFC development. The latest fuel cell news in Japan,December 2007.
325. Homma T. Trends in SOFC development. The latest fuel cell news in Japan, FCDIC, February 2007.
326. Someya Y, Koide H, Ando M, Yoshida (Tonen Corporation) T. The performance and reliability of 29 cm square planar SOFC. In: Dokiya M, Yamamoto O, Tagawa H, Singhal SC, editors. *Solid oxide fuel cells IV.* The Electrochemical Society; 1995.
327. Someya Y, Koide H, Ando M, Yoshida T (Tonen Corp.). The performance and reliability of 20cm square planar SOFC, Solid oxide fuel cells (SOFC-IV), Yokohama, Japan, 18–23 June 1995.
328. Ando M, Nakata K, Wakayama S (Tonen Corp.). Study on durability for thermal cycle of planar SOFC, Fuel Cell Seminar, Kissimmee, FL, 17–20 November 1996.
329. Yoshioka I (Petroleum Energy Center, Japan). Technology Development of Utilization of Light Petroleum Fraction for Fuel Cells. Proceedings of International Fuel Cell Conference, NEDO, Makuhari Messe Convention Hall, Chiba, Japan, March 1992.
330. Yamaji K, Fabrication of Anode Supported SOFC by EPD Method, AIST, Energy, Fuel Cells, <unit.aist.go.jp/energy/fuelcells-mate/english/material/epd-2e.html, no date [accessed March 2010].
331. Electric power technology development survey, FY2002 Result Report, Techno Research, NEDO, March 2003.
332. TOTO: Realized the World's Highest Level Efficiency in SOFCs. News release, http://www.toto.co.jp/company/press/2004/09/17.htm, TOTO, 17 September 2004.
333. Aikawa S, Kuroishi M, Saito T, Abe T, Takeuchi H, Fujinaga K, et al. Development of 10 kW class SOFC System, Nippon Kikai Gakkai Kankyo Kogaku Sogo Shinpojiumu Koen Ronbunshu, vol.13, Nippon Kikai Gakkai, Japan, 2003.
334. New Sunshine Project: Fuel Cell Power Generation Technology R&D (SOFC), Pre-Final Evaluation Report, METI, Industrial Technology Council, May 2000.
335. TOTO to realize the world's highest level electric efficiency under NEDO Contract, Plans to Put into Practical Use in 2007. News release, http://www.toto.co.jp/company/press/2004/09/17.htm, TOTO, 17 September 2004.
336. TOTO's SOFCs achieved 55 percent power efficiency and 3000 hour operation, Nikkei Sangyo Shimbun, 17 September 2004.
337. Minamizono T. Research on SOFC, Techno Report, Kyushu Electric, http://www.kyuden.co.jp/library/pdf/company/report/report27.pdf, March 2007.
338. Fuel cell technologies Ltd. and TOTO Ltd. sign memorandum of understanding for development of a 2-3 kW solid oxide fuel cell system, PR News wire, 12 January 2005.
339. TOTO to Develop the World's First SOFC Operable at 500°C, 50W SOFC with Start-up Time in Five Minutes

Possible. News release, http://www.toto.co.jp/company/press/2005/10/06_1.htm, TOTO, 6 October 2005.
340. TOTO to Participate Fuel Cell Field Test, Selected by NEF's SOFC Demonstration Research. News release, http://www.toto.co.jp/company/press/2007/07/27_2.htm, TOTO, 27 July 2007.
341. Homma T. Hitachi and TOTO, Latest News, FCDIC, No.165, 28 January 2010.
342. Ponthieu E. SOFC support programmes by the European commission. In: Stimming U, Singhal SC, Tagawa H, Lennert W, editors. *Solid oxide fuel cells V.* The Electrochemical Society; 1997.
343. Thematic network on solid oxide fuel cell technology (S O F C n e t). http://ec.europa.eu/research/energy/pdf/efchp_fuelcell36.pdf.
344. *Fuel cells: powering the future, sustainable power for the European Union.* Brussels: European Commission; 2000.
345. Future export adventure? El &Energi (energisektorens nyhedsblad), http://www.Risøe.dtu.dk/~/media/Risøe_dk/about_Risøe/abf/documents/el%20og%20energi.ashx,Risø, Denmark, 17 October 2007.
346. Danish solid oxide fuel cell project: DK-SOFC 1997-1999, Risø National Laboratory, Denmark, April 2001.
347. Risø annual report 1996, Risø National Laboratory, Denmark, April 1997.
348. Improving durability of SOFC stacks, Advanced Search, http://cordis.europa.eu/search/index.cfm?fuseaction=proj.document&PJ_LANG=EN&PJ_RCN=2360281&pid=0&q=1560FDEBDC1C16E0ACAEEDBEBB8456C9&type=adv, CORDIS Services, no date [accessed June 2010].
349. Linderoth S, Smith A (Risø National Laboratory). Worldwide SOFC OVERVIEW from a scandinavian and a European perspective, Advances in Solid Oxide Fuel Cells III, The American Ceramic Society, 14 December 2009.
350. History of SOFC Research at Risø; Fuel Cells at Risøe, http://www.Risoe.dk/Risoe_dk/Home/About_Risoe/research_departments/ABF/Fuel_cells/recent_history.aspx,Risoe. DTU, National Laboratory for Sustainable Energy, 18 April 2007, [accessed May 2010].
351. Topsoe Milestone, About Us, Haldor Topsoe, Denmark, no date [accessed June 2010].
352. Christiansen N, Hansen JB, Kristensen S, Holm-Larsen H, Linderoth S, Hendriksen PV, et al. (Haldor Topsøeand Risø national laboratory), SOFC development program at haldor Topsøe/Risø national laboratory—ProgressPresentation, http://iea-hia-annex18.sharepointsite.net/Public/National%20Documents/Denmark/DK9-20SOFC-20development-20FuelCellSem2004.pdf, 2004 Fuel Cell Seminar, San Antonio, Texas, 1–5 November 2004.
353. Richter A.B (Topsoe Fuel Cell). Topsoe fuel cell and our cooperation with Wärtsilä, Finish Fuel Cells Annual Seminar, Espoo, Finland, 14 September 2010.
354. Hawkes J, Centeck K, Drejer-Jensen M, Emerson S, Erikstrup N, Hale1 T. A power dense 2 kW logistics fueled solid oxide fuel cell system for mobile applications, 2009 Fuel Cell Seminar & Exposition, Palm Springs, CA, 16–19 Novmber 2009.
355. Topsoe Fuel Cell and Wärtsilä Corporation demonstrate operating results, News Release, Haldor Topsoe, Denmark, 4 September 2006.
356. February 2009, News, http://www.demo-sofc.eu/, demo SOFC, no date [accessed June 2010].
357. Topsoe Fuel Cell opens new manufacturing plant, News and Press releases, http://www.topsoefuelcell.com/sitecore/content/Topsoe_fuel_cell/news_and_info/press_releases/280409.aspx, Topsoe Fuel Cells, 28 April 2009.
358. Holm-Larsen H, Jørgensen MJ, Christiansen N, Hendriksen P. V, Hagen A, Ramousse S, and et al. Recent progress in development and manufacture of solid oxide fuel cell at Topsoe Fuel Cell A/S and Risø DTU, 2009 Fuel Cell Seminar & Exposition, Palm Springs, CA, 16–19 November 2009.
359. Emspak J. Danish company builds home fuel cells, article http://hken.ibtimes.com/articles/87709/20101201/danish-company-building-home-fuel-cells.htm, International Business Times, 1 December 2010.
360. Project Fact Sheet: LARGE-SOFC, Toward a Large SOFC Power Plant, Sixth Framework Programme, CORDIS, http://cordis.europa.eu/fetch?CALLER=FP6_PROJ&ACTION=D&DOC=13&CAT=PROJ&QUERY=011f3bd549f8:02be:4a85aab1&RCN=81194, last updated 15 January 2010, [accessed June 2010].
361. 24 stacks delivered to Wärtsilä. Press release, Topsoe Fuel Cell, Denmark, 12 February 2010.
362. Topsoe Fuel Cell SOFC stacks in successful demonstration. Press release, Topsoe Fuel Cell, Denmark, 24 February 2010.
363. Wärtsilä has been highly successful in operating its groundbreaking fuel cell unit. Press release, Wärtsilä Corporation, Finland, 22 February 2010.
364. Danish Micro Combined Heat and Power project, The Project. http://www.dk-mchp.eu/project.html, Denmark, no date [accessed June 2010].
365. Pedersen A.H, Balslev P. Demonstration of micro chip based on Danish fuel cells (Phase 2), 2009 Fuel Cell Seminar & Exposition, Palm Springs, CA, 16–19 Novmber 2009.

366. Christiansena N, Hansena J.B, Holm-Larsena H, Linderothb S, Larsenb P.H, Hendriksenb P.V, et al. (Topsoe Fuel Cell A/S and Risø National Laboratory), Solid oxide fuel cell development at Topsoe Fuel Cell A/S and Risø National Laboratory, ECS Transactions, 7(1) 31-38 (2007) 10.1149/1.2729070, The Electrochemical Society, 2007.
367. Topsoe Fuel Cell: Rethinking Energy, Brochure, Topsoe Fuel Cells, http://www.topsoefuelcell.com/news_and_info/~/media/Fuel_cell/pdf/topsoe_fuel_cell_company_brochure.ashx, no date.
368. Butler J, Dehamna A. Survey of Denmark, Fuel Cell Today, August 2009.
369. Sponsors, About us, http://www.topsoefuelcell.com/about_us/sponsors.aspx, Topsoe Fuel Cell, no date, [accessed June 2010].
370. Topsoe and Risø DTU Receive Grant for on-Going Fuel Cell Development, News Article, http://www.fuelcelltoday.com/online/news/articles/2010-06/topsoe-Risø-dtu-additional-grant, Fuel Cell Today, 28 June 2010.
371. Fontell E, FINSOFC Päätösseminaari, Teollisuuden näkökulma, WÄRTSILÄ CORPORATION, 5 April 2007.
372. Kiviaho J, Halinen M, Noponen M, Saarinen J, Rosenberg R. SOFC system development in VTT, http://www.h2cu.com/public/H2CU/SOFC%20System%20Development%20in%20VTT.pdf,, EFC2005, 14 December 2005.
373. Fuel cell technology, Products & Services,< http://www.Wartsila.com/, en, productsservices,0, product,E5BF1B55-B877-4139-8E2E-F4667EC39BD9,5-CA49D21-7B6C-45CC-BA6B-D894FB826B5C, htm, Wärtsilä, last modified 16 February 2010.
374. Wärtsilä and Haldor Topsøe start co-operation in fuel cell development, Press Release, Wärtsilä Corporation, 16 August, 2002.
375. Fontell E, Hottinen T, Eklund K. *Status of the solid oxide fuel cell system development at Wärtsilä.* Palm Springs, CA: 2009 Fuel Cell Seminar & Exposition; 16–19 Novmber 2009.
376. METHAPU, Leaflet, http://www.methapu.eu/File.ashx?mode=file&file=///Downloadables/Leaflet+METHAPU+2009.pdf, Wärtsilä Corporation, 2009.
377. Wärtsilä installs fuel cell unit on vessel – unique SOFC technology provides power to Wallenius' car-carrier 'Undine', Press release, Wärtsilä Corporation, Jun 11, 2010.
378. LARGE-SOFC – Toward a Large SOFC Power Plant, Homepage, http://www.vtt.fi/proj/largesofc/?lang=en, VTT.fi, no date, [accessed June 2010].
379. Rosenberg RJ, Kiviaho J, Göös J, Jansson P, Tarnowski OC, Jacobsen J. *Large-SOFC, toward a large SOFC power plant.* Palm Springs, CA: 2009 Fuel Cell Seminar & Exposition; 16–19 November 2009.
380. Matsuyama K. Hitachi Zosen to Develop Fuel Cells With Wärtsilä, Nikkei Says. http://www.bloomberg.com/apps/news?pid=20601101&sid=aTt6_UqjHdWk, Bloomberg.com, 6 June 2010.
381. Wärtsilä and Versa to jointly develop fuel cell technology, http://gcaptain.com/Wartsila;-versa-jointly-develop/?26374, Captain, 6 June 2011.
382. Sandström C-E, Noponen M, Hottinen T (Wärtsilä). WFC20 biogas unit operation, Biogas Workshop, Lund, 21 June 2011.
383. Wärtsilä Fuel cell Development Program, Presentation, http://www.tekes.fi/fi/gateway/PTARGS_0_201_403_994_2095_43/http%3B/tekes-ali1%3B7087/publishedcontent/publish/programmes/polttokennot/documents/seminaariaineistot/files/wartsila_fuel_cell_program_stationary_ws_290311.pdf, Tekes Workshop: Stationary Fuel Cells, Espoo, Innopoli, Finland, 29 March, 2011.
384. Development of an advanced 1kW solid oxide fuel cell prototype, Advanced Search, http://cordis.europa.eu/search/index.cfm?fuseaction=proj.document&PJ_LANG=EN&PJ_RCN=250745&pid=33&q=FE76EBD5ABFBD96EA4C3704029F90323&type=adv. CORDIS Services, no date, [accessed March 2010].
385. Wetzko M, Belzner A, Rohr FJ, F. Harbach (ABB Corporate Research). Solid oxide fuel cell stacks using extruded honeycomb type elements, Journal of Power Sources, 83, Elsevier, http://www.docin.com/p-42421868.html, 1999.
386. Michael P. D, Maguire J. European Fuel Cells R&D Review Final Report, ANL-94/46, Prepared for Argonne National Laboratory, September, 1994.
387. New manufacturing technologies for advanced solid oxide fuel cells, Advanced Search, http://cordis.europa.eu/search/index.cfm?fuseaction=proj.document&PJ_LANG=EN&PJ_RCN=251748&pid=0&q=FE76EBD5ABFBD96EA4C3704029F90323&type=adv, CORDIS, no date, [accessed March 2010].
388. Delphi and BMW Announce Development of Fuel Cell Auxiliary Power Unit, PR Newswire, http://www.accessmylibrary.com/article-1G1-54469769/delphi-and-bmw-announce.html, AcccessMyLibrary, 26 April 1999.
389. Crawley G, Solid Oxide Fuel Cells (SOFC), http://www.fuelcelltoday.com/media/pdf/archive/Article_1153_SOFC%202007.pdf, FUEL CELL TODAY, January 2007.

390. IEF-3 Report 2007, From Basic Principles to Complete System, vol. 70, Forschungszentrum Jülich, 2007.
391. Schiller G, Franco T, Lang M, Metzger P, Stormer (German Aerospace Center and BMW) AO. Recent results of the SOFC APU development ad DLR. In: Singhal SC, Mizusaki J, editors. *Solid oxide fuel cells IX (SOFC IX). Cells, stacks, and systems*, vol. 1. The Electrochemical Society; 2005.
392. Integration techniques for up-scaled SOFC modules (SOFC integration), Advanced Search, http://cordis.europa.eu/search/index.cfm?fuseaction=proj.document&PJ_LANG=EN&PJ_RCN=261390&pid=3&q=DB43C1544D5B7D4597948DCB2DD7D34C&type=adv, CORDIS Services, no date, [access March 2010].
393. Modeling of the mass and energy balances of SOFC modules, Advanced Search, http://cordis.europa.eu/search/index.cfm?fuseaction=proj.document&PJ_LANG=EN&PJ_RCN=319917&pid=0&q=DB43C1544D5B7D4597948DCB2DD7D34C&type=adv, CORDIS Services, no date, [accessed March 2010].
394. Operational tests of SOFC modules and use of SOFC as chemical reactors, Advanced Search, http://cordis.europa.eu/search/index.cfm?fuseaction=proj.document&PJ_LANG=EN&PJ_RCN=250759&pid=2&q=DB43C1544D5B7D4597948DCB2DD7D34C&type=adv, CORDIS Services, no date, [accessed March 2010].
395. Blum L, Medenberg WA, Nabielek H.,Steinbcrgcr-Wilckens R (Jülich), Worldwide SOFC technology overview and benchmark, Advances in Solid Oxide Fuel Cells, http://ceramics.org/wp-content/uploads/2009/05/cesp-volume-26-issue-5-solid-oxide-fule-cells-chapter-1-blum-et-al1.pdf, the American Ceramic Society, 2005.
396. Development of Materials and Processes for the Solid Oxide Fuel Cell (SOFC), Brochure, Forschungszentrum Jülich, Germany, May 2004; Ceramic Fuel Cells Limited Signs Partnership Agreement with Germany's Forschungszentrum Jülich, Press Release, Ceramic Fuel Cells Limited, 17 April 2007.
397. Ceramic Fuel Cells Limited Signs Partnership Agreement with Germany's Forschungszentrum Jülich, Press Release, Ceramic Fuel Cells Limited, 17 April 2007.
398. Die Festoxid-Brennstoffzelle (SOFC, Solid Oxide Fuel Cell), Brennstoffzellenseiten, SOFC, http://www.fz-juelich.de/ief/ief-3/brennstoffzellen/entwicklungsthemen/sofc/, Institut fur Energieforschung (IEF), Forschungszentrum Jülich, last updated 12 July 2003, [accessed May 2001].
399. 1000-Watt-Schallmauer' durchbrochen, Meilenstein der Brennstoffzellen-Entwicklung erreicht, Press Release, Forschungszentrum Jülich, Germany, 31 January 2001.
400. New milestone in fuel cell development, Dual world record in SOFC, Press Release, Forschungszentrum Jülich, Germany, 6 June 2002.
401. Weltrekord bei Hochtemperatur-Brennstoffzellen, Jülicher Forscher machen viel versprechende Energietechnologie immer leistungsfähiger, Press Release, Forschungszentrum Jülich, 9 June 2004.
402. Steinberger-Wilckens R. Overview over Mission and Activities at Forschungszentrum Jülich and in the area of high temperature fuel cell. Projekt Brennstoffzelle, Forschungszentrum Jülich, 12 July 2006.
403. Crofer 22 APU, High Temperature Alloy, Material Data Sheet No. 4046, http://thyssenkruppvdm.de/_pdf/Crofer22APU_e.pdf, ThyssenKrupp VDM GmbH, Germany, January 2010.
404. Forschungszentrum Jülich entwickelt neuen Werkstoff für effiziente Brennstoffzellen, ThyssenKrupp stellt die Metalllegierung bereits her – BMW erprobt das Material in Prototypen, Press release, Forschungszentrum Jülich, 30 October 2003.
405. Crofer 22 APU: new material for fuel cells, Trade press release, http://www.thyssenkrupp.com/en/presse/art_detail.html&eid=tk_pnid930, ThyssenKrupp Steel AG, Sep. 09, 2003.
406. Executive summary: 2. Fuel cells, efficient energy conversion, http://www.fz-juelich.de/scientific-report-2004/index.php?item=35&lang=en, Scientific Report 2004, Forschungszentrum Jülich, 2004.
407. Highlight: SOFC, scientific report 2006, http://www.fz-juelich.de/fs/index.php?index=431, Forschungszentrum Jülich, 2006.
408. Highlights: fuel cells, scientific report 2007, http://www.fz-juelich.de/fs/index.php?index=537, Forschungszentrum Jülich, 2007.
409. Rosenberg R. *The publishable executive summary of the 1st periodic activity report: M1–M12: LARGE-SOFC toward a large SOFC power plant, project no. 019739*. VTT Technical Research Center of Finland; 8 May 2008.
410. Steinberger-Wilckens R (Forschungszentrum Jülich). Executive summary: realising reliable, durable, energy efficient and cost effective SOFC systems (Real-SOFC), Real-SOFC, http://www.real-sofc.org/datapool/page/390/Real-SOFC_PublicExecSummary_M37-M48_coordinator_final.pdf, 31 March 2008.
411. Feck T (Forschungszentrum Jülich). Real-SOFC project successfully terminated, News, http://www.real-sofc.org/news_extended/news-12-2008,Real SOFC, 4 December 2008.

412. Hannover Messe: fuel cells with world record-life; Over 20 000 hours of operation for fuel cell Jülich, Numerous interactive exhibits at the Jülich Stand, Press Release, http://www.fz-juelich.de/portal/index.php?cmd=show&mid=766&index=163, Forschungszentrum Jülich, 19 April 2010.
413. Forschungszentrum Jülich erzielt erneut Weltrekord mit Brennstoffzellen, Lange Lebensdauer bei hoher Leistung nachgewiesen / Großer Schritt in Richtung kommerzieller Einsatz, Press Release, Forschungszentrum Jülich, Germany, 15 May 2009.
414. Component reliability of solid oxide fuel cell systems for commercial operation (CORE-SOFC), Advanced Search, http://cordis.europa.eu/search/index.cfm?fuseaction=proj.document&PJ_LANG=EN&PJ_RCN=4938439&pid=0&q=37581187467EE500933ADA54D2E5A20B&type=adv, Cordis Services, no date, [accessed May 2010].
415. Koch S, Hendriksen PV, Mogensen M (Risø National Laboratory, Denmark), Nico Dekker and Bert Rietveld (ECN), Bert de Haart and Frank Tietz (Forschungszentrum Jülich, Germany). Solid Oxide Fuel Cell Performance Under Severe Operating Conditions, the 6th European Solid Oxide Fuel Cell Forum, Lucerne, Switzerland, 28 June–2 July 2004.
416. Gen FC Project, Generic Fuel Cell Modeling Environment. http://www.genfc.org/index.php?id=27; no date [accessed May 2010].
417. Lim T-H, Song R-H, Shin D-R, Yang J-I, Jung H, Vinke I (Korea Institute of Energy Research, Daejeon, Korea, and FZ Juelich Gmbh, Juelich, Germany). Development of 5 kW class SOFC power generation system for gas turbine-fuel cell hybrid system, 2006 Fuel Cell Seminar, Honolulu, HI, 13–17 November 2006.
418. Jülich und Korea verstärken ihre Zusammenarbeit in der Brennstoffzellenforschung, Partnerschaftsvertrag auf der Hannovermesse unterschrieben, Press Release, Forschungszentrum Jülich, Germany, 20 April 2009.
419. The Future of Fuel Cells. Energy Science Infosite. http://cdsweb.cern.ch/record/479405/files/sis-2000-521.html, Ben Wiens, 27 June 2000.
420. Steinberger-Wilckens R, Blum L, Buchkremer H-P, de Haart L, Nabielek H, Quadakkers J, et al. (Forschungszentrum Juelich, Juelich, Germany), Results of solid oxide fuel cell development at forschungszentrum juelich, 2006 Fuel Cell Seminar, Honolulu, HI, 13–17 November 2006.
421. Scientific Report 2006: Highlight. http://www.fz-juelich.de/fs/index.php?index=431 Forschungszentrum Jülich, Germany, 2006.
422. Lang M, Auer C, Eismann A, Franco T, Lachenmann C, Schiller G, et al. Development of solid oxide fuel cells and short stacks for mobile application, 7th European Solid Oxide Fuel Cell Forum, Luzern, Switzerland, 3–7 July 2006.
423. Revolt Newsletter 116, Press releases, http://www.revolt.co.uk/news116.html, Revolt, UK, 6 March 2002.
424. IEF-3 Report 2009: Basic Research for Applications, Forschungszentrum Jülich, Germany, 2009.
425. History, About Staxera. http://www.staxera.de/History.329.0.html?&L=1, Staxera, no date, [accessed June 2010].
426. Otterstedt R, Rietveld G, Huiberts RC. Indec b.v. on the road toward commercialization. In: Singhal SC, Mizusaki J, editors. *Solid oxide fuel cells IX (SOFC-IX). Cells, stacks, and systems* 2005;vol. 1.
427. InDEC Ltd, organisation, industry directory, http://www.fuelcelltoday.com/online/industry-directory/organisations/in/InDEC-Ltd, Fuel Cell Today, 20 April 2010.
428. Enerday GmbH. Brief Portrait. http://www.enerday.de/en/enerday.html; no date [accessed June 2010].
429. Frequently asked questions (FAQ), About Staxera, http://www.staxera.de/FAQ.327.0.html?&L=1, Staxera, no date, [accessed June 2010].
430. Ouweltjes JP, van Berkel FPF, Nammensmam P, Christie GM. Development of 2nd generation, supported electrolyte, flat plat SOFC components at ECN Solid Oxide Fuel Cells IV (SOFC VI). In: Singhal SC, Dokiya M, editors. *The electrotechnical society*; 1999.
431. Ceramic Fuel Cells (ASX: CFU) partners with H.C. Starck to secure volume supply of fuel cell components, 2008 Announcements, http://www.cfcl.com.au/Assets/Files/Starck_final_Eng.ASX.pdf. Ceramic Fuel Cells Limited, 29 January 2008.
432. A New Trendsetting Fuel Cell Alliance, News Releases, http://www.prnewswire.com/news-releases/a-new-trendsetting-fuel-cell-alliance-97056964.html, PR Newswire, 24 June 2010.
433. Wunderlich (Staxera) C. *Staxera SOFC stacks in series production*. San Antonio, TX: 2007 Fuel Cell Seminar; 10 October 2007.
434. Staxera's new higher power 1.4 kW SOFC integrated stack module (ISM) is now available to all customers. The 1.4d kW ISM has a performance guarantee of 1.000 hours and 20 thermal cycles. http://fuelcellsworks.com/news/2010/04/20/staxera-releases-higher-power-and-increase-lifetime-guarantee-on-its-1-1-kw-sofc-integrated-stack-module/, FuelCellWorks, 20 April 2010.
435. Mai BE, Heller T, Schimanke D, Lawrence J, Wunderlich C (Staxera). Influence of Operating Conditions on the Reliable Performance of Stacks and Integrated Stack Modules, B10 -Solid Oxide Fuel Cells, Eleventh International Symposium (SOFC-XI), Vienna, Austria, 4-9 October 2009.

436. Mk200-10.000 hours of operation. News, http://www.staxera.de/announcement.105+M5e3d89b983a.0.html?&L=1, Staxera, 21 September 2009.
437. Target of 150 thermal cycles achieved. News, http://www.staxera.de/announcement.105+M56914e11a0e.0.html?&L=1, Staxera, 23 October 2009.
438. Staxera releases higher power ISM. News. http://www.staxera.de/announcement.105+M501407651e8.0.html?&L=1, Staxera, 19 April 2010.
439. Integrated Stack Module, 1.4 kW. Products. http://www.staxera.de/ISM-1-4-kW.455.0.html?&L=1, Staxera, no date, [accessed June 2010].
440. Staxera and EBZ cooperate. News. http://www.staxera.de/announcement.105+M51753a73e1a.0.html?&L=1, Staxera, Germany, 17 November 2009.
441. Saxera GmbH exhibiting at Hannover Messe 2010. News. http://www.staxera.de/announcement.105+M5014c597929.0.html?&L=1, Staxera, 13 April 2010.
442. New staxera products at the Hannover Fair 2011. News. http://www.staxera.de/announcement.105+M568b747b25d.0.html?&L=1, Staxera, 9 March 2–11.
443. Integrated Stack Module 1.7 kW, Products. http://www.staxera.de/ISM-1-7-kW.800.0.html?&L=1, no date, [accessed March 2012].
444. Product roadmap, http://translate.google.com/translate?hl=en&sl=de&u=http://www.staxera.de/&ei=fCytT57sMKb10gGwnd2xDA&sa=X&oi=translate&ct=result&resnum=1&sqi=2&ved=0CFYQ7gEwAA&prev=/search%3Fq%3Dstaxera%26hl%3Den%26biw%3D1024%26bih%3D601%26prmd%3Dimvns, Staxera, no date, [accessed May 2012].
445. Enerday GmbH, Home. http://www.enerday.de/en/enerday.html, Enerday, no date, [accessed June 2010].
446. Successful presentation of the product strategy of Enerday at Hannover fair 2009. News. http://www.enerday.de/en/news.html, Enerday, 19 May 2009.
447. Highest Efficiency of a NG SOFC system with dry CPOX reformer. News. http://www.enerday.de/en/news.html, Enerday, 6 July 2009.
448. Customer use and advantages. Corporate Strategy, http://www.enerday.de/en/efficienct.html, Enerday, no date, [accessed June 2010].
449. What the Future brings, Corporate Strategy. http://www.enerday.de/en/future.html, Enerday, no date, [accessed June 2010].
450. Advances in technology and product development. News. http://translate.google.com/translate?hl=en&sl=de&u=http://www.new-enerday.com/&ei=VDatT7WICMLw0gH09IC1DA&sa=X&oi=translate&ct=result&resnum=1&ved=0CGIQ7gEwAA&prev=/search%3Fq%3Dnew-enerday%26hl%3Den%26biw%3D993%26bih%3D583%26prmd%3Dimvnsb, New Enerday, 10 June 2011.
451. 2 units of the EN500 test field generation in continuous operation, News. http://translate.google.com/translate?hl=en&sl=de&u=http://www.new-enerday.com/&ei=VDatT7WICMLw0gH09IC1DA&sa=X&oi=translate&ct=result&resnum=1&ved=0CGIQ7gEwAA&prev=/search%3Fq%3Dnew-enerday%26hl%3Den%26biw%3D993%26bih%3D583%26prmd%3Dimvnsb, New Enerday, 1 April 2012.
452. Development of ceramic oxide fuel cell (SOFC) for power, Advanced Search, http://cordis.europa.eu/search/index.cfm?fuseaction=proj.document&PJ_LANG=EN&PJ_RCN=319908&pid=1&q=382145D2A228F39E63A7AC01A3A91F6A&type=adv., CORDIS Services, no date, [accessed March 2010].
453. Development of a solid oxide fuel cell (SOFC), Advanced Search, http://cordis.europa.eu/search/index.cfm?fuseaction=proj.document&PJ_LANG=EN&PJ_RCN=250805&pid=0&q=1DDA7B15440979C1C496A7CE81D7CEB9&type=adv, CORDIS Services, no date, [accessed March 2010].
454. Development of SOFC flat plate reactor with metal separator plate of alternative materials for low temperature SOFC operation, Advanced Search, http://cordis.europa.eu/search/index.cfm?fuseaction=proj.document&PJ_LANG=EN&PJ_RCN=263012&pid=0&q=4DE66F766EF6FC8C40FEC037A9AE8E8A&type=adv, CORDIS Services, no date, [accessed March 2010].
455. Development of co fired ceramic cells for low temperature 800øC solid oxide fuel cell (SOFC) operation, Advanced Search, SA http://cordis.europa.eu/search/index.cfm?fuseaction=proj.document&PJ_LANG=EN&PJ_RCN=352238&pid=2&q=FE76EBD5ABFBD96EA4C3704029F90323&type=adv, CORDIS, no date, [accessed March 2010].
456. Development of 50kw class SOFC system and components, Advanced Search, http://cordis.europa.eu/search/index.cfm?fuseaction=proj.document&PJ_LANG=EN&PJ_RCN=2360144&pid=10&q=DE666C5C56D761518914033B5D9BF325&type=adv, CORDIS Services, no date, [accessed March 2010].
457. Membrane-based thin layer SOFC technology, Advanced Search, http://cordis.europa.eu/search/index.cfm?fuseaction=proj.document&PJ_LANG=EN&PJ_RCN=319920&pid=0&q=8001FDE47051020C7152DE2C87A1DCE6&type=adv, Cordis Services, no date, [accessed April 2010].
458. Ponthieu E. Status of sofc development in Europe. In: Singhal SC, Dokiya M, editors. *Proceedings of the sixth international solid oxide fuel cells (SOFC VI)*. The Electrochemical Society; 1999.

459. Author Huiberts R.C, Huijsmans J.P.P. Pilot production of planar SOFC components at InDEC Ltd, ECN-RX−01-007, ECN, The Netherlands, 1 January 2001.
460. H.C. Starck acquires majority holding in InDEC, ECN. http://www.ecn.nl/news/item/date/2003/04/02/hc-starck-acquires-majority-holding-in-indec/, 2 April, 2003.
461. École Polytechnique Fédérale de Lausanne, École Polytechnique Fédérale de Lausanne. Wikipedia, last modified on 21 June 2010.
462. SOFCpower has purchased 100% of HTceramix' shares, Press Release, http://www.htceramix.ch/upload/documents/SOFCpower%20purchase%20of%20all%20HTc%20shares%20final.pdf, HTceramix, 9 May 2007.
463. Intellectual Property based on HTc' Technology, Technology, http://www.htceramix.ch/technology.php, HTceramix, no date [accessed June 2010].
464. HTceramix, Industry Directory, Fuel Cell Today, 21 June 2010.
465. Company profile, About Us. http://www.htceramix.ch/company-profile.php, HTceramix, no date [accessed June 2010].
466. Selection of Completed Projects, Projects/Completed. http://www.htceramix.ch/completed.php, HTceramix, no date [accessed June 2010].
467. Current Projects, Projects. http://www.htceramix.ch/current-projects.php, HTceramix, no date [accessed June 2010].
468. European Projects, Projects. http://www.htceramix.ch/european.php, HTceramix, no date [accessed June 2010].
469. Production facilities, Technology / Production.< http://www.htceramix.ch/production.php, HTceramix, no date [accessed June 2010].
470. A solid oxide fuel cell (SOFC) development platform demonstrated by fideris Inc. and HTceramix SA, Turnkey system reduces barrier to entry to financially promising SOFC development field. Press release, http://www.htceramix.ch/upload/documents/Dantherm-HTceramix-SOFC-partnership_final.pdf, HTceramix, 13 April 2005.
471. Dantherm AS and HTceramix SA announce successful integration of a Solid Oxide Fuel Cell (SOFC) stack into system producing net electricity and hot water proof of concept SOFC system successfully tested. Press release, http://www.htceramix.ch/upload/documents/Dantherm-HTceramix-SOFC-partnership_final.pdf, HTceramix, 15 April 2005.
472. HTceramix Exhibit at the Fuel Cell Seminar. Press release, http://www.htceramix.ch/upload/documents/press%20release%20-%20%20palm%20springs.pdf, HTceramix, 14 November 2005.
473. HTceramix launches new portable SOFC system − the Lilith 025. Press release, http://www.htceramix.ch/upload/SOFC%20stack%20exceeds%20expectations.pdf, HTceramix, 18 November 2005.
474. HTceramix partners SOFCpower for fuel cells, News Article. http://www.fuelcelltoday.com/online/news/articles/2007-02/HTceramix-partners-SOFCpower-for, Fuel Cell Today, 5 February 2007.
475. About Us, Company. http://www.sofcpower.com/index.php?page=about-us&hl=en_GB, SOFCpower, no date, [accessed June 2010].
476. 1000 h self-sustaining operation of SOFCpower module HoTbox, NEWS, http://www.sofcpower.com/index.php?page=agg-news&mact=News,cntnt01, detail,0&cntnt01articleid=11&cntnt01detailtemplate=DettaglioNewsInInterne&cntnt01lang=en_GB&cntnt01returnid=36&hl=en_GB, SOFCpower SpA, 1 December 2008.
477. Validation of pilot manufacturing line. Press release, http://www.htceramix.ch/upload/Validation%20of%20pilot%20manufacturing%20line.pdf, HTceramix, 26 March 2009.
478. ASC-700, ASC-800, Products. http://www.sofcpower.com/uploads/File/products/asc-700_asc-800.pdf, SOFC power, no date, [accessed December 2010].
479. Milestone in the development of large-SOFC-Stacks achieved. Press release, http://www.htceramix.ch/upload/Milestone%20in%20the%20development%20of%20large%20%20SOFC.pdf, HTceramix, 19 May 2009.
480. SOFC stack exceeds expectations, SOFCpower /HTceramix operate their new, larger size SOFC stack platform stably at 53% electrical efficiency. Press release, http://www.htceramix.ch/upload/SOFC%20stack%20exceeds%20expectations.pdf, 11 August 2009.
481. 65% electric efficiency achieved on SOFCpower's SOFC stack, Efficiency development target exceeded. Press release, http://www.htceramix.ch/upload/Efficiency%20development%20target%20exceeded_.pdf, HTceramix SA, 9 February 2010.
482. Zahid M (EDF). EDF measured HTceramix stack over 3000 hours of operation with a degradation rate of only 1% per 1000 hours. News. http://www.real-sofc.org/news_extended/news_04-2008, Real-SOFC. Real-SOFC Org, 30 April 2008.
483. van Herle J (EMPA-EPFL). HTceramix (S-Design) stack reached 5000 hours of operation. News. http://www.real-sofc.org/news_extended/news_04-2008, Real-SOFC.Org, 23 April 2008.
484. Products. Home. http://www.dantherm-power.com/Products/Micro_combined_heat_and_power_units.aspx, Dantherm Power, 2009, [accessed June 2010].

485. SOFC micro CHP plants to be climate-friendly power stations in homes. Archives, http://fuelcellsworks.com/news/2010/11/29/sofc-micro-chp-plants-to-be-climate-friendly-power-stations-in-homes/, FuelCellWorks, 29 November 2010.
486. Diethelm R. The Sulzer HEXIS SOFC System the report on international fuel cells, NEDO,Japan, March 1992.
487. Sulzer Hexis AG, industry directory, Fuel Cell Today, 17 June 2010.
488. Ponthieu E. Status of sofc development in Europe. In: S.C. Singhal, M. Dokiya, editors. *Solid oxide fuel cells IV (SOFC VI)*. The Electrotechnical Society; 1999.
489. Bove R. 11. Solid oxide fuel cells: principles, designs and state-of-the-art in industries. In: Basu S, editor. *Recent trends in fuel cell science and technology*. New Delhi, India: Anamaya Publishers; 2007.
490. Assessment of Planar Solid Oxide Fuel Cell Technology, Report to DOE FETC. http://www.netl.doe.gov/technologies/coalpower/fuelcells/publications/assessmentofplanarsofctechnology.pdf, Arthur D. Little, Inc., October 1999.
491. OZAWA & SARL, A Report on EU Hydrogen and Fuel Cell R&D Trend, NEDOP0219, a FY2002 Survey Report, NEDO Paris, September 2003.
492. Hoogers G, Fuel cell technology handbook, ISBN: 0-8493-0877-1, CRC Press, Boca Raton, London, New York, Washington, D.C., 2002.
493. Sulzer Hexis and the Dutch Energy Research Center ECN sign co-operation agreement. Media release, Sulzer Hexis, 27 June 2002.
494. Sulzer Hexis—the next generation of fuel cell system. Technical news, http://www.sulzer.com/en/desktopdefault.aspx/tabid-243/297_read-701/. Sulzer, 11 April 2005.
495. Orders in first half of 2005: Up 19% from previous year. Media release, http://www.sulzer.com/en/desktopdefault.aspx/tabid-242/295_read-522/257_view-242,195/295_page-17/. Sulzer, 14 July 2005.
496. Investments in Sulzer Hexis to discontinue. Media release, http://www.sulzer.com/en/desktopdefault.aspx/tabid-242/295_read-524/. Sulzer, 25 August 2005.
497. Decentralized electricity and heat gain. Home, http://www.hexis.com/, Hexis, no date [accessed June 2010].
498. Advances in the development of fuel cell technology, Swiss company Hexis tests fuel cell system in the field. Media release, http://www.hexis.com/downloads/MM_2007_07_11Hexis_e.pdf, Hexis AG, 11 July 2007. Volker Nerlich, Alexander Schuler, Andreas Mai, and Thomas Doerk (Hexis AG), Cogeneration in single family homes with fuel cell heating appliances. Associated File, http://www.fuelcellmarkets.com/3,1,629,1, 28621.html, Fuel Cell Markets, no date [accessed 2010].
499. Heading for series production. Fuel Cell, http://www.hexis.com/brennstoffzelle_e.htm, Hexis, no date [accessed December 2010].
500. Data «Galileo 1000 N». Products, http://www.hexis.com/brennstoffzelle_daten_e.htm, Hexis, no date [accessed December 2010].
501. Swiss company Hexis tests fuel cell system in the field. Media release, Hexis AG, Winterthur, Switzerland, 11 July 2007.
502. Mai A, Iwanschitz B, Weissen U, Denzler R, Haberstock D, Nerlich V, et al. Status of Hexis' SOFC stack development and the Galileo 1000 N micro-CHP system. Abstracts, http://www.electrochem.org/meetings/scheduler/abstracts/216/1219.pdf. The 216th ECS meeting, the Electrochemical Society, Vienna, Austria, 4–9 October 2009
503. Fuel cell system Galileo: advantages that convince. Galileo 1000 N, http://translate.google.com/translate?hl=en&sl=de&u=http://www.hexis.ch/&ei=uLhXT_6eLufq0QGS7qjGDw&sa=X&oi=translate&ct=result&resnum=9&sqi=2&ved=0CH0Q7gEwCA&prev=/search%3Fq%3Dhexis%26hl%3Den%26biw%3D1342%26bih%3D677%26prmd%3Dimvnsa. Hexis, no date [accessed March 2012].
504. Public events. News, http://translate.google.com/translate?hl=en&sl=de&tl=en&u=http%3A%2F%2Fwww.hexis.ch%2F; no date [accessed March 2012].
505. Leading expertise in fuel cell technology. Hexis AG, http://translate.google.com/translate?hl=en&sl=de&tl=en&u=http%3A%2F%2Fwww.hexis.ch%2F; no date [accessed March 2012].
506. Alstom. industry directory, http://www.fuelcelltoday.com/online/industry-directory/organisations/al/Alstom, Fuel Cell Today, 8 August 2010.
507. Pyke S. H, Howard PJ, Leah RT. Planar SOFC technology: stack design and development for lower cost and manufacturability (ETSU F/01/00194/REP, DTI/Pub URN 02/1350). DTI, UK, 2002.
508. Pyke S.H, Burnett A.J, Leah R.T. Systems development for planar SOFC based power plant (ETSU F/01/00195/REP, DTI/Pub URN 02/868). DTI, UK, 2002.
509. Distributed power generation plant based on planar SOFC, proof of concept (PROCON). European fuel cell projects 1995–2000, EUR 19368, European Commission, 2000.
510. Ceres power. Organisation, http://www.fuelcelltoday.com/online/industry-directory/organisations/ce/CERES-Power-Ltd, Fuel Cell Today, 30 March 2010.

511. Kanellos M. Making fuel cells from car parts. http://www.greentechmedia.com/articles/read/making-fuel-cells-from-car-parts/. Greentech Media, 21 January 2010.
512. The ceres cell. Technology, http://www.cerespower.com/Technology/TheCeresCell/. Ceres Power, no date [accessed March 2010].
513. Baron S (Imperial College). Intermediate temperature (500−850 °C) Solid oxide fuel cells (IT-SOFCs) explained. Fuel Cell Today, 21 January 2004.
514. Development of innovative metal-supported IT-SOFC technology, project summary, NO PS210, DTI, UK, January 2005.
515. BOC trial uses LPG to generate electricity with Ceres fuel cells. Press release, http://www.cerespower.com/store/files/55-BOC%20trial.pdf. Ceres Power, October 5, 2004.
516. Ceres takes lead role in £1.9 million government contract. Press release, http://www.cerespower.com/store/files/52-DTi%20award.pdf. Ceres Power, 23 February 2005.
517. Larger fuel cells cheaper to produce and easier to assemble. Press release, http://www.cerespower.com/store/files/50-Cell%20Size%20May%2005.pdf, Ceres Power, 23 May 2005.
518. Ceres' fuel cell sets new world standards in performance tests. Press release, < http://www.cerespower.com/store/files/48-Ceres%20Power%20performance%20milestones%20-%2013%20May%202005%20_3_.pdf. Ceres Power, 16 May 2005.
519. Key technical milestone: compact, lightweight 1 kW fuel cell. Press release, http://www.cerespower.com/store/files/31-March%2006%201%20kW%20Stack%20announcement.pdf. Ceres Power. 13 March 2006.
520. Ceres Power secures £2.7 million programme with British Gas to accelerate the introduction of fuel cells into UK homes. Press release, http://www.cerespower.com/store/files/30-March%2006%202.7%20million%20programme%20with%20BG.pdf. Ceres Power. 20 March 2006.
521. Ceres Power announces wall-mountable system design under British Gas CHP programme. Press release, http://www.cerespower.com/store/files/29-May%2006%20wall%20mountable%20system%20design.pdf. Ceres Power, 25 May 2006.
522. Ceres secures new 'product facility' for assembly of complete products and also plans a 'mother plant' for volume manufacturing of its core fuel cell components, Press release, http://www.cerespower.com/store/files/23-Sept%2006%20Product%20Facility.pdf. Ceres Power, 25 September 2006.
523. Technical update: key performance and durability targets achieved. Press release, http://www.cerespower.com/store/files/22-Dec%2006%20Technical%20Update.pdf. Ceres Power, 1 December 2006.
524. Ceres Power and British Gas enter development, supply and distribution agreement for UK residential CHP. Press release, http://www.cerespower.com/store/files/12-BG%20development%20supply%20and%20distribution%20agreement.pdf. Ceres Power, 14 January 2008.
525. Fuel cell module meets alpha phase target on schedule. Press release, http://www.cerespower.com/store/files/75-Fuel%20Cell%20Module%20RNS%20FINAL%2029%2009%2008.pdf. Ceres Power. 30 September 2008.
526. Mother plant facility secured. Press release, http://www.cerespower.com/store/files/87-Ceres%20Power%20Motherplant%20secured%20Mar%2009.pdf. Ceres Power, 23 March 2009.
527. British Gas confirms Alpha CHP product testing successfully completed. Press release, http://www.cerespower.com/store/files/89-CHP%20Alpha%20sign%20off%20Final%2026.05.09.pdf. Ceres Power, 26 May 2009.
528. Fletcher N. Ceres Power falls on increased losses and delayed trial. Market Forces Blog. http://www.guardian.co.uk/business/marketforceslive/2010/mar/25/2, guardian.co.uk, 25 March 2010.
529. Feed-in tariff announced for residential CHP. Press release, http://www.cerespower.com/store/files/151-FiT%20Announcement%20Feb%202010.pdf. Ceres Power, 2 February 2010.
530. Ceres Power Holdings plc (or the Company) Result of AGM, Annual general meeting held on Friday 2nd, Ceres Power, 2 December 2011. Chief Executive Officer's Address.
531. Developing new technologies for the energy sector. Innovation for the future, http://www.rolls-royce.com/reports/environment_report_07/innovation/energy.html. Rolls Royce, 2007 [accessed April 2010].
532. Koehler T.M, Jarrell DB, Bond L.J. High temperature ceramic fuel cell measurement and diagnostics for application to solid oxide fuel cell systems. PNNL-13716, Prepared by Pacific Northwest National Laboratory for the U.S. Department of Energy under Contract DE-AC06-76RL01830, October 2001.
533. Hart N, Wright G, Pashley M, Lapena-Rey N, Dean E. Improving the manufacture and durability of multi-cell modules for the IP-SOFC (ETSU F/01/00179/REP, DTI/Pub URN 01/1293). Rolls Royce, 2001.
534. Collins R, Pashley M, Day M, Oakley J. Development of A 1 kW IP-SOFC stack. ETSU F/01/00144/REP. Rolls-Royce Plc, 2001.

535. Hart NT. Scale-up to multi-tens of kW for the IP-SOFC, ETSU Report F/01/00197/REP. Rolls Royce, 2004.
536. Pashley M, Collins R, Oakley J, Dean E. Design & build a 5 kW thermally self-sustaining IP-SOFC stack, ETSU F/01/00199/REP, DTI/Pub URN 01/1161. Rolls Royce, 2001.
537. Pressurisation of IP-SOFC technology for second generation hybrid application. Project Summary NO PS214, DTI, UK, March 2005.
538. Pressurization of IP-SOFC Technology for second Generation Hybrid Application PIP-SOFC, NNE5-2001-00791, http://ec.europa.eu/energy/efficiency/industry/doc/pipsofc.pdf, Rolls-Royce Fuel Cell Systems Ltd.,no date [accessed April 2010].
539. High temperature fuel cells use for the next ten years. Advanced Search, http://cordis.europa.eu/search/index.cfm?fuseaction=proj.document&PJ_LANG=EN&PJ_RCN=3758701&pid=10&q=09B2E51E9986F478CEF F29997EBAF67A&type=adv, CORDIS Services, no date [accessed April 2010].
540. Integrated modeling study of fuel cell/gas turbine hybrids (IM-SOFC-GT), Advanced Search, http://cordis.europa.eu/search/index.cfm?fuseaction=proj.document&PJ_LANG=EN&PJ_RCN=4938374&pid=3&q=E7BC973961F0AAB35B12A6F865E1352D&type=adv, CORDIS Services, no date, [accessed March 2010].
541. Pressurized IP-SOFC: a path to successful SOFC/Hybrids (PIP-SOFC). European fuel cell and hydrogen projects, 1999–2002, EUR 20718, European Commission, 2003.
542. Integrated modeling study of FC/GT hybrids (IM-SOFC-GT). European fuel cell and hydrogen projects, 1999–2002, EUR 20718, European Commission, 2003.
543. Liste des subventions accordées en 2003 par la Direction générale de l'énergie et des transports, http://ec.europa.eu/dgs/energy/tenders/subsidies/doc/2003_dg_tren_subsidies.pdf. The European Commission, 2003.
544. Goettler R. Overview of the rolls-royce SOFC technology and SECA program. In: 10th annual solid state energy conversion alliance (SECA) workshop, http://www.netl.doe.gov/publications/proceedings/09/seca/presentations/Goettler_Presentation.pdf. DOE, NETL, Pittsburgh, PA, 14–16 July, 2009.
545. European Union awards $5.8m grant for the development of solid oxide fuel cell-based power plants, http://www.azom.com/News.asp?NewsID=7670. AZo Materials, 31 January 2007.
546. Dr Dicks AL, Dr Bradley J, Alexander D.T (Department of Chemical Engineering, the University of Queensland). Australian hydrogen activity. Australian Government Department of Resources, Energy and Tourism, 2008.
547. Milestones, Company History, http://www.cfcl.com.au/Company_History/, Ceramic Fuel Cell Limited, 2010 [accessed July 2010].
548. Why solid oxide fuel cells? Fuel cell facts, No. 4, http://www.cfcl.com.au/Assets/Files/4._Why_Solid_Oxide_Fuel_Cells_-_Rev.4_2009.pdf, Ceramic Fuel Cell Limited, 4 January 2009.
549. Dinsdale J, Coduit A, Foger K. Development of a residential combined heat and power (CHP) system at ceramic fuel cell limited. Ceramic fuel cell limited, 2004 Fuel Cell Seminar, San Antonio, Texas, 1–5 November 2004.
550. Company introducton. About, http://www.cfcl.com.au/Company_Introducton/, Ceramic Fuel Cell Limited, 2010 [accessed July 2010].
551. Open Briefing®; Ceramic fuel cells. Plans for Zircon Powder Plant, Attention ASX Company announcements platform, Lodgment of Open Briefing®. http://www.cfcl.com.au/Assets/Files/20060807%20CFCL%20Open%20Briefing%20-%20Powder%20Plant.pdf, Ceramic Fuel Cell Limited, 7 August 2006.
552. Dow B. Ceramic fuel cells manufacturing plant, Fuel Cell Foundry, Heinsberg, Germany. Presentation. http://www.cfcl.com.au/Assets/Files/20061213%20CFCL%20Heinsberg.pdf, Ceramic Fuel Cell Limited, 12 December 2006.
553. TD-1. Into Center, http://www.cfcl.com.au/Content.aspx?ID=209, Ceramic Fuel Cell Limited, 2010 [accessed July 2010].
554. CHP1. Into Center, http://www.cfcl.com.au/Content.aspx?ID=210, Ceramic Fuel Cell Limited, 2010, [accessed July 2010].
555. CHP2. Into Center, http://www.cfcl.com.au/Content.aspx?ID=211. Ceramic Fuel Cell Limited, 2010 [accessed July 2010].
556. CHP field trials – emonstration units installed internationally. Smart Power, vol. 4, http://www.cfcl.com.au/Assets/Files/Smart_Power_Newsletter_06-02.pdf>. Ceramic fuel cell limited, February 2006.
557. 10,000 hours & counting. Smart Power, vol. 5, Ceramic fuel cell limited, April 2006. http://www.cfcl.com.au/Assets/Files/Smart_Power_Newsletter_06-04.pdf>.
558. Net~Gen launched in London. Smart Power, vol. 3, http://www.cfcl.com.au/Assets/Files/Smart_Power_Newsletter_05-10.pdf. Ceramic Fuel Cell Limited, October 2005.
559. CFCL's Net Gen unit on show at global fuel cell conference. 2005 Announcements, http://www.cfcl.com.au/Assets/Files/20051118%20CFCL%

20Net~Gen%20at%20Palm%20Springs.pdf, Ceramic Fuel Cell Limited, 18 November 2005.

560. Ceramic fuel cells achieves European safety approval for microgeneration fuel cell system. 2007 Announcements, http://www.cfcl.com.au/Assets/Files/NetGenPlus%20CE%20Nov07%20final.pdf, Ceramic Fuel Cell Limited, 22 November 2007.
561. Ceramic Fuel Cells Ltd (ASX: CFU), 12.4M investment in manufacturing facility and commercial order from nuon. 2008 Announcements, http://www.cfcl.com.au/Assets/Files/20080227_CFCL_Capex_investment_and_Nuon_order.pdf, Ceramic Fuel Cells Limited, 27 February 2008.
562. Ceramic fuel cells achieves world-best 60% efficiency for its electricity generator units. Company Announcement, http://www.cfcl.com.au/Assets/Files/20090219_CFCL_Announcement_60_percent_Efficiency.pdf, Ceramic Fuel Cells Limited, 19 February 2009.
563. Payne R, Love J, Kah M (Ceramic Fuel Cells Limited). Generating electricity at 60% electrical efficiency from 1 to2 k we SOFC products. 216th ECS Meeting, Vienna, Austria, 4–9 October 2009, http://www.electrochem.org/meetings/scheduler/abstracts/216/1300.pdf.
564. Dow B. Product development agreement, fuel cell powered Micro CHP units. Presentation, http://www.cfcl.com.au/Assets/Files/20061215_GdF_DDT_Development_Agreement_Presentation.pdf, Ceramic Fuel Cell Limited, 15 December 2006.
565. Ceramic Fuel Cells moves ahead with development of combined heat and power units for French market. 2009 Announcements, http://www.cfcl.com.au/Assets/Files/20090729_CFCL_GDF_SUEZ_Beta%20Announce_July09_071309_modif_GDF_SUEZ.pdf. Ceramic Fuel Cells Limited, 29 July 2009.
566. Ceramic fuel cells limited signs product development agreement for a german residential combined heat and power system. 2007 Announcements, http://www.cfcl.com.au/Assets/Files/20070315_CFCL_Bruns_PDA_(English)_final.pdf. Ceramic Fuel Cell Limited, 15 March 2007.
567. Ceramic fuel cells limited Ships a Net Gen Plus unit to EWE. 2007 Announcements, http://www.cfcl.com.au/Assets/Files/20071218_NetGen_EWE.pdf, Ceramic Fuel Cell Limited, 18 December 2007.
568. Ceramic Fuel Cells, Nuon and Remeha Collaborate on Microgeneration for Enelux Region. Press release, http://www.cfcl.com.au/Assets/Files/20070724_CFCL_Nuon_Remeha_collaboration.pdf, Ceramic Fuel Cell Limited, 24 July 2007.
569. E.ON UK and ceramic fuel cells to collaborate on development of new greener microgeneration in the home. Press release, http://www.cfcl.com.au/Assets/Files/E%20ON%20and%20CFCL%20press%20release.ASX.pdf. Ceramic Fuel Cell Limited, 9 July 2007.
570. Gledhill selected as appliance partner for Ceramic Fuel Cells & E.ON UK collaboration working for greener home energy. Press release, http://www.cfcl.com.au/Assets/Files/20070830_CFCL_and_Gledhill.pdf. Ceramic Fuel Cell Limited, 30 August 2007.
571. Ceramic fuel cells limited completes first shipments to European partners, 2008 Announcements, http://www.cfcl.com.au/Assets/Files/20080110_CFCL_Net_Gen_plus_shipments_v2.pdf. Ceramic Fuel Cell Limited, 10 January 2008.
572. E.ON and Ceramic Fuel Cells (CFCL) extend microgeneration collaboration, Funded product development, Proposed volume order for commercial products, 2009 Announcements, http://www.cfcl.com.au/Assets/Files/20090202_CFCL_EON_and_CFCL_Announcement.pdf, Ceramic Fuel Cells Limited, 2 February 2009.
573. Ceramic fuel cells (ASX: CFU) and paloma industries to collaborate on fuel cell CHP products for Japan. 2008 Announcements, http://www.cfcl.com.au/Assets/Files/20080117_CFCL_Paloma_Collaboration_Announcement.pdf, Ceramic Fuel Cell Limited, 17 January 2008.
574. Ceramic Fuel Cells Limited Completes Commissioning of UK Ceramic Powder Plant 2007 Announcements, http://www.cfcl.com.au/Assets/Files/20071029_CFCL_Powder_Plant.pdf, Ceramic Fuel Cell Limited, 29 October 2007.
575. Ceramic fuel cells launches modular generator product, Signs MoU with VicUrban. 2009 Announcements, http://www.cfcl.com.au/Assets/Files/20090522_CFCL_Product_Launch_Announcement.pdf, Ceramic Fuel Cells Limited, 22 May 2009.
576. BlueGen, Products, http://www.cfcl.com.au/BlueGen/, Ceramic Fuel Cell Limited, 2010 [accessed July 2010].
577. Ceramic Fuel Cells, Technical advances result in 50% increase in cell power density and longer lifetime. 2008 Announcements, http://www.cfcl.com.au/Assets/Files/20080701_CFCL_Lucerne_1July08_final_ASX.pdf. Ceramic Fuel Cells Limited, 1 July 2008.
578. Annual Report 2009. Ceramic Fuel Cell Limited, 2009.
579. Introducing: blue gen, modular generator clean power + heat. Presentation, Ceramic Fuel Cell Limited, 22 May 2009.
580. *World fuel cells, an industry profile with market prospects to 2010*. Elsevier; 2002.
581. FCT, SWPC sign joint development agreement. News article, Fuel Cell Today, 21 July 2003.
582. Stannard JH, Allen G, Varma V. Commercialization of combined heat and power SOFC systems. 2005 Fuel

Cell Seminar, Palm Springs, CA, 14–18 November 2005.

583. *Fuel cell technologies Ltd. system enters demonstration phase.* In: *Fuel cell catalyst, the U.S. fuel cell council, national energy technology laboratory*, vol. 2. national fuel cell research center; Summer 2002. No. 4.

584. Natural gas fuel cell. About Us, http://www.fngas.com/fuelcell.html. Fairbanks, Natural Gas, 2002–2005, [accessed July 2010].

585. Worldwide fuel cell installations. http://www.fuelcells.org/FCInstallationChart.pdf. Fuel Cells 2000, updated May 2010, [accessed July 2010].

586. SOFC: FCT ships four SOFCs in first quarter. Fuel Cell Technology News, AllBusiness.com, 1 April 2005.

587. Fuel cell technologies to undergo restructuring. News Article, http://www.fuelcelltoday.com/online/news/articles/2006-02/Fuel-Cell-Technologies-to-Underg. Fuel Cell Today, 2 February 2006.

588. Borglum B, Fan J, Neary E (Global Thermoelectric Inc.). Following the critical path: an update on global thermoelectric's SOFC technology and product development. 203rd meeting, the electrochemical society, Paris, France, 27 April–2 May 2003.

589. Global Thermoelectric Inc. Industry directory, http://www.fuelcelltoday.com/online/industry-directory/organisations/gl/Global-Thermoelectric-Inc-. Fuel Cell Today, 9 July 2010.

590. Quantum to Acquire Global Thermoelectric Inc. http://www.encyclopedia.com/doc/1G1-131728426.html. PR Newswire,9 April 2003.

591. Quantum receives agreement from a major shareholder in global thermoelectric to support transaction. Press release, http://www.qtww.com/files/qtww_press/030602%20QT%20Receives%20Agreement%20from%20a%20Major%20Shareholder%20in%20G.pdf. Quantum Technologies, 2 June 2003.

592. FuelCell energy to acquire global thermoelectric. http://www.thefreelibrary.com/FuelCell+Energy+to+Acquire+Global+Thermoelectric-a0106403812. The Free Library, 7 August 2003.

593. Korean fuel cell technology. ATIP05.041, Asian Technology Information Program (ATIP), Tokyo, Japan, 24 August 2005.

594. Shin S-C (Korea Institute of Energy Research). Innovation in fuel cell technology in Korea. OECD conference on the innovation of energy technologies, in collaboration with The National Academy of Sciences, Washington, D.C., 29–30 September 2003.

595. Current status of fuel cell development in South Korea. Hydrogen, fuel cell special issue, http://www.nedo.go.jp/kankobutsu/report/981/981-12.pdf. NEDO Overseas Report, No. 981, 5 July 2006.

596. Park J-K, Yoo Y-S, Yang S-Y, Oh J-M, Park J-W, Kim (Korea Electric Power Research Institute) J-H. Performance of solid oxide fuel cells with controlled microstructure of anode substrate. In: Singhal SC, Mizusaki J, editors. *Solid oxide fuel cells IX (SOFC IX). Cells, stacks, and systems*, vol. 1. The Electrochemical Society; 2005.

597. Lee T, Park J-K, Oh J-M, Yoo Y-S (Korea Electric Power Research Institute). Development of 1 kW class SOFC system at KEPRI. 10th International Symposium on Solid Oxide Fuel Cells (SOFC-X), vol. 7(1), Nara, Japan, 3–8 June 2007.

598. Lee T, Choi JH, Park T-S, Yoo Y-S. Development of planar type SOFC stacks and systems at KEPRI. 2008 Fuel Cell Seminar & Exhibition, 27–30 October 2008.

599. Yoo YS, Lee T, Choi J, Yang SY, Park TS. Operation of 1 kW class SOFC stack and system for residential power generation application. Korea Electric Power Research Institute, 2007 Fuel Cell Seminar & Exposition, San Antonio, TX, 15–19 October 2007.

600. Lee T, Choi JH, Choi M-H, Yoo Y-S (Korea Electric Power Research Institute). Development of kw class planar type SOFC stacks and a 5 kw class cogeneration system. ASME 2010 Eighth International Fuel Cell Science, Engineering and Technology Conference (FuelCell2010), Brooklyn, New York, 14–16 June 2010.

601. Message from the president. About Kier, http://www.kier.re.kr/open_content/eng/kier/message.jsp. Korea Institute of Energy Research, no date [accessed July 2010].

602. Song R-H. R & D status of solid oxide fuel cell at KIER. Fuel Cell Conference, 10 March 2010.

603. Song R-H, Kim J-H, Son H-J, Lee G-Y, Yu S-H, Jung D-H, et al. (KIER). Development of anode-supported flat tube cell stack. 2003 Fuel Cell Seminar, Miami Beach, FL, 3–7 November 2003.

604. Hyoung Lim T, Hyun Song R, Hee Kim J, Dong Un S, You Chung D, Hyun Peck D, Jung D-H, Shin (Korea Institute of Energy Research) DR. Development of anode-supported tubular SOFC stack. In: Singhal SC, Mizusaki J, editors. *Solid Oxide Fuel Cells IX (SOFC IX). Cells, stacks, and systems*, vol. 1. The Electrochemical Society; 2005.

605. Lim T-H, Song R-H, Shin D-R, Yang J-IL, Jung H, Vinke I.C. Operating characteristics of a 5 kW class anode-supported planar SOFC stack for a fuel cell/gas turbine hybrid system, International Journal of Hydrogen Energy, vol. 33(3), Elsevier, 2008.

606. Song R-H, Kim J-H, Chung D-Y, Son H-J, Yun K-S, Lim T-H, et al. (KIER). Development of 500 W-class anode-supported flat-tube solid oxide fuel cell stack in KIER. 7th European SOFC Forum, Lucerne, Switzerland, 3–7 July 2006.

607. Dong S-K, Yun B-T, Lee D-K (Korea Institute of Energy Research). Design and numerical study for 1 kW tubular SOFC APU System. 2007 Fuel Cell Seminar and Exposition, San Antonio, TX, 15–19 October 2007.
608. Lim T-H, Yun U-J, Lee J-W, Lee S-B, Park S-J, Song R-H, et al. Fabrication of tubular segmented-in-series SOFC unit bundle.2010 Fuel Cell Seminar & Exposition, San Antonio, TX, 18–20 October 2010.
609. Lim T-H, Park S-J, Song R-H (Korea Institute of Energy Research). Operation characteristics of 1.3 kW class anode-supported flat tubular SOFC stack. ASME 2010 8th International Conference on Fuel Cell Science, engineering and technology (FUELCELL2010), Brooklyn, New York, 14–16 June 2010.
610. Fuel cell development in China. Energy & Fuels, http://www.ecoworld.com/energy-fuels/fuel-cell-development-in-china.html, Eco World Magazine, no date [accessed July 2010].
611. The first SOFC manufacturing line in China. Research, http://english.nimte.cas.cn/rh/rp/200908/t20090804_26980.html. Ningbo Institute of Material Technology & Engineering, Chinese Academy of Sciences, no date [accessed July 2010].
612. Wang WG, Guan WB, Li HM, XueYJ, WanJX, Wu YN, et al. SOFC research and development at NIMTE. B0404, Abstract 236, Oral presentation, Session B04, Innovative applications and designs, Ningbo Institute of Material Technology and Engineering, 1 July 2008.
613. Planar SOFC single cells with highest flexural strength have been developed in NIMTE, research progress, http://english.nimte.cas.cn/rh/rp/200909/t20090921_39553.html, Division of Fuel Cell & Energy Technology, Ningbo Institute of Materials Technology & Engineering, Chinese Academy of Sciences, 21 September 2009.
614. Guan W, Zhai H, Li F, Li Z, Xu C, Wang WG. (Ningbo Institute of Material Technology and Engineering, Chinese Academy of Science), Development and performance of planar SOFC stacks,ECS Transactions. *Electrochem Soc* 2009;**25**(2):485–8.
615. SOFC R&D Group in NIMTE obtain an important development in SOFC stack. News, http://www.sofc.com.cn/news3.html. Division of Fuel Cell & Energy Technology, Ningbo Institute of Materials Technology & Engineering, Chinese Academy of Sciences, 29 April 2009.
616. New advances in SOFC stack module development at NIMTE. News, http://www.sofc.com.cn/news11.html. Ningbo Institute of Materials Technology & Engineering, Chinese Academy of Sciences, 23 December 2010.
617. Improving power density of solid oxide fuel cell stacks using a new contacting. News, http://www.sofc.com.cn/news17.html. Ningbo Institute of Material Technology & Engineering, 23 April 2012.
618. Solid oxide fuel cells. Research, <ahttp://english.sic.cas.cn/rh/RD/em/so/>. Shanghai Institute Of Ceramics, Chinese Academy of Sciences, no date [accessed July 2010].
619. Yang YJ, Tu HY, Lu ZY, Yang JH, Wen (Shanghai Institute of Ceramics) TL. Properties of (La0.7Sr0.3) 0.9Mn1-yCryO3-δ cathode materials for solid oxide fuel cell. *J Australian Ceram Soc* 1998;**34**(1).
620. Wen TL, Wang D, Chen M, Tu H, Lu Z, Zhang Z (Shanghai Institute of Ceramics). Material research for planar SOFC stack. Available Online 15 March 2002, Solid State Ionics, vol. 148(3–4), 2 June 2002.
621. Liu RZ, Wang SR, Huang B, Zhao CH, Li JL, WanZR. (Shanghai Institute of Ceramics), Dip-coating and co-sintering technologies for fabricating tubular solid oxide fuel cells, Received 29 October 2008, Published online 20 December 2008, vol. 13, Number 12, Journal of Solid State Electrochemistry, Springer Berlin, Heidelberg, December, 2008.
622. Zhao C, Liu R, Wang S, Wen T (Shanghai Institute of Ceramics). Improvement on the performance of cathode supported tubular solid oxide fuel cells fabricated by dip-coating and co-sintering techniques, http://www.electrochem.org/meetings/scheduler/abstracts/216/1514.pdf. ECS 216th Meeting, Vienna, Austria, 4–9 October 2009.
623. Liu X, Zhan Z, Meng X, Huang W, Wang S, Wen T. Enabling catalysis of Ru–CeO_2 for propane oxidation in low temperature solid oxide fuel cells, J Power Sources, Elsevier 199():138–141(2012) vol. 199 February 1, 2012.
624. Welcome to FCET. Home, http://www.sofc.com.cn/. Division of fuel cell & energy technology, Ningbo Institute of Materials Technology & Engineering, Chinese Academy of Sciences, Ningbo City, Zhejiang Province, no date [accessed July 2010].
625. Batra V.S, Bali S. Status of fuel cell technology. Energy technology news, issue 2 and 3, Organizations for the promotion of energy technologies, India, January and April 2001.
626. Status of planar SOFC technology development at CGCRI, Kolkata, India.2nd German-Indian conference on research for sustainability, United Nations University, Bonn, 27–28 April, 2009.
627. History/Overview of CGCRI. About CGCRI, http://webserver.cgcri.res.in/page.php?id=2. Central Glass and Ceramic Research Institute, Updated on 1 July 2010.

628. Overview of fuel cell and battery division. Overview, Central Glass and Ceramic Research Institute (CGCRI), http://webserver.cgcri.res.in/page.php?id=39, 22 May 2010.
629. Solid oxide fuel cell. CGCRI products, http://www.cgcri.res.in/page.php?id=282. Central Glass and Ceramic Research Institute (CGCRI), 2 July 2010.
630. Homma T. Research and development of SOFC. The latest fuel cell news in Japan, FCDIC, July 2001.
631. Fuel cell breaks 40,000-hour mark. News, http://www.rdmag.com/News/2012/03/Industries-Energy-Electricity-Fuel-cell-breaks-40000-hour-mark/, Advantage Business Media, 29 March 2012.
632. Steinberger-Wilckens R, Blum L, BuchkremerH-P, de Haart B, Malzbender J, Pap M (Forschungszentrum Jülich). Recent Results in Solid Oxide Fuel Cell Development at Forschungszentrum Juelich. ECS Transactions, 35(1) 53–60. The Electrochemical Society, 2011.
633. Tang E, Martell F, Brulé R, Marcotte K, Borglum B (Global Thermoelectric Inc.). Global thermoelectric's integrated cell manufacturing of planar SOFCs. The 203rd Meeting, the electrochemical society, Paris, France, 27 April–2 May 2003.
634. Enbridge and global thermoelectric establish fuel cell evelopment Alliance. News release, ENBRIDGE INC., 31 July 2000.
635. Weaver G. *In DEC pilot production BV, world fuel cells: an industry profile with market prospects to 2010*. Elsevier; 2002.
636. HoTbox. Products, http://www.htceramix.ch/hotbox.php, HTceramix, 2006, [accessed June 2010].

CHAPTER

7

History of Proton Exchange Membrane Fuel Cells and Direct Methanol Fuel Cells

OUTLINE

Fuel Cells
http://dx.doi.org/10.1016/B978-0-444-56325-5.00007-7

7.1. INTRODUCTION

The modern history of the proton exchange membrane fuel cell (PEMFC)[i] started in the early 1950s, when a number of companies and institutions began PEMFC development. They included Aerojet Company, General Electric (GE), National Carbon Company, the Patterson-Moos Division of the Universal Winding Company, and Pittsburgh Consolidation Coal.[1]

GE had the most ambitious program and in 1954 GE researchers Willard T. Grubb and Leonard Niedrach started research on the "ion-exchange membrane cell." They invented the first PEMFC in June 1955,[2] which generated 0.02 W.[1] GE's PEMFC was simpler in design and operation than Grove's original fuel cell and seemed almost elegant.[3] While Grove's cell used a corrosive sulfuric acid bath as the electrolyte, Grubb's fuel cell was made of all solid materials. Three years later, Niedrach devised a way of depositing platinum onto the fuel cell,[4] which accelerated electrochemical reactions and resulted in a significantly higher efficiency. The improvements made through the combined efforts of Grubb and Niedrach produced the "Grubb-Niedrach fuel cell."

- GE used the PEMFC in research programs with the US Navy's Bureau of Ships (Electronics Division) and the US Army Signal Corps.[5] Although the cell was compact and portable, it was highly expensive to produce.

7.2. US NATIONAL AERONAUTICS AND SPACE ADMINISTRATION BOOSTS GE's PEMFC R&D IN THE LATE 1950s

A lucky break came to GE's PEMFC effort in the late 1950s when the US National Aeronautics and Space Administration (NASA) asked GE to play a major role in the US manned space program. NASA's Mercury Program (1958–1963) flew men into space in a capsule that was battery powered.[6] NASA needed options other than heavy batteries for its second manned program, Gemini (1962–1966), which would fly men into space for much longer periods.[7] Gemini was designed to test man and equipment for space flights up to 2 weeks in preparation for the third critical manned spaceflight program, the Apollo Lunar-Landing Mission (1963–1972). Relying on solar energy was still too expensive, and nuclear energy was considered too risky.

- NASA believed fuel cells were the best option, and so a massive infusion of funds was directed at developing space-qualified fuel cells. Annual funding reached more than $20 million (in 1990 dollars) during the mid-1960s.[8]
- NASA awarded two contracts to develop fuel cells to power the Gemini spacecraft, one to GE and one to United Technologies Corporation (UTC). They were selected to develop PEMFCs and alkaline fuel cells (AFCs), respectively.[9] In 1963, GE produced the first prototype PEMFC for NASA using the "Grubb-Niedrach fuel cell."[10] The 1 kW power plant consisted of 96 cells in three parallel groups of 32 cells. Two 1 kW modules were to provide primary power for each of the 12 Gemini missions.

7.2.1. GE PEMFCs Fail Manned Space Mission in the late 1960s

Power density was a low 38 mW/cm^2 at a voltage of 0.83 V per cell. Water was removed with a wick in each cell, which was not very

[i] Proton exchange membrane fuel cells (PEMFCs) go by many names. General Electric (GE), the original developer, referred to them as ion-exchange membrane fuel cells in the early stages. Later, GE named them solid polymer electrolyte fuel cells. There have been a few more names used, including solid polymer fuel cells (SPFCs) and polymer electrolyte fuel cells (PEFCs).

reliable. Gemini missions 1 through 4 flew with batteries instead of fuel cells.[12] GE redesigned their cells, and the new cells, despite malfunctions and poor performance on Gemini 5, adequately served for the remaining Gemini missions. Nevertheless, performance and lifetimes of the Gemini fuel cells were limited by the sulfonated polystyrene membrane used at that time.

- NASA decided to use AFCs developed by UTC for both the command and lunar modules for the Apollo missions, as did designers of the Space Shuttle a decade later.

7.2.2. New Nafion Membrane Increases PEMFC Efficiency…But GE Abandons PEMFC Effort in 1984

GE maintained a high level of inhouse investment in PEMFC R&D even after the Gemini project was over. GE reportedly spent \$8.5 million of its own money over the 1960s and 1970s.[13] In 1968, the EI DuPont de Nemours and Company (DuPont) developed an advanced membrane, Nafion, which had excellent thermal, chemical, oxidative, and mechanical stability, and seemed perfect for fuel cells.[14] In 1969, GE used Nafion in its second PEMFC system, a 350 W module that flew in three Biosatellite spacecraft (1966–1969), which NASA built to assess the effects of spaceflight on living organisms.[15] GE next developed a 3 kW flight-qualified module for a Navy high-altitude Balloon program. This unit demonstrated a power density more than five times greater than GE's Gemini fuel cells. GE by the mid-1970s developed PEM water electrolysis technology and began work on the U.S. Navy Oxygen Generating Plant.[5] The British Navy adopted this technology in the 1980s for its submarine fleet.

- As DuPont continued to make improvement in Nafion, GE made significant progress in improving the performance of PEMFC units. In 21 years between 1959 and 1980, GE's cell performance increased 160-fold from 5 to 800 W/ft^2 (from 5.4 to 861 mW/cm^2) (Table 7.1).

In 1984, nonetheless, GE apparently made a strategic decision to leave the fuel cell business. During 1983–1984, it licensed the basic know-how for PEMFCs to Siemens AG in Germany. In the mid-1980s, GE transferred PEMFC and electrolysis technology to UTC-Hamilton Standard.

7.3. CANADIAN GOVERNMENT DECIDES TO FOSTER DOMESTIC PEMFC CAPABILITIES IN THE EARLY 1980s

In the early 1980s, the Government of Canada was thinking about developing domestic

TABLE 7.1 General Electric: Power Densities Attained for PEMFC

Time	Membrane	Power density (W/ft^2)
1959–1961	Phenol sulfonic	5–10
1962–1965	Polystyrene sulfonic	40–60
1966–1967	Polytrifluorostyrene sulfonic	75–80
1968–1970	Nafion experimental	80–100
1971–1980	Nafion production	600–800

Source: Leo J. M. J. Blomen & Michael N. Mugerwa, "Fuel Cell Systems," Springer, 1993.

PEMFC capabilities. Canada clearly believed that PEMFC technology could be further improved and decided to have a Canadian firm improve the cost and performance of PEMFC technology and develop a PEMFC capability to tolerate impure hydrogen as fuel. Sometime in the early 1980s, Ottawa's Department of National Defense (DND) purchased from NASA the PEMFC technology developed by GE that had lain dormant for some time.

7.3.1. Canada's Defense Department Commissions Ballard to Advance GE's PEMFC Technology in 1983

In 1983, the DND awarded a contract to a small start-up firm to develop PEMFCs, according to the then Chairman and CEO of the firm.[16] The DND wanted the small firm to produce a low-cost PEM fuel cell that could run on impure hydrogen produced by reforming a liquid fuel like methanol.[17] The small start-up was the Vancouver-based Ballard Power Systems, which was soon to make history in successfully advancing PEMFC technology and applying it for practical applications.

- Ballard was founded in 1979 under the name Ballard Research to conduct research and development in high-energy lithium batteries.[18] Under the DND contract, Ballard began PEMFC research in 1984.[19]

Ballard quickly demonstrated outstanding technical progress. In 2 years, Ballard succeeded in developing an 8-cell stack that produced 130 W, considerably higher than the 50–100 W specified in the DND contract.[17] By mid-1986, it had succeeded in creating a 12-stack version capable of producing 280 W.

7.3.1.1. *Second Contract Makes Marked Advances toward Practical Applications of PEMFCs*

In view of the successful completion of its first contract, the DND awarded Ballard a second contract to support further development.[17] The contract work also was remarkably successful. Over the next several years, Ballard developed ever more powerful and smaller PEMFC stacks, establishing itself as the world leader in PEMFC technology. In 1985, Ballard demonstrated the ability to run PEMFCs on fuel gas by using a selective oxidation process after the fuel reformer to convert CO to CO_2 external to the cell. Ballard constructed several 72 W demonstration stacks in 1986.[1] In 1988, it made a two-and-one-half to threefold improvement in maximum power density of PEMFCs, from 800 to 2000–2500 W/ft^2

- The program went so well that in 1988 several 2 kW stacks were delivered to military and civilian customers.[1]
- In July 1989, Perry Energy successfully incorporated Ballard's 2 kW power plant into the first commercial fuel cell–powered submarine.[20] The boat was a two-man observation submersible.
- In 1990, Ballard demonstrated an independent methanol-air PEMFC brassboard power plant. The 4 kW, 28 V DC methanol-air power plant was developed for the Canadian DND as a portable field generator.

There were two external factors that likely contributed to Ballard's accomplishments:

- In the 1980s, researchers at Los Alamos and Texas A&M showed that the amount of platinum required for the fuel cell could be reduced by a factor of 10. Their work likely convinced Ballard that low-cost PEM fuel cells were possible.
- Also in the early 1980s, Dow Chemical was experimenting with a sulfonic polymer.[19] In 1987 and 1988, Ballard tested Dow's new membrane, which allowed the cell to produce four times the current compared to that of Nafion at the same cell voltage. Cell

durability of over 10,000 h was also demonstrated.[19] Dow's new membrane was the most significant development since the introduction of Nafion by DuPont in 1968. Dow Chemical's rights to manufacture this membrane were subsequently transferred to DuPont,[19] and DuPont has continued to improve durability and performance of its membranes.

7.3.2. Canada's Energy Department Sponsors Ballard to Develop PEM Fuel Cell Bus in 1990

In 1990, Canada's Federal Department of Energy Mines and Resources, the Canadian Province of British Columbia, and the Federal Department of Energy Mines and Resources implemented a government-industry cost-share program to develop a PEMFC-powered bus. Ballard was chosen as the prime contractor and it started to develop PEMFC stacks up to 5 kW.[1] The bus was to seat 20 passengers, and the Canadian government reportedly invested a total of $4.84 million.[21]

- In 1993, Ballard introduced the world's first PEMFC bus built by New Flyer, which drove on the streets of Vancouver.[17] The 32 ft, low-floor transit bus, *P1*, was powered by 90 kW Ballard stacks.[19]
- Two years later, in June 1995, Ballard introduced its Phase Two *P2* bus, a full-size prototype ZEV powered by a 205 kW fuel cell engine. It had a range of 250 miles before requiring refueling.[19] The P2 bus was based on a 40 ft New Flyer Model 40 1F bus.[22]

7.3.3. Ballard Becomes the Global PEMFC Leader

By the early 1990s, Ballard had become the global PEMFC leader. It was viewed worldwide as having established PEMFCs as the predominant fuel cell technology that would finally deliver the fuel cell's unfulfilled promise. Inspired by Ballard's developments, nearly every major automaker in the world launched a program to build a PEM fuel cell vehicle (FCV).[23] The automotive world was beating a path to Ballard's door.[19] All of the former "big three" US automotive manufacturers—General Motors, Chrysler, and Ford—were early customers of Ballard.

In Europe, Daimler-Benz clearly was the most enthusiastic believer in Ballard's PEMFC capabilities. In 1993, Daimler-Benz[ii] and Ballard signed a 4-year joint venture agreement, to which Daimler committed $35 million.[24] In April 1994, Daimler surprised the world with its first New Electric Car (NECAR) incorporating an improved version of the Ballard stack that was used in the P2 bus.[25] The NECAR, which was subsequently named the ***NECAR 1*** to distinguish itself from succeeding NECARs, was a boxy cargo van, based on the MB-180 van. The van was powered by 12 Ballard fuel cells stacks, which collectively generated 50 kW and produced 60 hp.[17]

- With this vehicle, Daimler proved to the world audience the basic suitability of the fuel cell technology as an electric vehicle propulsion system.[25]

Ballard continued to make progress. In May 1996, Daimler and Ballard demonstrated the

[ii] In 1998, Daimler-Benz of Stuttgart, Germany, merged with Chrysler of the United States and renamed itself Daimler-Chrysler. In 2007, DaimlerChrysler sold Chrysler Division to Cerberus Capital Management of New York, a private equity firm and changed its name to Daimler. This book will use the three names during the corresponding time periods; Daimler-Benz (or Daimler for short) for the period prior to 1998, DaimlerChrysler between 1996 and 2007, and Daimler for the period after 2007.

world's first passenger car with PEMFC drive, the ***NECAR 2***.[25] The six-passenger car, based on the V-Class minivan, had two 25 kW stacks and achieved a maximum speed of 68 mph and a range of 155 miles. NECAR 2's fuel cell system weighed only about 270 kg (600 lbs) and was much smaller in size and volume, about one-third of its predecessor.

- The world felt the era of the fuel cell cars was around the corner.

7.3.3.1. Ballard/Daimler-Benz/Ford Fuel Cell Car Alliance Formed in 1997

Throughout 1997, the world was kept in awe by a series of astonishing developments at Ballard. In April 1997, Daimler said it was investing US$145 million to buy a one-quarter interest in Ballard, and announced it was investing $150 million in a joint venture with Ballard to create a new vehicle fuel cell engine company.[22] Daimler also announced that beginning in 2005, the new company would produce 100,000 fuel cell engines annually. This was a remarkable figure, considering that the company, the world's 15th largest auto manufacturer, made only 700,000 passenger cars a year at that time. Moreover, 8 months later, Ford Motor Company announced that it would join the alliance and invest $420 million in the venture.[26]

- The Ballard/Daimler/Ford alliance established three jointly owned companies: DBB Fuel Cell Engines (later named Xcellsis), Ecostar Electric Drive Systems, and Ballard Automotive.[19]

The alliance would soon grow larger. Daimler and Ford brought in their affiliates. Ford brought in Mazda, Volvo, and Th!nk Nordic, a Norwegian electric car developer which it acquired in 1999. Daimler would bring in Chrysler, then US's third largest automaker, with which it merged in 1998.[27] The automakers would use Ballard fuel cell stacks to power their fuel cell cars, creating a powerful international fuel cell car alliance.

- Meanwhile, Ballard's first 110,000 ft^2 manufacturing plant for large-scale production of portable and automotive fuel cells was opened in Burnaby, British Columbia, in December 2000.[19]

At the end of 2001, however, the alliance announced that, in a transaction worth $547 million, Ballard had entirely taken over Xcellsis and Ecostar. At the same time, DaimlerChrysler and Ford raised their stakes in Ballard.[19]

7.4. A GLOBAL FUEL CELL RACE BEGINS

By the early 2000s, the global fuel cell race had begun in earnest. The Daimler/Ford/Ballard alliance virtually dominated media attention by their spectacular fuel cell development activities. They undoubtedly challenged and inspired others to enter the race. While it is not clear exactly when the global fuel cell race started, the establishment of the alliance between Daimler, Ford, and Ballard likely marked its beginning, and it grew in intensity until 2005. In the first few years, a large number of major automakers and bus manufacturers began FCV development and announced that they would commercialize their cars and buses in a few years. They made aggressive, large-scale investments in PEMFC technology R&D, energetically built up alliances, and unveiled their concept cars and prototypes with fanfare, creating major media around the world.

Moreover, the governments of major industrialized countries, including Japan, the United States, and European countries, started national fuel cell development programs to develop not only PEMFCs for transport applications but also stationary applications. They started anew or reinforced their existing measures to subsidize fuel cell development and nurture the business environment for the fuel cell industry. The governments and the fuel cell industries thus worked in tandem for a common goal of achieving global competitiveness and economic growth. The broader, technically informed public was also excited with the belief that PEMFC technology had come of age at last and they would be able to drive their FCVs soon. The enthusiasm by major global automakers, major industrialized governments, and the general public all propelled global excitement over FCVs, intensifying the global fuel cell race.

7.4.1. Major Global Automakers and Bus Manufacturers Join the Race

In the early 2000, virtually every major automaker entered the fuel cell race. They rapidly unveiled increasingly sophisticated FCVs and buses and demonstrated them on public roads. Some automakers were announcing their near-term market entry dates. Others were somewhat more modest in their goals; they perhaps had less investment capital or more limited technical capabilities. Nonetheless, the participants seemed enthusiastic and committed.

There also were a large number of fuel cell bus development projects worldwide. Transit buses were viewed as well suited for fuel cell applications.[28] They had fixed routes, centralized fueling and maintenance infrastructures, and dedicated maintenance personnel. They were large enough to install fuel cell systems and hydrogen fuel tanks. Diesel transit buses were noisy and polluting, providing an opportunity to show off the advantages of fuel cells. There was even speculation that fuel cell buses might enter the commercial market much earlier than fuel cell passenger vehicles.

The following sections are a compendium of the activities of the various competitors in the global fuel cell race. They are listed in alphabetical order (Table 7.2).

7.4.1.1. Beijing LN Green Power Company (China)

The Beijing LN Green Power Company (Beijing LN Power Sources), established as the LN Research Institute in 1998, worked to transfer their experience in electric vehicles to fuel cells.[29] Within a few years, LN Green Power unveiled one bus, one taxi, and two passenger cars. LN Green Power rapidly achieved international recognition for its robust developmental activities.

- In 2001, Beijing LN Green unveiled a prototype fuel cell bus ***Green Power FC Bus China.***[30] The 6 m long bus was powered by a 18 kW Lu Neng PEMFC and 36 kW AC induction motor and fueled by compressed hydrogen (25 MPa).[31] Tsinghua University was reported to have worked with LN Green Power to develop the fuel cell bus.[32] The bus had a top speed of 45 mph and a driving range of 102 miles.
- In 2001, LN Green Power unveiled a ***PEM fuel cell powered taxi*** with a range of 93 miles and a top speed of around 47 mph.[29] LN Green Power developed the fuel cell taxi in collaboration with the Electric Vehicle R&D Centre of the Beijing Institute of Technology (part of the Chinese Academy of Science).
- Also in 2001, LN Green Power presented the ***Green Power FC Passenger Car.***[33] The FCV was developed in collaboration with the Tsinghua University and the Beijing Institute of Technology. The FCV was powered by six 5 kW Chinese PEMFCs with an electric motor and ran on compressed hydrogen. It had a top speed of 14 mph.
- In 2002, LN Green Power introduced the ***Green Power FC Kombi.***[34] The FCV was powered by an 18 kW Chinese PEMFC and an electric motor and ran on compressed hydrogen. No detail is available.

The world recognized that China's growing fuel cell industry clearly had become a potentially powerful competitor in the global fuel cell race.

7.4.1.2. Daihatsu

Daihatsu, Japan's leader in electric vehicles (EV), became interested in fuel cell technology

Company	Prototype/Demonstration model (fuel, fuel cell size, fuel cell maker if not made inhouse) year	Planned commercialization date
Ballard	Ballard P1/New Flyer Fuel Cell Transit Bus (hydrogen) 1993 Ballard P2/New Flyer Heavy-Duty Fuel Cell Transit Bus (hydrogen) 1995 Ballard P3/New Flyer Fuel Cell Transit Bus in Chicago (hydrogen) 1997 Ballard P3/New Flyer British Columbia Transport Bus (hydrogen,) 1998 Fraunhofer and CWA CCN AUTOTRAM (hydrogen) 2005 University of Delaware Ebus Plug-in FC Hybrid Bus (hydrogen, fuel cell as range extender) 2007 University of Texas Ebus Plug-in FC Hybrid Bus (hydrogen, fuel cell as range extender) 2007 UNDP/GEF Tuttotrasporti/Ballard FC Bus for Sao Paulo (hydrogen) 2009 University of Delaware Ebus Plug-in FC Hybrid Bus (hydrogen, fuel cell as range extender) 2009 CHIC Project: ISE/Wrightbus FC Bus London (hydrogen, Ballard) 2010 APTS Phileas Fuel Cell Hybrid (hydrogen, Ballard) 2010	2002
BMW	BMW 1 Series Fuel Cell Hybrid (gasoline, UTC fuel cell range extender) 2010	
China	Buses: Beijing LN Green Power Company (Beijing LN Power Sources): Green Power FC Bus China (hydrogen, Lu Neng) 2001 Dalian Institute of Chemistry and Physics (DICP): Dalian University Fuel Cell Bus China (hydrogen) 2001 Dalian University Fuel Cell City Bus (hydrogen) 2002 Dalian University FC-Bus 2 (hydrogen) 2003 Foton Motor: Foton Motor FC Low Floor City Bus (hydrogen, fuel cell maker unknown) 2008 Higer Bus: Higer U90B fuel cell Shanghai (hydrogen, Shanghai Shen Li High Tech Company (Shen Li) 2006 SAIC Shanghai Automotive Industry Corporation (SAIC): SAIC Fuel Cell Bus Expo 2010 (Hydrogen, Ballard) 2010 Shanghai Shen Li High Tech Company (Shen Li): Shen Li SLFC-Tourist Car 2 (hydrogen) 2004 Shen Li Shanghai Shenwo Fuel Cell Bus (hydrogen) 2007 Shen Li Shen-Kai No.1 Fuel Cell Bus (hydrogen) 2007 Shen Li No.1 Fuel Cell Bus (hydrogen) 2007 Shen Li No.2 Fuel Cell Bus (hydrogen) 2008 Tsinghua University: Qing Neng I (hydrogen, Shen Li) 2002 Tsinghua University & DICP FCB 60 kW (hydrogen, Shen Li) 2003 Tsinghua University Fuel Cell City Bus 863 (hydrogen, Shen Li) 2004 Tsinghua University FC Low Floor City Bus (hydrogen, Shen Li) 2006	

(Continued)

TABLE 7.2 Major PEM Fuel Cell Vehicle Development* (*cont'd*)

Company	Prototype/Demonstration model (fuel, fuel cell size, fuel cell maker if not made inhouse) year	Planned commercialization date
	Tsinghua University FC Low Floor City Bus 2 Generation (hydrogen, fuel cell manufacture unknown) 2008Tsinghua University + Nanyang Technological University Fuel Cell Bus (hydrogen, fuel cell developer unknown) 2010	
	Wuhan University: Wuhan University FC Midibus (hydrogen) 2006	
	Yutong Bus Company: Yutong Low Floor Fuel Cell Range Extender EV City Bus (hydrogen, Shen Li) 2010	
	Cars: Beijing LN Green Power Company (Beijing LN Power Sources): Fuel Cell Taxi (no details available) 2001 Green Power FC Passenger Car (hydrogen, fuel cell developer unknown) 2001 Green Power FC Kombi (hydrogen, fuel cell developer unknown) 2002	
	Chang'an Automobile: Chang'an Z-SHINE FCV (hydrogen, Shen Li) 2010	
	Chery Automobile Company: Chery Eastar FCV (hydrogen, probably Chen-Li) 2007 Chery Eastar FCEV (hydrogen, Shen Li) 2010	
	Dongfeng Motor: Wuhan University Chu Tian 1 (hydrogen, Wuhan University) 2004	
	FAW Group: FAW Besturn B70 FCV (hydrogen, Shen Li) 2010	
	Geely Automobile: Geely MK Fuel Cell (hydrogen, fuel cell developer unknown) 2008	
	Pan Asia Technical Automotive Center Co: Phoenix Fuel Cell Minivan (hydrogen, GM) 2001	
	SAIC: SAIC Roewe 750 FC (hydrogen, fuel cell developer unknown) 2007 Shanghai Fuel Cell Car (hydrogen, Shen Li) 2007 SAIC Roewe 750 Plug-in FC (hydrogen, fuel cell developer unknown) 2008 SAIC Roewe 750 GM-Fuel Cell (hydrogen, GM) 2009 Shanghai Brand Plug-in Fuel Cell Car (hydrogen) 2011	2015

	Shanghai Volkswagen: Tongji University Fuel Cell PASSAT Lingyu (hydrogen, Shen Li) 2007 Tongji University Fuel Cell PASSAT New (hydrogen, Shen Li) 2010 Shen Li: Shen Li Technical Company FC SUV (hydrogen) 2008 Tongji University: Chunhui (hydrogen, fuel cell developer unknown) 2002 Tongji University Chao Yue I (hydrogen, Shen Li) 2003 Tongji University Chao Yue II (hydrogen, Shen Li) 2004 Tongji University Chao Yue III • Tongji University Chao Yue III S3000 (hydrogen, Shen Li) 2005 • Tongji University Chao Yue III MPV (hydrogen and gasoline, Shen Li) 2005 • Tongji University Chao Yue III Eastar (hydrogen, Shen Li) 2005	
Chrysler	EcoVoyager (hydrogen, fuel cell manufacturer unknown, used as a range extender) 2008	2003, 2004
Daimler (DaimlerChrysler)	NECAR 1 (hydrogen, Ballard) 1994 NECAR 2 (hydrogen, Ballard) 1996 NECAR 3 (methanol, Ballard) 1997 DaimlerChrysler Jeep Commander (hydrogen, Ballard) 1998 NECAR 4 (hydrogen, Ballard) 1999 NECAR 5 (methanol, Ballard) 2000 NECAR 4 Advanced (hydrogen, Ballard) 2000 DaimlerChrysler Jeep Commander 2 (methanol, Ballard) 2001 NECAR 5.2 (methanol, Ballard) 2001 DaimlerChrysler Sprinter (hydrogen, Ballard) 2001 DaimlerChrysler Natrium (sodium borohydride (NaBH4), Ballard) 2001 Mercedes-Benz Mercedes F-Cell (hydrogen, Ballard) 2002 Mercedes-Benz Sprinter UPS (hydrogen, Ballard) 2003 Jeep Treo (chemical hydride, Ballard) 2003 DaimlerChrysler Mercedes F600 Hygenius (hydrogen, Ballard) 2005 Daimler B-Class F-Cell Tourer (hydrogen, Ballard) 2005 Mercedes-Benz Mojave Runner (hydrogen, fuel cell manufacturer unknown, fuel cell used as a power generator) 2006 Daimler F-Cell Roadster (hydrogen, Ballard) 2009 Mercedes BlueZERO F-Cell (hydrogen, Ballard) 2009 Mercedes-Benz B-Class F-Cell (hydrogen, Ballard) 2009	2004, 2009, 2012–2013, 2015, 2014, 2002, 2015

(Continued)

TABLE 7.2 Major PEM Fuel Cell Vehicle Development* (*cont'd*)

Company	Prototype/Demonstration model (fuel, fuel cell size, fuel cell maker if not made inhouse) year	Planned commercialization date
	Mercedes-Benz F800 Style (hydrogen, Ballard) 2010 • The F800 Style with a plug-in V6 gasoline and electric engine hybrid • The F800 Style with a 100 kW/136 hp fuel cell Mercedes-Benz B-Class F-Cell (hydrogen, Ballard) 2011 Benz F125! F-CELL plug-in Hybrid (hydrogen, Ballard) 2011	
	Daimler—EvoBus: NEBUS (hydrogen, Ballard) 1997 ZEBUS (hydrogen, Ballard) 1999 Citaro Bus (hydrogen, Ballard) 2001 DaimlerChrysler Citaro Bus (hydrogen, Ballard) 2002 Mercedes Citaro FuelCELL Hybrid (hydrogen, Ballard) 2009	
Fiat	Fiat: Seicento Elettra H2 Fuel Cell Car (hydrogen, Nuvera) 2001 Fiat Seicento (hydrogen, Nuvera) 2003 Fiat Panda (hydrogen, Nuvera) 2004 Fiat Panda Hydrogen Concept Car (hydrogen, Nuvera) 2006 Fiat Phyllis Electric and Fuel Cell (hydrogen, fuel cell manufacturer unknown, fuel cell perhaps used as a range extender) 2008 Fiat Panda HyTRAN (hydrogen, Nuvera) 2009 Serenergy Fiat Scudo Fuel Cell (methanol, Serenergy fuel cell as a range extender) 2009 Fiat Panda Hybrid Fuel Cell/Battery-Powered Vehicle (hydrogen, Nuvera) 2010	
	Iris/Iveco Bus: Irisbus CityClass Hybrid Bus for Turin (hydrogen, UTC) 2001 Irisbus CityClass Hybrid Bus for Madrid (hydrogen, UTC) 2003 Irisbus CityCell Bus (hydrogen, UTC) 2003 Hybrid FC Europolis (hydrogen, UTC) 2006	
Ford	P2000 FC EV (hydrogen, Ballard) 1999 Ford Focus FC5 (methanol, Ballard) 2000 Ford Th!nk FCV (methanol, Ballard) 2000 Th!nk FC5 (methanol, Ballard) 2000 • Ford Th!nk FCV (methanol, Ballard) 2000 • Ford Th!nk FCV (hydrogen, Ballard) 2001 Ford FCV Hybrid (hydrogen, Ballard) 2002 Ford Explorer Hydrogen SUV (hydrogen, Ballard) 2006 Ford Edge with HySeries Drive Plug-In Hybrid (hydrogen, Ballard fuel cell as onboard recharger) 2007 Ford Airstream Concept (hydrogen, Ballard fuel cell as onboard recharger) 2007 Ford Fusion Hydrogen 999 (hydrogen, Ballard) 2007	2005, 2015

General Motors	General Motors: passenger car FC EV 1 (methanol, fuel cell manufacturer unknown) 1998 Opel GM Zafira (methanol, Ballard) 1998 Opel/GM Passenger Car Sintra (hydrogen, fuel cell manufacturer unknown) 1998 GM Passenger Car Precept (hydrogen) 2000 Opel/GM Passenger Car HydroGen1 (hydrogen) 2000 GM Chevrolet S-10 FCEV (gasoline) 2001 GM HydroGen3 (liquid hydrogen) 2001 • Opel HydroGen3 compressed 700 (hydrogen) 2002) • Opel HydroGen3 Liquid (hydrogen) 2004 GM Shanghai—Phoenix (hydrogen) 2001 GM AUTOnomy concept car (hydrogen) 2002 GM HyWire (hydrogen) 2002 Chevrolet Sequel (hydrogen) 2005 GM Chevrolet Equinox FCV (hydrogen, 115 kW) 2006 • GM HydroGen4 (hydrogen) 2008 Chevrolet Volt (hydrogen) 2007 Cadillac Provoq (hydrogen) 2008	2004, 2005, 2010, 2011–2012, 2015, 2025
Gillig	Gillig Ballard Bus (hydrogen, Ballard) 2005	
Honda	FCX-V1 (metal hydride; Ballard fuel cells) 1999 FCX-V2 (methanol), 1999 FCX-V3 (hydrogen; Ballard fuel cells) 2001 • Some of the FCX-V3 equipped with Honda fuel cells were put to road tests FCX-V4 (hydrogen; Ballard fuel cells) 2001 Honda FCX (hydrogen, Ballard) 2002 • Second-Generation FCX (hydrogen) 2004 Kiwami FCV Concept Car (hydrogen) 2003 FCX Concept (hydrogen) 2006 • Second Generation FCX Concept (hydrogen) 2007 Honda FCX Clarity (hydrogen) 2007 • Honda FCX Clarity Japan Model (hydrogen) 2008	2003, 2002, 2008, 2009–2010, 2020
Hydrogenics	Cars: GEM(TM) FC "neighborhood" vehicle (hydrogen) 2004 Buses: Hawaii Hickham Air Force/University of Hawaii Fuel Cell Hybrid Bus (hydrogen, fuel cell used as a range extender) 2004 Hydrogenics FC Hybrid Tecnobus Midibus owned by Hydrogenics and used for exhibition use in Europe (hydrogen, fuel cell used as a range extender) 2005 Winnipeg ISE/Siemens/New Flyer FC Bus (hydrogen) 2006 Ostseebus Heliocentris/NEOPLAN FC Midibus in Barth, Germany (hydrogen, fuel cell as a range extender) 2006 Rheinbahn Rheinische Bahngesellschaft Tecnobus Gulliver Fuel Cell Hybrid Midibus (hydrogen, fuel	

(Continued)

TABLE 7.2 Major PEM Fuel Cell Vehicle Development* (*cont'd*)

Company	Prototype/Demonstration model (fuel, fuel cell size, fuel cell maker if not made inhouse) year	Planned commercialization date
	cell used as a range extender) 2006 HyChain–Minitran/NRW/Tecnobus Gulliver Minibus (hydrogen, fuel cell used as a range extender) 2006 University of Rome "La Sapienza" Tecnobus FC Hybrid (hydrogen, fuel cell used as a range extender) 2006 University of Glamorgan IVECO Daily FC Hybrid (hydrogen, fuel cell as a range extender) 2007 ECObus for Arizona Public Service (APS) Hydrogen Park (hydrogen) 2007 City of Barth (Mecklenburg–West Pomerania) Hybrid Midibus (hydrogen, fuel cell as a range extender) 2008 Expo Zaragoza Tecnobus Gulliver FC Hybrid Bus (hydrogen, fuel cell used as a range extender) 2008 Vestische Strassenbahnen/Tecnobus Electric FC Midibus (hydrogen, fuel cell used as a range extender) 2009 University Clinic–Hamburg Eppendorf/Tecnobus FC Hybrid Bus (hydrogen, fuel cell used as a range extender) 2009 Hydrogenics Rampini FC Midibus (hydrogen, fuel cell used as a range extender) 2010	
Hyundai-Kia	Hyundai Cars: Santa Fe FCEV SUV (hydrogen, UTC) 2000 Santa Fe FCHEV (hydrogen, UTC) 2002, 2001 Hyundai Tucson FCEV (hydrogen, UTC) 2004 Hyundai i-blue (hydrogen) 2007 Hyundai Tucson ix35 FCEV (hydrogen) 2010 Hyundai Blue2 (hydrogen) 2011 Hyundai Buses: Hyundai FC bus (hydrogen) 2006 Hyundai FC-Bus Second Generation (hydrogen) 2009 Kia: Kia Sportage FCEV (hydrogen, UTC) 2004 Kia Borrego FCEV (Hydrogen, Hyundai) 2008	2004, 2010, 2012, 2015 2012
MAN Truck and Bus (Germany)	MAN Fuel Cell City Bus (hydrogen, Siemens) 2000 ARGEMUC–MAN fuel cell bus (hydrogen, Ballard) 2004 Hybrid APU for H2 ICE Bus (hydrogen, Hydrogenics fuel cell used as an APU) 2008	
Mazda	Demio FCEV (metal hydride) 1997 Demio FCEV (metal hydride, Ballard) 1999 Premacy FCEV (methanol, Ballard) 2001	

Mitsubishi Motors	Mitsubishi Fuel Cell Concept Vehicle (methanol, Mitsubishi Heavy Industries) 1999 Mitsubishi Space Liner FCV (hydrogen, Ballard) 2001 Mitsubishi FCV Grandis (hydrogen, Ballard) 2003	2005, 2010–2020
NEOPLAN	NEOPLAN Midi FC Bus (hydrogen, Nuvera) 1999 Proton Motors Bayern Bus II (hydrogen, Proton Motor) 2000	
New Flyer	ISE Corporation/Ballard/New Flyer FC Bus for the winter Olympic, Whistler, Canada (hydrogen, Ballard) 2009 SunLine Transit/CARB/FTA National FC Bus Program (hydrogen, UTC) 2010	
Nissan	R'NESSA recreational vehicle (methanol, Ballard) 1999 Xterra FCV (hydrogen, Ballard) 2000 • Xterra FCV (methanol, Ballard) 2000 X-Trail '02 (hydrogen, UTC) 2002 X-Trail '03 (hydrogen, UTC) 2003 Effis FCV (hydrogen, UTC) 2003 Nissan X-Trail FCV (hydrogen) 2004 X-Trail '05 model (hydrogen) 2005 Nissan X-Trail FCV Advanced (hydrogen) 2009	2005, 2002,
Nova Bus	Georgetown University Methanol Transit Bus (methanol, Ballard) 2000	
Proterra	Proterra HFC35 Plug-in Hybrid FC Bus/FTA FC Bus for Burbank, CA (hydrogen, Hydrogenics fuel cell used as a range extender) 2008 Proterra Plug-in Hybrid FC Bus for demonstration in Columbia, SC, Austin, TX, Hartford, CT, and Victoria, BC, Canada (hydrogen, Hydrogenics fuel cell used as a range extender) 2009 Proterra Fuel Cell Hybrid Bus for DoD (hydrogen, Hydrogenics fuel cell used as a range extender) 2009 Burbank CARB AFIP/Proterra FC Battery Plug-in Hybrid (hydrogen, Hydrogenics fuel cell used as a range extender) 2010	
PSA Peugeot Citroen	Peugeot Hydro-Gen (hydrogen, Nuvera) 2001 Peugeot Taxi PAC (hydrogen, H-Power) 2001 Peugeot H2O Fire Prevention Vehicle (chemical hydride, H-Power fuel cell as auxiliary power unit) 2002 Quark (hydrogen, CEA fuel cell as a range extender) 2004 Peugeot 207 ePURE (hydrogen, CEA fuel cell as a range extender) 2006 Peugeot Flux (hydrogen, fuel cell manufacturer unknown) 2007 H2Origin Light Utility Urban Vehicle (hydrogen, Intelligent Energy fuel cell as range extender) 2008 FiSyPAC Demonstrator (hydrogen, CEA fuel cell as range extender) 2009	2010–2020
Renault	FEVER Laguna FCV (hydrogen, Nuvera) 1997 Renault Scenic ZEV H2 (hydrogen, Nissan) 2008	
Scania	Scania, Air Liquide FC Midi City Bus (hydrogen, Nuvera) 2001	

(*Continued*)

TABLE 7.2 Major PEM Fuel Cell Vehicle Development* (*cont'd*)

Company	Prototype/Demonstration model (fuel, fuel cell size, fuel cell maker if not made inhouse) year	Planned commercialization date
Suzuki	Wagon R-FCV 35 Mpa (hydrogen, GM) 2003 MR Wagon-FCV 35 Mpa (hydrogen, GM) 2003 Mobile Terrace (hydrogen, GM) 2003 • Improved MR Wagon-FCV (hydrogen, GM) 2004 Suzuki Ionis (hydrogen, GM) 2005 Suzuki SX4-FCV (hydrogen, GM) 2008 Suzuki Crosscage Motorcycle (hydrogen, Intelligent Energy) 2007 Fuel Cell Senior Car MIO (methanol, Intelligent Energy) 2008 Burgman Fuel-Cell Scooter (hydrogen; Intelligent Energy) 2009	2010
Thor Industries	Thor ThunderPower Bus (hydrogen, UTC) 2001 AC Transit Hyroad FC Bus (hydrogen, UTC) 2003	
Toyota Motor	Toyota: RAV4 FCEV SUV (metal hydride, fuel cell manufacturer unknown) 1996 RAV4 FCEV (methanol, fuel cell manufacturer unknown) 1997 FCHV-3 (metal hydride, fuel cell manufacturer unknown) 2001 FCHV-4 (hydrogen, fuel cell manufacturer unknown) 2001 FCHV-5 (gasoline, fuel cell manufacturer unknown) 2001 Toyota FCHV 2002 (hydrogen) 2002 Fine-N FCV (hydrogen) 2003 Fine-X FCV (hydrogen) 2005 FCHV-Adv (hydrogen) 2008 Toyota FCV-R (hydrogen) 2011 Daihatsu: Daihatsu Hijet FCEV (methanol, fuel cell manufacturer unknown) 1972 Move EV-FC (methanol, Toyota) 1999 Move FCV-K-2 (hydrogen, Toyota) 2001 Tanto FCHV (hydrogen, Toyota 2005 Hino: Toyota-Hino FCHV-BUS1 (hydrogen, Toyota) 2001 Toyota-Hino FCHV-BUS2 (hydrogen, Toyota) 2002 • Toyota-Hino FCHV-BUS2 Improved (hydrogen, Toyota) 2005 Toyota–Hino FCHV—BUS (hydrogen, Toyota) 2006 Toyota–Hino FCHV–Bus New (hydrogen, Toyota) 2010	2004, 2003, 2002, 2015

Van Hool	AC Transit Hyroad Fuel Cell Hybrid (hydrogen, UTC) 2006 SunLine Transit/CARB/DOE Technology Validation (hydrogen, UTC) 2006 UTC/Van Hool Hybrid FC Bus (hydrogen, UTC) 2007 Marin Transit/Van Hool A300L FC Bus (hydrogen, UTC) 2009 AC Transit ZEBA Advanced FC Hybrid Bus (hydrogen, UTC) 2010 Connecticut Transit/DOE Technology Validation Program (hydrogen, UTC) 2010
Volkswagen	Volkswagen: Volkswagen CAPRI (methanol, Ballard) 2000 Volkswagen Bora HyMotion (hydrogen, UTC) 2000 Volkswagen Bora HyPower (hydrogen, Paul Scherrer Institute) 2002 Volkswagen Touran HyMotion (hydrogen, Ballard) 2004 Volkswagen Space UP! Blue ( ) 2007 Volkswagen Caddy Maxi HyMotion (hydrogen) 2009 New Volkswagen HY.POWER Bora (hydrogen, Paul Scherrer Institute) 2010 Audi: Audi A2 (hydrogen, Ballard) 2004 Audi Q5 HFC (hydrogen, fuel cell manufacturer unknown) 2009 2012 Audi Q5 Hybrid Fuel Cell Concept (hydrogen, Volkswagen) 2011
Miscellaneous Other Car and Bus Demonstrations	Buses: Ansaldo Ricerche CityBus (hydrogen, Nuvera) 1997 European Regional Development Fund/Czech Republic Skoda TriHyBus (hydrogen, Proton Motor) 2009 Connecticut Transit Nutmeg/FTA National FC Bus Program (hydrogen, UTC) 2010 HAN Automotive HyMove Bus (hydrogen, NedStack) 2010 COPPE-UFRJ Onibus a hidrogenio (hydrogen, COPPE UFRJ) 2010 Cars: Billings LaserCel 1 (hydrogen, Billings) 1991 Energy Partners Inc.: passenger car "Green Car" (hydrogen, perhaps Energy Partners) 1993 Kewet EL-Jet -NEV FC (hydrogen, fuel cell manufacturer unknown) 1998 New Jersey—Venturer (hydrogen, H Power) 1999 Proton Motor VW E2G (hydrogen, Proton Motor) 1999 Hypercar Revolution (hydrogen, fuel cell manufacturer unknown) 2000 ESORO HyCar (hydrogen, ESORO HyCar) 2001 Virginia Tech University—ZEburban (hydrogen, Honeywell) 2001 Millennium Cell New Jersey Genesis (hydrogen, H Power) 2001 DMFC Go-Cart (methanol, fuel cell manufacturer unknown) 2001 Clean Urban Vehicle (CUV) (hydrogen, Anuvu) 2001 Virginia Tech University—Magellan (Future Truck 3) (hydrogen, Honeywell) 2002 Pac-CAR I (hydrogen, Paul Scherrer Institute) 2003 HYSUN3000 (hydrogen, Ballard) 2004

(*Continued*)

TABLE 7.2 Major PEM Fuel Cell Vehicle Development* (*cont'd*)

Company	Prototype/Demonstration model (fuel, fuel cell size, fuel cell maker if not made inhouse) year	Planned commercialization date
	Quantum "AMV" (Alternative Mobility Vehicle)—"Aggressor"—for the U.S. Army (hydrogen, Hydrogenics) 2004	
	Michelin Hy-Light (Hydrogen, Paul Scherrer Institute) 2004	
	Formula Zero go-kart (hydrogen, Hydrogenics) 2005	
	ETH PAC-CAR II (hydrogen, Paul Scherrer Institute) 2005	
	Purolator FC Hybrid van for Canada's Transportation FC Alliance (hydrogen, Hydrogenics) 2005	
	Agnis's REAL FCV (hydrogen, Agni) 2005	
	REVA FCV (hydrogen, Hydrogenics) 2005	
	Th!nk's hydrogen city car (hydrogen, Hydrogenics) 2005	
	Microcab H4 (hydrogen, Intelligent Energy) 2005	
	Hywet (methanol, fuel cell manufacturer unknown) 2007	
	Hochschule Esslingen Hydro-Smart (hydrogen, Ballard) 2008	
	Melbourne TAXI (hydrogen, fuel cell manufacturer unknown) 2009	
	Riversimple (hydrogen, Horizon Fuel Cell) 2009	
	Lotus Engineering Zero Emissions Black Cab (hydrogen, Intelligent Energy) 2010	
	MicroCab H2EV (hydrogen, Serenergy) 2011	

* *Major global automakers and bus manufacturers are listed individually; Chinese automakers, however, are grouped together in the table because they appear to follow a set of unified strategies, plans, projects, and fuel cell technologies set by the Chinese government, instead of their own independent policy and business decisions.*

relatively early. In April 1971, Daihatsu engaged in FCV development as a participant in the "Light Vehicle EV" project, which was part of the large-scale EV research and development project sponsored by Japan's Ministry of Economy, Trade, and Industry (METI).[35] In 1972, Daihatsu developed Japan's first FCEV, based on the Daihatsu Hijet light truck, in cooperation with the Osaka National Research Institute (now AIST Kansai, part of the National Institute of Advanced Industrial Science and Technology (AIST)).[36]

In October 1999, Daihatsu unveiled the ***Move EV-FC*** at the Tokyo Motor Show. The four-passenger hybrid FCV used a methanol reformer and nickel–metal hydride (NiMH) batteries.[37] The vehicle was based on Daihatsu's popular light car Move model and was the world's first minicar, light car (Japanese categorization), or city car (European categorization) powered by fuel cells.[38] The fuel cell stack and the methanol reformer were developed by Daihatsu, in cooperation with the AIST Kansai.[39]

In October 2001, Daihatsu unveiled the ***Move FCV-K-2*** at the Tokyo Motor Show.[40] The Move FCV-K-2, developed jointly with Toyota (Daihatsu's parent company since 1998 and a technical collaborator in fuel cell technology since 1999) was a refined version of the Move EV-FC shown in 1999. The FCV integrated a 30 kW Toyota fuel cell stack in a hybrid configuration with Daihatsu's NiMH batteries. The entire fuel cell system, including fuel cells, high-pressure hydrogen tank, and other fuel supply components, was incorporated into a single cartridge-type frame. This modular system was positioned under the rear floor panel to leave enough room for four passengers in a minicar.[37] The Move FCV-K-2 had a maximum speed of 65 mph and a driving range of 75 miles.[40]

- Daihatsu was ambitious enough to announce that it would commercialize FCVs "as soon as possible."[28] Daihatsu likely was planning for market entry in 2004, focusing on the minicar market.

7.4.1.3. *Daimler-Benz (DaimlerChrysler)*[iii]

Daimler-Benz (Daimler for short), established by the merger of Daimler Motoren and Benz in 1926, is the manufacturer of well-known Mercedes-Benz. Daimler began a serious assessment of fuel cell technologies in the 1990, as part of an evaluation of hydrogen as an alternative to conventional internal combustion engines.[28]

- Shortly after establishing its joint venture with Ballard, Daimler in 1994 unveiled its first FCV, the ***NECAR 1 (New Electric Car).*** The fuel cell van was powered by a 50 kW Ballard fuel cell fueled by compressed hydrogen. Daimler testdrove the NECAR 1 for 1250 miles, and its performance convinced the company that fuel cells deserved serious research investment.[28] Daimler set up a separate research facility in 1996, called Fuel Cell House, and focused on methanol as the hydrogen carrier.

With the formation of the Ballard/Daimler/Ford/Mazda alliance in 1997, Daimler took a 25-percent equity stake in Ballard Power Systems and a 67-percent stake in a joint venture with Ballard called Daimler-Benz–Ballard Fuel Cell Engines.[28]

- Just prior to forming its alliance with Ballard, Daimler had already presented the world with two impressive fuel cell passenger vehicles, the ***NECAR 1 and 2***, establishing its reputation as the world's foremost fuel cell automaker.

[iii] Daimler developed fuel cell buses at EvoBus, part of Daimler. EvoBus' fuel cell bus development activities will be discussed in a separate section in this book.

In September 1997, Daimler unveiled the ***NECAR 3*** in Frankfurt, which demonstrated for the first time that an FCV could be powered by liquid methanol. NECAR 3 used an onboard methanol reformer and a 50 kW Ballard fuel cell system.[272] NECAR 3 had a maximum speed of 75 mph, delivering 90 percent of its maximum output in just 2 s, the power equivalent of an internal combustion engine.[25] NECAR 3 had two PEMFC stacks, each with 150 fuel cells. A 10-gallon tank of methanol provided a range of 186 miles.

In March 1999, DaimlerChrysler (Daimler) demonstrated the hydrogen-fueled ***NECAR 4*** in Washington, D.C. [25] It had space for five passengers plus luggage, a driving range of over 280 miles, and a top speed of 90 mph. The two PEMFC stacks in the NECAR 4 each consisted of 160 individual, tightly packed fuel cells that attained an output of 70 kW, about 40 percent more than its predecessor, the NECAR 3. The stacks fitted together into a suitcase-sized box under the floor of the compact car. Cold liquid hydrogen (−423°F) was located in the rear in a cylindrical tank with 26.5 gallons or 11 lbs capacity.

- In November 2000, the ***NECAR 4a*** ran on pressurized hydrogen. The optimized electric motor generated 55 kW/75 hp.[41]

In October 2000, Daimler unveiled the ***Jeep Commander 2*** methanol FCV at George Washington's northernmost farm on the Potomac River.[42] The Jeep Commander 2 was the first drivable, methanol-powered vehicle in the North America.[42] It was virtually pollution-free and achieved double the fuel efficiency of a conventional SUV.[43]

- The Jeep Commander 2 was developed by Daimler's Chrysler Division. Chrysler's FCV program prior to the merger with Daimler was relatively modest.[28] Its pre-merger FCV development was undertaken in the context of the Partnership for a New Generation of Vehicles (PNGV), funded by the US federal government.

In November 2000, Daimler presented the ***NECAR 5*** in Berlin.[25] Based on the Mercedes-Benz A-Class, NECAR 5 was viewed as the world's most advanced fuel cell car. Hydrogen fuel was supplied from methanol with the aid of an onboard reformer, which was half the size used in the NECAR 3. The vehicle was 660 lbs lighter than NECAR 3. NECAR 5 housed the entire fuel cell system, including the reformer, in the vehicle floor using a sandwich construction design. The car's interior was entirely set aside for passengers and luggage. Top speed was over 90 mph and the 12-gallon tank enabled the car to cover a distance of more than 250 miles before refueling.[25]

In July 2001, Daimler introduced the ***Mercedes-Benz Sprinter Van*** with fuel cell drive.[41] The Sprinter Van was powered by an 85 kW Ballard fuel cell and an 80 kW electric motor. It was fueled by pressurized hydrogen (at 350 bar), had a top speed of 80 mph, and a driving range of 155 miles.[44]

- The Sprinter Van was tested in everyday operations in collaboration with the Hermes delivery service company.[41] The prototype was initially deployed to the greater Stuttgart area, and later demonstrated in Hamburg.
- In May 2003, Daimler and the U.S. Environmental Protection Agency, and UPS (United Parcel Service) launched a demonstration project for practical testing of fuel cell vehicles for daily commercial delivery services.[44]

In December 2001, Daimler unveiled the ***Town & Country Natrium,*** a minivan with a fuel cell battery hybrid configuration incorporating a sodium borohydride ($NaBH_4$) hydrogen storage system developed by Millennium Cell.[45] The fuel and the fuel cell system gave Chrysler's Natrium a range of

300 miles, comparable to a gasoline-powered vehicle and significantly greater than any FCV to date.[46]

After having developed five generations of Mercedes-Benz research NECAR models, Daimler was now ready to unveil a new fuel cell car suitable for day-to-day operations and extensive field testing. In October 2002, DaimlerChrysler unveiled the Mercedes-Benz "*F-Cell*" A-Class, the world's first small-scale FCV production series (60 cars).[47] The Mercedes-Benz A-Class "F-Cell" marked a significant milestone in fuel cell car development. The entire fuel cell drive system was sandwiched in the floor of the long-wheelbase Mercedes-Benz A-Class "F-Cell". Two fuel tanks contained enough compressed hydrogen (at 350 bar) to give the "F-Cell" a 93 mile range. The electric motor, which generated 65 kW/88 hp, accelerated the A-Class from 0 to 62 mph in 16 s and to a maximum speed of 87 mph—enough energy to handle day-to-day driving conditions.

- Daimler set a target of 2004 for the commercialization of fuel cell cars while targeting 2002 for fuel cell buses.[26]

7.4.1.4. Dalian Institute of Chemical Physics (DCIP) (China)

Dalian Institute of Chemical Physics (DICP), the Chinese Academies of Sciences, began PEMFC R&D in the 1990s.[48] By the early 2000s, DICP was China's foremost PEMFC research institute. It developed fuel cells and also built fuel cell buses.

- In January 2001, DICP unveiled its first fuel cell bus (FCB), the ***Dalian University Fuel Cell Bus China.***[49] The FCB was powered by a 25 kW DICP PEMFC stack combined with a 27 kW AC induction motor. The fuel cell bus was fueled with compressed hydrogen (25 MPa).[50] The bus was 7 m long, had a top speed of 37 mph, and accelerated from 0 to 25 mph in 22.1 s.
- In 2002, DCIP unveiled the ***Dalian University Fuel Cell City Bus.***[51] The FCB was based on an 11 m Jinghua Coach Company City Bus, powered by a 40 kW DICP PEMFC, lithium-ion batteries (100 kWh), and a 102 kW AC induction motor. It ran on compressed hydrogen (25 MPa), reached a top speed of 40 mph, and accelerated from 0 to 31 mph in 40 s.
- In 2003, ***Dalian University FC-Bus 2*** was unveiled.[52] The fuel cell bus was powered by a 40 kW DICP fuel cell and an electric motor and fueled with compressed hydrogen. Little information is available.

7.4.1.5. EvoBus[iv]

EvoBus, a bus manufacturer created in 1995 to amalgamate Mercedes-Benz buses and Setra buses, is part of Daimler.[53] It is the largest European manufacturer of buses[54] and has developed three generations of fuel cell buses—the NEBUS, the ZEBUS, and the Citaro Bus—since 1997.

In May 1997, EvoBus presented its first fuel cell bus, the ***NEBUS ('New Electric Bus')*** in Stuttgart.[25] The bus was based on EvoBus's Mercedes-Benz 0405 N regular-service city bus that accommodated 34 seated and 24 standing passengers. The NEBUS was powered by ten 25 kW Ballard fuel cell stacks, the same fuel cell system used in the Ballard P3 bus. The 250 kW output provided enough power for driving and for the auxiliaries such as the onboard electrical system and the air-conditioning unit. No additional onboard battery power was needed. The NEBUS used

[iv] EvoBus is part of Daimler Group (or Daimler-Benz or DaimlerChrysler). EvoBus' fuel cell bus development is presented separately from Daimler's passenger fuel cell vehicle development in this book.

pressurized hydrogen, had a range of 155 miles, and a top speed of 50 mph.[55]

In October 1999, EvoBus unveiled the ***ZEBUS*** (Zero Emission Bus). The hydrogen-fueled ZEBUS used the Ballard P4 205 kW fuel cell system, which included ten Md700 Ballard stacks. It had a range of about 186 miles.[56]

- The ZEBUS concluded a 13-month testing program with the Sun Line Transit Agency in California in September 2001, covering a total of 14,850 miles.

In March 2001, two major demonstration projects were unveiled.[57] They were the European Union Fifth Framework Program (FP5) project, "Clean Urban Transport for Europe (CUTE)," and a related project—the "Ecological City TranspOrt System (ECTOS)." Both projects aimed to operate hydrogen buses in inner cities. The projects were due to start in late 2002 and run for 2 years.

- CUTE would introduce three hydrogen buses to each of nine cities (Amsterdam, Barcelona, Hamburg, London, Luxembourg, Madrid, Porto, Stockholm, and Stuttgart) for public transportation.[58] It had $22 million in EU funding.
- ECTOS was to operate three buses in the Icelandic capital of Reykjavik.[57]

In support of these projects, EvoBus unveiled the ***Citaro Bus*** in Vancouver in November 2001.[59] The 12 m bus carried 69 passengers and was powered by the 250 kW *P5* Ballard fuel cell system and an electric motor.[60] The P5 engine was jointly developed by Ballard and Daimler for the FP5 project. Eight compressed hydrogen gas tanks provided a driving range of 124 miles. The Citaro had a top speed of 50 mph. Its price tag was about $1.2 million.

- By 2003, 30 Mercedes-Benz Citaro city buses with fuel cell engines were put in regular service on the streets of 10 European cities. As of mid-2005, the buses had been driven more than 940,000 km and carried over 3 million passengers. [47]
- Another three buses were operated in Perth, Australia, and later also were to be tested in Beijing, China.

7.4.1.6. *Fiat*

Fiat, Italy's largest carmaker and the world's ninth largest carmaker, had long focused on electric cars. It produced a few electric vehicle prototypes in the 1970s, sold a few electric vans in the 1980s, and by 1990 was producing a 22-passenger electric minibus.[61] Fiat began to focus on FCV development in 1999, with special emphasis on developing a small fuel cell–powered city car.

The Italian Ministry of the Environment provided encouragement and financial support to promote the effort.[62] Around that time, diesel and gas vehicles were banned on certain days and were considered too smoggy in Italy.[28] The Italian Ministry of the Environment[62] undoubtedly hoped that Fiat would develop an environmentally friendly FCV that could replace diesel and gas vehicles in Milan. The Italian Government also likely hoped that Fiat would become a major FCV developer that could compete in the global fuel cell race.

In February 2001, Fiat unveiled its first prototype fuel cell car—the ***Seicento Elettra H2 Fuel Cell Car (or Fiat 600 Elettra Fuel Cell Vehicle)***.[63,19] The two-passenger hydrogen-powered fuel cell battery hybrid was based on Fiat's smallest city car—the Seicento Elettra—an electric car produced 2 years earlier.[64] All the power required by the electric drive motor was delivered by the battery pack. The fuel cell system, which used Nuvera's 7 kW PEMFC stack, was designed to charge the batteries and power the fuel cell system auxiliaries, including compressor, water pumps, and fan.[65] Elettra H2 shared several components with the Seiceno

Elettra, including the 30 kW asynchronous three-phase motor and the battery pack (in a different size).[62] Seicento Elettra H2 had a top speed of 60 mph and a driving range of 85 miles.[19] The FCV was driven without degradation for more than 1800 miles.[63]

In October 2003, Fiat unveiled two FCV prototypes at a ceremony in Milan: a prototype of the *FIAT Seicento* employing a 40 kW Nuvera fuel cell stack and a prototype of FIAT's next-generation FCV, the ***Panda Hydrogen***,[66] which would feature an advanced 60 kW Nuvera Andromeda PEMFC fuel cell stack. Both Fiat prototypes used compressed hydrogen.[67] The Fiat Seicento had a driving range of 136 miles.[68] Its top speed was 80 mph. The Fiat Panda accelerated from 0 to 31 mph in 5 s.

7.4.1.7. Ford

Ford Motor Company has long had an explicit commitment to build vehicles that would operate on alternative fuels; its fuel cell program was among the most ambitious.[28]

- As one of the Ballard fuel cell car alliance members since 1997, Ford invested hundreds of millions or perhaps billions of dollars to create and preserve a relationship with Ballard as part of its core strategy for commercialization of fuel cell cars.
- Ford also invested in a Norwegian electric car manufacturer, Th!nk Nordic, which was founded in 1991.[69] Around 1999, Ford purchased 51 percent of its shares, later purchasing the other 49 percent and forming the Th!nk Group, which would lead to the creation of the Th!nk FC5.[70]

In January 1999, Ford unveiled its first FCV, the ***P2000 FC EV***, at the Detroit Auto Show.[71] The prototype was a concept car but was likely on the road by June 1999 when it was presented in Europe. The P2000 FC EV was based on the P2000 vehicle platform, which was comparable to a "Taurus." It was powered by three 25 kW Ballard PEM fuel cell stacks with a 76 kW traction drive. The FCV was fueled with compressed hydrogen, had a driving range of about 100 miles, and a top speed of 90 mph.

- The prototype was believed to be the world's first full-size, full-performance, five-passenger car powered by fuel cells. Ford hoped to achieve weight reduction, build a test fleet of the FCVs in 2001, and reach production ready status by 2004.[71]

In January 2000, Ford unveiled the ***Th!nk FC5*** concept car at the Detroit Motor Show.[71] The FC5 was based on the 2000 Ford Focus four-door family sedan with a Ballard Mark 9000 fuel cell system and an electric drive train from the Ford EV Ranger. The FC5 was fueled by methanol with a reformer.

- In November 2000, Ford announced the Ford ***Th!nk FCV***,[72] a drivable methanol-powered version of the Th!nk FC5 concept car. The FCV was also based on a 2000 Ford Focus, powered by a 75 kW Ballard PEM fuel cell system, and fueled with methanol via a reformer. The FCV had a driving range of 100 miles and a top speed of 80 mph.
- In October 2001, Ford presented the Ford Th!nk FCV in Berlin. The FCV was fueled by pressurized hydrogen (35 MPa).

In June 2002, Ford unveiled the Ford ***Focus FCV Hybrid***, one of the industry's first hybrid FCVs.[73] The Ford Focus FCV Hybrid was powered by a new Ballard Mark 902 fuel cell stack matched with a 300 V Sanyo battery pack and a brake-by-wire electrohydraulic regenerative braking system. The Focus FCV also was equipped with a more advanced hydrogen storage tank, which could handle 5000 pounds per square inch (psi) of hydrogen—compared to 3600 psi in the previous version. As a result, the four-passenger Focus FCV had a driving range between 160 and 200 miles and a top speed of 80 mph. The new Focus FCV was

Ford's third-generation drivable FCV. The first, the P2000 FCV, was introduced just 4 years earlier.

- Ford believed that the latest technology would bring the company one step closer to making FCVs viable for consumers. Ford planned to produce five Focus FCVs in 2002 for testing and demonstration, leading to low-volume customer production by 2004.

7.4.1.8. General Motors

General Motors (GM), the world's largest automaker for many years and the world's second largest automaker as of March 2012,[74] was the first automaker to demonstrate an FCV. In the 1960, the automaker demonstrated the Electrovan powered by hydrogen-fueled alkaline fuel cells developed by Union Carbide. GM test-drove the van on its campus, but did not continue FCV development efforts. It was not until the late 1990s that GM decided to develop FCVs, undoubtedly prompted by the successes shown by Ballard and its allied automakers.

GM first embarked on building up its in-house R&D infrastructure. In late 1997, GM established the Global Alternative Propulsion Center (GAPC)[28] by integrating its fuel cell development activities at Warren, Michigan, and Rochester, New York with Opel fuel cell development activities at Mainz-Kastel, Germany.[56] The GAPC team in Warren would be responsible for the basic system research while the team in Rochester would focus on fuel cell and component development. A team of about 200 employees at GAPC's Mainz-Kastel facility would deal with system integration. Total staff was about 500.

Next, GM aggressively moved to establish alliances and partnerships with expert fuel cell developers to secure technological expertise and capabilities.[75] Its partners included Toyota, Quantum Fuel System Technologies Worldwide of Irvine, California, Giner Electrochemical Systems of Newton, Massachusetts, General Hydrogen of Vancouver, British Columbia, Hydrogenics of Toronto, Canada, and Suzuki Motors.

In September 1998, GM unveiled its first PEM FCV at the Paris Motor Show, the ***Opel Zafira FCEV***, a concept car based on the Opel Zafira minivan.[76] The Opel Zafira FCEV was powered by two 25 kW fuel cell stacks manufactured by Ballard.[77] The specific power of the fuel cell system was 0.31 kW/kg.[19] It used methanol and an onboard reformer to create hydrogen fuel. The vehicle achieved a maximum speed of 75 mph and a range of 300 miles. It had a battery that provided transient power and storage of electricity gained from regenerative braking.

- Zafira FCEV was GM's first methanol FCV and also its last. GM subsequently abandoned the methanol reformer.[19]

Starting in 2000, GM demonstrated a series of FCVs using its own PEMFC systems.

In January 2000, the ***Precept FCEV*** was unveiled at the Detroit Auto Show. The Precept FCEV hybrid used a 100 kW fuel cell system developed by GM and a hydride hydrogen storage system that could deliver up to 500 miles range.[78] Owing to its lightweight aluminum body, the five-passenger Precept FCEV was able to accelerate from 0 to 60 mph in just over 9 s and achieve 108 miles per gallon equivalent (MPGe).[79]

- GM developed the Precept model under the Partnership for a New Generation of Vehicles (PNGV) implemented in 1993. The PNGV was a cooperative research program between the US government and three major US automakers that aimed to bring to market by 2003 vehicles with high fuel efficiencies—up to 80 mpg. The PNGV ended in 2001 without meeting this goal.

In June 2000, GM unveiled ***HydroGen1***, powered by the first generation of its mainstay

PEMFC system, which was codeveloped with Hydrogenics.[77] The HydroGen1 was based on the same five-passenger Opel Zafira compact van as the Opel Zafira FCEV.[19] The car used an 80 kW fuel cell system that ran on pure hydrogen. The fuel cell system comprised 200 individual fuel cells, had stack dimensions of 590 × 270 × 500 mm (length × width × height), and a specific power of 0.47 kW/kg. The front-wheel drive vehicle reached a top speed of nearly 90 mph and had a range of about 250 miles per tank of hydrogen. The stainless steel tank installed under the rear seat held 16 gallons of liquid hydrogen.

In August 2001, GM unveiled the world's first onboard gasoline fuel processor for fuel cell propulsion on the ***Chevrolet S-10 pickup***.[19] The gasoline processor, packaged in a Chevrolet S-10 pickup, reformed gasoline on board, extracting a stream of hydrogen to send to the 25 kW fuel cell stack. It achieved an efficiency of 40 mpg equivalent.[80] The gasoline reformer created a major sensation among automakers and oil companies alike. It had long been believed that gasoline-powered FCVs would be much more practical, in terms of using the existing fuel delivery infrastructure, than hydrogen FCVs. After all, there was no hydrogen or methanol supply infrastructure to fuel and service FCVs for the foreseeable future.

- GM was clearly proud of the gasoline reformer, but subsequently abandoned it. The gasoline reformer, not unlike a small-scale oil refinery, was a difficult technology to master.

In September 2001, GM unveiled the ***HydroGen3*** at the Frankfurt Motor Show.[80] HydroGen3 was also based on the Opel Zafira and was powered by liquid hydrogen. The HydroGen3 incorporated a more advanced GM fuel cell system—the Stack2000—consisting of 200 cells and achieving a power output of 94 kW, with a specific power of 0.94 kW/kg. It used an electronic motor with a maximum output of 60 kW, enabling the fuel cell—powered Zafira to accelerate from 0 to 60 mph in 16 s and to reach a top speed of 100 mph.[81] It also had a driving range of 250 miles.[77] The fuel cell system was built by the German-based GM research facility in Mainz-Kastel.[82]

- The HydroGen3 also was the first prototype to be used for fleet demonstrations. The FCV was certified by the Japanese government for driving on public roads.[83] In addition, HydroGen3's 250 mile driving range was the highest of any FCV approved for public roads in Japan.[84]

In 2002, GM introduced the ***Advanced HydroGen3***, fitted with a 10,000 psi high pressure hydrogen storage tank.[77] The vehicle otherwise was the same as the HydroGen3. It incorporated the same 94 kW PEMFC system as the HydroGen3. The advanced HydroGen3 achieved a top speed of 100 mph and a range of 170 miles.

- The HydroGen3 was used in many demonstration programs. In a 2-year program, Members of Congress had the chance to test drive HydroGen3 in Washington DC. The US Postal Service also used HygroGen3 minivans to deliver mail in and around the Washington DC area[83] and in Irvine, California.[85]

The HydroGen3 likely marked a critical milestone for GM; it succeeded in developing an environment-friendly, resource-saving FCV that also had the performance and endurance for day-to-day operations. For example, test drives conducted by an international group of expert journalists at the Grand-Prix course in Monte Carlo in December 2002 likely convinced GM that the HydroGen3 had brought mass production of hydrogen-fueled automobiles within reach.[86]

- In July 2002, GM announced that the company would produce prototype units by late 2003 or early 2004 and have fuel cell–powered electric generators commercially available by 2005.[87]

7.4.1.9. *Gillig*

Gillig (USA), the second largest producer of transit buses in North America,[88] was selected to build three fuel cell buses that would operate in the Zero-Emission Bus Demonstration Program in Santa Clara, California. The $18.5 million multiyear program was partially funded by the Federal Transit Administration (FTA).[89] It was one of several US fuel cell bus demonstration programs to be implemented through partnerships with local agencies in the early 2000s.

- In December 2000, Santa Clara Valley Transportation Authority adopted a low-emission strategy as its approach for complying with the California Air Resources Board regulations and decided to demonstrate hydrogen fuel-cell technology in a joint effort with the neighboring San Mateo County Transit District.[89] As part of the program, Gillig was contracted to build three 40 ft, low-floor, hydrogen fuel-cell buses. The buses would be powered by 205 kW Ballard fuel cells using compressed hydrogen.[28] The total program budget was $18.5 million, which would include $10.6 million for the buses. Of the total, FTA would subsidize $5.1 million, with the rest paid by local government agencies, transit services, and local sales taxes.[89]
- Gillig was expected to deliver the buses in August 2004 and the transit service was to begin in February 2005[89] and continue for 2 years.[28]

7.4.1.10. *Hino*

Hino Motors, a manufacturer of diesel trucks, buses, and other vehicles, based in Hino, Tokyo, was founded in 1942. It merged with Toyota Motor in 1966,[90] which holds a 50.1 percent stake in Hino. Since 1973, Hino has been the leading producer of medium and heavy-duty diesel trucks in Japan.

In June 2001, Hino and Toyota jointly developed the ***FCHV-BUS1***, a fuel cell bus powered by a 90 kW Toyota PEM fuel cell feeding two 80 kW permanent magnet motors.[91] The fuel cell bus used a hybrid system, including nickel–metal hydride secondary batteries to recover and store energy from braking. The FCHV-BUS1 was fueled with high-pressure hydrogen stored at 25 MPa in roof-mounted tanks. The prototype had a cruising range of over 186 miles and reached a top speed of 50 mph. The fuel cell bus was based on a Hino low-floor 10.5 m city bus model that could hold 63 passengers.[92]

In September 2002, Toyota and Hino announced that the ***FCHV-BUS2***, an advanced follow-on to the FCHV-BUS1, was developed and approved by the government for use on Japanese public roads.[93] The FCHV-BUS2, like the FCHV-BUS1, stored its hydrogen in roof-mounted hydrogen storage tanks, but the FCHV-BUS2 had two high-performance Toyota fuel cell stacks, each with an output of 90 kW. This gave it double the performance of the FCHV-BUS1, which had only one stack. The FCHV-BUS2 also had two NiMH batteries with a maximum output of 80 kW each and maximum torque of 260 Nm each.[94] The fuel cell bus was based on a 10.5 m Hino Blue Ribbon City Bus that carried 60 passengers. The fuel cell bus was fueled with compressed hydrogen at 35 MPa and had a driving range of 217 miles.

- The FCHV-BUS2 was the first fuel cell powered bus to be certified for use on Japanese public roads. Four FCHV-BUS2 units, including a 60-passenger municipal model, were test-driven in Tokyo starting in 2002.[95]
- The FCHV-BUS2 municipal model became the first fuel cell bus to go into revenue service as part of a municipal fleet in Tokyo starting in August 2003.[96]

7.4.1.11. *Honda*

Honda began to explore FCV technology as early as 1989. The company devised a two-pronged approach; it worked on developing its own PEMFC technology while deploying Ballard fuel cell stacks to power its concept, prototype, and demonstration FCVs. Thus, it maintained a long-term supply relationship with Ballard. According to Ballard, Honda's purchases of Ballard fuel cells and related equipment amounted to over $2.6 million in October 1999, $1.5 million in August 2000, and $1.3 million in January 2001.[19]

- Honda reportedly had a contract to buy Ballard engines through 2005.[97]

In September 1999, Honda unveiled its first two fuel cell prototypes: ***FCX-V1*** and ***FCX-V2***. Both FCVs were built on the chassis of the discontinued Honda EV Plus battery electric vehicle, but with a different and more aerodynamic body.[98] The two-passenger FCX V1 contained a hydrogen-powered 60 kW fuel cell system manufactured by Ballard and a metal hydride alloy developed by Honda to store hydrogen.[97] The vehicle had a maximum speed of 78 mph and a driving range of 110 miles.[97] The second version—the FCX-V2—had a 60 kW PEMFC system and a methanol processor, both developed by Honda. Although little information is available on the FCX-V2, the Honda fuel cell system likely was not mature. It was the only methanol fuel vehicle Honda built. Honda subsequently settled on hydrogen as its fuel of choice.

- As early as 1999, Honda had been already contemplating its FCV market entry. The company indicated it had been planning to make FCVs commercially available by the year 2003[99] (or by the Japanese fiscal year of 2003, starting in April 2003).

In September 2000, Honda announced the ***FCX-V3***, a four-passenger FCV that was equipped with a Ballard fuel cell system. Honda managed to reduce the size of the fuel cell system in the FCX V3 by incorporating Honda's ultracapacitor with the Ballard fuel cell, creating a fuel cell hybrid.[97] The vehicle was powered by pressurized hydrogen. It had the maximum speed of 78 mph and a driving range of 108 miles.

- In March 2001, Honda outfitted the FCX-V3 with a newly developed Honda fuel cell stack and started to operate a fleet of vehicles in the California Fuel Cell Partnership's (CaFCP) public highway demonstration program.[100] In July 2001, public road testing of the FCX-V3 with the 70 kW Honda fuel cells began in Japan.[101] Honda likely hoped that the on-road evaluation of the latest FCX-V3 hydrogen-fueled prototype would help confirm performance and dependability of Honda's own fuel stack design.[100] Maximum speed of the vehicle was 78 mph, and the driving range was 109 miles.[19]

In October 2002, Honda surprised the world by unveiling a new prototype ***Honda FCX*** and announcing that it would begin leasing the vehicles by December 2002.[102] This was the world's first production fuel cell electric vehicle. Moreover, Honda advanced the commercialization date of April 2003 by a few months. The company likely impressed global automakers and consumers with its technological prowess and its bold business decision. The 4-passenger Honda FCX was equipped with a 78 kW Ballard fuel cell stack powered by compressed hydrogen gas, which reached a maximum speed of 93 mph and had a driving range of 220 miles.

- In November 2002, the FCX was granted approval for driving on public roads in Japan. In December 2002, Honda started limited leasing of Honda FCXs, with plans to lease about 30 FCXs in the US and Japan within 2–3 years. It would lease the FCVs to institutional customers first.[103]

Since then, Honda FCX cars have been used as fleet vehicles by such agencies as the City of Los Angeles, City of San Francisco, and California's South Coast Air Quality Management District, which has made daily use of the vehicles.[104]

7.4.1.12. Hyundai

Hyundai, South Korea's biggest automaker, became interested in fuel cells rather late, but once it started, the vigor of its developmental effort was astonishing. In December 1999, the Korean government formed a cooperative program with Ballard so that Hyundai would have access to Ballard fuel cell technology.[105] Under the agreement, the Korean government purchased $391,000 worth of Ballard fuel cells, related equipment, and support services. Immediately following the purchase, Hyundai and its Kia Motors affiliate decided to make a major investment in fuel cell development. In February 2000, Hyundai and Kia announced they would spend about US$9 million on fuel cells that year.[106] Although it was modest compared to the spending of other major global automakers, it was a major investment for Korea at that time. Hyundai also established a fuel cell task force at its research and development center in Yongin, South Korea, in 2000.[107]

Of particular note, Hyundai in April 2000 signed a $40 million agreement with UTC to develop FCVs.[108] The agreement specified that UTC would equip between two and four sports utility vehicles (SUVs) with its 75 kW ambient pressure PEMFC stacks. Little information is available, but it is clear that Hyundai did not use Ballard fuel cells.

- The company also embarked on ambitious plans to produce a prototype fuel cell car by November 2000. It intended to make a decision on whether mass produce fuel cells under its own brand or purchase fuel cells from abroad by 2003.[106]

As announced earlier, Hyundai indeed unveiled its first prototype FCV—the ***Santa Fe FCEV***—in October 2000, only 6 months after it had signed an agreement with UTC to supply the fuel cell stacks.[107] The Santa Fe FCEV was also equipped with a 5000 psi hydrogen storage tank from IMPCO Technologies, Irvine, California.[109] The fuel tank supported a driving range of 100 miles. Electric drive train, motor, and control unit were from Enova systems of Torrance, California. Top speed was 77 mph and acceleration from 0 to 60 mph took 13.4 s. Vehicle start-up time was 10 s.[110] The Santa Fe FCEV was one of the fuel cell cars and buses that were showcased at the inaugural ceremony of the newly constructed CaFCP headquarters facility on November 1, 2000.

- In 2002, Hyundai introduced the hybrid Santa Fe FCHEV, which had a 250-mile range.[28]

UTC and Hyundai proudly announced that they had achieved tremendous results in demonstrating the viability of FCVs and confirming the performance of UTC's fuel cell design.[111] Indeed, a Hyundai Santa Fe scored best in class in efficiency and noise at the 2001 Michelin Challenge Bibendum, an annual event where new automotive technologies were evaluated by independent judges. In the 2001 Challenge Bibendum, the Santa Fe FCEV earned gold medals in both the emissions and noise tests, and received silver medals for fuel efficiency, the highest grades given to fuel cell cars that were entered in those categories.[19]

- A total of six hydrogen-powered FCVs from various automakers competed in the Michelin event.[19] These were Ford, DaimlerChrysler, GM, Toyota, and Honda.
- It was an amazing achievement for a relatively small automaker to produce its first FCV just 1 year before winning the top prize.

By mid-2003, a total of six Hyundai fuel cell Santa Fe prototypes were produced and powered by the UTC Fuel Cells power plant.[111] The Santa Fe FCEV's solid performance since

2001 clearly had given Hyundai the confidence that its FCVs were ready for real-world applications.

In June 2003, Hyundai and UTC signed an agreement to jointly develop a new fuel cell power plant capable of operating under sub-freezing conditions.[112] The agreement called for Hyundai and UTC Fuel Cells to work jointly to develop the fuel cell power plant and integrate it into a Hyundai SUV platform.

- Hyundai then announced its plans to make FCVs available to US fleet operators as early as 2004, with limited consumer availability planned for 2010.[111]

7.4.1.13. *Irisbus*

Irisbus—a major bus manufacturer based in Lyon, France—was established in January 1999 with the merger of the Fiat-Iveco Group and the Renault Group.[61] Since 2001, Irisbus has been 100 percent owned by Fiat-Iveco.[113]

Irisbus started to develop fuel cell urban buses in a national program in 1999[56] or 2000. In 2000, Irisbus signed a partnership agreement with UTC, in which UTC was to provide Irisbus with UTC PEMFC fuel cell systems for buses.[114] The two buses were to be built for passenger service in Madrid, Spain, and Torino, Italy, respectively, under national projects. The projects were funded equally by private and public sectors; the public sponsors included the Ministry of the Environment, the Turin Transit Authority, and the City of Turin.[115] Private partners included Air Liquide, Alstom, Ansaldo, Axane, ENEA, Fiat, SAPIO, and Total.[116] Two different prototype buses were built:

- In January 2001, Irisbus unveiled the first ***CityClass Irisbus***,[117] which went into passenger service in Turin, Italy.[118] The bus was based on a 77-passenger Iveco City Class bus powered by a 60 kW UTC PEMFC fuel cell. It had a top speed of 60 kmph.
- In April 2003, Irisbus presented the second ***CityClass Irisbus***,[119] which went into service in Madrid.[120]

Irisbus also participated in the EU FP program to develop fuel cell buses. In January 2002, Irisbus participated in a 60-month FP5 project, "Fuel cell energy systems standardized for large transport, busses, and stationary applications" (FEBUSS) (January 2002–December 2006).[121] The project was to develop two standardized 100 kW PEMFC prototype power modules for heavy-duty vehicle and stationary applications, culminating in a 2-year demonstration test of the two prototypes.[122] The total cost of the project was €7.8 million, and a large consortium of 17 companies and universities and research institutions participated.[123]

- The goal of the project was to achieve a mass-production cost target of €300 Euro/kW for the fuel cell plus hydrogen storage.[121] It is not clear if the project was completed.

In spring 2003, Irisbus participated in another FP5 project, "CityCell—fuel cell energy in cities," a 48-month, €11-million project to develop and demonstrate five fuel cell buses in four European cities: Berlin, Madrid, Paris, and Turin.[124] The participants included Air Liquide, Ansaldo, ENEA, SAPIO, and Total. The CityCell project—somewhat more complex than other FP projects—would fund the development and demonstration of two new CityCell Irisbuses as well as fund the demonstration of three CityClass Hybrid Irisbuses that Irisbus had been developing with national and local government funding.

- Two CityCell Irisbuses to be demonstrated in Berlin would be newly funded under the FP5 for both development and demonstration.
- One bus each was to be demonstrated in Madrid and Turin (these vehicles had already been developed under national programs). The FP5 would fund operation of the two buses.

- The fifth bus, to be demonstrated in Paris, was being developed under a national program; the EU would fund the demonstration portion only.[116]

It was expected that the five CityClass and CityCell Irisbuses would show how fuel cell technology could be introduced and applied in urban public transport systems, thus increasing the general public's exposure to the technology and promoting market introduction of the technology in the near term.

In 2006, Fiat/Iveco manufactured a fuel cell hybrid bus powered by a 60 kW UTC fuel cell hybridized with UTC batteries, regenerative braking, and an electric motor. The bus was based on a 7.4-m, 40-passenger, Iveco Europolis. It had a driving range of 200 km. Fleet trials were planned for several Italian cities between 2008 and 2010.[125] It is not clear if the trial operation took place.

7.4.1.14. Man Nutzfahrzeuge (MAN Trucks and Bus)[v]

MAN Truck & Bus (formerly MAN Nutzfahrzeuge), based in Munich, Germany, is the biggest company of the MAN Group and a top-ranking international supplier of commercial vehicles.[126] MAN Nutzfahrzeuge acquired NEOPLAN in 2001, another German bus company that manufactures fuel cell buses. MAN Nutzfahrzeuge changed its name to MAN Truck and Bus in January 2011. It produces trucks and buses under the brand name MAN and VIP class buses under the brand name NEOPLAN.[127]

In 1994, MAN Nutzfahrzeuge participated in a feasibility study on fuel cell urban buses coordinated by Ludwig-Bölkow-Systemtechnik (LBS) and sponsored by the Bavarian State Ministry of Economics, Transport, and Technology.[128] The project was cofinanced by participating Bavarian companies, including Daimler-Benz, Erlanger Stadtwerke, Gottlob Auwarter, Linde, Magnet Motor, MAN Nutzfahrzeuge, Siemens, and Solar-Wasserstoff-Bayern.

- In January 1996, the development of two bus prototypes—one by MAN and another by NEOPLAN— began with 50/50 cofinancing of the entire activity by the Bavarian and Federal governments and industry participants. [128]

In May 2000 at the Munich "Fuel Cell Day," MAN unveiled the ***MAN Fuel Cell City Bus***.[129] The fuel cell bus was based on a 12 m long, low-floor MAN Nutzfahrzeuge NL 263 "Bavaria I" city bus. Siemens developed the PEM fuel cell system.[130] Four fuel cell modules delivered a net electrical output of 120 kW to two electric motors (2 × 75 kW), which were linked via a gearbox manufactured by the Siemens. The fuel cell bus was fueled by compressed hydrogen, which provided a driving range of around 155 miles. Linde developed the hydrogen auxiliary systems.

- Total project cost was DM13.5 million; the Bavarian Ministry contributed 50 percent of the total.[129]
- The Nuremberg traffic authority conducted public operations of the bus in the cities of Nuremberg, Erlangen, and Fürth between October 2000 and April 2001.[130] Passenger response was overwhelmingly positive.[28]

MAN was also involved in two consecutive FP5 projects: "Fuel Cell Bus for Berlin, Copenhagen, Lisbon" (project period unclear) and "Fuel Cell Bus for Berlin, Copenhagen, Lisbon II" (March 2000—February 2004).[131] It appears that the first project ran out of funds and only the initial phase of the project—development of the fuel cell bus—was completed. The

[v] Although NEOPLAN and MAN are now one company, MAN and NEOPAN each develop buses with distinctly different features and thus are presented separately in this book.

subsequent FP5 project was aimed at conducting demonstrations of the bus in three European cities: Berlin, Copenhagen, and Lisbon.

MAN unveiled the ***Fuel Cell Bus*** in 2001 to be operated by transport operators in three European cities for demonstration purposes.[132] The 12 m, low-floor bus based on MAN NL 223 bus was propelled by three PEMFC stacks, which had an output of 40 kW each and were developed by De Nora (now Nuvera). The electric drive had a maximum power of 150 kW and was manufactured by Siemens. The bus was fueled by liquid hydrogen, stored in two 350 l hydrogen thermo-insulated bottles manufactured by Messer Griesheim, and provided a driving range of 186 miles. Total project cost is not clear. The first project amounted to at least DM11 million,[133] and the second project was €5.23 million.

- Other partners included Air Liquide, Berlin Senate, Berlin Transport Company, Copenhagen Transport, CARRIS (Portugal), Instituto Superior Técnico (Portugal), and MVV InnoTec (Germany).[131]
- A 1 year demonstration was planned to take place in Berlin, Copenhagen, and Lisbon.[130]

7.4.1.15. Mazda

Mazda has viewed PEMFCs as one of the most promising alternative fuel sources for vehicles and began its studies of the technology in 1991,[134] when it received a Ballard fuel cell on loan.[19] In early 1992, it incorporated a Ballard fuel cell stack into its first FCV with a golf cart-type chassis.[134] Subsequently, Mazda apparently has attempted to develop its own PEM fuel cell stacks. In 1998, however, it joined the Ballard/Daimler/Ford fuel cell car alliance and started using Ballard fuel cell stacks.

In December 1997, Mazda exhibited a prototype FCV, the ***Demio FCEV*** at Eco Japan 1997.[135] The FCEV was based on Mazda's small passenger car, the basic Demio model. The FCEV was powered by a 20 kW PEM fuel cell stack developed in-house, with a 20 kW ultra capacitor and an AC synchronized motor with a maximum power output of 40 kW. The vehicle was fueled with hydrogen stored in hydrogen-absorbing metal alloys. The Demio FCEV achieved a top speed of 55 mph and had a driving range of approximately 100 miles using compressed hydrogen gas.

- Mazda received financial support from AIST and METI for development of the Demio FCEV.[135]

In November 1999, Mazda unveiled the ***Second Generation Demio FCEV*** at the Tokyo Motor Show.[135] The FCEV was based on the same basic Demio model as the first-generation FCEV, but was powered by newly developed 50 kW Ballard fuel cell stacks and drive system. In April 1998, Mazda participated in the fuel cell alliance established by Ford, Daimler-Chrysler, and Ballard, through its close relationship with Ford.[134] The second-generation Demio FCEV also was equipped with a metal hydride hydrogen storage tank. The volume of the fuel cell system has decreased by 20 percent compared to the first-generation Demio FCEV.

In February 2001, Mazda unveiled the ***Premacy FCEV*** vehicle,[134] the latest FCEV, which was based on the five-passenger Mazda Premacy.[136] The Premacy FCEV was developed in cooperation with Mazda's partners. The new ECEV incorporated 75 kW Ballard Mark 900 PEM fuel cells and a vehicle control system developed by Ford's TH!NK group.

- In February 2001, Japan gave its first approval for an experimental FCV to be driven on public streets.[137] Mazda's Premacy FCEV would soon run on Hiroshima streets while Daimler's NECAR 5 would run in Tokyo and Yokohama for feasibility tests. Both models used methanol as fuel.
- Mazda planned to start marketing fuel cell cars around 2005.[138]

7.4.1.16. *Mitsubishi Motors*

Mitsubishi Motors was spun off from Mitsubishi Heavy Industries (MHI) in 1970.[139] It is the sixth largest automaker in Japan and the 17th in the world[140] with a full line of automotive products, including automobiles, trucks, buses, parts, and powertrains.[19] Mitsubishi Motors has alliance relationships with a number of major automakers, including Chrysler, Daimler-Benz, Hyundai, PSA Peugeot Citroën, and Volvo, but its alliance with Daimler-Benz is the most noteworthy.

- In 1987, Mitsubishi Motors signed a basic partnership agreement with Daimler-Benz to sell Mercedes-Benz in Japan and produce commercial vehicles in Europe.[139]
- In 2000, Mitsubishi Motors formed a capital alliance with Daimler.[139] Since then, Daimler has had a varying stake in Mitsubishi Motors, from 34 percent (2000–2001) to 37.3 percent (2001–2005), and 12.4 percent (2005), when Daimler sold off its entire Mitsubishi stock.[141] The alliance with Daimler clearly helped Mitsubishi Motors advance its fuel cell technology.

In 1998, Mitsubishi Motors began the development of FCVs.[142] Initially, Mitsubishi Motors focused on fuel cell electric system powered by methanol and worked to develop a proprietary fuel cell system and reformer in collaboration with other Mitsubishi Group members, including MHI and Nippon Mitsubishi Oil.[143]

In October 1999, Mitsubishi unveiled a methanol powered fuel cell concept vehicle at the Tokyo Motor Show, incorporating a fuel cell stack developed by Mitsubishi Heavy Industries.[19] The stacks reportedly used DuPont PEM membranes. Mitsubishi released a new fuel cell minivan in 2003 featuring a Daimler–Ballard fuel cell under the body of a new Mitsubishi model called the Grandis. It underwent road tests in Japan,[28] but little information is available on the concept vehicle.

In summer 2001, Mitsubishi Motors was working to develop fuel cells for vehicles in cooperation with DaimlerChrysler, its capital alliance partner and the holder of a controlling stake in the company.[19] In October 2001, the company unveiled the ***Mitsubishi Space Liner FCV***. The FCV used a 75 kW fuel cell stack from Daimler's NeCar series developed by Ballard and a drive-by-wire technology.[144] Nippon Mitsubishi Oil, which was working to develop a liquid fuel that could be used in fuel cells instead of gasoline, was also involved in the development of the FCV.[19] The Mitsubishi Space Liner FCV used the body of the Mitsubishi Space Liner, a futuristic looking, monobox four-seat vehicle that was also exhibited for the first time at the 2001 Tokyo Motor Show. Little additional information is available on the Space Liner FCV.

In 2002, Mitsubishi Motors started the development of a new FCV in close collaboration with Daimler.[142] To maximize the benefits of the Daimler–Mitsubishi alliance, the companies reached an agreement whereby Mitsubishi Motors would use Daimler's F-Cell technologies in the development of the ***Mitsubishi FCV***, ***MFCV***, or the ***Mitsubishi FCV Grandis***.[145] The FCV, based on a Mitsubishi Grandis model, was a hydrogen-powered hybrid electric vehicle using a 68 kW Ballard fuel cell stack fueled with compressed hydrogen and nickel–metal hydride rechargeable batteries.[142] Final assembly of the Mitsubishi FCV was made in May 2003 at the Daimler facility in Nabern, Germany, and the MFCV's basic performance was verified there. Driving tests of the FCV started in August 2003 at Mitsubishi Motors test course in Okazaki, Japan.

- Mitsubishi Motors exhibited the Mitsubishi FCV Grandis at the October 2003 Tokyo Motor Show as the core element of its pro-environmental strategy.[142]

Also in October 2003, the Mitsubishi FCV received a vehicle type certification that would allow it to operate on public roads.[146] Subsequently, Mitsubishi Motors entered the FCV in various tests and events in the Okazaki and Kanto areas.[142] In 2003, the FCV was used in the Japan Hydrogen & Fuel Cell Demonstration (JHFC) Project sponsored by METI, Japan's first large-scale fuel cell demonstration project.

- Mitsubishi Motors stated in November 2003 that the company would use its alliance with Daimler to further the FCV development program and work to make FCVs a more common sight on Japan"s roads.[147]
- Backed by its alliance partner Daimler and due to the success of its FCVs, Mitsubishi Motors clearly was buoyant about achieving its goal of commercializing its FCVs in 2005.

7.4.1.17. NEOPLAN[vi]

NEOPLAN was founded in 1935 in Stuttgart, Germany, as a manufacturer of bodywork for buses and truck chassis.[148] In 2001, MAN Nutzfahrzeuge acquired NEOPLAN, and in January 2011, MAN Nutzfahrzeuge changed its name to MAN Truck and Bus.[149] MAN produces trucks and buses under the MAN brand name and VIP buses under the NEOPLAN brand.[127]

In 1994, NEOPLAN (formerly Gottlob Auwarter) along with MAN participated in a feasibility study on fuel cell urban buses coordinated by LBS and sponsored by the Bavarian State Ministry of Economics, Transport, Technology.[128]

- In January 1996, NEOPLAN and MAN each started to develop their own fuel cell bus prototypes.

In October 1999, NEOPLAN unveiled its first fuel cell bus, the ***NEOPLAN Midi FC Bus***.[150] The fuel cell bus was based on a Midibus N8008 FC, which was an 8 m, 47-passenger bus designed for cities and resorts. The propulsion system consisted of a 40 kW Nuvera PEM fuel cell system and nickel–metal hydride batteries that had a total energy storage capacity of 21 kW, with a maximum output of about 100 kW. The bus was fueled with compressed hydrogen that provided a driving range of 372 miles.

- Total cost of the bus itself was estimated to be about DM2 million, and the cost of the entire project was about DM10 million.[150]

In May 2000 at Munich's "Fuel Cell Day," NEOPLAN and Proton Motor displayed another fuel cell bus, the ***Bayern-Bus II***.[151] The fuel cell bus was based on NEOPLAN's low-floor N8012 bus, which was 10 m long and carried 33 passengers. The bus was powered by an 80 kW Proton Motor PEM fuel cell system, with four 70 kW electric motors (one at each wheel) with regenerative braking. It was fueled by compressed hydrogen and had a driving range of up to 155 miles.

7.4.1.18. New Flyer Industries

New Flyer Industries, headquartered in Winnipeg, Manitoba, Canada, began in 1930 by building truck and bus bodies. New Flyer is the largest transit bus manufacturer in North America today[152] and is famous for its low-floor transit fuel cell buses that incorporate Ballard PEM fuel cells.

Subsequent to the development of the Phase one ***P1 bus*** and Phase two ***P2 buses*** in 1993 and 1995, respectively, the *Ballard P3* was developed and three buses each were delivered to the Chicago Transit Authority (CTA)[153] and TransLink (formerly BC Transit, Vancouver) in 1997–1998. The P3 was based on a 60-passenger New Flyer 40 LF model and was powered by a 260-kW Ballard PEM fuel cell system, fueled by compressed hydrogen at 25 MPa. It had a driving range of 248 miles.

[vi] NEOPLAN and MAN are one company, but their fuel cell bus development activities are presented separately in this book.

- Two-year tests were successful; they revealed some problems with specialized components and the system layout, but overall public acceptance of the buses was good.[19]

In 1999, the next-generation ***P4 ZEbus*** (zero emission bus) was unveiled at the International Public Transportation Exposition in Orlando, Florida.[19] The P4 ZEbus also was a standard, low-floor New Flyer 40 LF transit bus.[154] Almost 4000 lb lighter than its predecessor, the 205 kW P4 engine was less complex than the P3, incorporating more off-the-shelf components.[19] The P4 ZEbus operated in SunLine's 13-month demonstration program (from August 2000 to September 2001) in cooperation with CaFCP.[155] SunLine Transit Agency, AC Transit, and Santa Clara Valley Transportation Authority were associate members of the Partnership and served as the initial testing sites for the bus program. The new P4 engine was to be used in approximately 25 buses under the CaFCP program "Driving for the Future."

In late 2002, New Flyer had an opportunity to work with another fuel cell developer, Hydrogenics. In December 2002, Natural Resources Canada (NRCan) selected Hydrogenics to develop a next-generation hybrid fuel cell bus.[156] The project aimed to integrate Hydrogenics' new proprietary vehicle-to-grid technology into a 12.5 m New Flyer Invero 40i transit bus. Rather than using a single 180 kW PEMFC power module, Hydrogenics would incorporate a distributed array of three discrete modules, providing flexibility for both grid and road power. This vehicle-to-grid capability would enable a vehicle's power supply to provide off-board power and power to the electrical grid while the vehicle was sitting idle. Another distinct feature was the use of ultra capacitors and regenerative braking for peak power demands. Hydrogen fuel would be stored in high-pressure storage tanks (35 MPa).

- Other project participants included Dynetek Industries Ltd., ISE Research, Maxwell Technologies, and the Province of Manitoba.
- NRCan contributed CDN$3 million toward the project, which was scheduled for completion in March 2005.

7.4.1.19. Nissan

Nissan, Japan's third largest car manufacturer after Toyota and Honda, had an interest in fuel cell technology early on and was a long-time customer of Ballard.[19] Nissan purchased its first fuel systems from Ballard in 1991. Nissan began a full-scale FCV development in 1996.[157] Ballard received further orders from Nissan in March 1997 (worth $1.6 million), in January 1999 ($2.3 million), and in March 2001 ($2.2 million).[19] A delivery of Mark 900 fuel cell power modules worth $1.6 million was made to Nissan in January 2001.

Nissan presented its first fuel cell concept vehicle at the 1997 Tokyo Motor Show and unveiled its first FCV prototype, the ***R'nessa*** SUV, also called Nissan Altra EV, 2 years later.[158] In May 1999, Nissan joined CaFCP[159] and started driving tests of the R'nessa SUV, a fuel cell battery hybrid FCV. The vehicle was equipped with a methanol reformer to generate hydrogen on board and a10 kW Ballard PEM fuel cell stack.[19] The vehicle was propelled by a neodymium magnet synchronous traction motor combined with lithium ion batteries, using a control system that enabled the vehicle to switch between a fuel cell–powered driving mode and a battery-powered driving mode depending on operating conditions. The methanol reformer, however, took 10–20 min to warm up before generating hydrogen, and the car reached a top speed of only 44 mph.

In October 2000, Nissan unveiled another fuel cell prototype, the ***Xterra FCV***. The FCV, which was based on Nissan's Xterra SIV, was powered by a 75 kW (Mark 900) Ballard fuel cell.[19] In April 2001, public road testing of the Xterra FCV, powered by high-pressure hydrogen, began under CaFCP. Driving tests of the Xterra FCV powered by reformed methanol began in May 2001. Efficiency and maximum speed were 45 percent and 75 mph, respectively.

In March 1999, Renault acquired a 36.8 percent equity in Nissan,[19] marking the beginning of the two companies' long collaborative relationship in FCV development. In April 2001, Nissan entered into a 5 year, ¥850 million[159] (about $7 million at the 2001 exchange rate of $1/¥125.57) alliance with Renault to develop FCVs that run on gasoline.[158]

In February 2002, Nissan signed a joint fuel cell system development agreement with UTC,[160] marking a change in Nissan's fuel cell strategy, which was previously exclusively associated with Ballard hardware. Renault also participated in these fuel cell projects.[19] UTC provided fuel cells for both the 2002 and 2003 ***Nissan X-TRAIL*** fuel cell hybrid vehicles.[160]

In November 2002, the Nissan X-TRAIL FCV high-pressure hydrogen-powered FCV received approval from the Japanese Minister of Land, Infrastructure, and Transport (the MLIT) to start public road testing in Japan.[159] (That month, Nissan also began road testing the Xterra FCV in Japan).

After receiving the 2002 MLIT approval, Nissan began driving tests on public roads in Japan. The company clearly made major improvements to the basic X-Trail FCV and announced its plans to begin limited leasing of FCVs by March 2003, 2 years ahead of its original commercialization date of 2005.[159]

- In December 2003, Nissan unveiled the X-Trail 2003 model that used the Neo FC System.[159] The X-Trail FCV '03 had the top speed of 90 mph and an extended cruising range of more than 217 miles, 1.75 times longer than that of the 2002 Model. The FCV was powered by a 63 kW UTC fuel cell stack and lithium-ion batteries, and fueled by compressed hydrogen.
- In December 2003, a limited number of X-Trail '03 FCVs were made available for lease. X-Trail '03s were delivered to Cosmo Oil in March 2004 and the Kanagawa prefectural government and the Yokohama city government in April 2004.[159]

In October 2003, Nissan presented a newly developed fuel cell urban vehicle—***Effis***—during the 37th Tokyo Motor Show. The company was upbeat and announced that it would lease the Effis to selected partners by March 2004.[158]

7.4.1.20. Nova Bus

Nova Bus, owned by the Volvo Bus Corporation, is a bus manufacturer headquartered in Quebec, Canada.[161] In 1993, the company became involved in a fuel cell bus development program administered by Georgetown University (GU).

- GU administered a feasibility study on the application of fuel cells to buses for the U.S. Department of Transportation in 1983.[162] The study was led by Los Alamos National Lab and was completed in 1984. It concluded that urban transit buses were ideally suited for fuel cell power.
- In 1987, GU began the development of two methanol-powered 25 kW PAFC brassboard systems to demonstrate the technology, which was successfully completed in 1990.[56] One system was developed by Engelhard teamed with Fuji Electric, and the second by Energy Research Corporation (now FuelCell Energy).[162] In 1991, the Federal Transit Administration (FTA) awarded GU to manage the ***Generation I Bus Program***, in which three methanol-fed 30 ft buses were to be built and powered by a 50-kW PAFC system. Fuji Electric was contracted to build them because by that time Engelhard had abandoned PAFC development and licensed its technology to Fuji. The (now defunct) Bus Manufacturing USA, Inc. (BMI) of Santa

Barbara[163] built the three buses, and rolled them out in 1994 and 1995.[162]

In 1993, Nova Bus was contracted by the FTA to build two methanol-powered, 40 ft hybrid ***Generation II Buses***. The project was aimed at demonstrating the commercial viability of fuel cell buses.[28]

- Nova Bus built the first in May 1998. The bus used a 100 kW PAFC system built by UTC Power.
- Nova Bus rolled out the second bus in 2000. By that time, the PEMFC technology was believed to be the most suitable power source for buses and so this bus used a 100 kW PEMFC system developed by Ballard.[164]

GU 's subsequent ***Generation III Program*** was planned to begin in the near term in partnership with EPRI, ZSW, and NuCellSys. The Generation III Buses were also planned to use PEMFC systems, further fueling excitement over PEMFC technology.

- The GU Generation III Program concluded in July 2011 when federal funding ran out.[165] A methanol fuel cell system design and prototype hardware had been completed by then, along with extensive testing.

7.4.1.21. PSA Peugeot Citroen

PSA Peugeot Citroen is the second largest automaker in Europe and the sixth largest in the world.[166] PSA started FCV development in the mid-1990s under the sponsorship of the French government and the EU FP programs. The company also had a number of alliances with research institutes and corporations that helped advance its FCV capabilities.

In January 1996, PSA began the development of the ***Hydro-Gen FCV*** sponsored by European Union's 63-month long (January 1996–March 2001) Fourth Framework Program (the FP4). PSA was the coordinator of the Hydro-Gen project, working with Ansaldo (Italy), Commissariat a l'Energie Atomique (CEA), De Nora Fuel Cells (now Nuvera Fuel Cells) (Italy), Renault (France), Solvay (Belgium), and Sorapec (France).[167]

- In June 2001, PSA unveiled the Hydro-Gen FCV,[168] based on a Berlingo van.[169] The Hydro-Gen was a hybrid electric hydrogen FCV propelled by a 20 kW electric motor (210 N·m maximum torque) powered by a 30 kW (70 V) De Nora (now Nuvera) fuel cell and a 162 V battery for energy recovery. With 3.3 kg of hydrogen stored at 35 MPa in a pressure tank (140 l) developed by CEA, the vehicle could run 186 miles and reached a maximum speed of 59 mph.

Also in June 2001, PSA presented the Peugeot Fuel Cell Cab, the ***Taxi PAC***, [168] conceived as a "London taxi for the 21st century."[170] The Taxi PAC demonstrator was based on an electric vehicle, the Peugeot Partner, powered by its original electric motor (rated at 22 kW) and equipped with a small, 5.5 kW H Power fuel cell, nickel–metal hydride batteries and an interchangeable rack of hydrogen bottles.[171] The hydrogen was stored on board in the form of gas compressed at 300 bar.[170] An empty rack could be quickly replaced with a full one and the hydrogen bottles were easily refilled when off the vehicle. In the Taxi PAC, the fuel cell served as a range extender, which recharged the batteries, and which delivered electricity to the traction motor as needed. The vehicle could drive between 124 and 186 miles.[19]

In September 2002, PSA unveiled the ***H_2O Fire Prevention Vehicle*** at the Paris Motor Show.[172] The two-seater H_2O demonstrator was a battery-powered fire truck equipped with an electric generator with a 5.5 Millennium Cell fuel cell.[173] The H_2O represented a new application for fuel cells as an auxiliary power unit (APU).[174] It recharged the batteries, which delivered electricity to the drive motor.[170] In the H_2O, hydrogen was produced on board using an aqueous sodium borohydride solution and a catalyst. The vehicle therefore never carried more than 2.5 g of hydrogen on board, the energy equivalent of a glass of gasoline. The H_2O had all its capabilities in an anaerobic (oxygen-free) environment, enabling it to fight fires in a tunnel or an underground parking lot. The oxygen required to power the vehicle was supplied from two onboard canisters. Ultracompact breathing systems were also included in the passenger compartment.

PSA appeared to be taking a long-term view of fuel cell development and commercialization from early on.[175] In the 1990s and early 2000s, the company believed that FCVs would be gradually introduced after 2010. Entering the 2000s, the company changed its mind and assessed that it would be in about 2020 when mass production of FCV with onboard hydrogen storage would start.

7.4.1.22. *Renault*

Renault, another major French automaker in addition to PSA Peugeot Citroen, has been more modest in supporting fuel cell development. Nonetheless, Renault has, attracted close attention from automakers around the world.

- Renault has a 44.3 percent stake in Nissan, and Nissan holds a 15 percent stake in Renault. Combined vehicle sales by Renault and Nissan in 2008 reached 6.9 million, making the Renault–Nissan Alliance the world's third largest automaker.[176]

In August 1997, Renault presented the ***FEVER Renault Laguna Estate*** in Grimstad, Norway. The development of the FCV was sponsored under the EU FP3 project titled "Fuel Cell Powered Electric Vehicle for Efficiency and Range (FEVER)" (January 1994–May 1998).[177] Renault designed, built, and tested the liquid hydrogen-fueled, two-seated FCV, based on the Renault Laguna Nevada station wagon. The partners included Italy's De Nora (now Nuvera) (the fuel cell supplier) and Ansaldo (assembly of secondary systems), Air Liquide of France (the hydrogen tank), and Volvo of Sweden (simulations).[169] The FEVER was powered by three 10 kW Nuvera PEM fuel cells (115 cells), weighing 320 kg and taking up a volume of 225 l, assisted by a NiMH battery with an energy reserve of 2.8 kWh. It also took advantage of regenerative braking.[19] The 115 kg Air Liquide hydrogen tank contained 8 kg or 120 l of liquid hydrogen. The 2200 kg vehicle could drive 312 miles and reached a maximum speed of 75 mph.

Since then, Renault has had a few other partnerships to develop fuel cells and fuel reformers:

- From January 1996 to March 2001, Renault participated in the FP4 HYDRO–GEN project to develop an FCV. Peugeot led the effort to develop a hydrogen-fueled PEMFC vehicle using a Berlingo van, in partnership with Ansaldo, CEA, Nuvera, DeNora, and Air Liquide.[178]
- In May 1999, it finalized a research partnership with BMW and Delphi on solid-oxide fuel cell APUs for diesel cars and trucks.[179] Plans called for prototypes by 2008 and commercial products by 2015.[28]
- From July 2000 to June 2004, Renault engaged in an FP5 project— the "Fuel cell systems and components general research for vehicle applications (FUERO)."[180] The €4.5-million project aimed to identify the operational requirements for the operation of fuel cell systems in vehicles, including studies of energy sources and efficiency. Other major participating automakers included Peugeot, Volkswagen, and Volvo.
- In April 2001, Renault launched a $7 million gasoline reformer program jointly with its partner, Nissan.[28] Renault and Nissan said they intended to develop gasoline powered FCVs, with a market date as early as 2005.
- In February 2002, Nissan announced an agreement with UTC Fuel Cells, in which Nissan would develop fuel cells and fuel cell

components for vehicles with UTC.[19] Renault, a partner and shareholder in Nissan, also participated in the project.

- In July 2002, Renault signed a multiyear agreement with Nuvera to develop multifuel processing technologies for FCVs.[181] The joint program sought to deliver by 2004 a fuel processor that could be adapted to onboard vehicle requirements for Renault. Also in April 2003, Nuvera entered into an agreement with TotalFinaElf, a global oil company, to study the effects of gasoline on fuel processors and fuel cell stacks in fuel cell cars, as well as to identify the constituents of reformate that would adversely affect PEM (proton exchange membrane) fuel cell life.[182]

In the early 2000s, Renault seemed to be planning to commercialize its FCVs by about 2010.[183] To prepare for the move, the company established alliances with key fuel cell players. It cultivated partnerships with Nissan and Nuvera to develop an onboard gasoline reformer. It also clearly counted on Nuvera, UTC, and Nissan for the development of fuel cell stacks.

7.4.1.23. *Scania*

Scania, a Swedish automotive industry manufacturer of heavy trucks and buses founded in 1891, is the world's fourth largest bus manufacturer.[184] MAN has a 17.37-percent voting stake in Scania.[185]

In 1996, Scania participated in the FP5 project, "FC Bus Fuel Cell Midi City Bus," that aimed to build a hybrid fuel cell bus.[186] The hybrid fuel cell bus was based on a Scania Midi City hybrid Bus developed by Scania and ZF (ZF Friedrichshafen of Germany, a global supplier of driveline and chassis technology) because the model already had all the necessary hybrid subsystems. Other participants included Air Liquide, CEA, Genoa University, Nuvera, and SAR.[186] The overall project budget was funded at €4.3 million.

- In May 2001, Scania presented the ***Scania MIDI City Bus*** at the UITP (L'Union internationale des transports publics) Mobility and City Transport Exhibition in London.[187] The bus was designed for inner-city operations; it was 9.2 m long and had an extremely low flat-floor bus that optimized accessibility. The fuel cell hybrid bus was equipped with a hybrid electric propulsion system, integrating two 30 kW Nuvera fuel cell stacks, two wheel hub motors each with 50 kW peak power, driveline electronics, a 600 V battery, subsystems for steering and brakes, and a 24 V onboard power supply. The fuel cell bus was fueled with pressured hydrogen (at 20 MPa) and had a driving range of 155 miles with a top speed of 50 mph. It accelerated from 0 to 30 mph in 18 s.

7.4.1.24. *Suzuki*

Suzuki Motor, Japan's second largest subcompact car maker next to Daihatsu and the ninth largest automaker in the world, specializes in manufacturing compact automobiles and 4 × 4 vehicles, a full range of motorcycles, all-terrain vehicles (ATVs), outboard marine engines, wheelchairs, and a variety of other small engines.[188]

- In 1981, Suzuki formed a partnership alliance with GM, which made a 5.3 percent capital investment in Suzuki. GM subsequently increased its stake in Suzuki to 10 percent in September 1998[189] and to 20 percent in September 2000.[190]

In October 2001, Suzuki established an agreement with GM to collaborate on FCVs as part of GM's FCV development effort.[191] The main goal of the collaboration with Suzuki was the miniaturization, weight reduction, and integration of fuel cell systems into compact and small-sized cars.[19]

In October 2003, Suzuki unveiled its first FCVs at the Tokyo Motor Show. In fact, it presented three FCVs, which included the following:

- The ***Suzuki Wagon R-FCV***. The FCV was powered by a 50 kW fuel cell system developed by GM. It used two 35 MPa compressed hydrogen tanks.[192] It had a 33 kW AC synchronous electric motor, but there was no secondary battery.[193] The FCV had a driving range of 80 miles and a top speed of 68 mph.
- The ***MR Wagon-FCV***, a mini FCV that seated four adults.[28] The FCV had exactly the same specifications as the Suzuki Wagon R-FCV.
- The ***Mobile Terrace***, a six-passenger concept car with a novel design. The FCV's sliding doors on both sides and the gullwing roof opened wide, and the floor extended outward. Three rows of seats rotated around a table, which slid out from the instrument panel.[194] The 4-m-long body featured an underfloor fuel cell system based on GM's "Hy-wire" platform and optimized by Suzuki to fit in small cars.

In October 2003, the Wagon R-FCV and the MR Wagon-FCV obtained ministry certification to drive on public roads and Suzuki started field tests that month with the Japan Hydrogen and Fuel Cell Demonstration Project (JHFC), sponsored by Japan's METI.[28]

- In December 2004, Suzuki presented the ***improved MR Wagon-FCV***, which used 70 MPA compressed hydrogen storage tanks.[195] That month, the FCV obtained ministry approval for public roads, the first ministerial approval for FCVs that used hydrogen compressed at that pressure. The FCV was to start field tests in January 2005. The MR Wagon-FCV had a driving range of 124 miles and a top speed of 68 mph.
- Suzuki had three MR Wagon-FCV vehicles with ministerial permits for public road driving.

Suzuki had a much longer-term commercialization strategy than GM regarding FCVs. GM stated that it planned its FCV commercialization to begin between 2004 and 2005.

- In October 2003, Suzuki announced that its commercialization would occur by 2010.[196]

7.4.1.25. *Thor Industries (ThunderPower)*

Thor Industries, founded in 1980 and headquartered in Jackson Center, Ohio, builds mid-sized buses and RVs.[28]

In June 2000, Thor announced that it would build a zero-emission, high-performance, low-cost bus in partnership with UTC and ISE Research.[197] The company hoped the fuel cell bus would be the world's first commercially viable, fuel cell transit bus. Funding partners for this project were the U.S. Department of Transportation, Sacramento Municipal Utility District (SMUD), CALSTART, and South Coast Air Quality Management District.[198]

By the end of 2001, the ***ThunderPower Bus*** was built.[199] The bus was based on a low-floor E-Z Rider transit bus built by Eldorado National, a Thor company. The ThunderPower Bus was powered by a 75 kW UTC PEM fuel cell, which produced electricity both to propel the bus and charge the batteries. The bus used a drive system by ISE, featuring a dual motor/controller set that had a continuous power rating of 170 kW and a peak power rating of 288 kW. It was fueled with compressed hydrogen and had a driving range of 175–200 miles and carried 26 passengers.

In November 2002, the Thor's 30 ft bus went into operation with the Sun Line Transit Agency, the first fuel cell hybrid bus to enter into passenger service in California.[199] The DOE National Renewable Energy Laboratory evaluated the performance of the ThunderPower Bus while it operated at the Sun Line.

- ThunderPower, LLC, was under contract to demonstrate the ThunderPower bus with at least on other transit agency in California, probably the AC Transit in the San Francisco Bay Area.

Following the demonstrations, commercial products were expected to follow within 12 months.[200]

7.4.1.26. Toyota

Toyota began fuel cell development work in 1992. In the early years, it likely relied on Ballard's PEMFC stacks, but by 1996 when it presented its first FCVs to the public, Toyota had already developed its own PEMFCs. In October of that year, Toyota displayed its first ***RAV4L FCEV*** at the 13th International Electric Vehicle Symposium in Osaka.[19] The RAV4L FCEV was a hybrid SUV powered by a 20 kW PEMFC stack (developed inhouse) and a backup battery. It ran on hydrogen stored in a metal hydride system. The unit stored 2 kg of hydrogen, which was about 2 percent of the weight of the 100 kg metal hydride. The car had a top speed of 60 mph and a driving range of 109 miles.

- Toyota's hydrogen storage system stirred global excitement; metal hydride could carry hydrogen in a far less combustible form.

In September 1997, Toyota unveiled the ***RAV4 FCEV***, which featured a methanol reformer.[19] The methanol-powered RAV4 FCEV had a 25 kW PEMFC system and a NiMH battery with a regenerative braking system. The FCEV had the top speed of 80 mph and a driving distance of 310 miles.

In March 2001, Toyota announced development of the ***FCHV-3*** (Fuel Cell Hybrid Vehicle) at the International Symposium on FCVs in Tokyo.[201] The vehicle was based on Toyota's Highlander SUV. The powertrain consisted of a 90 kW fuel cell system fueled by onboard compressed hydrogen and a nickel–metal hydride accumulator battery. The FCHC-3 had a secondary battery for storing energy created during braking and had other features to ensure high-efficiency driving, such as precise control of the charge and discharge of the secondary battery and the accumulator battery.[19]

In June 2001, Toyota presented the ***FCHV-4*** at the Fourth Toyota Environmental Forum in Tokyo. The FCHV-4 was Toyota's first production FCV[202] and was powered by hydrogen stored in four high-pressure (25 MPa) tanks fitted under the rear passenger compartment.[19] The FCHV-4 had a range of 156 miles—slightly less than the FCHV-3's range—in stop-and-go city driving. Each of the hydrogen tanks held the equivalent of 11.2 l of gasoline. Toyota hoped to raise tank pressure to 35 MPa in 2002 and 50 MPa in 2004, increasing the vehicle's driving range to 217 miles and 310 miles, respectively. All major components—stack, compressor, motor, and power control unit—were fitted into the front engine compartment. Also, the FCHV-4 had a newly developed heat pump air conditioning system that used CO_2 as the refrigerant instead of hydrofluorocarbons (HFCs).

- Later in the year, Toyota began test driving the FCHV-4 on public roads in Japan and in the United States in cooperation with the CaFCP.[203]
- Also, in 2001, Toyota announced plans to start selling fuel cell cars based on Toyota's FCHV-4 by 2004[vii] in Japan.[19]

In October 2001, Toyota presented the gasoline-powered ***FCHV-5*** at the 35th Tokyo Motor Show,[19] 2 months after GM unveiled its Chevrolet S-10 pickup equipped with a gasoline reformer. The FCHV-5 was a fuel cell hybrid

[vii] Japanese corporations typically use the fiscal year and Toyota referred to fiscal year in its announcement; a marketing date for fiscal year 2003 would mean the year starting in April 2003 and ending in March 2004.

vehicle that generated electricity from hydrogen derived from Clean Hydrocarbon Fuel (CHF),[viii] using Toyota's CHF reformer. The CHF reformer was packaged with a newly developed catalyst and heat exchanger. Main system components in the new FCHV-5 included a 120 l reformer (positioned under the rear seat), a 35 l fuel tank (located under the passenger compartment floor), a power control unit, a permanent magnet synchronous motor, and the Toyota-developed fuel cell stack, all fitted in the front engine compartment. A rechargeable nickel–metal hydride battery was fitted above the rear axle.[19] The reformer module included a mixer, evaporator, combustor, heat exchanger, high- and low-temperature shift catalyzers, and carbon-monoxide remover. The battery pack, which operated at 288 V, was supplied by Panasonic EV Energy, a joint venture of Toyota, Matsushita Electric Industrial, and Matsushita Battery Industrial. Like other Toyota hybrids, FCHV-5 employed energy regenerative technology whereby kinetic energy during deceleration and braking was channeled into the vehicle's battery for reuse during acceleration.

- GM and Toyota likely jointly developed the gasoline reformer under their cooperative agreement.

In early 2002, Toyota was confident that it could soon start marketing its FCVs in Japan and in the United States. In January 2002, Toyota integrated all of its FC researchers into one unit and established the FC Development Center[204] to strengthen FCV development as well as its service capabilities. In July 2002, the company advanced the FCV commercialization target date from March 2004 to March 2003.

- Toyota started commercial leasing of the ***Toyota FCHV*** in January 2003, 3 months earlier than the revised commercialization date of March 2003.[204]

7.4.1.27. Tongji University (China)

The Center of Automotive Engineering at the Tongji University has had long experience in vehicle development and has worked closely with the Chinese government to develop fuel cell and hybrid vehicles.[205] In 2000, China launched a major national program to promote FCVs, including cars and buses, under the 10th 5-year 863 High Technology Development Program (named after its start date in March 1986). It directed Tongji to take charge of fuel cell car development. Specifically, Tongji was to develop a special fuel cell electric vehicle and three generations of fuel cell powertrains that could be used in various vehicle types. In 2001, Tongji's Automotive Engineering Center spun off the Shanghai Fuel Cell Vehicle Powertrain, a joint venture company established with Shanghai Automobile Industry Corporation (SAIC). Tongji also partnered with Shanghai Shen Li High Tech Company (Shen Li), China's leading PEMFC developer established in 1998, to be its PEMFC supplier. Subsequently, the team of four would demonstrate to the world a remarkable ability to increase the power of its fuel cell stacks and advance the performance of its fuel cell cars:

- In August 2003, Tongji presented its first-generation fuel cell powertrain platforms, the ***Tongji University Chao Yue I***.[206] The fuel cell powertrain consisted of a 30 kW Shanghai Shen Li (Shen Li) PEMFC, 30 ampere-hour (Ah) lithium-ion batteries, and a 50 kW traction motor that was integrated into a Volkswagen Santana 2000 model.[207] It ran on compressed hydrogen (10 MPa) and had a driving range of 130 miles and a top speed of 68 mph.[207] It accelerated from 0 to 50 mph

[viii] CHF is a liquid fuel that can be derived from crude oil, coal or natural gas, has a clean sulfur content, and can be dispensed from current gasoline pumps.

in 15.4 s. Its fuel efficiency reportedly was 1.4 kgH_2/62 miles.[208]

- In 2004, Tongji unveiled the second-generation ***Tongji University Chao Yue II*** at Challenge Bibendum in Shanghai.[209] The Chao Yue II was based on a VW Santana 3000 model. The FCV was powered by a 35 kW Shen Li PEMFC, 15 Ah lithium-ion batteries, and a 60 kW traction motor.[208] It ran on compressed hydrogen,[206] had a top speed of 71 mph, and accelerated from 0 to 62 mph in 26.7 s.[208] It had a driving range of 135 miles. The fuel efficiency was 1.13 kgH_2/62 miles. Two vehicles of this kind were built.
- In 2005, Tongji developed its third-generation fuel cell powertrain, the ***Tongji University Chao Yue III***. The powertrain system consisted of a 40 kW Shen Li PEMFC, 15 Ah lithium-ion batteries, and a 65 kW permanent magnet synchronous motor.[208] It had a top speed of 76 mph and a driving range of 143 miles. It accelerated from 0 to 62 mph in 19 s and had the fuel efficiency of 1.12 kgH_2/62 miles. The powertrain platform was used in two passenger vehicle models by two different automakers.
    - Tongji University built the ***Chao Yue III S3000*** in partnership with Shanghai Volkswagen, using the Santana 3000 as the base vehicle.[210]
    - The ***Chao Yue III MPV*** was built in partnership with Chery Automobile, a state-owned automaker. The vehicle was based on a Chery Easter MPV model, which incorporated a third-generation Chao Yue powertrain with modifications to accommodate both compressed hydrogen and gasoline as fuel.[211]
    - The ***Chao Yue III Eastar***[212] used the same vehicle model and powertrain as the Chao Yue MPV. It also ran on both hydrogen and gasoline.

Ten Chao Yue III vehicles were manufactured,[206] including eight Chao Yue III C3000 vehicles, one MPV, and one Eastar.

- According to an assessment of reliability and endurance, the 10 Chao Yue III vehicles drove a cumulative 63,342 miles, and a single FCV operated for 18,009 miles without repair.[213]
- Of the three models, the Chao Yue III S3000 model was perhaps the most successful. Eight Chao Yue III C3000 vehicles were built,[206] and two of them finished in 14th and 15th places in the 2006 Bibendum Challenge in Paris in June.[214]

In addition to the three FCV powertrains, Tongji University developed a fuel cell mini-sedan, the ***Chunhui No.1***.[215] In November 2004, Tongji unveiled the Chunhui at the Shanghai Industrial Fair. The Chunhui was one of the special electric vehicle projects under the 863 program during the 10th 5 year period.[216] The FCV was a PEMFC and battery hybrid and won an Innovation Award at the Shanghai exhibition.

7.4.1.28. Tsinghua University (China)

Tsinghua University worked on PEM fuel cells, fuel cell engines, and technologies to make hydrogen from ethanol. In 1999, Tsinghua demonstrated its ability by building a 5 kW PEM fuel cell-powered golf cart. In 2000, China assigned Tsinghua University to develop three generations of fuel cell powertrain platforms for buses under the 863 program. The first generation would be powered by a 50 kW PEMFC, the second generation by a 60 kW PEMFC, and the third by a 100 kW PEMFC.[208] These FCBs were seen as the first steps toward accomplishing China's ambitious plan to operate 100 fuel cell buses during the 2008 Olympic Games in Beijing.[48]

- In December 2002, Tsinghua unveiled its first-generation fuel cell bus, the ***Qing Neng I***, which was an 11-m-long vehicle powered by a 50 kW Shen Li fuel cell system.[206] It had a 42 percent fuel efficiency.

- In September 2003, the second-generation powertrain, the ***Tsinghua University & DICP FCB 60 kW***, was unveiled.[217] The FCB was based on an 11-m-long Jinghua Coach Company City Bus and was powered by a 50 kW Shen Li PEMFC, lithium-ion batteries, and a 120 kW Enova AC induction motor (650 Nm). It ran on compressed hydrogen (15 MPa) stored in 900 l pressure vessels carrying 9.6 kg of hydrogen. It had a top speed of 53 mph and accelerated from 0 to 31 mph in 30.9 s. During the first year, the FCB operated for over 3000 miles.

In May 2004, Tsinghua presented another fuel cell bus powered by a 60 kW PEMFC, the ***Tsinghua University Fuel Cell City Bus 863***, at the International Hydrogen Forum in Beijing.[218] The FCB was developed with partial support from the US FTA. The bus was based on the 11 m Jinghua Coach bus and was powered by a 60 kW Shen Li PEMFC, nickel–metal hydride batteries, and a 120 kW Enova AC induction motor (650 Nm). It used compressed hydrogen (15 MPa). It had a top speed of 43 mph and accelerated from 0 to 31 mph in 25 s.[217] The bus carried the same amount of compressed hydrogen as the second-generation FCB.

- In 2006, Tsinghua University presented its third-generation ***Tsinghua University FC Low Floor City Bus*** (2006).[219] The bus was based on the 12 m Jinghua Coach low-floor city bus. It was powered by a 100 kW Shen Li High Tech PEMFC, nickel–metal hydride batteries, and 120 kW Evova AC induction motor (650 Nm). The bus had a top speed of 53 mph and accelerated from 0 to 31 mph in 25.5 s.

7.4.1.29. Van Hool

Van Hool, founded in 1947, is a Belgian manufacturer of buses, coaches, trolleybuses, and trailers.[220] Van Hool became involved with fuel cell buses starting in the late1980s.

- In 1988, Van Hool participated in the Eureka[ix] project to develop a fuel cell bus.[221] In November and December 1995, the ***Elenco EUREKA City Bus*** was demonstrated at Brussels. The fuel cell bus was based on a Van Hool-articulated city bus and powered by 78 kW Elenco Alkaline fuel cells, nickel–hydride batteries, and a 200 kW hybrid electric drive.[222]

In the early 2000s, Van Hool participated in the development of fuel cell buses in the United States. At the end of 2001, AC Transit started discussions with ISE on developing a new fuel cell bus design. In April 2002, negotiations were started with ISE and UTC with plans to use Van Hool buses.[223] In April 2003, a contract to develop the fuel cell buses was signed. This order included three new Van Hool fuel cell buses for AC Transit and one new Van Hool fuel cell bus for SunLine. Van Hool designed and constructed an entirely new chassis based on the existing A330 bus design so that it could more effectively accommodate fuel cell and hybrid system components.

- The Van Hool fuel cell bus was to be powered by a UTC fuel cell system integrated with an ISE hybrid electric drive systems.
- Funding for this project included more than $21 million from public agencies and private sector companies, including the State of California ($8 million), the California Air Resources Board ($2.8 million), Bay Area Air Quality Management District ($1 million), Federal Transit Administration ($1 million), California Energy Commission ($1 million),

[ix] EUREKA, a pan-European research and development funding and coordination organization established in 1985, consists of European member states and the European Commission.

DOE ($700,000 plus partial funding for the hydrogen station construction and operating expenses), AC Transit matching funds ($1.2 million), CALSTART-WestStart ($200,000), Chevron Technology Ventures ($2 million), and miscellaneous contributions (more than $3 million).[223]

- The buses were scheduled to be delivered in September 2005 and the demonstrations were to be conducted as part of the CaFCP.[224]

7.4.1.30. *Volkswagen*

Volkswagen (VW) started developing fuel cells in January 1996, when it first served as the coordinator of the EU FP4 project, "Car Autothermal Process Reactor Initiative (*CAPRI*)."[225] The CAPRI project (January 1996–December 2000) aimed to develop a fuel cell battery hybrid vehicle powered by reformed methanol. The total project budget was about €4 million (about $3.12 million at the 1996 exchange rate of $1/€0.78).[226] VW modified a VW Golf station wagon to incorporate a 15 kW Ballard PEMFC stack and a methanol reformer developed by Johnson Matthey (UK), with the Netherlands' Energy Research Foundation (ECN) responsible for the systems integration, and Volvo for fuel cell system modeling, testing, and the air compressor/expander development.[169] Little information on the other Volkswagen CAPRI project is available.[227]

In November 2000, VW unveiled its first fuel cell car, the ***Volkswagen Bora HyMotion***, at the grand opening of CaFCP.[228] The Bora HyMotion was based on the VW Jetta and featured a 75 kW PEM fuel cell of unknown German origin, which powered a 75 kW asynchronous motor, giving a top speed of 87 mph. The vehicle tank held 50 l of liquid hydrogen, giving the Jetta a driving range of about 217 miles. It accelerated from 0 to 62 mph in 12.6 s reaching a top speed of 87 mph.

In January 2002, VW unveiled the ***Bora HyPower***, incorporating a PEM fuel cell stack developed by the Paul Scherrer Institute.[19] With a 28 kW fuel cell and supercapacitors for brake energy recovery, it could drive approximately 93 miles on a tank of compressed hydrogen (at 350 bar).[233] Top speed was 72 mph.[19] In 2002, the vehicle was tested over a Swiss mountain pass along with an advanced diesel-powered vehicle operating on a synthetic liquid hydrocarbon. The FCV's performance was comparable to that of the diesel.[28]

- VW made it clear that it was more enthusiastic about developing advanced diesel engines and other high-efficiency combustion engines than fuel cell cars.

7.4.2. Many Governments Join the Race to Boost Domestic PEMFC Capabilities

Undoubtedly encouraged by the wide range of fuel cell development activities by the Ballard alliance and other global automobile and bus manufacturers, many governments began to put in place major policy and financial support programs to promote their nations' PEMFC capabilities, not only for transport applications but also for stationary and portable applications. The Japanese government was the most aggressive in this regard:

- Japan increased its budget for PEMFCs eightfold from ¥1 billion (about $8.8 million at the 1999 exchange rate of $1/¥114) in 1999 to about ¥4 billion ($37 million at the 2000 exchange rate of $1/¥108) in 2000, and quadrupled it to ¥21 billion ($181 million at the 2003 exchange rate of $1/¥116) by 2003.
- The Unites States, which maintained a small PEMFC R&D budget since 1993 under the Partnership for a New Generation of Vehicles (PNGV), implemented a sweeping FreedomCar Initiative in 2002 to develop FCVs and a Hydrogen Fuel Initiative in 2003 to develop a hydrogen fuel supply infrastructure. As a result, the US PEMFC budget increased to $54 million in 2003 and to $64 million in 2004.

- Europe, including the European Commission and European governments, invested a total of about $60 million per year in fuel cells from 1999 to 2001, with about 60 percent devoted to PEMFC and hydrogen technology R&D. The EU planned to double the budget to $120 million in the following years.

7.4.2.1. *Japanese Government Begins Modest R&D Investment in 1992*

Taking note of the development of PEMFC technology in Canada, Japan's Ministry of Economy, Trade, and Industry (METI, formerly MITI) implemented a modest 8-year PEMFC system technology development project in 1992 in two phases: Phase I (1992–1995) focused on the development of 1 kW modules for stationary and portable applications,[229] and Phase II (1996–2000) focused on the development of 5–10 kW potable, residential, and distributed power generation systems. The Phase I budget was ¥663 million (about $5.8 million at the average exchange rate $1/¥115 for the period 1992–1995),[230] and the Phase II budget was ¥3272 million (about $28.7 million at the average exchange rate of $1/¥114 during the period 1996–2000). Participants included Aisin, Asahi Chemical, Asahi Glass, Mitsubishi Electric, Sanyo, and Toshiba.[229]

- Both phases of the project were viewed as highly successful.[17] All power generation systems operated at low temperatures, were compact in size, and had excellent start-up characteristics.

7.4.2.1.1. JAPAN MAKES DRAMATIC STRATEGIC MOVES IN 2000

A METI committee in August 1999 reviewed the outcome of the 8-year R&D program and concluded that Japanese PEMFC capabilities had caught up with those of their foreign counterparts.[229] The committee decided, however, that long-term R&D would be necessary to bring the technology to where PEMFCs could be made into products and commercialized. The committee proposed that a next generation PEMFC project be implemented to nurture Japan's PEMFC capabilities.

- The committee proposed that Japan establish support mechanisms to promote commercialization, including subsidies, tax incentives, safety standards, certification systems, and systemization standards common to electric power systems.[229]

In view of the accelerating pace of global PEMFC development, Japan decided to move quickly. In December 1999, the Japanese Prime Minister implemented the Millennium Project that funded PEMFC development for FCVs and residential Combined Heat and Power (CHP) applications[231] and which would serve to boost METI's PEMFC R&D. The Millennium Project aimed for fuel cell cars and CHP systems to be on the market by FY2005 and it promoted the establishment of safety standards, standardization, and certification systems and hydrogen technology and infrastructure to support PEMFC commercialization. It also called for the establishment of a fuel cell commercialization promotion council, an industry association, to facilitate fuel cell development and commercialization activities.

The Millennium Project plans were acted on immediately:

- In December 1999, METI's Agency of Natural Resources and Energy (ANRE) established the Fuel Cell Commercialization Strategic Study Group, comprised of government and private fuel cell experts, to serve as an advisory group for the director of ANRE.[232]
- In December 1999, METI initiated a 4-year PEMFC system development program (2000–2004), which was to serve as a Phase III project, to facilitate PEMFC commercialization.[232]
- In March 2001, the Japanese cabinet adopted a Science and Technology Basic Plan that

placed a high priority on fuel cell R&D in Japan's S&T policy.[234]

- In March 2001, METI established the Fuel Cell Commercialization Conference of Japan (FCCJ), an industry association. The FCCJ would provide a forum for the private sector to cooperate with the government in the commercialization of fuel cells.[235] The FCCJ would study how to facilitate commercialization, coordinate regulatory reforms and standardization, help formulate the government's fuel cell policy, and promote fuel cell demonstrations.[236]
- In August and December 2001, METI adopted Japan's strategic fuel cell roadmaps to define the pathway to fuel cell commercialization.[233]
- In June 2002, the Japanese cabinet adopted the Energy Policy Basic Law that defined fuel cell R&D as essential for ensuring energy security for the nation.[233]
- In May 2004, METI laid out a mid- to long-term economic growth scenario, the New Industry Creation Strategy, which laid out the schedule for development of fuel cells and hydrogen technology (Table 7.3).[234]

In conjunction with these plans, Japan rapidly increased its PEMFC R&D funding. Starting at ¥430 million (about $3.3 million at the 1998 exchange rate of $1/¥ 132) in 1998, METI's PEMFC R&D budget had grown 60-fold within 5 years, to ¥21.02 billion by 2003 ($181.4 million at the 2003 exchange rate of $1/¥ 115.9) (Fig. 7.1).[237]

Throughout the decade of 2000s, Japan vigorously hammered out policies and programs to promote fuel cell development. In March 2006, the third Science and Technology Basic Plan defined fuel cell and hydrogen technology as a priority strategic science and technology

TABLE 7.3 Japan: Promotion Schemes and Time Schedules for Fuel Cell and Hydrogen Development

Time period	Developmental stage and goals	Actions
2000–2005	Base Building and Technology Validation	• Creation and implementation of technology development strategy • Implementation of Millennium Project to build the system foundation (standards and standardization) • Validation testing (collection of data on operation characteristics, public education) • Establishment of fuel cell fuel quality and standard
2005–2010	Introduction to Market Goal: 50,000 fuel cell cars, 2.1 GW stationary power-generation units	• Step-by-step creation of fuel supply system in preparation for product introduction starting about 2003 • Initial product purchases by government and public organizations and fuel cell companies
2010–2020	Product Diffusion Goal: 5 million fuel cell cars, 10 GW stationary power generation units	• Establishment of fuel cell infrastructure, cost reduction, market expansion and development with no subsidy • Product distribution in the public sector

Source: "Heisei 17, Post Evaluation of Research and development Policy, Solid Polymer Fuel Cell/Hydrogen Energy Use Program," Evaluation Report, http://www.meti.go.jp/policy/policy_management/17fy-jigo-hyoka/kotaisuiso/17fy-jigo-kotaisuiso.pdf, METI, Japan, June 2006.

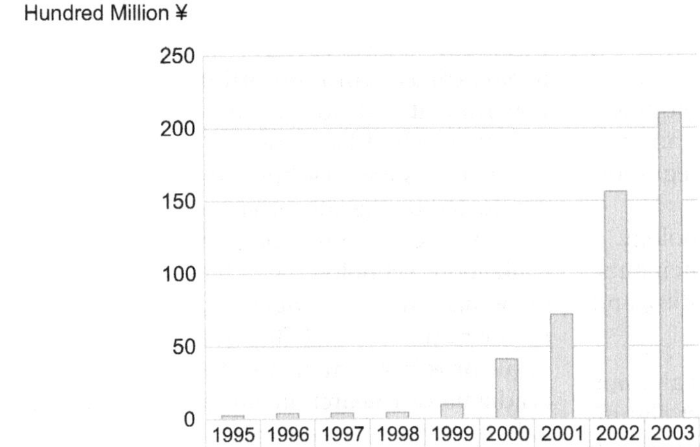

	1995	1996	1997	1998	1999	2000	2001	2002	2003
METI PEMFC Budget	2.4	3.8	3.8	4.3	10	41.03	71.89	156.2	210.2

FIGURE 7.1 METI fuel cell R&D budget: 1995–2003* (hundered million yen). Japanese METI fuel cell budget is not always clearly distinguished by fuel cell types. Between 1995 and 2003, however, the PEMFC budget was easily identifiable due to METI's strong budgetary focus on it. The budget here is presented in hundred millions of yen; an exchange rate of $1/¥100, although not precise, would help make a rough estimate of the Japanese budget in terms of the US dollar. (For the color version of this figure, the reader is referred to the online version of this book.)

area. In March 2007, the Energy Basic Law reasserted the importance of fuel cell technology for next generation automobiles. In April 2008, METI implemented the Innovation Program Basic Plan, encompassing the promotion of seven innovation programs, including energy.[238] Under the associated Energy Innovation Program, METI began implementation of several ambitious PEMFC R&D projects.

7.4.2.1.1.1. MINISTRY OF ECONOMY, TRADE, AND INDUSTRY IMPLEMENTS MAJOR PEMFC R&D PROJECTS METI's NEDO implemented three notable PEMFC R&D projects starting in 2000. They included two stationary PEMFC R&D projects, administered consecutively (in 2000–2004 and 2005–2009) and one portable Direct Methanol Fuel Cell (DMFC) R&D project (from 2003 to 2005).

- The research themes and approaches in these projects have changed over the years, and most interestingly, research emphasis has migrated from product development back toward basic research, and the research approach has become more collaborative and cooperative.
- NEDO might have realized that gaining a deep understanding of PEMFC technology required basic research by a consortium of collaborating researchers from various disciplines. Conversely, commissioning a few companies to make incremental improvements on pre-commercial technology might be counterproductive. As a result, Japan never implemented R&D programs for PEM FCVs.

7.4.2.1.1.1.1. PEMFC System Technology Development Project (2000–2004) NEDO implemented a PEMFC project in 2000 that was considered to be Phase III of the earlier PEMFC development project. It was meant to be the last government support phase, which was expected to be able to transition into a commercialization phase. The Phase III PEMFC system technology development project merged with another program in 2001 and by 2002 had evolved into a large-scale PEMFC System Technology Development Project, which continued until 2004.[239]

- The total budget was ¥19 billion ($176 million at the 2004 exchange rate of $1/¥108), an average ¥4.75 billion (about $44 million) per year, which was an extraordinary sum for an R&D project in Japan at that time. The expectation that this project would commercialize PEMFC technology, however, was never realized.

- The PEMFC system technology development project included 56 research topics, ranging from the development of systems integration, component technologies, and BoP technologies to manufacturing technology. The project included at least 44 participants, including 25 companies, 15 universities, and 4 research institutions: [239]

In September 2005, a post-project evaluation report stated that the Phase III project significantly advanced Japan's PEMFC capabilities in efficiency, durability, reliability, and in lowering costs. It also stated that these advances would contribute to creating the world's first residential-use PEMFC market in Japan.

- The committee, however, stated that successful commercialization of PEMFCs would require significant breakthroughs to achieve the needed efficiency, durability, and low cost.
- The committee then proposed that METI support further R&D and conduct it in a vertical, consortium structure, in which project leaders would lead collaborative, joint research activities. (METI R&D projects typically commission several corporations; each to develop a pre-commercial prototype of their own in parallel and in competition with each other). The committee also proposed that the project should include young talent from yet unexplored, different disciplines from universities and national laboratories.

7.4.2.1.1.1.2. Strategic Technology Development for PEMFC Commercialization Project (2005–2009) In 2005, NEDO implemented a 5-year PEMFC R&D project (2005–2009).[240] The total budget for the project was a whopping ¥25.86 billion ($276 million at the exchange rate of $1/¥93.7). The project had four areas of research. The basic and universal issue area had six research themes, which were to be carried out in a consortium with 100 percent government subsidies. The component technology area had 14 themes, all of which received 100 percent subsidies, but half of these were to be conducted by the consortia and the other half by groups of competing companies. The commercialization technology area had nine themes, all of which were conducted in a competitive manner and received 50 percent subsidies. Half of all project research was led by leaders from academia or research institutions, and the other half by corporate researchers. There also was one additional research area, the next generation technology development area. This segment received a 100 percent government subsidy.

Research themes were selected every year, and a total of 98 research themes were carried out during the 5 years. Over the life of the 5 year project, there were at least 42 participants, including 21 companies, 16 universities, and 5 research institutions. The proportion of participating universities and research institutions was greater than for the Phase III project, indicating the nature of research was becoming more basic and less focused on product development.[240]

- As part of the 2005–2009 project, NEDO implemented two additional projects.
  - The Fuel Cell Cutting-Edge Center Technology Research Project (FC-Cubic Project), which was to start up a center of excellence (COE) for fundamental PEMFC development as part of AIST.[241] The center would focus young scientists on investigation of electrolyte catalysts, interface mass transfer, and electrolyte membranes. The total budget was planned at ¥3.2 billion ($27 million at the 2005 exchange rate of $1/¥110) for the 2005–2009 period. The COE FC-Cubic would be located at AIST,Tokyo, and AIST, Tsukuba.[241]

- The High Performance Fuel Cell (HiPer-FC) Project, which was launched in April 2008 to continue until 2014, would focus on understanding degradation mechanisms and the creation of new, high-efficiency cells using nanotechnology for FCV use.[242] The research would be conducted in a consortium led by a project leader—a distinguished professor of physical chemistry from the University of Yamanashi—with participation of corporations and universities. The project included the start-up of a COE for Fuel Cell Nanomaterials, also headed by the same project leader. The total budget for 2008 and 2009, including the start-up of the COE, was ¥4.098 million ($39.6 million at the 2008 exchange rate of $1/¥103.4), which represented a major investment.[243] In July 2007, NEDO announced the successful visualization of oxygen distribution in fuel cells enabled by the Yamanashi University professor. Project participants included Waseda University, Shimadzu, Fuji Electric, and Hitachi.[244]

Total budget for the 5-year project and the two additional projects for the 2005–2009 period amounted to ¥23.5 million (about $250 million at the exchange rate of $1/¥93.7).

A post-project review committee stated that the project resulted in significant advancements, including major contributions to advancing the performance of stationary PEMFC CHP systems.[243] It stated, however, that progress fell short of making PEMFCs commercially competitive. The committee, however, proposed that METI continue to support PEMFC development. The committee's assessment seemed a little like asking someone to move on without knowing how far he/she must go. There had been some significant progress, but because the progress was not sufficient, more needed to be done. Going further on the current path would not guarantee achieving the ultimate goal, but at this point, no one appeared to be arguing that leadership should change course (Table 7.4).

7.4.2.1.1.1.3. Portable Fuel Cells for Electronic Devices—A Successful Development Project METI started a modest development effort on portable DMFCs in 2003. METI likely had thought that Japanese DMFC capabilities were mature enough to quickly achieve commercialization with a small government R&D support. METI implemented a 3 year project to develop fuel cells for portable information devices (2003–2005) budgeted at ¥1561 million ($14.2 million at the 2005 exchange rate

TABLE 7.4 Strategic Technology Development for PEMFC Commercialization Project (2005–2009): Target Goals to be Achieved by 2009

PEM Fuel Car System	
Efficiency	System efficiency of about 60 percent (LHV)
Durability	3000 h
Cost	¥10,000/kW (stack cost)
Stationary PEM CHP System	
Efficiency	Electrical efficiency of about 40 percent (HHV)
Durability	40,000 h
Cost	¥250,000/kW

of \$1/¥110).[245] The project aimed to develop compact, high-energy density DMFCs that would power portable electronic devices as well as help establish international standards and promote necessary regulatory changes. Participants included NEC, Hitachi, Hitachi Cable, Hitachi Maxell, AIST, and the Japan Electrical Manufacturers' Association.

- The post-project review committee assessed that the project had fostered first-in-class DMFC technology. The project progressed from component development to prototype development, and to field demonstration, satisfying virtually all project goals as well as demonstrating to the public Japan's leading position in the technology.[245] The project also helped create a system in Japan to deal with international standards and regulations. The effort subsequently led to wide acceptance and approval of the use of DMFC cartridges aboard aircraft.
- The committee, however, pointed out that current DMFC energy density still remained less than half the theoretical value.[245] Moreover, the committee noted that the current DMFC energy density was merely about the same or fell a little short of the energy density of lithium batteries.
- The committee, thus, proposed that further government support be provided to maintain Japan's global leadership in DMFC technology. The committee stated that there were problems that could not be solved by mere improvements in the current technology, and suggested the need for a research consortium consisting of new talent from industry and academia with different, diverse backgrounds.[245]

7.4.2.1.1.2. METI IMPLEMENTS FUEL CELL CAR AND RESIDENTIAL POWER SYSTEM DEMONSTRATION PROJECTS METI sponsored two fuel cell car demonstration projects, consecutively implemented over 9 years (2002–2005, 2006–2010), and one large-scale residential CHP system demonstration project (2005–2008).

7.4.2.1.1.2.1. Japan Hydrogen and Fuel Cell Demonstration Project, JHFC I (2002–2005) In March 2001, METI agreed to establish the Fuel Cell Commercialization Conference of Japan (FCCJ), a fuel cell industry association to promote cooperation between the government and industry for commercialization and diffusion of FCVs and stationary fuel cell systems.[235] Up until that point, Japan had left its automakers largely on their own to develop PEMFC technology. In 2002, at the urging of the FCCJ METI implemented a 4 year FCV demonstration project, the Japan Hydrogen and Fuel Cell Demonstration Project (JHFC) (2002–2005). The project examined the feasibility of PEMFC FCVs, hydrogen-fueling infrastructure, and its commercial potential.[246] The total budget was ¥11.17 billion (about \$101.5 million at the 2005 exchange rate of \$1/¥110).[247] The project was subsequently named the JHFC I to distinguish it from the second demonstration project, JHFC II.

- Seven automakers and one bus manufacturer participated in the program, including DaimlerChrysler, GM, Honda, Mitsubishi Motors, Nissan, Suzuki, Toyota, and Hino Motors.[246] They all supplied test FCVs to the project, which were used to verify fuel economy on regularly traveled routes as well as to conduct demonstration events.[17] In addition, one bus was operated as a city bus on a normal service line in Tokyo for 15 months, and eight shuttle buses were operated daily during the Aichi Expo to connect two sites for 6 months in 2005. These demonstrations effectively showed the new technology to the public. Two industry associations, Japan Automobile Research Institute (JARI) and the Engineering Advancement Association of Japan, managed the project.[246] JARI organized a working group of carmakers to plan and execute vehicle tests, data analysis, and vehicle performance evaluation.

- The participation of foreign corporations in Japanese national projects is uncommon. Japan was perhaps concerned about criticism from its trading partners and hoped the inclusion of foreign automakers in the project would preclude the appearance of giving Japanese automakers an unfair advantage over foreign competitors.

The project installed and operated ten hydrogen-fueling stations in the Tokyo Metropolitan area, all adopting different reforming and supply modes[246]:

- Groups of oil and fuel retail companies operated ten hydrogen-refueling stations: Babcock Hitachi (city gas reforming), Cosmo Oil (gasoline reforming), Idemitsu (kerosene reforming), Japan Air Gases (methanol reforming), Kurita, Shinanen and Ito-Chu Ene (alkaline electrolysis), Nippon Oil (naphtha reforming), Nippon Sanso and Tokyo Gas (LPG reforming), Nippon Steel Corporation (LH production from COG), Showa Shell Sekiyu and Iwatani International (LH storage), and Tsurumi Soda (off-site H_2 storage).[17]
- Further, two hydrogen stations were built in 2005 in Aichi Prefecture by Toho Gas and Nippon Steel, which participated in bus operations at the Aichi Expo.[17] One of these stations reformed hydrogen from city gas onsite. The other was operated by a steel company, providing pure hydrogen separated from COG (coke-oven gas) by PSA (pressure swing adsorption) at the steel plant and transported to the station in compressed hydrogen containers.[17]

Participating FCVs were routinely driven on 12 pre-determined routes and refueled at the hydrogen stations installed along the routes. Data on fuel performance, fuel efficiency, hydrogen consumptions, and hydrogen production efficiency were gathered. The data were used to investigate problems and to establish necessary standards and regulations, as well as to inform and educate the public[246]:

- The project review report in October 2006 indicated that the fuel cell cars demonstrated a high average energy efficiency, ranging from 37.5 percent to 49.5 percent and a high potential for well-to-wheel efficiency. The project also indicated that shuttle bus operations and hydrogen station operations were trouble free.
- The report, however, stated that there were many problems to be solved. FCVs would need further development to reach technical levels equal to their conventional counterparts. It stated that understanding fuel cell degradation mechanisms and development of new generations of component technologies to ensure reliability and durability were necessary.

7.4.2.1.1.2.2. Japan Hydrogen and Fuel Cell Demonstration Project II, JHFC II (2006–2010) In 2006, METI implemented JHFC Phase II, the second 5-year FCV demonstration project (2006–2010) to gain more field tests, technology development, and operational experience.[246] The Phase II demonstration was to address remaining problems as they occurred in real-life operating conditions, to conduct fleet tests via third parties rather than by automakers themselves, and to extend the test areas to three metropolitan areas (Tokyo, Nagoya, and Osaka). In the Tokyo area, FCVs were used in company fleets; third parties operated more than two vehicles per company on daily runs.[248] After 3 years, FCVs were used as taxis.[248] In Nagoya, a hydrogen station used in Aichi World Exposition was relocated to Chubu International Airport, and two FC buses were used as airport shuttles.[248] In the Osaka area, a hybrid hydrogen station was installed for refueling FCVs as well as stationary power generation systems. In addition, fuel cell–powered wheelchairs, FC carts, and FC bicycles were operated for demonstration.[246] The total budget was ¥5973 million (about $65.8 million at the 2010 exchange rate of $1/¥90.8).[249]

- Participating automakers included GM, Hino, Honda, Mazda, Mercedes-Benz, Nissan, Suzuki, and Toyota. About 60 FCVs participated in all.

In 2010, a METI committee evaluated the project. Four of the six committee members recommended a total re-evaluation of METI policy on FCV demonstrations in the future; one member suggested that the project be dropped: and one suggested that the project be improved.[250] The majority expressed that in the future, the auto industry should take the lead in FCV demonstrations and that the Japanese government should play no more than a minor role.

Also by 2010, Japan clearly had decided to postpone commercialization of FCVs and the establishment of a hydrogen infrastructure from 2010 to 2015.[251] NEDO revised its milestones for FCVs in its roadmap. For 2010, it raised vehicle efficiency from 45 percent (HHV) and 55 percent (LHV) to 61 percent (HHV) and 60 percent (LHV), respectively; it increased durability from 2000 h in 2010 to 5000 h. For 2015, it reduced the FC system target cost from more than ¥10 million (more than $110,000 at the 2010 exchange rate of $1/¥90.8) to ¥1 million ($11,000 at the 2010 exchange rate of $1/¥90.8).

7.4.2.1.1.2.3. Large-Scale Stationary PEMFC Demonstration Project (2005–2009) By 2005, Japan had felt PEMFC CHP technology was ready for field testing and validation of prototypes. In April 2005, METI implemented a 4-year stationary PEMFC power generation system demonstration project (2005–2008). The project would sponsor the demonstration of 1 kW class PEMFC CHP systems for residential use.[252] The scope of the project would cover field testing of a total of 3307 CHP units developed by five PEMFC developers. The manufacturers included ENEOS Celltech (a joint venture of Nippon Oil and Sanyo established in 2008), Ebara, Toshiba, Panasonic (the new name Matsushita Electric Industrial adopted in October 2008), and Toyota. Japan budgeted a total of ¥13.3 billion (about $128.6 million at the 2008 exchange rate of $1/¥103.4) for the project (Table 7.5).

In view of the fact that the PEMFC technology continued to advance during the 4-year demonstration project period, Japan decided to assist Japanese PEMFC developers to enter commercial markets upon completion of the

TABLE 7.5 Large-Scale Stationary Fuel Cell Demonstration Program (2005–2008)

Fuel cell manufacture	Type of fuel			
	LPG	Natural gas	Kerosene	Total
ENEOS Celltech	1062	191	0	1253
Ebara	0	396	314	710
Toshiba	554	194	0	748
Panasonic	0	520	0	520
Toyota	0	76	0	76
Total	1616	1377	314	3307

demonstration project. Field test data showed that the average system electric efficiency (HHV) improved from 29.4 percent in 2005 to 31.5 percent in 2008.[253] Those systems installed in later years demonstrated considerably better performance; units installed in 2006 operated for 2 years with virtually no change in efficiency and with little deterioration. The average unit cost also declined. Costs were cut by more than half from ¥7.7 million in 2005 ($70,000 at the 2005 exchange rate of $1/¥110) to ¥3.29 million (about $32,000 at the 2008 exchange rate of $1/¥ 103.4) in 2008. Unscheduled outages dropped to 0.5 times per unit per year in 2008.

Feeling confident, the Japanese government and the fuel cell industry set May 2009 as the date for market entry.[253] The industry association FCCJ announced in June 2008 that the name of the residential-use PEMFC unit would be ***Ene Farm***.[253] In January 2009, six PEMFC developers jointly announced that they would start selling Ene Farm units in May 2009. The unit cost, including tax, was ¥3.465 million ($33,510) for the Panasonic unit, ¥3.255 million (¥31,480) for the Toshiba unit, and ¥3.2 million (¥30,950) for the ENEOS Celltech unit.

- METI set up a 5-year subsidy program (2009–2013) to promote the sale of the residential CHP systems.[254] The 2009 budget for the subsidy was ¥6.07 billion ($64.78 million at the 2009 exchange rate of $1/¥93.7). The government subsidy per unit was set at a maximum ¥1.4 million (¥13,540), a little less than half the cost of a unit.[253]

7.4.2.1.1.3. METI LAUNCHES "BACK TO BASICS" PEMFC TECHNOLOGY R&D PROJECT (2010–2014) In 2010, NEDO implemented another 5-year project, called Technology Development for Promotion of PEMFC Commercialization (2010–2014).[255] Contrary to what the title suggests, the project was more akin to basic research than product development. The budget for 2010 was ¥5.1 billion ($56.2 million at the 2010 exchange rate of $1/¥90.7). The project had three segments: basic technology development, commercial technology development, and next-generation technology development.

- The basic technology development segment was the largest.[256] It was allocated ¥3.4 billion ($37.5 million).[256] Also, the High Performance (HiPer-FC) Project on degradation and high-performance cells that was already underway was merged with the basic technology segment. Research topics included degradation mechanism analysis and materials for high-efficiency cells using nanotechnology, which would be carried out under the HiPer-FC project; high efficiency: low-cost MEAs; lowering the use of platinum; carbon alloy catalysts; non-noble metal oxide catalysts; structure, reaction, and mass transfer analysis of MEA materials; developing a common technology base for cell evaluation and analysis; and international standardization work. Research would be conducted in a consortia led by program leaders, each of whom had considerable academic standing.
- The commercialization technology development segment and the next generation technology development segment were allocated ¥1.6 billion ($17.6 million).[257] The commercialization segment included the development of technology that would enable the use of diverse fuels, the development of multipurpose, high value-added PEMFC systems, cost reduction technologies, and safety technology. The next generation technology development segment explored innovative technologies, excluding the current technology, which could dramatically increase reliability and reduce cost.

This METI 2010–2014 project clearly was a landmark event. The project deployed more academic scientists and physicists than applied researchers. NEDO announced in July 2010 that at least 32 universities, 30 companies, and 3 research institutions were participating.[256] It was unprecedented that more universities than companies would be participating in a METI or NEDO project. The project also used a collaborative R&D approach centering on COEs and the consortia.

- The HiPer-FC project, which merged with the 5-year project, was being conducted at a center of excellence (COE) established at University of Yamanashi in April 2008. The COE would conduct basic scientific research to understand reaction and degradation mechanisms and develop innovations to create high-performance fuel cells using nanotechnology.
- The 5-year project would also have another noteworthy Fuel Cell Innovation Research Center.[258] The Center, which opened at the University of Electro-Communications (UEC) in Tokyo in September 2010, was headed by one of Japan's most renowned scientists in the field of solid surface science. He would also serve as the senior project leader of the 5 year project.
- Of note, the research center at UEC was to have a research annex at the world's most powerful third-generation synchrotron radiation facility, SPring-8 in Hyogo, Japan. The center also planned to build the world's most advanced real-time catalyst structure and reaction measurement XAFS beamline at the SPring8 facility. The new facility was to be built by 2012 with a budget of ¥1.3 billion (about $13 million).[259] The XAFS beamline facility will use synchrotron X-rays from the Spring-8 facility to study and understand electrochemical reaction mechanisms and changes in catalyst surface and structure, according to UEC,[258] thereby hopefully leading to dramatically different fuel cells of high efficiency, high durability, and low cost.

It is remarkable as well as unprecedented for METI to be involved in such basic scientific research, which typically falls within the domains of the Ministry of Education, Culture, Sports, Science, and Technology and the Science and Technology Agency. Moreover, there are no institutions outside of Japan that have supported or encouraged basic scientific research on PEMFCs at this scale. The project is a ground-breaking achievement for Japan that has set it apart from United States and Europe. For a Japanese bureaucracy to take such a bold step is highly commendable. It calls to mind the Japanese proverb that "extreme times require extreme measures."[258]

7.4.2.1.1.3.1. ....But Small Budget and Overarching Goal of Commercialization Might Impede Basic Research Effort Despite the promise that this project holds, there remain factors that cause concern. The budget of the project is not large enough to create a critical mass for achieving technical breakthroughs. It might be adequate for applied research, but not for basic research that is intended to elucidate how fuel cells work at the atomic level. Also of concern is that all the individual projects are under the umbrella of a 5-year project that is governed by the overarching goal of PEMFC commercialization. This could thwart the basic 2010 research effort prematurely in favor of more expedient product development. Nonetheless, the intention of the project is worthy and a successful outcome would have monumental significance.

- If this effort is successful, Japan will be the first one in the world to master fuel cell technology for commercialization.

7.4.2.2. US Government Launches Ambitious Fuel Cell Car and Hydrogen Technology Initiatives in 2002

Throughout the 1990s, the US DOE supported small PEMFC R&D efforts under the PNGV, but this budget consistently was much greater than that of the Japanese PEMFC program. That all changed in 2000 when the Japanese PEMFC budget surpassed that of the US. The surprising turn of events may have prompted the US President to support major fuel cell and hydrogen initiatives in 2002 and 2003. The President initiated two ambitious PEMFC and hydrogen technology development programs. In January 2002, the US Secretary of Energy announced the implementation of the FreedomCAR Initiative to develop FCVs that would run on hydrogen fuel. In the following year, the President in his State-of the-Union address in January 2003 announced another major initiative—the Hydrogen Fuel Initiative—which was aimed at developing hydrogen technologies and infrastructure to produce, store, and distribute hydrogen to fuel FCVs. The President's goal was ambitious; it was to reverse America's growing dependence on foreign oil by developing the technology for hydrogen and hydrogen-powered FCVs by 2015.[260] He proposed allocating a total of $1.7 billion over 5 years to fund the FreedomCAR Initiative and the Hydrogen Fuel Initiative.[261] The Hydrogen Fuel Initiative included $720 million in new funding.

- The two initiatives created a greater sense of competition among fuel cell developers and intensified the global fuel cell race.

7.4.2.2.1. QUESTIONABLE PROGRAM GOVERNANCE

Unfortunately, the new US initiatives ran aground because of implementation problems. The two initiatives were governed by a government–industry partnership, in which government and industry had equal responsibility for policy making and implementation. In January 2001, the governing body, the FreedomCAR Partnership was created, consisting of the DOE and the USCAR (United States Council for Automotive Research), which included the three US automakers, GM, Daimler-Chrysler (subsequently Chrysler), and Ford.[262] In 2003, when the Hydrogen Fuel Initiative was implemented, the Partnership was expanded to include five more partners who would share the responsibility of governing the new hydrogen initiative. With the participation of the five—BP America, Chevron, ConocoPhillips, Exxon Mobil, and Shell Hydrogen—the newly expanded partnership was called the FreedomCAR and Fuel Partnership.[263]

- It is unusual that the government and the industry would share an equal responsibility for all aspects of policy and decision-making activities in a government program. Moreover, the Partnership also served as the communication mechanism for all parties involved, including the government, the private sector, the national laboratories, universities, the public, and others.[264] The Partnership's executive steering group included the DOE Assistant Secretary for Energy Efficiency and Renewable Energy (EERE) and vice presidents of the Partnership companies. [264] The automakers and oil companies apparently neither contributed nor received government funds, reducing their incentive to work hard and contribute to the Partnership and the program.
- It is also questionable that those with a direct business interest in competing technologies should be involved in making policy and budgetary decisions for this activity. The automakers had commitments and interest in conventional automotive technologies, making it difficult for them to focus on long-term, high-risk fuel cell R&D projects, particularly if near-term combustion engine technologies were to compete for the same pool of funding resources. Partnership activities included conventional vehicle technologies. In reality, the automakers' inherent interest in conventional vehicles ultimately prevailed over any longer-term interest in fuel cells.

7.4.2.2.2. FREEDOMCAR AND FUEL INITIATIVE BECOMES VIRTUAL EXTENSION OF THE PARTNERSHIP FOR NEW GENERATION OF VEHICLES (PNGVS)

It is noteworthy that the FreedomCAR and Fuel Partnership turned out to be virtually an extension of the Partnership for a New Generation of Vehicles (PNGV),[264] a cooperative partnership formed back in 1993 between the federal government and the USCAR to develop more fuel-efficient vehicles. The PNGV had six technical areas of R&D, aimed at developing fuel-efficient vehicles, including conventional vehicle technologies and fuel cell technologies. These same six areas would continue to be funded under the new FreedomCar and Fuel Initiative.

- The only difference between the PNGV and the FreedomCar and Fuel Partnership was the addition of the Fuel Tech Teams that focused on hydrogen technology and the Joint Tech Teams that focused on codes and standards for hydrogen storage (Fig. 7.2)[264]

The immediate good news following the implementation of the 2003 FreedomCAR and Fuel Partnership initiative was that the fuel cell R&D budget started to increase. The PEMFC budget under the PNGV program was in the low \$20 millions in the 1990s and increased gradually to \$41.5 million in 2001. In 2003, the DOE PEMFC R&D budget increased by close to 30 percent over the previous year to \$53.9 million. In 2004, the PEMFC R&D budget was \$63.78 million, an 18 percent increase from the year before.

- However, the fuel cell budget soon started to decline; it dropped to \$55 million in 2005 and to \$34 million in 2006. Inexplicably, the fuel cell budget started to climb again and reached a peak of \$83 million in 2009. Since then it has declined to \$43.6 million in 2012.
- The US fuel cell budget under the Partnership was dwarfed by Japan's PEMFC fuel cell R&D. In 2003, the Japanese budget for PEMFC R&D was \$181.4 million, more

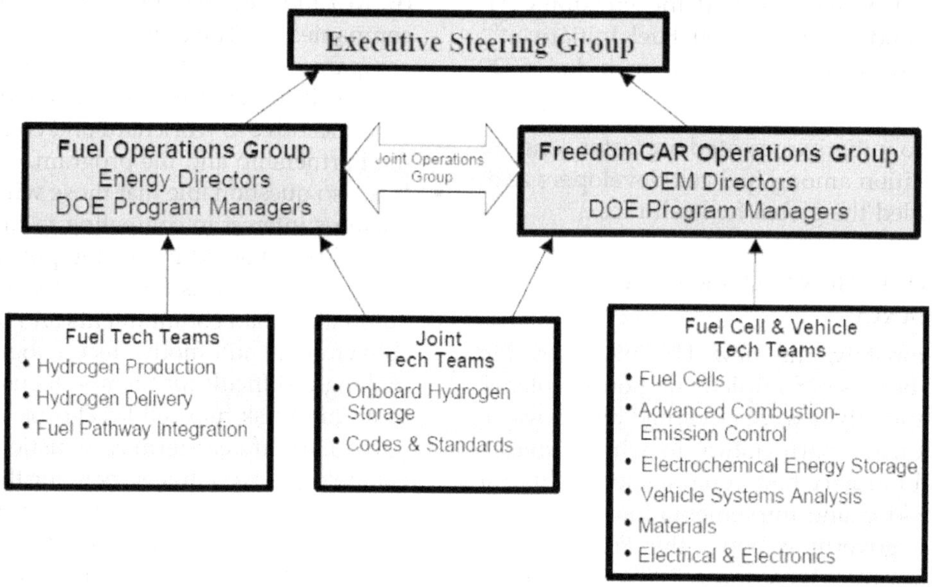

FIGURE 7.2 FreedomCAR and fuel partnership organizational structure. *Source: "Partnership Plan, FreedomCar & Fuel Partnership," http://www1.eere.energy.gov/vehiclesandfuels/pdfs/program/fc_fuel_partnership_plan.pdf, FreedomCar & Fuel Partnership, Energy Efficiency and Renewable Energy, US Department of Energy, March 2006.* (For the color version of this figure, the reader is referred to the online version of this book.)

than three times the US budget. In 2004, the Japanese budget was not broken down into comparable categories—such as PEMFCs and hydrogen—making comparison difficult. Nonetheless, Japan's fuel cell budget in 2004 was ¥32.9 billion (about $307.5 million at the 2004 exchange rate of $1/¥107), and even if only half was allocated to PEMFC R&D, this would still be more than twice that of the US PEMFC and hydrogen budgets (Fig. 7.3).

7.4.2.2.3. FUEL CELL R&D GETS INCREASINGLY LESSER FOCUS

The US budget for the FreedomCAR and Fuel Partnership exhibits two troubling characteristics. As we noted earlier, one was the inexplicable ups and downs of the budget that undoubtedly made long-term research planning difficult or impossible. Second, fuel cell R&D seemed to be virtually always receiving a smaller share of funds than hydrogen or conventional vehicle technologies in the total program budget. Moreover, the budget for conventional vehicle technologies always received a dramatically greater budget each year, with no ups and downs.

- A possible explanation is that in and around 2005, the automaker partners in the Partnership perhaps became less interested in fuel cell R&D and more interested in the R&D of conventional vehicles. Funds for conventional technology likely had top budget priority and were protected, leaving an ever-diminishing remainder of funds for fuel cell R&D (Fig. 7.4).

7.4.2.2.4. EFFECTIVENESS OF THE FUEL CELL PROGRAM QUESTIONABLE

Was the US fuel cell R&D program under the Partnership effective? The program yielded some incremental improvements. The National Research Council's (NRC) Committee on Review of the FreedomCAR and Fuel Research Program published three review reports, one each in 2005, 2008, and 2010, and praised the program's achievements:

- In the 2010 report, for example, the NRC committee commended the DOE program for achieving significant improvements in performance.[265] The achievements included an increase in demonstrated stack lifetimes in on-road vehicles from approximately 1250 h to 1977 h. NRC also praised the single-cell and short-stack tests at the laboratory scale, which demonstrated (using accelerated test protocols) longer run times (3M Company, 7200 h).
- Generally, NRC seemed to be satisfied with these achievements, though it apparently was

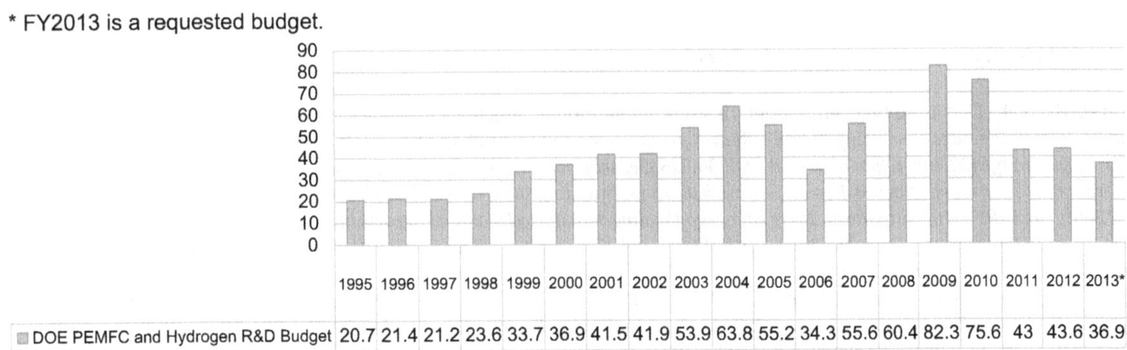

FIGURE 7.3 US PEMFC budget: 1995–2013*. *The figure is created based on various sources, including "DOE's Budget Justifications and Supporting Documents," from FY2002 to FY2013, http://www.cfo.doe.gov/crorg/cf30.htm, Office of Budget - CF-30, US Department of Energy, no date, accessed May 2012.* (For the color version of this figure, the reader is referred to the online version of this book.)

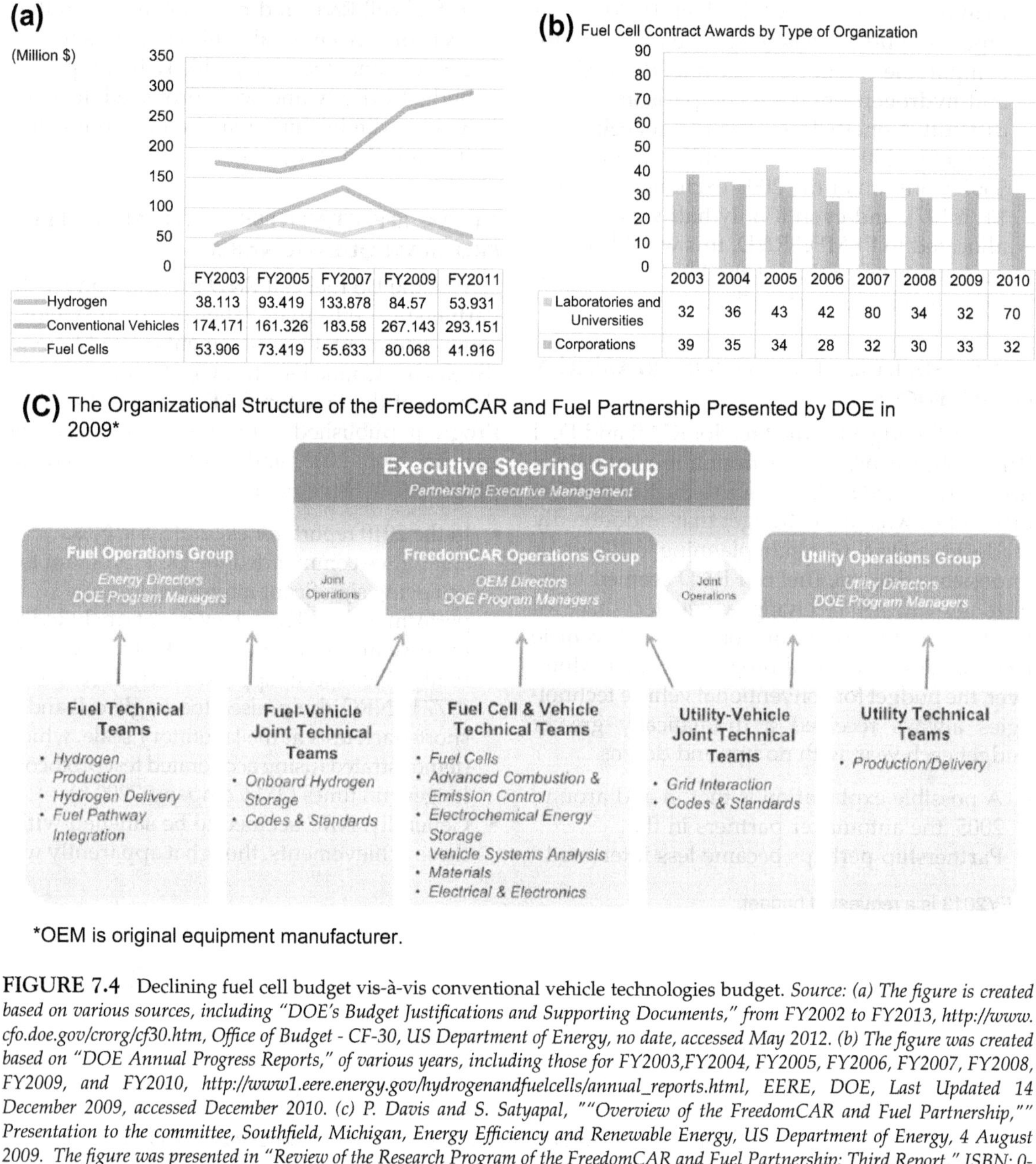

FIGURE 7.4 Declining fuel cell budget vis-à-vis conventional vehicle technologies budget. *Source: (a) The figure is created based on various sources, including "DOE's Budget Justifications and Supporting Documents," from FY2002 to FY2013, http://www.cfo.doe.gov/crorg/cf30.htm, Office of Budget - CF-30, US Department of Energy, no date, accessed May 2012. (b) The figure was created based on "DOE Annual Progress Reports," of various years, including those for FY2003,FY2004, FY2005, FY2006, FY2007, FY2008, FY2009, and FY2010, http://www1.eere.energy.gov/hydrogenandfuelcells/annual_reports.html, EERE, DOE, Last Updated 14 December 2009, accessed December 2010. (c) P. Davis and S. Satyapal, ""Overview of the FreedomCAR and Fuel Partnership,"" Presentation to the committee, Southfield, Michigan, Energy Efficiency and Renewable Energy, US Department of Energy, 4 August 2009. The figure was presented in "Review of the Research Program of the FreedomCAR and Fuel Partnership: Third Report," ISBN: 0-309-15684-X, http://www.nap.edu/catalog/12939.html, Committee on Review of the FreedomCAR and Fuel Research Program, Phase 3; National Research Council, The National Academies Press, Washington, D.C., 2010.* (For the color version of this figure, the reader is referred to the online version of this book).

disappointed at the fact that no single fuel cell technology achieved a performance and cost point that would make it competitive with conventional technologies.[265] In any case, the NRC never challenged the DOE program managers or advised DOE to fundamentally and radically change its status quo research approach to make the program more productive and effective.

It is true that NRC made one excellent recommendation. It advised DOE to provide more funding for basic scientific research at universities and national laboratories. Unfortunately, the DOE program managers have yet to do so.

- DOE annual progress reports from 2003 to 2010 showed that DOE perhaps followed the NRC's advice only twice. The number of contracts awarded to universities and laboratories increased two times, although it is not clear if the contract awards also increased. Nonetheless, the changes were not enough to encourage basic research to flourish.

It appears as if the Partnership program managers were not focused on pursuing a competitive, pioneering, consequential scientific research effort. Rather, it seems they were administering the program as if it were a form of corporate welfare. The program's primary goal appeared to be maximizinge the distribution of funds across as many different research projects as possible.

- The program's solicitations attracted a large number of R&D project proposals, and an extraordinarily large number of small contract awards were granted, resulting in a large number of minor, uncoordinated activities.
- As of May 2012, 10 years after the current Partnership was formed, there have been no breakthroughs. Moreover, there is no evidence that the DOE leadership in the Partnership has asked itself why the program has not achieved any breakthroughs in technology, whether the approach needed to be re-thought, or what could have been done to change program outcomes.

What is clear is that the DOE program has made few, if any, significant technological contributions to the US auto industry's fuel cell capabilities. Ironically, the automakers appear to have achieved considerable technological improvements independent of the DOE R&D program.[265] According to the NRC committee, the automakers, using their own proprietary engineering, made significant on-road progress toward the 5000 hour reliability and durability target. They also achieved increased power density for the fuel cell stack and BoP, while at the same time packaging and operating modes became quite sophisticated.

- The automakers probably lost interest in the Partnership due to the lack of benefits. Indeed, the Executive Steering Group, of which the US automakers were part, has not met in more than 2 years, according to the 2010 NRC committee review.[265]
- Conversely, if the automakers did not use DOE's publicly funded program, what did the more $500 million of public funds actually accomplish? That question remains unanswered.

7.4.2.2.5. FUEL CELL R&D BUDGET STARTS DECLINING IN 2009...FUTURE US PEMFC DEVELOPMENT UNCERTAIN

DOE's raison d'être for fuel cell R&D has become increasingly fuzzy and ill-defined over the years. In 2008, in an almost fatal blow to the fuel cell R&D program, DOE's interest shifted to plug-in hybrid electric vehicle (PHEV) and battery electric vehicle (BEV) technologies. These were viewed to have more near-term feasibility than FCVs. PHEVs and BEVs were designed to be plugged into the electric grid to charge an onboard battery. DOE became interested in incorporating the electric power utility sector into the

Partnership.[265] Two electric utility companies—DTE Energy (Detroit) and Southern California Edison— joined the Partnership as its third group, the new Utility Operations Group.[265] Accordingly, the focus on fuel cells became even weaker.

- The organization chart for the Partnership updated in 2009 (Fig. 7.4c on Page 482) showed that fuel cell activity became one of the 14 technical areas in the 2009 chart instead of one of the 11 in the 2004 chart (Fig. 7.2 on Page 480).
- The funding for fuel cell system R&D declined to $43 million in FY2011 and to $36.9 million in the FY2012 request. The outlook for future PEMFC R&D in the United States looks dim.

7.4.2.3. *Europe Recognizes Global Fuel Cell Challenge in Late 1990s*

Europe has had an early interest in PEMFC technology and supported a range of important projects in the 1990s.

- For example, the EU provided €4.8 million (about $4.36 million at the 1994 exchange rate of $1/€1.1) for the 1990–1994 Joule II Program in the EU Third Framework Program (FP3) to develop a small-to-medium PEMFC vehicle.

It is known that many European governments (or EU Member States (MS)), including France, Germany, Italy, and UK, contributed R&D funds in amounts equal to FP funding.[266] Europe thus likely spent about €9.6 million ($8.72 million) during the FP3 period for PEMFCs.

EC stepped up its effort to increase PEMFC R&D funding for FP4 (1994–1998),[267] which according to the EC document, was about €30 million ($27.27 million at the 1998 exchange rate of $1/€1.1).[267] With equal funding from individual countries, Europe likely invested a total of €60 million ($54.54 million) for PEMFC R&D during the FP4 period.

In the late 1990s, European R&D organizations noted the many PEMFC activities going on elsewhere in the world and argued for increased funding for FP5 (1998–2002). FP5's budget rose to €45.68 million (about $43.1 million at the 2002 exchange rate of $1/€1.06),[268] about 50 percent greater than it was for FP4. With EU MS contributions, total European PEMFC R&D budgets likely reached €91.36 million ($86.2 million), or an average annual European PEMFC funding of €22.84 million ($21.5 million).

- Despite the increase, the FP5 funding still was much smaller than US and Japanese PEMFC budgets, which were rapidly rising. In 2002, the United States spent $41.9 million and Japan's funding was ¥15.62 billion (about $134.66 million at the 2002 exchange rate of $1/¥116).

7.4.2.3.1. EUROPEAN COMMISSION TAKES CHARGE IN 2000

Undoubtedly shocked by the widening funding gaps between Europe and its foreign competitors, the European Commission (EC) made a noteworthy decision in 2000. It decided to improve the European fuel cell and hydrogen R&D policy and management structure.

In January 2000, EC proposed the creation of a European Research Area (ERA) to overcome the fragmentation of research in Europe and eliminate national and institutional barriers. It was believed that Europe's many stove-piped R&D activities had prevented it from fulfilling its research and innovation potential, at a huge cost to Europeans taxpayers, consumers, and citizens.[269] The ERA was launched in March 2000 at the Lisbon European Council.

In October 2002, the EC established the High Level Group (HLG) for Hydrogen and Fuel cell Technologies to coordinate efforts between EC and the EU Member States and compile a strategic research agenda that would speed up the introduction of fuel cells and hydrogen.[270] In

June 2003, the HLG recommended the creation of a "Hydrogen and Fuel Cell Technology Partnership," to quick-start the fuel cell and hydrogen technology commercialization.[271]

- In response, the EC launched in January 2004 the "European Hydrogen and Fuel Cell Technology Platform (HFP)" with an advisory council consisting of key stakeholders in the European hydrogen and fuel cell sector.[271] At the January 2004 inaugural session, the HFP unveiled an ambitious vision aimed at facilitating and accelerating the R&D and deployment of cost-competitive, world-class European hydrogen and fuel cell-based energy systems and component technologies for applications in transport, stationary, and portable power.[273]

In 2004, based on the vision unveiled by the HFP, the EC started to make major changes in fuel cell R&D programs. Although FP6 (2002–2006) was already in progress, the EC increased the fuel cell budget by a large margin. According to an EC document, the fuel cell and hydrogen budget increased from some €100 million[274] (about $125 million at the 2004 exchange rate of $1/€0.8) to €279.61 million[268] ($349.51 million), nearly tripling FP5 funding. Some €153.92 million ($192.4 million) was allocated to fuel cells and €125.69 million ($162.11 million) for hydrogen.

- FP6 allocated €115.35 million ($144.18 million) specifically for PEMFC R&D and demonstrations, 75.4 percent of the total fuel cell funding and 41.25 percent of the combined fuel cell and hydrogen funding.[268] Adding MS contributions, total European spending on PEMFC projects during the FP6 period amounted to at least €230.7 million ($288.38 million). The average annual European PEMFC R&D budget was €57.67 million ($ 72.09 million).
- With this additional funding, the European PEMFC budget was now on a par with or slightly higher than the US budget, although it was still lower than the Japanese budget.
- While the EC's effort to raise the FP funding level was successful, change was slow, and some goals were not met. The EU had no central policy management structure, and any major change required time-consuming coordination and consensus building. The policy change it wanted in 2000 had only a limited effect as funding increased in 2004 under FP6 (2002–2006). The EC's ultimate goal to facilitate and accelerate the R&D and deployment of cost-competitive, world-class European hydrogen and fuel cell technology was still far away.

7.4.2.3.1.1. EC STILL HAPPY WITH R&D UNDER FP5 (1998–2002) While the EC was struggling to change policy in the early 2000s, three major PEMFC R&D and demonstration projects already were being implemented under FP5 (1998–2002) with considerable success (Table 7.6). While many projects under FP5 were selected late and ran beyond 2002, the EC was clearly satisfied with the project outcomes. It stated that projects from FP5 were well advanced, with many entering the critical phase of exploiting and disseminating results. EC PEMFC funding was €45.68 million (about $51.32 million at the 2002 exchange rate of $1/€0.89) with likely MS contributions of another €45.68 million. EC funding for the three projects constituted the bulk of the FP5 PEMFC budget.

The three major projects under FP5 were FUERO, CUTE, and ECTOS:

1. ***The FUERO project (2000–2004)***. The "Fuel Cell Systems and Components General Research for Vehicle Applications (FUERO)" project was a large-scale effort to bring the European fuel cell technology one step closer to market introduction. The project consisted of a cluster of eight projects— PROFUEL, BIO-H_2, DREAMCAR, AMFC, PEM-ED, ASTOR,

TABLE 7.6 EU Fifth Framework Program (FP5): Major PEMFC R&D and Demonstration Projects

Project title (Acronym)—(Project description)	Coordinator; Other major participants	Program	Total budget; EU funding	Official project period
Fuel cell systems and components general research for vehicle applications (FUERO)—(the identification of the relevant operational requirements for the operation of fuel cell systems in vehicles with regard to components and sub-systems)	Aachen University Of Technology (Germany); Centro Ricerche Fiat, Volkswagen, Institut Francais du Petrole, Peugeot Citroën, Volvo,	FP5	€21.4 million	July 2000–June 2004
FUERO includes the following eight projects:				
Onboard Gasoline Reforming for Fuel cell Vehicles (PROFUEL)	FEV Motorentechnik (Germany); Johnson Matthey, Centre Ricerche Fiat, ECN, Ansaldo Ricerche, Volvo	FP5 FUERO	Included in FEURO funding	36 months
Producing Clean Hydrogen for Fuel Cells by Reforming of Bioethanol (BIO-H_2)	Centro Ricerche Fiat, ENEA, Peugeot Citroen, The Queen's University of Belfast (UK), University of Patras (Greece), ENC			36 months
Direct methanol fuel cell system for car applications (DREAMCAR)	Thales Engineering & Consulting (France); Israel Plastics And Rubber, SOLVAY, Centro Ricerche Fiat			February 2001–January 2005
Advanced methanol fuel cells for vehicle propulsion (AMFC)	VOLVO; Technical University of Denmark, Norwegian University of Science And Technology, University of Newcastle upon Tyne (UK), STATOIL (Norway)			January 2001–August 2004
Proton exchange membranes for application in medium temperature electrochemical devices (PEM-ED)	FUMA-TECH (Germany); Centre National de la Recherche Scientifique (France), Consiglio Nazionale delle Ricerche (Italy), University of Strathclyde (UK), Univercita di Perugia (Italy), Dlectricite de France, SAFAR (Switzerland), Nuvera Fuel Cells			April 2000–September 2004

Assessment and testing of advanced energy storage systems for propulsion and other electrical systems in passenger cars (ASTOR)	Volkswagen; DaimlerChrysler, Ecole D'ingnieurs En Genie Des Systemes Industriels (France), Renault Recherche et Innovation, Volvo, BMW, Adam Opel, Centro Ricerche Fiat			April 2001–September 2004
Miniaturized gasoline fuel processor for fuel cell vehicle applications (MINIREF)	Peugeot Citroën; Universite Paul Sabatier De Toulouse III (France), Institut Fuer Mikrotechnik Mainz (Germany), Politecnico Di Torino, Infragas (Italy)			February 2002–January 2005
Ammonia cracking for clean electric power technology (ACCEPT)	VITO (Belgium)			January 2002–June 2005
Clean urban transport for Europe (CUTE)—(Demonstration operations of three Citaro buses each in nine European cities)	EvoBus; Sydkraft (Sweden), PE Product Engineering (Germany), Hamburger Hochbahn Aktiengesellschaft, Instituto Superior Technico (Portugal), EMPRESA Municipal de Transportes (Spain), BP International (UK), Hamburgische Elektrizitätswerke, Sociedade de Transportes Colectivos do Porto (Portugal), Statkraft (Norway), Transports de Barcelona, Universitaet Stuttgart, Shell Hydrogen (the Netherland), City of Stockholm, Fédération Luxembourgeoise des Exploitants d'Autobus et d'Autocars, First Group (UK), AB Storstockholms Lokaltrafik (Sweden), POLIS IASBL (European Cities And Regions Networking)(Belgium), PLANET Planungsgruppe Energie und Technik (Germany), Norsk Hydro (Norway), London Bus Services, BUSSLINK (Sweden), MVV Verkehr (Germany), DaimlerChrysler, Autobus de la Ville de Luxembourg, Gemeentevervoerbedrijf Amsterdam, Milieudienst Amsterdam, Stuttgarter Strassenbahn	FP5	€18,551,209	November 2001–June 2006
Ecological City Transport System (ECTOS)—(Demonstration operations of three Citaro buses in Reykjavik, Iceland)	Iceland New Energy	FP5	€2,857,173	March 2001–August 2005

MINIREF, and ACCEPT[275]—mobilizing most major European automakers, fuel cell developers, systems manufacturers, research institutes, and universities. It aimed to identify operational requirements of fuel cell systems in vehicles and make available advanced systems and components that would facilitate market commercialization of FCVs.[276] It studied energy sources, evaluated the well-to-wheel efficiency, and analyzed components available with respect to their performance, production, and recycling aspects. EC funding for the entire FUERO project was €21.4 million.[275]

2. ***The CUTE project (2001–2006).*** "The Clean Urban Transport for Europe (CUTE)" was another major FP5 undertaking, which aimed to realize a unique public transport system and the necessary hydrogen supply infrastructure to support it. The project was supported by a large consortium of transport operators, hydrogen infrastructure and fuel cell developers, universities, and city authorities. The project aimed to deliver 27 Citaro fuel cell buses (developed by Daimler and EvoBus) to nine European cities and operate the buses over a period of 2 years to prepare the fuel cell technology for series transport application.[277]
3. ***The ECTOS project (2001–2005).*** "The Ecological City Transport System (ECTOS)" was an extension of the CUTE project that was implemented in the Icelandic capital of Reykjavik. The Icelandic project received EC funding of €2.87 million. Major fuel and infrastructure partners included BP, Norsk Hydro, and Shell Hydrogen.[58]

7.4.2.3.1.1.1. EC Programs Emulated Elsewhere in the World Undoubtedly impressed with the CUTE and ECTOS projects, the Government of Western Australia's Department of Planning and Infrastructure decided to implement its own project similar to CUTE and ECTOS in the city Perth, the capital of Western Australia[278]:

The Sustainable Transport Energy for Perth (STEP) project (2001–2005). The Australian project was not funded under FP; it received $2.5 million from the Department of the Environment and Heritage and the Australian Greenhouse Office. It was endorsed by the United Nations Environment Program and the United Nations Industrial Development Organization.[278] Three Citaro buses were delivered to Perth, called EcoBuses in Australia, and began service in September 2004. BP would produce the hydrogen as a by-product at its Kwinana oil refinery, 31.1 miles south of Perth. The hydrogen would then be transported by road in specially designed road tankers to a bus depot in the northern suburbs of Perth.

- Perth's buses achieved greater reliability and better fuel economy than in any other city in the CUTE and ECTOS project.[278] By June 2005, the Perth buses had covered more than 37,282 miles and completed almost 3000 operational hours. Almost 60,000 passengers had used the service.

7.4.2.3.1.2. BUT POST-FP6 REVIEW IN 2007 REVEALS EUROPE 5 YEARS BEHIND THE UNITED STATES AND JAPAN In 2007, the EC conducted a review of FP6 (2002–2006) at the European Hydrogen and Fuel Cell Review Days held on October 10–11 in Brussels and found the outcome of FP6 to be less than satisfactory.[279] FP6 supported at least five PEMFC R&D projects, two PEM FCV fleet demonstration projects, five projects to develop PEMFCs for transport applications, one project for PEMFCs for stationary applications, and one project to develop DMFCs (or direct ethanol fuel cell) for portable applications (Table 7.7).

- ***Membrane and catalyst R&D projects.*** Four of the five PEMFC R&D projects, (FURIM, APPOLLON-B, AUTOBARNE, and FRANODE) focused on the development of low-cost membranes and non-noble catalysts. The fifth project, CARISMA, aimed to

TABLE 7.7 EU Sixth Framework Program (FP6): PEMFC R&D and Demonstration Projects

Project Title (Acronym)—(Project description)	Coordinator; Other participants	Program	EU funding	Official project period
Further Improvement and System Integration of High-Temperature Polymer Electrolyte Membrane Fuel Cells (FURIM)	Danmarks Tekniske Universitet; 12 participants	FP6	€3,999,000	April 2004–March 2008
Polymer Electrolytes and Non-noble Metal Electrocatalysts for High-Temperature PEM Fuel Cells (APOLLON-B)	Foundation Of Research And Technology Hellas; 10 participants	FP6	€1,800,000	October 2006–September 2009
Automotive high-temperature fuel cell membranes (AUTOBRANE)	Daimler; 29 participants	FP6	€8,299,693	November 2005–October 2009
Coordination action of research on intermediate and high temperature specialized membrane electrode assemblies (CARISMA)	Centre National De La Recherche Scientifique (CNRS); 37 participants	FP6	€560,400	January 2007–December 2008
Non-noble catalysts for proton exchange membrane fuel cell anodes (FCANODE)	Danmarks Tekniske Universitet (Technical University Of Denmark); 6 participants	FP6	€1,492,866	February 2007–January 2010
Hydrogen for clean urban transport in Europe (HYFLEET: CUTE)—(the continued operation of the FC-fleet from the former CUTE and ECTOS projects)	DaimlerChrysler; 30 participants	FP6	€18,986,145	January 2006–September 2009
Deployment of innovative low-power fuel cell vehicle fleets to initiate an early market for hydrogen as an alternative fuel in Europe (HYCHAIN); The demonstration of fleets of innovative fuel cell vehicles in France, Germany, Spain, and Italy (MINI-TRANS)	Air Liquide; 23 participants	FP6	€17,000,000	January 2006–January 2011
Hydrogen and Fuel Cell Technologies for Road Transport (HYTRAN)—the development of an 80 kW PEM system for an automobile and a 10 kW APU	Volvo; at least 18 participants	FP6	€8,811,143	January 2004–November 2009

(Continued)

TABLE 7.7 EU Sixth Framework Program (FP6): PEMFC R&D and Demonstration Projects (*cont'd*)

Project Title (Acronym)—(Project description)	Coordinator; Other participants	Program	EU funding	Official project period
Fuel Cell Hybrid Vehicle System Component Development (HYSYS)—research on of low-cost components for fuel cell systems and electric drive systems	Daimler; 27 participants	FP6	€11,197,201	December 2005–November 2010
High-density power electronics for FC- and ICE-Hybrid Electric Vehicle Powertrains (HOPE)	Siemens; 14 participants	FP6	€2,398,000	January 2006–December 2008
Fuel cell power trains and clustering in heavy-duty transports (FELICITAS)—the development of PEMFC and SOFC APUs for trucks, buses, and light and heavy rail and marine applications	Fraunhofer; 14 participants	FP6	€7,943,597	April 2005–March 2008
Fuel cell application in a new configured aircraft (CELINA)—the investigation of a 65 kW kerosene reformate PEMFC APU as emergency power supply for airplanes	Airbus; 18 participants	FP6	€4,497,400	January 2005–June 2008
The next generation of stationary fuel cells (NextGenCell)—a joint EU–US effort to develop 1–5 kW residential microCHP for commercialization	Vaillant; 8 participants	FP6	€2,498,928	December 2006–March 2009
Compact direct methanol fuel cell for portable application (MOREPOWER)	Forschungszentrum Geesthacht; 8 participants	FP6	€2,147,784	February 2004–April 2007
Total EU Funding			€91,632,157	

network research activities in Europe on high-temperature PEMFC membrane electrode assemblies and their components.[279] The EC called these projects basic research, but they were closer to applied research. According to the EC review, the R&D projects made little progress. Results were either described in too little detail to be assessable, were late, or had not resulted in promising outcomes.

- ***PEM FCV fleet demonstration projects.*** The HYFLEET-CUTE project included the continued operation of the FC fleet from the former CUTE and ECTOS projects.[280] It also included the development and demonstration of a new FC hybrid pre-prototype aimed at reducing fuel consumption 20 percent below that of a comparable diesel bus, the construction and demonstration of hydrogen infrastructure for operating a fleet of 14 buses in Berlin, and the development and production of four buses with naturally aspirated hydrogen ICE. Another fleet demonstration project was the HYCHAIN MINI-TRANS, which deployed fleets of 180 FCVs to four regions in Europe (France, Germany, Spain, and Italy) to assess costs and overcome major cross-region barriers.[281] Although the demonstrations were successful, the EC was not certain as to how these could lead to early market entry.[279] The EC felt that it was unlikely that spontaneous replication would occur based on the demonstration sites.
- ***PEMFC projects in transport applications.*** The HYTRAN project was to improve the performance and cost of fuel cells by developing an 80 kW PEMFC system, consisting of innovative stack and BoP and integrating it in a vehicle for validation.[279] The project also included the development of a 5 kW PEMFC APU system, including a steam reformer, clean-up reactor, stack, and BoP. The HYSIS projects aimed to develop low-cost components for PEMFC FC systems and electric hybrid drive systems. The components were tested in validated vehicles and delivery vans. The HOPE projects were to achieve cost reduction, meet reliability requirements, and reduce volume and weight of PEMFC and ICE-hybrid vehicles.[282] The CELINA project was to investigate the PEMFC system under aircraft-operating conditions and identify requirements of an airworthy design.[279] The FELICITAS project was to develop 200 kW PEMFC and SOFC APU and propulsion systems for trucks, buses, light and heavy rail, and marine applications.[283] The EC stated that the projects were well balanced, but that the progress in PEMFC stack and BoP technology seemed insufficient to bring Europe into a leading position or to close the gap with the state-of-the art PEMFC in North America and Japan.[279]
- ***PEMFC stationary application project.*** The NextGenCell project was a joint EU and US collaborative effort, aimed at the development of a 1–5 kW PEMFC for residential CHP system. [279] Three major hurdles identified were costs, reliability, and the need to increase the system temperature.[284] The EC stated indicated that Europe-based fuel cell developers were preparing for technology validation and demonstration efforts during the FP7 study period (2007–2013). Without substantial increase in member states' support, according to the EC, it was likely that the benefits of Europe's expertise would diffuse outside the EU toward the larger support programs available in Japan and the United States.[279]
- ***DMFC portable application project.*** The MOREPOWER project aimed to develop a low-cost, low-temperature, portable DMFC (or direct ethanol fuel cell) of modular design with an output of 500 W, operating at 40 A and 12.5 V and at temperatures up to 60°C. Applications envisaged included weather stations, medical devices, signal units, APUs,

gas sensors, and security cameras.[279] The project appeared successful. The EC indicated that the project delivered a 350 W fuel cell stack operating with 1 molar methanol solution at 60 °C.

The EC review report stated as follows: "The EU is five years behind Japan and the United States. The US and Japanese programs are strategically managed by the central government. Central management is not possible within the EU political structure. It is unrealistic, at this stage, to expect national governments to put substantial funding behind the FP projects. The dispersed European effort has evident disadvantages, although it may also have some advantages in that it allows a more diverse offering of technical solutions. What is important is not so much central management, but central coordination of strategy, supported by a central validation and assessment exercise from which all can learn."[279]

7.4.2.3.1.3. EC MAKES A SURPRISE MOVE—ENTRUSTS FP7 FUEL CELL R&D MANAGEMENT TO PUBLIC-PRIVATE PARTNERSHIP At the October 2007 Brussels Review meeting, EC announced dramatic changes.[279] First, the EC would allocate €470 million ($361.3 million at the 2007 exchange rate of $1/€1.3) for fuel cell and hydrogen development projects under FP7 (2007–2013), which would run for 7 years from 2007 to 2013.[285] Second, the EC would establish a new mechanism, the Fuel Cells and Hydrogen Joint Technology Initiatives (FCH JTI), that could provide a more sustained and focused program for fuel cells and hydrogen research than that could be achieved within the proposal-driven structure of FP7. Third, the JTI would be managed by a public-private partnership, the Fuel Cells and Hydrogen Joint Undertaking (FCH JU).

- The establishment of the FCH JU was extraordinary; the FCH JU was virtually a carbon copy of the US Freedom CAR and Fuel Partnership. It consisted of the EC representatives, fuel cell and hydrogen industries represented by the New Industry Grouping NEW-IG, and the research community represented by Research Grouping N.ERGHY. The JU was to run from 2008 to 2017 and would receive the bulk of the FP7 fuel cell and hydrogen R&D budget minus €20 million (about $15.4 million), to fund the JTI projects under FP7 (2006–2013).[286] In fact, the EC would outsource management responsibilities and pay €20 million for what it had been doing at no cost. The costs for the basic administration of the JTI program, estimated at about €20 million, would be shared: 50 percent from the Industry Grouping and 50 percent from the other member(s) of the JTI.[287]
- There was some unease in the EC to entrusting projects to the FCT JU, which was so strongly oriented toward industry interests.[279] One concern was a possible tendency to focus on immediate commercial options over basic research and training that would be important in the long term. As in the case with the US FreedomCar and Hydrogen Partnership, those were legitimate concerns.

7.4.2.3.1.3.1. But Little Increase in Fuel Cell R&D Funding Despite the expectations for greater FP funding, however, the EU's FP7 funding increased only modestly. As a result, the EC PEMFC budget was slightly lower than that of the United States and far lower than that of Japan's. Because there is no breakdown of the FP7 fuel cell funding, we estimate that the total FP7 PEMFC R&D and demonstration funding might be about €193.9 million (about $149.15 million). This estimate is based on the following analysis:

- An EC document indicates that the FP6 funding for PEMFC R&D and demonstration was 41.25 percent of the combined fuel cell and hydrogen funding.[268] Using this as a benchmark, 41.25 percent of the combined

FP7 fuel cell and hydrogen funding of $470 million would be about €193.9 million for PEMFC R&D funding. Assuming an equal amount is expected from Member States, total European PEMFC R&D spending during the FP7 period would be about €387.8 million (about $298.3 million); over the 7 year FP7 period this would amount to about €55.4 million per year (about $42.6 million).

- The European funding level is likely comparable to that of the United States, but much lower than that of Japan.

7.4.2.3.1.4. FP7 MAY NOT ACHIEVE NEAR-TERM COMMERCIAL SUCCESS Although the JTIs under FP7 (2007–2013) were to cover basic science as well as large-scale demonstrations, the focus on basic science could be weakened under the new FCT JU management. Without much basic research, the near-term outlook for PEMFC commercialization would be less promising (Fig. 7.5).

7.4.2.3.2. EUROPEAN MEMBER STATE GOVERNMENTS CONDUCT MODEST PEMFC R&D

Each of the European countries has different interests in different fuel cell technologies and make R&D investments based on different priorities. Key countries researching hydrogen and fuel cell technologies in Europe are Germany, the UK, Italy, and France.[268]

7.4.2.3.2.1. FRANCE France started PEMFC R&D in 1987. Like other European countries, France abandoned fuel cell research in the late 1970s. CEA (Atomic Energy Commission), however, was developing water electrolysis and in 1987 started PEMFC research by adapting this technology.[288]

7.4.2.3.2.1.1. France Establishes Core Hydrogen and PEMFC R&D Group In a few years, France established a group of PEMFC and hydrogen developers that served as the core of French hydrogen and PEMFC R&D efforts.

- In 1989, the car manufacturer PSA Peugeot Citroen prepared a development program including hydrogen and fuel cells for FCVs.[288] Renault joined PSA in this effort. In November 1989, they jointly proposed to the CEA a collaboration on PEMFC development for transport applications.
- In 1990, the French government launched the national VPE (Véhicule automobile Propre et Econome en énergie—Clean Energy-Saving Automobile Vehicle) program, which

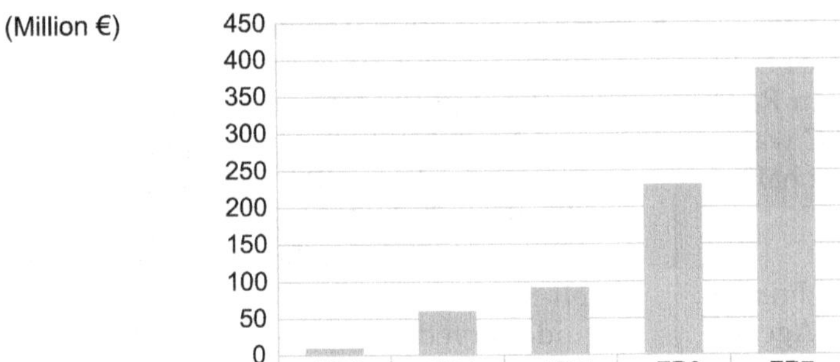

	FP3 (1990-1994)	FP4 (1994-1998)	FP5 (1998-2002)	FP6 (2002-2006)	FP7 (2006-2013)
Member State Funding	4.8	30	45.68	115.35	193.9
EU Funding	4.8	30	45.68	115.35	193.9

FIGURE 7.5 European PEMFC R&D budget: EU framework program funding and member state R&D budget, FP4, FP5, FP6, and FP7 (million euors). *The figure was created based on various EC documents, including "The State and Prospects of European Energy Research, Comparison of Commission, Member and Non-Member States' R&D Portfolios," EUR 22397, European Communities, 2006.* (For the color version of this figure, the reader is referred to the online version of this book.)

included technologies of the future such as hydrogen and fuel cells.[288] The program included PSA, Renault, CEA, as well as CNRS (French National Center for Scientific Research), thus making them the core of the French PRMFC development team.

7.4.2.3.2.1.2. France Established Public Funding Channel France also established a number of funding bodies that would channel public funds to fuel cell (PEMFC and SOFC technology) and hydrogen R&D. In 1999, the Ministry of Research, in association with the Ministry in charge of Industry, created the Fuel Cell Technological Research Network (PACo network).[288] The PACo network was to foster development of fuel cell and hydrogen technology in France through public-private partnerships and interdisciplinary cooperation. Other funding sources included ADEME (French Environment and Energy Management Agency), PREDIT (Programme de Recherche sur le Développement des Transports Terrestres), ANVAR (Agence Nationale de Valorisation de la Recherche), the DGA (Délégation Générale pour l'Armement), and French regions (via Regional Councils), departments (via Departmental Councils), and Districts[288]:

- The PACo network provided nearly 80 percent of the public financing for fuel cell and hydrogen R&D in France. The public funding for PACo was about €10 million ($12.5 million at the 2004 exchange rate of $1/€0.8) for PEMFC, SOFC, and hydrogen R&D projects per year. Funding for PEMFC was about €5 million ($6.25 million), about half of the total.

In 2005, France established the French National Research Agency (ANR) (l'Agence Nationale de la Recherche), a funding agency that would provide French public research funding to both public research institutions and industries based on calls for proposals and peer review selection processes.[289]

- Also in 2005, France launched the PAN-H program (2005–2008) to be administered under ANR that would fund fuel cell and hydrogen R&D projects in place of PACo. Unlike PACo, which funded both PEMFCs, SOFCs, and hydrogen, PAN-H funded primarily PEMFC and hydrogen technology to promote transport applications. Between 2005 and 2008, PAN-H funded a total of 73 projects involving 420 organizations for a total budget of €83 million,[290] €27.7 million per year, fivefold more than that of the budget under PACo.
- In 2009, France implemented the H-PAC program (2009–2010) under ANR to fund fuel cell and hydrogen projects. The program has areas of funding priorities, including low- and high-temperature electrolysis, hydrogen storage, PEMFC stationary applications, prototype development for early markets, and mobile applications including transport.[290]

7.4.2.3.2.1.3. Some Success in Building PEMFC Capabilities, but Falls Short of a Critical Mass for Developers France achieved some successes in persuading a number of developers to engage in PEMFC development and commercialization. For example, it succeeded in incubating two new start-up PEMFC developers: Axane and Helion.

- Axane was a wholly owned subsidiary of Air Liquide, which was created in 2001.[291] It specialized in PEMFCs in a power range from 0.2 to 5 kW fuel cells for portable, stationary, backup, UPS, and plug-in power applications.[292] In 2006, Axane participated in the 5-year, €37.6-million FP6 HYCHAIN MINI-TRANS project (2006–2011). Air Liquide served as coordinator.[293] Axane provided 0.25–0.5 kW fuel cell modules for wheelchairs and 3 kW detachable generators for auxiliary mobile power. In May 2007, Axane provided a 5 kW power generator to a hydrogen station installed by TOTAL in

Berlin and refueled MAN fuel buses operated by Berlin Transportation Company BVG.[294] Also in March 2009, Axane developed and tested a 2.5 kW fuel cell generator to light a cross-country ski trail at the Autrans (Alps, France) resort.[295] An AXANE fuel cell of this type had already supplied 7000 h of energy to a mobile telephony relay station that was isolated from the electricity network.[291]

- Helion was created in 2001 as an R&D unit by Areva, a French multinational industrial conglomerate headquartered in Paris specializing in nuclear power.[296] Helion's purpose was to develop an alternative to nuclear power for naval propulsion.[297] In 2007, Helion was attached to the business unit of Areva Renewable. Helion, filiale d'AREVA spécialisée dans l'hydrogène énergie et les piles à Helion, presented its first prototype of a 20 kW power unit in November 2005 and installed its first 20 kW UPS prototype at the inauguration of CEA's new headquarters in July 2006.[298] The system operated to the satisfaction of the customer for 10 months. In May 2007, Helion commissioned its first commercial 30 kW UPS power unit at the headquarters of the CEA in Saclay.

Nonetheless, two major global automakers in France—PSA and Renault—were seemingly unenthusiastic about FCV developments. In 2003, PSA indicated that it would not make large investments in FCVs. Renault presented no new FCVs since the FP3 FEVER program in 1997. The two automakers also had no interest in developing their own PEMFC capabilities; they used PEMFCs developed by others.

France, thus, failed to create a critical mass of domestic PEMFC developers. France is heavily dependent on nuclear energy and does not have a critical need for hydrogen. This may have handicapped any progress on PEMFC capabilities in France. If the government had little interest in hydrogen and hydrogen-fueling infrastructure, corporations likely would not be encouraged to develop PEM FCVs or PEMFC products that would depend on that infrastructure.

7.4.2.3.2.2. GERMANY Germany started to support PEMFC R&D relatively late in the game. Since 1974, Germany had supported fuel cell R&D under its National Program for Energy Research and Technologies, but the bulk of the funding went to MCFCs and SOFCs. The total annual budget was around €8 million.[299, x,xi] In 1998, Germany implemented a 4-year project (1998–2001) under the Fourth National Energy Program (1996–2004), which for the first time included some PEMFC R&D for transport applications, along with SOFCs and MCFCs for decentralized CHP applications[299]:

- The government budget for the 4 year fuel cell project was roughly €4 million (about $4.4 million at the 1998 exchange rate of $1/€0.91) per year; €3 million ($3.3 million) went to SOFCs. Perhaps only a fraction of the budget was allocated to PEMFC R&D. The program was cofinanced by industries that invested another €4 million per year.
- Germany perhaps never intended to develop domestic PEMFC or DMFC capabilities. The country's major automaker Daimler was investing heavily in Ballard of Canada. Since 2000, a small start-up, SFC Smart Fuel Cell

[x] In addition to the Federal Republic of Germany, there are PEMFC development programs sponsored by the governments of German federal states. Those programs are not discussed in this book. Similarly, programs sponsored by prefectural governments of Japan and state governments in the United States are not discussed.

[xi] Due to lack of exchange-rate information, no dollar equivalent is provided.

(now SFC Energy), has been developing DMFCs for small portable and off-grid applications.

7.4.2.3.2.2.1. Germany Implements PEMFC Demonstration Programs in the 2000s In the beginning of the 2000s, Germany began to promote fuel cell demonstrations and establishment of a hydrogen infrastructure to facilitate commercialization of fuel cell programs. The German government likely believed that its support for fuel cell R&D was unnecessary. There were only a few PEMFC and DMFC companies in Germany, and they seemed to be equipped with market-ready fuel cell technologies. Daimler was supported by Ballard's PEMFC technology and SFC Energy appeared to be self-sufficient and using its own DMFC capability.

7.4.2.3.2.2.2. The Program on Investment into the Future (2001–2005) In 2001, Germany started to fund the Program on Investment into the Future (ZIP), in addition to the fourth Energy Research Program.[300] ZIP aimed at demonstration and market stimulation of fuel cells, including PEMFCs, MCFCs, and SOFCs. ZIP, a public–private partnership program, received a total of €60 million[301] ($53.6 million at the 2001 exchange rate of $1/ €1.12) in German government funding, with the private sector financing another €60 million. German government funding for PEMFCs perhaps amounted to about €29 million ($25.9 million), an average of €5.8 million ($5.2 million) per year.

- ZIP included development and demonstration of several 2–5 kW PEMFC systems for residential applications. Other PEMFC projects included APUs and portable applications and further development of components for automotive applications, onboard reforming of fuels, and demonstration of fuel cell buses.[301]
- ZIP also cofinanced the FP5 project Fuel Cell Buses for Berlin, Copenhagen, and Lisbon and the FP5 CUTE project to demonstrate the DaimlerChrysler fuel cell bus NEBUS/Citaro and hydrogen infrastructure.[302]

7.4.2.3.2.2.3. Clean Energy Partnership (2002–2007) In May 2002, Germany launched an ambitious PEM FCV demonstration project, Clean Energy Partnership (CEP). The project was to demonstrate hydrogen production, supply, and refueling technologies and test-operate FCVs and hydrogen combustion vehicles in Berlin.[303] The CEP aimed to develop hydrogen and fuel cell technologies along the same lines as the FP5 CUTE project and create a path toward hydrogen and FCV commercialization via cooperative action between the government, energy companies, and automakers. The project was run by a public-private partnership, but the private sector clearly took the leading role. The CEP partnership consisted of the German federal government represented by the Ministry of Transport, Building, and Urban Affairs (BMVBW) and nine corporate partners, including Aral, BMW, the Berlin Transit Agency, Daimler, Ford, GM, Hydro/GHW, Linde, and Vattenfall Europe. The corporate partners supplied the infrastructure, technology, and vehicles for the 5-year project.[552]

- Total cost of the CEP project was €33 million ($31.1 million at the 2002 exchange rate of $1/ €1.06).[301] BMVBW provided subsidies to the tune of €5 million ($4.7 million).[303]

The CEP (2002–2007) demonstrated a fleet of 17 hydrogen cars. BMW showed four hydrogen combustion vehicles, DaimlerChrysler eight FCVs, Ford four FCVs, and GM/Opel one FCV.[44]

- In November 2004, the CEP opened the demonstration part of the project with a hydrogen-fueling station fully integrated into a public Aral-fueling station on Messedamm in Berlin. The station supplied both liquid and gaseous hydrogen.[303]

Individual consumers were allowed to test drive the FCVs until the end of the program in December 2007.[44]

- In March 2006, the CEP opened a second hydrogen-fueling station in Berlin–Spandau. The hydrogen equipment, which was integrated into a conventional TOTAL fueling station, provided liquid and gaseous hydrogen for cars and buses operated by the Berlin Public Transit System.[303]
- In June 2006, the first two buses in the FP HYFLEET–CUTE project refueled with hydrogen at the TOTAL fueling station and provided shuttle services for journalists during the 2006 FIFA Soccer World Cup as part of FIFA's Green Goal Initiative. Volkswagen joined the project in July 2006.[303] The two hydrogen buses were taken over by the Berlin Public Transit System and put into regular service in August 2006. Two more hydrogen buses were added in October 2006 and refueled at the TOTAL fueling station.[44]

After successfully demonstrating that hydrogen could be used safely in road traffic situations, CEP moved into a second phase (2008–2010), which concentrated on the validation of technology under everyday driving conditions and further development of technologies essential for market entry of hydrogen fueled vehicles.[303]

- This is to be followed by a third phase for CEP (2011–2016), which would be dedicated to preparing for market entry of hydrogen in the traffic sector.[303]

7.4.2.3.2.2.4. National Innovation Program for Hydrogen and Fuel Cell Technology (2007–2016) In 2007, Germany launched the National Innovation Program for Hydrogen and Fuel Cell Technology (NIP) (2007–2016)[304] to accelerate R&D and commercialization of internationally competitive hydrogen and fuel cell technology products made in Germany. NIP is a major strategic alliance between the German government, industry, and academia. The overall project budget was €1.4 billion, €200 million to be funded by the Ministry of Economics and Technology (BMWi) focusing on R&D, €500 million funded by the Ministry of Transport, Building, and Urban Affairs (BMVBS), and €700 million cofinanced by industry. The level of German government funding was significant and unprecedented. Indeed, government funding amounted to €700 million over 10 years, making the project the biggest R&D and demonstration program of its kind in Europe.[305]

- Under this program, fuel cell R&D would be conducted by researchers from dozens of German universities and state-funded research institutes, such as German Aerospace Center (DLR), Fraunhofer, and Max Planck, which specialized in fuel cell system integration, fuel cell system, fuel cell stack, stack component, balance of plant (bop)/electrolyzer/reformers, and manufacturing technology.[306] The goal was to reduce cost and increase lifetimes, reliability, and efficiency.[306] The program appears to be focused on applied research. According to a NIP-prepared document, R&D activities should be orientated toward the entire system as a whole and to the requirements of the users.[307]
- The demonstration projects under NIP aimed to validate systems and methods devised during the course of R&D activities and to prepare for commercialization. NIP planned for an intensive exchange of data between R&D projects and demonstrations. In addition, NIP envisioned the operation of progressively larger-scale demonstrations in conjunction with the required hydrogen infrastructure.

NIP is coordinated and managed by the National Organization for Hydrogen and Fuel Cell Technology (NOW), which was established in February 2008.[308] NOW, the 100 percent-federally owned organization, brings together researchers from Germany's universities and state-funded research institutes with engineers from companies like Daimler and Siemens, as well as other players in the supply chain, to forge cooperative projects.[309] The program included more than 180 industry partners.

A key focus of NIP is to link together R&D and field activities for projects that demonstrate hydrogen fuel cell technology.[309] NIP consists of three segments:

- ***Transportation.*** The second phase of CEP (2008–2010) would play a major role in NIP transportation demonstration.[305] In NIP, 50 percent of the total budget would be devoted to transportation.[310] CEP demonstrations were to be expanded to include Berlin and Hamburg and include more than 30 FCVs and bus fleets and several hydrogen-fueling stations.[304]
- ***Stationary Applications.*** The CALLUX Lighthouse Project (2008–2015) would install 800 fuel cell power generation plants for demonstration in five regions.[304] This would include PEMFC micro-CHPs for residential use, and industrial SOFC and MCFC generator sets for CHP and trigeneration, and some marine applications.[553] Of the total budget, 29 percent would be devoted to stationary applications.[310]
- ***Special Markets.*** This segment of the project would focus on fuel cells for critical electricity supply, information technology, telecommunications, APUs, propulsion for logistic, leisure, and tourist markets (boats, bikes, caravans, etc.), and special vehicles (e.g., forklifts).[304] Twenty percent of the total budget would be dedicated to this segment.[310]

7.4.2.3.2.2.5. H2 Mobility Initiative (2009–2015) Germany launched the 6-year H2 Mobility Initiative (2009–2015) in September 2009.[311] The initiative would evaluate and expand the hydrogen infrastructure in Germany to support the commercial production of FCVs. Daimler and leading energy companies—EnBW, Linde, OMV, Shell, Total, Vattenfall—and NOW (the management of NIP) signed a Memorandum of Understanding in Berlin under the auspices of the German Ministry of Transport.

- The H2 Mobility launch came one day after leading automakers signed a Letter of Understanding (LOU) to commercialize and start series production of FCVs in 2015.[311] Noting the importance of a hydrogen infrastructure with sufficient density, the automakers—Daimler, Ford, GM/Opel, Honda, Hyundai, Kia, Renault/Nissan Alliance, and Toyota—all strongly supported building a hydrogen infrastructure in Europe, with Germany to be the regional starting point.

The H2 Mobility Initiative has two phases[311]:

- Phase One includes the evaluation of options for an area-wide roll-out of hydrogen-fueling stations in Germany. The partners would leverage plans to install new hydrogen-fueling stations by 2011. This would take place within the framework of the German economic stimulus package (Konjunkturpaket II) and other national and state programs to jointly address standardization and cost-reduction issues.
- Subject to a positive and satisfactory outcome of Phase One, the partners would implement an action plan in Phase Two. The nation-wide roll-out of hydrogen-fueling stations would be continued in parallel with the start of series production around 2015.

The German government would also provide a package of financial incentives starting in 2012 to support the production and sale of 100,000 FCVs annually.[311] The plan envisaged that around one million FCVs would be on German roads by 2020.

7.4.2.3.2.2.6. But Success of FCV Commercialization in Near Term Remains Uncertain Germany should be able to expand its hydrogen infrastructure. It can build upon its leading in Europe, with initial hydrogen centers having been established in urban areas such as Berlin and Hamburg, and with many of the current 30 hydrogen-fueling stations in Germany already integrated into public gas stations.[311] German fuel cell programs—including ZIP, CEP, H2 Mobility, and NIP—would highly likely bolster this infrastructure base.

Nonetheless, the prerequisite for successful FCV commercialization is PEMFC technology that is efficient, durable, and cost competitive. Germany may still be able to develop such PEMFC technology. NIP includes a substantial R&D program, with one-third of the total budget devoted to fuel cell research. NIP R&D, however, leans toward applied research rather than basic research. Moreover, NIP R&D activities are programmed to make improvements of those fuel cell systems that are already being used in demonstration programs. There are no plans to conduct basic research to achieve breakthroughs in PEMFC performance.

7.4.2.3.2.3. ITALY Italy began fuel cell R&D in the late 1980s.[299] Hydrogen and fuel cell R&D was promoted by four organizations, including Italy's Ministry of Education, University, and Research, the Ministry of Environment, local governments, and the FP Program.[44] ENEA (the Italian National agency for new technologies, Energy, and sustainable economic development) was primarily responsible for Italian fuel cell development. It conducted research and implemented research programs in close collaboration with Italian Ministries, the European Union, industry, academic institutions, and research organizations.[312] ENEA also acted as a source of government funding to Italian industry and universities for fuel cell R&D and demonstrations.[313]

7.4.2.3.2.3.1. Italy provides Relatively Small PEMFC R&D Funding Italy has provided modest funding to PEMFC R&D. Although information is incomplete, various reports suggest funding for PEMFC R&D could have been as much as €3 million per year.

- In the 1980s, Italy's annual fuel cell R&D was around €5 million, to which both government and industry contributed.[299] The main focus was on the development of MCFCs for distributed cogeneration and PEMFCs for transport applications. The government budget for PEFCs likely was less than €1 million per year.
- In the mid-1990s, Italy implemented a 3-year program that would spend about $19 million on fuel cells for transport applications.[313] Industry (and other sources) were to provide $20 million toward this program, for a total of $39 million. Of this, perhaps $20 million was for R&D on PEFCs for transportation. Assuming the government's share was half ($10 million), this would imply an annual government budget for PEMFC R&D of $3.3 million.
- An EC document indicated that Italy provided about €5 million to MCFC, PEMFC, and SOFC R&D during the 1994–1999 period.[314] This implies that an annual PEMFC R&D perhaps was about €1.5–2 million.

7.4.2.3.2.3.2. Italy Focuses on PEM FCVs in Early 2000s In November 2000, the Italian government decided to support a new research program with a budget of €7 million over the following 3 years to support the fuel cell effort.[299]

- In January 2001, the first hydrogen-powered urban fuel cell bus was developed by Iveco/Irisbus for the municipal transport authority of Turin.[302] The bus was powered by a 60 kW fuel cell system supplied by UTC. It had a hybrid configuration equipped with a battery system and was fueled with hydrogen.
- In February 2001, Fiat introduced its first prototype fuel cell car, the ***Seicento Elettra***

H2 Fuel Cell.[19] The development of the FCV was partly funded by the Italian Ministry of Environmental Affairs. A more advanced prototype of the same car was presented in 2003.[302] Both Fiat prototypes used compressed hydrogen as fuel and fuel cell stacks supplied by Nuvera.

7.4.2.3.2.3.3. Lack of Government Will and Absence of Domestic PEMFC Capabilities Hinders Commercial Success The Italian government has not fostered a critical mass of PEMFC technology and it lacks a strong, long-term policy that would incentivize the industry to develop and commercialize PEMFC products. Italy has only implemented a series of short-term PEMFC development.[315] So far, the country has fostered only two PEMFC developers: Nuvera and Acrotronics. Nuvera has produced 5–25 kW fuel cell systems for material-handling vehicles, 4 kW systems for CHP applications, and 75–120 kW for distributed power generation and FCV applications.[44] Arcotronics Fuel Cells has produced PEMFC systems ranging from 500 W to 50 kW. It marketed 3–7 kW systems for CHP applications. Arcotronics (now Exergy Fuel Cells) was acquired by Morbid Group of Sweden in 2007.[316] Nuvera's ultimate success would likely depend on the success of the system integrator, Fiat. It is not clear if Fiat is committed to FCV development, however.

7.4.2.3.2.4. UNITED KINGDOM The UK government started to support fuel cell R&D in 1992. Its funding effort has been modest compared to Germany, primarily focusing on SOFC and PEMFC technologies.[299]

7.4.2.3.2.4.1. UK Implements National Plans and Strategies to Promote PEMFC R&D In 2002 and 2003, United Kingdom made an energetic effort to formulate a national plan and strategy to promote fuel cells as a means to strengthen UK industrial competitiveness as well as advance clean energy policy.[317] One key goal of the United Kingdom's plan was to encourage national suppliers to develop PEMFCs suitable for stationary and transportation. The United Kingdom, like the other European countries, was undoubtedly aware that it had not been able to implement important fuel cell projects because they did not have a domestically based fuel cell maker.

7.4.2.3.2.4.2. UK Deploys Three R&D Funding Channels To incubate domestic PEMFC capabilities, the United Kingdom created three funding channels. It provided funding to industries through the Department of Trade and Industry (DTI) and to academic institutions through the Engineering and Physical Science Research Council (EPSRC).[318] The country also established in 2001 the Carbon Trust, a non-profit corporation, to support industries engaged in the development of fuel cells and other new emerging low-carbon technologies.[319]

The DTI budget for fuel cell R&D funding was relatively modest, around £2 million per year, in the early 2000s ($3 million at the 2000 exchange rate of $1/£0.66).[317] Nonetheless, DTI support appears to have produced one major success—it likely helped foster one PEMFC start-up.

- According to a survey conducted in 2003, DTI provided the largest subsidy to Intelligent Energy, a start-up PEMFC developer that was spun off from Loughborough University's research work.[317] DTI provided a total of £3.84 million ($5.8 million) to Intelligent Energy between 2000 and 2003 for the development of a 50 kW PEMFC system for APUs for Boeing's single-engine aircraft. DTI provided Johnson Matthey £0.9 million for MEA development, £0.5 million to ITM Power for polymer research, and £0.3 to £0.4 million to Morgan Fuel Cells, a subsidiary of the US company Morgan, for bipolar development.

ESPRC focuses on funding academic research and receives its funding primarily from the UK Department for Business, Innovation, and Skills (BIS). In 2007/2008, ESPRC provided a total of £507.3 million in research grants,[320] but it is not clear how much was allocated to PEMFC R&D. Between 2000 and 2003, several universities—Imperial College, St. Andrew's College, the University of Surrey, and University of Southampton—received half of ESPRC's total fuel cell R&D budget.[317] Most funding was for SOFC research, however.

The Carbon Trust was established in 2001 to support the development and commercialization of low-carbon technologies, including fuel cells.[321] Its 3-year budget was £75 M. The Trust helped build a solid, robust, commercially vibrant industry sector, focusing on those elements of the emerging fuel cell business where the United Kingdom could compete successfully.

7.4.2.3.2.4.3. Commercial Success in Near Term Uncertain The one major successful outcome of the UK fuel cell funding effort was the emergence of Intelligent Energy as a global PEMFC developer. It is not certain, however, if United Kingdom would be able to foster more start-up companies like Intelligent Energy and thereby create a critical mass of PEMFC capabilities in United Kingdom.

- The United Kingdom lacks national R&D programs to advance PEMFC technology and likely will not implement any in the near term. The UK seems to depend on the FP projects to leverage the growth of its domestic PEMFC capabilities. UK companies, including Intelligent Energy and Johnson Matthey, and as well as UK universities have regularly participated in FP projects throughout the 2000s. However, UK participation for the most part has been that of a project member rather than that of a coordinator or project leader. The United Kingdom also participated in some of FP fuel cell bus demonstration projects. With support from the non-profit UK Energy Savings Trust and the Greater London Authority, the government of the City of London,[302] three fuel cell buses operating on hydrogen were demonstrated in central London as part of the FP CUTE project (2001–2006) and the FP6 HYFLEET–CUTE project (2006–2010). These projects, however, did not employ UK PEMFC capabilities; the buses used Ballard's PEMFCs.
- The United Kingdom also perhaps has lost some of its earlier enthusiasm for hydrogen and fuel cell development. It recently reduced its funding support for organizations that directly and indirectly affect the development of hydrogen and fuel cell technologies. The Carbon Trust announced a 40-percent cut in its funding from the government.[554] A few weeks prior to that, the Energy Saving Trust announced a 50-percent reduction in its grant income.

7.4.2.4. *Other Governments*

7.4.2.4.1. CHINA

As noted earlier, China started to promote PEMFC R&D in the mid-1990s[323] under its 863 Program, which is the government's multiyear high technology development program. China's Ministry of Science and Technology (MOST) funded a number of research institutions, universities, and companies to promote fuel cell R&D. Its funding on fuel cell research during the 1990s, however, was small and thinly spread among a large number of competing recipients.

- Those who received MOST funding for fuel cell research during the period included Beijing Fuyuan Century Fuel Cell Power, Beijing LN Green Power Company, the Beijing University of Science and Technology, the Changchun Institute of Applied Chemistry (Chinese Academy of Sciences), the Dalian Institute of Chemical Physics (DICP) (Chinese Academy of Sciences), Fudan University, Shanghai University (in cooperation with Beijing Petroleum University), the South China University of Technology, Tianjin Institute of Power Sources, Tianjin University, and Tsinghua University.[324]

By the end of the 1990s, some of the research had yielded positive results. The Changchun Institute of Applied Chemistry developed a 100 W PEMFC prototype.[323] DICP built and tested some 20 PEMFC stacks ranging in size from 1 to 5 kW.[324] In 2000, the Beijing Green Energy Institute made a 15 kW PEMFC. Also, the Beijing Fuyuan imported Nafion membranes and built a 5 kW PEMFC stack, and the Tsinghua University Automotive Engineering Department built a 5 kW PEMFC golf cart.

- Likely encouraged by these results, MOST began to investigate the integration of fuel cells into vehicles under a large-scale development program. The Chinese FCV development program began to take shape in 1998.[208]

7.4.2.4.1.1. CHINA LAUNCHES AGGRESSIVE PEM FCV DEVELOPMENT PLAN IN THE EARLY 2000s China pursued an ambitious plan to establish PEMFC technology for transport applications, particularly for passenger cars and buses, during the 2000s. To achieve its goals, China took full advantage of the nation's major high technology development programs and increased and refocused their resources on PEMFC and FCV development. China's motivation was clear. FCVs would help China reduce imported oil.[325] The Chinese economy would benefit; China could leap-frog foreign counterparts and the fuel cell technology was still under development, giving time for China to be a global fuel cell player.

China chose to boost the budget of three major national programs to promote PEMFC and FCV technologies. In effect, this refocused the goals of the 863 Program, the 973 Program (named after its start date in March 1997), and the Torch Program. China's Ministry of Science and Technology (MOST) ensured that many eligible, promising institutions, universities, and companies received funding under one or more of these programs throughout the 2000s[48]:

- ***Two consecutive 5-year plans under the 863 Program.*** China placed major emphasis on the development of PEM fuel cell cars and buses under the 10th 5 year plan (2001–2005) and the 11th 5 year plan (2006–2010), primarily through the 863 program. The 10th 5 year plan allocated $106 million for fuel cell development, the majority of which was provided for PEMFCs.[48] The 11th 5 year plan further stepped up funding. A total of $172 million was allocated for energy technology development, the majority of which would go to fuel cells and hydrogen.[326]
- ***The National Hydrogen Energy Project 973.*** With a budget of approximately $3 million from 2000 to 2005, Project 973 promoted basic research to strengthen fundamental R&D, notably with regard to the production, storage, and distribution of hydrogen and fuel cells.[208] Much of the research and development under this project was conducted by Dalian Institute of Chemical Physics (DICP).
- ***The Torch Program.*** Torch was China's most important high-technology industry program. Launched in August 1988, it aimed to promote the development of high-technology products that would have high economic potential in domestic and foreign markets.[327] As such, the program would fund individual organizations and their efforts to develop and commercialize new high-technology products, including fuel cells and FCVs.
    - Some of the projects funded under the program were: the 2001 project to

support the development of the Phoenix Fuel Cell Vehicle by GM and SAIC; the 2002 project to develop a 5 kW PEMFC by DICP and Sunrise Power; the 2004 project to develop a PEMFC vehicle engine by Shanghai Shen Li (Shen Li); the 2007 project to establish the China Vehicle Energy Onboard Technology Center with GM, SAIC, and Tsinghua University; the 2007 project to develop the Shanghai Volkswagen Passat Fuel Cell Vehicle by SAIC and Shanghai Volkswagen and demonstrate at the Beijing 2008 Olympics.[328]

  - Also included were most FCV projects that involved SAIC, Cherry Motor, Shanghai Volkswagen, Dongfeng, Beijing Foton, Changan, Tsinghua University, Tongji University, Shen Li, Dalian Sunrise Co., Beijing Tsingneng Huatong S&T Company, and Beijing Inpower Company.[328]

In addition to fuel cell R&D, the three national programs funded demonstration of FCVs, development of domestic FCV brands, establishment and expansion of research facilities and factories, international collaboration, joint ventures, and mergers and acquisitions. Moreover, China assigned local governments, especially the Beijing and Shanghai municipal governments, to fund PEMFC R&D, FCV development, and demonstrations.[325] Although little information is available, local governments reportedly allocated significantly more for FCV development than the central government.[325] Thus, the Chinese government not only had flexibility and broad administrative authority, but also was able to marshal ample financial resources[208]:

- The core of China's capacity-building strategy was a conscious effort to build a domestic research capacity at Chinese universities and companies, to network local enterprises, research institutes, and universities, and to stimulate learning through partnerships with foreign automobile, fuel cell, and oil companies. Not surprisingly, most of these activities were eligible for central and local government funding.
- MOST funded national and university research institutes for PEMFC R&D as well as industries for fuel cell and FCV-related business activities.
- The central government also directed the municipal governments of Beijing and Shanghai to cofund FCV technology projects in the period 2002–2005 and the provincial government in Hubei province to support FCV development projects over the period 2005–2007.

In sum, the Chinese effort to develop PEMFC technology and FCVs was massive, systematic, and broad. Any notable research institution, university, or company with fuel cell development activities was provided with long-term funding by the government. Indeed, in view of the impressive lineup of fuel cell cars and buses developed during the 2000s, it is not difficult to see that massive funding must have been given to PEMFC and FCV developers, including DICP, SAIC, Shanghai Fuel Cell Vehicle Power Train Company, Shen Li, Tongji University, and Tsinghua University.

7.4.2.4.1.1.1. China Provides Administrative Guidance for FCV Development China also provided specific direction and guidance to fuel cell developers, which they were obligated to implement under the national programs. While this direction was logical and reasonable, it would be highly unusual if any Western government were to apply such management tools to private-sector industries.

Some aspects of China's guidance were plainly stated; some were not. In any case, the guidance could be inferred from demonstrated activities. Seven different guidelines are identified and presented here.

First, China directed the number and specifications of fuel cell cars and buses that should be developed under the 863 Program as part of the 10th 5 year plan. The guidance specified that three prototype fuel cell cars and three fuel cell bus prototypes were to be built by 2005.[208]

Second, China directed Tongji University and Shanghai Fuel Cell Vehicle Powertrain Company (SFCV) to lead development of fuel cell cars.[213] China assigned Tsinghua University and Beijing Sinohytec Company to head the fuel cell bus development.

Third, China directed that specialized powertrain or platforms were to be built, including a third-generation set for fuel cell cars and for fuel cell buses. The powertrains for fuel cell cars were to be designed to accommodate various passenger car models. Similarly, the powertrains for fuel cell buses were to be designed to accommodate various bus vehicle models.

Fourth, China directed that when the powertrain technology was ready, the universities should transfer the technology to automakers and bus manufacturers.

Fifth, China instructed Chinese companies to engage in joint ventures with global automakers and fuel cell developers. Joint ventures would speed up the learning process and advance Chinese fuel cell capabilities faster. With strong financial incentives from the Chinese government and potential access to a large Chinese market, many foreign automakers agreed to form joint ventures and joint FCV development projects with Chinese companies.

Sixth, China directed that a number of major FCV demonstrations were to be held. China solicited support of local municipal governments, international organizations, and other foreign governments. Demonstrations of FCVs were to offer practical experience as well as showcase Chinese fuel cell prowess to the world.

Seventh, China directed that domestic FCV brands should be created. China would provide resources to Chinese companies to engage in acquisition of foreign automakers with recognized brand names, joint ventures, and other collaborative ventures.

7.4.2.4.1.2. CHINA UNVEILS A SERIES OF FCVS In 2000, China assigned FCV development efforts to Tongji and Tsinghua universities[32]:

- Tongji University was assigned to develop three generations of fuel cell passenger cars. Tongji would work with its close associates: Shanghai Fuel Cell Vehicle Powertrain (SFCV), SAIC, and Shanghai Shen Li High Tech Company (Shen Li). SFCV was a spin-off from Tongji's Center of Automotive Engineering and was established in 2001 as a joint venture with SAIC. SFCV was also one of the main grantees of the government's 863 program. SAIC, a state-owned automaker, was then China's second largest automaker next to China First Auto Works (China FAW), which SAIC would overtake in 2009. Shen Li was China's leading PEMFC company, established in 1998.
- Tsinghua University was assigned to lead the fuel cell bus development effort. Tsinghua would work with DICP, Shen Li, and Sunrise Power (a joint-stock company owned by the DICP) to develop three generations of fuel cell buses. Tsinghua had worked on developing PEM fuel cells, fuel cell engines, and making hydrogen from ethanol. In 1999, Tsinghua built a fuel cell golf cart with a 5 kW PEMFC developed by Beijing Fuyuan. Together with Beijing LN Green Power, Tsinghua University introduced various vehicles in 2001, one of which was a small 12-seat bus.

In addition to these two team efforts, all or most of the other FCV development efforts undertaken in China during the 2000s likely were funded under one program or another.

- In total, at least 23 fuel cell cars were developed and at least 19 fuel cell buses were developed by the end of the 2000s.

7.4.2.4.1.2.1. Fuel Cell Cars China began unveiling fuel cell cars in 2001. FCVs developed in the early 2000s were generally the result of the fuel cell activities funded during the 1990s. Four FCVs were introduced. Three of them were introduced by LN Green Power, including the ***PEM Fuel Cell Powered Taxi*** and ***the Green Power FC Passenger Car*** in 2001, and ***the Green Power FC Kombi*** in 2002. LN Green Power probably used its own PEMFC stacks and built the FCVs in partnership with the Beijing Institute of Technology and Tsinghua University. The sudden emergence of LN Green Power as an FCV developer and its rapid introduction of prototypes created a major sensation in the global fuel cell community.

The fourth FCV unveiled in China in 2001 was an even greater sensation. In October 2001, GM and SAIC presented the ***GM Shanghai Phoenix Minivan*** in Shanghai. The Shanghai Phoenix FCV was a product of Pan Asia Technical Automotive Center (PATAC) Company, a $50 million, 50–50 joint venture between GM and SAIC, established in June 1997. The FCV in reality was a GM product; it was based on a 2.5-ton GM Buick GL8 XEA1 minivan with room for eight passengers, powered by a 35 kW GM PEM fuel cell, and built at Shanghai GM, another GM-SAIC joint venture.[48] The Shanghai Phoenix had a driving range of 125 miles and a top speed of around 70 mph.[329] It ran on compressed hydrogen. The FCV had already met the government's minimum requirements for an FCV. The project was funded by the Chinese government.[328]

Starting in the early 2000s, Tongji established itself as China's pre-eminent FCV developer. Over the next 5 years, Tongji would successfully develop increasingly powerful powertrains in three successive generations: the ***Chao Yue I, II,*** and ***III***.

- The PEMFC stack had better performance in each generation. The first-generation stack was 30 kW, the next was 35 kW, and the third was 40 kW, providing a power increase of 33 percent from the first generation to the third. Driving speed increased from 65.7 mph to 75.7 mph, about a 15-percent increase, from the first generation to the third. Acceleration dramatically improved. The first generation accelerated from 0 to 62 mph in 36.6 s, the second in 26.7 s, and the third in only 19 s.

The third-generation Chao Yue III was undoubtedly viewed by Chinese policymakers as ready for transition to automakers. Indeed, the Chao Yue III powertrain was incorporated into two existing vehicle types: the Volkswagen Santana 3000 model and the Chery Automobile's Eastar MPV model. Two automakers—Shanghai Volkswagen and Chery Automobile—likely participated in the development effort. Ten Chao Yue III vehicles were built, eight Chao Yue III S3000 vehicles, one Chao Yue III MPV, and one Chao Yue III Eastar. The 10 Chao Yue III vehicles drove a cumulative total of 63,342 miles. The longest-running vehicle traveled a total distance of 18,009 miles with no repair.[213] Two Chao Yue III S3000 vehicles demonstrated excellent performance at the 2006 Bibendum Challenge in Paris.[214]

Beyond Tongji's powertrain successes, there was another FCV development effort, which perhaps was not too successful:

- In August 2004, Dongfeng Motor presented the ***Wuhan University Chu Tian 1***.[330] The FCV was based on a Citroen Fukang with a capacity of five passengers.[331] It was powered by a 25 kW Wuhan University PEMFC, batteries, and a motor and ran on compressed hydrogen. It had a top speed of 65 mph and accelerated from 0 to 50 mph in 22 s. Dongfeng, China's second largest automaker and a public (not a state owned) company, unveiled the Chu Tian 1 about the same time as Tongji's Chao Yue II. But, the Chu Tian 1 was less powerful and the performance was less impressive. The Chu

Tian 1 was never followed by a second model. Perhaps, Dongfeng and Wuhan University gave up their fuel cell car effort.

Amid the introduction of a number of domestically produced FCV prototypes in the mid-2000s, China's research institutions nonetheless reached out to Ballard of Canada for assistance:

- China had been soliciting assistance from Ballard for over a year. In the fall of 2005, China's President visited Ballard at his express request, the only company he visited during his 10-day September visit to Canada.[214] Ballard indicated that the President's interest in visiting the company was first conveyed in August when a delegation from China's Ministry of Science and Technology visited Ballard, which was the most recent of a series of contacts between Ballard and the ministry, as well as with China's Tsinghua and Tongji Universities.[332]
- In July 2006, Shanghai Fuel Cell Vehicle Powertrain Company (SFCV)—a member of the Chao Yue development team—and Ballard of Canada signed a memorandum of understanding and fuel cell supply agreement to cooperate on the development of FCVs for demonstration and field trial programs in China.[333] As a first step in the relationship, Ballard and SFCV signed an agreement for Ballard to supply to SFCV up to 20 Mark 902 automotive fuel cells. Ballard would ship two Mark 902 automotive fuel cells later in 2006.[334] Ballard would also provide related applications engineering and field-service support. Upon successful integration of the fuel cells into vehicles, SFCV would purchase up to an additional 18 Mark 902 fuel cells for the program in 2007.[335]
- The Mark 902 platform, the fourth generation of fuel cell technology from Ballard, was scalable from 10 kW to 300 kW of output depending upon customer requirements and applications.[333] Typical power output for transportation applications was 85 kW for passenger vehicles and 300 kW for transit bus applications.

It was Chinese government policy to encourage Chinese companies to form collaborative relationships with global fuel cell players. But the purchase agreement with Ballard's PEMFC products after the seemingly successful development of the Chao Yue III was peculiar. Moreover, two Chao Yue III S3000 vehicles had just demonstrated excellent performance at the 2006 Bibendum Challenge in Paris a month earlier.

- It was possible that China was not fully satisfied with the Chao Yue III powertrain technology and wanted some other options.

In 2006, Tongji, SFCV, and SAIC started to build a new-generation platform, reportedly changing the configuration and using improved parts.[208] By the end of 2006, the team built the ***Chao Yue IV***.[336] As with the Chao Yue I, II, and III, the powertrain platform would be shared by various vehicle types.

In 2007, three FCVs were built for demonstration at the Shanghai New International Expo Center in 2007: the ***Tongji University Fuel Cell Passat*** based on the VW Passat Lingyu, the ***SAIC Roewe 750 FC*** based on SAIC Roewe 750, and the ***Chery Eastar FCV*** based on Chery Automobile's Eastar.[337] The SAIC Roewe 750 was built based on the Rover 75 design, which SAIC had purchased in early 2005. With the completion of these Chao Yue IV FCVs, all the automakers involved seemed to have full confidence in their FCVs and started to discuss commercial production dates:

- Shanghai Volkswagen envisioned mass production of the Fuel Cell Passat sometime in 2010.[338]
- The Roewe 750 FC technology was transferred to SAIC to tool up for production in 2008.[208]

- Chery Automobile envisioned commercialization of Chery Eastar FCV in the next 5–10 years.[339]

The FCVs built with the Chao Yue IV powertrain showed striking improvement over those with the Chao Yue III. The fuel cells were smaller and more powerful.[208] The Chao Yue IV powertrain used a 50 kW fuel cell, enabling a top speed of 93 mph, a driving range of 186 miles, and a fuel efficiency of 1.2 kgH_2/62 mile. It accelerated from 0 to 62 mph in 15 s.

- Twenty Tongji University Fuel Cell Passat vehicles were built and successfully demonstrated during the summer Olympic games 2008 in August in Beijing.[32]
- The Chao Yue IV vehicles likely were powered by Ballard fuel cells (Table 7.8).

In 2008, separate from the Chao Yue IV effort, there would be two other FCV development projects, which never came to fruition:

- Geely Automobile, China's tenth largest automaker and a non–state-owned company, presented the ***Geely MK Fuel Cell***.[340] The FCV was a fuel cell hybrid with a capacity for five passengers. It ran on compressed hydrogen. Little information is available.
- Shanghai Shen Li presented the ***Shen Li High Tech Company FC SUV***.[341] It integrated a 35 kW Shen Li PEMFC, supercapacitors, and a 301 W brushless DC motor. It ran on compressed hydrogen (20 MPa), which was carried in 120 l vessels. It had a top speed of 75 mph and accelerated from 0 to 37 mph in 15 s.

In 2008, SAIC began to show concrete results from its development efforts. Starting in 2008, SAIC unveiled three new FCVs—one nearly every year. SAIC, however, still seemed to be in search of the right fuel cell for its FCV. For the three FCVs unveiled in 2008, 2009, and 2011, SAIC used one fuel cell stack by an unidentified manufacturer, a GM stack, and an SAIC stack that it developed for the first time, respectively.

- In 2008, SAIC presented the ***SAIC Roewe 750 Plug-In FC***.[342] The plug-in FC hybrid was based on a SAIC Roewe 750E CSA7001FCEV. It was powered by a 25 kW PEMFC, Johnson Control & Saft lithium-ion batteries, and a 42 kW Shanghai Electric Drive permanent

TABLE 7.8 Progress in Developing Fuel Cell Cars in China

Model	Chao Yue I	Chao Yue II	Chao Yue III	Chao Yue IV
Fuel Cell/Battery Power	30 kW/30 Ah	35 kW/15 Ah	40 kW/15 Ah	50 kW
Traction Motor	50 kW	60 kW	65 kW	
Maximum Speed (mile/h)	65.7	71	75.7	93
Acceleration (0–62 mile/h), Seconds	36.6	26.7	19	15
Range (mile)	143.4	134.8	142.8	186.3
Fuel Efficiency (kgH_2/mile)	1.394	1.132	1.12	1.2

Source: Wei Guo WANG (Ningbo Institute of Material Technology and Engineering), "After the Beijing Olympics: Fuel Cells in China," PPT Presentation, http://www.climate-change-solutions.co.uk/pictures/content473/1-2_wei_gho_wang_-_20100322_birmingham_wei_guo_wang.pdf, Birmingham, UK, March 15, 2010; Ming Pingwen, Lun Jingguang and Lynn Mytelka, "15. Hydrogen and fuel-cell activities in China," in Making Choices About Hydrogen, Transport Issues for Developing Countries, Edited by Lynn K. Mytelka and Grant Boyle, http://www.idrc.ca/en/ev-132182-201-1-DO_TOPIC.html, the International Development Research Centre, 2008.

magnet synchronous motor. The manufacturer of the fuel cell stack was not identified. The plug-in FC ran on compressed hydrogen (35 MPa). It had a top speed of 93 mph.

- In 2009, SAIC presented the ***SAIC Roewe 750 GM-Fuel Cell***.[343] The FCV was based on SAIC Roewe 750 E CSA7002FCEV and powered by a 55 kW GM PEMFC, Johnson Control & Saft nickel–metal hydride batteries, and a 45 kW three-phase synchronized motor. It ran on compressed hydrogen (70 MPa). The FCV had a top speed of 93 mph.
- In April 2011, SAIC unveiled the ***Shanghai Brand Plug-In Fuel Cell Car***.[344] It used a battery as main power source and a small pressurized fuel cell to provide auxiliary power if the battery ran out.[345] The plug-in fuel cell car was developed by SAIC itself. The battery would run the car the first 12 miles, but for longer-distance driving or when the battery ran out of charge, the system could be shifted into a hybrid mode, where the fuel cells would generate electricity to power the car and recharge the battery.[344]

In April 2010, four FCVs were unveiled in China all at the same time, which was unusual. Three models used an identical powertrain: a 55 kW Shen Li PEMFC, Suzhou Star Constant Power lithium–manganese batteries, and a 42 kW Shanghai Electric Drive permanent magnet synchronous motor.[346]. The fourth (listed below) curiously used a different set of components, consisting of a 40 kW Shen Li PEMFC, lithium-ion batteries, and a 42 kW Shanghai Drive permanent magnet synchronous motor.[347]

- Chang'an Automobile, China's fourth largest automaker and a state-owned corporation, presented the ***Chang'an Z-SHINE FCV*** at the Auto China 2010 in Beijing.[346] The FCV was based on a Chang'An Automobile Z-SHINE SC7003EV. It ran on compressed hydrogen and had a top speed of 93 mph. It was funded under the 863 Program.
- Chery Automobile Company presented the ***Chery Eastar FCEV*** at Expo 2010 in Shanghai.[348] The FCV was based on a Chery Eastar SQR700FEB11. It ran on compressed hydrogen. No more details are available.
- FAW Group, China's first automaker established in 1958 and a state-owned corporation, presented the ***FAW Besturn B70 FCV*** at Expo 2010 in Shanghai.[349] The FCV was based on a Besturn B70 CA7904FC. It ran on compressed hydrogen and had the top speed of 75 mph.
- Tongji University introduced the ***Tongji University Fuel Cell PASSAT New*** at Expo 2010 in Shanghai.[347] The FCV was based on a Volkswagen Passat Lingyu New 2010 SVW7553FCV. It incorporated a 40 kW Shen Li PEMFC, lithium-ion batteries, and a 42 kW Shanghai Drive permanent magnet synchronous motor. It ran on compressed hydrogen and had the top speed of 79 mph.

It is clear that the 863 program had fostered the development of several fuel cell car power-train generations (and perhaps a fifth-generation unit that powered the three vehicle models unveiled in April 2010). Although the extent of government success in promoting FCV technology is still not yet known, the transfer of the powertrain technology to automakers clearly encouraged several automakers to offer their vehicles for FCV use and participate in the testing of FCV viability.

7.4.2.4.1.2.2. Fuel Cell Buses In the early 2000s, the Chinese Fuel Cell Buses (FCB) development program was launched, centering on DICP, Shen Li, and Tsinghua University. Shen Li and DICP worked on fuel cells for buses that were to be assembled by Tsinghua, but they also developed FCBs of their own. Their PEMFC research and FCV development efforts were funded by MOST and the Beijing municipal government. Demonstrations were funded by MOST, the Beijing municipality, the Global Environment Fund (GET), the UNDP, and other foreign governments.[328] Tsinghua was expected to develop three generations of prototype buses.

- In addition to DICP, Shen Li, Tsinghua, and six other universities and companies developed fuel cell buses. All of them likely were funded by the government.

DICP, the largest beneficiary of the 973 program in fuel cell R&D,[208] developed fuel cells while building at least three FCBs on its own. It unveiled its first FCB in 2001 using a 25 kW PEMFC, based on the research from the 1990s. The second and third FCBs were unveiled in 2002 and 2003, respectively, each incorporating a 40 kW PEMFC. Since 2003, however, there has been no further reporting regarding PEMFCs made by DICP for use in any fuel cell buses.

Tsinghua, a major beneficiary of the 863 program, developed at least five FCBs, including one that used a third-generation powertrain that had been commissioned by the government. Tsinghua unveiled the first FCB in 2002, which was powered by a 50 kW PEMFC; a second-generation FCB, powered by a 60 kW PEMFC was unveiled in 2003; the third-generation FCB with a 100 kW PEMFC was shown in 2006.[208,219]

Tsinghua built another FCB in May 2004 with the same specifications as its second-generation bus. Called the ***Tsinghua University Fuel Cell City Bus 863***, it was presented at the International Hydrogen Forum in Beijing.[218] This likely was a special assignment by the government. It was funded under the 863 program as well as with US FTA funding.[218]

After completing three generations of FCBs and 1 second-generation duplicate, Tsinghua tried to develop another FCB with an unidentified PEMFC instead of the usual Shen Li units. This suggests that Tsinghua was not satisfied with the Shen Li PEMFC:

- In December 2008, Tsinghua University unveiled the ***Tsinghua University FC Low Floor City Bus 2 Generation*** in Suzhou.[350] The bus was based on a 12 m long, Higer low-floor city bus KLQ6129GQH2 and was powered by a 80 kW PEMFC made by an unidentified manufacturer. It also had nickel–metal hydride batteries and a 100 kW electric motor (531 Nm). It carried 9 pressure vessels storing 28.1 kg of compressed hydrogen (35 MPa), providing a driving range of 248 miles. It had a top speed of 53 mph and accelerated from 0 to 31 mph in 22 s.
- The bus seemed to have reliable performance, suggesting that Tsinghua perhaps used a Ballard PEMFC.

In 2010, the Tsinghua University FC Low Floor City Bus 2 Generation was modified and exported to Singapore.[351] Tsinghua and Nanyang Technology University of Singapore altered the bus for right-hand drive and to operate in the high humidity in Singapore. After the modifications, the bus was called the ***Tsinghua University + Nanyang Technological University Fuel Cell Bus***. The fuel cell bus was Singapore's first eco-friendly bus, which was used to transport athletes and officials around the Youth Olympic Village located at the Yunnan Garden Campus of the Nanyang Technology University from August 14–16, 2010.

Another major Chinese FCB developer was Shen Li, which, of course, also developed PEMFCs. Shen Li presented at least five fuel cell buses from 2004 to 2008.[352] The data shows that Shen Li's PEMFC technology leapfrogged from 8 kW output in 2004 to 100 kW in 2007.

- In October 2004, the ***Shen Li Technical Company SLFC-Tourist Car 2*** was unveiled.[353] The 8-passenger tourist bus was 4 m long. It was powered by four 8 kW Shen Li PEMFC and a 5 kW electric motor. The bus used hydrogen stored in metal hydride.
- In 2007, Shen Li presented the ***Shen Li Technical Company Shanghai Shenwo Fuel Cell Bus***.[354] Shen Li collaborated with SAIC.

The bus was based on a 11.4 m, SAIC low-floor city bus and had a capacity of 66 passengers. The bus was powered by a 100 kW Shen Li PEMFC and electric motor and ran on compressed hydrogen. It had a top speed of 48 mph and accelerated from 0 to 31 mph in 22 s. The bus operated for demonstration purposes in Shanghai.

- In 2007, Shen Li unveiled the ***Shen Li Technical Company Shen-Kai No.1 Fuel Cell Bus***.[355] The bus was powered by a 100 kW Shen Li PEMFC and electric motor and ran on compressed hydrogen. Little information is available.
- In 2007, Shen Li again presented another fuel cell bus, the ***Shen Li No.1 Fuel Cell Bus***.[355] The bus was based on a 10.5 m long, Shanghai Shenlong Bus SLK6103 SunLong City Bus and had a capacity of 55 passengers. The bus was powered by a 100 kW Shen Li PEMFC, battery, and an electric motor. The FCB carried 23 kg of compressed hydrogen (35 MPa) stored in 990 l vessels. It had a top speed of 53 mph and accelerated from 0 to 31 mph in 21 s.
- In 2008, Shen Li presented the ***Shen Li No.2 Fuel Cell Bus***.[356] The bus was based on a 12 m long, Shanghai Shenlong Bus SLK6123 SunLong City Bus. The bus was integrated with a 100 kW Shen Li PEMFC, battery, and an electric motor. It was fueled in the same way as the Shen Li No.2 FCB. It had a top speed of 53 mph and accelerated from 0 to 62 mph in 21 s.

In addition to the three major FCB developers, there had been at least six others that in the 2000s built at least one FCB each:

- In April 2001, Beijing LN Green Power Company (LN Green Power) unveiled a prototype FCB ***Green Power FC Bus China***.[30] The 6-m-long bus was powered by an 18 kW Lu Neng PEMFC and 36 kW AC induction motor and was fueled by compressed hydrogen (25 MPa).[31] The bus had a top speed of 45 mph and a driving range of 102 miles. It is not certain, but the Lu Neng PEMFC perhaps was made by LN Green Power.
- In 2006, Higer Bus, a bus manufacturer established in 1998, unveiled the ***Higer U90B fuel cell Shanghai (2006)***.[357] The 11.5 m bus was based on a Higer U90B bus and was integrated with a 100 kW Shen Li PEMFC, batteries, and electric motors. It was equipped with nine pressure vessels storing compressed hydrogen totaling 990 l. It had a top speed of 53 mph and accelerated from 0 to 31 mph in 21 s.
- In 2006, Wuhan University presented the ***Wuhan University FC Midibus***.[358] The bus was based on the Dongfeng EQ6700FCV East Wind model and powered by a 50 kW Wuhan University PEMFC, nickel–metal hydride batteries, and a 60 kW reluctance motor. It ran on compressed hydrogen. No further information is available.
- In 2008, Foton Motor introduced the ***Foton Motor FC Low Floor City Bus***.[359] The bus was based on a 12 m long, Foton Motor BJ6123C6N4D low-floor city bus model with a capacity of 49 passengers. The bus was integrated with an unidentified PEMFC, Taizhou Chunlan nickel–metal hydride batteries, and a 100 kW Zhuzhou Electric Locomotive Research induction motor. It ran on compressed hydrogen (35 MPa). It had a top speed of 50 mph. It was demonstrated during the Olympic Summer Games 2008. The FCB perhaps used a Ballard fuel cell to showcase its performance.
- In April 2010, SAIC presented the ***SAIC Fuel Cell Bus Expo 2010*** for demonstration at Expo 2010.[360] The bus was based on a 12 m long, SAIC low-floor city bus SWB6129FC with a passenger capacity of 49. The bus was powered by an 80 kW Ballard PEMFC, Xianyang Limited lithium-ion phosphate batteries, and a 100 kW Zhuzhou Electric Locomotive Research induction motor. It ran

on compressed hydrogen and had a top speed of 43 mph. The bus was demonstrated at Expo 2010 in Shanghai with UNDP–GEF funding.

- In May 2010, Yutong Bus Company presented the ***Yutong Low Floor Fuel Cell Range Extender EV City Bus*** at Busworld 2010 in Shanghai.[361] The bus was based on a 12 m long, Yutong low-floor city bus ZK6125EGAA model with a passenger capacity of 62. It incorporated a 20 kW Shanghai Shen Li High Tech PEMFC, lithium-ion battery (169 kWh), and a 110 kW AC induction motor (778 Nm). It ran on compressed hydrogen (16 MPa) stored in four pressure vessels. The bus had a driving range of 186 miles and had a top speed of 56 mph. No further information is available.

7.4.2.4.1.2.3. FCV Performance Falls Short of Expectations China had an intense desire to successfully demonstrate the results of its national FCV development plan at various public events, such as the Beijing Olympics and Shanghai Expo. The data from these demonstrations, however, were either insufficient to draw any conclusion or fell short of expectations. For example:

- China demonstrated 20 Passat Lingyu FCVs and three fuel cell buses during the 2008 Olympic Games in Beijing.[32] The 20 FCVs operated for 68 days from July 14, 2008 –to September 19, 2008.[213] The FCVs ran a total of more than 970 driving trips for a total of 47,196 miles during this period, an average of 2360 miles per vehicle. The three FCBs were likely those unveiled in 2008 by Foton, Shen Li, and Tsinghua.[32] Little information on their performance has been reported, however.
- China demonstrated 16 Passat Lingyu FCVs for 6 months in California from January 1 to June 30, 2009.[213] The FCVs drove a total of 22,977 miles or an average of 1436 miles per vehicle.
- China demonstrated 100 fuel cell sightseeing cars, 70 fuel cell passenger cars from five different manufacturers, and six fuel cell buses for 6 months from May 1 to October 31, 2010 during Expo 2010 Shanghai China.[213] The sightseeing cars at the Expo 2010 Shanghai closely resembled the small 4-seat, no-door, Chunhui fuel cell hybrid mini-sedan unveiled by Tongji University in 2004. According to an assessment made in September 2010, Chinese fuel cell technology was not sufficiently mature for commercialization.[213] The power density of PEMFCs was 250 W/kg, which fell well short of the 1 kW/kg power densities of traditional internal combustion engines. Life span was less than 2000 h, which was short of the 5000 h target for commercial applications. The assessment concluded that China's fuel cell engines needed to be improved in power density, life span, low-temperature start capability, and reliability.

7.4.2.4.1.3. COMMERCIAL SUCCESS IN THE NEAR TERM UNLIKELY Over the past decade, China has been highly supportive of PEMFC and FCV technology development and pursued a focused approach while funding highly diversified programs. The result has been a range of remarkably dynamic FCV development activities.

Nonetheless, the PEMFC programs clearly have not achieved what China had hoped. They likely fell short of the capabilities achieved by many fuel cell–developing countries, including Japan, Germany, the United States, and Korea. Although it is difficult to be precise, China's PEMFC technology development likely had run into major barriers around the mid-2000s, when progress slowed substantially. DICP, China's foremost PEMFC research institute, by 2003 was no longer announcing the achievement of new milestones in performance. Shen Li's PEMFC seemed to have lost its vitality in the mid-2000s; about the same time Tongji

University and Tsinghua University, China's leading FCV integrators, appeared to have turned to Ballard fuel cells instead of using Shen Li fuel cells.

7.4.2.4.2. SOUTH KOREA

South Korea started a relatively modest PEMFC development program in 1988. It allocated $50 million[362] to support fuel cell R&D during the 1988–2003 period, about $3.3 million per year. Korea's Ministry of Education, Science, and Technology promoted basic research with a long-term perspective and the Ministry of Knowledge Economy (MKE) promoted applied research and product development for commercialization of fuel cells in the near term.

7.4.2.4.2.1. KOREA INITIATES AGGRESSIVE 10-YEAR FUEL CELL DEVELOPMENT PROGRAM In 2003, the South Korean government initiated an aggressive 10-year program for the development and commercialization of new and renewable technology.[363] In the program, fuel cells were identified as one of South Korea's three high-priority technologies to be developed. Thus, in 2004, Korea dramatically increased its budget for fuel cell R&D. Korea funded $37.1 million for hydrogen and fuel cell R&D in 2004,[362] a more than ten-fold budget increase over the previous 1988–2003 period. Korea kept up its aggressive funding and allocated $46 million in 2005 and $66 million in 2006. Funding declined a little in 2007, but still was an impressive $60.5 million in 2007. It appears that Korea continued to sustain a similar level of funding for subsequent years, with an impressive total of perhaps about $500 million for the 10-year program.

The fuel cell goals to be achieved under the 10 year program were also ambitious[362]:

- ***Fuel cell vehicles***. South Korea would develop commercially viable fuel cell cars and buses by 2012, when Korea would demonstrate and disseminate 200 fuel cell buses and 3200 fuel cell passenger cars on the roads. To achieve these goals, Korea chose *Hyundai-Kia Motors* to play a leading role. Hyundai has been developing FCVs since 1999 in collaboration with Ballard and UTC. Hyundai was to lead three projects:
  - A $33.4 million (2004–2008) project to develop an 80 kW PEMFC passenger car.
  - A $49 million (2005–2009) project to develop a 200 kW PEMFC bus.
  - A $48 million project (2006–2008) to demonstrate the operation of 30 fuel cell cars, four fuel cell buses, and 10 hydrogen-refueling stations.
- ***Residential fuel cell power generation***. Korea aimed at developing prototype residential CHP plants by 2006. Korea would distribute 10,000 residential CHP units at a price of $10,000/kW by 2012. To achieve these goals, Korea chose GS Fuel Cell and Fuel Cell Power (FCP) to lead a CHP development project and KOGAS, the national LNG wholesale monopoly distributor and the second largest LNG importer in the world, to lead the demonstration and dissemination project.[362] Details included:
  - A 2004–2006 project to develop 3 kW prototype residential power generation unit.
  - A 2006–2012 CHP demonstration and dissemination project. Korea would purchase CHP units from manufacturers and distribute the units with a government subsidy. It would try to reduce the price of CHP systems year by year until the ultimate target price was achieved.
- ***Portable devices***. Korea would commercialize *both PEMFC and direct methanol fuel cell DMFC technology* for portable devices, such as notebooks, PDAs, mobile phones, and digital media broadcasting (DMB) devices by 2008. Korea chose Samsung SDI, Samsung SEM, and LG Chem, to lead three portable fuel cell development projects.[362]

7.4.2.4.2.2. BUT PLANS YET TO DELIVER RESULTS The South Korean PEMFC R&D effort has not delivered on the planned goals for FCVs. Similarly, residential fuel cell units or portable devices have not yet been successfully commercialized.

7.5. THE GLOBAL FUEL CELL RACE SO FAR FAILS TO ATTAIN COMMERCIAL SUCCESS

There were immense hopes and dreams that PEMFC technology would change the world by the early twenty-first century. The fuel cell industry believed that PEMFCs would bring commercial success to the industry, and governments believed that efficient and inexpensive fuel cells would propel cars, light homes, and operate laptop computers. The dream did not come true.

Nonetheless, the PEMFC industry remains the most vibrant segment of all the fuel cell technologies. A large number of companies have entered the fuel cell race and developed a variety of applications and products.

7.5.1. Transportation Applications

Transportation applications include motive power for cars, buses, and other fuel cell passenger vehicles, specialty vehicles, material-handling vehicles (e.g., forklifts) and APUs for highway and off-road vehicles. Some transportation applications, such as aircraft, locomotives, trucks, mining vehicles, cranes, marine vessels, golf carts, bikes, scooters, and water taxis, are still in an early stage of development and will not be discussed in this chapter.

7.5.1.1. Passenger Cars

The level of enthusiasm for fuel cells among automakers dramatically downshifted over the past several years. Based on their fuel cell development profiles, five groups of automakers have emerged. One group is still focused on fuel cell development and remains in the fuel cell race. Another group was once active and has now abandoned fuel cell activities. A third has always shown limited interest in fuel cells and experimented with a few FCV prototypes or tried out auxiliary applications of fuel cells, such as range extenders. A fourth is unsure as to what their next move should be. Finally, a fifth group has taken a long-term view and a measured approach, because it believes fuel cells will be an important commercial product one day, but not yet. Automakers in each of these five groups are described in the following pages.

7.5.1.1.1. SOME AUTOMAKERS REMAIN FOCUSED... BUT WITH A WAVERING MIND

Some automakers believe that fuel cells are the ultimate clean car technology and continue to invest in fuel cell car development, demonstrations, and commercialization. They include Daimler, GM, Honda, Hyundai-Kia, Nissan, and Toyota. There are, however, some differences in the degree of determination, commitment, and investment. Some appear more committed or less conflicted than others. Nonetheless, they all hope to achieve near-term commercial success. They also have focused on the use of fuel cells for primary transport missions, while using them for auxiliary operations.

7.5.1.1.1.1. DAIMLER Daimler unquestionably has been the global fuel cell car leader and trend setter since 1994. It has unveiled more than 20 new fuel cell cars and has rarely shown lack of confidence in its plans for fuel cell commercialization. The company built the world's first fuel cell car and has introduced one new fuel cell car nearly every year between 1996 and 2002. In each successive model, fuel cells were smaller in size and better in performance. In 2002, Daimler unveiled the world's first series-produced fuel cell car, the ***Mercedes-Benz F-Cell A-Class***. The car had a range of 100 miles, with a top speed of 82 mph. A fleet of 60 vehicles were produced and leased to customers for daily use in Germany, Japan, Singapore, and in the United States.[555] With this car, the company felt confident enough to announce that it would start mass marketing of fuel cell cars in 2004. However, no commercialization has materialized, but Daimler continues to introduce new fuel cell cars.

In September 2009, Daimler presented a significantly advanced ***Mercedes-Benz B-Class F-Cell***, which had a range of 248 miles and a top speed of 106 mph.[364] Daimler felt quite confident in the car's market appeal. In early 2009, it announced plans to begin commercial FCV production in 2009 with the aim of 100,000 vehicles in 2012–2013.[365] Before long, however, plans for commercialization were postponed to 2015. At the beginning of 2010, Daimler started to lease up to a total of 200 B-Class F-Cell vehicles.[557] In June 2011, three B-Class F-Cell vehicles completed a 5-month trip around the globe, covering 18,641 miles,[366] which convinced Daimler that the technology was now ready for mass production.

- In a surprise announcement in the same month, the Daimler chairman stated that the company would introduce a production hydrogen fuel cell car into showrooms as early as 2014, 1 year earlier than the other major fuel cell car companies.[367] According to Daimler, current plans are to produce it in volume by 2014 at a price comparable to that of a diesel-powered car.[368]
- In September 2011, Daimler introduced the ***Mercedes-Benz F125!***, an F-CELL plug-in Hybrid, another indicator of the company's commitment to F-Cell.

7.5.1.1.1.2. GM GM has been more committed to FCVs than any other US manufacturer and remains the only US automaker pursuing the near-term commercialization of fuel cell cars. Ford likely is no longer committed to investing in FCV development,[369] and Chrysler, with its new parent company Fiat, likely views FCV as a technology for the distant future.

GM, with the development of the HydroGen3 in 2001, undoubtedly felt confident in the commercial viability of FCVs and it announced plans to begin commercialization by 2005. GM unveiled two futuristic-looking concept cars— the ***GM AUTOnomy*** and the ***HyWire***—in 2002. The AUTOnomy was a sleek, spacious drive-by-wire 2-seater vehicle with a fuel cell propulsion system and three hydrogen fuel tanks that were packaged in a 6-inch thick skateboard platform.[370] The HyWire was a drivable proof-of-concept vehicle derived from the-AUTOnomy model using HydroGen3 technology. The skateboard grew in thickness to 11 inches, but the vehicle still was controlled by wire.[82] It had a top speed of 100 mph.[371] The fuel cell system was capable of providing 100 kW of continuous power. Using 2 kg of hydrogen stored in tanks at 5000 psi, it had a driving range of 100 miles.[371]

- But plans for commercialization of the HyWire in 2005 did not materialize.

Still, in 2005, GM unveiled the ***Chevrolet Sequel*** at the Detroit auto show[372] and in September 2006, GM unveiled two Chevrolet Sequels that were as drivable as regular cars. The vehicles were controlled by wire, giving the driver simulated driving feedback[370] and used HydroGen3 technology[372] along with a 65 kW lithium-ion battery pack.[370] Three high-pressure storage tanks contained 8 kg at 10,000 psi, giving the Sequel a 300 mile driving range. The lithium-ion battery served to capture energy from regenerative braking in the same manner as a hybrid vehicle. It accelerated from 0 to 60 mph in 10 s and had a top speed of 90 mph.

- In February 2007, it was reported that GM was planning to begin production of the fuel cell Chevrolet Sequel by 2010.[373]

In September 2006, GM introduced the ***GM Chevrolet Equinox Fuel Cell Vehicle*** as part of a major field research project, called Project Driveway, which would place more than 100 Equinox vehicles with customers across the United States, China, Japan, and South Korea.[374] The FCV was powered by GM's fourth-generation fuel cell propulsion system and was able to start and operate in sub-freezing temperatures. The FCV had a driving range of about 200 miles and a top speed of 99 mph, and accelerated from 0 to 62 mph in 12 s.[375] In September 2007, GM introduced the ***GM HydroGen4*** in Germany,[376] which was the European version of the Chevrolet Equinox Fuel Cell.[377] From mid-2008, a total of 10 HydroGen4 vehicles took part in day-to-day testing in the CEP in Berlin.[376]

In April 2007, GM unveiled the ***Chevrolet Volt FC Hybrid*** at the Shanghai Motor Show, which was a fuel cell variant of the Chevrolet Volt under the E-Flex electric drive family.[378] The Chevrolet Volt Fuel Cell Hybrid was powered by a new 80 kW fifth-generation GM fuel cell stack with an 8 kWh (50 kW peak power) lithium-ion battery to provide up to 300 miles of electric driving. The Volt fuel cell variant was plug-in capable, adding up to 20 additional miles (34 km) of range each time it was charged.

- In November 2007, GM indicated that it hoped to deliver production fuel cell cars as early as 2011, most likely in the form of the fuel cell Chevy Volt.[379]
- Also that month, a GM vice president indicated that the company aimed to be the first automaker to produce one million FCVs.[380]

In January 2008, GM unveiled the ***Cadillac Provoq*** at the Consumer Electronics Show in Las Vegas, Nevada.[381] GM's E-Flex propulsion system initially was designed for the Chevrolet Volt, but was also incorporated in the Cadillac Provoq. The vehicle achieved a 300 mile range by packing two high-pressure hydrogen storage tanks beneath its rear cargo floor. It combined an 88 kW fuel cell with a 60 kW lithium-ion battery pack that could store up to 9 kWh of electrical energy. It featured a 70 kW coaxial motor drive system for the front wheels and a 40 kW motor in each of the rear wheel hubs, accelerating the vehicle to 60 mph in about 8.5 s. The vehicle had a top speed of 100 mph.

- However, GM's plans for near-term fuel cell commercialization were wavering. In March 2008, GM's vice chairman at the Geneva Motor Show suggested that GM's fuel cell priorities had shifted to plug-in electric vehicles.[382]
- As of October 2009, General Motors CEO stated that GM had reduced its hydrogen program because the cost of building hydrogen cars was too high.[365] In the same month, GM's vice president for research and development, who led the company's fuel cell development for 30 years, elected to retire.[383] It seemed as if GM had abandoned its fuel cell effort after all.

But the story of GM's interest in FCVs might not yet be over. In September 2009, Chevrolet Equinox fuel cell electric vehicles passed the one million mile mark in the Project Driveway road test.[384] At the same time, Opel, a European subsidiary of GM, announced in June 2011 that its HydroGen4 fleet, the European versions of the Chevrolet Equinox, passed the two-million mile mark.[385] The HydroGen4 vehicles had been part of a test program with the partners such as ADAC, Allianz, Bild/Axel Springer AG, Coca-Cola, Enertrag, Hilton, IKEA, Linde, Pace, and Total to test different applications for FCVs.[386]

- Contrary to pessimistic reports of a year or two ago, GM in March 2010 was reported to be testing a production-class hydrogen fuel cell system for a passenger vehicle that would be ready for commercial production in 2015.[387] The new fuel cell system was GM's fifth-generation fuel cell system.[388] The powertrain would be half the size, 220 pounds lighter, and use about one-third of the platinum compared to the Chevrolet Equinox Fuel Cell electric vehicles used in Project Driveway.[387]

- It was also reported that Opel revealed that the company had been looking toward launching a FCV by 2015[389] or 2016.[390]

In August 2011, however, GM seemed to back away from its commercialization goals. A top GM executive was unsure if GM would be able to reduce the cost of fuel cell cars to a practical level even after 2020.[391] GM might have encountered difficulty with its fifth-generation fuel cell system. But, it is also possible that confidence might return if it achieves some success in reducing costs.

7.5.1.1.1.3. HONDA Honda is one of the most, if not the most, committed FCV developers in the world. In 2004, Honda reached a major milestone in developing its own PEMFC technology when it unveiled a second-generation Honda FCX built with the company's own fuel cells. Honda had previously been using Ballard fuel cells[392]:

- The Honda fuel cell stack was the world's first fuel cell capable of starting in temperatures as low as −20°C (−4°F). The stack delivered increased horsepower and a longer range than that of the previous ones. It also had half the number of components than previous fuel cell stacks.

In November 2006, a fully functional Honda ***FCX Concept fuel cell vehicle*** was displayed to the general public at the Los Angeles Auto Show,[393] and in June 2007, Honda's next-generation FCX Concept FCV made its European driving debut at the Gotland Ring in Sweden. The vehicle had an energy efficiency of around 60 percent—approximately three times that of a gasoline-powered vehicle and twice that of a hybrid vehicle.[394] The FCX Concept was equipped with a Honda V Flow fuel cell system. The new system featured a vertical flow design, allowing gravity to facilitate water drainage, which was a key to high-efficiency fuel stack performance. The fuel cell stack was 20 percent smaller and 30 percent lighter than the current FCX FC stack, and power output was 14 kW greater. Low-temperature start-up was also improved, enabling cold-weather starting at temperatures 10°C lower than the current FCX—as low as minus 30°C.

In November 2007, Honda unveiled the ***FCX Clarity FCV*** at the Los Angeles Auto Show and announced plans to begin limited retail marketing of the vehicle in the US, starting in summer 2008.[395] In July 2008, Honda unveiled the Japan model of its FCX Clarity FCV powered by a Honda 100 kW V Flow fuel cell stack, a lithium-ion battery pack, a 95 kW electric motor, and using a 5000 psi compressed hydrogen gas storage tank.[396] It could achieve a top speed of 99 mph and had a cruising range of 240 miles. Leasing in Japan was scheduled to begin November 2008. The Clarity was limited to small-scale production of FCVs, but had its own automated assembly.[397] Honda hoped to sign 200 leases over the next 3 years.[397]

Honda reaffirmed in 2009 that it would continue to put resources into hydrogen fuel cell development, which it saw as "a better long-term bet than batteries and plug-in vehicles."[365]

- Honda stated that it would start mass producing vehicles based on the FCX concept by the year 2020.[365]
- In May 2011, Honda announced that it would become the newest member of the CEP in Germany.[398]

7.5.1.1.1.4. HYUNDAI-KIA Supported by the robust backing of the South Korean government, Hyundai has been aggressively pursuing fuel cell development. In March 2004, Hyundai presented the ***Hyundai Tucson FCEV*** at the Geneva Motor Show.[399] The Tucson FCEV, Hyundai's second-generation FCV, achieved remarkable improvements over the first-generation Santa Fe FCEV. The Tucson FCEV had a driving range of 186 miles and was equipped with 40-gallon hydrogen storage tanks, double that of the Santa Fe FCEV. In a major technology breakthrough, the Tucson FCEV was capable of starting in freezing temperatures. Testing proved that the vehicle was able to start after being subjected to −20°C temperatures for five days. The Tucson FCEV was powered by an 80 kW UTC PEMFC fuel cell, 5 kW greater and marginally lighter than its predecessor. Its maximum speed was 93 mph compared to the Santa Fe's 77 mph.

Also in 2004, Kia, Hyundai's subsidiary, unveiled the ***Kia Sportage SUV*** at the Paris Motor Show.[400] The fuel cell propulsion system in the Kia Sportage SUV appears to be the same as that used in the Tucson FCEV. The Kia Sportage SUV was powered by a 80 kW UTC PEMFC fuel cell and had a driving range of 186 miles and a top speed of 93 mph.

In 2005, Hyundai succeeded in developing its own fuel cell stacks, which it improved significantly over the next few years.[401] In December 2007, Hyundai introduced an all-new, purpose-built FCEV concept, the ***Hyundai i-Blue***, incorporating Hyundai's own—but its third-generation—fuel cell technology.[402] The Hyundai i-Blue was powered by a 100 kW Hyundai PEMFC fuel cell. It had a projected range of about 370 miles before refueling. It also had an estimated top speed of 100 mph.

In 2008, Kia unveiled the ***Kia Borrego FCV*** at the Los Angeles Auto Show.[397] The FCV, which incorporated Hyundai's fourth-generation fuel cell system, more than doubled the cruising range of the group's 2006 Sportage FCEV to 460 miles, and had an expected operational life of about 10 years, more than three times that of the Sportage.[401] It was powered by a 115 kW fuel cell stack, with 100 kW of electrical assistance from a supercapacitor, and had a 700 bar hydrogen tank.

- In 2008, Hyundai was clearly confident of the commercial viability of its PEMFC fuel cell technology and announced its plans to produce 500 FCVs by 2010 and to start mass production of its FC vehicles in 2012.[365]

In 2010, Hyundai unveiled the ***Hyundai Tucson ix35 FCEV***, first in a cutaway version in March 2010 and then as a drivable car at the Geneva Motor Show in December 2010.[403] The Tucson ix35 FCEV was powered by a 100 kW Hyundai PEMFC with a 100 kW AC induction motor (300 Nm) and 21 kW lithium-polymer batteries. The vehicle carried 5.6 kg of hydrogen in two pressure vessels, which gave it a range of 403 miles.[403] It had a top speed of 100 mph.

- According to Hyundai, by 2012 it had planned to begin manufacturing of FCEVs in the low thousands and delivering them to fleet customers in Korea.[404] According to another source in June 2010, the plan was to produce up to 2000 cars a year for the next few years— to be leased to companies and governments for evaluation—before high-scale production of 10,000-plus starts by 2015.[405]

In March 2011, Hyundai unveiled the ***Hyundai Blue2*** at the Seoul Motor Show in South Korea.[406] The Hyundai Blue2 was the company's first sedan-style fuel cell electric vehicle powered by hydrogen.[407] The concept vehicle used a 90 kW fuel cell to deliver a stack power of about 120 horsepower[408] and an 82 mph fuel efficiency rating.[406]

- Hyundai clearly hoped to become a leader in the global fuel cell race. Hyundai president indicated that Toyota had expected to price their FCVs at $50,000, but that Hyundai could beat it.[401]

7.5.1.1.1.5. NISSAN Nissan appears determined to keep pace in FCV development and has unveiled an average of almost one FCV per year since 1999. In 2005, it introduced the ***X-Trail '05 Model***, powered by a 90 kW Nissan PEMFC, a 90 kW electric motor (280 Nm), and a Li-ion battery.[409] It had a driving range of 229 miles with compressed hydrogen at 35 MPa or 310 miles at 70 MPa. It started leasing the FCV in December 2005 and planned a field trial in Canada for February 2006. After that, however, Nissan presented no new FCVs for 3 years. It appears that Nissan was struggling with major challenges at that time; it was unable to put enough fuel cells together to create durable stacks greater than 400 cells.[410] Finally in February 2009, Nissan presented the ***Nissan X-trail FCV Advanced***, an improved version of the Nissan X-Trail FCV with good cold-start capability.[411] The FCV was powered by a 90 kW Nissan PEMFC, a Nissan lithium-ion battery, and a 90 kW coaxial motor integrated with reduction gears (280 Nm).[412] It was fueled by hydrogen compressed to 70 MPa.

- It appeared at that time, however, that Nissan had finally given up on FCVs. In 2009, French Renault-Nissan announced that it was canceling its hydrogen car R&D efforts.[365]

Nonetheless, according to Nissan engineers, the company was neither discouraged nor had given up.[410] Nissan now claims that it has made great strides in a number of critical areas: power density, cost, and advanced stacking techniques. Nissan's 2011-model fuel cell stack reportedly delivers a power density of 2.5 kW/l, 2.5 times better than the 2005 model. The stack thus is much smaller, packing 85 kW of power in a 34 l volume, according to Nissan. It also brought stack production costs down by 85 percent—close to meeting the U.S. Department of Energy cost target for 2010, a widely referenced benchmark—by reducing the need for platinum by 75 percent. According to Nissan, the membrane electrode assembly (MEA) comprises 80 percent of the stack's cost, and platinum is half the cost of an MEA, so this was a huge step forward. Moreover, Nissan resolved the earlier problem of creating stacks larger than 400 cells. In a breakthrough concept, it used molded plastic around the MEA to create insulating frames between each of the more than 400 layers in the stack, thereby ensuring the layers did not short out or come loose.

- Nissan says it is ready to go to market any time. The only remaining hurdle is hydrogen distribution, it says.

7.5.1.1.1.6. TOYOTA In December 2002, Toyota began limited commercial leasing of FCVs in the United States and Japan.[413] A total of 20 first-generation fuel cell hybrid vehicles (FCHV) were in service in California with universities, corporations, and government agencies. Since then, Toyota's fuel cell technology has advanced at an impressive pace. Toyota improved vehicle range, durability, and efficiency through improvements in the fuel cell stacks and the high-pressure hydrogen storage system, while achieving significant cost reductions in materials and manufacturing[353]:

- ***Toyota*** presented the ***FCHV-adv*** to the public in June 2008. In driving tests conducted by the U.S. Department of Energy, Savannah River National Laboratory, and the National Renewable Energy Laboratory, the FCHV-adv averaged the equivalent of 68 mpg and achieved an estimated driving range of 431 miles on compressed hydrogen. This was an increase in range of more than 150 percent compared to the first-generation FCHV.[414]
- Also in late 2007, prior to public introduction, technology improvements implemented in the FCHV-adv were road tested in extreme conditions on a 2300 mile trek from Fairbanks, Alaska to Vancouver, British Columbia, along the Alaska–Canadian (ALCAN) highway. The seven-day trip confirmed that substantial progress had been made in reliability and durability under cold-weather conditions, as well as in long-range driving.

Nonetheless, even Toyota appeared to have uncertainty about the future of fuel cell cars. At the Geneva Motor Show in 2008, for example, Toyota president expressed concern about the high costs of fuel cells and the lack of an infrastructure to produce and distribute hydrogen fuel to a wide range of consumers.[382] He reportedly said that it would be difficult to see the spread of fuel cells in 10 years' time.

- By February 2009, however, Toyota appeared to have regained its determination to commercialize FCVs. A Toyota executive at the 2009 Washington DC Auto Show stated that although cost reduction and the lack of hydrogen infrastructure remained major obstacles, Toyota was committed to bring FCVs to global markets in 2015.[415]
- In January 2010, Toyota announced that it had planned to place more than 100 Toyota FCHV-adv vehicles in a nationwide demonstration program over the next 3 years.[413]

In May 2010, Toyota said that it had cut the cost of making fuel-cell vehicles by about 90 percent since the mid-2000s and might be able to price its first retail hydrogen model at about $50,000.[416]

- The automaker displayed its hydrogen-powered FCV-R concept car in December 2011 at the Tokyo motor show and in March 2012 at the Geneva motor show, where it reiterated its intention to sell FCVs by 2015.[417]

7.5.1.1.1.7. BUT COMMERCIAL SUCCESS IN THE NEAR TERM UNCERTAIN Significant financial resources have been invested in fuel cell technology over the last 20 years and remarkable advances have been demonstrated during the past several years. Still, barriers to mass market commercialization of FCVs persist, including cost, durability and reliability, system size, air, thermal, and water management.[418] In particular, cost and durability are the most critical challenges to FCV commercialization.

7.5.1.1.2. SOME SHIFT FOCUS AWAY FROM FUEL CELLS

There are automakers that were once active in fuel cell car development, but who have shifted their focus away from FCVs. They include Chrysler, Ford, Mazda, Mitsubishi, and Renault. These companies have gradually de-emphasized their fuel cell car activities, have stopped unveiling new fuel cell cars, or have developed other types of vehicles.

7.5.1.1.2.1. CHRYSLER As part of DaimlerChrysler, Chrysler unveiled three prototype FCVs. In October 2001, Chrysler unveiled the ***Jeep Commander 2***, a Jeep Grand Cherokee powered by a 50 kW fuel cell and nickel–metal hydride battery.[419] It was fueled by methanol with an onboard reformer. In 2001, it presented the DaimlerChrysler ***Natrium***, a Chrysler Town&Country powered by a 75 kW Ballard fuel cell with Li-ion battery and with hydrogen stored in sodium borohydride.[420] It had a top speed of 90 mph and a driving range of 310 miles. In 2003 Chrysler presented the ***Jeep Treo*** during the 37th Tokyo Motor Show.[421] The Treo was based on an all-new platform designed to utilize drive-by-wire technology and a zero-emission fuel cell. After splitting off from Daimler in 2007, Chrysler unveiled the ***ecoVoyager*** in January 2008, a four-passenger, four-door fuel cell and hybrid concept car.[422] The hybrid car used a 45 kW fuel cell stack as a range extender and 268 hp (200 kW) electric motor plus "range-extended electric vehicle technology" to deliver a 300-mile range.

- Since early 2008, however, Chrysler has not presented any fuel cell cars. It reportedly put fuel cell development on the back burner.[397]
- On June 10, 2009, Chrysler emerged from a Chapter 11 reorganization bankruptcy and was sold to the Italian automaker Fiat.[423] On June 3, 2011, Fiat bought out the remaining US Treasury's stake in Chrysler for $500 million, thereby increasing its ownership of

the automaker to 52 percent. It is not certain how Chrysler's new affiliation with Fiat will change its future fuel cell strategy.

7.5.1.1.2.2. FORD Ford appears to have abandoned its fuel cell effort. Since it unveiled the ***Ford FCV Hybrid*** in 2002, its public announcements on fuel cell cars had been few and far between. Between 2002 and 2006, Ford developed no fuel cell cars but instead worked on vehicles with a hydrogen-fueled internal combustion engine, the Ford H2RV (2003), Ford U SUV (2003), and Ford Focus C-MAX (2004). At the 2006 Greater Los Angeles Auto Show, Ford did showcase a fuel cell version of its ***Ford Explorer SUV.***[424] The Fuel cell Explorer had a combined output of 174 hp (130 kW). The vehicle reached a notable range of 350 miles, the farthest of any FCV at that time. The fuel cell Explorer was the first in a series of prototypes partly funded by the United States Department of Energy to expand efforts to determine the feasibility of hydrogen-powered vehicles. In 2007, Ford presented three cars using fuel cells. One was the ***Ford Fusion Hydrogen 999.***[425] The FCV was powered by a 400 kW Ballard fuel cell and developed as a high-speed vehicle jointly with the Ohio State University. It established a speed record of 207 mph. The other two cars developed in 2007 were the ***Ford Edge*** and ***Ford Airstream,*** both of which were plug-in battery hybrids (336 V Li-ion) with hydrogen fuel cells as onboard rechargers.[426]

- In 2009, Ford announced that they were shifting resources away from fuel cells and into battery vehicle development.[397]

7.5.1.1.2.3. MAZDA Mazda was long interested in FCVs and developed its first FCV, the ***Demio FCEV,*** in 1997. The Demio FCV was powered by a 20 kW PEMFC,[427] which had been developed in-house. Owing to Ford's alliance with Ballard, Mazda started to use Ballard's fuel cells in 1998. Mazda developed the ***Demio FCEV*** in 1999 and the ***Premacy FCEV*** in 2001, both powered by Ballard fuel cells. The Premacy FCEV was given approval for driving on public roads, the first such approval given to an experimental FCV in Japan.[137] Mazda apparently planned to start marketing fuel cell cars around 2005.[138] It developed two hydrogen internal combustion vehicles—the ***Mazda Premacy Hydrogen RE Hybrid*** and the ***Mazda 5 Hydrogen RE Hybrid***—in 2007 and in 2009, respectively.

- Mazda has not unveiled any fuel cell car since 2010. Mazda is no longer focused on fuel cells, but shifted its focus to hydrogen internal combustion hybrid systems and electric hybrid systems.[428]

7.5.1.1.2.4. MITSUBISHI MOTORS Mitsubishi has not developed another FCV since unveiling the Mitsubishi FCV Grandis in October 2003. Nonetheless, as of March 2006, it still believed that FCV commercialization and distribution would occur in 2010–2020 and that the company would be active in the market.[103] The company envisioned that the Mitsubishi FCVs would be a compact car in the 50–60 kW range and priced at or less than hybrid electric vehicles.

- On October 4, 2006, however, it decided to abandon development of FCVs and to concentrate its efforts on development of EV and diesel engine cars.[429]

7.5.1.1.2.5. RENAULT Renault always had limited interest in FCVs. It developed the ***FEVER Laguna FCV*** in 1997 under the FP sponsorship and the ***Renault Scenic ZEV H2*** in 2008. Renault also showed no interest in developing its own fuel cell technology and instead used Nuvera fuel cells.[430]

- Like many other major automotive companies, Renault once planned to have a FCV in mass production by 2010.[28]

In April 2011, Daimler and Renault along with its subsidiary Nissan repeatedly planned to cooperate on the development of battery and fuel cell technology.[431] It is not clear, however, if Renault would pursue fuel cell commercialization in the near term.

- It is likely that Renault would largely depend on Nissan for a future course of action regarding FCVs.

7.5.1.1.3. SOME AUTOMAKERS SHIFT TOWARD USING FUEL CELLS FOR AUXILIARY POWER USE

There are some automakers that have increasingly become more interested in the use of fuel cells as a source of auxiliary power than for primary power. These automakers include BMW, Lotus Engineering/Intelligent Energy, PSA Peugeot Citroen, Suzuki, and Volvo.

7.5.1.1.3.1. BMW BMW has a long development record of hydrogen internal combustion engines and bivalent hydrogen/gasoline ICEs, including the BMW Hydrogen 7 (2006) with a hydrogen/gasoline bivalent ICE, the BMW Record Vehicle 'H2R' (2004), a high-speed racing car with a hydrogen ICE, the BMW 7 series 6.0 l V12 (2003) with a bivalent ICE for hydrogen and gasoline, the BMW Mini Hydrogen (2001), a mock-up with a hydrogen/gasoline ICE, the BMW 745h 4.4 l V8 (2001) with a hydrogen/gasoline ICE, the BMW 750hL 5.4 l V12 (1999) with a hydrogen/gasoline ICE, the BMW 728 hL (1998) with a hydrogen ICE, the BMW 520h (1979) with a hydrogen/gasoline ICE, and the BMW 7 series (1978) with a hydrogen/gasoline ICE.

In March 2010, BMW finally unveiled a gasoline/fuel cell hybrid, the ***BMW 1 Series Fuel Cell Hybrid***, which had a 88 kW BMW gasoline 4-cycle in-line ICE for front-wheel drive and a 5 kW UTC PEMFC, supercapacitors, and a 82 kW electric motor for rear-wheel drive.[432]

- The 5 kW fuel cell was used as a range extender. The car was equipped with hydrogen for the fuel cell and gasoline in a small tank for the internal combustion engine. The fuel cell would serve well for city traffic, and in combination with the ICE would provide fast acceleration. The ICE would be used for long-distance driving.

7.5.1.1.3.2. LOTUS ENGINEERING/INTELLIGENT ENERGY In June 2010, Lotus Engineering, LTI Vehicles, TRW Conekt, and Intelligent Energy unveiled the Lotus Engineering ***Zero Emissions Black Cab***, a full performance, zero-emissions fuel cell hybrid London taxi developed with funding from the UK Government's Technology Strategy Board.[433] It had a top speed of over 80 mph and a range of 248 miles on a full tank of hydrogen, accelerated from 0 to 62 mph in 14 s, and produced no emissions other than water vapor.[434] The Black Cab was powered by a 30 kW Intelligent Energy PEMFC fueled by compressed hydrogen, hybridized with 14 kW lithium polymer batteries, and an electric motor. The fuel cell acted as a range extender for the electric drive,[435] solving the range limitations of battery propulsion and eliminating battery recharging time.

- In March 2011, the Fuel Cell Black Cabs were driven on the roads of London for the first time.[436] A fleet of the taxis is planned for delivery in time for the London 2012 Olympics.

7.5.1.1.3.3. PSA PEUGEOT CITROEN PSA has taken a long-term view of fuel cell technology and is moving rather slowly on FCV development. From early on, it has indicated that the commercialization of fuel cells would take place between 2010 and 2020. It views using fuel cells as range extenders or as auxiliary power units. In 2001, it introduced the ***Taxi PAC*** in 2001, with a 5.5 kW H Power fuel cell as a range extender. In 2002, it presented the ***H_2O Firefighter***, which used fuel cells for an auxiliary power unit that charged the batteries. In 2004, PSA developed a four-wheel vehicle, the ***Quark***, with a 1.5 kW fuel cell developed by MES-DEA, a Swiss energy company.[437] The fuel cell also served as a range extender. In October 2006, PSA unveiled the ***Peugeot 207 ePURE***, powered by a 50 kW lithium-ion battery, CEA's 20 kW GENEPAC PEMFC, and an electric motor.[438] The fuel cell was used as a range extender. The Peugeot 207 ePURE had five compressed hydrogen tanks (70 MPa) containing 3 kg of hydrogen. In 2007, the company unveiled the Peugeot Flux, a futuristic design car, at the Frankfurt Motor Show.[439] The car had a hydrogen internal combustion engine. In 2008, PSA developed the ***H_2Origin Light Utility Urban Vehicle***, powered by a 10 kW Intelligent Energy PEMFC, a 180 V Panasonic nickel−metal hydride battery pack, and an electric motor.[440] The fuel cell also was used as a range extender. The H2Origin had a top speed of around 60 mph and a driving range of 186 miles.[441] In 2009, PSA unveiled the ***Peugeot 307 CC FiSyPAC FCV hybrid***, a fully electric vehicle, based on a Peugeot 307 coupe cabriolet.[442] The FCV hybrid was powered by a 17 kW GENEPAC fuel cell stack, developed by the French Atomic Energy Commission (CEA), and a 13 kW Li-ion battery pack (with energy densities of 100 Wh/L and 88 Wh/kg). The fuel cell served as a range extender, giving the EV demonstrator a range of 310 miles at 62 mpg.

- PSA clearly has made considerable use of fuel cells as range extenders in electric vehicles. The FiSyPAC demonstrator ranked among the world's top performers, needing less than 1 kg of hydrogen per 62 miles.[443] PSA also successfully quadrupled the fuel cell's lifespan and increased its efficiency by nearly 20 percent since 2006.
- It still has the view that fuel cells have a number of unresolved challenges, including cost, lifespan, the need for lithium-ion batteries, and the need for an infrastructure to market hydrogen to the general public.[443] Nonetheless, PSA believes that it would be able to mass market vehicles with fuel cell range extenders in the 2020−2025 period.

7.5.1.1.3.4. SUZUKI Suzuki has participated in GM's fuel cell car development strategy. Suzuki developed a series of fuel cell cars using GM fuel cells, including the ***Wagon R-FCV 35 MPa*** in 2003, ***MR Wagon-FCV 35 MPa*** in 2003, ***Mobile Terrace*** in 2003, ***Improved MR Wagon-FCV*** in 2004, ***Suzuki Ionis*** in 2005, and the ***Suzuki SX-4-FCV*** in 2008. In 2006 and 2008, the company bought back the stock that GM had held in Suzuki since 1981[444]:

- Although there was an agreement that the two companies would continue to work on hybrid and fuel cell cars, Suzuki's interest in developing fuel cell cars appears to have tapered off.

During the past few years, Suzuki seems to be more interested in developing fuel cell motorcycles, scooters, and wheel chairs. It developed the ***Suzuki Crosscage Motorcycle*** in 2008, ***Fuel Cell Senior Car MIO*** in 2008, and ***the Burgman Fuel Cell Scooter*** in 2009. These products used fuel cell stacks made by Intelligent Energy of United Kingdom.

In January 2010, Suzuki announced that it would no longer collaborate with GM on hybrid and hydrogen fuel cell technologies.[445] Suzuki shifted its alliance to VW. In early January 2010, VW purchased a 19.9-percent stake in Suzuki, becoming the automaker's largest shareholder. It is not clear how the new alliance with VW has affected Suzuki's future fuel cell strategy.

- In February 2012, Suzuki announced that it would establish a joint venture company—SMILE FC—with Intelligent

Energy to develop and manufacture fuel cell systems, including cars and motorcycles.[446]

7.5.1.1.3.5. VOLVO Volvo was one of the few global automakers that have not shown much interest in developing fuel cell cars. The company is now initiating the development of a fuel cell that can extend an electric car's operating range.[447] It planned to develop two prototype chassis based on the Volvo C30 DRIVe Electric for testing in everyday traffic in 2012. Volvo is working with Powercell Sweden on this project, which is owned by Volvo Technology Transfer. The work is supported by the Swedish Energy Agency. A preliminary study will be conducted on a range extender that would be expected to increase the electric car's operating range up to 155 miles. In a second phase, Volvo and Powercell would produce two test cars based on the current Volvo C30 DRIVe Electric.

7.5.1.1.4. SOME HAVE UNCERTAINTY ABOUT FUEL CELL FUTURE

China's government and its fuel cell car developers may belong to this group. Since 2000, the government has orchestrated the development of fuel cell cars and provided a massive infusion of resources. China's automotive industry and research institutions have churned out a large number of FCV prototypes (see section 7.4.2.4.1.2 for details). Recently, however, the government and industry have indicated that they have not achieved what they had hoped and are uncertain as to what to do next.

7.5.1.1.4.1. SHANGHAI AUTOMOTIVE INDUSTRY CORPORATION SAIC has emerged as the most dynamic fuel cell car developer in China. The automaker, however, seems to be in search of the optimum application of fuel cells. SAIC used three different fuel cell stacks for its three most recently unveiled FCVs, for example. In its 2011 plug-in FCV hybrid, SAIC used fuel cells as an auxiliary power unit to extend driving range. SAIC announced that it would commercialize a plug-in fuel cell vehicle by 2015[448]:

- The SAIC plug-in fuel cell hybrid vehicle has been reported to cost around $80,000 and the Chinese government would subsidize half the cost.

7.5.1.1.4.2. OTHER CHINESE FUEL CELL CAR DEVELOPERS Several Chinese automakers built FCVs incorporating the fuel cell powertrain platforms developed by Tongji University. The automakers included Chang'an Automobile, Chery Automobile, SAIC, and Shanghai Volkswagen.

- Shanghai Volkswagen's fuel cell cars clearly performed better than the FCVs built by other automakers. Despite this success, it has not announced any plans for near-term FCV commercialization in China.

The Chinese government as well as the fuel cell industry likely recognized that the Chinese fuel cell technology was not mature for near-term commercialization. They also might have felt that China's large investment in fuel cells during the past decade might have focused on the wrong target—building FCV prototypes vice conducting basic research. China's strategy of "import-absorb-digest-innovate," which enabled it to successfully acquire intellectual property and know-how through licensed use of foreign technology, also might have created a habit of ignoring basic research.[449]

Moreover, in the case of PEMFC technology, China would not have been able to acquire foreign intellectual properties as easily as with other technologies. Ballard would be unlikely to provide China with its coveted technology without the strictest of conditions, for example.

- China has the resources, however, to support basic research and make a breakthrough on its own, though this would take time.

7.5.1.1.5. SOME HAVE A LONG-TERM PERSPECTIVE

A few automakers believe FCVs will only be practical after much more R&D and development have been conducted. In their view, fuel cells will be ready for commercialization in 10–20 years. These automakers include Fiat and Volkswagen–Audi.

7.5.1.1.5.1. FIAT According to a Fiat presentation made at an International Partnership for Hydrogen and Fuel Cells in the Economy (IPHE) meeting in November 2007, the company suggested that it might consider engaging in commercialization of FCVs around 2020.[450] Fiat viewed zero emissions and sustainable energy to be the company's long-term goals and that FCVs and hydrogen from renewable resources were key technologies for achieving these goals. It also thought that it would require 10 to 15 years to solve the remaining problems, including cost, durability, the robustness to endure extreme operating conditions, and viable onboard hydrogen storage, as well as the development of satisfactory hydrogen-fueling infrastructure. Although the Italian government likely was pressuring Fiat to speed its effort, it appears that Fiat was taking its time and proceeding with FCV development at its own pace. Over the last decade, it has unveiled half a dozen prototype and concept cars that used fuel cells for primary or secondary power.

In October 2003, Fiat presented the ***Fiat Seicento.***[68] The FCV was powered by an improved 40 kW Nuvera PEMFC stack and had a driving range of 136 miles.[38] It had a top speed of 80 mph. In 2004, Fiat introduced the ***Fiat Panda Hydrogen,*** based on Fiat's new supermini Panda.[451] The Panda Hydrogen was Fiat's first fuel cell car with performance and usability comparable to a conventional car and thus was intended for commercial fleet trials.[452] The Panda Hydrogen used a hybrid architecture, with the fuel cell directly connected to the traction motor for primary power, and the battery pack providing supplementary power to the motor during start-up and acceleration. The batteries were recharged by regenerative braking. The 40 kW PEM stack was fed by hydrogen stored at 350 bar in two tanks under the floorpan. The Panda Hydrogen had a top speed of 80 mph, it accelerated from 0 to 31 mph in less than 7 s, and had a range of more than 124 miles.

In 2006, Fiat presented the ***Fiat Panda Hydrogen Concept Car,*** featuring an advanced 60 kW Nuvera Andromeda II PEMFC stack.[453] The vehicle used a fuel cell full power system in which the fuel cell system produced sufficient electricity to power high-torque electrical motors. The Fiat Panda Hydrogen Concept Car accelerated from 0 to 30 mph in 5 s and reached a top speed of about 78 mph. It also had a travel distance of more than 120 miles. In February 2006, the Fiat Panda Concept Hydrogen Concept Car was shown off at the first Hydrogen Olympics in Turin, Italy, which coincided with the 2006 Winter Olympics in the same city.

- In September 2007, Fiat presented three Fiat Panda Hydrogen Concept Cars to the municipality of Mantova as part of the Zero Regio project.[454] The Fiat Panda appeared to be powered by a 70 kW Nuvera PEMFC system. It ran on compressed hydrogen (350 bar), had a top speed of 80 mph, a driving range of 186 miles, and accelerated from 0 to 31 mph in 5 s.[455]

In May 2008, Fiat presented the ***Fiat Phylla (or Phyllis) Fuel Cell,*** a fuel cell–type hydrogen concept vehicle, in the International Motor Show Bologna, Italy.[456] The car, which did not yet exist in drivable form, was designed by Fiat and the Ecole Polytechnic of Turin with the support of the Piedmont Region of Italy.[457] Fiat Phylla was equipped with electric propulsion and batteries (150 kg), with a driving range from 90 miles (with lithium-ion batteries) to 136 miles (with lithium polymer batteries).[458] The full recharge time was 5 hours. It also had a 1 kW PEM fuel cell with hydrogen, which served as an onboard power generator.[459] Its roof was covered with solar photovoltaic cells supplying 340 W to its electric motors, giving the Fiat Phylla Fuel Cell an additional range of 10–18 miles a day. It accelerated from 0 to 31 mph in 6 s. Its maximum speed was 80 mph.

In 2009, Fiat unveiled the ***Fiat Panda HyTRAN***.[460] The Fiat Panda HyTRAN was developed under the €16.8-million FP6 HyTRAN project (2004–2009) coordinated by Volvo.[461] The Panda HyTRAN was fully powered by fuel cells. It used an 80 kW Nuvera fuel cell, had a top speed of 80 mph, and a driving range of 124 miles.

Also in 2009, Fiat presented the ***Serenergy Fiat Scudo Fuel Cell***, which was powered by a 6.4 kW Serenergy PEMFC fuel cell hybrid and 44 kWh Lithium Balance lithium-ion batteries. The fuel cells were used as a range extender.[462] It ran on methanol using an onboard autothermal reformer to produce hydrogen. It had a driving range of 310 miles with the fuel cells and 62 miles only on batteries. It had a top speed of 68 mph.

- Fiat was clearly impressed with fuel cells as a range extender. Fiat's brochure stated that the capability of electric commercial vehicles could be dramatically extended by onboard fuel cells.[463] Used as a range extender, the fuel cells functioned as an onboard charger for the battery pack continually feeding energy during the working day, thereby extending the range between recharges by anything up to 400 percent.

In October 2010, Fiat unveiled the ***Fiat Panda Hybrid Fuel Cell/Battery Powered Vehicle*** at the Battery Show, San Jose, California.[556] The Panda prototype incorporated an 80 kW Nuvera fuel cell and a 0.88 kWh Quallion battery pack. The vehicle accelerated from 0 to 31 mph in 5 s, had a maximum speed of 87 mph, and a range of 155 miles.[464] The prototype was developed under the sponsorship of two FP6 projects: HyTRAN and HySYS. HyTRAN (2004–1999) aimed to develop a PEM fuel cell system to power a passenger car (see details in Section 7.4.2.3.1.2).[465] HySYS (2005–2010) was to develop low-cost components for fuel cell systems and electric drive for future hybridized FCV and ICE vehicles.[466]

Fiat is maintaining its steady pace of developing new FCVs. Fiat has used each new FCV to assess the best applications for fuel cells, whether as an auxiliary power unit for charging batteries, as a primary propulsion power, as a hybrid application, or as a range extender. Fiat has also been experimenting with hydrogen to fuel internal combustion engines (ICE), hydrogen mixed with natural gas, or bivalent gasoline/hydrogen fuels.

- Over the longer term, Fiat will consider the most viable cost-effective applications, and that likely is when the company will decide on its market entry.

7.5.1.1.5.2. VOLKSWAGEN–AUDI VW also envisions FCV commercialization as a longer-term proposition. It has indicated that it might be able to mass produce a FCV with a sufficiently durable and affordable fuel cell drive by 2020, after a complete hydrogen infrastructure had been established.[467] Owing to this longer-term perspective, VW's FCV development has proceeded without a sense of urgency. Since it developed the ***Bora HyPower*** and test drove the vehicle over a Swiss mountain pass in 2002, VW has developed only two new FCVs:

- In September 2004, VW introduced the ***Touran HyMotion*** at the California Fuel Cell Partnership Rally.[468] The FCV was powered by an 80 kW Ballard fuel cell system with Ni-MH batteries and an 80 kW electric motor. It accelerated from 0 to 62 mph in 14 s, had a top speed of 87 mph, and a driving range of 99 miles. The FCV's storage tank held 1.9 kg of compressed hydrogen at 350 bar. After undergoing testing in the United States, the FCV was unveiled in Berlin in July 2006.

- In 2009, VW presented the ***Volkswagen Caddy Maxi HyMotion***.[469] The car was a fuel cell/battery hybrid vehicle, powered by an 85 kW Volkswagen PEM fuel cell, 22 kW lithium-ion batteries, and an 80 kW asynchronous motor (230 Nm). It was fueled by compressed hydrogen at 70 MPa.

In addition to VW's FCV program at its headquarters, VW also had a fuel cell program in China at Shanghai Volkswagen (see Section 7.4.2.4.1.2.1 for details). Shanghai VW and Tongji University developed the ***Tongji University Fuel Cell Passat Lingyu***, in 2007, which was powered by a 55 kW fuel cell system developed on the Tongji campus. The FCV ran on compressed hydrogen (350 bar), accelerated from 0 to 62 mph in 15 s, had a top speed of 87 mph, and a range of 146 miles.

- It was reported that the Fuel Cell Passat Lingyu was slower and noisier than the VW Touran HyMotion.[470] The compressor was loud, and the exhaust constantly spat water on the road surface. In LA's fast moving traffic, acceleration was not enough to keep up, and the Passat bobbed and weaved like an underpowered ship in a storm. China's fuel cell technology clearly was not satisfactory.

In 2010, Shanghai VW built the ***Tongji University Fuel Cell Passat New***, which used Tongji's fifth-generation fuel cell system. According to a Chinese assessment in September 2010, the Chinese FCV technology was not ready for commercialization (see Section 7.4.2.4.1.2.3 for details).[213]

In an interesting development, VW has been developing another type of fuel cell, the phosphoric acid fuel cell (PAFC), for transport applications. VW began the R&D program in 1999 and had solved the electrode problem for the most part and obtained very promising results in the autumn of 2006.[471] VW has developed three new FCVs powered by PAFC propulsion systems:

- In 2007, it presented the Volkswagen Tiguan HyMotion.[472] The Tiguan HyMotion was powered by VW's 80 kW PAFC and a 22 kW hybrid battery system, which delivered a total output of 100 kW.[473] The FCV accelerated from 0 to 60 mph in about 14 s and had a top speed of 93 mph.[474]
- In November 2007, VW introduced the ***Space UP***! Blue at the Los Angeles Auto Show.[475] The vehicle seemed to feature a high-temperature fuel cell (the world's first, according to VW) combined with a lithium-ion battery array to power a rear-mounted electric motor.[476] A solar panel on the roof also helped charge the onboard batteries. On battery power alone, the vehicle had a range of 65 miles with a full charge. When the fuel cell was added into the mix, the maximum range increased to 220 miles. VW claimed that its high temperature fuel cell (FC-HT) was more practical for mass production than competing fuel cell technology because it was lighter, more geared to everyday use, and less costly to produce.
- In August 2010, VW announced that it had successfully completed its first long-range test drive of the ***New Volkswagen HyPower Bora***,[477] a new FCV featuring a PAFC membrane developed by the Paul Scherrer Institute (PSI) in Zurich, Switzerland. The FCV was tested in the depths of winter and driven over the demanding 6578 ft high Simplon Pass that connects Switzerland and Italy. Together with PSI, VW developed a low-cost hydrogen fuel cell with ultra high-performance supercapacitors that could store a fuel cell engine's electrical energy for use during strenuous driving, such as passing on upgrades. These "supercaps" eliminated the need for heavy storage batteries that were

a problem with many other experimental hydrogen vehicles. Although little additional information is available, the PAFC membrane seemed to perform better and was cheaper to produce.

Audi entered the fuel cell development relatively late and has developed a few fuel cell cars. It unveiled the ***Audi A2* (or *Audi A2H2*)** in Hannover Fair, Germany, in April 2004.[478] The A2H2 was based on an Audi A2 alloy body and was powered by a 66 kW Ballard PEM fuel cell and a 38 kW NiMH Battery. It had a driving range of 137 miles and a top speed of 109 mph. It ran on compressed hydrogen and accelerated from 0 to 62 mph in 10 s. Audi's second fuel cell car—the *Audi Q5 FCEV*—was unveiled in October 2009 at CEP Berlin.[479] It ran on compressed hydrogen (70 MPa) and was powered by an 85 kW Volkswagen PEMFC, a 22 kW lithium-ion battery pack, and an 80 kW asynchronous motor (230 Nm). In May 2011, Audi unveiled its third fuel cell car—the ***2012 Audi Q5 Hybrid Fuel Cell Concept***—in the Michelin Challenge Bibendum event in Berlin. The 2012 Audi Q5 Hybrid contained two 10,000 psi compressed hydrogen tanks for extended driving range. The FCV was powered by a 98 kW fuel cell, two 90 kW electric motors, and a 1.3 kW lithium-ion battery pack.

- Audi stated that despite many problems that had yet to be solved, the group was committed to continued work toward the development and testing of hydrogen-powered fuel cell cars.[480]
- Both VW and Audi likely are moving along with their own long-range plans.

7.5.1.2. Buses

The development of fuel cell buses has been less vigorous than that of fuel cell passenger cars. The global bus industry is smaller than the passenger car industry and has fewer players. There have been about a dozen or more fuel cell bus manufacturers and about 10 fuel cell stack developers. Those currently producing PEMFCs for buses likely include Ballard, COPPE UFRJ, Hydrogenics, Hyundai, Nedstack, Nuvera, Proton Motor, Toyota, Tsinghua University, and UTC.

7.5.1.2.1. GOVERNMENT PROGRAMS LED FUEL CELL BUS DEVELOPMENT

During the 18 years from 1993 to December 2010, at least 86 new fuel cell buses were built. It appears that there were three peaks in fuel cell bus development during this time, which occurred in 2001, 2006, and 2010. For the first several years until 2001, only a few new fuel cell buses were developed each year. In 2001, six new fuel cell buses were introduced, the highest number to date. This was followed by 5 years of rather slow, deliberate development. Then, in 2006, 11 new fuel cell buses were built, nearly double the 2001 peak. Following another period of slow development, 14 new fuel cell buses were introduced in 2010, the highest number so far (Fig. 7.6).

None of these peaks can be attributed to market demand. More likely, the increased rate of new bus introductions was due to major government bus demonstration projects that occurred in Europe, the United States, and elsewhere around 2001, 2006, and 2010, including the following:

In 2001, the EU started two FP5 bus demonstration projects: the Clean Urban Transit for Europe (CUTE) for nine European cities and the Ecological City Transport System (ECTOS) to be implemented in Reykjavik, Iceland.

- In 2006, the EU implemented the HyFLEET—CUTE ("Hydrogen for CUTE"), the follow-on to CUTE. The project continued to operate the existing fleet of 21 Mercedes Citaro buses over a period of 12 months in seven European cities and also operated another six Citaro buses in China and

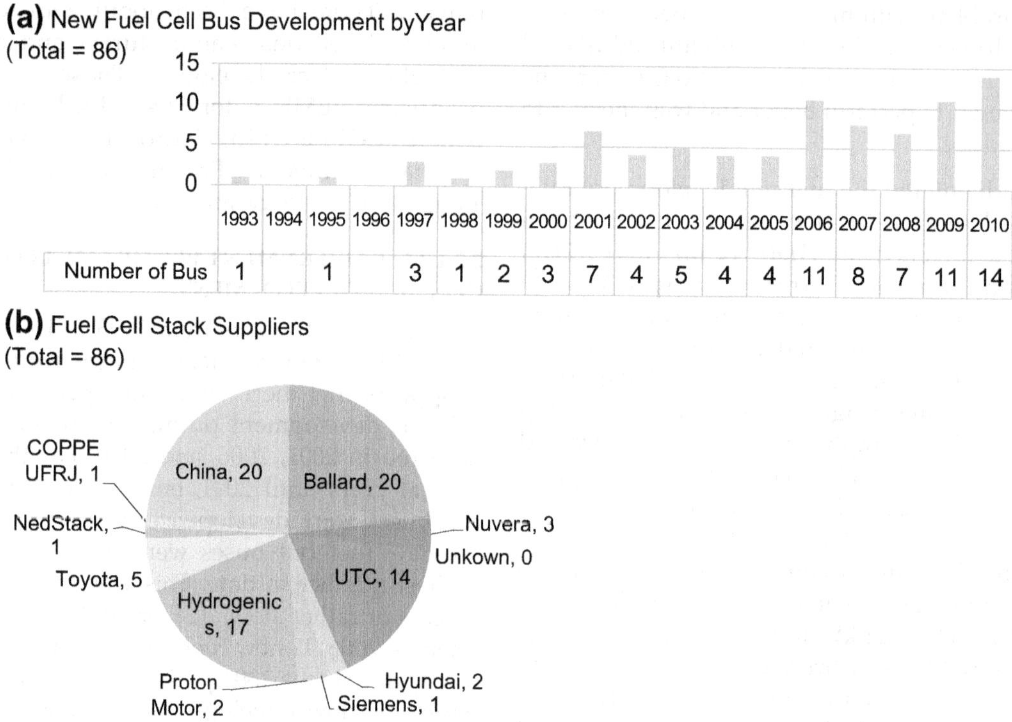

FIGURE 7.6 New fuel cell bus development by year: 1993–2010. (For the color version of this figure, the reader is referred to the online version of this book.)

Western Australia until 2009. In 2005, the US Federal Transit Authority (FTA) implemented its new National Fuel Cell Bus Program (NFCBP), which authorized an initial budget of $49 million[481] to develop and demonstrate a number of fuel cell buses through 2010.

- In 2010, The US FTA provided an additional $13.5 million to continue the NFCBP.[481] In addition, Alameda-Contra Costa Transit District (AC Transit) demonstrated 12 Van Hool buses powered by a 120 kW UTC PEMFC.[482] British Columbia, Canada, also deployed 20 fuel cell buses for the 2010 Winter Olympics. Furthermore, the EU implemented the CHIC Program in 2010 to demonstrate 26 fuel cell buses in five European cities thru 2016.[483] The project would include at least three different bus manufacturers, funded at €26 million.

7.5.1.2.2. TOP THREE LEADERS IN FUEL CELL BUS DEVELOPMENT: BALLARD, HYDROGENICS, AND UTC POWER

Of the fuel cell stack developers, ***Ballard*** has been the most favored by bus manufacturers and suppliers. Since it built the world's first PEMFC bus—the P1 New Flyer fuel cell transit bus—in 1993, Ballard has supplied fuel cells for at least 20 new fuel cell buses as of December 2010. Here, each new generation of fuel cell bus is counted as one, even if some buses were produced in volume for the FP5 CUTE project.

Hydrogenics was the second most popular fuel cell supplier next to Ballard. The company

(c) By Fuel Cell Stack Suppliers by Year (Total = 86)

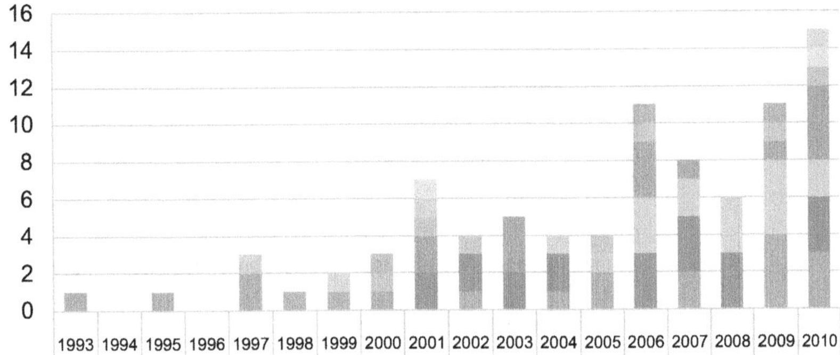

	1993	1994	1995	1996	1997	1998	1999	2000	2001	2002	2003	2004	2005	2006	2007	2008	2009	2010
Unknown									1									
COPPE UFRJ																		1
Siemens								1										
NedStack																		1
Hyundai														1			1	
Proton Motor								1									1	
Nuvera					1		1		1									
Toyota									1	1			1	1				1
UTC									2		3			3	1		1	4
Hydrogenics												1	1	3	2	3	4	2
China									2	2	2	2		3	3	3		3
Ballard	1		1		2	1	1	1		1		1	2		2		4	3

(d) Top Three Global Fuel Cell Developers: Number of Fuel Cell Buses Developed by Year

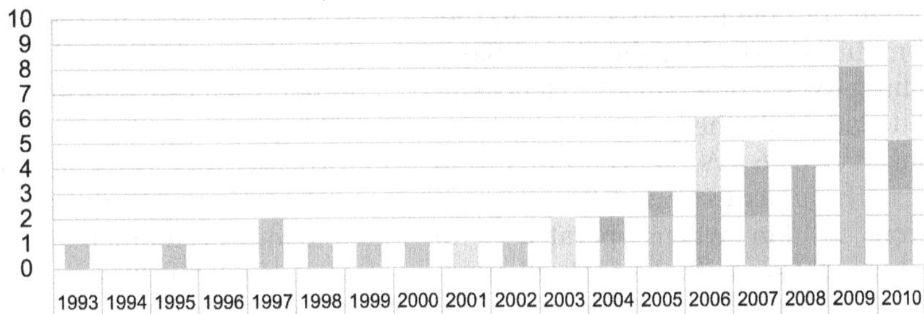

	1993	1994	1995	1996	1997	1998	1999	2000	2001	2002	2003	2004	2005	2006	2007	2008	2009	2010
UTC									1		2			3	1		1	4
Hydrogenics												1	1	3	2	4	4	2
Ballard	1		1		2	1	1	1		1		1	2		2		4	3

FIGURE 7.6 (*continued*).

(e) Change in the Use of Fuel Cells: From Primary Power Source to Auxiliary Power Unit

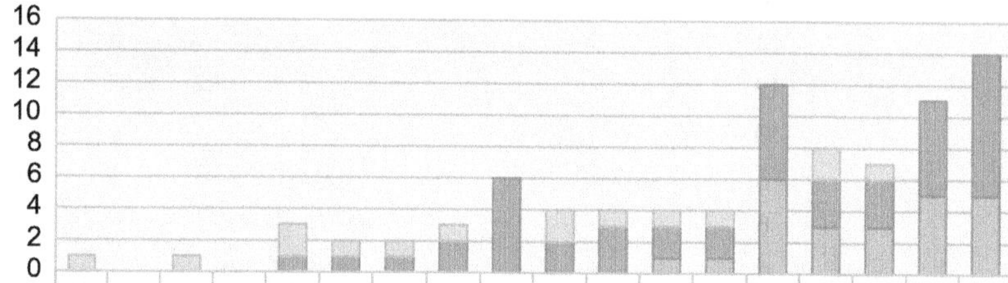

	1993	1994	1995	1996	1997	1998	1999	2000	2001	2002	2003	2004	2005	2006	2007	2008	2009	2010
Primarily Fuel Cell-Powered Bus	1		1		2	1	1	1		2	1	1	1		2	1		
Fuel Cell/Battery Hybrid					1	1	1	2	6	2	3	2	2	6	3	3	6	9
Battery Hybrid with Smaller Fuel Cell as Range Extender												1	1	6	3	3	5	5

has been a long-term, close associate of GM in the fuel cell car development business. By 2004, it had emerged as an independent fuel cell stack supplier for buses. Hydrogenics appears to have a close relationship with Heliocentris, a leading fuel cell educational system manufacturer and an emerging systems integrator of stationary fuel cell backup power and auxiliary power units for transport applications. In April 2008, Heliocentris delivered a 32 kW fuel cell system for operation in a battery/fuel cell hybrid midibus owned by the city of Barth in Mecklenburg-West Pomerania.[484] Heliocentris designed and built the power integration unit that linked two 16 kW fuel cell units from Hydrogenics. The units extended the range of the bus from 20 km when operated with batteries only to 200 km with the fuel cell auxiliary power-range extender.[485] Also in August 2010, Heliocentris and Hydrogenics, in collaboration with FAUN Umwelttechnik, a major garbage collection and road sweeper vehicle manufacturer, unveiled a hybrid waste disposal vehicle for BSR (Berliner Stadtreinigung), the largest municipal disposal company in Germany.[486] The prototype was a diesel/fuel cell hybrid where the diesel engine could be shut off during garbage collection and the fuel cell would drive the electrical propulsion system during these periods.

- Hydrogenics has supplied its fuel cells for 17 new fuel cell buses as of December 2010. It is not clear how many fuel cell buses it has built with Heliocentris.

The third most popular bus manufacturer—***UTC***—started to supply PEMFCs for buses in 2001 and has used its fuel cells in 12 new fuel cell buses.

Other fuel cell suppliers have been much smaller players in the bus field. Toyota started to provide fuel cells to Hino, Toyota's bus manufacturing subsidiary, in 2001. It contributed to the development of five new generations of Hino fuel cell buses as of December 2010. Nuvera, an Italian fuel cell developer, started supplying fuel cells for buses in 1997. It has supported the development of four new fuel cell buses. Hyundai produced fuel cells for a new fuel cell bus in 2006 and another in 2009. Proton Motor, a German fuel cell developer, produced fuel cells for two new fuel cell buses, one in 2000 and another in 2009. Nedstack, the Netherlands fuel cell developer, provided fuel cells to one bus project in 2010. COPPE UFRJ, a teaching and research center at the Federal University of Rio de Janeiro, Brazil, worked on one new fuel cell bus in 2010. One unidentified fuel cell manufacturer supported an Irisbus/IVECO bus project in 2001.

Siemens abandoned its PEMFC fuel cell development after it supplied fuel cell stacks for a NEOPLAN fuel cell bus in 2000.

Regarding China's fuel cell bus development effort, it is unclear who supplied some of the fuel cell stacks used in the buses and how many generations of fuel cell buses have been built. In 2001, China designated Tsinghua University to be its leading fuel cell developer, which incentivized the use of indigenously produced fuel cell stacks.

- It is likely that some fuel cell buses might have been built with non-Tsinghua but made-in-China fuel cells, particularly prior to Tsinghua's introduction of its first fuel cell powertrain in 2002. It is also possible that Ballard fuel cells were used in some of the fuel cell buses after Tsinghua started developing fuel cells.
- Due to many ambiguities, Chinese fuel cell bus development cannot be assessed.

7.5.1.2.3. EMERGING TREND: FUEL CELLS AS RANGE EXTENDERS

Over the past 18 years, a noteworthy trend has emerged in the use of fuel cells in buses. In earlier years, fuel cells were used for primary power to propel the buses, but recently fuel cells have been used more and more with batteries in a hybrid configuration. At the same time, fuel cell stacks in hybrid systems have become smaller and have functioned as an auxiliary power unit or a range extender, while batteries are used to provide the primary vehicle propulsion. This shift in function is undoubtedly because current fuel cells are not robust enough to support the frequent load changes in vehicle engine operations.[487] Many fuel cells, moreover, still, have a long start-up time and problems of lifespan and reliability.

- Reflecting this reality, of the 86 fuel cell buses developed since 1993, only 15 buses were built with primary power supplied by fuel cells (data is as of December 2010). Another 47 buses were built with hybrid systems. In 2004, a new breed of fuel cell bus emerged; Hydrogenics and Proterra built a bus powered by a fuel cell/battery hybrid system, in which a small 10 kW class fuel cell system functioned as a range extender.[487] Since 2004, at least 24 such new buses have been developed, as of December 2010.

In a successful collaboration, Hydrogenics and Proterra have emerged as global leaders in building plug-in battery-electric hybrid buses using fuel cells as range extenders. Proterra is a new US start-up founded in 2004, specializing in plug-in battery electric commercial buses. Since 2004, they have received a number of developmental grants[488]:

- In 2007, the FTA NFCBP funded Proterra to design and integrate a hybrid electric fuel cell bus using Hydrogenics fuel cells as range extenders. The battery is recharged by an APU that consists of two small 16 kW Hydrogenics PEMFC fuel cells. The bus is fueled by pressurized hydrogen in tanks on the bus roof. The bus started demonstration operations in Burbank, California in 2008. Without the fuel cell APU, the Proterra buses could operate in a charge-depleting (CD) battery- electric mode (BEB) for up to 20 miles, but with the fuel cell APU range is extended up to 250 miles and/or 20 h of operation. The batteries can be recharged in 5–10 min at bus stops.
- In 2009, Hydrogenics and Proterra were funded to develop two plug-in fuel cell battery electric hybrid buses, one funded by the FTC for demonstrations in Columbia, South Carolina, and another funded by DoD

to be deployed in Fort Lewis, Washington.[489] In both buses, the fuel cells were used as range extenders.

- Also, Hydrogenics participated in two other FTA NFCBP projects in 2010 that was to use the Hydrogenics fuel cell auxiliary power unit and a lithium-ion battery for onboard rechargeable energy storage.

7.5.1.2.4. BUT NO NEAR-TERM COMMERCIAL SUCCESS

The use of fuel cells as range extenders for transport application is a trend that is growing popular. Clearly, it has merits. Nonetheless, the cost of this capability makes it difficult to say whether this fuel cell application will be competitively viable in the marketplace. The DOE National Renewable Energy Laboratory (NREL) indicated that fuel cell buses were in the early development cycle and that it would probably be another 4 to 6 years before transit authorities put in large orders.[488] Meanwhile, according to the NREL, there were other alternative propulsion technologies—such as diesel hybrid electric and compressed natural gas—that were gaining ground vis-à-vis fuel cell buses.

- The Proterra bus with a fuel cell range extender is costly, at about $1.2 million. This is about half the cost of a fuel cell bus, but twice the cost of a compressed natural gas–powered bus.[488]

7.5.1.3. *Material Handling Vehicles*

In recent years, PEMFC technology has been making rapid inroads into the material handling market, which includes forklifts, fork trucks, and other warehouse vehicles (so-called forklifts). These applications tend to be indoors where emissions must be controlled. The growth in deployment of forklifts probably is due to tangible economic and operational benefits over traditional technologies. The material handling equipment sector appears to be a niche where the advantages of fuel cells serve user needs so well that the cost disadvantages can be tolerated. Market growth is only occurring in North America, however, and is barely discernible elsewhere.

7.5.1.3.1. GROWING FUEL CELL MATERIAL HANDLING MARKET IN THE UNITED STATES

The first four fuel cell–powered forklifts were deployed at a warehouse of Ozburn-Hessey Logistics in Smyrna, Tennessee, in 2004. Since then, fuel cell–powered material handling vehicles have become increasingly popular at warehouses, distribution centers, and factory floors in the United States and Canada. Fuel cell–powered equipment have some attractive characteristics compared to the traditional material-handling equipment, which generally use propane and diesel internal combustion engines (ICE) or lead–acid batteries. Fuel cell material handling vehicles operate silently, with zero- or near zero-emissions, and offer faster refueling as well as significantly longer run times compared to their conventional counterparts.[490] Fuel cells also do not require lengthy recharging time or need large amounts of plant space for recharging equipment, as compared to battery-powered vehicles.

- Material handling is an attractive market for the fuel cell industry. It is a sizable market, with around 600,000 battery electric material handling vehicles (mostly lead acid batteries) sold worldwide each year.[490] If ICE material handing vehicles are included, the market size is some 3 million units per year.

During the past few years, there has been significant growth in the North American market for fuel cell–powered material handling vehicles, which is a source of optimism in the fuel cell industry. According to the chart by Fuel Cell 2000, as of July 2011 at least 2220 forklifts or other material handling vehicles have been deployed in the North America since 2004.[491] Although the number was insignificant until 2008, it grew to 386 in 2009, 16 times greater than the number in 2008. In 2010, 646 forklifts were deployed, nearly double that of the previous year. In 2011, 1003 were deployed or are planned to be deployed; a 55 percent increase over 2010 (Fig. 7.7).

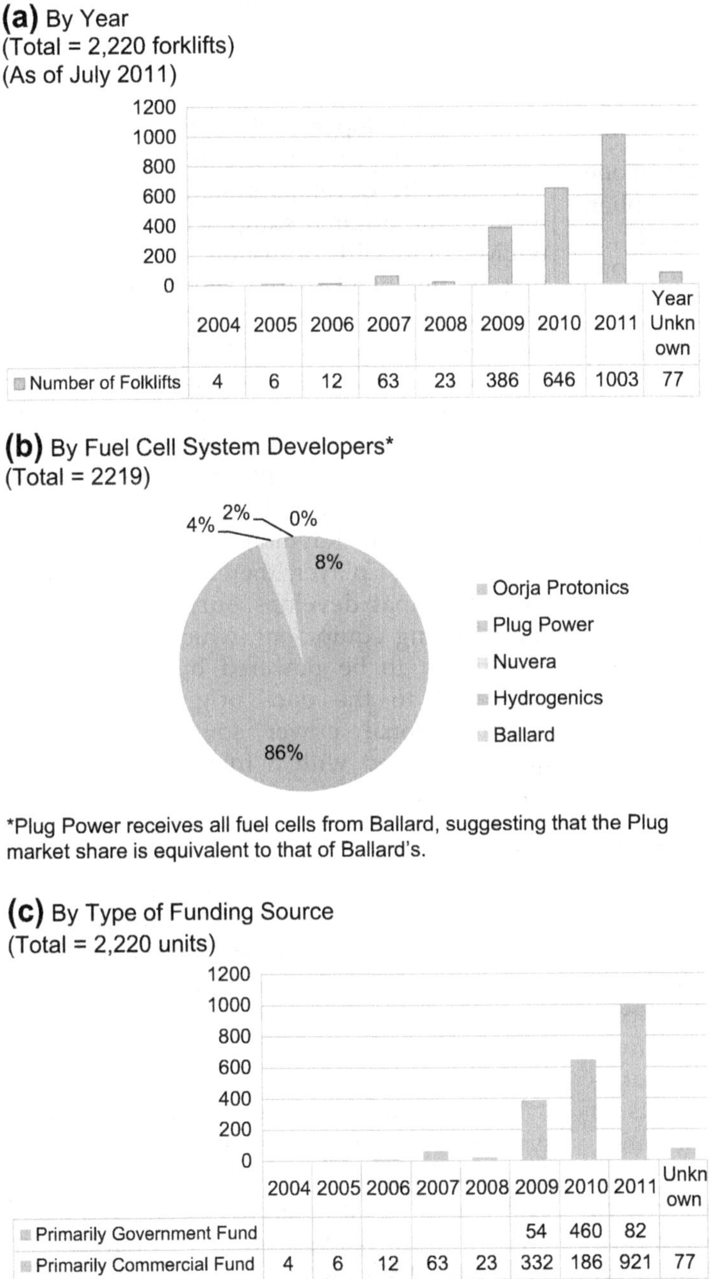

FIGURE 7.7 Deployment or planned deployment of forklifts in North America. (For the color version of this figure, the reader is referred to the online version of this book.)

According to the same Fuel Cell 2000 chart, about 36 companies deployed forklifts at sites in North America as of July 2011[491]:

- Notable deployment examples span diverse industries, including grocery distribution centers, beverage handling facilities, and automotive manufacturing operations. Central Grocers in Illinois operate solely on fuel cell–powered material handling equipment with 220 fuel cell lift trucks at one of its distribution centers. One food supplier, Sysco, deployed or plans to deploy fleets totaling over 334 lift trucks for its operations in Virginia, Texas, Pennsylvania, and Michigan.[491] Walmart deployed 151 fuel cell material handling equipment for grocery and food handling operations both in the United States and Canada, and Wegmans acquired 136 forklifts in Pennsylvania in 2010.

The growing numbers of lift-truck installations reportedly have demonstrated economic, environmental, and performance advantages of fuel cells. The DOE preliminary analysis indicates that fuel cells incurred 1.5 times lower maintenance costs, 8 times lower refueling/recharging labor costs, and 2 times lower net present value for total system cost.[492] It seems that fuel cells could be cost competitive with batteries on a lifecycle basis (Table 7.9).

7.5.1.3.2. EMERGING LEADERS IN MATERIAL HANDLING

The fuel cell–powered material-handling sector has been dominated by only a few major players. Of the 2220 forklifts that were deployed or are planned to be deployed in the North America as of July 2011, 86 percent or 1904 forklifts used PEMFC fuel cell systems manufactured by Plug Power, 8 percent or 185 units by Oorja Protonics, a manufacturer of methanol-powered DMFC stacks, 4 percent or 77 units by Nuvera, an Italian PEM fuel cell developer, and 3 percent or 49 units by Hydrogenics, a Canadian PEM fuel cell developer. Plug Power purchases fuel cells exclusively from Ballard of Canada. Ballard supplies other customers in addition to Plug Power, which makes Ballard the top player in this sector.

There are three types of players in the fuel cell–powered material handling market sector. The first is the fuel cell and stack developer, such as Ballard, which supplies stacks to a system developer, such as Plug Power. The system developer is the second type of player, and typically develops systems and supplies them to the original equipment manufacturer (OEMs) of forklifts or installs them on forklifts that are deployed at user sites. System developers often are fuel cell and stack developers, such as Hydrogenics. The third type of player is the OEM, such as Crown Equipment, which manufactures conventional material handling vehicles and is expanding into fuel cell–powered vehicles. In addition, some fuel cell developers, such as Plug Power, and some OEMs, such as Raymond, are trying to be a new kind of player; they hope to be a one-stop shop that develops purpose-built material handling equipment, which are designed specifically to be powered by fuel cells, as compared to the ones originally built for a conventional power source and subsequently refitted with a fuel cell power unit (Table 7.10).

7.5.1.3.2.1. BALLARD Ballard clearly is the leading supplier of PEM stacks and systems in the material handling vehicle market. As noted, Ballard's major customer in the material handing sector is Plug Power. In July 2011, Ballard received a purchase order commitment from Plug Power for a minimum of 3250 fuel cell stacks over the next 18 months. These units will be used in Plug Power's line of GenDrive systems for various forklift trucks in the North American material handling market.[493] The deal would run through 2014 and use Ballard fuel cell stacks with power levels ranging from 1.5 kW to 20 kW each.

TABLE 7.9 Deployment of Fuel Cell–Powered Material Handling Vehicles in North America* ** (Total = 2219) (as of July 2011)

Year deployed	Customer	Location	Fuel cell developer	OEM	Number of units
2004	Ozburn–Hessey Logistics	Smyrna, TN	Ballard		4
2005	FedEx	Toronto, Canada	Hydrogenics	NACCO Materials Handling Group	1
	GM	Oshawa, Canada	Hydrogenics	NACCO Materials Handling Group	1
	Walmart	MO	Plug Power		4
2006	Walmart	OH	Plug Power		12
2007	Walmart	Washington Court House, OH	Plug Power		55
	Four companies deployed two forklifts each in 2 week trials, including ISOLA Laminates, Leigh Fibers, Michelin, and PBR	SC	Hydrogenics	LiftOne, Linde Material Handling of Summerville, SC	8
	New United Motor Manufacturing, Inc. (NUMMI) (joint-owned by GM and Toyota	Fremont, CA	Oorja Protonics		N/a
2008	Bridgestone–Firestone	Aiken County, SC	Plug Power		23
2009	Bridgestone–Firestone	Aiken County, SC	Plug Power		20
	Central Grocers	Joliet, IL	Plug Power		220
	Defense Logistics Agency, U.S. Department of Defense	Susquehanna, PA	Plug Power		20
	Defense Logistics Agency, U.S. Department of Defense	Susquehanna, PA	Nuvera		20
	H-E-B	San Antonio, TX	Nuvera,		14
	LPC	Lodi, CA	Oorja Protonics		N/a ("entire fleet")
	Nestlé Waters	Dallas, TX	Plug Power		32
	Nissan North America	Smyrna, TN	Oorja Protonics		60
	Super Store Industries	Lathrop, CA	Oorja Protonics		N/a

(Continued)

TABLE 7.9 Deployment of Fuel Cell–Powered Material Handling Vehicles in North America* ** (Total = 2219) (as of July 2011) (*cont'd*)

Year deployed	Customer	Location	Fuel cell developer	OEM	Number of units
Deployed or order placed in 2010	BMW Manufacturing Company	Spartanburg, SC	Plug Power		86
	Defense Logistics Agency, U.S. Department of Defense	Susquehanna, PA	Nuvera		15
	Defense Logistics Agency, U.S. Department of Defense	Warner Robins, GA	Hydrogenics		20
	FedEx	Springfield, MO	Plug Power		35
	Martin-Brower	Stockton, CA	Oorja Protonics		15
	Sysco	Houston, TX	Plug Power		98
	Sysco	Philadelphia, PA	Plug Power		95
	Testa Produce	Chicago, IL	Oorja Protonics		20
	United Natural Foods, Inc. (UNFI)	Sarasota, FL	Plug Power		65
	Wegmans	Pottsville, PA	Plug Power		136
	Whole Foods Market	Landover, MD	Plug Power		61
Deployed or order placed in 2011	Coca-Cola	Charlotte, NC	Plug Power		40
	Coca-Cola	San Leandro, CA	Plug Power		37
	Defense Logistics Agency, U.S. Department of Defense	Susquehanna, PA	Nuvera		18
	Defense Logistics Agency, U.S. Department of Defense	San Joaquin, CA	Plug Power		20
	Defense Logistics Agency, U.S. Department of Defense	Fort Lewis, WA	Plug Power		19
	EARP Distribution	Kansas City, KS	Oorja Protonics		24
	Golden State Foods	Lemont, IL	Oorja Protonics		20
	Kimberly-Clark	Graniteville, SC	Plug Power		25
	Kroger Company	Compton, CA	Plug Power		161
	Sysco	Front Royal, VA	Plug Power		100
	Unnamed Food Distributor	Boston, MA	Plug Power		160

TABLE 7.9 *Deployment of Fuel Cell–Powered Material Handling Vehicles in North America* ** (Total = 2219) (as of July 2011) (cont'd)*

Year deployed	Customer	Location	Fuel cell developer	OEM	Number of units
	Unnamed Food Distributor	San Antonio, TX	Plug Power		113
	Unnamed Food Distributor	Long Island, NY	Plug Power		42
	U.S. Foodservice	Livermore, CA	Oorja Protonics		40
	WinCo Foods, LLC	Modesto, CA	Plug Power		184
Year of deployment unknown	Ace Hardware	N/a	Oorja Protonics		6
	East Penn Manufacturing	Topton, PA	Nuvera		11
	GM	Oshawa, ON, Canada	Hydrogenics		19
	Sysco	Canton, MI	Plug Power		30
	Sysco	Grand Rapids, MI	Plug Power		11
	Unified Grocers	N/a	Oorja Protonics		N/a

* *This table is a composite of data found in various sources and "Fuel Cell-Powered Forklifts in North America," Charts, http://www.fuelcells.org/info/charts/forklifts.pdf, Fuel Cells 2000, updated July 2011.*
** *Those deployment cases in dark gray background were supported by government funding, either by the DoD funds or by the DOE funds from the American Recovery and Reinvestment Act of 2009 (Recovery Act or ARRA).*

TABLE 7.10 Players in Fuel Cell–Powered Forklift Manufacturing

	OEMs	Fuel cell system developers	Fuel cell stack developers
North America market	Crown equipment, LiftOne, Raymond, Toyota, Yale Equipment	Eat Penn manufacturing, Plug power, H2 logic (Denmark)	Ballard (Canada), Nuvera (Italy)
		Hydrogenics (Canada), Nuvera (Italy), Oorja Protonics, Plug Power	
European Market	Jungheinrich (Germany), Still (Germany)	H2 Logic, Hoppecke, Julich Forschungszentrum (Germany), Proton Motor (Germany)	Nuvera, Ballard
	Linde Material Handling	Hydrogenics	

*7.5.1.3.2.2. **PLUG POWER*** Plug Power is the dominant player in the material handling market. Plug was formed in June 1997, headquartered in Latham, New York.[494] It acquired competitor H Power in November 2002 and became a leading developer of PEMFC systems for stationary power generation systems, backup power systems, and automotive applications.[322] Plug Power developed the popular line of GenDrive systems, which were integrated into material handling vehicles manufactured by OEMs. As a skillful strategic move, Plug Power in 2007 acquired capabilities to develop fuel cells and material handling vehicles. It acquired Cellex of Canada, a leading developer of fuel cell lift trucks and other material handling equipment, in April,[495] and General Hydrogen, a Canadian fuel cell system developer in May 2007.[496] Plug Power's 2009 deal with wholesaler Central Grocers was a breakthrough that flowed directly from this acquisition. Central Grocers opened a new warehouse in Illinois in May with an all new fuel cell fleet of 220 vehicles, including pallet trucks, reach trucks, stand-up counterbalanced trucks, and sit-down trucks.[497] Plug Power supplied all the fuel cell systems, taking advantage of its ability to offer a range of forklift products after acquiring Cellex and General Hydrogen in 2007. Forklift OEM Yale received the fuel cell maintenance contract.

- Plug Power has decided to focus on the material handling business.[498] In October 2010, it sold to Oregon-based IdaTech its GenSys LPG off-grid and GenCore backup power stationary fuel cell product lines.

Plug Power became the biggest winner in the American Recovery and Reinvestment Act of 2009 (ARRA) stimulus package forklift awards. Plug Power was allocated $9.7 million[499] to support the shipment of 483 fuel cell forklifts.[500] It appears that Plug Power received 97 percent of the entire award allocated to fuel cell forklifts, with the remaining 3 percent allocated to Nuvera, which deployed 14 fuel cell forklifts to the H.E.B. Grocery Company's distribution facility in San Antonio in 2009. In the award package for Plug Power, the company supplied or planned to soon deploy 350 material-handling units to GENCO Supply Chain Solutions (North America's second largest third-party logistics provider), 35 warehouse trucks to FedEx, and 72 Raymond pallet trucks and 26 Raymond fork trucks to Sysco.[501] Under the GENCO package, Plug Power deployed in 2010 or planned soon to deploy 61 GenDrive-powered Crown forklifts to Whole Foods,[502] 50 units to Wegmans, 25 GenDrive®-powered Toyota electric lift trucks to move product at Kimberly-Clark's distribution facility in Graniteville, South Carolina,[503] and 95 units to Sysco.

In addition to the ARRA award, Plug Power made two major fleet sales in 2010. In July 2010, United Natural Foods, Sarasota, Florida, deployed 65 GenDrive-powered forklifts.[504] Of the fleet, 29 were new Raymond lift trucks, and 36 were existing Raymond lift trucks retrofitted with Plug Power GenDrive fuel cell systems. Also in October 2010, BMW deployed 86 GenDrive-powered lift trucks manufactured by Raymond and Crown.[505]

7.5.1.3.2.3. OORJA PROTONICS Oorja, a DMFC developer headquartered in Fremont, California, specializes in a unique onboard DMFC fuel cell charger, the ***OorjaPac,***[497] which serves as an APU for the material handling vehicle's existing lead–acid battery. The Oorjapac DMFC unit continuously charges the battery, thus eliminating the need for a change-out battery for recharging. It also uses methanol and eliminates the hydrogen infrastructure.

Oorja, a privately held company, was founded in 2005.[506] It appears that Oorja deployed a number of forklifts to New United Motor Manufacturing (NUMMI) in Fremont, California, in 2007 for customer trials. In 2009, Oorja started to receive major purchase orders. Nissan in Smyrna, Tennessee, which had conducted an 18-month field trial of OorjaPac units, ordered Oorja to install the units on 60 forklifts. LPC placed an order to replace LPC's entire fleet of propane forklifts with electric forklifts retrofitted with methanol-fuel cells as onboard battery chargers at its warehouse in Lodi, California. The fleet size was not reported.[507] In 2010, Oorja installed 20 OorjaPac units to Testa Products in Chicago and 15 units to Martin-Brower in Stockton, California.[491] Martin-Brower placed an add-on order to convert the entire Stockton pellet jack fleet to fuel cell power in 2011. In 2011, Oorja also sold 40 units to US Foodservice, Livermore, California, 24 to EARP Distribution, Kansas, and 20 to Golden State Foods, Illinois.

7.5.1.3.2.4. NUVERA Nuvera has a 4 percent share of the North-American forklift market. Nuvera (Italy) was established in April 2000 and focused on the development of multifuel processing and fuel cell technology for industrial utility vehicle and transport applications.[508] The company started to develop fuel cells for material handling vehicles in 2003. Initially, Nuvera was interested in developing a fuel cell and battery hybrid power system. In 2003 and 2004, Nuvera started joint development of such units with Fiamm and East Penn Manufacturing, both major lead–acid battery manufacturers. In 2008, Nuvera and East Penn developed the ***Deka Nuvera ReadyPower*** hybrid fuel cell power system,[509] which was interchangeable with standard lead–acid forklift batteries. It appears that East Penn immediately deployed 11 forklifts powered by the ReadyPower system.[510] In February 2009, Nuvera deployed 20 ReadyPower-powered forklifts to the Defense Logistics Agency (DLA).[511] In February 2009, Nuvera decided to end the joint development terms with East Penn and pursue its own development of a forklift power system,[512] while continuing to supply East Penn with fuel cells to produce ReadyPower systems. In March 2009, Nuvera announced a new line of hybrid fuel cell power products for forklift trucks called ***PowerEdge***. PowerEdge, which would also replace standard lead–acid batteries, consisted of Nuvera's PowerFlow hydrogen fuel cell system, sealed maintenance-free batteries, compressed hydrogen storage, and a computerized control system.[513] It appears that the PowerEdge system is more popular than the ReadyPower system. In June 2009, Nuvera retrofitted East Penn's fleet of 11 forklifts with the new PowerEdge system and deployed 14 PowerEdge-powered forklifts to an HEB distribution center with ARRA funds.[510] Nuvera also deployed 15 PowerEdge-powered forklifts in 2010 and additional 18 in 2011 at the DLA distribution depot at Susquehanna, Pennsylvania.[514]

7.5.1.3.2.5. HYDROGENICS Hydrogenics has a 2 percent share in the North-American forklift market and deployed at least 49 forklifts as of July 2011. In 2005, Hydrogenics developed a fuel cell hybrid power pack, replaced the lead–acid batteries in two seat-rider forklifts built by NACCO Material Handling Group, and conducted 3-month field trials of the forklifts at GM of Canada's car assembly plant in Oshawa and FedEx Canada's hub at Toronto International Airport.[515] In 2007, a number of 2 week field evaluation projects were planned in South Carolina.[516] Under the sponsorship of the Greater Columbia Fuel Cell Challenge, LiftOne installed Hydrogenics' HyPX Fuel Cell Power Packs in electric lift trucks manufactured by Linde Materials Handling of Summerville, SC. Two forklifts each were deployed at four different sites in South Carolina for 2 weeks, including ISOLA Lainates in Ridgeway, Leigh Fibers in Spartanburg, Michelin in Columbia, and PBR in West Columbia.[491] LiftOne, the leading supplier of lift trucks for South Carolina, replaced the electric engine of the forklift trucks with a Hydrogenics-manufactured HyPX Fuel Cell Power Packs.[517] In 2010, Hydrogenics received a major purchase order; the DLA's distribution depot in Warner Robins, Georgia, deployed 20 forklifts. Hydrogenics also deployed 19 forklifts at GM in Oshawa, Canada, but the year of deployment is not known.[491]

7.5.1.3.2.6. H2 LOGIC Danish company H2 Logic is one of the several manufacturers showing interest in capturing this market sector in Europe. Since 2006, H2 Logic and the forklift manufacturer Dantruck have been collaborating to develop fuel cell–powered forklifts.[518] In

2011, they developed a 2.5 ton forklift for heavy-duty applications with an H2 Logic H2Drive fuel cell system. The forklift was presented at the CeMAT logistics fair in Hannover, Germany, in May 2011. The forklift was to be demonstrated for 2 years as part of the EU project HyLIFT-DEMO, co-funded by the European Joint Undertaking for Fuel Cells and Hydrogen FCH JU.[518] The HyLift-DEMO would demonstrate a total of thirty 2.5–3.5 ton fuel cell–powered forklifts at 30 user sites for 2 years with the hope that this would kick off commercial market deployment by the end of 2013. In January 2008, H2 Logic signed a strategic supply agreement with Ballard, and under the 2 year agreement, Ballard would supply fuel cell stacks for H2 Logic fuel cell hybrid systems for the material handling market.[519]

7.5.1.3.2.7. PROTON MOTOR Proton Motor, a German fuel cell system developer, has been a manufacturer of hydrogen fuel cells and fuel cell systems since 1994.[520] Proton Motor develops fuel cell systems that can power forklift trucks, city buses, light-duty vehicles, as well as stationary and maritime vehicles. In 2003, Proton in collaboration with German forklift truck producer Still developed a fuel cell–powered forklift.[521] The vehicle was presented at a "Hydrogen Day" press event at Munich Airport and also shown at the H2Expo international trade fair in Hamburg. The logistics company Cargogate, the subsidiary of Flughafen München GmbH, deployed the prototype as the first fuel cell–powered vehicle to be used at Munich airport.[522] In September 2007, Proton introduced the world's first triple-hybrid fuel cell forklift truck during the inauguration celebration of its new facility in Puchheim near Munich.[523] The system combined fuel cells, batteries, and supercapacitors and provided more than 30 kW of peak power.[524] The system was devised as a plug and drive, which would replace the standard battery package for the Class 1 forklift with no additional modifications. The fuel cell system allowed the forklift truck to work for an 8-hour shift. It could be refueled with hydrogen in 1 min.

7.5.1.3.3. LEADING ORIGINAL EQUIPMENT MANUFACTURERS OF FUEL CELL MATERIAL HANDLING VEHICLES

A number of US and European original equipment manufacturers (OEMs) of material handling vehicles have entered the fuel cell–powered forklift market. They were initially interested in fuel cell power systems as a replacement for diesel or propane ICE engines or batteries in their existing material handling vehicles. Recently, they have become more interested in designing and building totally new fuel cell forklifts.

7.5.1.3.3.1. CROWN EQUIPMENT Crown, one of the world's leading forklift manufacturers, has been interested in using fuel cells in forklifts. In 2008, Crown was awarded \$1 million from the state of Ohio to address the technical and commercial barriers to using fuel cell power systems in lift trucks.[525] Crown worked with PEM suppliers Plug, Hydrogenics, Nuvera, and DMFC supplier Oorja Protonics,[497] and established the industry's first qualification program for safe and efficient fuel cell packs for lift trucks.[526] It qualified 20 of its electric forklift models to operate with various fuel cells and offered 29 qualified combinations of fuel cell packs and trucks. In 2009, the company was awarded another \$1 million in Ohio to develop an integrated fuel cell forklift over 2 years.[527] The project aimed to identify and partner with fuel cell suppliers, construct a bench-top model, and define specifications for creating a prototype forklift. In August 2011, Crown introduced a new version of its Crown ***PC 4500 Series center control pallet truck*** designed to accommodate Plug Power GenDrive fuel cells.[528] It was reported to be the industry's first pallet truck that was designed specifically for fuel cells.

7.5.1.3.3.2. LINDE MATERIAL HANDLING Since 2008, Linde has been developing fuel cell–powered forklifts with its long-term partner, Hydrogenics, the Canadian fuel cell manufacturer.[529] In January 2010, Linde delivered two fuel cell–powered fuel cell trucks to the Linde Gases Division, part of the technology company. At the same time, Linde announced that it would incorporate fuel cell trucks into its product range, the first industrial truck manufacturer in Europe to do so. The fuel cell–powered forklifts would be offered as standard and as a customized option. The fuel cell forklifts were based on the Linde E30 three-ton electric counterbalanced forklift.[530] In place of the usual 80 volt battery, the forklifts had a fuel cell and a tank which stored 1.6 kg of hydrogen gas at 350 bar. Linde's contractual partner, Gruma Nutzfahrzeuge, based in Garching near Munich, was involved in the project from the beginning, and is responsible for forklift service and maintenance.

7.5.1.3.3.3. RAYMOND CORPORATION/TOYOTA Raymond, a leading North American manufacturer of material handling vehicles, was founded in 1922 in Greene, New York, and was acquired by Toyota in 2000. It started investigating fuel cells for forklifts in 2004.[531] In January 2007, it began a fuel cell forklift test laboratory after receiving funds from the New York State Energy Research & Development Authority.[532] Raymond tested several fuel cell forklifts and pallet trucks in a real world test environment. Fuel cell products were supplied by Ballard, Hydrogenics, Nuvera, and Plug Power. In April 2008, Raymond signed an agreement with Ballard to jointly develop fuel cell–powered lift trucks.[533] The goal was to develop a prototype using the next generation of Ballard's Mark9 SSL fuel cell system to power Raymond lift trucks. In March 2010, Raymond became an independent distributor and service provider of Plug Power GenDrive fuel cells.[534] As an independent distributor, Raymond and its network of authorized sales and service centers will sell, rent, and lease Plug Power's GenDrive fuel cell units to material handling customers in North America.[535] In August 2010, Raymond and Plug participated in a ribbon-cutting event at United Natural Food, Inc. (UNFI) in Sarasota, Florida, and demonstrated the seamless integration of Plug Power's GenDrive fuel cell unit into UNFI's full fleet of Raymond electric lift trucks.[536]

- As of June 2011, Raymond had deployed or would soon deploy at least 447 fuel cell–powered lift trucks in the United States.[531]

Toyota, a parent company of Raymond, had developed a fuel cell hybrid forklift, ***FCHV-F***, in 2005 and exhibited it at CeMAT in Hannover in October 2005.[537] The forklift truck was powered by a 30 kW Toyota PEMFC hybrid with supercapacitors. Toyota presented the prototype a couple of times in 2007 with no changes in vehicle specifications.[538] Press releases in 2007 indicated that Toyota would bring a fuel cell lift truck to market in the next few years. It is possible that Toyota is depending upon Raymond to pursue fuel cell forklift commercialization.

7.5.1.3.4. HOPES FOR COMMERCIAL SUCCESS RISING

It was recently reported that the atmosphere was electric at Plug Power.[539] Plug Power, which struggled long in search of a niche for its fuel cells appears to have found one in warehouse and distribution centers across North America with its GenDrive fuel cell. To focus on this market, Plug Power decided to shed its initial focus on residential CHP systems. Demand was too low to turn a profit in the sector. Reportedly, the change in focus served Plug Power well, and the company should be able to achieve profitability by 2012.

Commercial success perhaps is finally within reach for fuel cell–powered material handling vehicles. Fuel cell forklifts were sold in large numbers between 2009 and 2010 due to ARRA and owing to the DoD purchases. After all, the ARRA incentives were highly attractive and virtually paid half the deployment cost. In 2011 (as of June 2011), however, 1003 forklifts were sold, and 921 units (92 percent) were sold without subsidies. In the aggregate, out of a total 2230 forklifts sold since 2004 thru June 2011, 1628 (73 percent) were sold independently of the ARRA subsidies.

- This perhaps means that the fuel cell forkliflts could be sold without subsidies.

Still, there are concerns. Some industry experts have expressed doubt about the fuel cell technology becoming a power source of choice for forklifts in the near term. An executive of Exide Technologies, one of the world's largest producers and recyclers of lead–acid batteries in United Kingdom, indicated in September 2010 that the lead–acid forklift battery would still be relevant in the forklift market.[540] An executive of Jungheinrich, a major European forklift manufacturer in Germany, also indicated in 2010 that for at least 10 years, fuel cell–powered forklifts would not be deployed in large numbers in Europe. He stated that in the near term, lithium-ion batteries that would provide for rapid charging, long running times, and compact dimensions would be available. Both companies have had experience in developing fuel cell–powered forklifts. In 2007, Exide Technologies teamed with Ballard to develop fuel cell hybrid power systems for forklift trucks.[541] In 2008, Jungheinrich exhibited the world's first DMFC–powered forklift truck at the CeMAT 2008, developed jointly with Forschungszentrum Julich.

- Both Plug Power and Ballard have indicated reduced sale activities for fuel cell forklifts since June 2011.

Moreover, with its record-breaking sales performance in the past few years, one might expect that Plug Power and Ballard, the two dominant players in the fuel cell-powered material handling vehicle sector, should have made some profits. But neither of them made any. Our figures, which are based on the financial records posted on company websites, show a puzzling trend. On the one hand, their losses are shrinking, but on the other hand, the more they sell, the more losses they incur.

- Their products perhaps are more expensive to make than the price at which they can be sold. The companies might be hoping for massive sales and mass production to achieve cost reduction. If the cost of the material itself is high, however, mass production would not achieve the cost reduction goal.
- Today, it is yet not clear if fuel cell forklifts are viable products that would prevail over conventional forklift products in the long run. Only time will tell (Fig. 7.8).

7.5.1.4. Other Transport Applications (Scooters, Bikes, Trains, Marine Vessels, and Aircraft)... Perhaps No Near-Term Commercial Success

In addition to passenger cars, buses, and material handling vehicles, fuel cells are used in bikes, scooters, three-wheel vehicles, trains, marine vessels, and manned or unmanned aircraft. This book, however, does not discuss these applications in detail. Some of these applications are in an experimental stage or are being field tested. Some of them are on the commercial market.

- ***Bikes, scooters, and wheelchairs***. There are a number of bikes, scooters, and wheelchairs that are powered by fuel cells. For example, Intelligent Energy has unveiled some.
- *ENV*. In February 2005, Intelligent Energy presented the world's first purpose-built fuel cell motorbike ENV.[542] The ENV was powered by a 6 kW 48 V motor, with energy supplied by Intelligent Energy's 1 kW PEMFC fuel cell.[543] It could be refilled with

(a) Plug Power, 2005-2012* **

*The 2012 figure shows that of the first quarter of 2012.

(Millions of $)

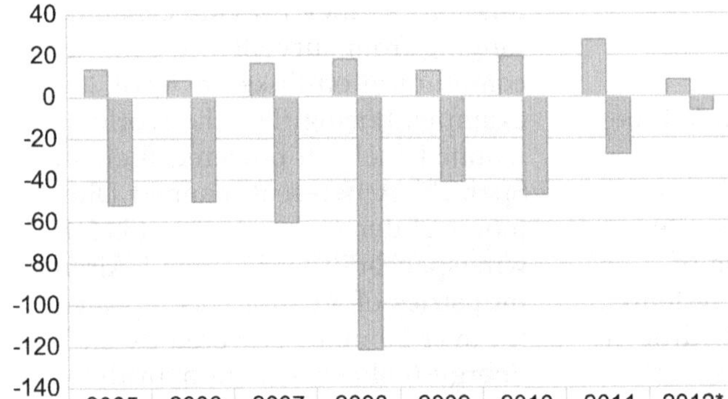

	2005	2006	2007	2008	2009	2010	2011	2012*
Revenues	13.5	7.8	16.27	17.9	12.3	19.5	27.5	7.8
Net Loss	-51.7	-50.31	-60.57	-121.7	-40.7	-47	-27.6	-6.6

The figure was creased based on Plug Power's annual financial reports

(b) Ballard Power System, 2005-2012*

*The 2012 figure shows that of the first three-months as of 31 March 2012.

(Millions of $)

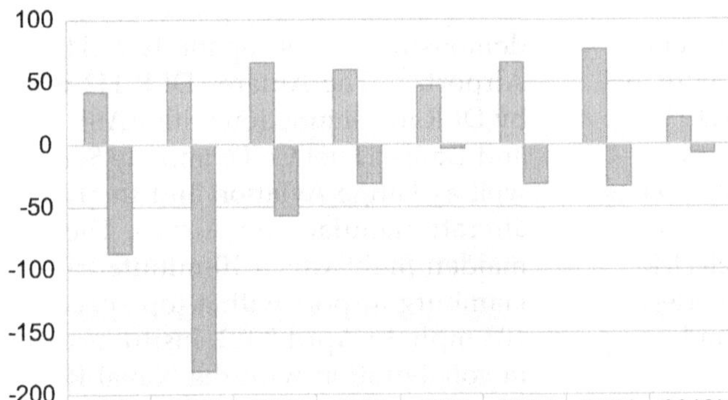

	2005	2006	2007	2008	2009	2010	2011	2012*
Revenues	42.25	49.82	65.53	59.58	46.72	65.019	76.009	21
Net Loss	-86.98	-181.14	-57.3	-31.46	-3.258	-31.532	-33.42	-7.3

The figure was creased based on Ballard's annual financial reports

FIGURE 7.8 Financial performance: Plug Power and Ballard (million US dollars). The figure displays both revenues and net loss to illustrate two sides of one picture. The net loss is the residual that the revenues could not offset. For example, Plug Power had revenues of $13.5 million in 2005, which cancelled out the same amount of loss it had, leaving $51.7 million as the net loss. (For the color version of this figure, the reader is referred to the online version of this book.)

hydrogen in less than 5 min, made virtually no noise, had a top speed of 50 mph, and a range of 100 miles.[544]

- ***Suzuki crosscage motorbike.*** In November 2007, Intelligent Energy and Suzuki unveiled the Suzuki Crosscage Hydrogen Concept bike at the Tokyo Motor show.[543] The motorbike was powered by Intelligent Energy's PEMFC pack hybridized with an electric battery and had a range of 62 miles on a tank of compressed hydrogen.[545]
- ***Suzuki senior car MIO.*** In September 2006, Intelligent Energy and Suzuki presented a fuel cell senior car MIO.[546] The prototype electric

wheelchair was powered by a DMFC fueled by methanol and ran about 24 miles with a tank of methanol.[547] Suzuki presented an updated MIO at the 2009 Tokyo Motor Show.[548]

- ***Suzuki Burgman fuel cell scooter.*** In November 2009 Tokyo Motor show, Intelligent Energy and Suzuki presented the Suzuki Burgman fuel cell scooter.[544] Improved on the Crosscage fuel cell motorbike, the Burgman Fuel Cell Scooter was equipped with the latest version of Intelligent Energy's fuel cell power systems. Fuel was provided via a cylinder of hydrogen and gave a riding range of 217 miles, comparable to a conventional Burgman scooter. In March 2011, the fuel cell scooter obtained Whole Vehicle Type Approval, which certified the scooter design as safe to use on public roads without having to be inspected and tested individually.[549]
- ***Ships and boats***. In August 2008, Hamburg launched the FCS Alsterwasser, a fuel cell–powered 100-passenger inland waterway tourist vessel.[550] The vessel suffered a fire in 2010, but was back in service in June 2011.[551] Cause of the fire allegedly was traced to incorrectly connected batteries. The vessel has two 50 kW (67 hp) proton motor fuel cells powering a 100 kW (134 hp) hybrid electric propulsion system with lead acid batteries. Using fuel cells as propulsion for inland waterway craft such as Hamburg's Alsterwasser is a logical first step leading toward more deployments. However, the application of fuel cells for propulsion in ships in the open ocean would require further technological maturity and development of capabilities that could withstand rough open-ocean conditions.
- ***Manned Aircraft and UAVs***. Fuel cell aircraft are gaining increasing popularity. In October 2009, the Naval Research Laboratory's Ion Tiger, an unmanned aerial vehicle (UAV) powered by a 550 W Protonex fuel cell, flew 23 h and 17 min, setting an unofficial flight endurance record for a fuel cell–powered flight.[558] Small UAVs are growing in importance for naval missions as they provide capabilities ranging from surveillance collection to communication links. In April 2008, for example, Boeing flew the world's first manned fuel cell powered flights in Spain.[559] A two-seat manned Dimona light motor-glider with a 16.3 m (53.5 ft) wingspan built by Diamond Aircraft Industries of Austria was repeatedly flown for over 20 miles powered by an Intelligent Energy fuel cell system hybridized with lithium-ion batteries to power an electric motor coupled to a conventional propeller.[560] The aircraft used the lithium-ion batteries for takeoff, but switched to using the PEM fuel cell system for flight cruise. In July 2009, the Antares DLR-H2, the world's first piloted aircraft capable of taking off using only power from fuel cells, demonstrated this capability at Hamburg Airport.[561] The Antares DLR-H2 was built by DLR in conjunction with BASF Fuel Cells and Danish Fuel Cell company Serenergy as well as Lange Aviation and the European aircraft manufacturer Airbus. The aircraft's maiden flight was a 10-minute trip from Hamburg airport with a top speed of 105 mph. In April 2012, Insitu Incorporated, in collaboration with the Naval Research Laboratory, flew a UAV, the ScanEagle, for 2 h, which was powered by a single 1.5 kW hydrogen fuel cell provided by UTC.[562]
- ***Trains***. The benefits of using fuel cells onboard trains are unclear, but a few countries have made attempts to develop a fuel cell rail system. In 2003, Japan Railway Technical Research Institute with government support tested a 100 kW fuel cell rail system made by Nuvera fuel cells following 2 years of development.[563] The project was mothballed or stalled through lack of interest or funds.[564]

7.5.2. Stationary Applications Shore up Only Two Notable Markets

Stationary power includes those operated at a fixed location, either for primary or for backup power, or for CHP.[565] At present, there are only two notable markets for this application. They are small residential CHP systems and backup power systems for telecommunications industries. The market for residential CHP units is emerging in Japan, perhaps facilitated by government subsidy programs. The market for backup power products appears to be emerging in the United States through popular demand owing to the products' technology advantages and cost competitiveness. Both markets are at an early stage of development, and only a handful of fuel cell developers are playing in each. None of the fuel cell developers appear to be making profits.

7.5.2.1. *Small Residential Combined Heat and Power Market in Japan Sustained by Government Subsidies*

Small stationary units include several watts to 10 kW. Units with outputs of about 0.7–1 kW are most commonly deployed in Japan. The Japanese market for residential CHP is not self-sustaining; it is supported by the government, which subsidizes nearly 50 percent of the capital and installation cost of the systems.

Japan has implemented aggressive policies to develop, demonstrate, and commercialize PEM CHP technology. As a result, Japan has the world's largest residential PEMFC CHP market today. Since 1992 until the end of FY2011, the government had spent an estimated $1 billion for R&D, demonstration, and subsidies for promotion of residential PEMFC CHP.

Since 1992, Japan had spent at least $ 615.1 million for stationary PEMFC technology R&D and demonstrations alone:

- From 1992 to 2000, the government spent at least $34.5 million for R&D, consisting of $5.8 million for phase I (1992–1995) and $28.7 million for phase II (1996–2000).
- From 2000 to 2009, the government spent at least $452 million for R&D, including a $176-million R&D program (2000–2004) and a $276-million R&D program (2005–2009).
- From 2005 to 2009, it implemented a 4-year $128.6 million residential CHP demonstration program. A total of 3307 1 kW CHP units were deployed for demonstration. Five PEMFC CHP developers participated, including ENEOS Celltech (a joint venture of Nippon Oil and Sanyo established in 2008), Ebara, Toshiba, Panasonic (formerly Matsushita Electric Industrial), and Toyota.[566]

In 2009, the Japanese government started a 5-year subsidy program (FY2009–2013) to spur the commercialization of residential CHP systems, which will likely cost the government at least $485.4 million.

- METI had spent at least $263.4 million from FY2009 to the end of FY2011. METI allocated a budget of ¥6.07 billion (about $64.8 million at a 2009 exchange rate of $1/¥93.7) to subsidize CHP unit purchasers (typically new home buyers) for FY2009, ¥8 billion (about $88 million at a 2010 exchange rate of $1/¥90.76) for FY2010, and ¥9.07 billion (about $110.6 million at a 2011 exchange rate of $1/¥82) for FY2011.
- The subsidy program will continue another 2 years until the end of FY2013 (ending in March 2014), which would add perhaps another ¥20 billion, an estimated $222 million at an estimated exchange rate of $1/¥90.

Three CHP system developers entered the initial commercial market, including ENEOS Celltech, Panasonic, and Toshiba. Ebara and Toyota dropped out.

- The subsidy in 2009 would disburse about half the CHP system and installation cost

after subtracting the cost of a traditional hot-water heater (¥230,000 or $2455).[567] The subsidy was expected to decrease as the capital cost was also expected to decrease.
- The subsidies were appropriated to pay for 5030 units in FY2009 (ending March 2010), but only 5006 applications were filed. In 2010, a total of 7084 units were sold,[568] despite a limit on subsidies of 5127 units. In 2011, the applications reached the allocated number of 8133 units as of July 7, 2011, when the program ended for the fiscal year.[569] Total units sold reached 20,223 as of that point. The March 2011 earthquake and tsunami increased the CHP demand, and the expected number of CHP units to be sold in FY2011 was 11,490, leading to a likely total of 23,580.

METI implemented in 2009 a Feed-In-Tariff (FIT) system to encourage residential PEMFC CHP installation. Electric utilities would purchase from homeowners the extra power generated from fuel cell and/or solar power generation. There was an existing FIT system for solar power, but the new program doubled the tariff to ¥48/kWh ($0.51). The new system includes the purchase of combined solar and fuel cell power at ¥39/kWh ($0.42) (Fig. 7.9).

7.5.2.1.1. PANASONIC (FORMERLY MATSUSHITA ELECTRIC INDUSTRIAL)

Panasonic is Japan's largest residential PEMFC CHP system developer today. Formerly known as Matsushita Electric Industrial, the company was founded in 1918. Panasonic is a leading Japanese electric machinery and consumer electronics manufacturer headquartered in Osaka with its main electronics products under the brand names of Panasonic and Technics.[570] Panasonic started to develop residential 1 kW PEMFC CHP systems in 1999.[571] In December 2004, it announced that it would start marketing through Tokyo Gas the world's first commercial residential PEMFC CHP system.[572] Also, Panahome, Panasonic group's home-building operation, started to market houses with these units pre-installed.[573] During 2006, its system was awarded the Good Design Award for Ecology Design.[574] The PEM units were rated at 33 percent HHV electrical efficiency and 45 percent HHV thermal efficiency.

- In April 2005, Panasonic participated in METI's large-scale stationary PEMFC demonstration project (2005–2009).

In April 2008, the company developed a residential PEMFC CHP system that had the world's highest power-generating efficiency of up to 39 percent LHV.[575] The system had a durability of 40,000 operation hours and 4000 start-stop times, and a predicted lifetime of over 10 years. In FY2008, Panasonic suddenly became the second largest manufacturer in terms of the number of deployed CHP systems in the METI demonstration project. Prior to FY2008, Panasonic had ranked fourth among the five participants in the program.

- In June 2008, Panasonic unveiled a mass production facility at its Kusatsu factory in Shiga Prefecture.[576] The initial annual production capacity was 1000–2000 units, which could be expanded to an annual production of 50,000 units by FY2010 and 100,000 units by FY2015.[577]

Today, Panasonic is the largest residential CHP manufacturer in Japan, ahead of Toshiba and ENEOS Celltech. Since the launch of the commercial market in May 2009 until January 2011, Panasonic sold a total of 5000 units, of which approximately 4000 were sold through Tokyo Gas.[578]

In February 2011, the company developed an improved and less expensive model.[579] The new model provided 750 W of rated output instead of 1 kW and had an electric efficiency of 40 percent LHV and a durability of about 50,000 h. It cost ¥2,761,500 (about $33,676 at a 2011 exchange rate of $1/¥82), which was ¥700,000 ($8536) less than the previous price.

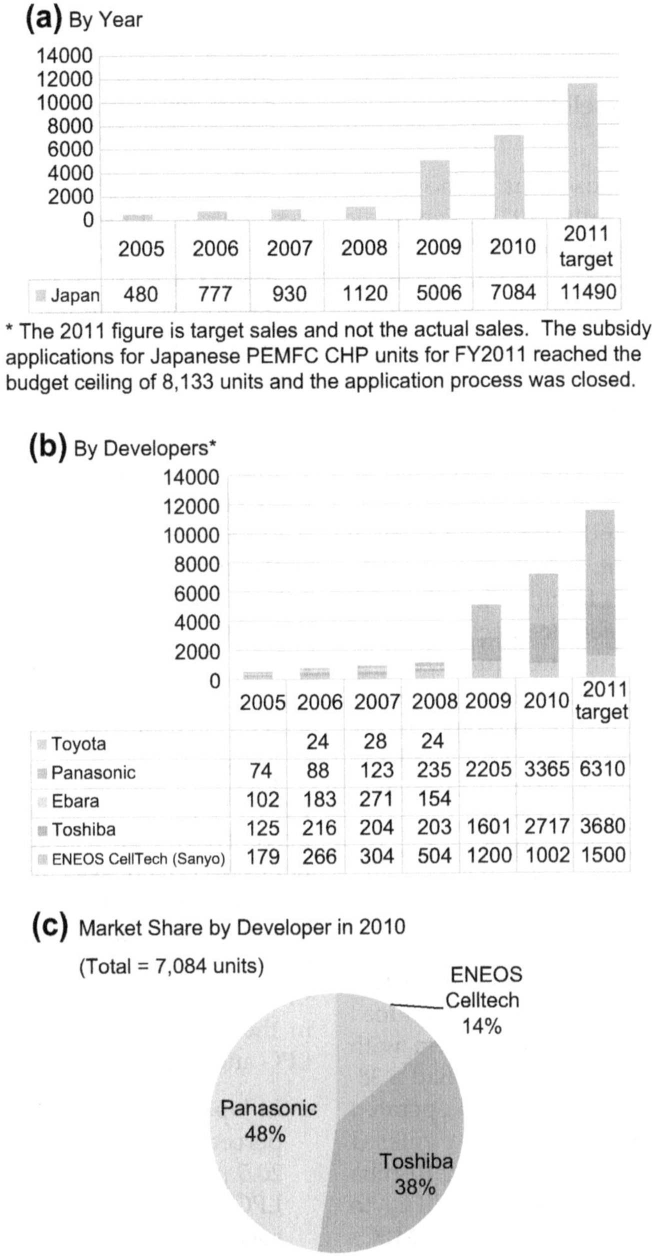

FIGURE 7.9 Deployment of residential PEMFC CHP units in Japan: 2005–2011*. *(b) The data was compiled based on various sources, including "On FY2010 Sales Record of ENE FARM Residential Fuel Cell Systems," Topics, http://www.digital-research.co.jp/enefarm_2010.pdf, Digital Research, 21 June 2011 and four residential fuel cell demonstration project reports for FY2005, FY2006, FY2007, and FY2008, presented at http://www.nef.or.jp/happyfc/pdf/17fc.pdf, http://www.nef.or.jp/happyfc/pdf/18fc.pdf, http://www.nef.or.jp/happyfc/pdf/19fc.pdf, and http://www.nef.or.jp/happyfc/pdf/20fc.pdf, New Energy Foundation, Japan.* (For the color version of this figure, the reader is referred to the online version of this book.)

The new model was to be on the market in April 2011.

- Panasonic planned to sell 6310 units in FY2011.

Panasonic has ambitious plans to move into overseas markets. In July 2011, Panasonic launched a fuel cell research center in Langen, Germany.[580] In April 2012, it revealed plans to establish another fuel cell research center in Cardiff, United Kingdom.[581] The company plans to market residential CHP systems in Western Europe, then the United States, and the developing world.[580]

7.5.2.1.2. TOSHIBA

Toshiba is the second largest residential PEMFC CHP system manufacturer in Japan today. The company was founded in 1938.[508] It is a leading Japanese electric machinery and electronics manufacturer headquartered in Tokyo. Toshiba first became interested in PAFCs and started R&D on the technology in the 1960s.[582] It built and tested 20 kW and 35 kW pressurized PAFC stacks in 1981 and formed with UTC of the United States a joint venture called International Fuel Cells (IFC) to develop PAFC plants for the Asian market in 1985.

In 1992, Toshiba started PEMFC development by participating in METI's early PEMFC development projects.[583] In 1999, Toshiba built a 1 kW fuel cell vending machine and in 2000 it introduced the first residential PEMFC CHP system with a rated output of 700 W.[584] The system had a 38-percent generating efficiency LHV and operated more than 6000 h.[585] In 2001, Toshiba entered another joint venture with UTC to form Toshiba International Fuel Cells Corporation (TIFC) to commercialize by 2008 residential PEMFCs fueled by LPG and natural gas.[574]

- In 2002, Toshiba developed a desulfurizer for kerosene and a kerosene reformer using catalysts developed by Idemitsu Kosan, a major Japanese petroleum refiner and retailer.[586] In March 2007, the company succeeded in building a PEMFC system fueled by kerosene and planned to commercialize a 1 kW-class kerosene-fueled PEMFC CHP system in FY2009.

In 2005, Toshiba participated in the METI residential PEMFC CHP demonstration project. During the following 4 years, Toshiba deployed a total of seven hundred and forty eight 700 W residential PEMFC CHP units, 22.6 percent of the 3307 total units that were demonstrated by the five participating PEMFC CHP manufacturers.

In 2009, Toshiba was one of the three CHP manufacturers that decided to enter the residential CHP commercial market, along with Panasonic and ENEOS Celltech. Toshiba sold 1601 units in FY2009 (about 32 percent of the total sales of 5006 units) and 2717 in FY2010 (38.4 percent) and planned to sell 3680 units in FY2011 (32 percent of the total sales target).

7.5.2.1.3. ENEOS CELLTECH

ENEOS Celltech was founded in 2008 by Nippon Oil (81 percent) and Sanyo Electric (19 percent) as the third largest developer of residential PEMFC CHP systems in Japan. ENEOS Celltech, however, is the only one in the world that has the capability to build PEMFC systems fueled by kerosene as well as natural gas and LPG. The other two Japanese PEMFC developers in the market—Panasonic and Toshiba—build LPG and natural-gas-fueled systems only.

- In Japanese households, as of 2008, 50.1 percent of all energy consumed is electricity, 20.5 percent is natural gas, 10.5 percent is LPG, 17.8 percent kerosene, and 1.1 percent solar and other energy sources.[587] These figures imply a considerable market for kerosene-powered CHP.

Nippon Oil, which became JX Nippon Oil and Energy by merging with Japan Energy in 2010, is a Mitsubishi Group company. It started PEMFC

R&D in 1986 focusing on catalysts that would reform petroleum fuels into hydrogen.[588] The company developed systems fueled by kerosene, LPG, natural gas, and naphtha with various developers, including Ebara Ballard, MHI, Sanyo, and Toshiba. The company placed LPG and kerosene-fueled residential CHP systems on the commercial market in 2005 and 2006.

- Nippon Oil developed a 5 kW naphtha-fueled power generation system using MHI stacks in 1999. Under the NEF stationary fuel cell demonstration project, the company conducted field testing of one system in 2003 in Shizuoka Prefecture and another in 2004 in Ishikawa Prefecture.[589]
- In 2000, Nippon Oil began the development of CHP systems fueled by LPG with Sanyo[590] and participated in METI's PEMFC systems development project (2000–2004) to develop a system fueled by gas-to-liquid fuels.[591] In February 2003, Nippon Oil installed twenty 1 kW LPG-fueled residential CHP systems jointly developed with Sanyo in Tokyo for field tests,[588] and in March 2005 started commercial sales of the LPG-fueled CHP systems.[592] In November 2007, Nippon Oil announced it would install a 1 kW LPG-fueled system at 22 homes in Ibaraki Prefecture. The contract fee per home was ¥60,000 a year (about $510 at the 2007 exchange rate of $1/¥117.7).[593]
- In 2003, Nippon Oil and Ebara Ballard started the development of a 1 kW-class kerosene-fueled residential CHP[594]and soon thereafter began the world's first field testing of a kerosene PEMFC unit.[595] In July 2006, Nippon Oil installed the first commercial kerosene-fueled residential CHP unit at a private residence in Aomori Prefecture.[596] The 1 kW system generated 950 W with an electric efficiency of 35 percent LHV and operated at temperatures below −10°C.[597] Nippon Oil was also developing a 10 kW kerosene-fueled unit with MHI and planned on commercializing the unit in 2007.[574] In June 2007, a 10 kW kerosene unit that was installed in 2005 at Hiroshima Diamond Hotel reached 10,000 h.[598]
- While engaging in commercial sales of residential CHP systems on its own, Nippon Oil in May 2005 also participated in METI's 4-year residential PEMFC demonstration project. Under the project, Nippon Oil installed a total of 1328 units (1062 LPG-fueled units and 191 natural gas-fueled units jointly developed with Sanyo, and 75 kerosene-fueled units jointly developed with Ebara Ballard).[599]

Sanyo, a leading Japanese electric machinery and battery manufacturer headquartered in Gunma, Japan, started PAFC R&D in the 1970s and developed a 220 kW PAFC power generation system for evaluation in 1987.[56] Sanyo also became interested in PEMFC technology in the early 1990s and participated in many METI PEMFC R&D projects. The company attempted and failed to commercialize a 1 kW PEMFC CHP system in 2005. It has remained a PEMFC supplier to Nippon Oil since then. In 2008, Nippon Oil bought out Sanyo's entire fuel cell business, which was incorporated into ENEOS Celltech with a minority investment from Sanyo.

- Sanyo participated in the METI's 8-year PEMFC system technology development project (1992–2000) and developed a 1 kW residential CHP system during Phase I (1992–1995) and a 5–10 kW portable power generation system during Phase II (1996–2000). In April 1998, Sanyo developed a 1 kW portable system for backup power, uninterruptible power sources, and auxiliary power for outdoor use.[600] The system generated 1 kW for 3 hours with a 10 l hydrogen container. In December 1999, the company developed a 1 kW residential CHP system fueled by natural gas.[601] In July 2002, Sanyo announced that it would develop within FY 2002 a 1 kW residential CHP system fueled by natural gas and planned to market it in 2005.[571] The system was to have an electric-generation efficiency of 35 percent and cost about¥500,000.[602] Sanyo, however, did not participate in METI's residential PEMFC CHP demonstration project, which started in 2005. Instead, it remained a supplier of stacks to Nippon Oil for 1 kW CHP units through 2008, when it sold its fuel cell business to the company.

In July 2010, Nippon Oil and Japan Energy merged and became JX Nippon Oil & Energy. JX has 34 percent of the Japanese oil market and has the goal of being a comprehensive energy provider.[306]

- Japan Energy and Toshiba jointly field tested 700 W LPG-fueled residential PEMFC CHP systems in Kanagawa Prefecture in January 2005.[604] In June 2005, Japan Energy participated in METI's 4-year residential PEMFC demonstration project (2005–2009) and deployed a total of 144 Toshiba LPG-fueled CHP units during the 4-year period.[599] The system had an electric efficiency of more than 30 percent HHV and a total efficiency of more than 65 percent HHV.[604]

In April 2009, ENEOS Celltech started building a mass-production plant for Ene Farm residential PEMFC CHP units.[605] The new plant was built at the Gunma Factory site of Sanyo Electric. It established an annual production capacity of 10,000 units in 2010, to ramp up to 40,000 units per year by 2015.[606] The company estimated that it would be able to produce 150,000 units from 2009 to 2015.[605]

7.5.2.1.4. NEAR-TERM SUCCESS OF JAPANESE SMALL RESIDENTIAL CHP NOT CERTAIN

The industry association, NEF, indicated that, for major market penetration CHP systems needed to be cheaper and more durable, indicating that Ene Farm's commercial success in the near term is not certain. It states that the cost must come down at least to ¥500,000-¥700,000 (about \$5500–\$7700 at the 2010 exchange rate of \$1/¥90.8) to compete with conventional power generation systems, and lifetime durability needs to get to 40,000–90,000 h.[253]

- An NEF survey showed that consumers had the same views. The most frequent critiques from the consumers were that the system was not reliable enough, too bulky, and not economical.[253]
- Current residential CHP consumers in Japan primarily consist of relatively wealthy homeowners with concerns about the environment and interest in advanced technology.[603]

7.5.2.2. Backup Power (UPS/Emergency Power/Remote Power) Market in the United States ...With Potential Success in the Near Term

During the past few years, a market for fuel cells for backup power applications is fast emerging in the United States. Backup power (or uninterruptible power) is a power source that is used in the event of a grid system failure. Backup power is essential for business for continuation of service and is especially important for telecommunications providers to ensure reliable, high-quality service with 24/7 availability.[607]

- There are a number of backup electrical power systems used for this purpose. The most common are batteries and generators operating on diesel, propane, or gasoline.[608] They have advantages, but the disadvantages are significant enough that service providers are searching for alternatives.[609]
- It appears that the market for backup power is vast. The worldwide market for data centre backup power is estimated to be in excess of \$6 billion and the market for telecom backup power is estimated to be \$2 to \$3 billion in the

United States alone, based on a complete replacement of existing products serving this market.[610]

For the past few years, fuel cells have demonstrated capabilities that could favorably compete with traditional technologies in backup power applications. One unique capability of fuel cells that far surpasses batteries is their long lifetimes when not in use. Batteries deteriorate and lose efficiency rather fast whether or not they are in use, but fuel cells do not deteriorate when not in operation and can remain in a ready-to-start mode indefinitely. Fuel cells also in general offer a longer service life than batteries even when in use. Today's commercially available PEM fuel cells perhaps are mature enough for low-power applications that require intermittent backup power.[608] Backup power requirements generally are less than 5 kW for a few hours to several days at a time, in contrast to the need for continuous operations for tens of thousands of hours, which is required for residential CHPs.

PEM fuel cell stacks in backup power applications today are expected to operate under real-world conditions for 5000–10,000 h.[603]

- With fewer moving parts, they require less maintenance than generators and batteries. They can also be monitored remotely, reducing maintenance time. They are quieter and have no emissions.

In addition, it has been indicated that PEM fuel cells could offer significant cost advantages over battery-generator and battery-only systems. One study, for example, indicated that fuel cell backup power was more cost competitive than conventional backup power sources, when short run-time applications were compared (up to three days). In a 2007 study for the US DOE, Battelle Memorial Institute examined lifecycle costs of emergency response radio towers, comparing fuel cells with 2 kW battery-only backup for 8 hours and 5 kW battery-generator backup for 52 hours, 72 hours, and 176 hours.[608] The study found that on a lifecycle basis, PEM fuel cells could provide service at substantially lower total cost than current technologies for service periods up to 72 h.

- When longer run times from 72 to 176 h are required, however, the fuel cell system incurred a higher cost. This was due to the cost of hydrogen storage tank rentals.

To alleviate the cost disadvantage, however, some fuel cell developers recently developed a new backup power system equipped with a fuel reformer that can reform methanol into hydrogen, eliminating the need for costly hydrogen and hydrogen storage.

- Nonetheless, a conclusive cost advantage has not been demonstrated. Few studies have been made that would compare fuel cell backup power with other new technologies that have been developed, including flywheels, ultra supercapacitors, and more advanced batteries (Table 7.11).

It is noteworthy that the market for backup power systems is growing fast in the United States. The market for backup power started around 2007 and has increased rapidly during the past several years. According to the annual reports published by Ballard, which likely is the global leader in the backup power market, the company shipped 200 fuel cell backup power units in 2007, 720 in 2008, 1225 units in 2009, and 1716 in 2010.[611]

- On the basis of various press reports and corporate annual reports, it appears that Ballard deployed a total of 3861 units by the end of 2010, ReliOn at least 1300 units, IdaTech 916 units, Altergy 700 units, and Hydrogenics 67 units. The total backup power sales for the past several years, thus, is estimated to be 6844 (Fig. 7.10).

TABLE 7.11 Net Present Value of Total Cost of Backup Power Systems for Emergency Response Radio Towers

	Outdoor installations			Indoor installations		
	Battery/ Generator	**PEM fuel cell with no tax incentive**	**PEM fuel cell with \$1 K/kW incentive**	**Battery only**	**PEM fuel cell with no tax incentive**	**PEM fuel cell with \$1 K/kW incentive**
8 hour run time				\$19,037	\$14,023	\$12,136
52 hour run time	\$61,082	\$61,326	\$56,609			
72 hour run time	\$47,318	\$33,901	\$32,014			
176 hour run time	\$75,575	\$100,209	\$95,491			

Notes:

1. Total cost includes capital costs and operations and maintenance costs.
2. As of October 2008, fuel cells are eligible for a tax credit of up to \$3,000/kW. As shown here, however, fuel cells can be cost competitive with the previous tax credit of \$1,000/kW. Calculations assume a 5 year battery replacement schedule.
3. Analysis of 3 year replacement schedules (for cold or harsh environments) indicates PEM fuel cells compare more favorably to traditional technologies.

Source: "Identification and Characterization of Near-term Direct Hydrogen Proton Exchange Membrane Fuel Cell Markets," Battelle Memorial Institute, April 2007.
Source: "Early Markets: Fuel Cells for Backup Power," Fuel Cell Technologies Program, http://www1.eere.energy.gov/hydrogenandfuelcells/education/pdfs/early_markets_backup_power.pdf, EERE, DOE, December 2010.

The rapid rise in demand for backup power has been triggered by three factors:

- First, the US power grid is becoming less reliable and this has created a demand for backup power. According to an academic study in 2006, the United States ranked toward the bottom among developed nations in terms of the reliability of electricity service.[612] The 2006 study indicated that customers in the Unites States lost power for an average of 214 min per year compared to 70 min in United Kingdom, 53 in France, 29 in the Netherlands, 6 in Japan, and only 2 min in Singapore. In Japan, the average customer lost power once every 20 years, and in the United States, it is once every 9 months, in addition to hurricanes and other strong storms.
- The growth in demand is also attributed to a recent US government mandate. In May 2007, the Federal Communications Commission (FCC) implemented recommendations of the FCC's Independent panel reviewing the impact of Hurricane Katrina on communications networks.[613] Specifically, the FCC mandated installation of emergency backup power for at least 8 hours for telephone service providers and commercial mobile service radio providers.
- Also in April 2009, DOE drew on ARRA to deploy fuel cell backup power.[614] The total DOE ARRA funding was \$41.9 million for all fuel cells, of which \$18.6 million was allocated for telecommunications backup power.[615] The \$18.6 million included \$2.7 million for Plug Power, \$7.3 million for Sprint Communications, and \$8.6 million for ReliOn. The number of backup power units

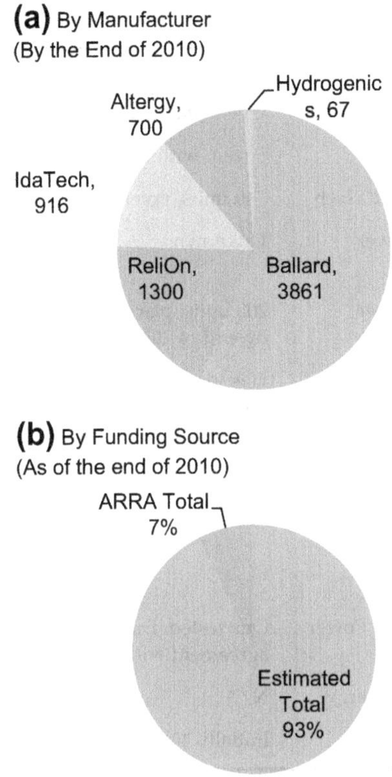

FIGURE 7.10 Estimated cumulative backup power deployment. (For the color version of this figure, the reader is referred to the online version of this book.)

deployed under the ARRA was estimated to be 539, including 20 GenSys units to be deployed by Plug Power,[616] 189 units by ReliOn,[617] and 330 units (260 new and 70 retrofit units) by Sprint Communications.[618]

Nonetheless, the proportion of backup power deployments under ARRA (539) compared to the cumulative total deployments at of the end of 2010 (6844) was a modest 7 percent.

After receiving major funding under the ARRA, however, Plug Power in October 2010 sold its backup power fuel cell business to Oregon-basedIdaTech, a leading developer and manufacturer of extended run backup power fuel cell products.[498]

In the rest of the world, there is only a modest demand for backup power where the national infrastructure lacks a modern power grid or landline telecommunications network, such as India, Indonesia, and South Africa (Table 7.12).[619]

7.5.2.2.1. BALLARD/DANTHERM

Ballard was founded in 1979 and is the world's leader in PEMFC technology. It has the largest gross revenues, the largest R&D expenditures, and the largest assets of any PEMFC technology company.[508] The company today focuses on backup power, distributed power generation, buses, and material handling.[620] The company had an interest in backup power for many years but only entered the market in 2007, a little later than other companies.

- As early as 2001, Ballard had developed a 60 kW engineering prototype stationary fuel cell power generator that used the Mark 900 automotive fuel cell stack.[19] The system was designed for backup and standby applications and has demonstrated synergies between stationary and automotive applications.
- In 2004, Ballard announced two new products that were in the laboratory demonstration phase.[494] They were a hydrogen-fueled 60 kW system for backup and standby applications and a 10 kW natural gas–fueled system designed for backup, light industrial and standby applications for telecommunications, and other applications. The products attracted little attention at the time.

In 2007, Ballard jumped back into the backup power systems market in a big way when it entered into a 2-year supply agreement with Dantherm, a European supplier of fuel cell systems for the telecommunications market.[621] Ballard supplied 646 kW of Mark 1020 ACS fuel cell products for integration into Dantherm's backup power products for European telecommunications customers.[622]

TABLE 7.12 Emerging Backup Power Markets in the Rest of the World: Players and Number of Units

Country	Partner Co.	Fuel cell Co.	# of units
Australia	Telstra	IdaTech	One installed at headquarters
Germany	O_2	P21 GmbH	Seven, with more planned
India	ACME Telepower	Ballard, IdaTech	310 units, potential for more
India	Telecommunications Consultants India Ltd. (TCIL)	Plug Power	Unknown
India	Wireless TT Info Services Limited (WTTIL)	Plug Power	200 units, planned to increase to 1000 by end of 2010
Indonesia	Working through Cascadiant, a network integrator, with several wireless companies, including PT Hutchinson CP Telecommunications	IdaTech	Has installed more than 100 systems around Indonesia, completed sale of 154 additional units
Indonesia	Working through Consistel, with several wireless companies, including PT Hutchinson CP Telecommunications and PT Telkomsel (Telekomunikasi Selular)	ReliOn	200 (PT Hutchinson)
Russia	Beeline	Dantherm Power	One tested. Dantherm signed agreement with Russian distributor
South Africa	Clean Energy Investments	Altergy Energy	N/A
South Africa	IST Holdings	Plug Power	Initially 400 with 120 added

Source: Sandra Curtin and Jennifer Gangi (Fuel Cells 2000), "The Business Case for Fuel Cells: Why Top Companies are Purchasing Fuel Cells Today," <http://www.fuelcells.org/BusinessCaseforFuelCells.pdf>, Fuel Cell 2000, September 2010.

- In 2007, Ballard/Dantherm shipped 200 fuel cell backup products.[621]

This agreement was followed in March 2008 by a 3 year supply agreement between Ballard andIdaTech, under which Ballard would provide its Mark1020 ACS product for integration into IdaTech's next generation 250 W iGen power supply, which was designed as an auxiliary power unit.[623] The agreement was expected to generate approximately 100 fuel cell product shipments from Ballard to IdaTech in the first year, with the potential for more than 1000 units over the full 3 year term of the agreement.[624] Also in October, 2008, Ballard signed a high-volume supply agreement with IdaTech and ACME Tele Power of India.[625] Subject to meeting product design and acceptance specifications, the agreement provided for the purchase of approximately 1000 fuel cell units in 2009 and 9000 units in 2010 for wireless base station backup power in India. The agreement with ACME of India, however, was terminated early. Ballard would continue its work with IdaTech as its primary indirect channel partner to develop backup power supply units for the telecommunications industry in India as well as other projects.[614]

- In 2008, Ballard shipped 720 backup power systems, a 260 percent increase over the previous year.[508]

In 2009, Ballard had another good year in the backup power business. In July 2009, Ballard announced that Motorola had deployed its

1.6 kW backup power system, which used the Ballard FCgen 1020 ACS stack, for Motorola's TETRA base stations throughout Denmark's SINE network.[626] Motorola was responsible for the establishment and operation of Denmark's nationwide SINE public safety network, which consisted of approximately 500 sites. Ballard fuel cells would provide backup power in 123 base stations across Denmark. Dantherm Power also installed a fuel cell backup power system for a municipal radio system in Columbia, South Carolina.[614]

- Ballard had a record shipment of 1225 backup power units in 2009.[627]

In 2010, Ballard exceeded its past sales record again. In view of the rapid growth in the backup power market in Europe, Ballard decided to establish a direct channel to the backup market by acquiring Dantherm Power, which had a well established access to the European market. [620] In 2010, Dantherm Power launched two new hydrogen-fueled systems utilizing Ballard's FCgen 1020ACS fuel cell product: the DBX 2000 (2 kW) and DBX 5000 (5 kW). In 2010, Ballard's indirect channel partner IdaTech also had a good year. IdaTech announced the sale of over 250 of their direct hydrogen products to Hutchison Telecom in Indonesia. IdaTech also launched a new 5 kW methanol fueled product that also contained a Ballard stack.

- In 2010, Ballard shipped a total of 1716 units,[627] a 40 percent increase over 2009.
- In April 2012, Ballard announced a 1 year extension of its equipment supply agreement with IdaTech.[628] It planned to support IdaTech's projected need for 2500 fuel cell stacks through the end of 2013 (Fig. 7.11).

Despite rapidly increasing sales of fuel cell backup power systems, the Ballard balance sheet has not yet shown profits. It is not certain when Ballard will become a major commercial success in backup power systems, at least in the near term.

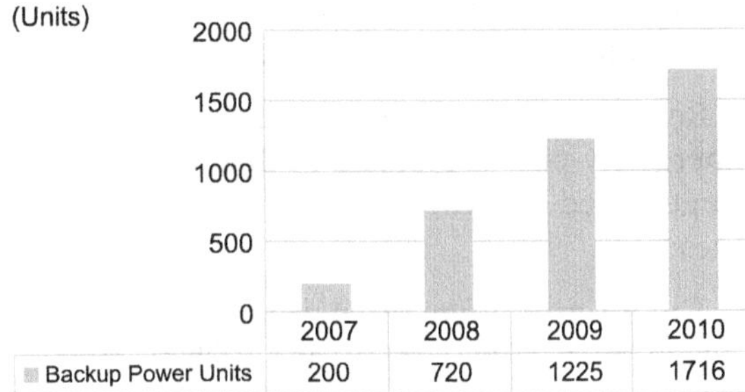

FIGURE 7.11 Ballard/Dantherm: Backup power shipments. (For the color version of this figure, the reader is referred to the online version of this book.)

7.5.2.2.2. RELION

ReliOn (formerly Avista Laboratories, Inc.) was established in 1995 as a subsidiary of Avista Corporation, formerly Washington Water Power of Spokane, Washington, to develop new energy technologies.[629] In November 1998, Avista began a 2.5 year project to pursue its unique technical approach toward development of a commercially viable PEM fuel cell.[13] By March 2000, the company obtained a comprehensive patent covering 162 claims for its PEMFC power system. In the summer of 2000 it deployed the first alpha test units to the field. In March 2003, the US Army Corps of Engineer's Construction Engineering Research Laboratory (CERL) successfully completed a 1 year demonstration of Avista's 3 kW fuel cell system in a base load application, achieving an availability rate of over 90 percent.[630] In 2003, the company started to sell commercial products.[630] In March 2004, Avista Labs changed its name to ReliOn to mark the transition from an R&D lab to a business entity.

- ReliOn's PEMFC technology features unique, modular cartridge construction, instead of a stack of cells, which serves as the basic building block.[13] The modular architecture is scalable to meet various design requirements. Cartridges also can be replaced without interrupting operation, eliminating potential single points of failure and offering easier servicing.[631]

ReliOn today specializes in backup and emergency power systems in the 50 W–20 kW range fueled by hydrogen.[632] It has two product lines:

- The E-series offers three products: the E-200 (175 W), E-1100 (1100 W) and E-2500 (2500 W) fuel cell systems. The E-series offers higher power in a smaller footprint.
- The T-series offers two products: the T-1000 (1200 W) and the T-2000 (2000 W). The T-series offers an increased feature set designed around hot-swappable maintenance.

ReliOn got a major commercial break in 2009. In April, it received an award from DOE's ARRA funding, totaling $8.6 million.[614] The ARRA funding was a cost share with two customers, AT&T and Pacific Gas & Electric (PG&E). Under the ARRA, ReliOn was expected to deploy a total of 385 units serving 270 sites at AT&A and PG&E sites,[617] which ReliOn fully completed by September 2011.[633]

- According to a news release on September 28, 2011, ReliOn had deployed a cumulative total of 1300 systems, serving sites in 38 US states and 28 countries.[634]
- The ARRA constituted 30 percent of ReliOn's cumulative total sales in terms of units, helping ReliOn to be a bigger player than IdaTech.

ReliOn's backup power systems might have a substantial advantage over some other systems. In 2010, ReliOn worked with hydrogen providers to roll out a bulk hydrogen refueling service for its fuel cell systems, providing ReliOn with a substantial advantage. Bulk hydrogen storage would give installations greater run time between refueling.

- Still, it is not certain if ReliOn would achieve major commercial success in the near term (Fig. 7.12).

FIGURE 7.12 ReliOn sales performance: 2003–2011. (For the color version of this figure, the reader is referred to the online version of this book.)

7.5.2.2.3. IDATECH

IdaTech specializes in extended-run time backup power fuel cell products for telecommunications.[635] It was founded 11 years ago as a unit of IDACORP, a Boise, Idaho-based energy and technology holding company. IdaTech was purchased in 2006 by Investec Group Investments Limited, a UK-based banking group that took the company public on the London Stock Exchange in 2007, and which holds 49.1 percent of company shares as of April 2011.[636]

IdaTech launched its line of commercial backup power products in 2007.[637] In October 2010, IdaTech acquired the assets of Plug Power's GenSys off-grid power systems and GenCore backup power fuel cell product lines.[498]

- IdaTech's main telecommunications backup power fuel cell product is the ElectraGen, which is available in both 3 kW and 5 kW units.[638] The system can be either fueled by hydrogen or can be equipped with a fuel reformer that converts methanol and water into hydrogen gas to power the system. The ElectraGen ME system, which has the methanol reformer, was commercially launched in 2010. By generating its own hydrogen, the ElectraGen ME does not require delivery or storage of bottled hydrogen for operations that would require backup power over long periods.[508]
- IdaTech also manufactures a 250 W iGen System designed to be integrated with

photovoltaic, wind, and grid power to recharge batteries in industrial applications, including traffic signaling and portable traffic signage.[638] The iGen System includes a fuel reformer that converts methanol and water into hydrogen.

In 2007, IdaTech installed 32 commercial backup systems; it more than doubled its sales to 83 units in 2008.[639] In 2009, IdaTech gained more customers, completed product certification tests, and sold 445 units, five times more than in 2008.[614] The company's goal in 2009 was to focus on achieving profitability in the near term. During 2010, IdaTech received orders for over 70 ElectraGen ME systems, which began shipping to customers in December.[508] Overall, the company sold a total of 350 commercial fuel cell systems in 2010. Since 2006, IdaTech has sold a total of 916 units.

- These backup systems were sold with no ARRA subsidies (Fig. 7.13).

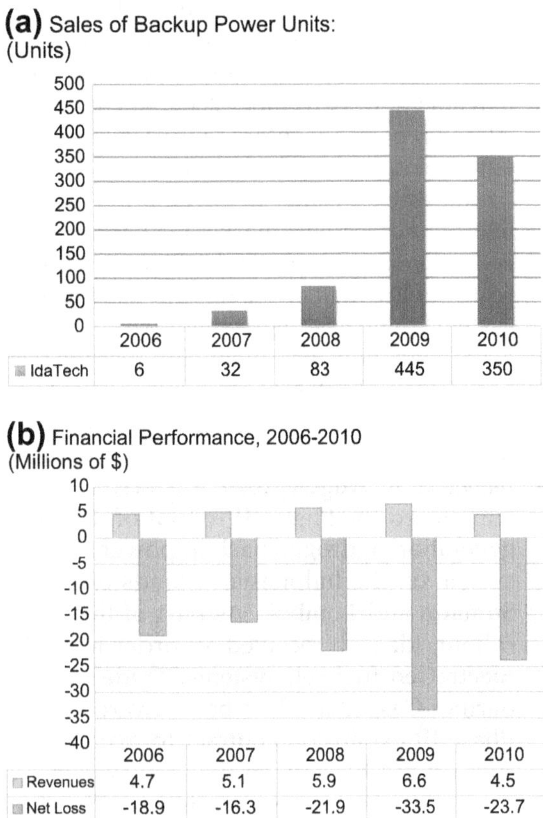

FIGURE 7.13 IdaTech financial performance: 2006–2010. Figure 7.13b is created based on IdaTech financial reports. (For the color version of this figure, the reader is referred to the online version of this book.)

IdaTech appears to have numerous market activities overseas, including Mexico, India, Indonesia, and Germany, but all of them are modest in size:

- *Mexico.* In October 2009, MICROM Electronics,IdaTech's Latin American partner, deployed 43IdaTech backup power units at Radiomovil DIPSA SA de CV (Telcel), Mexico's largest mobile phone carrier, in Mexico City.[640]
- *India.* It appeared thatIdaTech almost made a major inroad into India's telecommunications industry. In October 2008, the company signed a supply agreement to deliver up to 30,000 units to provide backup power for wireless communications towers in India.[641] The agreement called for IdaTech to supply 5 kW fuel cell systems (minimum order of 10,000) to the ACME Group, a leading supplier of passive infrastructure to the telecommunications industry in India. Initial orders for 310 hydrogen systems were scheduled for delivery in 2009. After completion of the core design of the natural-gas fuel cell system to be delivered under the agreement, deployment of 9690 systems was to commence in early 2010. Potential orders for 20,000 additional natural gas units were expected between 2011 and 2013. The agreement was a three-party contract betweenIdaTech, ACME, and Ballard, and Ballard was to supply the fuel cell stacks. The agreement was terminated in September 2009 without penalty, however.[642] It appears that

IdaTech was unable to deliver cost-competitive natural gas–powered backup systems in the specified timeline under the agreement. In November 2009, IdaTech and ACME reportedly entered into a distribution agreement, under which ACME and IdaTech would work together to develop the Indian telecommunications market for backup fuel cell systems, with particular focus on large-scale deployment of the natural gas fuel cell system.[643] No sales agreement was exchanged, however. It appears that IdaTech had delivered 310 hydrogen backup systems to ACME under the previous agreement, and no additional deployment has been made.

- *Indonesia.* In August 2010, IdaTech announced the Phase III rollout of its latest-generation hydrogen backup power fuel cell system on the Indonesian islands of Sumatra and Lombok. As part of this rollout, IdaTech received an order for 154 ElectraGen fuel cell systems.[644] IdaTech's partner Cascadiant had been awarded the Phase III expansion contract to install and support the ElectraGen fuel cell systems. Cascadiant had experience in installing fuel cell backup systems across Indonesia with several of the wireless operators. In March 2011, at a press conference attended by Indonesian Hutchison Telecommunications and Indonesia's Minister of Communication, it was announced that the backup systems at Hutchison Indonesia's network would use ElectraGen fuel cell systems instead of traditional diesel generators.[645]
- *Germany.* In July 2009, IdaTech received an order for 30 ElectraGen hydrogen fuel cell systems from its German OEM partner, b+w Electronic Systems for delivery in 2009 and early 2010.[646] These systems were to be deployed on the German TETRA network (Terrestrial Trunked Radio), telecom sites (indoor and container), as well in UPS applications. This order was in addition to an order for 30 systems received from b+w at the end of 2008. It appears that b+w Electronic Systems delivered most of the backup systems to the US Army in October 2009 in support of the Infrastructure Modernization program in Germany.[647] The first 10 were successfully installed in September 2009 in Grafenwoehr, Germany, and an additional 12 systems were to be deployed at US military sites throughout Germany before the end of 2009. In July 2009, the German National Hydrogen and Fuel Cell Technology Innovation Program (NIP) approved b+w fuel cell development projects as part of the NIP program,[646] in which IdaTech would participate through its partner b+w.[638] In September 2011, IdaTech announced that it had satisfied all requirements for European Conformity (CE) certification for its 5 kW hydrogen-based ElectraGen PEM fuel cell system.[648] The certification would now allow sales and operation ofIdaTech's ElectraGen and marked the opportunity for full-market deployment throughout Europe.

Although it is making a significant increase in commercial sales, IdaTech has incurred progressively larger losses each year.

7.5.2.2.4. ALTERGY

Altergy specializes in PEMFC backup power for telecommunications and UPS systems for institutions such as the banking industry. It is a privately held California corporation formed in 2001.[649]

Altergy began commercial delivery of fuel cell systems in 2005. In October, the company leased a new, large facility in Folsom, California, which is five times as large as the current facility, to house Altergy's corporate headquarters, sales, marketing, engineering, supply chain, and manufacturing operations.[650] By 2007, it had launched two major product lines and established the world's first automatic fuel cell assembly lines:

- In November 2005, Altergy announced that it had been producing two new commercial backup power systems: the 1 kW Freedom Power FPS-1 and the 5 kW Freedom Power FPS-5.[651] The units ran on gaseous hydrogen, could be used indoors or outdoors, and could be linked in series to deliver up to 30 kW.[574]
- In June 2007, Altergy established the world's first and only automated, high-volume fuel cell assembly line.[652] In November 2010, moreover, it expanded the capacity of its automated fuel cell assembly line.[653] With this expansion, Altergy had the capability to produce many thousands of units per year.[574]

Altergy established major domestic and international distribution networks by signing agreements with a number of US manufacturers to market its backup power systems:

- ***United States***. In September 2006, Altergy signed an agreement with Eaton Corporation, to incorporate Altergy's systems with Eaton's electrical control and power distribution products. Eaton would install, monitor, and service Altergy fuel cell products and systems at customer sites worldwide.[654] In March 2008, Altergy formed a global distribution and marketing alliance with EnerSys, the world's largest manufacturer of industrial batteries.[655] In December 2009, EnerSys invested an undisclosed amount in Altergy and also agreed to expand joint sales and development activity.[656] Under this expanded commercial alliance, EnerSys would be the exclusive distributor of the Extended Run Time Solution standby power system for specified markets and regions, including both telecommunications and uninterruptable power supplies.
- ***United Arab Emirates***. In April 2008, Altergy signed agreement with Spectrum Petroleum Services in the United Arab Emirates to market Altergy's PEMFC systems to businesses in the Middle East.[657]
- ***India***. In November 2010, the company entered into a collaboration agreement with TransAct Energy, which would distribute Altergy's hydrogen fuel cell backup power systems to the telecommunication industry in India.[658] TransAct would also have the option to manufacture its own hydrogen fuel cell generators using Altergy technology under a separate license agreement.
- ***Jamaica***. In December 2010, Altergy installed a 5 kW fuel cell system on a rooftop installation in Saint Andrew Plaza in Jamaica.[659] The system was designed and certified to provide uninterrupted backup power for 48 hours. In July 2011, the company announced that the system had been designed to be expanded to 10 kW if required by the client.[660]
- ***South Africa***. In August 2010, Altergy announced that Altergy, Anglo Platinum's Platinum Growth Metals Development Fund (PGM), and the government of the Republic of South Africa through its Department of Science and Technology (DST) had formed Clean Energy Investments, a company headquartered in South Africa whose principal objective was to manufacture and market Altergy fuel cell systems in the Republic of South Africa and other Sub-Saharan countries.[661] PGM and DST each invested undisclosed amounts in Clean Energy Investments, which would be jointly owned by Altergy, PGM, and the South African government. Clean Energy Investments initially would establish a distribution network for Altergy's fuel cell products throughout the sub-Saharan region.[662] Following the successful launch on the market, Clean Energy would establish a fuel cell manufacturing and assembly

plant in South Africa under license from Altergy. That facility would supply the sub-Saharan markets with Altergy fuel cell products manufactured locally.

Through these commercial and business networks, Altergy has made noteworthy sales, including the following:

- In April 2009, it announced that it had shipped more than a million watts of its Freedom Power Systems in the quarter ending March 31, 2009.[663] The majority of the sales were to Eaton.[574]
- In November 2009, Altergy announced the completion of deployment of more than 330 of its fuel cell backup systems in southern Florida for MetroPCS.[664] The deployment was believed to be the largest single deployment of fuel cells for the telecommunications market in US history.

Because Altergy is a private company and makes no public financial disclosures, it is difficult to assess the degree of its success. One foreign press report available as of April 2011 indicated that Altergy had installed 3.5 MW of its Freedom Power Fuel Cell Systems, ranging from 5 to 30 kW output at about 300 sites.[665]

- This report seems credible. The company had already sold more than 1 MW of its products by March 2009 and could have sold an additional 2.5 MW during the following 2 years. Indeed, major sales during the period included the deployment of 330 systems in Florida in November 2009.
- It is also possible that Altergy had installed up to 700 systems by April 2011. Its 5 kW systems were installed most frequently, and 3.5 MW worth of capacity would equal roughly 700 systems.

Altergy thus could have sold up to 700 systems since the launch of its commercial activities. This is a good record for a start-up company.

7.5.2.2.5. HYDROGENICS

Hydrogenics produces hydrogen-generation products based on water electrolysis technology and fuel cell power generation systems based on PEMFC technology. It was formed in 1988 and is headquartered in Mississauga, Canada.[666] In 1995, it entered the fuel cell technology development business and in 2000 changed its name from Traduction Militech to Hydrogenics. In 2005, Hydrogenics acquired Stuart Energy, a leading manufacturer of hydrogen-generation products based on alkaline electrolyte technology with experience in the field since 1948.[667]

- In 2007, the company narrowed the focus of its fuel cell activities by exiting the fuel cell testing business and concentrating on forklift power and backup power markets.[668]

In October 2001, Hydrogenics unveiled a regenerative backup power generator at the Las Vegas Telecom Show.[669] The Hydrogenics HyUPS fuel cell power generator was codeveloped by Hydrogenics and GM to replace diesel gensets and battery banks that were providing backup for telecom towers. The prototype unit was capable of delivering 25 kW of power and had regenerative-fueling capabilities. An integrated electrolyzer recharged the hydrogen supply, essentially providing a low-maintenance self-fueling system.

- In August 2002, one HyUPS system was installed at a trial site provided by Nextel Communications in northern California.[670] In November 2003, the company standardized the backup power system around a 10 kW HyPM fuel cell module, which according to the company, would enable it to manufacture the unit in greater numbers at lower cost.[671]
- The company also has 4, 8, and 12 kW HyPM systems, as well as a HyPM backup power systems ranging in output from 10 to 30 kW.[672]

In 2004, Hydrogenics appeared as if it had made early inroads in fuel cell backup power markets. In November 2004, Hydrogenics received an order for twenty five 10 kW HyPM modules from American Power Conversion (APC), a US manufacturer of uninterruptible power supplies, electronics peripherals, and data center products.[673] Hydrogenics modules were incorporated into its APC backup power products. It appears that the company delivered five units in 2004 and 20 units in 2005.[674] In August 2006, Hydrogenics and APC entered into a manufacturing and supply agreement.[675] Under the agreement, Hydrogenics would deliver up to 500 HyPM 12 kW modules for integration into APC products. The systems were scalable in increments of 10 kW up to 30 kW.

- In November 2006, Hydrogenics and APC jointly installed a 20 kW backup power unit for InterLink Connectivity in Toronto[676] and in August 2008 Hydrogenics announced that Schneider Electric (formerly APC) would soon market an extended-run backup power system powered by Hydrogenics' HyPM modules.[677]
- However, the deal with Schneider Electric apparently has not worked out well for Hydrogenics. In July 2009, Hydrogenics announced the termination of the agreement with Schneider.[678]

Nonetheless, Hydrogenics found another partner, CommScope, a multinational telecommunications company, with which it agreed to collaborate on backup power systems. In October 2008, Hydrogenics announced that it would deliver fuel cell power modules to CommScope for integration into CommScope's extended run and backup power systems for wireless and wireline operators.[679] In August 2010, Hydrogenics formed a strategic alliance with CommScope, in which the company would jointly develop next-generation fuel cell modules for telecom-related backup power applications.[680] Under the alliance, CommScope would invest $8.5 million in Hydrogenics as part of a joint product development program[610]:

- According to Hydrogenics, in 2010 the company was developing four new fuel cell backup power products with CommScope, which would be offered during 2011. The company reported that it had already tested the product in the United States, India, Mexico, Canada, and Europe.
- According to product literature, CommScope products incorporating Hydrogenics' fuel cell modules include the EcoPower cabinet, which is designed to generate up to 16 kW with two Hydrogenics 8 kW HyPM modules.[681] It could be configured with one 8 kW fuel cell unit in the factory and the second 8 kW unit could be added in the field.
- Hydrogenics expected commercial orders for these products in 2011.

Although it has been marketing backup power products since 2001, the company perhaps has not made many large sales. According to annual reports, while it might not be exhaustive, Hydrogenics sold at least 67 backup systems since 2001:

- In 2002, it installed one system at Nextel Communications in northern California.
- In 2004 and 2005, it delivered a total of 25 units.
- In 2006, it delivered one 8 kW fuel cell backup power generator to a Bell Canada telecommunication site in Burlington, Ontario,[682] and received orders for three 12 kW backup modules to be incorporated into a backup system by a leading mobile telecom company in Asia.[683]
- The company annual reports indicate that 13 units were delivered in 2008,[684] 18 units in 2009,[685] and seven units in 2010.[610]

Recently, Hydrogenics won a contract for a novel backup power applications. In October 2011, the company was asked to supply a hydrogen production, storage, and fuel cell system to the German city of Herten in North Rhine Westphalia.[686] Herten city had been searching for a means to manage intermittent electricity generation from wind-power farms, a key challenge for grid operators. City officials decided that this problem could be best addressed by using a Hydrogenics' HySTAT 30 hydrogen generator to electrolyze water, storing the resulting hydrogen and then converting it back to electricity using a Hydrogenics' HyPM 50 kW fuel cell power system when wind power was not being generated.

The company's performance in cutting losses over the last several years has been remarkable: it reportedly lost $130 million in 2006, $28 million in 2007, $14 million in 2008, $9 million in 2009, and $8.6 million in 2010. The company is now doing so well that it appears that it is on its way to commercial success. In view of the recent major business orders for new backup power applications, it is possible that Hydrogenics will make a profit in the near term. Nonetheless, it is also possible that the company's financial success could be attributable to other businesses, such as electrolyzers, hydrogen storage, and fueling stations rather than fuel cells themselves (Fig. 7.14).

7.5.3. Portable Fuel Cell Applications

Portable fuel cell applications are those that are designed to be moved.[687] These systems are small and grid independent and usually range from a few watts to a few kW. Portable applications include consumer electronic applications (cell phones, laptop computers, camcorders, and similar electronic devices), auxiliary power applications (APUs) for camping vans, mobile homes, boats, cars, trucks for long-distance hauling, and power systems for toy and educational applications.

- Fuel cell products for the military use are often categorized in this application area. These applications, for the most part, are ruggedized versions of the same portable fuel cell and APU products that are made for pre-commercial or commercial markets by the same manufacturers. Military applications are noted in this section where appropriate.

7.5.3.1. Consumer Electronic Devices Not Yet Commercially Viable

Over the past decade, portable fuel cell products—particularly for consumer electronic devices—have been viewed as the most promising fuel cell application to achieve an early

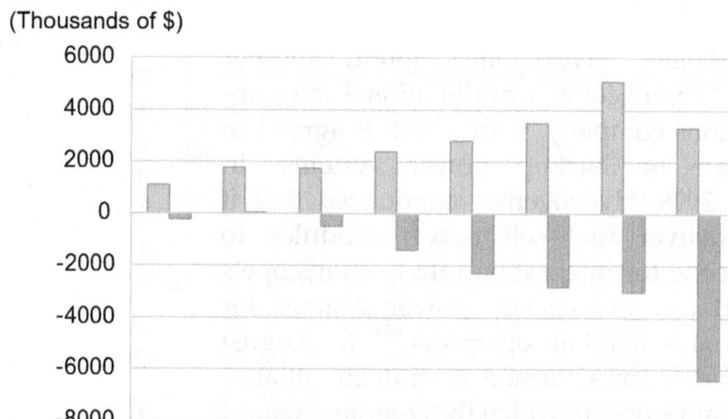

	2004	2005	2006	2007	2008	2009	2010	2011
Revenues	1095	1781	1768	2393	2814	3499	5102	3300
Net Loss or profit	-237	5	-480	-1392	-2293	-2806	-2998	-6400

FIGURE 7.14 Hydrogenics: Financial performance, 2004–2011. The figure is created based on Hydrogenics financial reports. (For the color version of this figure, the reader is referred to the online version of this book.)

market breakthrough. Portable electronics devices have progressed from primary (disposable) to secondary (rechargeable) batteries. The industry has been expected that fuel cells would soon start replacing lithium-ion batteries or other rechargeable batteries, since these battery systems are not well suited for high-power and long-lifespan portable devices.[688] In this application, fuel cells would be integrated in consumer electronic devices, such as mobile phones and laptops, as a replacement for batteries. As cell phones and laptops become more sophisticated and power-hungry, there is a need for power generation technologies that can keep up with the growing demand of these products.[373]

- For the past several years, many electronics manufacturers have been paying great attention to portable device applications. A number of prototypes have been developed, ranging from Casio's laptops, DoCoMo's mobile phone handsets, Hitachi's PDA, to Nokia's mobile phones. Nonetheless, few manufacturers have marketed a consumer electronic device that integrated a fuel cell power system. Some, however, plan to start selling portable fuel cells in the near term, including Lilliputian Systems.[689]
- It appears that product development has been blocked by major challenges. These include power density, miniaturization, system integration, fluid management, and cost.

While working to overcome these difficulties, many manufacturers are looking at perhaps slightly less-demanding alternatives. These include using fuel cells externally to an electronic device for charging the device's batteries. In this application, the fuel cell is used as a "range extender" or as an APU for the phone's battery to extend its run time.[373] Moving away from integrating fuel cells into electronic devices themselves and using fuel cells as external battery chargers appear to be the near-term strategy for both Toshiba and Sony.

- Still, few manufacturers have marketed fuel-cell battery chargers. The products perhaps are too expensive and not competitive with ever-improving batteries in the market.[19]

7.5.3.2. Major Success in Toys and Educational Systems … and Beyond

Fuel cell toys and educational kits are enjoying a remarkable market success today.[690] There is a secondary-effect market that stems from the need for education, demonstration, training, and hobbies.[19] Due to the large public interest in fuel cells, fuel cell toys and demonstration systems such as educational fuel cell kits are expected to create significant demand.

Indeed, fuel cell toys and educational kits have sold extraordinarily well. In 2008, of about 9000 portable fuel cell units that were shipped, most were toys and educational systems.[690] In 2010, shipments exceeded 100,000. Again, most of these were toys and educational kits.[691](Fig. 7.15)

- Toys and educational systems likely are sold without subsidies. This would be a clear commercial success and perhaps the only niche market that shows independent commercial strength. The net profitability of this market is not fully known, however.

There are two manufacturers of fuel cell toy and educational systems: Heliocentris and Horizon. Heliocentris generally makes demonstration kits for educational purposes, and Horizon makes fuel cell toys. However, both companies have made intriguing decisions, exploiting the strengths of fuel cells while bypassing the weaknesses.

7.5.3.2.1. HELIOCENTRIS

This Berlin-based start-up was founded in 1995[692] as one of the first providers of fuel cell educational systems with a worldwide distribution network.[19]

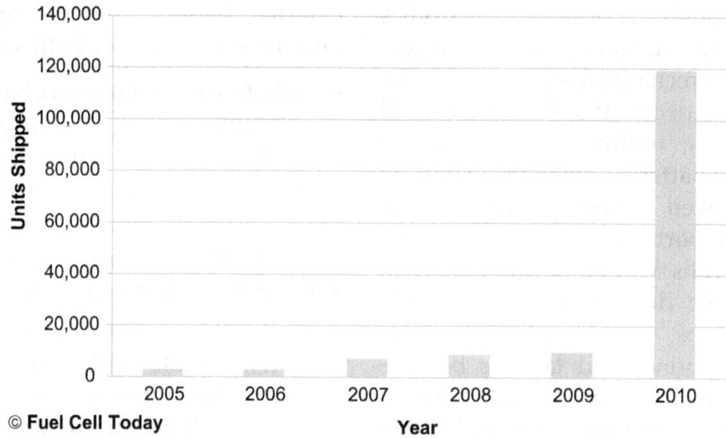

FIGURE 7.15 Portable fuel cell unit shipments including educational kits and toys: 2005–2010. *Source: "2010 Portable Fuel Cell Analysis, Snapshot Summary," Fuel Cell Today, 2010.* (For the color version of this figure, the reader is referred to the online version of this book).

7.5.3.2.1.1. EDUCATIONAL SYSTEMS In its early days, Heliocentris developed methanol-powered toy cars.[19] In 1999, it developed its first fuel cell learning system for universities.[693] At the 2000 Hannover fair, it presented a CHP demonstrator that incorporated a PEMFC powered by hydrogen, generating electricity and heat.[19] By 2002, the company was achieving annual sales exceeding €1 million.[693] Its export sales also increased fast, exceeding domestic sales in Germany for the first time in 2003 and accelerating the internationalization of the company. In 2004, Heliocentris made a major strategic decision; it ended in-house fuel cell stack development and transformed itself into a system integrator. It formed a partnership with Ballard to receive fuel cell stack supplies. In 2005, Heliocentris achieved profitability, with sales of €1.78 million and a net gain of €0.5 million. This, however, has not been duplicated since then. In 2009, the company moved to further expand its education system business. It reached an OEM-agreement with Lucas-Nulle, a global leader in the training systems industry, to enter the worldwide higher-education market. In 2011, it made an OEM-agreement with LabVolt, a global leader in manufacturing training laboratories for academia, industry, and the military, to access the worldwide technical education market. In October 2011, Heliocentris reported that it had won an order from Qatar University to supply a complete teaching laboratory for energy generation, storage, and management.[694] The Heliocentris system would combine energy generation from solar, wind, and fuel cells with energy storage in batteries and hydrogen technology to simulate an autonomous hybrid energy system. The price of the educational system was in excess of €100,000 (about $135,000 at a 2011 exchange rate of 1$/€0.74). The German-based company is also opening an office in Dubai to boost its presence in the Middle East.

In 2007, Heliocentris made a major strategic move; it decided to diversify its businesses beyond educational systems. The decision was insightful and the business direction it took was exactly where the fuel cell industry is heading. Although some applications are non-portable, they are discussed here because they are noteworthy.

7.5.3.2.1.2. STATIONARY UPS/BACKUP POWER APPLICATIONS In 2007, Heliocentris signed a partnership agreement with Hydrogenics and SMA Solar Technologie.[695] Under the agreement with Hydrogenics, Heliocentris would incorporate Hydrogenics' fuel cell power modules into stationary backup power systems. The products would have a power range from 4 kW to 16 kW. In September 2007, in collaboration with SMA Solar Technologie, Heliocentris developed a photovoltaic (PV) system with a fuel cell/battery hybrid backup power system.[696] The system would provide users of a PV installation, which feeds its electricity into the public network, with a stand-alone backup power supply in the event of a power failure. The system was based on the already successfully introduced battery-operated Sunny backup system from SMA, which was combined with a fuel cell system from Heliocentris. In December 2010, Heliocentris announced that it would also start an industrial field test of the "Nexa 1200" fuel cell system incorporating Ballard fuel cells.[697] The 1.2 kW Heliocentris Nexa would be integrated into different systems, ranging from indoor to outdoor stationary applications, including off-grid power, primary, emergency backup, and uninterruptible power supply units. The output ranged from a couple of hundred watts up to 10 kW and would be achieved through cascading the system's capability.

7.5.3.2.1.3. AUXILIARY POWER FOR TRANSPORT APPLICATIONS Heliocentris also incorporated Hydrogenics fuel cells for auxiliary power unit applications on-board vehicles. In April 2008, Heliocentris delivered a 32 kW fuel cell system to the city of Barth in Mecklenburg-West Pomerania to operate its battery/fuel cell hybrid midibus.[484] Also in August 2010, Heliocentris and Hydrogenics, in collaboration with FAUN Umwelttechnik, a major garbage collection and road sweeper vehicle manufacturer, developed a prototype hybrid waste disposal vehicle for BSR (Berliner Stadtreinigung), the largest municipal disposal company in Germany.[486]

- The company had a small profit in 2005, but losses since then. In view of the booming business activities, the reason for increasing loss could be attributed to high cost of materials and production (Fig. 7.16).

7.5.3.2.2. HORIZON FUEL CELL TECHNOLOGIES[508]

Horizon Fuel Cell Technologies, which was founded in 2003 and registered in Singapore,

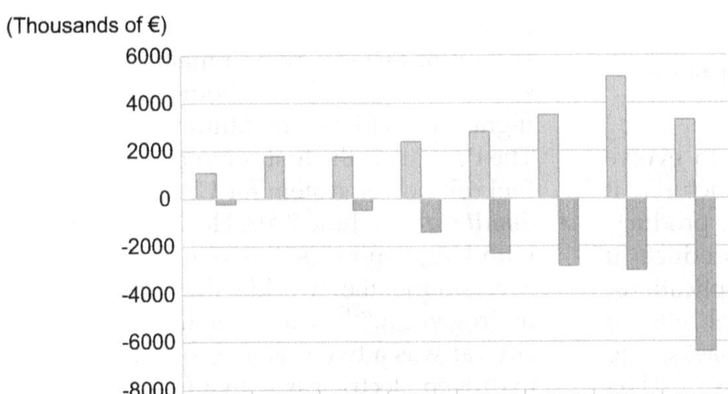

	2004	2005	2006	2007	2008	2009	2010	2011
Gross Revenue	1095	1781	1768	2393	2814	3499	5102	3300
Net Loss or profit	-237	5	-480	-1392	-2293	-2806	-2998	-6400

FIGURE 7.16 Heliocentris financial performance: 2004–2010. The figure is created based on Heliocentris financial reports. (For the color version of this figure, the reader is referred to the online version of this book.)

is a developer of small PEMFC systems based in Shanghai, China.[698] Like Heliocentris, Horizon is unique; it introduced a viable fuel cell technology into the market early and diversified into areas that took advantage of the strengths of fuel cells. Horizon also entered non-portable applications, which are discussed here because of their noteworthiness.

7.5.3.2.2.1. TOYS AND EDUCATIONAL KITS In June 2006, Horizon launched its "***H-Racer***" fuel cell toy car.[508] In November 2006, TIME magazine called the H-Racer the world's smallest hydrogen fuel cell car and one of the best inventions of the year.[699] In December 2006, Horizon unveiled the H-Cell, a power kit for 1:10 scale remote-controlled hobby cars, and has developed more than a dozen fuel cell and clean energy science kits.[508] Horizon has shipped various micro-fuel cell powered toys in the hundreds of thousands to over 60 countries. In February 2010, it introduced the world's first fuel cell hobby-grade radio-controlled vehicles, which could be refueled by model-scale hydrogen stations.[700] In March 2009, Horizon also introduced a new renewable energy tool kit.[701] The new kits include small wind-turbines, solar panels, and zero-emission vehicle construction sets.[701]

- Horizon claims that it is the world's largest volume producer of commercial PEM fuel cell products.[702]

Horizon also implemented plans to evolve beyond being a fuel cell toy manufacturer. It started to move from small and simple products to develop larger and more complex products. It now produces fuel cells for various applications, ranging from 0.5 W to 5 W micro fuel cells, 10 W−10 kW industrial grade stacks, to 10 W−2 kW aerospace grade systems.[614] They include consumer electronics, portable emergency power, stationary backup power, aerospace systems, and light vehicles.[703]

7.5.3.2.2.2. TRANSPORT APPLICATIONS INCLUDING UNMANNED AERIAL VEHICLES (UAVS) AND SMALL CARS Horizon diversified into transport applications, specializing in low-power applications for relatively short duration use.

- ***UAVs***. In April 2007, the German Air & Space Center (DLR) and a number of international partners, including Horizon, successfully developed the ***Hyfish***, a 5 kg[614] unmanned aerial vehicle (UAV), which flew over Bern, Switzerland.[704] The world's first hydrogen fuel cell jet performed vertical climbs, loops, and other aerial acrobatics at speeds reaching 200 km/h. In November 2007, Horizon announced that a new Horizon hydrogen fuel cell propulsion system incorporating Millennium Cell Hydrogen on Demand flew a micro UAV for 78 miles (128 km) in Lancaster, California.[705] The micro UAV was designed by two leading US aerospace research laboratories and supported by NASA, the Dryden Flight Research Center, the US Air Force Office of Scientific Research and the National Science Foundation. It set a new micro-UAV flight distance record; the previous record was 28 miles (40 km) set in 2006 in Estonia. In August 2010, Israel Aerospace Industries (IAI) and Horizon announced that Horizon's new PeroPak flew IAI's ***Bird Eye 650*** mini-unmanned aerial system (mini-UAS) for 6 hours, twice the flight time achieved by lithium batteries.[706] The PeroPak is the first commercial hydrogen fuel cell power system for UAVs.[707]
- ***Small Cars.*** In June 2009, Horizon developed with Hugo Spowers, the founder of Riversimple, the "world's first affordable hydrogen car."[708] The Riversimple small city car was a two-seater zero emission hydrogen-electric car with a fuel consumption equivalent to 250 miles per gallon (US). It integrated a 6 kW Horizon fuel cell with ultra super-capacitors and could

travel 240 miles (390 km) on one small tank of hydrogen weighing only 2.2 lbs (1 kg). The car could be rented for $315 per month.

7.5.3.2.2.3. PORTABLE AUXILIARY POWER APPLICATIONS FOR EMERGENCY/RECREATIONAL USES AND FOR POCKET-SIZE POWER CHARGERS In January 2008, Horizon unveiled a pre-production prototype, the HydroPak portable power generator combined with the Millennium Cell's Hydrogen on Demand storage system, which was designed to charge or operate low-power devices such as lighting, notebook computers, portable televisions, and ad hoc communication networks for more than 14 continuous hours.[709] In February 2008, it announced it would market the HydroPak in Japan within the year.[710] In June 2010, the company introduced the pocket-size MiniPak, which delivered 1.5–2 W of continuous power to recharge batteries with up to 12 Wh of net energy.[711] It was designed to extend the charge of small portable electronics, such as cell phones, smart phones, MP3/MP4, and GPS. The MiniPak would cost $99, but according to Horizon, at higher production levels, prices would be expected to drop to $29.

It has been reported that Horizon is making genuine commercial progress and generating revenue.[690] The fuel cell toy market sector may be a niche where market success perhaps is possible. Toys only require low power, intermittent and brief operations, and do not require high precision, efficiency, or durability.

7.5.3.3. SFC Energy: The Global Leader in Portable Auxiliary Power Unit Applications

The German manufacturer SFC Energy AG is the world leader in DMFC portable APU applications, which have become highly popular for supplying continuous power for battery recharging. SFC's APU systems are small, mobile, and portable, but not necessarily limited to portable applications only. SFC systems have been successfully commercialized for many applications, including the leisure sector (camping vans, cabins, and marine use), the industrial sector (power for remote locations including gas/oil well sites, communication towers, security, weather stations), and in the military sector.[712]

The success of SFC Energy is undoubtedly due to the company's insightful strategy. The company chose a specific fuel cell technology, DMFC, and developed it so that would exploit the strengths of the fuel cell while circumventing its weaknesses. No other company has taken this approach. SFC, thus, has no competitors.

- DMFCs are well suited for small applications that require continuous, low power for mobile and portable electrical devices.[603] In this role, SFC's DMFCs do not directly power the devices, but serve as an auxiliary power that recharges the battery that operates the electrical devices. SFC DMFC power output is small, ranging from 25 W to 90 W. It is SFC's intention to stay below the 1 kW boundary, where they see a "sweet spot" for this technology. As an auxiliary unit, SFC DMFC for the most part works in combination with a battery or a solar generator. The battery stores and dispenses energy as needed, including start-up and peak loads, while the DMFC continually recharges the battery, keeping it fully charged all the time.[713] In tandem with solar modules, the fuel cell would kick in if the solar module fails to produce enough power due to poor weather conditions.
- SFC chose DMFCs because it uses methanol and does not require hydrogen refueling. Methanol is a liquid fuel, ubiquitously available and easy to handle. Operated off-grid with 5–28 l methanol-fuel cartridges, DMFCs provide weeks of uninterrupted power without requiring user intervention.[714] One single 5 l cartridge provides enough fuel to run a surveillance

video camera system for 4 weeks—a major advantage over batteries, which have to be replaced every 10 hours. Methanol saves on refueling costs. In contrast, hydrogen is difficult to handle, is normally stored in heavy metal hydride cartridges, and manufacturing and refilling these cartridges is highly costly.[713]

It is interesting to note that the use of fuel cells as an APU has emerged in all three applications and that these APUs perhaps are the most commercially successful products today. Examples include APU range extenders in automobiles and buses (transport applications), backup power/UPS in the telecommunications industry (stationary applications), and portable APUs such as SFC's DMFC products (portable applications).

- SFC Energy likely is the only company in the world that has successfully commercialized DMFC APUs in many applications.

Because SFC is unique and few other companies have achieved its level of success with fuel cell products, other SFC applications are discussed below.

7.5.3.3.1. SFC ENERGY FOUNDED IN FEBRUARY 2000 NEAR MUNICH, GERMANY[715]

The company has had major sales successes since its founding. It reportedly sold over 21,000 systems and logged more than 9 million operating hours in end-user environments.[715] SFC products are offered as standard or optional equipment for many mobile applications, including motor homes and yachts globally.[508]

7.5.3.3.2. SFC COMMERCIALIZES ITS FIRST PORTABLE APU DMFC IN 2004

SFC developed and commercialized its products remarkably fast. In October 2001, a little more than a year and half after its establishment, SFC unveiled its first DMFC product prototype, the SFC A25. In August 2004, it started to sell the SFC A25, the first commercially available DMFC suitable for mobile, stationary, and portable applications.[716] The SFC A25 could deliver up to 25 W, enabling batteries to operate electrical equipment for daily use onboard mobile homes, including auxiliary heating, lighting, GPS, TV, and other digital systems. The SFC A25 could operate for four entire days with only one 2.2 kg methanol fuel cartridge. In August 2005, SFC introduced the SFC A50, a new DMFC power generator for mobile homes, at the Caravan Salon Dusseldorf, which is the leading trade fair for mobile homes. The SFC A50 could deliver 50 W and run without attention for several 1000 h.[717]

- The SFC A50 is now a standard option for the Hymer "S Class" motor home. It is also used as a backup power system for photovoltaic power-generation installations.

7.5.3.3.3. SFC INTRODUCES NEW FUEL CELL FAMILY EFOY DMFC IN 2006—WITH REMARKABLE MARKET SUCCESS

In May 2006, SFC introduced the EFOY 600, 1200, and 1600, which delivered charging capacities ranging from 600 to 1600 watt hours/day.[718] These units have had considerable success. The EFOY fuel cell is being offered as standard or optional equipment by 50 international motor home manufacturers.[508] In 2010, two yacht manufacturers—Arcona and Ridas Yachts—started to offer the EFOY as a standard option on all models. Rapido SAS, one of Europe's biggest motorhome manufacturers, is now offering EFOY as a standard option on its Series 10 models, and Notin, a French manufacturer, is pre-wiring all models for EFOY and will install one for no additional cost. As of May 2011, SFC had sold more than 20,000 units with no subsidies, mandates, or other government support.[603] SFC has a dealer infrastructure that sells SFC fuel cartridges at nearly 1000 sales points all over the world.[719] As of May 2012,

SFC has sold more than 24,000 units and has 1500 sales outlets in Europe.[720]

SFC's success has been recognized globally. In September 2010, Panasonic, which is a global DMFC developer, decided to incorporate SFC's DMFCs in its Toughbook laptop computer products.[721] Panasonic Toughbook computers are battery-operated laptops with various levels of ruggedness. The SFC DMFC would serve to extend the Toughbook battery runtime beyond its normal life. Panasonic had been developing DMFCs to power its laptops since late 1990s.[722] In January 2006, it unveiled a prototype DMFC that generated an average 13 W with a maximum power of 20 W. The system was used as a primary power source and could operate the laptop for 20 continuous hours with 200 cc of methanol. Panasonic was hoping to market the DMFC in 2010. In view of its partnership with SFC, however, Panasonic likely is not ready to commercialize its own DFMC system.

- It is noteworthy that Panasonic is prepared to deploy SFC's DMFC technology, which suggests it recognizes the global leadership status of SFC Energy.

7.5.3.3.4. SFC MAKES A STRATEGIC MOVE TO WIDEN AREAS OF APPLICATION

Having firmly established the utility of its basic technology, SFC has moved to widen its application to include professional, government, and military uses. Basically, SFC modifying the EFOY DMFC units to increase power output, ruggedize them, and reduce size as needed[690]:

- In May 2006, SFC offered the EFOY Pro 2200 to meet the needs of professional and government users.[508] It has 38 percent more power than the EFOY Pro 1600 and a charging capacity of 2200 W per day. It is designed for use in remote mobile and stationary applications where the units can deliver power for traffic control, security, and surveillance. Since 2009, the German Federal Office for Goods Transport has been using EFOY Pro in their official vehicles, eliminating the need for the vehicles to return to a re-charging point to power onboard electronics. In March 2010, the company launched a new model, the EFOY 2200 Pro XT, which would provide week-long constant power without requiring any user intervention to operate traffic and security cameras and traffic measurement and early warning sensors.[723] Also in March 2010, Altec, a French security company, announced it had integrated EFOY power supplies into their Sentinelle visual and thermal camera systems, where it could provide 20 days of unattended constant power.[724]
- In February 2009, SFC offered a new product line for the defence market. It launched the JENNY portable fuel cell system; a compact, autonomous, silent power supply worn by a soldier, which would supply power for the soldier's electrical devices in remote terrain and under rough environmental conditions.[725] In October 2010, SFC launched the next-generation JENNY ND Terra, a portable fuel cell system that could also be used as a stand-alone power source, including underground, if required for specific operations.[726] The system includes a DMFC, a 2.5 l methanol tank, a power management system, and a battery. It has a capacity of 2750 W and runs up to 4500 h.
- In October 2009, SFC unveiled the Emily 2200 direct methanol fuel cell, a long-lasting, reliable power supply for on- and off-vehicle defense applications.[727] Integrated into tactical vehicles or in the field, the fuel cell operates as a ruggedized fuel cell power generator. As an APU for military vehicles, the Emily 2200 fuel cell keeps vehicle batteries charged automatically, reliably, quietly, and virtually emission free. In operation, the unit is undetectable by sound or smell, and generates power almost imperceptibly— eliminating the need to start

a vehicle's engine to charge its batteries. Off-vehicle, the Emily 2200 provides power for mobile and stationary military applications. It can be combined with other alternative power sources like solar or wind. The Emily 2200 continues to produce maintenance-free electrical power for several weeks.

7.5.3.3.5. BUT SFC IS NOT YET PROFITABLE

Despite extraordinary sales successes, SFC has not yet become profitable. The company reportedly has been successfully reducing platinum loading by 75 percent to 80 percent and lowering total system costs by 65 percent over the past 4 years.[603] Still, there are no indications that cost reductions have improved the company's profitability. It is suggested that one factor that would contribute to future cost reductions would be scaling up assembly and supply operations. SFC is still not in a mass production mode and expects its products to be largely hand assembled for the next 5 years or so. Today, it requires two-person hours to assemble one unit. At its current facility, it could produce 10,000 units per year with existing equipment. It has space to expand to 20,000 units per year.

- It is not clear how SFC will achieve economies of scale. Even with better economies of scale, however, it is not certain when SFC would become profitable in the near term.
- The good news is that when SFC went public in May 2007, it raised enough capital to have financial stability for the next 20 years (Fig. 7.17).

7.5.4. Conclusion: An Unexpected and Disconcerting Trend

While hopes for broad market success of fuel cells endure, an unexpected and disconcerting trend has emerged in all areas of PEMFC and DMFC applications. The primary markets for PEMFCs and DMFCs are not progressing satisfactorily and instead niche markets for unexpected applications have started to blossom. For example:

- ***In transport applications***, the role of fuel cells as a source of primary motive power is

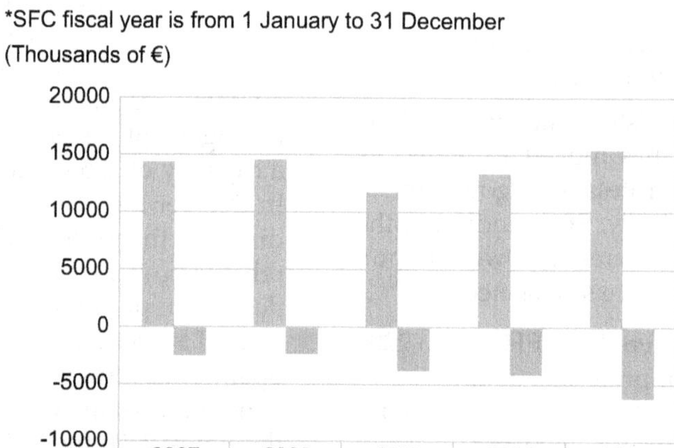

	2007	2008	2009	2010	2011
Revenues	14351	14554	11687	13330	15400
Net Loss	-2524	-2355	-3785	-4123	-6200

FIGURE 7.17 SFC energy financial performance: 2007–2011 (thousands of euros). The figure is created based on SFC Energy financial reports. (For the color version of this figure, the reader is referred to the online version of this book.)

looking less promising, and the use of fuel cells in an auxiliary power or range extender role for hybrid and plug-in hybrid cars and buses is becoming increasingly prevalent. Fuel cells also are becoming a popular choice to power material handing vehicles.

- ***In stationary applications***, fuel cells in CHP applications entered the commercial market in Japan in 2009, but have not become economically viable without government subsidies. On the other hand, fuel cells for backup stationary power applications appear to be fairly successful.
- ***In portable applications***, the introduction of consumer electronic devices powered by fuel cells has continued to be slow, and few fuel cell consumer electronic products are on the market today. On the other hand, fuel cell toys, educational systems, and auxiliary power units for recharging batteries in mobile homes, yachts, and vehicles are thriving markets.

7.5.4.1. Perhaps Current Fuel Cell Technology Is Only Adequate for Niche Market Applications

The trends suggest that current fuel cell technology is acceptable enough to meet power requirements that are called for in niche market applications. Fuel cells for these applications are only required to perform low-power operations for relatively short periods of time. There is great market potential, however, for toys, educational systems, and especially for portable APU applications.[687] Sales of PEMFC and DMFC units in this sector have outpaced all other fuel cell shipments over the past 5 years and are likely to continue to do so for the near future.

- The ability of fuel cells to compete successfully in the material handling and backup power markets also appears good. Near-term success in these markets, however, depends on achieving additional cost reductions and performance improvements.[687] Today, fuel cells in these markets are at a cost disadvantage as compared to established technologies, and the slim market shares achieved by fuel cells are highly dependent on government policies.[603]
- Another hopeful niche is an assortment of applications in the military market. The military continues to show a keen interest in experimenting with fuel cells because fuel cells have substantial operational value. Compared to batteries, fuel cells offer smaller package size and weight, longer running times, and a lower heat signature, making them more difficult to detect.[614] Fuel cells also deliver constant power output over time and can be refueled much more quickly than a battery can be recharged. The military has focused primarily on small PEMFC and DMFC in applications, including unmanned vehicles (both aerial and ground), soldier power, and material handling equipment, such as forklifts. Fuel cells for military use, for the most part, are ruggedized versions of the same fuel cell products that are made for commercial use. Near-term prospects for the military market are promising because the military, to a large extent, is more tolerant of cost and production time than is the commercial market.

7.5.4.2. And Not Mature Enough for Primary Market Applications

At the same time, the trend strongly suggests that PEMFCs and DMFCs are not sufficiently robust, efficient, and durable to serve as the primary power for cars and buses, CHPs, and consumer electronic devices. Today, they are unable to meet consumers' expectations for passenger cars' 100,000-mile operational life, which is the defacto standard for gasoline-powered automobiles.[11] Fuel cells are able to operate for 20,000 h in CHP applications in Japanese homes at present.[603] The durability for CHP applications must be at least 40,000 h to be economically viable. At the same time, fuel cells have not sufficiently evolved in terms of power density, miniaturization, and system integration to power consumer electric devices. Furthermore, current costs for fuel cells are still too high to be competitive in any of the primary market applications.

- These facts unfortunately suggest that current fuel cell technology is not sufficiently developed to compete in the prime-time commercial domain.

7.5.4.3. *Incremental Improvement Unlikely to Deliver Near-Term Commercial Success in Primary Markets*

In the long run, the biggest commercial opportunities for fuel cells would lie in the primary markets where fuel cells are used to provide primary power for cars and buses, CHPs, and consumer electronic devices, and not as APU, backup power, and battery chargers. Fuel cell manufacturers in general appear to believe that they would be able to reach these primary markets by making incremental technological advances as they have done in the past. They also appear to believe that they would be able to achieve cost reductions though economies of scale and cost savings in the supply chain.[603] Indeed, it has been reported that the fuel cell industry, worldwide, has made impressive progress in terms of both technology and cost reductions over the years.

- Nonetheless, viewing the slow rate of incremental improvement over the past decade, there is no assurance that technology and costs could be improved to the point of satisfying primary market needs in the next decade. Moreover, as long as the fuel cell industry relies on a strategy of incremental improvement, it would likely always have a need for government subsidies and other forms of support for survival. If that is the case, cutbacks in government support could well put the survival of the industry at risk.

References

1. Schaeffer GJ. *Fuel cells for the future*; 1998. ISBN 90 365 12 30 1.
2. Fuel cell United States Patent 2913511. <http://www.freepatentsonline.com/2913511.html>, FreePatentsOnline.com; no date [accessed August 2010].
3. Blomen JMJ, Mugerwa MN. *Fuel cell systems*. Springer; 1993.
4. Cleveland CJ (Lead Author), Lawrence T (Topic Editor). Grubb, Willard Thomas. <http://www.eoearth.org/article/Grubb,_Willard_Thomas>, Encyclopedia of earth; last revised August 23, 2008 [accessed 13 August 12, 2010].
5. Crawley G. Opening doors to fuel cell commercialization, Proton Exchange Membrane (PEM) fuel cells, <http://www.fuelcelltoday.com/media/pdf/archive/Article_1084_PEM%20article.pdf>, Fuel Cell Today, March 2006.
6. Project mercury. NASA. <http://science.ksc.nasa.gov/history/mercury/ercury-goals.txt> [accessed in August 2007].
7. Wand G. Fuel cell history part 1, Archive, <http://www.fuelcelltoday.com/media/pdf/archive/Article_1104_Fuel%20Cell%20History%20Part%201.pdf>, Fuel Cell Today, 15 June 2006.
8. Appleby AJ, Foulkes FR. *Fuel cell handbook*. New York: Van Nostrant Reinhold; 1989.
9. Fuel cells—the first 120 years, history of fuel cells, Fuel Cell Today, 24 July 2007
10. Erkan S. Synthesis of some metalophthalocyanines and their effects on the performance of PEM fuel cells. A thesis submitted to the graduate school of natural and applied sciences of middle east technical university, in partial fulfillment of the requirements for the degree of master of science in chemical engineering, September 2005.
11. Johnson RC. What happened to fuel cells? EE-Times Asia, <http://www.eetasia.com/ART_8800650716_765245_NT_10940dfc.HTM>, eMedia Asia; 6.September 2011.
12. PEM fuel cells, Collecting fuel cell history. <http://americanhistory.si.edu/fuelcells/pem/pemmain.htm>, Smithsonian Institution; no date [accessed August 2010].
13. Nail JM, Anderson G, Ceasar G, Hansen CJ. The role of the U.S. national innovation system in the development of the PEM Stationary Fuel Cell, NISTIR 7161, U.S. Department Of Commerce, February 2005.
14. Nafion, Article, <http://en.wikipedia.org/wiki/Nafionm>, Wikipedia, last modified on 9 August 2010 [accessed 13 August 2010]. Rajendran RG. Theme article—polymer electrolyte membrane technology for fuel cells, Archives, <http://www.mrs.

org/s_mrs/sec_subscribe.asp?CID=1823&DID=157685&action=detail>, Materials Research Society; 5 August 2005.
15. Biosatellite, Satellites & Space Probes, <http://www.daviddarling.info/encyclopedia/B/Biosatellite.html>, The World of David Darling; no date [accessed August 2010].
16. The global fuel cell race: can the United States win? Research paper, CENTRA technology, September 2001.
17. Mytelka ILK, Boyle G, Edited. Making choices about hydrogen: transport issues for developing countries, International Development Research Centre, United Nations University Press, 2008.
18. Company History, Corporate Information, <http://www.ballard.com/about_ballard/corporate_information/company_history.htm>, Ballard Power Systems, no date [accessed August 2010].
19. Hoogers G. *Fuel cell technology handbook.* CRC Press; 2002.
20. Ballard G (businessman), Article, <http://en.wikipedia.org/wiki/Geoffrey_Ballard_(businessman)>, Wikipedia, last modified on 25 July 2010 [accessed August 2010].
21. PEMFC development base survey, NEDO-P-9118, 1991 Survey Report, March 1992.
22. Ballard Power Systems: 40-foot heavy duty transit bus P2 (1995), Fahrzeug, <http://www.netinform.info/H2/H2Mobility/H2MobilityMain.aspx?ID=274&CATID=2>, TÜV SÜD Industrie Service, no date [accessed April 2011].
23. Leslie J. Dawn of the hydrogen age, <http://www.wired.com/wired/archive/5.10/hydrogen_pr.html>The Condé Nast Publications, 1993–2004, Wired Digital, 1994–2003.
24. Fuel cell industry assessment 2003. 2003 Fuel cell industry report, confidential through November 2005, Houston Advanced Research Center, November 2003.
25. Into the future with NECAR and NEBUS, press kit: history of fuel cell development at Mercedes-Benz, <http://media.daimler.com/dcmedia/0-921-882477-1-884208-1-0-0-884427-0-0-11702-614318-0-1-0-0-0-0-0.html>, Daimler, 25 June 2007.
26. Depalma A. *Ford joins in a global alliance to develop fuel-cell auto engines, Business Day.* The New York Times, <http://www.nytimes.com/1997/12/16/business/ford-joins-in-a-global-alliance-to-develop-fuel-cell-auto-engines.html?ref=ballard-power-systems-inc>;; 16 December 1997 .
27. Daimler AG, Article, <http://en.wikipedia.org/wiki/Daimler_AG>, Wikipedia, last modified on 18 August 2010 [accessed August 2010].
28. Fuel cell vehicle World survey 2003, prepared for US DOE, EERE, by Breakthrough Technologies Institute, Washington, DC, 20006, February 2004.
29. Beijing LN Green Power company, Industry Directory, <http://www.fuelcelltoday.com/online/industry-directory/organisations/be/Beijing-LN-Green-Power-Company>, Fuel Cell Today, 24 May 2011.
30. LUN Jingguang (Project Coordinator, GEF-UNDP-China Cooperation Project). Hydrogen-fuel cell vehicle development in China, Presentation, <http://www.un.org/esa/sustdev/csd/csd14/lc/presentation/hydrogen4.pdf> Beijing, China, 8 May 2006.
31. Green power FC bus China (2002), Fahrzeug, <http://www.netinform.info/H2/H2Mobility/H2MobilityMain.aspx?ID=489&CATID=2> TÜV SÜD Industrie Service; no date [accessed May 2011].
32. China, Partners. <http://www.iphe.net/partners/china.html> International partnership for hydrogen and fuel cells in the economy (IPHE); no date [accessed May 2011].
33. Green power FC Passenger Car (2001). Hydrogen and fuel cell vehicles worldwide, <http://www.netinform.de/H2/H2Mobility/Default.aspx?ID=127&CATID=1>, TÜV SÜD Industrie Service; no date [accessed May 2011].
34. Green power FC Kombi (2002). Wasserstoff-und Brennstoffzellen-Fahrzeuge weltweit, <http://www.netinform.de/H2/H2Mobility/Default.aspx?ID=486&CATID=1>, TÜV SÜD Industrie Service; no date [accessed May 2011].
35. 1970s, History, <http://www.daihatsu.com/company/outline/history/1970s.html>, Daihatsu; no date [accessed April 2011].
36. Environmental report 2001, <http://www.daihatsu.co.jp/company/csr/env_report/backno/pdf/2001/sec03.pdf>, Daihatsu, 2001.
37. World's first hybrid fuel cell minicar approved by Japan's ministry of land, infrastructure and transport, News, <http://www.daihatsu.com/news/n2003/03011501/>, Daihatsu, 15 January 2003.
38. Move EV-FC, 1999, <http://www.hydrogencarsnow.com/hydrogencars1999.htm>, Hydrogen fuel car now, no date [accessed March 2011].
39. Daihatsu cars unveiled at the 33rd Tokyo motor show, News Release, <http://www.daihatsu.co.jp/wn/991013-1f.htm>, Daihatsu, 13 October 1999.
40. Daihatsu motor, Industry Directory, <http://www.fuelcelltoday.com/online/industry-directory/organisations/da/Daihatsu-Motor>, Fuel Cell Today, 25 March 2011.
41. Timeline: alternative drive systems at Mercedes-Benz, alternative drive systems at Daimler AG for the mobility of the future, <http://www.emercedesbenz.

com/Nov07/21_Alternative_Drive_Systems_At_Daimler_AG_For_The_Mobility_Of_The_Future_8.html>, eMercedesBenz.com, 21 November 2007.
42. Daimler Chrysler unveiling of jeep commander 2, Fuel cells, <http://www.methanol.org/fuelcell/press/pr001026.cfm>, Methanol Institute 26 October 2000.
43. 2000 jeep commander 2, (Introduced October 23, 2000), the Gearheadz—concept jeeps, <http://mojeepin.com/concept_commander.html>, MoJeepIn.com, no date [accessed February 2011].
44. Sprinter F-CELL, Vehicles, <http://www.daimler.com/dccom/0-5-1228970-1-1230988-1-0-0-1228995-0-0-135-7165-0-0-0-0-0-0-0.html>, Daimler, no date [accessed May 2011].
45. *FY2006 FCV report*. Japan Automobile Research Institute; March 2007.
46. Chrysler group's town & country Natrium –fuel cell Minivan runs on unique, zero emission fuel, News, <www.media.chrysler.com/pdf.do?id=815>, Chrysler Group, Auburn Hills, Mich., 25 April 2002.
47. Fuel cells on track for series production, Fuel cell motor vehicles, <http://www.linvents.com/03_solar_fuel_producer/fuel_cell_motor_vehicles/fuel_cell_motor_vehicles.htm>, Linvents Century Labs, no date [accessed May 2011].
48. Geiger S. Fuel cells in China—a survey of current developments, Fuel Cell Today, 15 October 2003.
49. Activities related to hydrogen development in China, National statement, 2nd Steering Committee Meeting, International partnership for hydrogen and fuel cells in the economy (IPHE), Beijing, China, 26–28 May 2004. <http://www.iphe.net/docs/Meetings/China_5-04/China_Statement.pdf>
50. Dalian university fuel cell bus China (2002). Fahrzeug, <http://www.netinform.info/H2/H2Mobility/H2MobilityMain.aspx?ID=489&CATID=2>, TÜV SÜD Industrie Service, no date [accessed May 2011].
51. Dalian university fuel cell city bus (2002), Fahrzeug, <http://www.netinform.info/H2/H2Mobility/H2MobilityMain.aspx?ID=489&CATID=2>, TÜV SÜD Industrie Service; no date [accessed April 2011].
52. Dalian university FC-Bus 2 (2003), Fahrzeug, <http://www.netinform.info/H2/H2Mobility/H2MobilityMain.aspx?ID=489&CATID=2>, TÜV SÜD Industrie Service; no date. [accessed April 2011].
53. EvoBus, Article, < http://en.wikipedia.org/wiki/EvoBus>, Wikipedia, last modified on 18 January 2011 [accessed May 2011].
54. Discover Europe's largest range used buses, Used vehicles, <http://www2.mercedes-benz.co.uk/content/unitedkingdom/mpc/mpc_unitedkingdom_website/en/home_mpc/bus/home/consulting/brt.html>, Mercedes-Benz; no date [accessed May 2011].
55. Fuel cell technologies roadmap, Freedom car & fuel partnership, Department of Energy, 10 August 2005.
56. Weaver G, Reidy R. *World fuel cells: an industry profile with market prospects to 2010*. Elsevier; 2002.
57. News/Reykjavik, March 2, 2001-Hydrogen resources policy to start with three fuel cell buses, Archives, <http://www.fuel-cell-bus-club.com/index.php?module=pagesetter&func=viewpub&tid=1&pid=135>, Fuel Cell Bus Club, no date [accessed May 2011]. News/Amsterdam, March 20, 2001-Market launch as a further stage in Daimler Chrysler's pioneering initiative, Archives, <http://www.fuel-cell-bus-club.com/index.php?module=pagesetter&func=viewpub&tid=1&pid=134>, Fuel Cell Bus Club, no date [accessed May 2011].
58. CUTE–Clean Urban Transport for Europe, Industry Directory,<http://www.fuelcelltoday.com/online/industry-directory/organisations/cu/CUTE–-Clean-Urban-Transport-for> Fuel Cell Today, 7 May 2011.
59. News/Vancouver, November 14th, 2001: First public test drive of Citaro fuel cell bus prototype, Archives, <http://www.fuel-cell-bus-club.com/index.php?module=pagesetter&func=viewpub&tid=1&pid=133>, Fuel Cell Bus Club, no date[accessedMay 2011].
60. Daimler Chrysler: Citaro Bus (2002). Fahrzeug, <http://www.netinform.info/H2/H2Mobility/H2MobilityMain.aspx?ID=489&CATID=2>, ÜV SÜD Industrie Service, no date [accessed May 2011].
61. Fuel cell vehicle world survey 2003 by Breakthrough Technologies Institute, Washington, D.C. 20006, for Department of Energy, Energy Efficiency and Renewable Energy, February 2004.
62. Fiat Debuts Seicento Elettra H2, Battery & EV Technology, <http://goliath.ecnext.com/coms2/gi_0199-3809019/Fiat-Debuts-Seicento-Elettra-H2.html>, The Gale Group, 1 July 2001.
63. Kim C. Latest fuel cell fiat panda on display, Blog, <http://www.nuvera.com/blog/?p=1382>, Nuvera Fuel Cells, 5 October 2010.
64. Hydrogen and fuel cells: first FIAT (08.05.2001), Message, <http://www.netinform.de/Aktuelles_Detail.aspx?ID=2144>, TÜV SÜD Industrie Service, 8 May 2001.
65. Delzanno P, Zandiri S (Centro Richerche Fiat), Antonio Toro (Nuvera Fuel Cells Europe). Fuel cell in cars: development of an advanced fuel cell system suitable for high efficiency full performance urban vehicles, <http://www.nuvera.com/pdf/Fuel_Cell_in_Cars.pdf>, Nuvera, no date [accessed July 2011].
66. Nuvera fuel cell stack technology demonstrated on FIAT new generation hydrogen cars, Press Release, <http://www.nuvera.com/news/pr_crf.php>, Nuvera Fuel Cells, 29 October 2003.

67. Zoulias E, Tzanakus G, Lymberopoulos N (CRES), Hayter D (Intelligent Energy). Review of hydrogen community R&D agendas. ResearRoads2HyCom. <http://www.ika.rwth-aachen.de/r2h/images/6/64/Roads2HyCom_R2H6017PU_Review_of_Hydrogen_Community_R%26D_Agendas.pdf>, <www.roads2hy.com>; 22 February 2007.

68. Fiat Seicento (2003). Hydrogen and fuel cell vehicles worldwide, <http://www.netinform.de/H2/H2Mobility/Default.aspx?ID=103&CATID=1>, TÜV SÜD Industrie Service; no date [accessed July 2011]. Fiat Panda (2003). Hydrogen and fuel cell vehicles worldwide, <http://www.netinform.de/H2/H2Mobility/Default.aspx?ID=102&CATID=1>, TÜV SÜD Industrie Service; no date [accessed July 2011].

69. Think global, Article, <http://en.wikipedia.org/wiki/Think_Global>, Wikipedia, last modified on 2 May 2011 [accessed May 2011].

70. Think nordic hydrogen, hydrogen cars, <http://www.hydrogencarsnow.com/think-nordic-hydrogencar.htm>, Hydrogen Cars Now; no date accessed 2010.

71. Ford motor company: passenger car P2000 FC EV (1999), Fahrzeug, <http://www.netinform.info/H2/H2Mobility/H2MobilityMain.aspx?ID=186&CATID=1>, TÜV SÜD Industrie Service; no date [accessed May 2011].

72. Ford think FCV (2000), Fahrzeug, http://www.netinform.info/H2/H2Mobility/H2MobilityMain.aspx?ID=186&CATID=1, TÜV SÜD Industrie Service; no date [accessed May 2011].

73. Ford combines hybrid and fuel cell technology in all-new focus sedan, Article, <http://media.ford.com/article_display.cfm?article_id=12235>, Ford, 11 June 2002

74. Toyota motor overtakes general motors, regains number one spot, News, <http://economictimes.indiatimes.com/news/international-business/toyota-motor-overtakes-general-motors-regains-number-one-spot/articleshow/13396365.cms>, The Economic Times, 22 May 2012.

75. GM fuel cell partnerships and alliances, Partnerships, <http://www.gm.com/corporate/responsibility/partners/technology/index.jsp>, General Motors, no date [accessed February 2011].

76. Opel presents drivable fuel-cell powered Zafira, Opel, <http://www.gm-press.com/HU/site/article.php?aid=531>, GM Media Online, 29 September 1998.

77. Alternative energy materials, Tutorial, <http://www.sigmaaldrich.com/materials-science/alternative-energy-materials/tutorial.html>, Sigma-Aldrich Company, no date [accessed February 2011].

78. Ho J. *GM unveils next millennium fuel cell vehicle.* Auto World; 2001.

79. 2000, Hydrogen fuel cars. <http://www.hydrogencarsnow.com/hydrogencars2000.htm>, Hydrogen Fuel Cars Now, no date [accessed April 2011].

80. Hydrogen fuel cars 2001, History of hydrogen cars, <http://www.hydrogencarsnow.com/hydrogencars2001.htm>, Hydrogen Cars Now, no date [accessed October 2010].

81. GM HydroGen Minivan, Hydrogen cars, <http://www.hydrogencarsnow.com/gm-hydrogen3-minivan.htm>, Hydrogen Fuel Cars Now; no date [accessed March 2011].

82. General Motor Hy-wire Concept 2002, Car reviews, <http://www.autoconcept-reviews.com/cars_reviews/gm/gm-hiwire/cars_reviews-gm-hiwire.html>; no date [accessed March 2011].

83. GM HydroGen Minivan, <http://www.hydrogencarsnow.com/gm-hydrogen3-minivan.htm>, Hydrogen Fuel Cars Now; no date [accessed February 2011].

84. HydroGen3. Article, <http://en.wikipedia.org/wiki/HydroGen3>, Wikipedia, last modified on 8 July 2010 [accessed February 2011].

85. Magda M. Post office takes keys to GM fuel-cell delivery vehicle, <http://green.autoblog.com/2006/09/28/post-office-takes-keys-to-gm-fuel-cell-delivery-vehicle/>, <Autoblog.com>; 28 September 2006.

86. Winter U, Herrmann M. Hydrogen automobile heads toward series production HydroGen3 puts fuel cells on the road. *Fuel Cells* November, 2003;**3**(3). Wiley.

87. Walsh B, Moores P. Auto companies on fuel cells, Charts, <http://www.fuelcells.org/info/charts/AutoCompaniesonFuelCells.pdf>, Fuel Cells 2000, 27 April 2004.

88. Adamson KA. Fuel cell market survey: Buses, <http://www.accepth2.com/guidelines/docs/Paper_Fuel-Cell-Today_041201_Fuel-Cell-Bus-Market-Survey.pdf>, Fuel Cell Today, November 2004.

89. Zero-Emission bus demonstration program, Projects, studies & programs, <http://www.vta.org/projects/ZEBs.html>, VTA, no date [accessed April 2011].

90. Hino Motors. Article, <http://en.wikipedia.org/wiki/Hino_Motors>, Wikipedia, last modified on 14 April 2011 [accessed April 2011].

91. Toyota jointly develops fuel cell hybrid bus, the FCHV-BUS1, News, <http://www.toyota.co.jp/en/news/01/0618_3.html>, Toyota, 18 June 2001.

92. Hino motor. Industry directory, <http://www.fuelcelltoday.com/online/industry-directory/organisations/hi/Hino-Motor-Co-Ltd>, Fuel Cell Today, 24 April 2011.

93. Toyota/Hino fuel cell hybrid bus 1st to obtain ministry certification—FCHV-BUS2 to soon start

public road tests in Japan—, News, <http://www.toyota.co.jp/en/news/02/0927.html>, Toyota Motor Corporation and Hino Motors, Ltd., 27 September 2002.

94. Toyota Hino FCHV-Bus 2 (2002), Fahrzeug, <http://www.netinform.info/H2/H2Mobility/H2MobilityMain.aspx?ID=489&CATID=2>, TÜV SÜD Industrie Service; no date [accessed April 2011].
95. FCEV-BUS begins public road testing, Weekly News, <http://www.itsev.com/e/news.html>, EV Forum, 4 October 2002.
96. Toyota's role in developing fuel cell hybrid vehicles (FCHV), Quick Glance Fact Sheet, <http://www.cityofirvine.org/civica/filebank/blobdload.asp?BlobID=6272>, Toyota Motor Sales; no date [accessed April 2011].
97. Light-duty fuel cell vehicles. The fuel cell vehicle survey 2003, prepared by Breakthrough Technologies Institute, for EERE, DOE, February 2004.
98. Hordeski MF. *Alternative fuels: the future of hydrogen*. The Fairmont Press; 2008.
99. Honda introduces prototype fuel cell electric vehicles, News Release, <http://world.honda.com/news/1999/4990906b.html>, Honda, 6 September 1999.
100. FCX-V3 now powered with new honda fuel cell, News Release, <http://www.honda.com/newsandviews/article.aspx?id=443>, Honda, 11 March 2001.
101. Timeline, Environmental Leadership, <http://corporate.honda.com/environment/fuel_cells.aspx?id=fcx-concept>, Honda, no date [accessed February 2011].
102. Honda unveils the prototype of its FCX fuel cell vehicle, planned for commercial release this year, News Release, <http://world.honda.com/news/2002/4021022.html>, Honda, 22 October 2002.
103. FY2006 third report on overseas demonstration test-II, Japan automobile research institute, March 2007.
104. Honda FCX Clarity, Hydrogen cars, <http://www.hydrogencarsnow.com/honda-fcx-clarity.htm>, Hydrogen Cars Now; no date [accessed October 2010].
105. Ballard to make fuel cells for Hyundai, Electric Vehicle Online Today, <http://www.accessmylibrary.com/coms2/summary_0286-9336758_ITM>, AccessMyLibrary, 16 December 1999.
106. Korean carmakers become active on fuel cells, News, <http://www.h2mobility.org/1_cardata/c173.htm>, TÜV SÜD Industrie Service GmbH, 17 March 2000.
107. Hyundai motor company, Industry Directory, <http://www.fuelcelltoday.com/online/industry-directory/organisations/hy/Hyundai-Motor-Company>, Fuel Cell Today, 11 March 2011
108. Hydrogen and fuel cells: hyundai joins California fuel cell partnership, News, <http://www.h2mobility.org/1_cardata/c173.htm>, TÜV SÜD Industrie Service GmbH, 19 June 2000.
109. Hyundai motor company unveils fuel cell-powered santa fe suv at grand opening of California fuel cell partnership, Corporate News, <http://www.hyundainews.com/Corporate_News/Corporate/11_01_2000_1319.asp>, Hyundai, 1 November 2000.
110. McGrath D, Wing B. Partnership showcases record fuel cell vehicle Gathering, <http://www.evworld.com/article.cfm?storyid=124>, EV World; no date [accessed March 2011].
111. Hyundai's hydrogen fuel cell vehicle to hit the road next year, Car news, <http://www.carpages.co.uk/hyundai/hyundai_hydrogen_fuel_cell_vehicle_24_07_03.asp>, Car Pages, UK; 24 July 2003.
112. Hyundai and UTC fuel cells collaborate on all-weather fuel cell vehicle, CSR Press Release, <http://www.csrwire.com/press_releases/21118-Hyundai-and-UTC-Fuel-Cells-Collaborate-on-all-Weather-Fuel-Cell-Vehicle>, CSR Wire, 26 June 2003.
113. Irisbus. Article, <http://en.wikipedia.org/wiki/Irisbus>, Wikipedia, last modified on 15 April 2011 [accessed April 2011].
114. Fuel cells in buses, Fleet Vehicles, <http://www.utcpower.com/fs/com/bin/fs_com_Page/0, 11491, 0152,00.html>, UTC Power, no date [accessed April 2011].
115. The fiat group in 2001, report on operations, consolidated and statutory financial statements, <http://www.fiatspa.com/en-US/investor_relations/financial_reports/FiatDocuments/Bilanci/2001/BIL2001_Rel_Gest_ING.pdf>, Fiat Group, Turin, 27 March 2002.
116. City cell—fuel cell energy in cities, in European fuel cell and hydrogen projects, 1999–2002, EUR 20718, European Commission, 2003.
117. Irisbus (2001). Fahrzeug, <http://www.netinform.info/H2/H2Mobility/H2MobilityMain.aspx?ID=489&CATID=2>, TÜV SÜD Industrie Service; no date [accessed April 2011].
118. Irisbus, Industry Directory, <http://www.fuelcelltoday.com/online/industry-directory/organisations/ir/Irisbus>, Fuel Cell Today, 26 April 2011.
119. Irisbus City Cell Madrid (2003), Fahrzeug, <http://www.netinform.info/H2/H2Mobility/H2MobilityMain.aspx?ID=489&CATID=2>, TÜV SÜD Industrie Service, no date [accessed April 2011].
120. Fiat Irisbus shows fuel cell bus, E-Mail Newsletter, <http://www.hoover-ink.com/content_dev/michelin_issue3_eng.htm>, Michelin Challenge Bibendum, May 2003.

121. UNDP-GEF fuel-cell bus programme: update, GEF/C.25/Inf.8, prepared by the United Nations Development Programme, Global Environment Facility; May 2005.
122. Multiple end-user specified 100 kW FC power module (FEBUSS), European fuel cell and hydrogen projects 1999–2002, EUR 20718, European Commission, 2003.
123. Fuel cell energy systems standardised for large transport, busses and stationary applications (FEBUSS), simple search, record control number:60060, <http://cordis.europa.eu/search/index.cfm?fuseaction=proj.document&PJ_LANG=EN&PJ_RCN=5315764&pid=0&q=1CB592320658B846C32E66F5ACA63FA1&type=sim>, CORDIS, no date [accessed April 2011].
124. Irisbus, Industry Directory, <http://www.fuelcelltoday.com/online/industry-directory/organisations/ir/Irisbus>, Fuel Cell Today, 25 April 2011.
125. Hybrid FC Europolis (2006). Wasserstoff- und Brennstoffzellen-Fahrzeuge weltweit, <http://www.netinform.de/H2/H2Mobility/Default.aspx?ID=233&CATID=2>, TÜV SÜD Industrie Service; no date [accessed April 2011].
126. Man truck & bus—consistently efficient, Company, <http://www.mantruckandbus.com/en/Company/Company.jsp>, Man truck & bus; no date [accessed April 2011].
127. Brands, that make markets, Company, Article, <http://www.mantruckandbus.com/en/Company/Marken_/Marken_.jsp>, Man Truck & Bus; no date [accessed April 2011].
128. Solar hydrogen project at Neunburg vorm wald, Germany, <http://www.solarhydrogen.com/pdfs/eng/publications_e_10.pdf>, published in proceedings of the 11th World hydrogen energy conference,vol 2, Stuttgart, Germany: Solar-Wasserstoff-Bayern GmbH 23–28 June 1996.
129. Man: fuel cell city bus (2000). Fahrzeug, <http://www.netinform.info/H2/H2Mobility/H2MobilityMain.aspx?ID=489&CATID=2>, TÜV SÜD Industrie Service; no date [accessed April 2011].
130. Gruber C (Man Nutzfahrzeuge), Wurster R (L-B-Systemtechnik). Hydrogen-fueled buses: the Bavarian fuel cell bus project. <http://ieahia.org/pdfs/bavarian_proj.pdf>; no date [accessed April 2011].
131. European fuel cell projects, 1995–2000, EUR 19368, European Commission, 2000.
132. Man: low floor bus with FC and LH2 (2001), Fahrzeug, <http://www.netinform.info/H2/H2Mobility/H2MobilityMain.aspx?ID=489&CATID=2>, TÜV SÜD Industrie Service; no date [accessed April 2011].
133. Fuel cell bus for Berlin, Copenhagen, Liston, in European fuel cell projects, 1995–2000, EUR 19368, European Commission, 2000.
134. Mazda introduces new fuel cell electric vehicle, Premacy FC-EV Mazda, <http://www.autoweb.com.au/cms/newsarticle.html?&A=53580>, Autoweb 15 February 2001.
135. Mazda: small passenger car (1997). Fahrzeug, <http://www.netinform.info/H2/H2Mobility/H2MobilityMain.aspx?ID=165&CATID=1>TÜV SÜD Industrie Service; no date [accessed May 2011].
136. Mazda, Industry Directory, <http://www.fuelcelltoday.com/online/industry-directory/organisations/ma/Mazda>, Fuel Cell Today, 08 May 2011.
137. Green light for fuel cell cars' trial run on public roads, Messages, <http://groups.yahoo.com/group/nukenet/message/5197>, Yahoo! Group, 9 February 2001.
138. Fuel cell applications and case studies, FCIAC Team Report, <http://www.seco.cpa.state.tx.us/zzz_feulcell-initiative/fciac_applications.pdf>, State energy conservation office, Texas; no date [accessed May 2011].
139. History of Mitsubishi Motors, Overview of Mitsubishi Motors, <http://www.mitsubishi-motors.com/en/corporate/aboutus/history/1970/index.html>, Mitsubishi Motors; no date [accessed March 2011].
140. Mitsubishi Motors, Article, <http://en.wikipedia.org/wiki/Mitsubishi_Motors>, Wikipedia, last modified on 14 March 2011 [accessed March 2011].
141. Daimler Chrysler-Mitsubishi alliance, Article, <http://en.wikipedia.org/wiki/DaimlerChrysler-Mitsubishi_alliance>, Wikipedia, last modified on 19 December 2009 [accessed March 2011].
142. Tabo E, Kuzuoka N, Takada M, Yoshida H. Fuel cell vehicle technology trends and MMC initiatives, Mitsubishi Motors Technical Review, No.16, 2004.
143. Environmental report 1999, <http://www.mitsubishi-motors.com/corporate/environment/report/e/pdf/1999/1999e_06.pdf>, Mitsubishi Motors, 1999.
144. Hydrogen fuel cars 2001. History of hydrogen cars, <http://www.hydrogencarsnow.com/hydrogencars2001.htm>, Hydrogen fuel cars now; no date [accessed March 2011].
145. Mitsubishi FCV Grandis (2003). Hydrogen and fuel cell vehicles worldwide, <https://www.netinform.de/H2/H2Mobility/Default.aspx?ID=104&CATID=0>, TÜV SÜD Industrie Service, no date [accessed June 2011].
146. Mitsubishi FCV to run with 2004 Osaka International Ladies Marathon, Press Release, <http://www.

mitsubishi-motors.com/pressrelease/corporate/detail982.html>, Mitsubishi Motors, 22 January 2004.

147. Mitsubishi FCV receives vehicle type certification: Mitsubishi motors and Daimler Chrysler to share FCV maintenance facility, Press Release, <http://www.mitsubishi-motors.com/pressrelease/corporate/detail915.html>, Mitsubishi Motors, 5 November 2003.
148. NEOPLAN, Article, <http://en.wikipedia.org/wiki/NEOPLAN>, Wikipedia, last modified on 9 April 2011 [accessed April 2011].
149. Man truck & bus, Article, <http://en.wikipedia.org/wiki/MAN_Truck_%26_Bus>, Wikipedia, last modified on 21 April 2011 [accessed April 2011].
150. NEOPLAN: Midi FC bus (1999), Fahrzeug, <http://www.netinform.info/H2/H2Mobility/H2MobilityMain.aspx?ID=489&CATID=2>, TÜV SÜD Industrie Service, no date [accessed April 2011].
151. Proton Motors: Bayern-Bus II (2000), Fahrzeug, <http://www.netinform.info/H2/H2Mobility/H2MobilityMain.aspx?ID=489&CATID=2>; no date [accessed April 2011].
152. New York MTA awards new flyer order for up to 475 CNG buses, News, <http://www.newflyer.com/index/news-app/story.85>, New Flyer, 20 August 2010.
153. Ballard P3 (Chicago Transport Authority) (1997), Fahrzeug, <http://www.netinform.info/H2/H2Mobility/H2MobilityMain.aspx?ID=274&CATID=2>, TÜV SÜD Industrie Service; no date [accessed April 2011].
154. Ballard P4 ZEbus (1999), Fahrzeug, <http://www.netinform.info/H2/H2Mobility/H2MobilityMain.aspx?ID=274&CATID=2>, TÜV SÜD Industrie Service; no date [accessed April 2011].
155. Rips C (SunLine Transit Agency, Thousand Palms, CA), Clapper W, Johnston M (SunLine Services Group, Thousand Palms, CA), Elam C, Schucan TH (IEA Hydrogen Program). Sunline Clean Fuels Mall. <http://ieahia.org/pdfs/sunline.pdf>; no date [accessed February 2011].
156. Natural resources canada selects hydrogenics to lead development of next-generation hybrid fuel cell-powered bus, News Details, <http://www.hydrogenics.com/invest/News_Details.asp?RELEASEID=96761>, Hydrogenics, 6 December 2002.
157. Fuel-Cell vehicles, Environmental Activities, <http://www.nissan-global.com/EN/ENVIRONMENT/CAR/FUEL_BATTERY/DEVELOPMENT/FCV/index.html>, Nissan; no date [accessed March 2011].
158. Nissan motor, Industry Directory, <http://www.fuelcelltoday.com/online/industry-directory/organisations/ni/Nissan-Motor-Co-Ltd>, Fuel Cell Today, 1 April 2011.
159. History of NISSAN FCV technology development, Overviews, <http://www.nissan-global.com/EN/TECHNOLOGY/INTRODUCTION/DETAILS/XTRAIL-FCV/index.html>, Nissan; no date [accessed March 2011].
160. UTC fuel cells, Fuel Cell Suppliers, <http://www.transpowerusa.com/all_about_electric_drive/Hydrogen_Range_Extenders/utc_fuel_cells.shtml>, Transpower; no date [accessed April 2011].
161. Nova Bus. Article, <http://en.wikipedia.org/wiki/Nova_Bus>, Wikipedia, last modified on 2 April 2011 [accessed May 2011].
162. Generation I Bus General Information, TBB Technical Details, <http://fuelcellbus.georgetown.edu/gen1.cfm>, Georgetown University; no date [accessed April 2011].
163. Bus Manufacturing U.S.A., Inc. Santa Barbara, <http://www.corporationwiki.com/California/Santa-Barbara/bus-manufacturing-u-s-a-inc/41427105.aspx>, Corporation Wiki; no date [accessed April 2011].
164. Georgetown University: Methanol Transit Bus (2000), Wasserstoff- und Brennstoffzellen-Fahrzeuge weltweit, <http://www.netinform.de/H2/H2Mobility/Default.aspx?ID=259&CATID=2>, TÜV SÜD Industrie Service; no date [accessed July 2011]. Program Summary, Program Overview, <http://fuelcellbus.georgetown.edu/overview.cfm>, Georgetown University; no date [accessed April 2011].
165. Generation III fuel cell bus program, Georgetown University fuel cell bus program, <http://fuelcellbus.georgetown.edu/generation3.html>, Georgetown University; no date [accessed March 2012].
166. PSA Peugeot Citroën. Article, <http://en.wikipedia.org/wiki/PSA_Peugeot_Citro%C3%ABn>, Wikipedia, <http://en.wikipedia.org/wiki/PSA_Peugeot_Citro%C3%ABn>, last modified on 29 March 2011 [accessed April 2011].
167. Second generation PEM fuel cell working with hydrogen stored at high pressure for the electric vehicle. Record control number: 8945, Simple Search, <http://cordis.europa.eu/search/index.cfm?fuseaction=acro.document&AC_LANG=EN&AC_RCN=2360254&pid=0&q=D8C5AA34EAA6686AB8819F12C8877DE5&type=sim>, CORDIS; no date [accessed April 2011].
168. Goedgebuer JP. PSA Peugeot Citroen hydrogen and fuel cell research activities and perspectives, <http://www.storhy.net/finalevent/pdf/Welcome-Notes-and-Keynote-PSA.pdf>, StorHy Final Event, 4 June 2008.
169. Fuel cell report to congress, (ESECS EE-1973), <http://www1.eere.energy.gov/

hydrogenandfuelcells/pdfs/fc_report_congress_feb2003.pdf>, DOE, February 2003.

170. Fuel cell: energy for the future, Brochure, <http://www.psa-peugeot-citroen.com/modules/brochure_PAC_en.pdf>, PSA Peugeot Citroen, no date [accessed April 2011].

171. Peugeot Fuel Cell Cab (2001), Hydrogen and fuel cell vehicles worldwide, <https://www.netinform.de/H2/H2Mobility/Default.aspx?ID=150&CATID=1>, TÜV SÜD Industrie Service; no date [accessed June 2011].

172. Future of firefighting unveiled, Peugeot has designed a futuristic electric fire engine called the H2O, <http://www.electrifyingtimes.com/peugeot_h2o.html>, Electrifying times.com; no date[accessed April 2011].

173. Peugeot H2O Firefighter (2002), Hydrogen and fuel cell vehicles worldwide, <https://www.netinform.de/H2/H2Mobility/Default.aspx?ID=122&CATID=1>, TÜV SÜD Industrie Service; no date [accessed June 2011].

174. Millennium cell extends borohydride cooperation with France's Peugeot Citroen, Stories, <http://www.hfcletter.com/issues/XIX_2/stories/87-1.html>, vol. XIX, No. 2, Hydrogen and Fuel Cell Letter, February 2004.

175. Fuel cell: energy for the future, Brochure, http://www.psa-peugeot-citroen.com/modules/brochure_PAC_en.pdf, PSA Peugeot Citroen and intelligent energy; no date [accessed June 2011].

176. Renault, Article, <http://en.wikipedia.org/wiki/Renault>, Wikipedia, last modified on 4 April 2011 [accessed April 2011].

177. Fuel cell powered electric vehicle for efficiency and range, simple search, record control number: 17935, CORDIS; no date [accessed April 2011].

178. Second generation PEM fuel cell working with hydrogen stored at high pressure for the electric vehicle, Simple Search, Record Control Number: 37086, CORDIS; no date [accessed April 2011].

179. BMW, Renault, Delphi team up to develop fuel cell auxiliary power unit; MOU signed today, News Release, <http://www.prnewswire.com/news-releases/bmw-renault-delphi-team-up-to-develop-fuel-cell-auxiliary-power-unit-mou-signed-today-72929792.html>, Delphi automotive systems, 2 May 1999.

180. Fuel cell systems and components general research for vehicle applications (FUERO), Simple Search, Record control number: 52836, <http://cordis.europa.eu/search/index.cfm?fuseaction=proj.document&PJ_LANG=EN&PJ_RCN=4862739&pid=0&q=042F6445A6AE8F7D2CF5113C7C3F91DE&type=sim>; no date [accessed April 2011].

181. Nuvera, Renault forge fuel cell partnership, Press Release, <http://www.nuvera.com/news/pr_renlt.php>, Nuvera, 4 July 2002.

182. TotalFinaElf partners with Nuvera to optimize hydrocarbon fuels for fuel cells, Press Release, <http://www.nuvera.com/news/pr_tfe.php>, Nuvera, 9 April 2003.

183. Light-duty fuel cell vehicles, in the fuel cell vehicle survey 2003, <http://www1.eere.energy.gov/hydrogenandfuelcells/pdfs/ld_fc_vehicles.pdf>, Breakthrough Technologies Institute, February 2004.

184. Mikko K, Markus T. Electric and Hybrid Bus, Presentation, <http://www.tekes.fi/fi/gateway/PTARGS_0_201_403_994_2095_43/http%3B/tekes-ali1%3B7087/publishedcontent/publish/programmes/eve/documents/finpro_electric_hybrid_buses_overview_2011.pdf>, FINPRO, December 2010.

185. Scania AB, Article, http://en.wikipedia.org/wiki/Scania_AB, Wikipedia, last modified on 27 April 2011.

186. FC bus fuel cell midi city bus, in European fuel cell projects, 1995–2000, EUR 19368, European Commission, 2000.

187. Scania presents emission-free fuel-cell bus, Press Release, <http://www.scania.com/media/pressreleases/2001052317enen.aspx>, Scania, 23 May 2001.

188. Suzuki, Article, <http://en.wikipedia.org/wiki/Suzuki>, Wikipedia, last modified on 2 April 2011 [accessed April 2011].

189. 1990s, Corporate History, <http://www.suzuki.co.jp/about/development/1990.html>, Suzuki; no date [accessed April 2011].

190. 2000s, Corporate History, <http://www.suzuki.co.jp/about/development/2000.html>, Suzuki; no date [accessed April 2011].

191. 2001, Corporate History, <http://www.suzuki.co.jp/about/development/2001.html>, Suzuki; no date [accessed April 2011].

192. 2003, History of hydrogen cars, <http://www.hydrogencarsnow.com/hydrogencars2003.htm>, Hydrogen Fuel Cars Now; no date [accessed April 2011].

193. Suzuki 'wagonR-FCV' (2003), Hydrogen and fuel cell vehicles worldwide, <http://www.netinform.net/H2/H2Mobility/H2MobilityMain.aspx?ID=105&CATID=1>, TÜV SÜD Industrie Service; no date [accessed April 2011].

194. 2003 Suzuki mobile terrace, Suzuki Motors, Press Release, <http://www.seriouswheels.com/cars/top-2003-Suzuki-Mobile-Terrace.htm>, Serious Wheels; no date [accessed April 2011].

195. Suzuki to obtain ministerial permit on its third FCV, to start public test driving of FCV with 75 Mpa, Press Release, <http://www.suzuki.co.jp/release/d/d041203.htm> Suzuki, 3 December 2004.
196. Suzuki aims to launch fuel cell cars by 2010, Reuters, <http://www.simplemachines.org/about/copyright.php>, Simple Machines, LLC; 22 October 2003.
197. Thor to build world's first commercially viable, zero emissions fuel cell buses in alliance with International fuel cells, A United Technologies Company, and ISE Research, News Releases, <http://www.prnewswire.com/news-releases/thor-to-build-worlds-first-commercially-viable-zero-emissions-fuel-cell-buses-in-alliance-with-international-fuel-cells-a-united-technologies-company-and-ise-research-73432337.html>, PR Newswire, 5 June 2000.
198. Hydrogen commercialization: transportation fuel for the 21st century, Contract Number: DE FC36-96GO10139, <http://www.osti.gov/bridge/servlets/purl/928756-HWlkc7/928756.pdf>, Final Report to Department of Energy, <http://www.osti.gov/bridge/servlets/purl/928756-HWlkc7/928756.pdf>, SunLine Transit Agency; February 2003.
199. Chandler K (Battelle), Eudy L (National Renewable Energy Laboratory), ThunderPower bus evaluation at SunLine transit agency, A U.S. Department of Energy Transit Bus Evaluation Project, <http://www.nrel.gov/docs/fy04osti/34379.pdf>, Energy Efficiency and Renewabel Energy, DOE, November 2003.
200. Thor's Receives fuel cell grant from DOT, Business News, <http://goliath.ecnext.com/coms2/gi_0199-3797787/Thor-s-Receives-Fuel-Cell.html>, The Gale Group, 15 April 2001.
201. Toyota FCHV, Article, <http://en.wikipedia.org/wiki/Toyota_FCHV>, Wikipedia, last modified on 2 February 2011 [accessed February 2011].
202. Toyota FCHV, Article, <http://en.wikipedia.org/wiki/Toyota_FCHV>, Wikipedia, last modified on 2 February 2011 [accessed March 2011].
203. Toyota's role in developing fuel cell hybrid vehicles (FCHV), Quick glance fact sheet, http://www.cityofirvine.org/civica/filebank/blobdload.asp?BlobID=6272, Toyota; no date [accessed February 2011].
204. FY2005 report on the trends in fuel cell vehicle policies and technologies in the united states and europe, Japan Automobile Research Institute (JARI), March 2006.
205. Tongji University, Industry Directory, <http://www.fuelcelltoday.com/online/industry-directory/organisations/to/Tongji-University>, Fuel Cell Today, 7 June 2011.
206. WANG WG (Ningbo Institute of Material Technology and Engineering), After the Beijing Olympics: fuel cells in China, PPT Presentation, <http://www.climate-change-solutions.co.uk/pictures/content473/1-2_wei_gho_wang_-_20100322_birmingham_wei_guo_wang.pdf>, Birmingham, UK; 15 March 2010.
207. Tongji University Chao Yue I (2003), Wasserstoff- und Brennstoffzellen-Fahrzeuge weltweit, <http://www.netinform.de/H2/H2Mobility/Default.aspx?ID=100&CATID=1>, TÜV SÜD Industrie Service; no date [accessed May 2011].
208. Pingwen M, Jingguang L, Mytelka L. 15. Hydrogen and fuel-cell activities in China, in making choices about hydrogen, transport issues for developing countries. In: LK Mytelka, G, Boyle editors. <http://www.idrc.ca/en/ev-132182-201-1-DO_TOPIC.html>, The International Development Research Centre, 2008.
209. Tongji University Chao Yue II (2004), Wasserstoff- und Brennstoffzellen-Fahrzeuge weltweit, <http://www.netinform.de/H2/H2Mobility/Default.aspx?ID=89&CATID=1>, TÜV SÜD Industrie Service; no date [accessed May 2011].
210. Tongji University Chao Yue III S3000 (2005), Wasserstoff- und Brennstoffzellen-Fahrzeuge weltweit, <http://www.netinform.de/H2/H2Mobility/Default.aspx?ID=451&CATID=1>, TÜV SÜD Industrie Service; no date [accessed May 2011].
211. Tongji University Chao Yue III MPV (2005), Wasserstoff- und Brennstoffzellen-Fahrzeuge weltweit, <http://www.netinform.de/H2/H2Mobility/Default.aspx?ID=62&CATID=1>, TÜV SÜD Industrie Service; no date [accessed May 2011].
212. Tongji University Chao Yue III Eastar (2005), Wasserstoff- und Brennstoffzellen-Fahrzeuge weltweit, <http://www.netinform.de/H2/H2Mobility/Default.aspx?ID=450&CATID=1>, TÜV SÜD Industrie Service; no date [accessed May 2011].
213. Zhang T. The development of fuel cell vehicles in China, PPT Presentation, <http://www.iphe.net/docs/Events/China_9-10/1-2_FCV%20in%20China%20-%2020100921.pdf>, Clean Energy Automotive Engineering Center of Tongji University, Shanghai FCV Powertrain Co, Ltd., 21 September 2010.
214. Ballard signs with chinese group to provide up to 20 fuel cells for car trials, Issues, <http://www.hfcletter.com/issues/XXI_8/stories/720-1.html>, The Hydrogen and Fuel Cell Newsletter, August 2006.
215. Chunhui (2004), Hydrogen and fuel cell vehicles worldwide, <http://www.netinform.de/H2/H2Mobility/H2MobilityMain.aspx?ID=83&CATID=1>, TÜV SÜD Industrie Service, no date [accessed June 2011].
216. WAN Gang, Resume, 2004 Leadership Programme on Environment for Sustainable Development, <http://

www.rrcap.unep.org/uneptongji/resource/wan.htm>, UNEP-Tongji University, 2004.

217. Tsinghua University & DICP FCB 60 kW (2003). Fahrzeug, <http://www.netinform.info/H2/H2Mobility/H2MobilityMain.aspx?ID=489&CATID=2>, TÜV SÜD Industrie Service; no date [accessed May 2011].

218. Tsinghua University fuel cell city bus 863, Hydrogen and Fuel Cell Vehicles Worldwide, <http://www.netinform.net/h2/h2mobility/Default.aspx?ID=243&CATID=2>, TÜV SÜD Industrie Service; no date [accessed May 2011].

219. Tsinghua University FC low floor city bus (2006), Hydrogen and Fuel Cell Vehicles Worldwide, <http://www.netinform.net/h2/h2mobility/Default.aspx?ID=489&CATID=2>, TÜV SÜD Industrie Service; no date [accessed May 2011].

220. Hool V. Article, <http://en.wikipedia.org/wiki/Van_Hool>, Wikipedia, last modified on 12 April 2011 [accessed May 2011].

221. Hoffmann P. *Tomorrow's energy: hydrogen, fuel cells, and the prospects for a cleaner planet*. MIT Press; 2002.

222. Elenco EUREKA city bus (1995), Fahrzeug, <http://www.netinform.info/H2/H2Mobility/H2MobilityMain.aspx?ID=275&CATID=2>, TÜV SÜD Industrie Service; no date [accessed May 2011].

223. Chandler K (Battelle), Eudy L (National Renewable Energy Laboratory). Alameda-contra costa transit district (ac transit) fuel cell transit buses, Preliminary Evaluation Results, Technical Report NREL/TP-560-41041, <http://www.nrel.gov/hydrogen/pdfs/41001.pdf>, NREL, DOE, March 2007.

224. Fuel Cell Buses, Who Makes, <http://www.navc.org/whomakes.html>, NAVC; no date [accessed May 2011].

225. Car autothermal process reactor initiative, Simple Search, Record control number: 37124, <http://cordis.europa.eu/search/index.cfm?fuseaction=proj.document&PJ_LANG=FR&PJ_RCN=2360296&pid=0&q=F38532B93D3E7AB0FB9F567352979F01&type=sim>, CORDIS; no date [accessed April 2011].

226. Volkswagen AG: CAPRI (2000), Hydrogen Vehicles, <http://www.netinform.net/H2/H2Mobility/H2MobilityMain.aspx?ID=169&CATID=1>, TÜV SÜD Industrie Service GmbH; no date [accessed April 2011].

227. 2000, History Of Hydrogen Cars, <http://www.hydrogencarsnow.com/hydrogencars2000.htm>, Hydrogen Fuel Cars Now; no date [accessed April 2011].

228. Volkswagen unveils its first fuel-cell car at the grand opening of the California fuel cell partnership, Press Release, <http://media.vw.com/press_releases/volkswagen-unveils-its-first-fuel-cell-car-at-the-grand-opening-of-the-california-fuel-cell-partnership>, Volkswagen, 1 November 2000

229. A survey on electric power generation technology development. A report, FY2002 result report for NEDO. *Techno Research*; March 2003.

230. *Long-term energy technology strategy survey (new sunshine project systematic examination and ripple effect quantitative assessment survey) part I*. Heisei 12 Survey Report, NEDO-P-00-64-1. Techno Research Institute for NEDO; March 2001.

231. On millennium project, Decision by Prime Minister of Japanese Cabinet, <http://www.kantei.go.jp/jp/mille/991222millpro.pdf>; 19 December 1999.

232. Polymer electrolyte fuel cell system technology deployment project/polymer electrolyte fuel cell component technology development proje3ct, Mid-Term Evaluation Report; December 2003.

233. Dietrich P, Buchi F, Tsukada A, Bartschi M, Kotz R, Scherer GG, Rodatz P, Garcia O, Ruge M, Wollenberg M, Luck P, Wiartalla A, Schonfelder C, Schneuwly A, Barrade P. Hy.Power—a technology platform combining a fuel cell system and a supercapacitor. In: Vielstich W, Lamm A, Gasteiger HA, editors. *Part 11 in Handbook of fuel cells—fundamentals, technology and applications*, vol. 4. Chichester: John Wiley & Sons; 2003. ISBN: 0-471-49926-9.

234. *Heisei 17, post evaluation of research and development policy, solid polymer fuel cell/hydrogen energy use program*. Evaluation Report. Japan: METI, <http://www.meti.go.jp/policy/policy_management/17fy-jigo-hyoka/kotaisuiso/17fy-jigo-kotaisuiso.pdf>; June 2006.

235. Toward commercialization and wide use of fuel cells, Greetings, <http://fccj.jp/index_e.html>, Fuel cell Commercialization Conference of Japan; no date [accessed December 2010].

236. Greetings, <http://fccj.jp/index_e.html>, Fuel cell Commercialization Conference of Japan; no date [accessed December 2010].

237. Various sources, including METI's Natural resources and energy related budget outline published each year.

238. *On innovation program*. Material 4, the 23rd research and development subcommittee. Japan: METI, <http://www.meti.go.jp/committee/materials/downloadfiles/g80516a04j.pdf>; 16 May 2008.

239. *Polymer electrolyte fuel cell system technology development project post evaluation report*. NEDO, <http://www.nedo.go.jp/iinkai/gijutsu/hyouka/kotai/17h/jigo/page1-174.pdf>; September 2005.

240. Heisei 18, Implementation Guideline, New energy technology development program, Strategic

Technology Development for PEMFC Commercialization. Japan: NEDO; no date. <http://www.nedo.go.jp/activities/portal/gaiyou/p05011/h18jisshi.pdf> [accessed December 2010].

241. Fuel cell cutting cutting-edge center technology research project (FC-Cubic Project), Presentation, <http://www.meti.go.jp/policy/tech_evaluation/c00/C0000000H19/080215_fuel-cell2/fuel-cell2-A.pdf>, FC-Cubic Center, AIST, 28 December 2007.

242. *PEMFC technology development for promotion of commercialization, solicitation notice, the first round.* Japan: NEDO; 10 February 2010.

243. *Strategic technology development of pemfc commercialization/degradation mechanism elucidation and basic material research for high function cell using nanotechnology.* Mid-term Evaluation Report, Research Evaluation Committee. NEDO; February 2010.

244. Homma T. *Elementary technologies of PEFC and DMFC.* The Latest Fuel Cell News In Japan. FCDIC; July 2007.

245. *Portable information device use fuel cell technology development.* Post Project Evaluation Report. Japan: NEDO; September 2006.

246. *Polymer electrolyte fuel cell system demonstration research, phase I JHFC project.* Japan automobile research institute and engineering advancement association of Japan; October 2006.

247. Annual budget documents published each year by METI, various dates. <http://www.meti.go.jp/main/31.html>.

248. Homma T. *National policies.* The Latest Fuel Cell News In Japan. FCDIC; February 2006.

249. (Energy Innovation Program) fuel cell system demonstration research basic plan, P09013. Fuel cell hydrogen technology development division, <http://www.nedo.go.jp/activities/portal/gaiyou/p09013/kihon.pdf>, NEDO, no date [accessed December 2010].

250. Comments by project reviewers: project name, fuel cell system demonstration research, An unidentified committee. <http://www.meti.go.jp/committee/notice/2010a/20100514002com08.pdf>, METI, 2010.

251. Sato Y. Overview of scenario, roadmap and R&D projects of hydrogen and FCV in Japan, <http://www.hydrogen.energy.gov/pdfs/9_sato_0610.pdf>, NEDO, 4 June 2010.

252. *FY2008 stationary fuel cell large-scale demonstration project report workshop.* Japan: New Energy Foundation; 10 March 2009.

253. *Stationary fuel cell, Large-scale demonstration project report.* Japan: New Energy Foundation, <http://www.nef.or.jp/happyfc/pdf/h21b_report.pdf>; March 2010.

254. Kawahara M. *Japan's initiative to facilitate the diffusion of fuel cells.* FY2008 SOFC Demonstration Research Result Report Workshop. New Energy Foundation; 3 March 2009.

255. *(Energy Innovation Program) PEMFC practical use promotion technology development basic plan.* Japan: NEDO; 10 February 2010.

256. Heisei 22, PEMFC practical use promotion technology development, research development, item 1. Basic technology development, contractors list. <https://app3.infoc.nedo.go.jp/informations/koubo/koubo/FA/nedokouboplace.2010-06-14.0567338773/nedokoubo.2010-07-12.8038973727/(52256dfb1)59d48a174e885b9a51484e0089a7.pdf>; 14 July 2010.

257. List of contractors engaged in applied research related to PEMFC practical use promotion technology development/basic technology development/degradation mechanism analysis and basic material research for high efficiency cells using nanotechnology. <https://app3.infoc.nedo.go.jp/informations/koubo/koubo/FF/nedokouboplace.2010-08-30.7315445547/nedokoubo.2010-10-08.0070839540/301052256dfb130115171540c5b9f65bd4e885b9a51484e0089a7.pdf>, NEDO, Japan, 12 October 2010.

258. On the opening of fuel cell innovation research center, Media Release, <http://www.uec.ac.jp/about/publicity/media_release/pdf/100827.pdf>, University of Electro-Communication, Tokyo, Japan, 27 August 2010

259. Homma T. The latest news number 173, 2010, <http://www.fcdic.com/eng/news.html>, FCDIC, Japan, 24 September 2010.

260. Vehicle technologies program overview, <http://www.nrel.gov/programs/vehicles.html>, NREL, EERE, DOE, Last Updated, 22 September 2010 [accessed December 2010].

261. Hydrogen fuel: a clean and secure energy future, today's presidential action, For Immediate Release, <http://georgewbush-whitehouse.archives.gov/news/releases/2003/01/20030130-20.html>, The White House, 30 January 2003

262. Research and development: lessons learned from previous research could benefit FreedomCAR Initiative, GAO-02-810T. <http://www.thefreelibrary.com/Research+and+Development%3a+Lessons+Learned+from+Previous+Research.-a089019562>; 6 June 2002.

263. Freedom CAR and fuel partnership, USCAR; no date. <http://www.uscar.org/guest/partnership/1/freedomcar-fuel-partnership> [accessed December 2010].

264. Review of the research program of the freedom car and fuel partnership: first report, Committee on review of the FreedomCAR and fuel research

program, http://www.nap.edu/catalog/11406.html, National Research Council, 2005.

265. Review of the research program of the FreedomCAR and fuel partnership: third report, <http://www.nap.edu/catalog/12939.html>, National Research Council, 2010.

266. Antonio Paparella (European Commission). European prospects for fuel cells in the 6th framework programme, third annual solid state energy conversion alliance (SECA) workshop, Washington, DC; 21–22 March 2002. A presentation, made in Washington, 21 March 2002.

267. Assessing the international position of EU's Research and Technological Development and Demonstration (RTD&D) on hydrogen and fuel cells, EUR 21684, European Commission, 2005.

268. The state and prospects of European Energy Research, comparison of commission, member and non-member states' R&D portfolios, EUR 22397, European Communities, 2006.

269. Why do we need ERA? Understanding ERA. <http://ec.europa.eu/research/era/understanding/why/why_do_we_need_era_en.htm>, European commission, last updated, 7 June 2010 [accessed November 2011].

270. Lars Sjunnesson. High level group on hydrogen and fuel cells. PP Presentation, <http://www.iea.org/work/2003/hydrogen-coIEA/AG52-EUH.PDF> 2003.

271. A strong European technology partnership to move towards the hydrogen economy, IP/04/69, Press Release, <http://europa.eu/rapid/pressReleasesAction.do?reference=IP/04/69>, The European Commission, 20 January 2004.

272. Hydrogen fuel cars 1990–1998, History of hydrogen cars, <http://www.hydrogencarsnow.com/hydrogencars1990-1998.htm>, Hydrogen Cars Now, no date [accessed October 2010].

273. European hydrogen and fuel cell technology platform–HFP, http://www.certh.gr/EEC3F2F7.el.aspx; no date [accessed January 2011].

274. Hydrogen economy: new EU hydrogen and fuel cell Quick Start initiative, IP/04/363, Press Release, <http://europa.eu/rapid/pressReleasesAction.do?reference=IP/04/363&format=HTML&aged=0&language=EN&guiLanguage=en>, The European Commission, 18 March 2004.

275. Focused activities of the european vehicle manufacturers—The FUERO and FUEVA projects, Leaflet. European Commission, <ftp://ftp.cordis.europa.eu/pub/sustdev/docs/energy/sustdev_h2_posters_fuero-fueva.pdf>; no date [accessed January 2011].

276. European fuel cell and hydrogen projects, 1999–2002, EUR 20718, European Commission, 2003.

277. Clean urban transport for Europe, Record control number 57487, Simple Search. <http://cordis.europa.eu/search/index.cfm?fuseaction=proj.document&PJ_LANG=EN&;24&q=4E55C2EDF8D2781089F4B15B27B5ACC1&type=sim>, CORDIS, European Commission; no date [accessed January 2011].

278. Fuel cell bus club, Article, <http://en.wikipedia.org/wiki/Fuel_Cell_Bus_Club>, Wikipedia, last modified on 22 December 2009 [accessed January 2011].

279. European funded research on hydrogen and fuel cells, Review assessment, Future outlook. Hydrogen and Fuel Cell Review Days, Brussels, 10–11 October 2007.

280. Hydrogen for clean urban transport in Europe, CORDIS Search, http://cordis.europa.eu/newsearch/index.cfm?page=docview&collection=EN_PROJ&reference=85636&position=1&Highlights=HYFLEET,> CUTE, CORDI&prevPage=resultList&similarity_id=299129Quality validation date 20 March 2008 [accessed January 2011].

281. Deployment of innovative low power fuel cell vehicle fleets to initiate an early market for hydrogen as an alternative fuel in Europe, CORDIS Search, <http://cordis.europa.eu/newsearch/index.cfm?page=docview&collection=EN_PROJ&reference=86436&position=1&Highlights=HYCHAIN, MINITRAN, MINI, TRAN, CORDI&prevPage=resultList&similarity_id=300031>, European Commission, Quality validation date 30-04-2008 [accessed January 2011].

282. High density power electronics for FC- and ICE-Hybrid electric vehicle powertrains, CORDIS search, <http://cordis.europa.eu/newsearch/index.cfm?page=docview&collection=EN_PROJ&reference=78582&position=2&Highlights=HYSI, CORDI&prevPage=resultList&similarity_id=291895>, European Commission, 3 December 2009 [accessed January 2011].

283. Fuel cell power trains and clustering in heavy-duty transports, CORDIS Search, <http://cordis.europa.eu/newsearch/index.cfm?page=docview&collection=EN_PROJ&reference=75806&position=1&Highlights=FELICITA, CORDI&prevPage=resultList&similarity_id=290731>, European Commission, Quality validation date 22 January 2010 [accessed January 2011].

284. NextGenCell—The next generation of stationary fuel cells, Cordis search, <http://cordis.europa.eu/newsearch/index.cfm?page=docview&collection=EN_PROJ&reference=81407&position=1&Highlights=NEXTGENCEL, CORDI&prevPage=resultList&similarity_id=294799>, The European Commission,

Quality validation date: 03-12-2008 [accessed January 2011].

285. What is FP7? Frequently asked questions (FAQs), <http://cordis.europa.eu/fp7/faq_en.html>, CORDIS, European Commission, last updated 1 December 2011 [accessed May 2012].

286. Ralph-Uwe Dietrich (N.ERGHY). European Research to support fuel cell market introduction, PP Presentation, <http://www.fctedi.eu/symposia/effect/organizercontact/100713_I_6_European_Research_Dietrich.pdf>, International symposium on the introduction of fuel cell systems to early markets, Ulm; 13–14 July 2010.

287. Hydrogen & fuel cell joint technology initiative, JTI hydrogen & fuel cell, FP7 framework programme, <http://www.euresearch.ch/index.php?id=714>, Euresearch, Switzerland, last changed 20 December 2010 [accessed January 2011].

288. Hydrogen and fuel cell activities in France, report completed for the IEA hydrogen coordination group, 26/01/2004, <http://www.iphe.net/docs/Meetings/Germany_2-04/France_Statement.pdf>, ILC-023-04; 18 February 2004.

289. The French National Research Agency, <http://www.euronanomed.net/index.php?option=com_content&view=article&id=39>, The French National Research Agency; no date [accessedFebruary 2011].

290. Axel Strang. Fuel cell and hydrogen: from research dreams to industrial reality. Fuel cell and hydrogen joint undertaking, Stakeholders' General Assembly, Brussels; 9 November 2010.

291. Producing fuel cell, Air Liquide and Hydrogen, <http://www.hydrogen-planet.com/en/air-liquide-une-expertise-industrielle-reconnue-dans-l-hydrogene/ameliorer-les-performances-de-la-pile-a-combustible.html>, Air Liquide, 2008 [accessed May 2011].

292. Axane–fuel cell systems: hydrogen and fuel cells, tomorrow's energy, today, Brochure, <http://www.axane.net/docs/gb/documents/1-%20AXANE%20Fuel%20Cell.pdf>, Axane; no date [accessed May 2011].

293. Juan de Blas CEO (BESEL SA). HYCHAIN-MinitransLessons Learnt, PPT Presentation, <http://www.cessa.eu.com/sd_papers/madrid/CeSSA_Madrid_110_deBlas.pdf>, BASEL; no date [accessed May 2011].

294. Innovative hydrogen station uses fuel cells for power, News, <http://www.fuelcelltoday.com/online/news/articles/2007-05/Innovative-hydrogen-station-uses>, Fuel Cell Today, 20 May 2011.

295. Fuel cell developer axane lights up ski run, News, <http://www.fuelcelltoday.com/online/news/articles/2009-03/Fuel-Cell-Developer-Axane-Lights>, Fuel Cell Today, 18 Mar 2009.

296. Areva, Article, <http://en.wikipedia.org/wiki/Areva>, Wikipedia, last modified on 12 May 2011 [accessed May 2011].

297. Helion, About Helion, <http://translate.google.com/translate?hl=en&sl=fr&u=http://www.helion-hydrogen.com/&ei=rqDWTY25EsfXgQfzmJi9Bw&sa=X&oi=translate&ct=result&resnum=1&sqi=2&ved=0CCMQ7gEwAA&prev=/search%3Fq%3Dhelion%2Bfuel%2Bcell%26hl%3Den%26prmd%3Divns>, Areva; no date [accessed May 2011].

298. Une nouvelle étape dans le développement d'HELION"A new stage in the development of Helion, Press Release, <http://translate.googleusercontent.com/translate_c?hl=en&prev=/search%3Fq%3Dhelion%2Bfuel%2Bcell%26hl%3Den%26prmd%3Divns&rurl=translate.google.com&sl=fr&twu=1&u=http://www.helion-hydrogen.com/images/actualitesdyn/pdf70.pdf&usg=ALkJrhje-1VCazYQClwaGN4bLeXPFWnzDw>, Helion, 21 May 2007.

299. Gilles Lequeux. Status and perspectives of the European programme for stationary fuel cells, Workshop, Brennstoffzellensysteme Für Stationäre Anwendungen, Wirtschaftskammer Österreich, 30.Jänner 2001.

300. Wengel J (Flaunhoffer). Case study on innovation in the energy sector—the example of the fuel cell. OECD; December 2004.

301. Geipel H, Parker N. German hydrogen & fuel cell technology update. Vancouver: IPHE Steering Committee; 28 March 2006.

302. *Hydrogen & fuel cells: review of national R&D programs*. International Energy Agency, OECD; 2004.

303. CEP: report 2002-2007. Germany. <www.cep-berlin.de>; 2007.

304. Klaus Bonhoff (NOW). Electrification of future mobility—national programs and activities in Germany, Annual merit review & peer evaluating meeting| Arlington, Virginia; 18–22 May 2009.

305. *Energy storage & fuel cell systems, R&D Institutes in Germany (Hydrogen and fuel cell technology)*. Germany Trade & Invest GmbH, <http://www.gtai.com/fileadmin/user_upload/Downloads/Industries/Energy_Efficiency/Fuel_Cells/1_Englisch/FactSheet_RD-InstitutesGermany_EnergyStorageFuelCells_January2010_GTAI.pdf>; January 2011.

306. Klaus Bonhoff, Nilgün Parker. Fuel cell and hydrogen activities in Germany, Presentation, <http://www.iphe.net/docs/Meetings/Germany_2-08/Fuel_Cell_and_Hydrogen_Activities_in_Germany.pdf>, IPHE—The 9th Implementation- Liaison Committee Meeting, 20 February 2008.

307. Hydrogen and fuel cell technology innovation programme, National Development Plan Version 2.1, <http://www.nkj-ptj.de/datapool/page/3/300407%20NIP%20EP%202.1_english_final.pdf>, Strategoerat, Wasserstoff Brennstoffzellen, 30 April 2007.
308. National hydrogen and fuel cell technology innovation programme (NIP), Transport and Mobility, <http://www.bmvbs.de/SharedDocs/EN/Artikel/UI/national-hydrogen-and-fuel-cell-technology-innovation-programme-nip.html?nn=37150>, Federal Ministry of Transport, Building and Urban Development; no date [accessed January 2011].
309. Blau J, Germany leads Europe in hydrogen fuel-cell R&D, <http://findarticles.com/p/articles/mi_hb5716/is_201009/ai_n55487034/?tag=content;col1>, Research Technology Management. FindArticles.com; 26 January 2011.
310. Wolfgang Axthammer (NOW). *National Innovation programme Hydrogen and Fuel cell Technology (NIP), The situation in Germany focused on Special Markets*. Ulm: Joint Research Centre (FCTEDI); 13 July 2010.
311. *Germany Launches H2 mobility initiative to expand infrastructure for refueling hydrogen vehicles*. Green Car Congress, <http://www.greencarcongress.com/2009/09/h2-mobility-20090910.html>; 10 September 2009.
312. Brett D, Dornheim M, Dijkum PV, Lorenz H, Rufmann B, Stryk AV, Winkler W. FCTESTNET Fuel Cell Survey by region. Part 4 Europe, FCTESTNET ENG2-CT-2002-20657 03/2006. In: Winkler (leader of FCTESTNET WP 8) W, editor. FCTESTNET, <http://ie.jrc.ec.europa.eu/fctestnet/docs/Finals_WG_8/4_EUROPE.pdf>; March 2006.
313. Michael PD, Maguire J (ETSU). European fuel cells R&D review, AN L-94/46, final report, purchase order No. 062014, Prepared for Argonne National Laboratory; September 1994.
314. *Fuel Cells, Powering the Future, Sustainable Power for the European Union*. EUR 19367 EN. European Commission; 2000.
315. María Yetano Roche (IDMEC). HyWays, first phase of the project, profiling reports (deliverable D3.3), for the six involved member states: France, Germany, Greece, Italy, Norway and The Netherlands, <http://www.hyways.de/docs/deliverables/WP3/HyWays_D3.3_MS_Profiling_Reports_PhaseI_2005_Reedited_08JUN2007.pdf>, HyWays; March/April 2005.
316. Exergy fuel cells, Industry Directory, <http://www.fuelcelltoday.com/online/industry-directory/organisations/ex/Exergy-Inc>, Fuel Cell Today 17 May 2011.
317. OZAWA. Report on the survey of trends on european fuel cell and hydrogen, FY2004 survey report, NEDO Paris Office; March 2005.
318. Roy Williamson (UK Department of Trade and Industry). Fuel cells activity in the UK 2003, <http://www.oecd.org/dataoecd/12/10/31967829.pdf>, DTI, UK; September 2003.
319. The carbon trust, Article, <http://en.wikipedia.org/wiki/The_Carbon_Trust>, Wikipedia, last modified on 7 February 2011 [accessed February 2011].
320. Budget and finances, About Us, <http://www.epsrc.ac.uk/about/facts/Pages/budget.aspx>, Engineering and physical science research council (EPSRC); no date, accessed February 2011.
321. The carbon trust, Article, <http://en.wikipedia.org/wiki/The_Carbon_Trust>, Wikipedia, last modified on 19 April 2011 [accessed May 2011].
322. Plug power, Industry Directory, <http://www.fuelcelltoday.com/online/industry-directory/organisations/pl/Plug-Power>, Fuel Cell Today, 13 Jun 2011.
323. Jingjing Qian, Barbara Finamore, Tina Clegg. Fuel cell vehicle development in China, <https://www.nrdc.org/air/energy/chinadocs/fcv.pdf>, NRDC; no date, accessed May 2011.
324. CPR/01/G31: Demonstration for Fuel-Cell Bus Commercialization in China—Part I, (PIMS #764),<http://www.thegef.org/gef/sites/thegef.org/files/repository/CHINA-Fuel_Cell_Bus-full.pdf>, Global Environment Facility (GEF), United Nations Development Programme, 6 February 2002.
325. *Driven, Are China's car manufacturers ready to compete in the US and Europe? An economist intelligence unit briefing paper sponsored by roland berger strategy consultants*. Economist Intelligence Unit, <http://graphics.eiu.com/files/ad_pdfs/eiu_Roland_Berger_China_Cars.pdf>; 2006.
326. *WRI: 863 program's energy focus—11th five year plan, library & data*. World Resources Institute, <http://www.chinafaqs.org/library/wri-863-programs-energy-focus-11th-five-year-plan>; 1 March 2010.
327. Science and technology in the People's Republic of China, Article, <http://en.wikipedia.org/wiki/Science_and_technology_in_the_People's_Republic_of_China#Torch_Program>, Wikipedia, last modified on 2 May 2011 [accessed May 2011].
328. Hugh M (Fuel Cell Today), Shi Y (Department of Industrial Economic Research of the State Council of China). *Fuel cells in China 2008-policy and progress*. Fuel Cell Today; 2008.
329. Fuel cell vehicles (From Auto Manufacturers), Chart, <http://www.fuelcells.org/info/charts/carchart.pdf>, Fuel Cells 2000, Updated January 2011 [accessed May 2011].
330. Hydrogen fuel cell vehicles will become the second front their own brands, Latest News, <http://www.

hydrogenfuelcelluk.com/news-item.asp?id=171>, WPR Media; 30 March 2009.

331. Wuhan University Chu Tian 1 (2005). Hydrogen and fuel cell vehicles worldwide, <http://www.netinform.de/H2/H2Mobility/Default.aspx?ID=485&CATID=1>, TÜV SÜD Industrie Service; no date [accessed May 2011].

332. Fast Forward: Ballard/China, Stories, <http://www.hfcletter.com/issues/XX_10/stories/483-1.html>, The Fuel Cell And Hydrogen Newsletter, vol. XX, No. 10; October 2005.

333. Ballard power signs supply agreement with shanghai fuel cell vehicle powertrain, News, <http://www.asiapacific.ca/fr/news/ballard-power-signs-supply-agreement-shanghai-fuel-cell-vehi>, Fondation Asie Pacifique du Canada, 5 July 2006.

334. Ballard announces further progress in China, signs memorandum of understanding and fuel cell supply agreement with Shanghai fuel cell vehicle Powertrain Company, Ltd., Press Release, <http://www.marketwire.com/press-release/ballard-announces-further-progress-in-china-tsx-bld-602385.htm>, Ballard, July 05, 2006.

335. Ballard to partner with Shanghai powertrain to develop fuel-cell vehicles for China, Monthly Archives, <http://www.greencarcongress.com/2006/07/ballard_to_part.html>, Green Car Congress, 5 July 2006.

336. Vijayenthiran V. China's SAIC fuel cell Roewe 750, News, <http://www.motorauthority.com/news/1028073_chinas-saic-fuel-cell-roewe-750>, Motor Authority, 12 January 2007.

337. Muldoon D. China's biggest car manufacturer to debut fuel-cell vehicle at Shanghai show. Breaking News, <http://green.autoblog.com/2007/04/10/chinas-biggest-car-manufacturer-to-debut-fuel-cell-vehicle-at-s/>, Autobloggreen; 10 April 2007. Tongji University Fuel Cell PASSAT (2007), Hydrogen and Fuel Cell Vehicles Worldwide, <http://www.netinform.de/H2/H2Mobility/Default.aspx?ID=15&CATID=1>, TÜV SÜD Industrie Service; no date, [accessed June 2011].

338. Weber J. Shanghai-Volkswagen to showcase hydrogen fuel-cell Passat Lingyu at L.A. auto show, News, <http://www.motorauthority.com/news/1030875_shanghai-volkswagen-to-showcase-hydrogen-fuel-cell-passat-lingyu-at-l-a-auto-show> 18 November 2008.

339. Chery—Eastar FCV (2007). Hydrogen and fuel cell vehicles worldwide, <http://www.netinform.de/H2/H2Mobility/Default.aspx?ID=14&CATID=1>, TÜV SÜD Industrie Service; no date, [accessed May 2011].

340. Geely MK fuel cell (2008). Hydrogen and fuel cell vehicles worldwide, <http://www.netinform.de/H2/H2Mobility/Default.aspx?ID=474&CATID=1>, TÜV SÜD Industrie Service; no date, [accessed May 2011].

341. Shen Li Technical Company FC SUV (2008). Hydrogen and fuel cell vehicles worldwide, <http://www.netinform.de/H2/H2Mobility/Default.aspx?ID=456&CATID=1>, TÜV SÜD Industrie Service; no date, [accessed May 2011].

342. SAIC Roewe 750 plug-in FC (2008). Hydrogen and fuel cell vehicles worldwide, TÜV SÜD Industrie Service; no date, [accessed May 2011].

343. SAIC Roewe 750 GM-Fuel Cell (2009). Hydrogen and fuel cell vehicles worldwide, TÜV SÜD Industrie Service; <http://www.netinform.de/H2/H2Mobility/Default.aspx?ID=482&CATID=1>no date, [accessed May 2011].

344. SAIC Motor smartly displays its strength at Shanghai auto show 2011, News Bank, <http://www.saicmotor.com/english/xwzx/xwk/16437.shtml>, SAIC Motor, 19 April 2011.

345. SAIC China fuel cell vehicles plan, Fuel cells, <http://www.thehydrogenjournal.com/displaynews.php?NewsID=321&PHPSESSID=6qcn8cksl24ef5l6fh9p9fksr1>, The Hydrogen Journal, 30 November 2009.

346. Chang'an Z-SHINE FCV (2010). Hydrogen and fuel cell vehicles worldwide, <http://www.netinform.de/H2/H2Mobility/Default.aspx?ID=466&CATID=1>, TÜV SÜD Industrie Service; no date, [accessed May 2011].

347. Tongji University Fuel Cell PASSAT New (2010). Hydrogen and fuel cell vehicles worldwide, <http://www.netinform.de/H2/H2Mobility/Default.aspx?ID=480&CATID=1>TÜV SÜD Industrie Service; no date, [accessed May 2011].

348. Chery Eastar FCEV (2010). Hydrogen and fuel cell vehicles worldwide, <http://www.netinform.de/H2/H2Mobility/Default.aspx?ID=481&CATID=1>, TÜV SÜD Industrie Service; no date, [accessed May 2011].

349. FAW Besturn B70 FCV (2010). Hydrogen and fuel cell vehicles worldwide, <http://www.netinform.de/H2/H2Mobility/Default.aspx?ID=465&CATID=1>, TÜV SÜD Industrie Service; no date, [accessed May 2011].

350. Tsinghua University FC Low Floor City Bus 2. Generation. Wasserstoff- und Brennstoffzellen-Fahrzeuge weltweit, <http://www.netinform.de/H2/H2Mobility/Default.aspx?ID=406&CATID=2>, TÜV SÜD Industrie Service; no date, [accessed May 2011].

351. Tsinghua University + Nanyang Technological University Fuel Cell Bus (2010). Wasserstoff- und

Brennstoffzellen-Fahrzeuge weltweit, <http://www.netinform.de/H2/H2Mobility/Default.aspx?ID=476&CATID=2>, TÜV SÜD Industrie Service; no date [accessed May 2011].

352. Hydrogen and Fuel Cell Vehicles Worldwide. Buses, <http://www.netinform.de/H2/H2Mobility/Default.aspx>, TÜV SÜD Industrie Service; no date [accessed June 2011].

353. ShenLi Technical Company SLFC-Tourist Car 2 (2004). Fahrzeug, <http://www.netinform.info/H2/H2Mobility/H2MobilityMain.aspx?ID=489&CATID=2>, TÜV SÜD Industrie Service; no date [accessed May 2011].

354. ShenLi Technical Company Shanghai Shenwo Fuel Cell Bus (2007), Hydrogen and Fuel Cell Vehicles Worldwide, <http://www.netinform.net/h2/h2mobility/Default.aspx?ID=454&CATID=2>, TÜV SÜD Industrie Service; no date [accessed May 2011].

355. ShenLi Technical Company Shen-Kai No.1 Fuel Cell Bus (2007), Hydrogen and fuel cell vehicles worldwide, <http://www.netinform.net/h2/h2mobility/Default.aspx?ID=453&CATID=2>, TÜV SÜD Industrie Service; no date [accessed May 2011].

356. Shanghai Shenlong Bus SLK6123 SunLong City Bus, Wasserstoff- und Brennstoffzellen-Fahrzeuge weltweit, <http://www.netinform.de/H2/H2Mobility/Default.aspx?ID=455&CATID=2>, TÜV SÜD Industrie Service; no date [accessed May 2011].

357. Higer U90B fuel cell Shanghai (2006). Hydrogen and fuel cell vehicles worldwide, <http://www.netinform.net/h2/h2mobility/Default.aspx?ID=490&CATID=2>, TÜV SÜD Industrie Service; no date [accessed May 2011].

358. Wuhan University FC Midibus (2006). Hydrogen and fuel cell vehicles worldwide, <http://www.netinform.net/h2/h2mobility/Default.aspx?ID=484&CATID=2>, TÜV SÜD Industrie Service; no date [accessed May 2011].

359. Foton Motor FC low floor city bus (2008), Wasserstoff- und Brennstoffzellen-Fahrzeuge weltweit, <http://www.netinform.de/H2/H2Mobility/Default.aspx?ID=447&CATID=2>, TÜV SÜD Industrie Service; no date [accessed May 2011].

360. SAIC Shanghai Automotive Industry Corporation, Wasserstoff- und Brennstoffzellen-Fahrzeuge weltweit, <http://www.netinform.de/H2/H2Mobility/Default.aspx?ID=479&CATID=2>, TÜV SÜD Industrie Service; no date [accessed May 2011].

361. Yutong low floor fuel cell range extender EV city bus. Wasserstoff- und Brennstoffzellen-Fahrzeuge weltweit, <http://www.netinform.de/H2/H2Mobility/Default.aspx?ID=491&CATID=2>, TÜV SÜD Industrie Service; no date [accessed May 2011].

362. Tae-Hwan Lee (Team Leader of National RD&D Organization for Hydrogen and Fuel Cells, MKE, or Ministry of Knowledge Economy; KIST, or Korea Institute of Science and Technology), Hydrogen and fuel cell activities in Korea. Paper prepared for the 2008 Northeast Asia Energy Outlook Seminar, <http://www.keia.org/Publications/Other/LeeFINAL.pdf>, Korea Economic Institute Policy Forum, Washington, DC, 6 May 2008.

363. Butler J. *2010 Survey of Korea*. Fuel Cell Today; 2010.

364. Mercedes-Benz B-Class F-Cell (2009), Hydrogen and fuel cell vehicles worldwide, <http://www.netinform.de/H2/H2Mobility/Default.aspx?ID=423&CATID=0>, TÜV SÜD Industrie Service; no date [accessed June 2011].

365. Hydrogen vehicle.Article, <http://en.wikipedia.org/wiki/Hydrogen_vehicle>, Wikipedia, last modified on 8 September 2010, [accessed September 2010].

366. Ayapana E. Mercedes-Benz hydrogen fuel cell vehicles ready by 2014, according to Daimler Boss, Rumors, <http://rumors.automobilemag.com/mercedes-benz-hydrogen-fuel-cell-vehicles-ready-by-2014-according-to-daimler-boss-55305.html>, Automobile Magazine, 23 June 2011.

367. Mercedes bringing fuel cell cars to showrooms in 2014, Alternative Energy, <http://www.bearmarketinvestments.com/mercedes-bringing-fuel-cell-cars-to-showrooms-in-2014>, Bear Market Investments, 23June 2011.

368. Elliott H. Daimler's Zetsche: China, trucks, fuel cells will lead growth Autos, <http://blogs.forbes.com/hannahelliott/2011/06/21/daimlers-zetsche-china-trucks-fuel-cells-will-lead-growth/>, Forbes, 21 June 2011.

369. Hydrogen fuel cell vehicles (FCVs), Sustainability report 2010/2011, <http://www.corporate.ford.com/microsites/sustainability-report-2010-11/issues-climate-plan-migration-fcv>, Ford; no date [accessed July 2011].

370. Larry E. Hall, The Chevrolet sequel hydrogen fuel cell, Chapter 3 of GM's quest to 'reinvent the automobile' via hydrogen. Article, <http://editorial.autos.msn.com/article.aspx?cp-documentid=435658>, MSN Autos; no date [accessed March 2011].

371. Adrian Chernoff, General Motors HY-WIRE: Touring Sedan. <http://www.adrianchernoff.com/pages/HyWire.html>, AdrianChernoff.com; no date [accessed March 2011].

372. General Motors Sequel. Article, <http://en.wikipedia.org/wiki/General_Motors_Sequel>, Wikipedia, last modified on 25 May 2011, [accessed July 2011].

373. Crawley G. *2007 Light duty vehicle survey.* Fuel Cell Today; February 2007.
374. Joseph N. Project Driveway: GM launches largest ever fuel-cell fleet, Autoblog, <http://www.autoblog.com/2006/09/18/project-driveway-gm-launches-largest-ever-fuel-cell-fleet/> AOL, 18 September 2006.
375. GM HydroGen4, Article, <http://en.wikipedia.org/wiki/GM_HydroGen4>, Wikipedia, 30 June 2011.
376. Simona, 2007 GM HydroGen4, <http://www.topspeed.com/cars/chevrolet/2007-gm-hydrogen4-ar42774.html>, Car Reviews, 3 September 2007.
377. GM HydroGen4 (2008). Wasserstoff- und Brennstoffzellen-Fahrzeuge weltweit,<http://www.netinform.de/H2/H2Mobility/Default.aspx?ID=402&CATID=1>, TÜV SÜD Industrie Service; no date [accessed July 2011].
378. GM unveils second propulsion system for chevrolet volt: a fuel cell variant, Monthly Archives, <http://www.greencarcongress.com/2007/04/gm_unveils_seco.html>, Green Car Congress, 19 April 2007.
379. Healey JR. *GM pushes the pedal on hydrogen fuel-cell power.* Money; 5 November 2007.
380. Yan F. GM aims to be first to make 1 million hydrogen cars: exec, Article, <http://www.reuters.com/article/2007/11/14/us-gm-fuelcells-idUSSHA9988820071114>, Reuters, 14 November 2007.
381. Fuel-cell-powered cadillac provoq unveiled at electronics show, News, <http://www1.eere.energy.gov/vehiclesandfuels/news/m/news_detail.html?news_id=11527>, EERE, DOE, 16 January 2008.
382. Butler J. *2008 light duty vehicle survey.* Fuel Cell Today; May 2008.
383. Lawrence D. Burns. Distinguished Alumni, <http://www.kettering.edu/alumni/distinguished_alumni.jsp>, Kettering University; no date [accessed July 2011].
384. Merva J. Chevrolet equinox fuel cell passes 1 million miles, News, <http://www.gm-trucks.com/news/alt-energy/2009/09/11/chevrolet-equinox-fuel-cell-passes-1-million-miles/>, GM-Trucks.com, 11 September 2009.
385. Colter A. Opel hydrogen cars drive 2 million miles, Articles, <http://uk.ibtimes.com/articles/20110627/opel-hydrogen-cars-drivemillion-miles.htm>, IBTimes, 27 June 2011.
386. Udy J. Opel HydroGen4 helps set new GM fuel cell milestone, Wide Open Throttle, <http://wot.motortrend.com/opel-hydrogen4-helps-set-new-gm-fuel-cell-milestone-88449.html>, Motor Trend, 21 June 2011.
387. Chevrolet equinox fuel cell passes 1 million miles, GM News Center, <http://www.gm.com/vehicles/innovation/news/2010/fuel_cell_031610.jsp>, General Motors, 16 March 2010.
388. Mick J. GM to launch fifth generation fuel cells, commercial hydrogen vehicles in 2015, < http://www.dailytech.com/GM+to+Launch+Fifth+Generation+Fuel+Cells+Commercial+Hydrogen+Vehicles+in+2015/article16346.htm>, Daily Tech, September 25, 2009.
389. Cowan LK. Opel announces it will launch a fuel cell vehicle in 2015, Transportation, <http://inhabitat.com/opel-announces-it-will-launch-a-fuel-cell-vehicle-in-2015/>, inhabitat.com, 1 July 2011.
390. Eric Loveday. Opel looking to launch fuel cell vehicle by 2016, Green, <http://green.autoblog.com/2011/07/07/opel-looking-to-launch-fuel-cell-vehicle-by-2016/>, 7 July 2011.
391. General motors to postpone fuel cell technology, News, <http://www.autoevolution.com/news/general-motors-to-postopone-fuel-cell-technology-37594.html>, SoftNews NET, 1 August 2011.
392. 2005 Honda FCX with breakthrough honda FC stack, News Release, <http://world.honda.com/news/2004/4040901FCXFCStack/>, Honda, 1 September 2004.
393. Fully functional next-generation fuel cell vehicle makes first public debut in U.S., News Release, <http://world.honda.com/news/2006/4061129LAShowFCXConcept/> Honda, 29 November 2006.
394. FCX concept makes its European driving debut, News Release, <http://world.honda.com/news/2007/c070625FCXConcept/>, Honda, 25 June 2007.
395. Honda debuts all-new FCX clarity advanced fuel cell vehicle, dynamic styling and performance herald major advances for fuel cell vehicles; limited marketing to begin next summer, News Release, <http://world.honda.com/news/2007/4071114All-New-FCX/>, Honda, 14 November 2007.
396. Honda unveils Japan model of FCX clarity fuel cell vehicle—leasing in Japan to begin November 2008, News Release, <http://world.honda.com/news/2008/4080702FCX-Clarity/>, Honda, 2 July 2008.
397. Jerram LC. *2009 light duty vehicle survey.* Fuel Cell Today; May 2009.
398. Honda joins clean energy partnership with 2 FCX clarity vehicles, News Release, <http://world.honda.com/news/2011/c110519Clean-Energy-Partnership/index.html>, Honda, 19 May 2011.
399. Hyundai Tucson FCEV (2004), Wasserstoff- und Brennstoffzellen-Fahrzeuge weltweit, <http://www.netinform.de/H2/H2Mobility/Default.aspx?ID=97&CATID=0>, TÜV SÜD Industrie Service; no date [accessed July 2011].

400. Kia's new sportage fuel cell SUV, Monthly Archives, <http://www.greencarcongress.com/2004/09/kiarsquos_new_s.html>, Green Car Congress, 21 September 2004.
401. Hyundai-Kia's fuel cell push, News, <http://www.autocar.co.uk/News/NewsArticle/AllCars/250265/>, AutoCar, 4 June 2010.
402. Todd Kaho. Hyundai i-blue fuel-cell vehicle gets serious about hydrogen, Article, <http://www.greencar.com/articles/hyundai-blue-fuel-cell-vehicle-gets-serious-hydrogen.php>, Demand Media, 11 December 2007.
403. Hyundai Tucson ix35 FCEV (2010), Hydrogen and Fuel Cell Vehicles Worldwide, <http://www.netinform.de/H2/H2Mobility/Default.aspx?ID=431&CATID=0>, TÜV SÜD Industrie Service; no date [accessed July 2011].
404. Eon Youl Shin (Chief Representative, Hyundai Motor Company, Brussels office), Hyundai motor company contribution, European strategy on clean and energy-efficient vehicles Presentation, <http://ec.europa.eu/enterprise/sectors/automotive/files/pagesbackground/competitiveness/contribution_hyundai_en.pdf>, Hyundai, March 2010.
405. Hyundai/Kia—the hydrogen push, Electric Car News, <http://www.carsuk.net/hyundaikia-the-hydrogen-push/>, Cars UK, 6 June 2010.
406. Hyundai Blue2 FCEV unveiled for future drivers, Blog2, <http://www.hydrogencarsnow.com/blog2/index.php/hydrogen-cars/hyundai-blue2-fcev-unveiled-for-future-drivers/>, Hydrogen Fuel Cars Now, 31 March 2011.
407. Hyundai to showcase new thinking at the 2011 Seoul motor show, News, <http://worldwide.hyundai.com/company-overview/news-view.aspx?idx=353&&nCurPage=1&strSearchColumn=&strSearchWord=&ListNum=268>, Hyundai, 31 March 2011.
408. Steven J. Ewing.Hyundai Blue2 fuel cell concept brings 'Intersected Flow' design to Seoul <http://www.autoblog.com/2011/03/31/hyundai-blue2-fuel-cell-concept-brings-intersected-flow-design/>, AOL, 31 March 2011.
409. Nissan X-Trail FCV 2005 Model (2005), Wasserstoff- und Brennstoffzellen-Fahrzeuge weltweit, <http://www.netinform.de/H2/H2Mobility/Default.aspx?ID=54&CATID=1>, TÜV SÜD Industrie Service; no date [accessed June 2011].
410. Hanlon M. Nissan doubles power density with new fuel cell stack Automotive, <http://www.gizmag.com/nissan-doubles-power-density-with-new-fuel-cell-stack/20156/>, Gizmag, October 13, 2011.
411. FC EXPO 2009: 5th International hydrogen and fuel cell expo. Blog to H2 Motors, <http://www.hydrogen-motors.com/hydrogen/2009/03/fc-expo-2009-5th-international-exhibition-of-hydrogen-fuel-cell/>, Hydrogen Motors; March 2009. Nissan X-trail FCV advanced (2009), Wasserstoff- und Brennstoffzellen-Fahrzeuge weltweit, <http://www.netinform.de/H2/H2Mobility/Default.aspx?ID=444&CATID=1>, TÜV SÜD Industrie Service; no date [accessed June 2011]
412. Nissan X-trail FCV advanced (2009), Hydrogen and fuel cell vehicles worldwide, <http://www.netinform.de/H2/H2Mobility/Default.aspx?ID=444&CATID=1>, TÜV SÜD Industrie Service; no date [accessed July 2011].
413. Toyota fuel cell vehicle demonstration program expands, News Release, <http://pressroom.toyota.com/article_display.cfm?article_id=1840>, Toyota 11 January 2010.
414. Toyota FCHV-5 (2001), Wasserstoff-und Brennstoffzellen-Fahrzeuge weltweit, <http://www.netinform.de/H2/H2Mobility/Default.aspx?ID=139&CATID=0>, TÜV SÜD Industrie Service; no date [accessed July 2011].
415. 2009 Washington D.C. auto show, Irv Miller remarks, Washington DC auto show Executive Speeches, <http://pressroom.toyota.com/article_display.cfm?article_id=2360>, Toyota, 3 February 2009.
416. Ohnsman A. *Toyota targets $50,000 price for first hydrogen car (Update2)*. Bloomberg, Business Week; 6 May 2010.
417. Al Ohnsman. Toyota may make 'Tens of Thousands' of hydrogen cars by 2020s, Bloomberg News, <http://www.businessweek.com/news/2012-03-06/toyota-may-make-tens-of-thousands-of-hydrogen-cars-by-2020s>, Bloomberg, 6 March 2012.
418. Fuel cell technology challenges, Fuel Cells, <http://www1.eere.energy.gov/hydrogenandfuelcells/fuelcells/fc_challenges.html?m=1&> EERE, DOE, last Updated 8 March 2011.
419. Jeep Commander 2 (2001), Hydrogen and fuel cell vehicles worldwide, <http://www.netinform.de/H2/H2Mobility/Default.aspx?ID=134&CATID=1>, TÜV SÜD Industrie Service; no date [accessed June 2011].
420. DaimlerChrysler Natrium (2001), Hydrogen and Fuel Cell Vehicles Worldwide, <http://www.netinform.de/H2/H2Mobility/Default.aspx?ID=134&CATID=1>, TÜV SÜD Industrie Service; no date [accessed June 2011].
421. Jeep Treo, Image, <http://www.fuelcelltoday.com/reference/image-bank/Transport/Jeep-Treo>, Fuel Cell Today, 23 June 2011.
422. Chrysler ecoVoyager. Hydrogen Cars, <http://www.hydrogencarsnow.com/chrysler-ecovoyager.htm>,

Hydrogen Fuel Cars Now; no date [accessed June 2011].

423. Chrysler. Article, <http://en.wikipedia.org/wiki/Chrysler>, Wikipedia, last modified on 14 July 2011 [accessed July 2011].

424. Ford Motor Company, Article, <http://en.wikipedia.org/wiki/Ford_Motor_Company#cite_note-66>, Wikipedia, last modified on 2 October 2010 [accessed October 2010].

425. Ford Fusion Hydrogen 999 (2007), Wasserstoff- und Brennstoffzellen-Fahrzeuge weltweit, <http://www.netinform.de/H2/H2Mobility/Default.aspx?ID=13&CATID=1>, TÜV SÜD Industrie Service; no date [accessed June 2011].

426. Ford Airstream (2007), Hydrogen and fuel cell vehicles worldwide, <http://www.netinform.de/H2/H2Mobility/Default.aspx?ID=25&CATID=1>, TÜV SÜD Industrie Service; no date [accessed June 2011]. Ford Edge (2007). Hydrogen and fuel cell vehicles worldwide, <http://www.netinform.de/H2/H2Mobility/Default.aspx?ID=18&CATID=1>, TÜV SÜD Industrie Service; no date [accessed June 2011].

427. Mazda: small passenger car (1997), Wasserstoff-und Brennstoffzellen-Fahrzeuge weltweit, <http://www.netinform.de/H2/H2Mobility/Default.aspx?ID=195&CATID=1>,TÜV SÜD Industrie Service; no date [accessed June 2011].

428. Gearing up, going forward—resuming the offensive—MAZDA annual report 2010, <http://www.mazda.com/investors/library/annual/2010/pdf/ar10_03e.pdf>, Mazda, 2010. <http://www.mazda.com/investors/library/annual/2010/pdf/ar10_03e.pdf>.

429. Homma T. Forefront of development of FCV and other FC mobiles. The Latest Fuel Cell News in JAPAN, FCDIC; November 2006.

430. Renault, Article, <http://en.wikipedia.org/wiki/Renault>, Wikipedia, last modified on 18 June 2011 [accessed June 2011].

431. Daimler, Renault-Nissan in talks on fuel cell. Car News, <http://www.autoweek.com/article/20110419/carnews/110419849>, Autoweek, 19 April 2011.

432. BMW 1 Series Fuel Cell Hybrid (2010), Hydrogen and fuel cell vehicles worldwide, <http://www.netinform.de/H2/H2Mobility/Default.aspx?ID=437&CATID=1>, TÜV SÜD Industrie Service; no date [accessed June 2011].

433. Zero emissions fuel cell hybrid taxi unveiled at London's city hall, Company News, <http://www.intelligent-energy.com/news_events_and_press/news/8/>, Intelligent Energy, 7 June 2010.

434. Lotus engineering zero emissions black cab (2010). Hydrogen and fuel cell vehicles worldwide, <http://www.netinform.de/H2/H2Mobility/Default.aspx?ID=477&CATID=1>, TÜV SÜD Industrie Service; no date [accessed September 2011].

435. Zero emission fuel cell taxi. Fuel Cell Taxi, <http://www.lotuscars.com/engineering/en/fuel-cell-taxi#>, Lotus Engineering, Group Lotus; no date [accessed September 2011].

436. Road legal fuel cell black cabs make first appearance on London's roads, Company News, <http://www.intelligent-energy.com/news_events_and_press/news/86/>, Intelligent Energy, 28 March 2011.

437. Quark–FC range extender (2004), Wasserstoff- und Brennstoffzellen-Fahrzeuge weltweit, <https://www.netinform.de/H2/H2Mobility/Default.aspx?ID=335&CATID=3>, TÜV SÜD Industrie Service; no date [accessed June 2011].

438. Peugeot 207 ePURE (2006), Hydrogen and fuel cell vehicles worldwide, <http://www.netinform.de/H2/H2Mobility/H2MobilityMain.aspx?ID=30&CATID=1>, TÜV SÜD Industrie Service GmbH; no date [accessed June 2011].

439. Peugeot Flux (2007). Hydrogen and fuel cell vehicles worldwide, <http://www.netinform.de/H2/H2Mobility/Default.aspx?ID=12&CATID=1> TÜV SÜD Industrie Service; no date [accessed June 2011].

440. PSA Peugeot Citroen H2Origin (2008), Hydrogen and fuel cell vehicles worldwide, <https://www.netinform.de/H2/H2Mobility/Default.aspx?ID=34&CATID=1>, TÜV SÜD Industrie Service; no date [accessed June 2011].

441. Intelligent energy and PSA Peugeot Citroen deliver H2Origin zero emissions urban delivery vehicle, Company News, <http://www.intelligent-energy.com/news_events_and_press/news/archive/37/>, Intelligence Energy, 23 April 2008.

442. PSA Peugeot Citroën showing FiSyPAC fuel cell extended range electric vehicle demonstrator <http://www.greencarcongress.com/2009/12/fisypac-20091208.html>, Green Car Congress, 8 December 2009.

443. Parrish A. PSA Peugeot Citroën introduces fuel cell battery hybrid with 500 kilometer range, mass production seen in 2020–2025, Blog, <http://nanopatentsandinnovations.blogspot.com/2009/12/psa-peugeot-citroen-introduces-fuel.html>, Nano Patents and Innovations, 14 December 2009.

444. Suzuki, Page, <http://ja.wikipedia.org/wiki/%E3%82%B9%E3%82%BA%E3%82%AD_(%E4%BC%81%E6%A5%AD)>, Wikipedia, last updated 3 April 2011 [accessed April 2011].

445. Johnson D. Suzuki cuts ties with GM on hybrid, fuel-cell development, <http://www.leftlanenews.com/suzuki-cuts-ties-with-gm-on-hybrid-fuel-cell-

development.html>, Left Lane News, 18 January 2010.

446. Suzuki and IE to commercialize FC cars and bikes, <http://www.gizmag.com/suzuki-ev-scooter/21380/>, Gizmag, 8 February 2012.
447. Volvo starts development of fuel cells to improve operating range, Car News, <http://www.carpages.co.uk/volvo/volvo-27-10-10.asp>, Carpages, 27 October 2010.
448. Blackburn R. Hydrogen fuel-cell vehicle charge continues, Article, <http://theage.drive.com.au/green-motoring/hydrogen-fuelcell-vehicle-charge-continues-20110519-1etnu.html>, Fairfax Media, 19 May 2011.
449. The challenge of China's green technology policy Originally posted by perseverance, <http://forum.detik.com/kemajuan-china-t191567p164.html>, Digital Life Detik.com, 27 January 2011.
450. Delzanno P. Hydrogen fuel cell propulsione at Fiat, Presentation, <http://www.iphe.net/docs/Meetings/Italy_11-07/Hydrogen_Fuel_Cell_Proulsione_at_Fiat.pdf>, The IPHE meeting, November 2007.
451. 11.10.2004 All eyes will be on the fiat panda hydrogen concept as the 'hydrogen mobility research and industry embark on a path of convergence' conference opens in Rome. News, <http://www.italiaspeed.com/2004/cars/fiat/10/panda_hydrogen/panda_hydrogen.html>, <www.italiaspeed.com>; 11 October 2004 [accessed July 2011].
452. Fiat takes a first at monte carlo with hydrogen seicento, Monthly Archives, <http://www.greencarcongress.com/2005/04/fiat_takes_a_fi.html>, Green Car Congress, 8 April 2005
453. Fiat Panda hydrogen concept car. Hydrogen Cars, <http://www.hydrogencarsnow.com/fiat-panda-hydrogen-concept.htm>, Hydrogen Fuel Cars Now; no date [accessed July 2011].
454. CRF presents fuel cell powered-panda fleet, Press Release, <http://www.nuvera.com/news/press_release-33.php>, Nuvera Fuel Cells, 21 September 2007.
455. Demonstration of Fuel Cell Vehicles & Hydrogen Infrastructure, Project Zero Regio: November 2004–May 2010. Summary of Results <http://www.zeroregio.com/upload/reports/ZR_report_Abschluss.pdf>, Zero Regio Project Consortium; no date [accessed February 2011]
456. Yoney D. Video: Fiat Phylla powered by sunshine Green, <http://green.autoblog.com/2008/05/26/video-fiat-phylla-powered-by-sunshine/>, AOL, 26 May 2008.
457. Fiat Phyllis, Article, <http://en.wikipedia.org/wiki/Fiat_Phyllis>, Wikipedia, last modified on 30 December 2010 [accessed February 2011].
458. Bologna show 2008: Fiat Phyllis electric and fuel cell Blog of H2 Motors, <http://www.hydrogen-motors.com/hydrogen/2008/12/bologna-show-2008-fiat-phyllis-electric-and-fuel-cell/>, Hydrogen Motors, December 2008.
459. Fiat Phylla concept, Concept Cars, <http://www.carbodydesign.com/archive/2008/05/27-phylla-concept/>, Car Body Design 27 May 2008.
460. Fiat Panda HyTRAN (2009). Hydrogen and fuel cell vehicles worldwide, <http://www.netinform.de/H2/H2Mobility/Default.aspx?ID=461&CATID=0>, TÜV SÜD Industrie Service; no date [accessed July 2011].
461. Per Ekdunge, HyTRAN, Hydrogen and fuel cell technologies for road transport. Presentation, fuel cell and hydrogen joint undertaking stakeholders general assembly, Brussels, Belgium, October 2009, pp. 26–27.
462. Serenergy Fiat Scudo Fuel Cell (2009), Wasserstoff- und Brennstoffzellen-Fahrzeuge weltweit, <http://www.netinform.de/H2/H2Mobility/Default.aspx?ID=442&CATID=0>, TÜV SÜD Industrie Service; no date [accessed July 2011].
463. Electric Fiat Scudo 9-seat minibus with fuel cell range extender. Technical Information, Serenery and Lithium Balance; 15 February 2010.
464. Fiat Panda hybrid fuel cell/battery powered vehicle, Monthly Archives, <http://www.greencarcongress.com/2010/10/quallion-introduces-li-ion-anti-idling-hvac-solution-for-class-6-8-trucks-showcases-battery-pack-in-.html>, Green Car Congress, 9 October 2010.
465. Hydrogen and fuel cell technologies for road transport (HYTRAN), Record control number: 73959 Simple Search, <http://cordis.europa.eu/search/index.cfm?fuseaction=proj.document&PJ_LANG=EN&PJ_RCN=7495259&pid=0&q=1D10762EA2801749C99AFCE9F18B3470&type=sim>, CORDIS Services; no date [accessed February 2011].
466. Fuel cell hybrid vehicle system component development. Record control number: 78586, Simple Search, <http://cordis.europa.eu/search/index.cfm?fuseaction=proj.document&PJ_LANG=EN&PJ_RCN=8326914&pid=0&q=E71A2181935A48C33E27DA3F17E66CE5&type=sim>, CORDIS Services; no date [accessed February 2011].
467. Volkswagen launches fuel-cell vehicles on the road, News, <http://fuelcellsworks.com/news/2009/10/19/volkswagen-launches-fuel-cell-vehicles-on-the-road/>, Fuel Cells Works, 19 October 2009.
468. VW Touran 'Hy-Motion' (2004), Hydrogen and fuel cell vehicles worldwide, <http://www.netinform.

de/H2/H2Mobility/Default.aspx?ID=92&CATID=1> TÜV SÜD Industrie Service; no date [accessed July 2011].

469. Volkswagen Caddy Maxi HyMotion (2009), Hydrogen and fuel cell vehicles worldwide, <http://www.netinform.net/h2/h2mobility/Default.aspx?ID=448&CATID=0>, TÜV SÜD Industrie Service; no date [accessed July 2011].

470. VW fuel cells. ZEITSCHRIFT, <http://vwwatercooled.com/zeitschrift_files/2009-12%20December.pdf>, Volkswagen Group Australia, December 2009.

471. Volkswagen high temperature fuel cell, more efficient and cheaper, Press Release, <http://www.worldcarfans.com/10611025800/volkswagen-high-temperature-fuel-cell>, Volkswagen, 2 November 2006.

472. Volkswagen Tiguan HyMotion at 2007 Bibendum Challenge, Press Release, Volkswagen, <http://www.worldcarfans.com/10711163084/volkswagen-tiguan-hymotion-at-2007-bibendum-challenge/page2>, WorldCarFans.com, 16 November 2007.

473. Prototype of Tiguan HyMotion is the first fuel cell SUV by Volkswagen, Press Release. <http://www.zercustoms.com/news/Volkswagen-Tiguan-HyMotion.html>, Zer Customs.com; 14 November 2007.

474. Harley M. Volkswagen debuts zero-emission Tiguan HyMotion in the U.S., Auto News, <http://www.autoblog.com/2008/06/13/volkswagen-debuts-zero-emission-tiguan-hymotion-in-the-u-s/>, AOL, 13 June 2008.

475. Abuelsamid S. LA 2007: Volkswagen space up! blue, Green, <http://green.autoblog.com/2007/11/14/la-2007-volkswagen-space-up-blue/>, AOL 14 November 2007.

476. Nunez A. LA 2007: VW space up! blue, Auto News, <http://www.autoblog.com/2007/11/14/la-2007-vw-space-up-blue/>, AOL, 14 November 2007.

477. Volkswagen successfully tests its first hydrogen fuel-cell car, News, <http://fuelcellsworks.com/news/2010/08/18/volkswagen-successfully-tests-its-first-hydrogen-fuel-cell-car/>, Fuel Cell Works, 18 August 2010.

478. Audi A2 (2004), Hydrogen and fuel cell vehicles worldwide, <http://www.netinform.de/H2/H2Mobility/Default.aspx?ID=95&CATID=1>, TÜV SÜD Industrie Service; no date [accessed June 2011].

479. Audi Q5 FCEV (2009), Hydrogen and fuel cell vehicles worldwide, <http://www.netinform.de/H2/H2Mobility/H2MobilityMain.aspx?ID=439&CATID=1>, TÜV SÜD Industrie Service; no date [accessed June 2011].

480. Innovations for the future, Audi Environment Brochure, <http://www.audi.in/etc/medialib/ngw/company/environment/pdf/umwelt.Par.0009.File.pdf/pdf_par_0011_file.pdf>, Audi Ag, March 2008.

481. Gikakis C. FTA's National fuel cell bus program, accomplishments and future targets, Presentation, <http://www.ifcbus.com/agenda/Christina%20G.pdf>, Federal Transit Administration, U.S. Department of Transportation, 24 February 2011.

482. Bay area transit agencies propel fuel cell buses toward commercialization, Leaflet, <http://www.nrel.gov/hydrogen/pdfs/48115.pdf>, National Renewable Energy Laboratory (NREL), U.S. Department of Energy, Office of Energy Efficiency and Renewable Energy, July 2010.

483. Kentzler M (Daimler Buses, EvoBus), The CHIIC Project, Presentation, <http://www.fch-ju.eu/sites/default/files/documents/ga2010/monika_kentzler.pdf>, Daimler, 2010.

484. Heliocentris delivers 32 kW fuel cell system for bus project in Mecklenburg-West Pomerania, Press Release, <http://www.heliocentris.com/en/company/press/press-releases/new-details/article/heliocentris-delivers-32-kw-fuel-cell-system-for-bus-project-in-mecklenburg-west-pomerania.html>, Heriocentris, 1 April 2008.

485. Electric bus range extension, Reference System, <http://www.heliocentris.com/index.php?eID=tx_nawsecuredl&u=0&file=fileadmin/user_upload/11_Clean_Energy_Solutions/Datenbl%C3%A4tter/Referenz_Midi_bus_EN_1103.pdf&t=1318619952&hash=10d75057c53257bb701a3cf7f5523b4ce0e5ca44>, Heliocentris; no date [accessed October 2011].

486. "Heliocentris partners with FAUN to develop hybrid garbage truck for BSR, News, <http://www.fuelcelltoday.com/news-events/news-archive/2010/august/heliocentris-partners-with-faun-to-develop-hybrid-garbage-truck-for-bsr>, Fuel Cell Today, 6 August 2010.

487. Fuel cells rescued by batteries, News, Nanowerk, http://www.nanowerk.com/news/newsid=22241.php, Nanowerk; 27 July 2011.

488. Assessment of needs and research roadmaps for rechargeable energy storage system onboard electric drive buses, Report No. FTA-TRI-MA-26-7125-2011.1, <http://ntl.bts.gov/lib/35000/35700/35796/DOT-VNTSC-FTA-11-01.pdf>, Federal transit administration, US Department of Transportation, December 2010.

489. Hydrogenics provides fuel cell bus update, News Details, <http://www.hydrogenics.com/invest/News_Details.asp?RELEASEID=400883>, Hydrogenics, 4 August 2009.

490. Butler J. *2008 Niche transport survey.* Fuel Cell Today; 1 July 2008.

491. Fuel cell-powered forklifts in North America, Charts. <http://www.fuelcells.org/info/charts/forklifts.pdf>, Fuel Cells 2000, updated July 2011.
492. Early markets: fuel cells for material handling equipmentFuel Cell Technologies Program, <http://www1.eere.energy.gov/hydrogenandfuelcells/education/pdfs/early_markets_forklifts.pdf>, EERE, DOE, February 2011.
493. Gangi Je. *Fuel cells 2000's fuel cell technology update–August 2011*. Fuel Cell 2000; August 2011.
494. Fuel cell industry assessment, 2001 Industry Report, Houston Advanced Research Center (HARC), 19 November 2001, Released to Public 20 August 2004.
495. Plug power announces completion of cellex power acquisition, Press Release. <http://www.plugpower.com/News/PressReleases/07-04-04/PLUG_POWER_ANNOUNCES_COMPLETION_OF_CELLEX_POWER_ACQUISITION.aspx>, Plug Power; 4 April 2007.
496. Plug power announces acquisition of general hydrogen, News Release. <http://www.prnewswire.com/news-releases/plug-power-announces-acquisition-of-general-hydrogen-57974857.html>, Plug Power, 7 May 2007.
497. Adamson K-A, Jerrram LC. *2009 Niche Transport Survey*. Fuel Cell Today; August 2009.
498. IdaTech acquires Plug Power's LPG off-grid, backup power stationary product lines, News, <http://www.renewableenergyfocus.com/view/13470/IdaTech-acquires-plug-powers-lpg-offgrid-backup-power-stationary-product-lines/>, AltEnergyMag; 26 October 2010
499. Recovery act, <http://www1.eere.energy.gov/hydrogenandfuelcells/recovery.html>, Financial opportunities, EERE, DOE, last Updated 1 February 2011.
500. Recovery act projects funded for fuel cell market transformation, Financial Opportunities, <http://www1.eere.energy.gov/hydrogenandfuelcells/awards.html>, EERE, DOE, last Updated 27 January 2011 [accessed August 2011].
501. Sysco warehouse opens with fuel cell Forklifts, Palette trucks, Blog. <http://www.hydrogencarsnow.com/blog2/index.php/hydrogenforklifts/sysco-warehouse-opens-with-fuel-cell-forklifts-palette-trucks/>, Hydrogen fuel cars & vehicles; 21 June 2010.
502. Whole foods market to use plug power gendrive power solutions, delivering a healthy planet through fuel cell power, <http://www.plugpower.com/newsroom/pressreleases.aspx?action=details&newsid=304>, Plug Power, NY; 20 October 2009.
503. Plug power helps open hydrogen industrial park with GENCO and Kimberly-Clark Corporation. Press Release, <http://www.plugpower.com/news/pressreleases/11-02-10/PLUG_POWER_HELPS_OPEN_HYDROGEN_INDUSTRIAL_PARK_WITH_GENCO_AND_KIMBERLY-CLARK_CORPORATION.aspx>, Plug Power; 10 February 2011.
504. Plug power deploys 65 fuel cell forklifts in Florida Warehouse, Article, <http://www.fuelcelltoday.com/online/news/articles/2010-07/Plug-Power-Deploys-65-Fuel-Cell->, Fuel Cell Today; 13 July 2010
505. BMW deploys 86 GenDrive fuel cells from plug power, units to run material handling equipment at newly constructed manufacturing facility. Press Release, <http://www.marketwatch.com/story/bmw-deploys-86-gendrive-fuel-cells-from-plug-power-2010-10-12?reflink=MW_news_stmp>, Plug Power; 12 October 2010.
506. Oorja protonics unveils ultra-powerful OorjaPac™; an innovative onboard methanol-based fuel cell charger for material handling vehicles–announces customer success in pilot program with NUMMI. Business Wire, <http://findarticles.com/p/articles/mi_m0EIN/is_2008_March_19/ai_n24928631/?tag=content;col1>, Oorja Protonics; 19 March 2008.
507. Oorja protonics to replace LPC's propane forklifts with electric forklifts retrofitted with Methanol-fuel cell as onboard battery chargers. Press Release, <http://www.oorjaprotonics.com/news/press_releases/LPC_PR.pdf>Oorja Protonics; 11 May 2009.
508. 2010 Fuel cell technologies market report, <http://www.nrel.gov/docs/fy11osti/51551.pdf>, DOE, EERE, June 2011.
509. Deka-Nuvera ready power and power tap specification sheets released, Blog, <http://www.nuvera.com/blog/?p=76>, Nuvera; 7 May 2008.
510. Nuvera announces HEB and East penn contracts, Blog, <http://www.nuvera.com/blog/?p=724>, Nuvera; 11 June 2009
511. East Penn and Nuvera provide ready power units to defense logistics agency. Press Release. <http://www.nuvera.com/news/press_release-42.php>, Nuvera; 10 February 2009
512. Nuvera, East Penn end forklift-related project, News, <http://www.forkliftaction.com/news/forklifts_news_international/Nuvera_Fuel_Cells_Inc__East_Penn_Manufacturing_Co_Inc_6710.aspx>, Forkliftaction.com; 11 February 2009.
513. Nuvera fuel cells introduces power edge, News, <http://www.mhnetwork.com/news/nuvera-fuel-cells-introduces-poweredge/>, Material Handling Network; 12 March 2009.

514. DLA distribution Susquehanna, PA, DoD Alt Fuel Projects, <http://www.dlafuelcells.org/h2-projects/defense-depot-susquehanna-pa-new-cumberland-pa/>, DLA; 10 August 2011.
515. Hydrogenics completes Forklift pilot deployments at GM and FedEx, News Details, <http://www.hydrogenics.com/invest/News_Details.asp?RELEASEID=178333>, Hydrogenics; 1 November 2005.
516. Lift one, linde and hydrogenics deploy fuel cell forklift technology to South Carolina Companies, Company News, <http://www.forkliftnet.com/news/html/newshtml/news_121.html>, Forklitnet.com; 2 August 2007.
517. Fuel cell-powered forklift shown off in South Carolina, Article, <http://www.fuelcelltoday.com/online/news/articles/2007-12/Fuel-cell-powered-forklift-shown>, Fuel Cell Today; 18 December 2007.
518. H2 logic part of European fuel cell forklift demonstration project, FuelCellWorks 4 May 2011; . News, <http://fuelcellsworks.com/news/2011/05/04/h2-logic-part-of-european-fuel-cell-forklift-demonstration-project/>
519. H2 Logic A/S signs strategic supply agreement with Ballard power systems. Press Release, <http://www.h2logic.com/com/news/H2Logic_PR15012008_INT.pdf>, H2 Logic; 15 January 2008.
520. Cutting-edge in fuel cell technology, Home. <http://www.proton-motor.de/>, Proton Motor; no date [accessed August 2011].
521. *Fuel cell forklift from German partnership*. In: *Abstract*, 2003. Fuel Cells Bulletin, Elsevier. 12, <http://www.sciencedirect.com/science/article/pii/S1464285903000324>; December 2003.
522. Munich airport: forklift truck with fuel cell system, Applications, <http://www.proton-motor.de/?id=70>, Proton Motor; no date [accessed August 2011].
523. Proton power systems introduces triple-hybrid forklift: fuel cell, battery and supercaps, Monthly Archives, <http://www.greencarcongress.com/2007/09/proton-power-sy.html>, Green Car Congress; 26 September 2007.
524. Hydrogen and fuel cells: proton motor presents, News, <http://www.netinform.net/Aktuelles_Detail.aspx?ID=1457>, TÜV SÜD Industrie Service; 08 November 2007.
525. *Crown Awarded $1 million for fuel cell forklifts*. Fuel Cells 2000's Fuel Cell Technology Update; April 2008.
526. Crown equipment unveils unprecedented list Of 20 electric forklifts tested and qualified to operate with fuel cells, Archives, Current Article, <http://fuelcelldispatch.com/AutomotivePower/tabid/2704/articleType/ArticleView/articleId/229/Crown-Equipment-Unveils-Unprecedented-List-Of-20-Electric-Forklifts-Tested-And-Qualified-To-Operate-With-Fuel-Cells.aspx>, Fuelcelldispatch.com; 12 August 2011.
527. Ohio funds another crown fuel cell project, News Story, <http://www.forkliftaction.com/news/newsdisplay.aspx?nwid=7070>, Forkliftaction.com; 6 May 2009.
528. Crown unveils fuel cell-powered forklift with in-dash controls, gauges, Current Articles, <http://www.fuelcelldispatch.com/PortablePower/tabid/2705/articleType/ArticleView/articleId/208/Crown-unveils-fuel-cell-powered-forklift-with-in-dash-controls-gauges.aspx>, Fuelcelldispatch.com; 12 August 2011.
529. Linde MH delivers first trucks with fuel cells, News, <http://www.linde-mh.com/en/main_page/news/pressreleases/pressreleases_1_512.jsp>, Linde Material Handling; 27 January 2010.
530. Linde adds fuel cell forklifts to its product range, Article, <http://www.fuelcelltoday.com/online/news/articles/2010-02/Linde-adds-fuel-cell-forklifts-t>, Fuel Cell Today; 5 February 2010.
531. Medwin S. Fuel cells and electric forklift trucks, Presentation, <http://www.hydrogen.energy.gov/pdfs/htac_june2011_raymond.pdf>, The Raymond Corporation; 14 June 2011.
532. Hydrogen and fuel cell material handling vehicles. <http://www.ika.rwth-aachen.de/r2h/index.php?title=Hydrogen_and_Fuel_Cell_Material_Handling_Vehicles&oldid=6105>, Roads2HyCom.com, 9 August 2011.
533. Raymond corp and ballard power sign joint development agreement for fuel cell lift truck research and development, Articles, <http://business.highbeam.com/436102/article-1G1-178093954/raymond-corp-and-ballard-power-sign-joint-development>, HighBeam Business, 21 April 2008.
534. Plug power names Raymond as a distributor, News, <http://www.forkliftaction.com/news/newsdisplay.aspx?nwid=8265>, Forkliftaction.com; 18 March 2010.
535. Plug power announces Q1 2010 financial results, Online News. Fuel Cells Today; 11 May 2010.
536. Plug power and Raymond showcase fuel cell-powered electric lift trucks at UNFI distribution center, Article, <http://business.highbeam.com/438450/article-1G1-233547754/plug-power-and-raymond-showcase-fuel-cellpowered-electric>, HighBeam Business; 4 August 2010.
537. Toyota forklift truck (2005). Hydrogen and fuel cell vehicles worldwide, <http://www.netinform.de/

H2/H2Mobility/Default.aspx?ID=324&CATID=4>, TÜV SÜD Industrie Service; no date [accessed August 2011].

538. Toyota still showing off fuel-cell forklift, Green, <http://green.autoblog.com/2007/01/21/toyota-still-showing-off-fuel-cell-forklift/>, AOL; 21 January 2007.

539. Rulison L. Plugged in for profits, Business; <http://www.timesunion.com/business/article/Plugged-in-for-profits-811982.php>, Timesunion.com; 14 November 2010.

540. Hydrogen forklifts still a long way to go, News. <http://www.forkliftaction.com/news/newsprint.aspx?nwid=9028>, forkliftaction.com; 29 September 2010.

541. Exide technologies and Ballard power systems to develop fuel cell hybrid power systems for Forklift trucks, News Links Clippings, <http://www.green-energy-news.com/nwslnks/clips1007/oct07015.html>, Green Energy News; 9 October 2007

542. Hanlon M. Intelligent energy shows the ENV fuel-cell motorcycle prototype, Urban Transport, <http://www.gizmag.com/go/3857/>, Gizmag, 17 February 2005.

543. Hanlon M. Suzuki and intelligent energy commit to hydrogen motorcycle, Motorcycles, <http://www.gizmag.com/suzuki-and-intelligent-energy-commit-to-hydrogen-motorcycle/9058/>, Gizmag, 26 March 2008.

544. Motive power, markets, intelligent energy. <http://www.intelligent-energy.com/our_markets/motive_power/>; no date [accessed September 2011].

545. Suzuki Crossage (2007). Hydrogen and fuel cell vehicles worldwide, <http://www.netinform.de/H2/H2Mobility/Default.aspx?ID=279&CATID=3>, TÜV SÜD Industrie Service, no date [accessed September 2011].

546. New Suzuki MIO concept to debut at Tokyo auto show, Blog, <http://partmanpartcar.blogspot.com/2009/10/new-suzuki-mio-concept-to-debut-at.html>, Part Man Part Car, 7 October 2009.

547. Chen J. Suzuki shows off prototype fuel cell electric wheelchair, Gadgets, Gismodo, 27 September 2006.

548. Suzuki MIO fuel-cell senior car—Tokyo Motor Show 2009, MIO, <http://www.diginfo.tv/2009/11/04/09-0375-r-en.php>, Digitized Information, 4 November 2009.

549. Suzuki Burgman fuel cell scooter obtains whole vehicle type approval, Company News, <http://www.intelligent-energy.com/news_events_and_press/news/85/>, Intelligent Energy, 9 March 2011.

550. First fuel cell ship goes into service, the heart of the FCS Alsterwasser, the innovative hybrid fuel cell drive, was produced by Proton Motor, Press, <http://www.proton-motor.de/?id=102>, Proton Motor, 29 August 2008.

551. Henderson K. Fuel cell vessel back in service, News, <http://fuelcellsworks.com/news/2011/06/24/fuel-cell-vessel-back-in-service/>, Fuel Cells Works, 16 June 2011.

552. Fuel cell vehicle world survey 2003, by Breakthrough Technologies Institute, <http://www.fuelcells.org/info/charts/vehiclestudy.pdf>, EERE, DOE, Washington D.C., February 2003.

553. Kai Klinder (NOW). Callux and other NIP lighthouse projects in stationary fuel cell applications, FCH JU, Brussels, 26 October 2009.

554. Damian Carrington. Carbon trust funding cut by 40%, Environment, <http://www.guardian.co.uk/environment/2011/feb/14/carbon-trust-funding-cut>, Guandian News and Media, 14 February 2011.

555. Mercedes-Benz F-Cell. Article, <http://en.wikipedia.org/wiki/Mercedes-Benz_F-Cell>, Wikipedia, last modified on 13 June 2011, [accessed June 2011].

556. Panda fuel cell car on display at the battery show, Press Release, <http://www.nuvera.com/news/press_release-8.php>, Nuvera Fuel Cells, 5 October 2010.

557. Daimler teams with linde to build 20 hydrogen fuel stations, Latest, News, <http://www.alsetglobal.com/hydrogen-economy.php?pg=20>, Alset Technology, 5 June 2011.

558. Ion tiger fuel cell unmanned air vehicle completes 23-hour flight, Science News, <http://www.sciencedaily.com/releases/2009/10/091013123350.htm>, Science Daily, 15 October 2009.

559. Boeing successfully flies fuel cell-powered airplane, News Release, <http://boeing.com/news/releases/2008/q2/080403a_nr.html>, Boeing, 3 April 2008.

560. First fuel cell power system for manned flight, Aerospace Systems, <http://www.intelligent-energy.com/products_and_services/fuel_cells/power_systems/aerospace_systems/>, Intelligent Energy, no date [accessed September 2011].

561. DLR motor glider Antares takes off in Hamburg – powered by a fuel cell, News Archives, <http://www.dlr.de/en/desktopdefault.aspx/tabid-4819/8006_read-18278/>, German Aerospace Center (DLR), 7 July 2009.

562. Bergenson A. Insitu announces successful flight of hydrogen-powered UAV ScanEagle, Featured News, <http://www.hydrogenfuelnews.com/insitu-announces-successful-flight-of-hydrogen-powered-uav-scaneagle/852974/>, Hydrogen Fuel News, 8 April 2012.

563. Adamson KA. *2007 Niche transport survey (1)*. Fuel Cell Today; July 2007.

564. Jerram LC, Dehamna A, Adamson KA. 2010 transport fuel cell analysis, snapshot summary, Fuel Cell Today, 2010.
565. 2008 fuel cell technologies market report reveals latest trends, Reviews, <http://www.energyportal.eu/reviews/energy-savings/9062-2008-fuel-cell-technologies-market-report-reveals-latest-trends.html>, EnergyPortal.EU, 14 August 2010.
566. Our home's happy project: residential fuel cell system, FY2009, Pamphlet, <http://www.nef.or.jp/happyfc/pdf/happy_project.pdf>, New Energy Foundation, Japan, February 2010.
567. On subsidy, Outline, <http://www.fca-enefarm.org/subsidy/outline/page05.html>, Fuel Cell Association, 2011.
568. On FY2010 sales record of ENE FARM residential fuel cell systems, Topics, <http://www.digital-research.co.jp/enefarm_2010.pdf>, Digital Research, 21 June 2011.
569. On data, Subsidy, <http://www.fca-enefarm.org/subsidy/data/index.html>, Fuel cell association, 2011.
570. Panasonic Corporation, Article, <http://en.wikipedia.org/wiki/Panasonic_Corporation>, Wikipedia, last modified on 24 August 2011 [accessed August 2011].
571. Homma T. Shifts from PEMFC development to mass production, Latest News, No. 42, <http://www.fcdic.com/index.html>, FCDIC, 27 January 2000.
572. Commercial unit of residential fuel cell cogeneration systems launch into the market, Press Release, <http://www.tokyo-gas.co.jp/Press_e/20041206-2e.pdf>, Tokyo Gas, 6 December 2004.
573. Japan's fuel cell industry, Japan economic report, JETRO, April–May 2006.
574. Adamson KA. 2006 small stationary survey, Fuel Cell Today, November 2006.
575. Panasonic develops new fuel cell cogeneration system, Energy and Green Tech, <http://www.physorg.com/news127483336.html>, PhysOrg.com, 15 April 2008.
576. At the frontline of virtually zero CO_2 emissions, Global home, vol. 3, <http://panasonic.net/eco/feature/column_03/>, Panasonic, last updated 25 January 2010.
577. Okawara K. Public showing of Panasonic's residential fuel cell production facility, Column, <http://kaden.watch.impress.co.jp/cda/column/2008/06/17/2458.html>, Impress Watch Corporation, 17 June 2008.
578. Tokyo gas Panasonic Ene-Farm, High-Tech, <http://www.houseofjapan.com/electronics/tokyo-gas-panasonic-ene-farm>, House of Japan, 10 February 2011.
579. Tokyo gas and Panasonic start marketing new ENE-FARM product, <http://www.digital-research.co.jp/fuelcell_onl_2011_03.html#201103_16>, Digital Research Online, March 2011.
580. Panasonic to expand its Ene-Farm concept to Europe with a new R&D center in Germany, Company News, <http://theasiacareertimes.com/2011/07/panasonic-to-expand-it-ene-farm-concept-to-europe-with-a-new-rd-center-in-germany/>, The Asia Career Times, 4 July 2011.
581. Panasonic to heat up fuel cell market with £2m Cardiff plant, Business Green, 13 April 2012.
582. Toshiba history corner, Toshiba Science Museum, Toshiba Corporation, <http://kagakukan.toshiba.co.jp/history/1goki/1991plant/index.html> [accessed December 2008].
583. Okada O, Yokoyama K. Development of polymer electrolyte fuel cell cogeneration systems for residential applications, Review/Overview, No.1, <http://atmsp.whut.edu.cn/resource/pdf/4209.pdf>, Fuel Cells, January 2001.
584. Maekawa O (Toshiba). Residential fuel cell system of toshiba, <http://www.bids.cz/files/20090513-251-04%20II_1_2_Toshiba%20Fuel%20Cell%20Osamu%20MAEKAWA.pdf>, Toshiba, 4 May 2009.
585. Toshiba to reinforce fuel cell business for residential use in Japan, Press Release, <http://www.toshiba.co.jp/about/press/2004_11/pr1701.htm>, Toshiba, 17 November 2004.
586. Iwasaki W, Tanaka M, Isobe Y. 1kW-Class stationary fuel cell utilizing diversified raw fuels, Feature Articles, Toshiba Review, vol. 63, No 9, <http://www.toshiba.co.jp/tech/review/2008/09/63_09pdf/f02.pdf>, Toshiba, September 2008.
587. 2.1.2 Civilian sector energy consumption trend, Energy white paper 2010, <http://www.enecho.meti.go.jp/topics/hakusho/2010energyhtml/2-1-2.html>, Agency for Natural Resources, and Energy, METI, 2010.
588. ENEOS ECO LP-1 to be on commercial market, World's first LPG residential fuel cell systems will be installed at five celebrities residences, News Release, <http://www.noe.jx-group.co.jp/newsrelease/noc/2004/e71_copr20_20041220_02.html>, Nippon Oil, 20 December 2004.
589. Japan's fuel cell industry trend, Industrial Report, JETRO, 2006.
590. Watanabe N (ENEOS CELLTECH). For the 2nd year with EneFarm commercialization, Foreword, The Journal of Fuel Cell Technology; 9(4) <http://www.fcdic.com/index.html>, Spring 2010.
591. PEMFC system technology development project outline, Presentation, <http://www.nedo.go.jp/

content/100086736.pdf>, NEDO, Hydrogen Energy Technology Office, 6 November 2002.

592. Mohri M (Nippon Oil). Challenges for popularization of residential fuel cell system, Foreword, The Journal of Fuel Cell Technology; 5(4) <http://www.fcdic.com/index.html>, FCDIC, Spring 2006

593. Homma T. Demonstration and business deployment of home use PEFC, The Latest Fuel Cell News in Japan, December 2007.

594. Kawanishi T. Kerosene type home-use fuel cell co-generation system, New Products, Ebara Jiho, No. 217, <http://www.ebara.co.jp/business/tech/pdf/217/217_P41.pdf>, Ebara Ballard, October 2007.

595. Nippon oil to field-test 1kW kerosene fueled residential PEMFC system, News Release. <http://www.noe.jx-group.co.jp/newsrelease/noc/2004/e71_copr20_20040427_01.html>, Nippon Oil and Ebara Ballard, 27 April 2004.

596. Homma T. Development and demonstration of residential PEFC, Latest News, No. 124, <http://www.fcdic.com/index.html>, FCDIC, 18 August 2006.

597. Homma T. Development and commercialization of residential PEFC, Latest News, No. 120, <http://www.fcdic.com/index.html>, FCDIC, 7 April 2006.

598. Homma T. Demonstration results of business-use and home-use PEFC, The Latest Fuel Cell News in Japan, FCDIC, June 2007.

599. Project result data, <http://www.nef.or.jp/happyfc/progress.html>, NEF, no date [accessed August 2011].

600. Homma T. New PEMFC development, Latest News, No. 18, <http://www.fcdic.com/index.html>, FCDIC, 27 October 1997.

601. Homma T. Development and operation tests of residential and stationary PEFC, Latest News, No. 42, <http://www.fcdic.com/index.html>; 27 January 2000.

602. Homma T. Development and demonstration of residential FC, Latest News, No. 73, <http://www.fcdic.com/index.html>; 12 July 2002.

603 Greene DL (Oak Ridge National Laboratory), Duleep KG (ICF International), Upreti G (University of Tennessee). Status and Outlook for the U.S. Non-Automotive Fuel Cell Industry: Impacts of Government Policies and Assessment of Future Opportunities, ORNL/TM-2011/101, <http://www-cta.ornl.gov/cta/Publications/Reports/ORNL_TM2011_101_FINAL.pdf>, Oak Ridge National Laboratory, May 2011.

604. Japan energy to start operating LPG fueled stationary fuel cell, news release. <http://www.noe.jx-group.co.jp/newsrelease/je/2004/20050131_2.html>, Japan Energy, 31January 2005.

605. ENEOS CELLTECH completes new plant to mass produce fuel-cell cogeneration systems for home use, News, <http://fuelcellsworks.com/news/2009/04/28/eneos-celltech-completes-new-plant-to-mass-produce-fuel-cell-cogeneration-systems-for-home-use/>, Fuel Cell Works, 28 April 2009.

606. A world survey of major energy device production facility for secondary batteries, Solar Cells, and fuel cells, Report, <https://www.fuji-keizai.co.jp/market/10111.html>, Fuji-Sankei, 17 November 2010.

607. Telecommunications backup power, Permitting hydrogen facilities, <http://www.hydrogen.energy.gov/permitting/use_power.cfm>, Hydrogen Program, DOE, no date [accessed September 2011].

608. Early markets: fuel cells for backup power. Fuel cell technologies program, <http://www1.eere.energy.gov/hydrogenandfuelcells/education/pdfs/early_markets_backup_power.pdf>, EERE, DOE, December 2010.

609. Fuel cell in backup power applications, clean power generators—an alternative to conventional backup power sources. Technology Installation Review, DOE/EE-0310, <http://www1.eere.energy.gov/femp/pdfs/hydrogenfc_tir.pdf>, EERE, DOE, August 2005.

610. *2010 annual report*. Hydrogenics; 2010.

611. Annual reports, Ballard power systems, various years. <http://www.ballard.com/investors/financial-information/annual-reports.aspx> [accessed October 2011].

612. Apt J, Lave.LB, Morgan MG. Can the U.S. have reliable electricity?" Carnegie mellon electricity industry center working paper CEIC-06–02, <https://wpweb2.tepper.cmu.edu/ceic/pdfs/CEIC_06_02.pdf>, Carnegie Mellon University, February 2006.

613. FCC adopts order implementing katrina panel recommendations, FCC, <http://radiomagonline.com/fcc/fcc-katrina-panel-recommendations/>, NewBay Media, 19 June 2007.

614. 2009 fuel cell market report, <http://www.nrel.gov/docs/fy11osti/49492.pdf>, EERE, DOE, November 2010.

615. DOE announces ARRA-funded fuel cell initiatives (posted 4/17/09), Articles, <http://www.ceramicindustry.com/Articles/Industry_News/BNP_GUID_9-5-2006_A_10000000000000573227>, BNP Media, 17 April 2009.

616. Rohr D (Plug Power), Accelerating acceptance of fuel cell backup power systems, <http://www.hydrogen.energy.gov/pdfs/review11/h2ra007_rohr_2011_o.pdf, 2011>, Annual Merit Review Proceedings, May 2011.

617. Maxwell M, Cohen M (ReliOn). PEM fuel cell systems providing backup power to commercial cellular towers and an electric utility communications

network, <http://www.hydrogen.energy.gov/pdfs/review11/h2ra006_maxwell_2011_o.pdf>, 2011 Annual Merit Review Proceedings, May 2011.

618. Kenny K(Sprint-Nextel). Use of 72-hour hydrogen PEM fuel cell systems to support emergency communications, <http://www.hydrogen.energy.gov/pdfs/review11/h2ra012_kenny_2011_o.pdf>, Annual Merit Review Proceedings, May 2011.

619. Curtin S, Gangi J (Fuel Cells 2000). The business case for fuel cells: why top companies are purchasing fuel cells today, <http://www.fuelcells.org/BusinessCaseforFuelCells.pdf>, Fuel Cell 2000, September 2010.

620. Ballard power systems Inc., annual information form, <http://www.ballard.com/files/PDF/Investors/2011_Ballard_AIF.pdf>, Ballard Power Systems, 9 March 2011.

621. Ballard power systems Inc., notice of annual meeting, management proxy circular and 2007 annual report, <http://www.ballard.com/files/PDF/Investors/Annual_Reports/2007_AR_2008_ProxyCircular.pdf>, Ballard Power Systems, 2007.

622. Plunkett WJ. *Plunkett's engineering & research industry almanac*. E-Book, Prankett Research; 2008.

623. Ballard and IdaTech Sign Supply Agreement for Fuel Cell Backup Power Market, News Release. <http://phx.corporate-ir.net/phoenix.zhtml?c=76046&p=irol-newsArticle&ID=1122609&highlight=>, Ballard Power Systems, 27 March 2008.

624. *Fortnightly Newsletter*. Fuel Cell Today; 9 April 2008.

625. *Ballard power systems Inc., notice of annual meeting, management proxy circular and 2008 annual report*. Ballard Power Systems, <http://www.ballard.com/files/PDF/Investors/Annual_Reports/2008_Annual_Report.pdf>; 2008.

626. Ballard fuel cells to power telecom backup power units for Motorola, News Releases. <http://www.ballard.com/about-ballard/newsroom/news-releases/news07130901.aspx>, Ballard, July 13 2009.

627. Ballard power systems inc., notice of annual meeting, management proxy circular and 2010 annual report, <http://www.ballard.com/files/PDF/Investors/Annual_Reports/2010_Annual_Report___2011_Management_Proxy_Circular.pdf>, Ballard Power Systems, 2010.

628. Ballard and IdaTech extend fuel cell stack supply agreement for backup power through 2013, News Archives, <http://www.fuelcelltoday.com/news-events/news-archive/2012/april/ballard-and-idatech-extend-fuel-cell-stack-supply-agreement-for-backup-power-through-2013>, Fuel Cell Today, 4 April 2012.

629. About Us, ReliOn. <http://www.relion-inc.com/about.asp>; no date [accessed September 2011].

630. Avista Labs completes one year demonstration with army corps of engineers, Avista Labs fuel cell surpasses 90 percent availability objective, News Release. <http://www.relion-inc.com/pdf/04.03.LabsDoD.pdf>, Avista Labs, 21 April 21 2003.

631. The ReliOn Difference, ReliOn. <http://www.relion-inc.com/difference.asp>; no date [accessed September 2011].

632. Garbak J (Fuel Cell Technologies Program). Overview of hydrogen & fuel cell activities, Presentation, International Fuel Cell Bus Conference, EERE, DOE, 24 February 2011.

633. Blanchard J. A roadmap for tier 1 telecom carrier fuel cell installation, Presentation, OSP Expo 2010, October 2010.

634. ReliOn E-2500 fuel cell system chosen as hot product by association of public safety communications officials, News, <http://fuelcellsworks.com/news/2011/09/28/relion-e-2500-fuel-cell-system-chosen-as-%E2%80%9Chot-product%E2%80%9D-by-association-of-public-safety-communications-officials/>, Fuel Cells Works, 28 September 2011.

635. Fuel-cell maker plans to bring European success come to U.S. fuelcellworks.com, 24-Apr-2008.

636. Significant Share Transaction, News Release, IdaTech. <http://uk.finance.yahoo.com/news/ IdaTech-PLC-Significant-Share-afxcnf-3512511887.html?x=0&.v=1>, Yahoo!, 1 April 2011.

637. IdaTech Annual Report 2007. IdaTech. <http://www.IdaTech.com/uploadDocs/ IdaTech-RA07-WEB.pdf>; 2007.

638. Vincent B (Breakthrough Technologies Institute). 2008 Fuel cell technologies market report, <http://www.nrel.gov/docs/fy10osti/48219.pdf>, EERE, DOE, JUNE 2010.

639. IdaTech Annual Report 2008. IdaTech. <http://www.IdaTech.com/uploadDocs/ IdaTech_AR08.pdf>; 2008.

640. IdaTech's partner deploys 43 backup power fuel cell products in Mexico, Press Release. <http://www.IdaTech.com/uploadDocs/IDA_MICROM_Fv2_102109.pdf>, IdaTech, 21 October 2009.

641. Major Contract Win, Up to 30,000 fuel cell system supply agreement signed with ACME Group, Press Release. <http://www. IdaTech.com/uploadDocs/101008%20IDA_ACME_Supply_Agreement.pdf>, IdaTech, 10 October 2008.

642. Supply Agreement Update, Press Release. <http://www. IdaTech.com/uploadDocs/ACME%20Contract%20Update%20092309.pdf>, IdaTech, 23 September 2009.

643. New Distribution Agreement with ACME, Press Release. <http://www. IdaTech.com/uploadDocs/ACME%20Contract%20Update%20111609.pdf>, IdaTech, 16 November 2009.
644. IdaTech Announces Third Expansion of Advanced, for immediate release. <http://www. IdaTech.com/uploadDocs/080910_ IdaTech_Cascadiant.pdf>, IdaTech, 9 August 2010.
645. IdaTech Participates in Hutchison Indonesia Telecommunications Fuel Cell Press Conference, Press Release. <http://www. IdaTech.com/uploadDocs/HCPT_IdaTech%20Press%20Conf%20Final.pdf>, IdaTech, 31 March 2011.
646. Follow on Order for 30 Fuel Cell Systems, Press Release. <http://www. IdaTech.com/uploadDocs/NIP%20b+w%20PR%20072209.pdf>, IdaTech, 22 July 2009.
647. IdaTech's German partner, b+w electronic systems, supplies fuel cell systems to United States Army, News Release. <http://www. IdaTech.com/uploadDocs/b+w_IMOD_FV2.pdf>, IdaTech, 20 October 2009.
648. IdaTech obtains European conformity (CE) Certification of ElectraGen(TM), News, <http://www.prnewswire.co.uk/cgi/news/release?id=153856>, PRNewswire Europe, 16 September 2011.
649. Altergy fast facts, Investor, <http://www.altergy.com/investor/fastfacts.asp>, Altergy, no date [accessed August 2011].
650. Altergy systems signs lease for facility in Folsom, Announcements, <http://www.altergy.com/announcements/facility.asp>, Altergy, 17 October 2005.
651. Altergy introduces two new freedom Power™ systems, Announcements, <http://www.altergy.com/announcements/freedompower.asp>, Altergy, 4 November 2005.
652. Dedication of the World's first automated, high volume assembly line, Announcement, <http://www.altergy.com/announcements/default.asp>, Altergy, 4 June 2007.
653. Altergy increases production capacity, company experiencing growing global demand for fuel-cell system, Announcements, <http://www.altergy.com/announcements/increases_production_capacity.asp>, Altergy, 4 November 2010.
654. Eaton and altergy collaborate to provide fuel cell power solutions to telecommunications and other industries, worldwide, Announcements, <http://www.altergy.com/announcements/eaton.asp>, Altergy, 7 September 2006.
655. EnerSys and altergy collaborate to offer extended runtime power solutions, Announcements, <http://www.altergy.com/announcements/enersys_and_altergy_collaborate.asp>, Altergy, 18 March 2008.
656. EnerSys' Investment in altergy systems expands collaboration to bring green standby power technology to market, Announcements, <http://www.altergy.com/announcements/enersys_expands_with_altergy.asp>, Altergy, 17 December 2009.
657. Spectrum petroleum services signs agreement with altergy systems, Article, <http://www.fuelcelltoday.com/online/news/articles/2008-04/Spectrum-Petroleum-Services-sign>, Fuel Cell Today, 30 April 2008.
658. TransAct energy and altergy systems announce Indian collaboration, Article, <http://www.fuelcelltoday.com/online/news/articles/2010-11/Altergy-Moves-into-India>, Fuel Cell Today, 5 November 2010.
659. Altergy's zero emissions fuel cell systems now providing clean eco-friendly power in Jamaica, Announcements, <http://www.altergy.com/announcements/powering_jamaica.asp>, Altergy, 28 July 2011.
660. Further successful deployments for altergy's freedom poweArticle, <http://www.fuelcelltoday.com/online/news/articles/2011-08/Further-Deployments-for-Altergy>, Fuel Cell Today, 5 August 2011.
661. Altergy's freedom power™ fuel cell systems to power Africa, Announcements, <http://www.altergy.com/announcements/power_africa.asp>, Altergy, 27 August 2010.
662. Altergy, Anglo Pt and the SA Government form new venture to manufacture fuel cell systems, Article, <http://www.fuelcelltoday.com/online/news/articles/2010-09/Altergy-Forms-New-JV>, Fuel Cell Today, 1 September 2010.
663. Altergy ships over 1 million watts of freedom power systems, Announcements, <http://www.altergy.com/announcements/ships_over_million_watts.asp>, Altergy, April 5 2009.
664. Altergy completes largest known fuel cell deployment, Announcements, <http://www.altergy.com/announcements/completes_largest_fuel_cell.asp>, Altergy, 21 November 2009.
665. US Altergy's PEMFC units reached 5 million hour operation. <http://www.digital-research.co.jp/fuelcell_onl_2011_04.html#201104_08>, Digital Research Online, April 2011.
666. Hydrogenics Corporation, HYGS. <http://www.nasdaq.com/symbol/hygs>, NASDAQ, 30 September 2011.
667. Perkins D (Sole Panelist). Administrative panel decision, stuart energy systems corporation and

hydrogenics corporation v. Alexander Morozov, Case No. D2006-1533, <http://www.wipo.int/amc/en/domains/decisions/html/2006/d2006-1533.html>; 31 January 2007.

668. 2007 Annual Report, Hydrogenics, 2007.

669. Hydrogenics unveils regenerative back-up power generator at Las Vegas telecom show, News Release. <http://www.hydrogenics.com/invest/News_Details.asp?RELEASEID=65365>, Hydrogenics, 30 October 2001.

670. Hydrogenics announces successful installation of HyUPS(TM) at a telecommunications cell site in Northern California, News Release. <http://www.hydrogenics.com/invest/News_Details.asp?RELEASEID=87353>, Hydrogenics, 6 August 2002.

671. Hydrogenics announces commercialization of 10 kW power module, News Release. <http://www.hydrogenics.com/invest/News_Details.asp?RELEASEID=121578>, Hydrogenics; 3 November 2003

672. Forward shift: fully integrated hydrogenics power systems, Stationary Power Products. <http://www.hydrogenics.com/fuel/stationary>, Hydrogenics; no date [accessed September 2011].

673. Hydrogenics supplying fuel cell power modules for American Power conversion's new InfraStruXureTM with integrated fuel cells system, News Release. <http://www.hydrogenics.com/invest/News_Details.asp?RELEASEID=165912>, Hydrogenics, 13 June 2005.

674. Hydrogenics reports fourth quarter and full year 2004 results, Earnings Releases. <http://www.hydrogenics.com/invest/Earnings_Releases.asp?YEAR=2005>; 2 March 2005.

675. Hydrogenics awarded supply agreement from American power conversion to deliver up to 500 fuel cell power modules for backup power applications, News Release. <http://www.hydrogenics.com/invest/News_Details.asp?RELEASEID=207097>, Hydrogenics, 10 August 2006.

676. Hydrogenics and APC deliver fuel cell backup power system to Toronto internet service provider, News Release. <http://www.hydrogenics.com/invest/News_Details.asp?RELEASEID=218023>, Hydrogenics, 9 November 2006.

677. Hydrogenics to provide fuel cells to APC for extended run back-up power in data centers, News Release. <http://www.hydrogenics.com/invest/News_Details.asp?RELEASEID=328644>, Hydrogenics, 19 August 2008.

678. Hydrogenics announces termination of manufacturing and supply agreement With APC, News Release. <http://www.hydrogenics.com/invest/News_Details.asp?RELEASEID=398183>, Hydrogenics, 22 July 2009.

679. Hydrogenics to provide fuel cells to CommScope, Inc. for back-up power in telecom centers, News Release. <http://www.hydrogenics.com/invest/News_Details.asp?RELEASEID=342775>, Hydrogenics, 23 October 2008.

680. Hydrogenics announces strategic alliance with CommScope, News Release. <http://www.hydrogenics.com/invest/News_Details.asp?RELEASEID=497840>, Hydrogenics, 9 August 2010.

681. Hydrogenics, EcoPowerTM Cabinet, Product Literature, CommScope, March 2008.

682. Hydrogenics delivers fuel cell backup power generator to bell Canada telecommunication site, News Release. <http://www.hydrogenics.com/invest/News_Details.asp?RELEASEID=187832>, Hydrogenics, 16 February 2006.

683. Hydrogenics receives order for three 12 kW fuel cell power modules for asian telecom market, News Release. <http://www.hydrogenics.com/invest/News_Details.asp?RELEASEID=204928>, Hydrogenics, 25 July 2006.

684. 2008 Annual Report, Hydrogenics, 2008.

685. 2009 Annual Report, Hydrogenics, 2009.

686. Burger A. Intermittent wind power in Germany. <http://cleantechnica.com/2011/10/12/hydrogen-storage-fuel-cell-system-to-smooth-out-intermittent-wind-power-in-germany>, Clean Technica, 12 October 2011.

687. The fuel cell today, Industry Review 2011. <http://www.fuelcelltoday.com/media/1351623/industry_review_2011.pdf>, Fuel Cell Today, 2011.

688. Colpan CO, Dincer I, Hamdullahpur F. *Portable fuel cells—fundamentals, technologies and applications.* Mini-Micro Fuel Cells, Nato Science For Peace And Security Series C: Environmental Security. Springer, <http://www.springerlink.com/content/gw15077872j36740/>; 2008.

689. LaMonica M. Picket-sized fuel cell charges phones for two weeks, <http://asia.cnet.com/pocket-sized-fuel-cell-charges-phones-for-two-weeks-62215008.htm>, CNET, 9 May 2012.

690. Butler J. Portable fuel cell survey 2009, Fuel Cell Today, April 2009.

691. 2010 portable fuel cell analysis, snapshot summary, Fuel Cell Today, 2010.

692. Company profile. <http://www.heliocentris.com/en/company/about-us.html>, Heliocentris; no date [accessed October 2011].

693. Milestones, About Us. <http://www.heliocentris.com/en/company/about-us/milestones.html>, Heriocentris; no date [accessed October 2011].

694. Heliocentris wins energy lab order in Qatar, expands Middle East presence, News, <http://www.renewableenergyfocus.com/view/21233/heliocentris-wins-energy-lab-order-in-qatar-expands-middle-east-presence/>, Elsevier, 10 October 2011.
695. Hydrogenics signs distribution agreement with Heliocentris, Press Release. <http://www.marketwire.com/press-release/Hydrogenics-Signs-Distribution-Agreement-with-Heliocentris-630629.htm>, Hydrogenics, 15 January 2007.
696. Heliocentris and SMA unveil joint prototype of a fuel cell backup system for photovoltaic installations, Press Release. <http://www.heliocentris.com/en/company/press/press-releases/new-details/article/heliocentris-and-sma-unveil-joint-prototype-of-a-fuel-cell-backup-system-for-photovoltaic-installati.html>, Heliocentris, 24 September 2007.
697. Heliocentris starts industrial field test with the Nexa 1200 fuel cell system, News Archives, <http://www.fuelcelltoday.com/news-events/news-archive/2010/december/heliocentris-starts-industrial-field-test-with-the-nexa-1200-fuel-cell-system>, Fuel Cell Today, 3 December 2010.
698. Hugh M. Portable survey 2008, Fuel Cell Today, April 2008.
699. H-RACER named best invention of 2006 by time magazine, Press Release. <http://www.horizonfuelcell.com/file/Horizon_Best_Inventions.pdf>, Horizon, November 2006.
700. Hobby RC industry leaps into the future with hydrogen power, Press Release. <http://www.horizonfuelcell.com/files/HorizonpressreleasehcellFeb2010.pdf>, Horizon; 3 February 3 2010.
701. Preparing America for new jobs in renewable energy, horizon fuel cell develops science education kits for teachers and students, Press Release. <http://www.horizonfuelcell.com/file/renewablenergyeducation.pdf>, Horizon; 18 March 2009.
702. Horizon fuel cell technologies, fuel cell today, Industry Directory. <http://www.fuelcelltoday.com/industry/industry-directory/h/horizon-fuel-cell-technologies>; no date [accessed October 2011].
703. About Horizon, Horizon, About Us. <http://www.horizonfuelcell.com/about_1.htm>; no date [accessed October 2011].
704. World's first zero emission, hydrogen fuel cell jet, Press Release. <http://www.horizonfuelcell.com/file/hyfish.pdf>, Horizon; 6 April 2007.
705. Horizon fuel cell powers new world record in UAV Flight, Press Release. <http://www.horizonfuelcell.com/file/Pterosoardistancerecord.pdf>, Horizon; 1 November 2007.
706. Iai Malat equips BE650-LE unmanned aircraft with Horizon's AEROPAK fuel cell system, Press Release. <http://www.horizonfuelcell.com/news.htm>, Horizon; 23 August 2010.
707. First commercial hydrogen fuel cell power system for unmanned aircraft launched today, Press Release. <http://www.horizonfuelcell.com/news.htm>, Horizon; 11 August 2010.
708. Horizon's fuel cells power the World's first affordable hydrogen car, Press Release. <http://www.horizonfuelcell.com/file/RiversimpleHorizon.pdf>, Horizon; 16 June 2009.
709. Millennium cell and horizon fuel cell provide detailed plans for HydroPak™ portable power product. Press Release. <http://www.horizonfuelcell.com/file/HydroPak.pdf>, Horizon; 3 January 2008.
710. Power generator integrating a cartridge shape fuel cells will be marketed in Japan. Press Release. <http://www.horizonfuelcell.com/file/Horizon_HydroPak_Japan.pdf>, Horizon; 25February 2008.
711. Horizon launches pocket-size fuel cell power plant for portable electronics, Press Release. <http://www.horizonfuelcell.com/files/MinipakPR2010.pdf>, Horizon; 16 June 2010.
712. Fuel cell applications, Article. <http://en.wikipedia.org/wiki/Fuel_cell_applications>; Wikipedia, last modified on 4 August 2011 [accessed October 2011].
713. The technology behind EFOY fuel cells, Technology. <http://www.siriusintegrator.com/Sirius_Technology.html>, Sirius Integrator; no date [accessed October 2011].
714. SFC smart fuel cell, suppliers of fuel cell technologies, Suppliers. <http://www.security-technologynews.com/suppliers/sfc-smart-fuel-cell.html>; no date [accessed October 2011].
715. SFC energy AG–key facts, press kits, <http://www.sfc.com/images/stories/sfc/SFC/e_sfc_key_facts.pdf>, SFC Energy; no date [accessed October 2011].
716. Smart fuel cell A25, The first commercially available Direct-Methanol-Fuel-Cell (DMFC), Product Description. <http://texasrvtravel.com/admin/PDF/fuel_cell.pdf>, Smart Fuel Cell; no date [accessed October 2011].
717. Fuel Cell SFC A50, Products. <http://www.smart-powersystems.com/43_gb.html, M. Hauck/SPI GmbH>; no date [accessed October 2011].
718. SFC introduces new EFOY fuel cell generation for caravanning, Press Release. <http://www.sfc.com/en/sfc-press-release/10845-brennstoffzellen-marktfuehrer-sfc-startet-neue-efoy-brennstoffzellengeneration.html>, SFC Energy; 9 May 2006.

719. The SFC energy product line, Fuel Cells <http://www.sfc.com/en/sfc-fuel-cells-products.html>, SFC Energy; no date [accessed October, 2011].
720. EFOY comfort fuel cell once again in Top 3 of Motorhome Magazine reader survey, news archives. <http://www.fuelcelltoday.com/news-events/news-archive/2012/may/efoy-comfort-fuel-cell-once-again-in-top-3-of-motor-home-magazine-reader-survey>, Fuel Cell Today; 10 May 2012.
721. Panasonic is SFC energy approved partner of SFC. Press Release. <http://www.sfc.com/en/sfc-press-release/11204-panasonic-ist-sfc-energy-approved-partner-von-sfc.html>, SFC Energy, 27 September 2010.
722. Panasonic presents fuel cells for notebook personal computers. <http://www.digital-research.co.jp/fuelcell_onl_2006_01.htm#20060106_01>, Digital research online; January 2006.
723. SFC Energy launches EFOY Pro 2200 XT fuel cell for power generation of off-grid industrial applications, Press Release, <http://www.sfc.com/en/sfc-press-release/11152-sfc-smart-fuel-cell-launcht-efoy-pro-2200-xt-brennstoffzelle-fuer-stromversorgung-von-netzfernen-industrieanwendungen.html>, SFC Energy; 22 March 2010.
724. SFC EFOY fuel cell powers Altec mobile security and fire watch system, News, <http://www.renewableenergyfocus.com/view/7804/sfc-efoy-fuel-cell-powers-altec-mobile-security-and-fire-watch-system/>, Elsevier; 4 March 2010.
725. SFC energy launches JENNY portable fuel cell. Press Release, <http://www.sfc.com/en/sfc-press-release/11087-internationaler-verkaufsstart-der-tragbaren-jenny-brennstoffzelle.html>, SFC Energy; 9 February 2009.
726. SFC Jenny ND terra fuel-cell, Blog, <http://www.hightech-edge.com/sfc-jenny-nd-terra-fuel-cell-methanol-air-electrochemical-cells/9729/>, High-TechEdge.com, 21October 2010.
727. SFC smart fuel cell launches emily 2200 fuel cell, http://www.renewableenergyfocus.com/view/4363/sfc-smart-fuel-cell-launches-emily-2200-fuel-cell/ Elsevier; 5 October 2009.News,

CHAPTER

8

Strengths and Weaknesses of Major Government Fuel Cell R&D Programs: Europe, Japan, and the United States

OUTLINE

Three governments[i] stand out for their high-profile, long-term, large-scale fuel cell R&D policies and programs,[ii] and they are the EU (the European Union in combination with

[i] Three governments are the European Union (the EU in combination with individual European member state governments), the Japanese government, and the US government.

[ii] Combined fuel cell and hydrogen R&D is called fuel cell R&D.

http://dx.doi.org/10.1016/B978-0-444-56325-5.00008-9

individual European state governments), Japan, and the United States. Numerous others are engaged in fuel cell R&D, but these three are distinct leaders in terms of achievements, commitment, endurance, determination, and hard work.

The three governments have vigorously competed with each other for leadership in the field. They have different strengths and weaknesses, which have shaped their current status and which will likely influence their chances for success.

- Many factors have contributed to their respective competitive positions, including the size of their R&D budgets (including both government and industry investment), the government's consistency in implementing policies and programs, the soundness of program evaluations, robustness of the industry, and extent of patenting activities. We will assess each of these factors to appraise aggregate strengths and weaknesses of the three governments.

8.1. FUEL CELL R&D EXPENDITURE: JAPAN INVESTS THE MOST

Japan allocated the largest amount of funds for fuel cell development between 1995 and 2012.[iii] Europe and the United States were distant seconds.

- When government funding and private-sector expenditure are combined, Japan outspent the United States by a factor of two. It outspent Europe by a smaller margin.

8.1.1. Government R&D Funding: Japan Outspends the United States and Europe

Japan. Japan's Ministry of Economy, Trade, and Industry (METI) started to implement fuel cell R&D in the early 1980s and supported it for the past three decades. During the 1980s and 1990s, the Japanese fuel cell budget was fairly modest. In 2000, Japan boosted its budget, outstripping the United States, which was then the global fuel cell leader. Japan saw the global fuel cell race as intensifying and jumped in to compete. It believed that fuel cell technology was vital to its competitiveness and economic growth. Throughout the following decade, the budget for fuel cells eclipsed the rest of the world. By 2009, however, Japan believed that proton exchange membrane fuel cells (PEMFCs) were ready for commercialization, placed small residential PEMFC power generation units on the market, and gradually lowered its fuel cell budget. For the first time in over a decade, Japan's budget went below the US budget by about $30 million in 2010; Japan's budget in 2010 was ¥1.75 billion (about $188 million at a 2010 exchange rate of $1/¥93) and the US budget was $219 million.

It is worthy of note that, even while it was promoting fuel cell commercialization, Japan initiated two long-term basic fuel cell research projects, one for solid oxide fuel cells (SOFCs) and one for PEMFCs. Japan clearly recognized that it did not fully understand the fundamental electrochemistry of fuel cells and decided to go back to basics. The METI move

[iii] The budgets include R&D on fuel cells of all types and hydrogen technology. The US budget includes related programs within the Department of Energy's (DOE) Energy Efficiency and Renewable Energy (EERE) and Fossil Energy (FE) and does not include funding from the Department of Defense or the Department of Commerce. Japan's budget includes fuel cell—related activities of the Ministry of Energy, Trade, and Industry's (METI) budget. The European budget includes funding through the European Commission's Framework Program (EC FP) and funding by individual European countries.

stunned the world, but no other government followed suit.

If all the Japanese fuel cell budgets from 1995 to 2012 are summed, the total is ¥337.6 billion or roughly $3.38 billion at a hypothetical exchange rate of $1/¥100 (Fig. 8.1a). This represents the highest level of fuel cell spending in the world.

The United States. The US government, which has two separate fuel cell budgets, allocated a combined total of about $2.534 billion during the same 1995–2012 period (Fig. 8.1b).[1] It was the second largest next to Japan's budget and larger than the Europe's budget, except that the US budget was often subjected to congressionally earmarks, which funneled funds away from fuel cell R&D. Because of earmarks during the 1995–2012 period, appropriated fuel cell budgets were reduced by at least $132,928,000 to $2.401 billion.[iv]

- The US government fuel cell budget has been on a downward slope since 2009. The US administration requested less than $100 million for fuel cell R&D in 2011, though the Congress restored it to $170 million.

Europe. European spending on fuel cells consisted of European Commission's Framework Program (EC FP) funding and individual state member government spending, which totaled about €2.038 billion from 1990 to 2013 (Fig. 8.1c). The total from 1995 to 2012 was about €1.8106 billion, covering 3 years under the FP4 (1994–1998), FP5 (1998–2002), FP6 (2002–2006), and 6 years under the FP7 (2006–2013). At a hypothetical exchange rate of $1/€0.75, total spending amounted to about $2.414 billion, making it the world's second largest budget, albeit a virtual one, and equal to US cumulative spending (Fig. 8.1d).

- Because of the size and complexity of the EU, it took nearly ten years before Europe substantially increased fuel cell budgets. European efforts in overhauling its fuel cell policies were extensive:
  - *The European Research Area (ERA)* was launched in March 2000.
  - *The High Level Group (HLG) for Hydrogen and Fuel Cell Technologies* was established in October 2002 to coordinate the EC and the EU Member States and compile a strategic research agenda to speed up the introduction of fuel cells and hydrogen.
  - *The European Hydrogen and Fuel Cell Technology Platform (HFP)* was launched in January 2004 to facilitate and accelerate the development and deployment of cost-competitive, world-class European hydrogen and fuel cell technologies.
  - *The Fuel Cells and Hydrogen Joint Technology Initiative (JTI) and a Joint Undertaking on fuel cell and hydrogen (FCH JU)* was established in October 2007. The JTI was administered by the FCH JU, a public–private partnership.

Owing to these extraordinary initiatives, the European fuel cell budget started rising in 2004 and continued to increase throughout the decade. The European budget surpassed the declining US budget and became the world's second largest for the first time in 2011. Europe allocated about $179 million in 2011 (one seventh of the €470 million EC FP was

[iv] There have been a number of Congressional earmarks that directed approved funds to be spent on congressionally directed activities, which perhaps are not originally intended fuel cell and hydrogen R&D. A total earmarks amounted to $132,928,000. It included those directed at DOE FE fuel cell and hydrogen budgets to a total of $8,205,000—comprised of $53,000 in FY1992, $684,000 in FY1993, $689,000 in FY1994, $980,000 in FY1995, $1,227,000 in FY1996, $1,313,000 in FY1997, $1,054,000 in FY1998, $1,131,000 in FY1999, $1,126,000 in FY2000, and $1,426,000 in FY2001—and those directed at EERE fuel cell and hydrogen R&D budgets to a total of $124,723,000—comprised of $41,967,000 in FY2004, $40,236,000 in FY2005, and $42,520,000 in FY2006.

dedicated to fuel cell funding—€67.14 million or about $89.52 million—which doubled to $179 million with contributions from Member State governments).

- By comparison, the Japanese budget in 2011 was ¥16.8 billion or $204 million (at an exchange rate of $1/¥82 in 2011), the world's largest. The US budget was $145 million in 2011, the world's third largest.

8.1.2. Private-Sector R&D Investment: Japanese Government and Industry Together Outspend the United States by a Factor of Two; European Government and Industry Together Outspend the United States by 50 percent

While the government budget is a vital factor for the success of any science and technology

(a) Japan*

*The unit of hundred million ¥ is used for easy reference; although not accurate, an exchange rate of $1/¥100 would help make a quick estimate.

(Total = ¥337.6 billion)

(Hundred Million ¥)

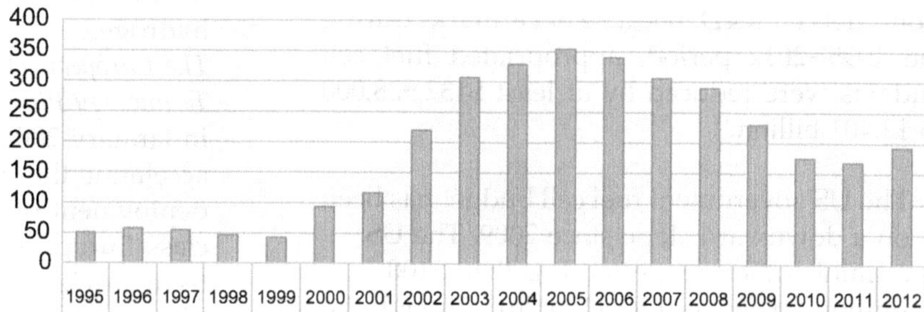

	1995	1996	1997	1998	1999	2000	2001	2002	2003	2004	2005	2006	2007	2008	2009	2010	2011	2012
Japanese Fuel Cell Budget	51	58	55	47	43	94	117	220	307	329	354	340	306	289	230	175	168	193

(b) United States*

(Total = $2.534 billion)

(Million $)

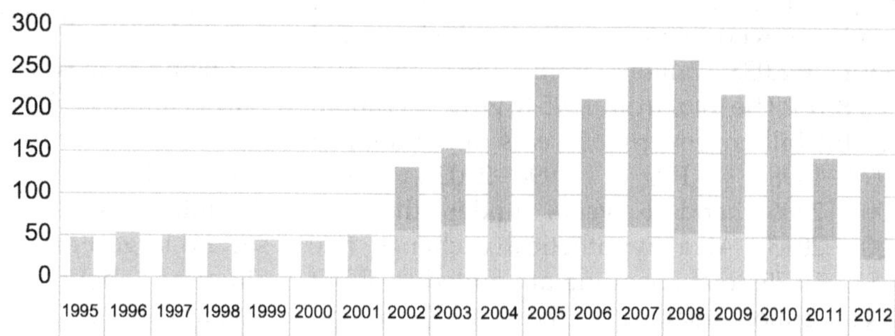

	1995	1996	1997	1998	1999	2000	2001	2002	2003	2004	2005	2006	2007	2008	2009	2010	2011	2012
DOE/EERE Fuel Cell Budget								76	92	144	167	153	190	206	165	170	96	104
DOE/Fossil Fuel Fuel Cell Budget	47	53	50	40	44	43	51	57	62	67	75	60	62	54	56	49	49	25

* During the 1995-2012, the appropriated fuel cell budgets were reduced by at least $132,928,000 to about $2.401 billion.

FIGURE 8.1 Fuel cell budget: 1995–2012*. (For the color version of this figure, the reader is referred to the online version of this book.)

(c) Europe
(Total = €2.038 billion)
(Million €)

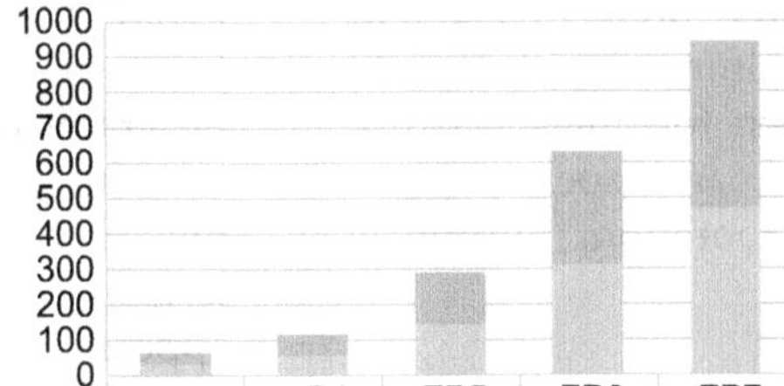

	FP3 (1990-1994)	FP4 (1994-1998)	FP5 (1998-2002)	FP6 (2002-2006)	FP7 (2006-2013)
Member State Funding	32	58	145	314	470
FP Funding	32	58	145	314	470

(d) Japan, US, and European Fuel Cell Spending: 1995-2012* **
(Estimated Billion $)

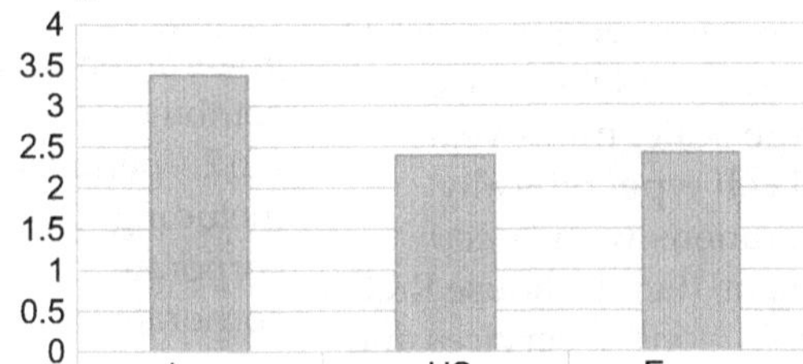

	Japan	US	Europe
1995-2012 Fuel Cell Spending	3.38	2.401	2.414

* European spending covers three years of FP4 (1994-1998), FP5, FP6, and 6 years of FP7 (2006-2013).

**A total of $132,928,000 of congressionally directed appropriations (earmarks) are deducted from the initial US total fuel cell and hydrogen R&D budgets of $2.534 billion, lowering it to $2.401billion.

(e) Estimated Combined Public and Private-Sector Fuel Cell and Hydrogen R&D Investment:
Japan, US, and Europe, 1995-2012
(Billion $)

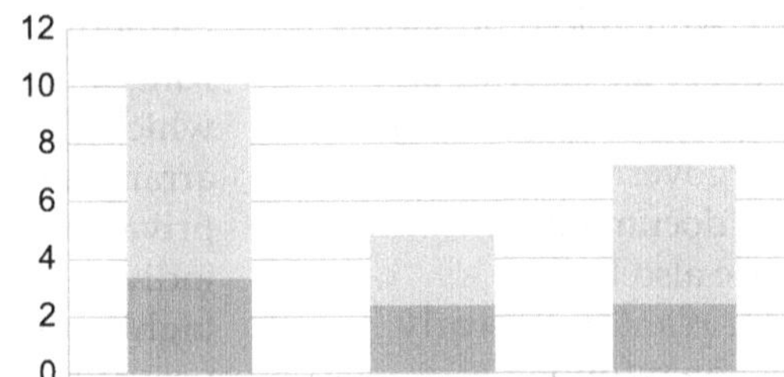

	Japan	United States	Europe
Private-Sector Investment	6.76	2.401	4.8
Government Budget	3.38	2.401	2.4

FIGURE 8.1 (*Continued*).

program, the private sector's willingness to collaborate with the government likely influences the level of corporate R&D investment and thus the program's ultimate outcome. In this regard, Japan has been fortunate because the government, industry, and society place high value on close collaboration. Europe is similar in some respects.

- *Japan.* Characteristically, the government promotes a high level of cooperation with industry, searching for the greatest leverage and return on its funding. The New Energy and Industrial Technology Development Organization (NEDO)—an independent government organization under METI (formerly MITI)—channels government R&D funding into the industry. NEDO focuses its funding on a few, large-scale, multi-year projects for each fuel cell type, typically involving large corporations with "deep pockets" and counting on their in-house R&D capabilities and contributions of services and facilities to leverage government funding. Japanese industry participants share the cost of the projects much like their US counterparts, but perhaps are more willing to spend more of their own funds on government projects than are their US counterparts.
- According to a Japanese industry document, METI allocated $163 million over 6 years for phosphoric acid fuel cell (PAFC) demonstration plants, with Tokyo providing $43 million and Toshiba, Fuji Electric, and participating utilities providing $120 million, nearly three times the government investment. The same document indicates that Mitsubishi Electric also built a demonstration PAFC plant in the early 1990s entirely with its own funds.[2]
- Also, according to diplomatic sources, a delegation of US government and industry officials visiting Japanese fuel cell companies and their plants as early as 1996 believed that, if total corporate cost-share as well as in-kind service contributions were combined, Japan outspent the United States in government-managed fuel cell R&D programs by a factor of more than three-to-one, and perhaps by as much as a factor of five.[3] The US delegation agreed that relative to the United States, Japanese companies appeared to have more trained people and more facilities. The delegation also saw several corporate laboratories with new and extensive (more than 40) test stands and equipment devoted exclusively to fuel cell testing and development.

We assess that Japanese corporations are willing to invest at least twice as much as the government and that Japan's cumulative combined fuel cell R&D spending from 1995 to 2012 perhaps amounted to $10.14 billion, including government funding ($3.38 billion), corporate cost-share ($3.38 billion), and additional corporate R&D spending ($3.38 billion).

- *Europe.* According to diplomatic sources, the EC FP program paid up to 50 percent of the cost of research but did not contribute to the payment of overhead or infrastructure construction, consigning a much greater financial burden to the private sector than in the United States. In this respect, it is similar to the Japanese approach.[4] At the same time, individual governments—including Denmark, France, Germany, Italy, the Netherlands, Switzerland, and the United Kingdom—have their own R&D efforts which also have similar cost-sharing arrangements. Under these circumstances, private-sector participants in EC FP and individual government R&D projects are highly committed to the government programs and are willing to spend twice as much as the public sector.

  We estimate that Europe's total combined government and private-sector spending for fuel cell and hydrogen R&D amounts to $7.2 billion, including EC FP budgets ($2.4 billion), individual European governments' R&D investment ($2.4 billion), and participants' co-share investments ($4.48 billion).

The United States. Private-sector participants in US projects are for the most part not in a strong financial position to invest more than their cost share. They are generally small venture-capital companies, start-ups, and subsidiaries with limited independent financial resources. Most major corporations in the United States may also have become a bit weary of being involved in government R&D projects. If they participate in a project, they typically establish fire walls between themselves and the corporate division involved in the government program, isolating it from in-house funding and resources. The end of government funding thus often results in a sell-off of the division or the end of the R&D activity. At the same time, companies do not wish to make large contributions to government projects. Allied Signal Aerospace started an SOFC division. When it merged with Honeywell, the division was sold to GE. The division was participating in the Department of Energy Solid Energy Conversion Alliance (DOE SECA) project as an industry team, a major role that required providing a major share of the costs. Not surprisingly, GE subsequently withdrew the division from the industry team. GE still remains in the SECA program as a nonindustry team member with no co-payment responsibility. In the past, some major corporations worked in DOE Fossil Energy's molten carbonate fuel cell (MCFC) and PAFC projects—including United Technologies Corporation (UTC) and Siemens Westinghouse—but they were the exception.

We estimate that the US cumulative combined total government and corporate investment in fuel cell and hydrogen R&D from 1995 to 2012 is about $4.8 billion—including government funding ($2.4 billion) and corporate cost-share ($2.4 billion).

We conclude that Japan outspent the United States in combined public and private fuel cell and hydrogen R&D by a factor of two between 1995 and 2012. Europe outspent the United States by about 50 percent (Fig. 8.1e).

8.2. CONSISTENCY IN POLICY AND PROGRAMS: JAPAN IS THE MOST CONSTANT AND STABLE

National R&D policies and programs are formed and shaped by the government. Europe, Japan, and the United States each have a unique style and approach in implementing and administering its fuel cell policies and programs.

Japan. Japanese fuel cell policies and programs have been highly consistent and stable. This has resulted in a steady, high level of funding throughout the past few decades. Moreover, Japan was highly motivated to succeed in fuel cell development. It lacks natural resources, is highly dependent on foreign oil, and is in dire need of new industries for economic growth. The Japanese administrative structure for fuel cell programs also was well-structured and coordinated.

- Japan's fuel cell R&D has been administered by three administrative entities. METI oversees the overall fuel cell R&D program and entrusts day-to-day management to the NEDO. NEDO works in tandem with the AIST, the National Institute of Advanced Industrial Science and Technology. AIST is Japan's largest public research organization with around 2400 researchers, comprising more than 40 autonomous research units under METI. AIST participates in NEDO's fuel cell R&D projects and works in collaboration with industrial participants. It also provides policy and programmatic inputs to METI. The three entities and their work are coherent, focused, and consistent.

The United States. The US fuel cell policy and programs have not been consistent over time. In 1993, the Clinton Administration launched the Partnership for a New Generation of Vehicles—a fuel-efficient, low-emission vehicle R&D program, which included modest fuel cell R&D. In 2003 and 2004, the Bush Administration implemented the FreedomCar and Hydrogen Fuel Initiatives—two major fuel cell and hydrogen vehicle R&D programs. In 2009, the Obama Administration showed little interest in fuel cells or continuing these programs.

- The lack of consistent, long-term focus by the government discourages industry from making long-term R&D plans and investments.

Moreover, the United States has duplicative fuel cell programs. Two different programs have been implemented under two different budgets in two different DOE components. DOE's Office of Fossil Energy (FE) started a fuel cell program in the 1970s and supported R&D on PAFCs, MCFCs, and SOFCs. DOE's Office of Energy Efficiency and Renewable Energy (EERE) started PEMFC R&D in 2003 and gradually expanded its R&D activities to include some SOFCs and other fuel cell types.

- Historically, the dual approach has been justified by the fact that FE and EERE have two different jurisdictions; FE administers fossil fuel and PAFC, MCFC, and SOFC R&D because those fuel cells use fossil fuels; EERE administers renewable energy and PEMFC R&D because PEMFC uses hydrogen.

Nonetheless, these two programs have led to duplicate administrative bureaucracies, two sets of policies and programs, two Congressional budget bills (Energy and Water Development Appropriations and Interior and Related Agencies Appropriations), all of which are not well coordinated, jointly managed, or in any way associated.

Europe. European R&D suffers from fragmentation among EC FP and individual European governments. Fragmentation prevented Europe from fulfilling its research and innovation potential, at a huge cost to Europeans as taxpayers, consumers, and citizens. Starting in 2000, Europe focused on overcoming the problem of fragmentation and launched a revolutionary policy to coordinate activities in 2007.

8.3. SOUNDNESS OF PROGRAM EVALUATION: US EVALUATION IS THE LEAST VALUABLE

European, Japanese, and the US fuel cell policies and programs have created numerous opportunities for reviews, evaluations, and assessments. In general, Japan encourages critical and objective evaluations that have resulted in major changes in policies and programs. Europe promotes such evaluations as well and these also have directly contributed to changes in fuel cell policies and programs. On the other hand, the United States tends to encourage conventional reviews and evaluations that maintain and reinforce the status quo.

- *Japan*. Japanese fuel cell projects typically are multi-year, multi-million dollar activities with major corporate participants. They are expensive and important and project evaluations are serious and critical. Japanese projects were originally modeled after DOE's FE fuel cell projects that started in the 1970s. Japan likely looked around the world and copied the DOE's National Energy Technology Laboratory (NETL) programs, including

PAFC, MCFC, and SOFC R&D, which were leading the world at the time. But the projects were evaluated in a manner that was unique to METI's philosophy and tradition. In general, METI would appoint a panel of committee members, often consisting of about a dozen prominent academicians, industrialists, policy analysts, and scientific commentators. The evaluation process was elaborate with multiple deliberate sessions encompassing a lengthy period of time, and final reports often carried significant recommendations. The reports often included candid comparisons between project outcomes with outcomes of projects of foreign counterparts, primarily in the United States. The reports tried to hammer out possible solutions to problems, again drawing on examples from the United States. In recent years, however, Japan appears to have outgrown this pattern of modeling and comparison; it has started to be more independent and inventive. Indeed, Japan has implemented pioneering R&D policies that neither the United States nor Europe has adopted.

- In 2008, a committee evaluation led METI to launch an SOFC "back to basics" research project. Also in 2010, METI launched a PEMFC basic research project. Both projects aimed to restart R&D activities from scratch. The launch of such projects after having invested in applied research and product development for 20–30 years is a bold undertaking as well as a politically difficult move.

- *Europe.* Major EC FP programs, such as FP5 or FP6 or European-wide fuel cell and hydrogen activities, are evaluated by reliable contractors or a panel of prominent EU policymakers and specialists. The evaluations are judicious, critical, and valuable. They have a global perspective and compare the status and the level of European achievements with those of Japan and the United States. The reviewing comments delve into the reasons for failures and explore future actions and solutions. These evaluations have led the EU to promote a number of significant policy and program changes.
  - It is noteworthy that when Europe searches for new options, it tends to look for US examples in forming new policies and programs.

In contrast, the reviews of individual European FP projects that are published on the EU website (http://cordis.europa.eu/newsearch/index.cfm?page=simpleSearch&language=en) are generally less useful. They often lack information on the outcomes of projects as well as critical comments. This problem has been well recognized in Europe. According to an EC publication, when formulating European-wide research objectives, the EC has to balance various priorities and objectives of individual countries and keep objectives at a general level to facilitate consensus.[5] Setting general objectives makes it difficult to assess the progress of individual research projects and take corrective action. Moreover, nonachievement of goals or project delays are not seriously criticized in Europe, according to the same source.

- *The United States.* DOE projects are reviewed by program managers, peers, and industry experts each year and by a prominent US academic institution every few years. As a general rule, large projects are evaluated by program managers, small projects are reviewed by peers, and a major program is reviewed by an academic institution.

DOE's FE, for the most part, has implemented major projects and awarded large contracts, and so program managers generally conducted evaluations. For example, FE launched PAFC and MCFC programs and commercialized them. FE subsequently launched an SOFC program focused on tubular technology. To FE's credit, it came to recognize shortfalls in the program and tried to mitigate these by instituting a dramatically new SOFC program, the SECA program. FE's new SECA program placed high priority on planar SOFCs or nontraditional tubular SOFC technology, which stunned and delighted the fuel cell industry and revitalized fuel cell R&D activities worldwide.

DOE's EERE, however, typically has implemented large numbers of small projects. Consequently, EERE reviews involve only a few peers with specific expertise who are assigned to evaluate each project following a brief project presentation. The evaluation process rarely has altered the course of the projects unless the outcome was a clear failure or alterations were required because of external events, such as the withdrawal of participants or budget cuts.

- Due to the vast number of projects and the brevity of the time to review them, EERE evaluations tended to be perfunctory and mechanical and included few new ideas that could change the course of the projects. For major programs, such as the FreedomCar and Fuel Partnership, which encompass all EERE fuel cell projects, the National Research Council (NRC) of the National Academy of Sciences has conducted evaluations every few years under contract with DOE. NRC evaluations, however, have provided little ground-breaking advice to the program. The NRC in December 2010, for example, provided its assessment that the program is "effective in progressing toward its goals and that there is evidence of solid progress in essentially all areas, even though substantial barriers remain." This is hardly a helpful evaluation that is worthy of an esteemed national research institution. Over the past decade, NRC evaluations did not recommend significant changes to the EERE fuel cell program.
- Indeed, NRC should have questioned the program's effectiveness, explored the reasons for the lack of progress, searched for innovative solutions, and helped the program work toward a breakthrough.

Another consideration that influences critical reviews of government fuel cell programs is the nature of long-standing government bureaucracy. In the United States, bureaucrats have greater power over policies and programs than do their foreign counterparts, which can be detrimental to program success if changes are warranted. This trend is fairly prominent with DOE's EERE. Government officials in the US civil service are hired to administer a specific program and generally remain in charge of the program for as long as it lasts, which is a fairly long time in most cases. In contrast, Japanese and European government officials are general officers who routinely are transferred to a new position every 2–3 years. It is a truism of bureaucracies that, by staying in a position over time, officials acquire power by building up their networks and relationships and by protecting the projects in which they have invested their time and energy. This often leads to institutions protecting a program's status quo and resisting any move to overhaul it.

- In 2009, for example, a new US administration decided to reduce a fuel cell program and make a smaller budget request. The bureaucrats, industry interests, and the Congress, however, worked to preserve the program by successfully restoring the 2009 budget to the level equal to the previous year.

While instability in government policies and programs can destabilize and retard US industry's R&D plans and investments, it is also true that long-in-the-tooth, uninspired bureaucratic leadership can lead to unsuccessful outcomes for R&D programs in technology areas that are complex and not well understood. Fuel cells are just such an area where solutions require revolutionary thinking.

8.4. RESILIENCE IN INDUSTRY: EUROPE IS THE LEAST STURDY

Fuel cell corporations and businesses of Europe, Japan, and the United States have managed R&D and product development activities differently for each of the five fuel cell types. We have assessed these differences to determine the resiliency of industries in each region. After many decades of hard work and billions of dollars of investment, commercial success in fuel cells remains elusive. Worldwide, numerous companies engaged in fuel cell R&D have failed and left the field, leaving only a handful of companies making efforts to commercialize the technology. None of these appear to be profitable, as yet.

- It is remarkable that the world has continued fuel cell R&D for so long and so persistently with so little commercial benefit.
- Despite this sad record, Japan and the United States remain resilient and determined in their quest for commercial fuel cell products. Europe, however, is falling behind.

8.4.1. Alkaline Fuel Cells

Francis Bacon of the UK built the world's first AFC in 1939. He later developed a 6 kW system and used it to power a forklift, a welding machine, and a circular saw, generating early interest in the novel source of power generation. Europe, Japan, and the United States are all engaged in AFC R&D, but despite advances that have been achieved AFCs face many major challenges before they can become commercially viable.

- *The United States.* Allis Chalmers, Union Carbide Corporation (UCC), and UTC began AFC development in the late 1950s. In 1959, UTC licensed Bacon's patents and successfully developed a practical AFC system that supported the NASA's Apollo moon-landing mission (1963–1972) and NASA's Space Shuttle mission (1981–2011). UCC and Allis Chalmers developed a number of AFC-powered prototype vehicles, including a van, an automobile, a motorbike, and a forklift. They abandoned AFC efforts around 1970; they believed that AFC's susceptibility to CO_2 poisoning was too difficult to overcome for terrestrial applications. UTC also froze its AFC technology in 1973 and has shown little interest in further developing the technology.
- *Europe.* More companies engaged AFC work in Europe than in the United States, starting in the 1950s and 1960s. Companies included Siemens and Varta of Germany, Alsthom, CGE, and IFP of France, Elenco of Belgium, and ASEA of Sweden. No companies, however, succeeded in developing commercially viable AFCs. All have left the field, mostly in the early 1970s.
- *Japan.* Several Japanese companies engaged in AFC work on a limited scale. Between 1961 and 1988, Fuji, Hitachi, Japan Storage Battery, Sanyo, and Matsushita Electric Industrial (now Panasonic) developed AFC systems of about 1–15 kW size. Fuji received government R&D support and had fairly significant R&D activities compared to the others. All of them abandoned AFCs by the late 1980s.

8.4.2. Phosphoric Acid Fuel Cells

The PAFC was considered to be the "first generation" fuel cell to be commercialized in the civilian markets. It has already proven its reliability. In the early 1960s, two US researchers G. V. Elmore and H. A. Tanner successfully experimented with PAFCs, triggering an unprecedented level of R&D activity and the staging of demonstration projects in the United States and Japan. Vigorous PAFC activities went on until the mid-1990s, followed by PAFC commercialization in both countries. Europe did not engage in PAFC R&D. PAFCs are inherently less powerful than other fuel cell types, can be heavy and bulky, and require expensive platinum catalysts and corrosion resistant materials. Today, many experts see this technology as declining because of these limitations.[6]

- *The United States.* The US military and public utilities started to sponsor PAFC R&D programs in the mid-1960s. In the mid-1970s, DOE started its R&D and field-test program. A number of US companies, including Engelhard, ERC (Energy Research Corporation), UTC, and Westinghouse, participated in the utility- and government-funded projects. In 1992, DOE decided that PAFC technology was sufficiently mature for commercialization and ended R&D funding. Just prior to the termination of the DOE funding, all the US corporations, except for UTC, abandoned their PAFC work. In 1993, UTC started marketing its commercial PAFC products and sold hundreds of units worldwide. These products likely have not yet become profitable. In May 2012, UTC reportedly was evaluating the future of its fuel cell business and could sell UTC Power, the subsidiary responsible for fuel cells.[7]
- *Japan.* Japanese utilities first got interested in PAFCs in the 1960s. Two gas companies participated in the US TARGET Program (1967–1976) and field-tested UTC PAFC power plants in Japan. Japanese electric machinery manufacturers—including Fuji Electric, Hitachi, Mitsubishi Electric, Toshiba, and Sanyo Electric—launched PAFC R&D in the 1970s, and METI started sponsoring a series of R&D projects programs in 1981. METI sponsorship of PAFCs continued until 1996, when Fuji Electric began the commercialization of PAFC systems. Fuji is still engaged in PAFC R&D and commercialization but has shown few signs of success.

8.4.3. Molten Carbonate Fuel Cells

MCFCs are more advanced than SOFCs for large-scale applications and the technology has been commercialized since 2003. MCFCs use low-cost materials, but typically have very limited life span due to the corrosive electrolyte. They also have the lowest power density of any of the major fuel cell types. There is no current research program to change the situation. The concept of an MCFC dates back to the 1950s when Dutch scientists G. H. J. Broers and J. A. A. Ketelaar began their fuel cell research at the University of Amsterdam. Subsequent to the end of the Dutch effort, a number of countries have shown interest in MCFC technology, including the United States, Japan, and eventually some European countries, including the Netherlands (again).

- *The United States.* Starting in the mid-1960s, the US government, utility companies, and manufacturers became active in MCFC development. The US Army first funded MCFC development in the mid-1960s. DOE started to get interested in the mid-1970s and funded the development and demonstration of MCFC power plants. US MCFC funding continued for nearly 25 years until 2000. Several companies undertook MCFC R&D, including Fuel Cell Energy (FCE), GE, M-C Power, and UTC. By 2000, FCE was the only MCFC developer left in the United States. The company has been engaged in MCFC marketing activities since 2003, but has made no profits as yet.
- *Japan.* The Japanese MCFC program was a major undertaking, nearly equal to that of the United States. In 1981, Japan's METI implemented a 24-year program that

continued until 2004, with a total budget of about \$470 million. Several companies participated, including Fuji Electric, Hitachi, IHI, Hitachi, Mitsubishi Electric, Sanyo Electric, and Toshiba. Most of them sought partnership with US MCFC developers to gain access to their expertise. When the METI project ended in 2004, IHI was the only remaining MCFC developer in Japan. The company started to sell MCFC products in 2005 but sold few. IHI likely is no longer engaged in the MCFC business.
- *Europe*. Some European countries, including the Netherlands, Italy, and Germany, revived MCFC development after the Chernobyl incident took place in April 1986. In 1987, the EC launched MCFC R&D under the FP2 (1986–1990). Europe worked to foster ECN of the Netherlands, Ansaldo Fuel Cells of Italy, and CFC Solutions of Germany. ECN sold its MCFC technology to FCE, United States, in 2005 and left the field. CFC Solutions was primarily interested in repackaging the US MCFC technology for the European market but went out of business in 2010. As of November 2011, Ansaldo is the only remaining company engaged in MCFC development.

8.4.4. Solid Oxide Fuel Cells

The technology for SOFCs took more than a century to develop. In 1899, Walther Hermann Nernst at Leipzig's Physicochemical Institute discovered the world's first SOFC, but it was not until the late 1950s that the idea of a high temperature fuel cell that would tolerate carbon monoxide caught the attention of researchers. Among them were scientists at Westinghouse, including J. Weissbart and R. Ruka, who built the first modern SOFCs in 1962 and D.H. Archer and his colleagues who demonstrated a multi-cell SOFC generator in 1965. The technical challenges for SOFCs remain stack reliability, thermal cycling resistance, and cost. SOFCs are not suitable for transport applications, but are considered useful for niche applications such as APUs for trucks.[6] SOFCs, like MCFCs, also are considered suitable for large-scale power applications (generating several megawatts), where the fuel cells could be linked with gas turbines in a combined cycle to achieve very high efficiencies. Today, numerous companies are engaged in SOFC R&D, and some have started marketing their products.

- *The United States*. In 1977, DOE started to support Westinghouse's tubular SOFC R&D. DOE sponsorship continued for nearly 30 years until 2005, with total cumulative funding of \$264 million. In the early 2000s, Siemens Westinghouse launched a large number of precommercial demonstration projects. None of these were successful, and the company abandoned the tubular SOFC. In 2001, DOE launched a new, 10-year, \$500 million SOFC R&D program—the Solid State Energy Conversion Alliance (SECA)—to develop planar and other SOFCs different from the traditional tubular type. The DOE's new SOFC program infused a sense of renewal in the fuel cell industry worldwide. Nonetheless, SECA's future is far from certain today, and the overall performance of US SOFCs has lagged behind that of Japan.
  - Bloom Energy has launched commercial activities and appears to be successful in marketing its products. Still, the company keeps its technology and financial information largely secret, and net profitability is unknown.
- *Japan*. METI launched a 13-year, 3-phase SOFC R&D program (1989–2001) in 1989. The goal was to develop planar SOFC technology using a wet method that was expected to produce inexpensive SOFCs. In

retrospect, METI thoroughly underestimated the difficulty of the technology; it has been sponsoring a series of SOFC R&D projects ever since. It is noteworthy that METI in 2008 launched a 5-year SOFC basic research program (2008–2012), which suggests that Japan recognized that continuous applied research and product development activities would not bring about needed breakthroughs. A greater number of companies and organizations are engaged in SOFC R&D in Japan than in Europe and the United States. The overall performance data indicate that Japan is the global leader in SOFC technology today. Still, no one has launched commercially successful SOFC products on the market.

- *Europe*. European SOFC development has been less resilient and less robust than in the United States and Japan. Europe started SOFC technology in the 1960s but dropped out in 1967. The reason appeared to be difficulties in raising the efficiency and lowering the cost of tubular SOFCs. Europe renewed its interest in SOFCs in 1986, triggered by the Chernobyl incident and a new hope for planar SOFCs. Several European countries began SOFC R&D, including Denmark, Finland, Germany, Italy, the Netherlands, Switzerland, and the United Kingdom. Nonetheless, few of them engaged in SOFC development from the ground up; they instead opted to acquire technology from other European countries. Finland outsourced SOFC stacks to Denmark and was not interested in developing its own technology. Italy indicated a desire to stand up a domestic SOFC industry but did not want to develop its own indigenous technology. Germany successfully attracted the Australian company Ceramic Fuel Cells Limited (CFCL) to build a major manufacturing base in Germany. CFCL appears to be highly successful, but as of November 2011, the company has no net profit.[8]

8.4.5. Proton Exchange Membrane Fuel Cells

PEMFCs are viewed as a new generation of fuel cells that are versatile and can power anything from small electronic devices to cars, buses, and submarines, as well as function in stationary power generation systems.[6] The modern history of PEMFCs started in the United States in the early 1950s, when a number of companies and institutions began basic PEMFC development. Among them, GE was the most noteworthy. In 1955, GE's Willard T. Grubb and Leonard Niedrach invented an innovative fuel cell, which was simpler in design and operation than Grove's. Three years later, Niedrach devised a way of depositing platinum onto the fuel cell, which accelerated electrochemical reactions and significantly improved fuel cell efficiency. PEMFCs were believed to have the largest potential markets, prompting many companies to begin PEMFC R&D. Some niche commercial markets are emerging in which PEMFCs are used as a small power source for materials handling vehicles or as an auxiliary power unit, range extender, backup power, or battery charger. On the other hand, there have been few bona fide commercial successes where PEMFCs are used as the primary power source for buses, cars, or for distributed power generation systems or consumer electronic devices. PEMFCs appear to have many outstanding technical barriers, including the membrane (sensitivity to water, to CO, operating temperature), electrode costs (cost of platinum), and the stability of the stack and the fuel flexibility (need for high purity hydrogen).

- *The United States*. In the late 1950s, NASA sponsored GE's PEMFC technology for deployment in a series of manned space missions. A massive infusion of funding was made for space-use fuel cell development, reaching a maximum of more than $20 million (in 1990 dollars) during the mid-1960s.

Nonetheless, GE's PEMFC was not selected for the manned space mission, ultimately leading GE to abandon PEMFC R&D in 1984. GE's PEMFC technology has remained dormant at NASA since the 1960s. In the early 1980s, it was licensed to Canada's Ballard, which became the world's leader in PEMFC technology. In the early 2000s, the United States launched a fuel cell car and hydrogen R&D program to develop PEMFCs to power automobiles, but no commercially viable fuel cell cars have been placed on the market yet. On the other hand, a number of companies have started to sell PEMFC products in niche markets. Plug Power and Oorja Protonics are marketing forklifts and materials handling products, and ReliOn, IdaTech, and Altergy are marketing backup power systems. It remains to be seen whether these few markets will be able to establish the roots needed to grow a large-scale global industry.

- *Japan*. Japanese government PEMFC R&D policy is focused primarily on stationary applications. In 1992, METI implemented a modest 8-year PEMFC system technology development project to develop a 1 kW module for stationary and portable applications (1992–1995) and 5–10 kW portable, residential, and distributed power generation systems (1996–2000). METI then followed it up with a series of PEMFC R&D projects for another 10 years, but METI clearly was not satisfied with the results. In 2010, METI launched a 5-year project (2010–2014) to conduct basic research on PEMFCs. Today, a large number of Japanese companies are engaged in PEMFC development, including Daihatsu, Eneos Celltech, Hino, Honda, Mazda, Mitsubishi Motors, Nissan, Panasonic, Suzuki, Toshiba, and Toyota. Three Japanese companies began commercial sales of CHP systems in 2009 but likely are making no profits. No other potentially viable PEMFC products are yet on the markets.
- *Europe*. Europe initially took a rather deliberate pace in promoting PEMFC development. The EU FP started to sponsor one PEMFC R&D project in the FP3 program (1990–1994) and several in the FP4 (1994–1998). Alarmed by the increasingly energetic PEMFC R&D activities in Japan and the United States, however, the EC made an extraordinary effort to increase R&D in subsequent FP programs, including FP5 (1998–2002) and FP6 (2002–2006). In 2007, the EC reviewed FP6 outcomes and found that European fuel cell and hydrogen capabilities were roughly 5 years behind that of the United States and Japan. As a result, Europe changed its fuel cell policy and priorities. In October 2007, the EU established the Fuel Cells and Hydrogen JTI—an ambitious industry-led integrated program of fuel cell and hydrogen R&D and demonstration activities—which would be administered through a unique public–private partnership, the FCH JU. This policy change was accompanied by little budget increase, however. At present, no commercially viable PEMFC products are on the primary markets. A handful of companies, including Dantherm (Ballard's subsidiary) of Denmark, H2Logic of the Netherlands, Nuvera of Italy, and Proton Motor and SFC Energy of Germany, are selling their products in niche markets for small power devices, auxiliary power, range extensions, and as battery chargers.

8.5. FUEL CELL PATENTING ACTIVITY

Patent activity is an indicator of trends in global interest in specific technologies. The increasingly energetic R&D on fuel cells can be seen in fuel cell patent activity over the past decade. Patents more than tripled from 2000 to

2010, well above what would be expected if nominal growth had occurred[9]:

- According to Fuel Cell Today—a UK research organization—US and European patent offices reported that a combined total of 403 patents were granted in 2000, 861 were granted in 2005, and 1801 in 2010 (Fig. 8.2). The number of patents has more than doubled every 5 years during the period, growing a spectacular 347 percent over the decade.[9]
- If the number of fuel cell patents had grown in line with the average growth in all patents, the number of fuel cell patents in 2005 would have been a mere 500 (compared to 861 actual) and only 620 in 2010, compared to 1801 actual.

8.5.1. Japan Grants the Largest Number of Fuel Cell Patents in 2010

Japan was the global leader in the number of fuel cell patents granted in 2010, according to a review of patents by country of origin published by Fuel Cell Today.[9] Although no specific numbers are presented, Fuel Cell Today indicates that Japan was far behind the United States in patent activities in 2000, was still behind the United States in 2005, but had surpassed the United States in 2010 (Fig. 8.3).[9] This indicates that a dramatic change has occurred in Japan's level of innovation and development of intellectual property with respect to fuel cell technology. The speed with which the change has occurred suggests this is a secular change that will ensure Japan remains in the lead for some time to come.

- Germany, France, the United Kingdom, Italy, and Switzerland were among the world's top 10, but were significantly behind Japan and the United States.

8.5.2. Japanese Corporations Expand Dominance in Fuel Cell Patent Activity During the Past Decade

The growing dominance of Japanese fuel cell activity can also be seen in Table 8.1 presented by Fuel Cell Today.[9] The table lists the top 10 companies and the number of patents they were granted in each of 3 years, 2000, 2005, and 2010. It shows that Japanese corporations in the top 10 had an increasingly greater number of patents granted during the decade compared with a decade earlier. Indeed, the number of patents granted to these companies in 2010 was 23 times as many as those in 2000. US corporations had about twice as many patents granted

* The number of average patents was calculated on the assumption that the 2000 baseline patents were = 403.

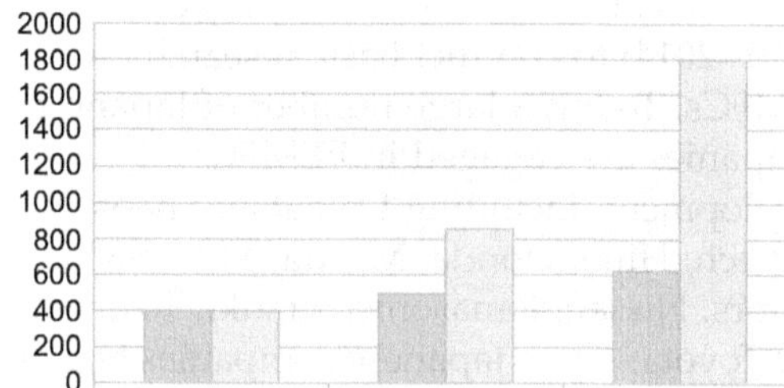

	2000	2005	2010
Projected patents based on average growth rates	403	494	627
Fuel cell patents granted	403	861	1801

FIGURE 8.2 Fuel cell patenting activity vis-à-vis average patenting activity: 2000, 2005, and 2010*. *Source: "The 2011 Fuel Cell Patent Review," http://www.fuelcelltoday.com/media/948977/2011-fuel-cell-patent-review.pdf, Fuel Cell Today, 2011.* (For the color version of this figure, the reader is referred to the online version of this book.)

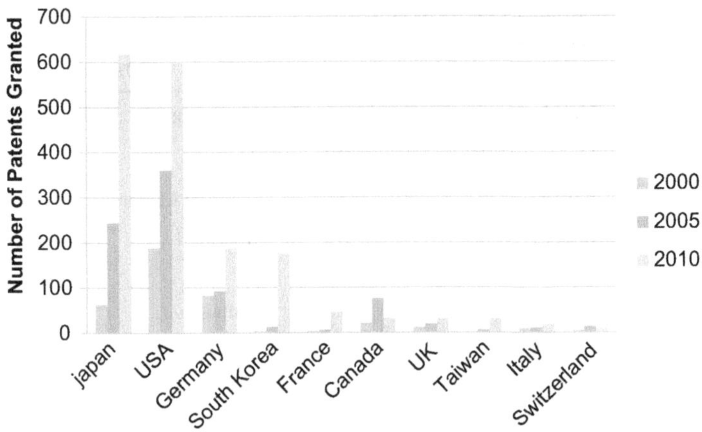

FIGURE 8.3 Annual fuel cell patents granted per country of origin (Top 10): 2000, 2005, and 2010. *Source: "The 2011 Fuel Cell Patent Review," http://www.fuelcelltoday.com/media/948977/2011-fuel-cell-patent-review.pdf, Fuel Cell Today, 2011.* (For the color version of this figure, the reader is referred to the online version of this book.)

in 2010 as in 2000. European corporations showed a decline in patent activity, and no European corporations were among the top ten in 2010. In contrast, other regions of the world were granted a growing number patents, an increase of sevenfold over the decade (Fig. 8.4).

It is noteworthy that none of the seven Japanese corporations listed in the top 10 in 2000, 2005, and 2010 have participated in any METI fuel cell R&D project nor received R&D subsidies, although they did participate in demonstration projects that allowed foreign participation. The seven companies include Honda, Toyota, Panasonic, Nissan, Hitachi, Toshiba, and Canon.

- This phenomenon may be attributable to long-term corporate commitment to invest in fuel cell R&D as a future energy source, even without government funding incentives. Nonetheless, the significant influence of government fuel cell policy and programs on Japan's overall R&D activities is unquestionable and probably indirectly influenced the R&D priorities of these seven corporations. Their fuel cell R&D activity would not have been as robust if the government had been indifferent to fuel cell technology.

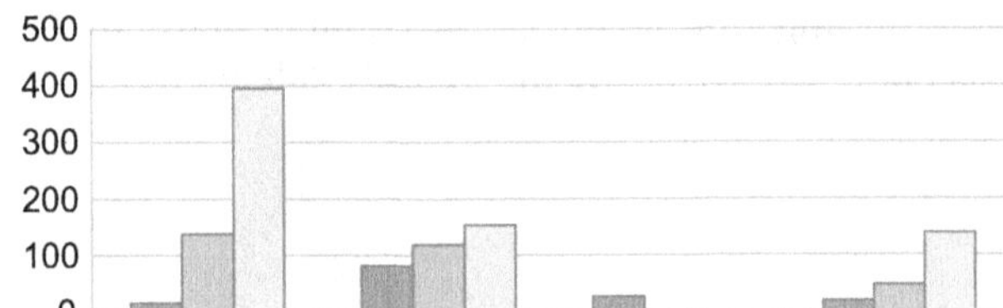

	Japan	United States	Europe	Others
2000	17	81	27	19
2005	138	119	0	48
2010	396	154	0	140

FIGURE 8.4 Top 10 assignees of fuel cell patents granted. *The figure was created based on Table I: Top ten assignees of fuel cell patents granted in each year, from "The 2011 Fuel Cell Patent Review," http://www.fuelcelltoday.com/media/948977/2011-fuel-cell-patent-review.pdf, Fuel Cell Today, 2011.* (For the color version of this figure, the reader is referred to the online version of this book.)

TABLE 8.1 Top 10 Assignees of Fuel Cell Patents Granted: 2000, 2005, and 2010

Rank	2000		2005		2010	
1	International Fuel Cells	29	Honda	68	Samsung	140
2	Siemens	23	General Motors	50	Honda	135
3	Ballard	19	Ballard	48	General Motors	130
4	Daimler-Benz-Ballard Fuel Cell Engines	12	Nissan	26	Toyota	90
5	Du Pont	10	UTC Fuel Cells	24	Panasonic	58
6	Honda	10	Panasonic	22	Nissan	46
7	Plug Power	10	Toyota	22	Hitachi	26
8	General Motors	9	Delphi	17	Delphi	24
9	Forschungszentrum Jülich	8	Hewlett Packard	16	Toshiba	21
10	De Nora/Toyota*	7	DaimlerChrysler	12	Canon	20

** De Nora and Toyota each had seven patents granted in 2000 and are therefore both ranked 10th.*

Source: "The 2011 Fuel Cell Patent Review," *http://www.fuelcelltoday.com/media/948977/2011-fuel-cell-patent-review.pdf, Johnson Matthey, 2011.*

8.6. THE GLOBAL FUEL CELL LEADER TODAY

Who is the current global leader in fuel cell technology and commercial applications?

The answer to this question objectively depends on a host of different characteristics. We examined programs over three decades in the United States, Europe, and Japan, including the level of government funding, the level of private funding, consistency of government policy and program direction over the long term, the soundness of program evaluations and critiques, the resiliency of the industry to deal with challenges and setbacks, and the quality and inventiveness of research results as measured by the numbers of patents granted. The overall assessment is shown in Fig. 8.5. When considering all these factors, Japan stands above the United States and Europe overall and is equal or better than the other regions in every single category.

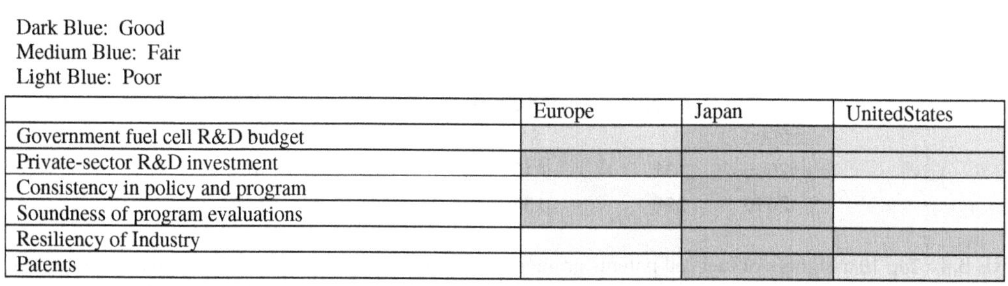

FIGURE 8.5 Strengths and weaknesses: Europe, Japan, and the United States. (For the color version of this figure, the reader is referred to the online version of this book.)

References

1. Our Budget. *Fossil Energy, US Department of Energy,* last updated on 14 February 2011. <http://fossil.energy.gov/aboutus/budget/index.html>; [accessed November 2011]; "Budgeting," Planning, Budgets, and Analysis, *Energy Efficiency and Renewable Energy, US Department of Energy,* last updated 14 February 2011. <http://www1.eere.energy.gov/ba/pba/budget_10.html>; [accessed November 2011].
2. *"Japanese technical literature Bulletin," The Japan technology program, issue No.* 22. Office of International Technology Policy, Technology Administration, US Department of Commerce; March 1994.
3. Japan. U.S. Department of State telegram, Message Reference No. Tokyo 010604, U.S. Department of State; 18 November 1996.
4. Europe. U.S. Department of State telegram, Message Reference No. Brussels 007087, U.S. Department of State; 20 December 2000.
5. *The state and prospects of European energy research: comparison of commission, member and non-member states' R&D portfolios.* European Commission Directorate-General for Research Directorate Energy, Brussels; 2006.
6. *Strengths, weaknesses, opportunities and threats in energy research.* EUR 21612. European Commission, Brussels; 2005.
7. *Stephen Singer (AP business writer), "united tech may sell fuel cell firm".* News. Advantage Business Media, <http://www.manufacturing.net/news/2012/05/united-tech-may-sell-fuel-cell-firm> US; 22 May 2012.
8. *Annual report 2011.* Ceramic Fuel Cell Limited, <http://www.cfcl.com.au/Assets/Files/20111019_CFCL_Annual_Report2011.pdf>, Australia; 2011.
9. *The 2011 fuel cell patent review,* <http://www.fuelcelltoday.com/media/948977/2011-fuel-cell-patent-review.pdf>; Fuel Cell Today, 2011.

CHAPTER

9

Policy Recommendations

OUTLINE

Fuel Cells
http://dx.doi.org/10.1016/B978-0-444-56325-5.00009-0

9.1. DIFFICULTIES OF PERFECTING FUEL CELL TECHNOLOGY NEVER UNDERSTOOD

Some technologies are always a few years away from commercialization, and the fuel cell is one of them. Governments and corporations continue to forecast an expected date for commercialization of fuel cell technology, but the dates come and go without results.[1] If products do make it to market, they are inefficient, unreliable, or expensive, and make little profit. It has been more than 170 years since the fuel cell was invented, but no truly commercially viable fuel cell products are on the market. Frustrated by so many years of optimistic forecasts and repeated failures, readers who have made it this far in this book may well ask "Why is it so difficult to make fuel cells?"

The answer is simple. Fuel cell technology is complex and difficult, and necessary research has not been done because few have appreciated the profound challenges that must be overcome to master the technology. History is full of stories in which fuel cell companies developed moderately functional fuel cells and rushed to commercialization. They subsequently found that their fuel cells did not work well. No matter how many times they encountered problems of cost or efficiency or reliability, they always believed things could be fixed with a little more time and a little more money. They were looking for tactical fixes when strategic solutions were needed.

- The difficulty of the technology is rooted in the complexities of how fuel cells work, which involve multiple chemical and physical interactions at the atomic level. Perhaps no advanced technology on the market today—including airplanes, computers, or even nuclear reactors—require the scale, magnitude, and range of scientific, physical, and engineering knowledge that fuel cell technology requires.
- It is only recently that fuel cell researchers started to recognize the magnitude of the challenges of fuel cell technology and how far they are from understanding it. One assessment of the state of knowledge of catalytic activity in the presence of electric

current makes the point: "Pessimism in the field of electrocatalysis may... . derive from the fact that science is still far from having attained one of the final aims of theoretical work in this field: that of providing a full quantum-mechanical foundation for electrocatalysis."[2]

9.2. UNTIL RECENTLY, SCIENCE AND PHYSICS TOO IMMATURE FOR FUNDAMENTAL UNDERSTANDING OF FUEL CELL

An unfortunate fact is that a working fuel cell device can easily be built, and so the fuel cell industry has not felt the need to know exactly how it works or why. When Grove unveiled his gas battery, the scientific community was trying to understand the basic principles of chemistry and electricity, matter, and energy. A few scientists, including Christian Schönbein (1799–1868), Johann Poggendorff (1796–1877), and Friedrich Wilhelm Ostwald (1853–1932) debated the question of exactly how Grove's gas battery worked.[3] Oswald, for one, experimentally determined the interconnected roles of the various components of the fuel cell in 1893. None of them, however, arrived at the theoretical understanding of how fuel cells work.

- Without understanding the basic physics and chemistry of the fuel cell, no matter how much money and time may be poured into product development, ultimate success will be doubtful. Fuel cells will remain always a few years away.

In the early days of Grove's invention, physics and chemistry were not mature enough to reveal fuel cell dynamics. In the twenty-first century, however, science has made many advances, which could provide the foundation for the groundbreaking discoveries that will be needed to master fuel cell technology. The time is ripe for undertaking this journey, as the following examples suggest:

- German researcher Gerhard Ertl, who received the Nobel Prize in Chemistry for 2007, carried out studies on surface chemistry. His research aimed to understand such processes as why iron rusts, how fuel cells function, and how the catalyst in cars works, according to a press release by the Royal Swedish Academy of Sciences.[4] Ertl's work was made possible by the modern technology of photoelectron microscopy, which captured for the first time the phenomenon of oscillatory reactions on platinum surfaces, how the reactions change surface structure, and the processes that occur during reactions.
- Researchers at the U.S. Department of Energy's Brookhaven National Laboratory recently unveiled details about a class of catalysts that could help improve the performance of fuel cells.[5] Using a technique called X-ray photoelectron spectroscopy at Brookhaven's National Synchrotron Light Source, as well as scanning tunneling microscopy, and calculations, the researchers established for the first time that two "next-generation" nanoscale catalysts—gold–cerium oxide and gold–titanium oxide—can efficiently convert nearly 100 percent of carbon monoxide into carbon dioxide in the "water–gas shift" reaction. Within a fuel cell, CO is a poison that deteriorates platinum-based catalysts. CO is often contained in hydrogen-rich fuels that feed the fuel cell, but up until now the removal of CO has been very difficult.
- Masahiro Watanabe of Yamanashi University, Japan, and his group of researchers from Waseda University, Shimadzu Corporation, Fuji Electric, and Hitachi have succeeded in imaging the oxygen distribution within a fuel cell stack for the first time in the world.[6] Watanabe used a chemical reagent that

- absorbs light and emits light of a specific wavelength where oxygen exists and captured an image of oxygen distribution with a charge-coupled device (CCD) camera. The visualization of the inner working of a fuel cell stack could reveal mechanisms of fuel cell deterioration. Watanabe is currently working on the simultaneous visualization of humidity, steam, water, temperature, and CO_2 within a fuel cell stack.

9.3. FUEL CELL KNOWLEDGE REQUIRES MULTIPLE SCIENTIFIC DISCIPLINES...BUT FEW INSTITUTIONS HAVE INTERDISCIPLINARY RESEARCH CAPABILITIES

The way scientific institutions are organized to conduct research provides some insight into why so little fuel cell basic research is being conducted today. Current academic, government, and corporate research institutions lack an adequate cross-discipline structure and process to conduct such investigations.

To understand the basic workings of fuel cells demands research that spans many scientific disciplines. Related disciplines include astrophysics, computational chemistry, molecular chemistry, surface chemistry, quantum chemistry, quantum electrodynamics, electromagnetism, mechanical engineering, metallurgical engineering, materials science, mathematics, molecular dynamics, quantum dynamics, thermodynamics, nanotechnology, quantum mechanics, statistical mechanics, atomic and molecular physics, experimental physics, computational physics, nuclear and particle physics, quantum physics, solid-state physics, theoretical physics, and many more.

- No single academic, government, and corporate research institution possesses such a vast spectrum of knowledge and expertise. Thus, no research institutes are in a position to lead, organize, and implement fuel cell basic research.

9.3.1. The United States Starts a Small Interdisciplinary PEMFC Basic Research Center in 2007

There has been a growing awareness that to advance fuel cell technology would first require substantial interdisciplinary basic research. The United States made perhaps the world's first attempt to create such an interdisciplinary center to engage in fuel cell basic research. In 2007, the National Science Foundation awarded a grant of $1.5 million to the University of Massachusetts in Amherst to a group of eight researchers, including chemists, polymer experts, and physicists, to study proton transfer in fuel cells for 3 years.[7] This team was to be housed in a newly established "Center for Fueling the Future" under the umbrella of MassCREST, the Massachusetts Center for Renewable Energy Science and Technology, at the University of Massachusetts.[8] MassCREST is a unique multi-interdisciplinary research center that brings together complementary expertise in molecular design and synthesis, physical characterizations, device fabrication, and theoretical modeling from at least five different departments within the University.

- The Center for Fueling the Future Chemical Innovation is one of seven NSF Chemical Bonding Centers in the nation and one of two that focus on clean, renewable energy research. The center aims to discover and apply the chemical design principles needed for making innovative organic charge-conducting materials with tunable transport properties, particularly for renewable energy applications. Such centers are designed to provide scientists opportunities to address "big questions" in basic chemical research that have potentially high scientific and societal impact.

- These initiatives provide strong evidence that the United States is still the world leader in originating new ideas and initiatives in sciences and technologies. Moreover, NSF officials say the University of Massachusetts Amherst is positioned to receive up to $30 million more over the next several years as fuel cell work progresses.

9.3.2. Japan Launches an Interdisciplinary PEMFC Basic Research Project in 2010

Japan has also recognized the need for basic interdisciplinary research to advance fuel cell technology. In 2010, Japan's METI implemented a 5-year basic research project, "Technology Development for Promotion of PEMFC Commercialization (2010–2014),"[9] designed to promote interdisciplinary fuel cell research. The project budget for 2010 was ¥5.1 billion ($56.2 million at the 2010 exchange rate of $1/¥90.7). The project has three segments, basic technology development, commercial technology development, and next generation technology development, with the largest amount of resources going to basic technology development.[10] Also, the High-Performance (HiPer-FC) Project, which has been studying the causes of fuel cell degradation since 2008, was merged with the basic technology segment of the project.

- The "Basic Technology Development" segment was allocated ¥3.4 billion ($37.5 million). Research topics include degradation mechanism analysis; materials for high efficiency cells using nanotechnology; high efficiency, low cost MEAs; lowering the use of platinum; carbon alloy catalysts; non-noble metal oxide catalysts; structure, reaction, and mass transfer analysis of MEA materials; development of a common technology base for cell evaluation and analysis; and international standards. Research would be conducted in consortia led by program leaders with high academic standing.

The "Commercialization Technology Development" segment and the "Next Generation Technology Development" segment were allocated ¥1.6 billion ($17.6 million).[11] The "commercialization technology development" segment would include the development of technology that would enable the use of diverse fuels, the development of multipurpose, high value added PEMFC systems, cost reduction technologies, and safety technology. The "next generation technology development" segment would foster exploratory, innovative technologies—excluding the current technology—that could dramatically increase reliability and reduce costs of fuel cells.

Although the project has not been fully defined, METI's 2010–2014 Technology Development for Promotion of PEMFC Commercialization is a landmark event:

- The basic technology development segment will conduct basic research. The project has more academic scientists and physicists participating than applied researchers, which is unprecedented. NEDO announced in July 2010 that at least 32 universities, 30 companies, and 3 research institutions were on the project.[10] The project also is using a collaborative R&D approach centering on COEs and consortia.
- The HiPer-FC project that merged with the 5-year project is being conducted at a center of excellence (COE) established at University of Yamanashi in April 2008. The project as well as the COE is headed by a distinguished physical chemistry professor of the university. The COE is conducting basic scientific research to understand the reaction and degradation mechanisms and, using nanotechnology, is working toward

innovative breakthroughs to create high-performance fuel cells.

- The 5-year project will also have a noteworthy Fuel Cell Innovation Research Center.[12] The center, opened at the University of Electro-Communications (UEC) in Tokyo in September 2010, is headed by a former Tokyo University professor and respected scientist in solid surface science who has developed techniques to elucidate the chemical reactions on solid surfaces at the atomic level. He would also serve as the senior project leader for the entire 5-year project. Of extraordinary note, the research center at UEC will use high-energy X-rays to study and understand electrochemical reaction mechanisms and changes in the catalyst surface and structure, thereby hopefully creating dramatically different fuel cells with high efficiency, durability, and low cost.

It is remarkable as well as unprecedented for a METI project to be involved in such a deeply basic, physical, and scientific domain. Also, there have been no foreign projects of this magnitude involved in PEMFC research at the basic research level. In this context, it is a truly groundbreaking achievement for Japan.[12]

9.3.3. Small Budgets, Short Deadlines, and Overarching Goal of Commercialization Might Inhibit Basic Research

There are factors that might limit the progress and outcome of both US and Japanese projects, including the following:

- *The budgets are too small.* A scientific breakthrough in understanding and elucidating the electrochemistry of fuel cells will not come cheap. Such projects should be provided with ample resources to support massive research activities.
- *Too few academic disciplines.* It is not clear whether a critical mass of scientific disciplines is involved. If the net is not cast widely enough and if the assembled expertise does not cover enough disciplines, the project might not be able to uncover the secrets of fuel cell electrochemistry.
- *Overly narrow research scope.* It is uncertain whether the scope of the 5-year Japanese project will be broad enough. In case of the University of Massachusetts, the research team there perhaps pursued research on proton transfer mechanisms only. To understand fuel cells, basic research must explore a wide range of fuel cell phenomena, not just one phenomenon.
- *Tight project deadline.* Basic research should not be bound by tactical deadlines. All the Japanese research activities were under the umbrella of a 5-year project, governed by the goal of commercializing PEMFCs in 5 years. The 5-year project undoubtedly will impede the basic research priorities in favor of product development. In case of the University of Massachusetts, the basic research should be funded for at least five years.

Nonetheless, the intentions of these projects are worthy and the progress these projects make is worth watching.

9.4. FUEL CELL DEVELOPMENT REQUIRES THREE LEVELS OF RESEARCH: BASIC RESEARCH SUPPORTED BY APPLIED RESEARCH AND PRODUCT DEVELOPMENT

Successful development of fuel cells requires three levels of research: basic research, applied research, and product development. Basic research must be supported by applied research and product development activities. However, the primary and foremost activity in fuel cell development today must be basic research. First, basic research needs to compile all fuel cell phenomena—what is known about how fuel cells work and behave—gathered from past and present applied research and product development activities. Second, basic research must examine those phenomena from the perspective of all possible scientific disciplines, drawing upon associated knowledge and theories to establish hypotheses that would give coherent explanations to why phenomena occur. Third, applied research and product development activities should support basic research by verifying and substantiating hypotheses. If a hypothesis fails to be validated, basic researchers need to scrutinize the failed hypothesis to develop another with better explanatory power. Through the repeated process of hypothesis generation, field verification, and data exchange, a body of individually verified theories would be constructed. If the construct leads to a coherent, unified explanation of fuel cell phenomena, it would lay the foundation for better, durable, and cheaper fuel cells.

- The repeated cycle of hypothesis, verification, and data exchange between basic research and applied research and product development must continue until the fuel cell industry evolves into a mature industry. Each step of the way, however, fuel cells will become more advanced.

9.4.1. Forschungszentrum Julich Assigned to Lead EU FP6 Real-SOFC Project to Address Degradation in 2004

Some governments have recognized the need for three levels of research to develop successful fuel cells. Indeed, the EU in 2004 implemented a research project combining all three levels of research to solve the SOFC degradation problem under the EU FP6.

The EU clearly was aware of Forschungszentrum Julich's global SOFC leadership and asked the institute to head the project. Julich took charge of the 4-year, €18 million (about $22.5 million at the 2004 exchange rate of $1/€0.8) project, entitled "Realizing Reliable, Durable, Energy Efficient and Cost Effective SOFC Systems" (Real-SOFC) Project (2004–2008). The goal was to solve the degradation problem and develop reliable, durable SOFCs. A consortium of 26 major European SOFC players, including corporations, research institutions, and academic institutions, were given access to Julich's SOFC cells and stacks as well as those of other major manufacturers and explored the degradation process and possible solutions to performance loss.[13] The consortium partners were allowed to use their findings to invent new cells and stacks—outside the project in order to secure intellectual property rights protection—which were then fed back into the project for testing and evaluation.[14]

9.4.2. AIST Assigned to Lead SOFC Basic Research Project to Address Degradation in 2008

Similarly, Japan's METI launched a 5-year program in 2008 called the "Development of System and Elemental Technology on Solid Oxide Fuel Cell Program" (2008–2012) to explore SOFC degradation. METI launched this effort because one of its program evaluation committees, in reviewing the 2004–2007 SOFC System Technology Development Program, recommended that fuel cell research go back to basics and explore SOFC degradation. It is noteworthy that METI swiftly launched the new project incorporating all three levels of research. METI undoubtedly saw the need for a change in fuel cell policy, because it recognized that the same kinds of projects it had implemented for the past several decades would not lead to successful fuel cell commercialization. The 5-year project included thermodynamic, chemical, and mechanical research to seek and clarify degradation mechanisms.[15]

- It is also remarkable that the program was welcomed by the fuel cell industry. Many fuel cell companies had invested billions of dollars in building manufacturing facilities and infrastructure, prototype and precommercial products, and creating intellectual property. Going back to basics was equivalent to abandoning what had been built and starting again from scratch. No corporation would agree to such drastic measures under normal circumstances, indicating that the government and industry had a common sense of desperation.

Another remarkable fact was that METI appointed a globally renowned Japanese SOFC researcher as the program leader and gave him the power to supervise, control, and manage the project. To identify a project leader by name was an extraordinary event in Japan, where anonymity is an accepted norm. This indicated METI's resolve to give the project prominence and prestige and ensure its success.

- When the 5-year basic research program was announced in March 2008, it drew immediate attention from the global fuel cell community. Scientists and engineers were amazed at the boldness with which the project was set up. There was a shared sentiment that this was exactly what was needed to get a breakthrough.

9.4.3. ...But Basic Research Limited by Serious Constraints

It is significant that both these two projects were launched to discover what causes fuel cell degradation and how to solve it. The Julich-led EU FP project was completed with good results, and the METI basic research project will continue until March 2013.

The EU FP6 Real-SOFC Project. The EU FP6 project led by Julich of Germany was completed in December 2008[16] and undeniably made many major advances in SOFC technology. The project resulted in the development of improved anode, cathode, and interconnect materials, which were demonstrated successfully in stacks made by HEXIS, HTceramix, Rolls-Royce Fuel Cell Systems, and Topsoe Fuel Cell. Julich indicated in 2009 that the EU FP6 Real-SOFC Project enabled the development of more efficient SOFC cells with thin functional layers, which, with the use of Julich's special Crofer steel (industrialized by ThyssenKrupp) and the EU project Real-SOFC qualified ITM steel (developed by Plansee) along with corrosion protective coatings, achieved an extended service life.[17] In April 2010, Julich announced that its SOFC stack developed under the EU FP6 project had been running for more than 20,000 h, or almost 3 years.[18] Moreover, Julich stated that the SOFC stack had shown significantly reduced degradation rates of around 10 mV (about 1 percent) per 1000 hours at 700 °C and 500 mA/cm^2 over tested stack lifetimes of over 30,000 h,[19] very close to the required 40,000 h standard for stationary fuel cells.

- Still, these successes were incremental improvements, short of a breakthrough. The EU FP6 project was not intended to be a basic research project in search of knowledge. Indeed, the project encouraged Julich to engage in applied research and product development; in effect, Julich taught the European fuel cell community how to build a better SOFC.

METI's Five-Year SOFC System and Elemental Technology Development Project (2008–2012). METI's 5-year SOFC project to address degradation became a major sensation, because by that time degradation had become widely known as the most difficult challenge of SOFC technology. The project included a basic research team consisting of AIST researchers and several applied research and product development teams from universities and fuel cell companies. Little information has become available on the status of the project, but the METI "basic research" project has exactly the same flaw that handicapped the EU FP6 Real-SOFC project—priorities that favored applied research and product development over basic research. In the METI project, research scientists were not solely dedicated to pursue basic research. The project's twin goals are to (1) conduct research on degradation and (2) support an SOFC stationary system demonstration project (2007–2010).[15] The demonstration project would provide SOFC degradation data to the "basic research" project, which would research and develop solutions, which in turn would deliver results to the demonstration project to improve SOFC system performance. The two projects had a committee and coordinating mechanisms for communication and data sharing,[20] constituting a compulsory process where developing solutions to improve the SOFC demonstration system became *defacto* the primary goal. The research activities that were most wanted in the project were focused on applied research and product development, not basic research.

- The project also was handicapped by a short deadline and small budget. The project had a term of 5 years and its budget was not only modest but also declining. The budget was ¥1.3 billion (about $12.5 million at the 2008 exchange rate of $1/¥104) for 2008, ¥1.2 billion (about $12 million at the 2009 exchange rate of $1/¥99) for 2009, and ¥800 million (about $8.8 million at the 2010 exchange rate of $1/¥90.8) for 2010. The declining budget indicated that the Japanese government did not truly appreciate the difficulty in finding solutions to degradation; it likely believed that the degradation challenge could be resolved in a short time and would not require a high level of funding for very long.

9.5. FUEL CELL TOO VALUABLE TO ABANDON: GO BACK TO BASICS NOW

Governments are faced with three options today: to abandon fuel cell development altogether; to provide assistance to sustain the temporary survival of the industry; or to implement the "National Fuel Cell Development Project" (NFCDP) for basic research that we propose here. Over the past decades, many governments have worked vigorously to help their fuel cell industry build technology capabilities but with little economic return. The fuel cell industry still is not commercially successful, however, and continues to ask for government assistance for survival, including demonstration projects, tax incentives, and subsidies for lowering purchase costs and building a hydrogen infrastructure. Today, governments are reluctant to stop fuel cell development because they have already invested too much in the industry to throw it all away. And yet, they cannot continue the same policy and subsidy program as before. It is expensive, unproductive, and never ending.

- It is our judgment that most governments would want to support a basic research strategy and implement the NFCDP concept. Once fuel cell technology is fully understood, the fuel cell industry would be able to produce commercially feasible fuel cell products virtually immediately because the industry already has well developed capabilities and infrastructure.

Governments should be aware of the extraordinary benefits of fuel cell technology, once it is successfully developed. Fuel cells are undeniably an extraordinary, revolutionary technology that can end the way in which the world depends on fossil fuel today. Fuel cells convert potential energy into power with higher efficiency than any internal combustion engine; and because fuel cells use hydrogen instead of fossil fuel, it will depend on a fuel supply that is virtually inexhaustible. When fossil fuel is depleted or becomes very costly to obtain, the world will be left with only four energy conversion options—the fuel cell, nuclear fission, nuclear fusion, and renewable energy—and the best option will be the fuel cell.

- Nuclear fission uses uranium, which like oil is a natural resource with high access cost. In addition, it generates long-lived, hazardous radioactive wastes.
- Nuclear fusion will not be ready for another 40 years. Fusion powered electricity generation was initially believed to be as achievable as fission power.[21] However, the extreme requirements for continuous reactions and plasma containment have led to projections of feasibility being extended out by several decades. In 2010, more than 60 years after the first attempts, commercial power production is still believed to be unlikely before 2050.
- Renewable energy depends on sunshine and wind or biofuels, which are either intermittent, unreliable, or environmentally unfriendly.

Over the medium term, however, while there are still fossil fuels available, fuel cells will serve as a far better energy conversion device than other conventional technologies. Fuel cells have high-energy conversion efficiency and can convert potential energy in a chemical fuel to electricity in a single step.

- An internal combustion engine requires multiple steps to generate electricity: converting chemical energy into thermal energy via combustion, using thermal energy to expand gases within the engine's cylinders, converting the high pressure gas into mechanical energy via the pistons and drive train, and then converting the mechanical energy to electricity. The process is inefficient, loses much energy through heat, and wastes excessive amounts of fossil fuel.

Because of its high efficiency, fuel cells would extend the life of fossil fuel reserves and can slow down their depletion. It has been reported recently that there are abundant supplies of natural gas in the United States and elsewhere in the world. Over the medium term, natural gas would be an economically viable source of hydrogen and might last long after oil becomes too costly for everyday use. Moreover, fuel cells create less carbon dioxide when operating on carbon-based fuels such as natural gas than combustion engines, and do not generate any other polluting emissions such as nitrogen oxide because fuel cells do not involve combustion.

In the long run, fuel cells should be able to operate on hydrogen that is generated from renewable sources. Fuel cells then would have no emissions whatsoever (except water). Thus, fuel cell technology can give an ordinary citizen the ability to generate clean electricity at home instead of buying electricity generated by polluting coal-fired power plants. Moreover, electricity today is delivered by an electrical grid system over long distances with high transmission losses, which are susceptible to frequent—sometimes catastrophic—grid failures. People also will be able to drive cars that operate on hydrogen instead of gasoline and help promote new industry and manufacturing sectors that will dramatically lower an individual's carbon footprint.

Fuel cells thus can spur the development of a new and better way of life, new energy supply-and-demand systems, transportation systems, and industrial and manufacturing systems, while generating trillions of dollars in new revenues. In addition, oil importing nations can reduce their use of oil, gain greater energy independence, slow the depletion of fossil fuel reserves, and thereby enhance their energy, economic, environmental, and national security.

9.6. LEARNING FROM PAST EXPERIENCE TO PLAN FUTURE COURSE OF ACTION

Launching a fuel cell basic research project is a serious undertaking and calls for careful assessment of what would be required. For fuel cell basic research to succeed, it will need unrestricted use of time, ample financial resources, and an extensive research infrastructure. It will also require a cadre of highly talented researchers from all related academic disciplines, practically and virtually collocated, and led by a strong project leader with direct access to the nation's top leadership. The project's leader must be a brilliant, enlightened, and open-minded thinker who can manage, fuse, and meld all the interdisciplinary concepts and knowledge that will be generated into a theoretical foundation that fully describes fuel cell dynamics.

Nonetheless, it is difficult in the abstract to envision all the prerequisites necessary to make up a fuel cell basic research project. Thus, we have examined some historical precedents:

- First, we looked at a few examples of past spending.
- Second, we examined the Manhattan Project as an exemplar of a major national research project implemented in a relatively short amount of time with high national priority. The project had very different objectives from the project we propose here. Nonetheless, both this exemplar and the NFCDP represent a desperate measure carried out in a desperate time to achieve a series of breakthrough solutions.

9.6.1. Past Spending

It is difficult to put a price tag on any activity that includes basic research, but funding shortfalls and short deadlines will handicap any serious undertaking. Three classes of spending give some indication of what level of funding would be required:

- The EU and individual European governments allocated a total of about $2.414 billion for fuel cell R&D from 1995 to 2012; Japan spent about $3.38 billion during the same period; the United States spent a total of about $2.401 billion (after earmarks). Since these efforts did not generate any significant commercial success in any of the three regions, the budget could be deemed insufficient. Taken together, the three regions invested a total of about $8.2 billion over 18 years or $456 million per year.
- The US government alone invests as much as this in its nuclear energy R&D. It invested a total of $8.3 billion to promote nuclear fission and fusion during the 13 years from 2000 to 2012, an annual average of about $640 million.[22] The nuclear energy budget in general has been increasing and rose above $1 billion in 2008 and 2009. Because nuclear fusion may not be possible for another 40 years, the United States will likely continue to allocate between $800 million and $1 billion per year for another 40 years on all forms of nuclear energy R&D. It is not out of all proportion if a similar budget is considered for fuel cell basic research. Fuel cells are potentially within reach, in just a quarter of the time currently projected for nuclear fusion.

If one thinks that fuel cell R&D project is expensive, a look at US spending on oil imports would change that view. The United States consumes more oil than any other nation in the world—about 18.7 million barrels of oil per day—according to the US Energy Information Administration's (EIA) short-term energy outlook.[23] Of the total, nine million barrels per day were imported.[24] EIA is currently forecasting crude oil prices to average about $95 per barrel for the rest of 2012,[25] indicating that the United States perhaps would pay $850 million per day or over $300 billion for oil imports for 2012. It is certain that the United States will continue to pay an exorbitant amount of money for the foreseeable future. Under these circumstances, it is not unreasonable to suggest that fuel cell basic research be allocated $2 billion a year—a fraction of the spending that goes to imported oil.

9.6.2. An Exemplar in History: The Manhattan Project

The Manhattan Project was a massive research and development effort during wartime, but it provides a blueprint that could be used for understanding what might be needed to achieve needed breakthroughs in fuel cell technology. The Manhattan Project exercised a number of potent measures to ensure successful project implementation.

9.6.2.1. Preceded by Nobel Prize-Class Nuclear Fission Basic Research in the 1930s

The basic research on nuclear energy conducted in the 1930s set the stage for the Manhattan Project of the 1940s. In 1932, James Chadwick, a British physicist, discovered the neutron, which was immediately recognized as a potential tool for nuclear experimentation because of its lack of an electric charge.[26] He was awarded the Nobel Prize for Physics in 1935.[27] Experimentation with the bombardment of materials with neutrons led Frederic and Irene Joliot-Curie, a French physicist and scientist couple, to discover induced radioactivity in 1934, which allowed the creation of radium-like elements at much lower cost than natural radium. In 1935, the couple was awarded the Nobel Prize in Chemistry. Further work by Enrico Fermi, an Italian-born, naturalized American physicist, focused on using slow neutrons to increase the effectiveness of induced radioactivity.[26] Fermi was awarded the Nobel Prize in Physics in 1938.

- Basic research on nuclear energy was not confined neatly to a single academic discipline but involved a wide range of different subjects, including quantum mechanics, nuclear and particle physics, and statistical mechanics. The same would be true for fuel cells; the broader the net that is cast, the greater the chances of achieving a breakthrough.
- Basic research leading up to the discovery of nuclear fission took about 6 years, starting with Chadwick's discovery of the neutron in 1932 to Fermi's work on induced radioactivity in 1938, and the confirmation of nuclear fission in 1939 by Otto Hahn and Lise Meitner. Although we cannot fathom the degree of difficulty that was involved in understanding the mechanism of nuclear fission, as compared to that of fuel cells, it would be prudent to allow for a longer time for fuel cell basic research.

9.6.2.2. The Manhattan Project—An Applied Research and Development Project—Began in 1939

The Manhattan Project started to take shape in 1939. In August 1939, prominent physicists, Leó Szilárdand Eugene Wigner, drafted the Einstein–Szilárd letter, which warned of the potential development of extremely powerful bombs of a new type and urged that the United States take steps to acquire stockpiles of uranium ore and accelerate the research of Enrico Fermi and others on nuclear chain reactions.[28] They had it signed by Albert Einstein—an ingenuous strategy to draw the highest level attention—and delivered it to President Franklin Roosevelt.

In response, President Roosevelt immediately called on Lyman Briggs of the National

Bureau of Standards to head the Advisory Committee on Uranium to investigate the issues raised by the letter. Briggs held a meeting in October 1939, which was attended by Szilárd, Wigner, and Edward Teller. The committee reported back to Roosevelt in November that uranium could provide a possible source of bombs with destructiveness vastly greater than anything then known.

9.6.2.2.1. PLACING PROJECT LEADERS DIRECTLY UNDER US PRESIDENT

President Roosevelt swiftly took a series of actions that led to launching the Manhattan Project:[29]

- In June 1941, the President created the Office of Scientific Research and Development (OSRD)—now the Office of Science and Technology Policy, part of the Executive Office of the President—to coordinate scientific research for military purposes, including the "S-1 Section," which later became the Manhattan Project that developed the first nuclear weapons.
- Next, the President appointed Vannevar Bush, former vice president and dean of engineering at MIT to head the OSRD. The President placed the OSRD directly under himself and directed that Vannevar Bush report only to him.
- On October 9, 1941, at a meeting between President Roosevelt, Vannevar Bush, and Vice President Henry A. Wallace, the President approved the atomic program.[28] To control it, he created a Top Policy Group consisting of himself—although he never attended a meeting—Wallace, Bush, President of Harvard University James Conant, Secretary of War Henry L. Stimson, and the Chief of Staff of the Army General George Marshall. Roosevelt chose the Army to run the project, as the Army had the most experience with management of large-scale construction projects.

9.6.2.2.2. PROVIDING VAST RESOURCES TO EXPLORE ALL POSSIBLE TECHNOLOGIES

President Roosevelt clearly gave the highest priority to the Manhattan Project. He ensured that the project received as many resources as needed to explore every possible technology option.[28] As a result, Arthur Compton, an American physicist and Nobel laureate in physics, in early 1942, brought together scientists working under his supervision at Columbia University and those at Princeton University to the University of Chicago, where he organized the Metallurgical Laboratory to study plutonium and nuclear reactors. On May 23, 1942, Lyman Briggs of the National Bureau of Standards, Compton, Ernest Lawrence (an American physicist and Nobel laureate at University of California, Berkeley), Eger Murphree (an American chemist), and Harold Urey (a Nobel laureate in chemistry) met to finalize the S-1 Committee recommendation, which called for five technologies to be pursued.

- Bush and Conant took the S-1 Committee's recommendation on nuclear reactor technologies to the Top Policy Group.[28] The budget proposal that accompanied the S-1 recommendation was $54 million for construction by the US Army Corps of Engineers, $31 million for research and development by OSRD, and $5 million for contingencies in fiscal year 1943, undoubtedly an unprecedented project with a budget of unprecedented scale. The Top Policy Group, in turn, sent recommendations to the President on June 17, 1942, who approved it by writing "OK FDR" on the document.

9.6.2.2.2.1. MOMENTOUS BUDGET The Manhattan Project, which started to take shape in 1939 and was officially funded in June 1942, ultimately cost nearly $2 billion in then-year dollars (roughly $24.4 billion in 2012 dollars),[28] which was a stunning amount of money. This budget likely covered the period from 1942 to

1946, implying an average annual budget some $6 billion a year (in 2012 dollars). Of the total amount, over 90 percent of the cost, or about $22 billion, was for building factories and producing fissionable materials, with about 10 percent or $2.4 billion (in 2012 dollars) allocated to the development and production of the weapons.

9.6.2.2.2.2. LIBERAL USE OF SILVER MINT Another episode demonstrates that no resources were spared for the project. Under the project, three different processes were considered for separating uranium-235 from uranium-238: electromagnetic, gaseous, and thermal.[54] Lawrence and his team at the University of California at Berkeley determined that of all the prospective processes, only electromagnetic separation appeared sufficiently advanced to be practical. However, the electromagnetic isotope separation process required 5000 tons of copper, which was in desperately short supply. But the scientists discovered that silver could be substituted.

- On August 3, 1942, Under Secretary of the Treasury Daniel W. Bell was approached and approved the transfer of 6000 tons of silver bullion from the West Point Mint Facility. Eventually, 14,700 tons were made into wire and wound into magnetic coils.
- After the war, all the machinery was dismantled and cleaned and the floorboards beneath the machinery were ripped up and burned to recover minute amounts of silver. In the end, only 1/3,600,000th was lost. The last silver was returned in May 1970.

9.6.2.2.3. TAKING WHATEVER ACTIONS NEEDED

The Manhattan Project required two very special individuals to bring the project to a successful conclusion. The project needed to acquire vast building sites, manage large-scale construction, as well as procure critical materials necessary for the project, all of which required an energetic, active, no-nonsense military officer. In addition, the project required a brilliant scientist and inspiring project leader who was intimately familiar with nuclear processes and bomb design concepts.

9.6.2.2.3.1. REPLACEMENT WITHIN THREE MONTHS OF THE INITIALLY APPOINTED COLONEL JAMES MARSHALL WITH A DYNAMIC GO-GETTER COLONEL LESLIE GROVES AND IMMEDIATE PROMOTION OF GROVES TO BRIGADIER GENERAL TO RAISE HIS STATURE Time was of the essence, and Bush took radical action to make sure that the Manhattan Project was executed speedily.

In June 1942, Colonel James C. Marshall was selected to head the Army's infrastructure construction part of the project.[28] Bush immediately became dissatisfied with Colonel Marshall; he felt that Marshall failed to move the project forward expeditiously. For example, when a survey team selected a site for production plants near Elza, Tennessee, Conant advised that it be acquired at once, and Styer agreed, but Marshall temporized. Marshall also failed to obtain a high enough Army priority rating for procuring scarce materials. Despite the need for Ratings AA-1 or AA-2, a lower priority AA-3 was offered, the same priority as a TNT plant. As a result, Bush replaced Colonel Marshall with Colonel Leslie Groves of the US Army Corps of Engineers, who took charge of the project in September 1942, scarcely 3 months after the initial appointment of Colonel Marshall. Groves was promoted to brigadier general as it was thought that the title "general" would hold more sway with the academic scientists working on the Manhattan Project. Groves' new rank placed him directly under General Somervell, with Colonel Marshall answerable to Groves.

Groves assumed command of the Manhattan Project on September 23, 1942 and took immediate action. On the same day, he attended a meeting called by Secretary of War Stimson, which established a Military Policy Committee, responsible to the Top Policy Group, consisting of Bush (with Conant as an alternate), General Styer, and Rear Admiral William Purnell. Richard Tolman, a professor of physical chemistry and mathematical physics at the California Institute of Technology and Conant were later appointed as Groves' scientific advisers. The following day, Groves took a train to Tennessee to inspect the proposed site there. On September 29, 1942, Under Secretary of War Robert P. Patterson authorized the Corps of Engineers to compulsorily acquire 56,000 acres (23,000 hectares) of land at a cost of $3.5 million. An additional 3000 acres (1200 hectares) was subsequently acquired. Also in September, Groves went to Donald Nelson, the chairman of the War Production Board, and asked for a higher priority rating (although Groves did not receive AA-1 authority until July 1, 1944).

9.6.2.2.3.2. APPOINTMENT OF DR. J. ROBERT OPPENHEIMER AS DIRECTOR OF PROJECT Y DESPITE SECURITY CONCERNS Groves' other vital responsibility was to find a director for Project Y, the group that would design and build the bomb. The obvious choice was one of the three laboratory heads, Urey, Lawrence, or Compton, but they could not be spared.[28] Compton recommended J. Robert Oppenheimer, who was already intimately familiar with bomb design concepts. However, Oppenheimer had little administrative experience, and, unlike Urey, Lawrence, or Compton, had not won a Nobel Prize, which many scientists felt was a requirement to head such an important project. There were also concerns about Oppenheimer's security status, as many of his associates were communists. In October 1942, Groves was convinced that Oppenheimer should be appointed as its director and personally waived the security requirements and issued Oppenheimer a clearance on July 20, 1943. Appointing an individual to such a sensitive position despite security concerns was a daring act, but it too illustrates that whatever action that was needed to be taken was taken.

9.6.2.2.4. ENLISTING NATION'S TOP QUALITY RESEARCH CADRE

The Manhattan Project undertook research and production at more than 30 sites, across the United States, the United Kingdom, and Canada.[54] The project began modestly in 1939 but grew to employ more than 130,000 people. In June 1944, the Manhattan Project employed some 129,000 workers, of whom 84,500 were construction workers, 40,500 were plant operators, and 1800 were military personnel. The project had about 2200 researchers engaged in applied research and development. Selection of candidate scientists and physicists was highly personalized and selective. Tolman and Conant, in their role as the project's scientific advisers, drew up a list of candidate scientists and physicists and had them rated by scientists already working on the project. Groves then sent a personal letter to the head of their university or company asking them to be released for essential war work.

- At the University of Wisconsin–Madison, Stanisław Ulam gave one of his students, Joan Hinton, nuclear physicist, an exam early so she could leave to do war work. A few weeks later, Ulam himself received a letter from theoretical physicist Hans Bethe, inviting him to join the project. Conant personally persuaded explosives expert George Kistiakowsky, a Ukrainian-American chemistry professor at Harvard, to join the project.

9.6.2.2.5. BUILDING MASSIVE RESEARCH INFRASTRUCTURE

Many national laboratories were needed for nuclear research and development under the Manhattan Project, including the Lawrence Berkeley National Laboratory (Berkeley Lab), Los Alamos National Laboratory (LANL), Oak Ridge National Laboratory (ORNL), Argonne National Laboratory, and Ames Laboratory. Two more laboratories were established as nuclear research facilities by Groves soon after the war, Brookhaven National Laboratory in Upton, New York, and Sandia National Laboratories in Albuquerque, New Mexico. Groves allocated $72 million for their research activities in fiscal year 1946–1947. The two labs eventually would be in the vanguard of the kind of large-scale research that Alvin Weinberg, the director of the ORNL, would call Big Science. This network of national laboratories is a valuable legacy left by the Manhattan Project.

9.6.2.2.5.1. LAWRENCE BERKELEY NATIONAL LABORATORY BERKELEY LAB Berkeley Lab, located on the grounds of the University of California at Berkeley, was founded in 1931 by Ernest Orlando Lawrence as a site for physics research centered around his new instrument, the cyclotron, a particle accelerator for which he won the Nobel Prize in Physics in 1939.[30] Lawrence was one of a few American physicists who were aware of the atomic bomb's potential, along with James B. Conant, Arthur Compton, and George Pegram, a physicist at Columbia University who was instrumental in bringing Enrico Fermi to the United States from Italy.[31] Throughout the 1930s, Lawrence worked to create larger and larger cyclotron machines for physics research.[30] The Manhattan Project adopted electromagnetic isotope separation using a cyclotron machine developed by Lawrence. Lawrence started to build a new 184-inch cyclotron and moved his lab to Charter Hill overlooking the Berkeley campus.[32] The 184-inch cyclotron was completed in November 1946.

From the 1950s through the present, Berkeley Lab has maintained its status as a major international center for physics research, and has also diversified its research program into almost every realm of scientific investigation.[30] The Laboratory's 14 scientific divisions are organized within the areas of computing sciences, general sciences, energy and environmental sciences, life sciences, and photon sciences. Many research projects are staffed and supported by multiple divisions, with computational and engineering activities integrated across the biosciences, general sciences, and energy sciences. The scientific divisions include earth sciences, genomics, life sciences, chemical sciences, environmental energy technologies, materials science, physical biosciences, computational research, accelerator and fusion research, engineering, nuclear science, nuclear medicine, and physics. Berkeley Lab has six main science thrusts: soft X-ray science, climate change and environmental sciences, matter and force in the universe, energy efficiency and sustainable energy, computational science and networking, and biological science for energy research. It was Lawrence's belief that scientific research is best done through teams of individuals with different fields of expertise, working together. His teamwork concept is a Berkeley Lab tradition that continues today. Additionally, Berkeley Lab is the host to six major national user facilities: the Advanced Light Source, the National Center for Electron Microscopy, National Energy Research Scientific Computing Center, the Energy Sciences Network, the Molecular Foundry, and the Joint Genome Institute (JGI).[i]

[i] This survey of national research laboratories covers those that were established in the course of the Manhattan Project. The Lawrence Livermore National Laboratory, which formally opened its doors in 1952 as a branch of the Lawrence Berkeley National Laboratory, was established after the Manhattan Project ended. At the time, the lab was focused on development of thermonuclear weapons that derive their energy mainly from the fusion of deuterium, an isotope of hydrogen. The lab today serves as a national resource for a broad range of science and engineering research, with national security remaining its core mission.

9.6.2.2.5.2. LOS ALAMOS NATIONAL LABORATORY. LANL, located in Los Alamos, New Mexico, was founded as a centralized facility to coordinate the scientific research of the Manhattan Project.[33] The site for the lab was authorized for acquisition in November 1942, and the construction of the lab was completed in November 1943.

LANL today is one of the largest science and technology institutions in the world, and conducts multidisciplinary research in fields such as space exploration, renewable energy, nanotechnology, and supercomputing. LANL has approximately 9000 employees and around 650 contractor personnel. Approximately one-third of the laboratory's technical staff members are physicists, one quarter are engineers, one-sixth are chemists and materials scientists, and the remainder work in mathematics and computational sciences, biology, geosciences, and other disciplines.[33] Professional scientists and students also come to Los Alamos as visitors to participate in scientific projects. The staffs collaborate with universities and industry in both basic and applied research to develop resources for the future. Its annual budget is approximately $2.2 billion.

9.6.2.2.5.3. OAK RIDGE NATIONAL LABORATORY ORNL, located in Oak Ridge, Tennessee, near Knoxville, is the DOE's largest science and energy laboratory.[34] ORNL was established as part of the Clinton Engineer Works for the Manhattan Project in 1943 to separate and produce uranium and plutonium for nuclear bombs. The site for the lab was acquired with compulsory wartime authorization in September 1942.

Today, ORNL conducts research and development that span a wide range of scientific disciplines, including high-performance computing, nanotechnology, basic energy science, materials science/advanced materials, chemical science, and electron microscopy. ORNL also is home to two of the most advanced neutron scattering research facilities in the world, the Spallation Neutron Source (SNS) and the High Flux Isotope Reactor (HFIR). SNS is an accelerator-based neutron source that provides the most intense pulsed neutron beams in the world and gives researchers detailed snapshots of smaller samples of physical and biological materials than were previously possible. HFIR provides one of the highest steady-state neutron fluxes of any research reactor in the world. The neutron scattering instruments at HFIR enable fundamental and applied research into the molecular and magnetic structures and behavior of materials.

9.6.2.2.5.4. ARGONNE NATIONAL LABORATORY ANL, located in DuPage Country, Illinois, is the first science and engineering research national laboratory in the United States. It received this designation on July 1, 1946.[35] When it was first established, it was known as the University of Chicago's *Metallurgical Laboratory (Met Lab)* previously located within Red Gate Woods, a forest reserve within the Palos Division of the Forest Preserve District of Cook County. The Met Lab was part of the Manhattan Project, where in December 1942 Enrico Fermi initiated the first self-sustaining nuclear chain reaction in an experimental reactor known as Chicago Pile-1.[54] In January 1943, Chicago Pile-1 was dismantled and reassembled at Argonne, as the operation of a reactor was regarded as too hazardous for a densely populated area.

In July 1946, the laboratory was formally chartered as Argonne National Laboratory to conduct "cooperative research in nucleonics."[35] Today, ANL is the largest national laboratory by size and scope in the Midwest. Its portfolio includes basic science research in physics, chemical sciences, and metallurgy. ANL is home to one of the world's largest high-energy light sources, the Advanced Photon Source (APS), a synchrotron X-ray research facility which produces the brightest X-ray beams in the Western Hemisphere. It also is equipped with one of the fastest computers for open science, the IBM Blue Gene/P supercomputer, and the Argonne Tandem Linac Accelerator System (ATLAS), the world's first superconducting particle accelerator for heavy ions at energies in the vicinity of the Coulomb barrier. It houses the Electron Microscopy Center (EMC), one of three DOE-supported scientific user facilities for electron beam micro characterization. The EMC conducts in situ studies of transformations and defect processes, ion beam modification and irradiation effects, superconductors, ferroelectrics, and interfaces. Its intermediate voltage electron microscope, which is coupled with an accelerator, is the only such system in the United States.

9.6.2.2.5.5. AMES LABORATORY Ames laboratory, Ames, Iowa, was established in 1947 as a DOE national laboratory.[36] The laboratory became involved in the Manhattan Project in 1942 when Iowa State College's Frank Spedding, an expert in the chemistry of rare earths, agreed to set up and direct a chemical research and development program to accompany the Manhattan Project's existing physics program. Its purpose was to produce high-purity uranium from uranium ores. The Ames Project developed an efficient process, called the Ames Process, to produce high-purity uranium metal in large quantities and produced more than 2 million pounds (1000 tons) of uranium for the Manhattan Project until industry took over the process in 1945.

In the 1950s and 1960s, Ames Laboratory took advantage of Iowa State University's synchrotron to pursue medium-energy physics research and served as a model for large-scale production of rare earth metals. Analytical chemistry efforts expanded to keep up with the need to analyze new materials. In the 1970s, Ames Laboratory put new emphasis on applied mathematics, solar power, fossil fuels, and pollution control. Innovative analytical techniques were developed to provide precise information from increasingly small samples. Among them was inductively coupled plasma-atomic emission spectroscopy, which could rapidly and simultaneously detect up to 40 different trace metals from a small sample. In the 1980s, high-performance computing research augmented the applied mathematics and solid-state physics programs. Ames Laboratory became a national leader in the fields of super conductivity and nondestructive evaluation. In addition, DOE established the Materials Preparation Center to provide public access to the development of new materials. In the 1990s, Ames Laboratory established the Scalable Computing Laboratory for computing, and researchers discovered the first noncarbon example of buckyballs, a new material important in the field of microelectronics. In recent years, Ames Laboratory has broadened the scope of its research into various areas of national concern, including energy resources, high-speed computer design, environmental cleanup and restoration, and the synthesis and study of new materials.

9.6.2.2.5.6. BROOKHAVEN NATIONAL LABORATORY Brookhaven Lab was established in 1946 at the site of Camp Upton, a former US Army base in Upton, New York. Its purpose was to serve as a regional "Big Science" research center for universities in the Northeast.[37] The Cold War led the United States to focus on high-energy physics to bolster scientific prowess.

Brookhaven Lab emphasized basic research in the physical, chemical, biological, and engineering aspects of the atomic sciences.[38] An equally important concept was the establishment of a national laboratory to design, construct, and operate large scientific machines that individual institutions could not afford to develop on their own. Thus, in April 1948, the Atomic Energy Commission approved a plan for a proton synchrotron to be built at

Brookhaven.[39] The new machine (called the Cosmotron) was the first accelerator in the world to accelerate particles to energies of a billion electron volts, or GeV, region, comparable to energies of cosmic rays (hence the name). The Cosmotron reached its full design energy of 3.3 GeV in January 1953. It ceased operation in 1966 and was dismantled in 1969. Knowledge gained from the Cosmotron led to the construction of Brookhaven's next big accelerator: the Alternating Gradient Synchrotron (AGS).[40] In addition, the lab is home to Relativistic Heavy Ion Collider (RHIC) designed to research quark-gluon plasma, the Center for Functional Nano materials (CFN), the National Synchrotron Light Source (NSLS), the Particle Beams Accelerator Test Facility, the Tandem Van de Graaff (once the world's largest electrostatic accelerator), and the New York Blue Gene supercomputer (an 18 rack Blue Gene/L and a 2 rack Blue Gene/P massively parallel supercomputer).[41] Today, Brookhaven has a staff of approximately 3000 scientists, engineers, technicians, and support staff and over 4000 guest researchers annually, and seven Nobel Prizes have been awarded for discoveries made at the Lab.[42]

9.6.2.2.5.7. SANDIA NATIONAL LABORATORIES Sandia National Laboratory, located on Kirtland Air Force Base in Albuquerque, New Mexico, has roots going back to the Manhattan Project.[43] In January 1941, construction began on the Albuquerque Army Air Base, which was later renamed Kirtland Field. During the war years, facilities were expanded and Kirtland Field served as a major Army Air Forces training installation. The Manhattan Project personnel in Los Alamos used the air base as a transportation center for work related to the testing of the atomic bomb.[44] In July 1945, the forerunner of Sandia Laboratory, known as "Z" Division, was established to handle future weapons development, testing, and bomb assembly for the Manhattan Project.[43]

Today, Sandia is a National Nuclear Security Administration Laboratory. Its mission is to maintain the reliability and surety of nuclear weapon systems, conduct research and development in arms control and nonproliferation technologies, and investigate methods for the disposal of the US nuclear weapons program's hazardous waste. Other missions include research and development in energy and environmental programs, as well as the surety of critical national infrastructure.[43] In addition, Sandia is home to a wide variety of research including computational biology, mathematics (through its Computer Science Research Institute), materials science, alternative energy, psychology, Micro Electro Mechanical Systems (MEMS) technology, and cognitive science initiatives. Sandia is also home to the Z Machine, the largest X-ray generator in the world designed to test materials in conditions of extreme temperature and pressure.

9.7. POLICY RECOMMENDATIONS: IMPLEMENTATION OF THE NATIONAL FUEL CELL DEVELOPMENT PROJECT

The foregoing discussion implicitly makes the case that major investments implemented with a sense of urgency by highly skilled scientists and engineers have achieved significant scientific breakthroughs and left a legacy of scientific institutions which contributed to the well-being of the United States for decades. In the same way, there is a compelling case for government to implement a major fuel cell basic research project at the earliest opportunity. We call this initiative the *National Fuel Cell Development Project (NFCDP).*

The NFCDP would be dedicated to obtain detailed knowledge of fuel cell electrochemical operations. This prime goal should be supported by an unrestricted use of time, ample financial resources, extensive research facilities, and a highly talented cadre of scientists and engineers from all related disciplines. It should be led by an accomplished and globally respected scientist and assisted by an experienced, hard-charging, nuts-and-bolts manager who can rapidly establish the needed infrastructure. The project should be able to bring together a critical mass of knowledge, intelligence, wisdom, expertise, and know-how that has never been assembled before for this purpose. It should seek as its ultimate goal, the discovery of breakthrough knowledge of how and why fuel cells work and ultimately establish a verifiable theory of fuel cell operations. This goal would lay the intellectual foundation for the development of commercially viable fuel cells.

9.7.1. Basic Research: A Central and Vital Mission

From the start, the NFCDP should set basic research as its central and primary mission. It should ensure that basic researchers be given the highest respect and priority throughout the life of the project. The project also needs to keep basic research independent and unencumbered from applied research and product development. Without this protection, basic research will not be able to compete with other research activities; it would be forced to abandon its knowledge-seeking mission and be compelled to conduct for-profit research that is normally carried out through applied research and product development activities.

- Few governments have experience in implementing a project such as the NFCDP. Even METI, an experienced industrial policymaker, failed to prevent its basic research projects from being diluted by competing goals. METI's SOFC basic research project was structured in such a way that it inadvertently compelled basic researchers to shift their time and attention to applied research and product development.
- The Manhattan project is one example of an effort that combined basic research with applied engineering to achieve a successful outcome. The NFCDP would have a different emphasis, however, in that nearly all of its activity would be focused on basic research.

It is also critical that fuel cell basic research be carried out in a multidisciplinary environment encompassing a wide range of scientific disciplines. Single discipline studies are unproductive when used to investigate complex, multidomain phenomena and as a result have generated few breakthroughs of consequence. In the NFCDP, basic research would begin by documenting and exploiting all known phenomena associated with the operation of fuel cells. Researchers would then undertake deep dives to understand the nature of these phenomena at the atomic level, employing all possible research instrumentation found at leading US, Japanese, and European research laboratories. Researchers would employ a host of scientific disciplines to generate new hypotheses and new theories to explain the characteristics of fuel cells as they exist and change under all operating conditions.

- A multidisciplinary approach would be the ultimate path to knowledge about fuel cells. As mentioned earlier, a professor from Yamanashi University and another from Waseda and their colleagues combined their knowledge from different disciplines to successfully depict the inner workings of fuel cells. In 2007, two professors met on a train by accident and were completely surprised to discover that they could help each other out through collaboration. Their collaboration led to the visualization of oxygen distribution within an operating stack for the first time in the world. The visualization of the fuel cell stack reveals details about the mechanism of fuel cell deterioration, thus leading to development of better and more efficient fuel cells.

For the NFCDP to be successful, the question of when to start applied research and product development would be crucial because a premature start could stunt basic research. When the details of fuel cell phenomena have been fully determined and codified in a hypothesis of fuel cell operations, only then would it be appropriate for the NFCDP to work with industry to begin applied research and product development activities. The role of applied research and product development would be to test and verify or refute each of the hypotheses and give empirical and experimental feedback. In any case, applied research and product development should always support basic research throughout the project.

- Any hypothesis that explains some facts and phenomena in a research environment may not apply under all conditions in a real life environment. Hypotheses are still hypothetical until a large body of facts has been accumulated and enough time has passed to confirm the theory through observation and experiment. Thus, feedback from applied research and product development need to be provided back to the basic researchers repeatedly. It is this process of timely feedback that would ultimately establish a true scientific theory that has the ability to accurately predict the behavior of fuel cell phenomena.

The establishment of a seamless relationship between the two activities—basic research on the one hand and applied research and product development on the other hand—might be difficult. Basic research and applied research and product development often are segregated from each other and researchers rarely exchange or share information across these boundaries. One-way flow of information occurs from basic research to the other side, but the reverse flow hardly ever occurs, making the job of the basic researcher all the more difficult. A fusion between the two research activities must occur. The following two examples illustrate the unfortunate consequences of not having a flow of information between the two activities and how frustrating it is to engage in basic research today:

- A 2006 survey compiled by a Japanese industry research association showed a total breakdown of communication between the Japanese fuel cell basic research community and the Japanese applied research and product development infrastructure. In the survey, a large number of professors engaged in basic research at Japanese universities complained that Japanese industry provided little or no feedback to them about their research findings.[45] The professors wished the industry were more frank, honest, cooperative, and solicitous so that they could know what they could do to help them. Because of the lack of feedback, professors were simply left out of the loop without knowing how their research was used, what the industry thought of their work, if any results were applied to solve practical problems, and what researchers might need to do in the future.
- A 2010 study by the National Research Council (NRC) illustrated a similar breakdown between the DOE fuel cell and hydrogen basic research activities and the US auto industry. The NRC study indicated that the DOE PEMFC R&D in the Freedom CAR and Fuel Research Program apparently did not yield results that could be used to advance the industry's fuel cell technology.[46] The automobile sector, on the other hand, achieved a high degree of technological sophistication in their fuel cell cars. Clearly, the automakers undertook proprietary research on their own. Moreover, the automakers showed little interest in the DOE program and did not even

attend the DOE R&D program steering group meetings, of which they were part, for more than 2 years. The automakers undoubtedly provided no feedback to the DOE program, did not help to improve or terminate it, and largely left DOE to guess how best to spend billions in public funds.

9.7.2. Budget and Research Period: $2 Billion a Year for 5 Years

The NFCDP should continue until it accomplishes its mission, and we believe that the mission can be accomplished in about 5 years. Even a physically massive project such as the Manhattan Project took only 10 years—including 5 years for precursor experimental basic research and 5 years for bomb development and deployment. With about 5 years of pure basic research, we believe that the NFCDP would be able to identify and understand almost all available empirical data on fuel cells, form multiple hypotheses, verify and validate most of them through a process of applied research and product development, and compile predictive theories that explain fuel cell operations.

In view of the critical importance of fuel cells to the national energy security and the complexity of the fuel cells' challenges, it is reasonable for a government to invest $2 billion a year for 5 years. The amount could be redirected from other sources. For example:

- Some of the existing renewable energy research and nuclear research effort at national laboratories could be refocused on the NFCDP.
- Some subsidies that are no longer imperative could be redirected to the NFCDP project. According to a recent editorial in *The New York Times*, the 45-cent-per-gallon tax credit for oil companies to blend ethanol into gasoline was allowed to expire on December 31, 2011.[47] The ethanol subsidy costs taxpayers $5 billion to $6 billion a year, according to the source. The subsidy likely is not required today because corn ethanol is now reportedly commercially viable without a subsidy. In addition, the same source suggests that the oil industry receives about $4 billion a year worth of tax breaks—including deductions for well depletion and intangible drilling costs—and subsidies.

9.7.3. Giving a New Mission to National and Industry R&D Labs

There are extensive networks of national laboratories and universities and well-established fuel cell industries in the world. The NFCDP would provide a new mission to selected national laboratories and universities and attract new research cadre to the fuel cell industry. Industry labs should be asked to lend a hand in fuel cell basic research so that, in return, industry will have the knowledge to produce commercially viable fuel cells.

9.7.3.1. Appointing Selected National Laboratories, Universities, and Their Research Cadre for Basic Research: At least 2000 Top-Notch Scientists and Physicists from All Related Disciplines

The existing national laboratories and universities would be able to offer the best resources to advance the NFCDP mission, and some should be appointed to engage in the project. In addition to the US national laboratories discussed in relation to the Manhattan Project, a number of outstanding national laboratories and universities in fuel cell basic research exist in other nations.

- Germany, for example, is noted for its SOFC leader Forschungszentrum Julich and a number of energetic fuel cell basic research centers, including the University of Duisburg-Essen, TU Munchen, University of Karlsruhe, and University of Stuttgart.

- Japan is equipped with the prominent AIST (the National Institute of Advanced Industrial Science and Technology) and many first-rate national universities, including Kyoto University, Kyushu University, Tokyo Institute of Technology, Tohoku University, Tokyo University, and Yamanashi University.

These national laboratories and universities collectively have remarkable research facilities and equipment and a cadre of noted Nobel-class scientists and physicists. They are equipped with many of scientific and engineering disciplines that would be directly relevant to the goals of the NFCDP. Rigorously selected facilities, infrastructure, and research cadre could be used to form the core of the NFCDP:

- Based on past experience, a cadre of at least 2000 top-notch scientists and physicists of every related discipline would be needed. They could be practically or virtually collocated for collaborating on fuel cell basic research goals.
- Selected research cadres from the fuel cell industry could be invited to join the NFCDP basic research effort.
- At the same time, the NFCDP should explore all types of fuel cells, including AFC, PAFC, MCFC, SOFC, and PEMFC, and examine all the similarities and differences in their operation, characteristics, behavior, and degradation phenomena.

There are a number of factors that may limit the progress of the NFCDP. For example, the national laboratories have an extensive established research focus so they may resist organizing their research around this effort. Moreover, national laboratory scientists and physicists may have concluded that fuel cells have inherent limitations that can never be overcome. According to a Japanese government document published in 2000, researchers of some companies engaged in PAFC R&D presented their papers at academic conferences but received few comments or questions from academic researchers.[48] The company researchers sensed the academics-regarded problems of fuel cell technology as engineering or manufacturing issues, which did not require basic research.

- It is likely, however, that national laboratory researchers recognize the gravity of the global energy situation today and understand that their foremost mission is to serve the country in times of need. Moreover, fuel cell technology presents a set of profoundly challenging intellectual problems to be solved.

9.7.3.2. Enlisting Fuel Cell Industry for Applied Research and Product Development

The NFCDP needs to make every effort to enlist the fuel cell industry to collaborate with and support the NFCDP. The fuel cell industry has invested as much as the government in fuel cells and established substantial research capabilities and infrastructure, not to mention massive manufacturing facilities. Their participation in the NFCDP would bring enormous experience and expertise to the effort.

Nonetheless, the fuel cell industry may not want to participate in a government basic research project of this magnitude just now. Industry leaders may think that they are closer to commercial success today than ever before. They may prefer the government provide subsidies that would lower the purchase price of fuel cell products and help build hydrogen infrastructure. The reality, however, is fuel cells are still inefficient and expensive and do not last long. It is in the industry's own interest not to repeat the mistakes of the past:

- According to a Japanese government document published in 2000, fuel cell companies indicated that they started producing fuel cells with little fundamental knowledge of the technology and regretted that they did not seek assistance from academic or government researchers engaged in basic research.[48]

If the fuel cell industry succeeds in getting government subsidies now, it may gain some time to make additional technological improvements. But the industry will not achieve much success. If only incremental improvements are pursued, it will take a very long time before fuel cell products reach a minimum level of acceptability without the use of subsidies. In the meantime, the industry likely will not reach profitability. Under those conditions, supporting industries that could serve as force multipliers would have little incentive to join up with fuel cell developers to create new products and services:

- BoP and other peripheral equipment manufacturing industries likely will not make much investment in their technologies and infrastructure due to lack of future prospects. Without viable BoP and other peripheral equipment, fuel cell products will not be commercially successful by themselves.
- Some fuel cell products, especially fuel cell cars, require hydrogen and hydrogen supply stations. The fuel cell industry's dependency on subsidies would not encourage the hydrogen industry to expand its hydrogen infrastructure.

On the other hand, if the NFCDP discovers principles of fuel cell operations that lead to solving the underlying causes of cell degradation or how catalytic performance could be enhanced with low-cost materials, the fuel cell industry would clearly become the direct beneficiary and could immediately apply these discoveries in their fuel cell products.

9.7.4. The NFCDP as Top National Energy Security Priority

The world's top leadership must come to champion fuel cells as the best, near-term option for dealing with our global energy crises. Fuel cells are the best option because they are pollution-free and use hydrogen which is present in abundance on earth. To ensure fuel cells are reliable and affordable, top national leaders should give the NFCDP the highest national energy security priority.

- Furthermore, the world's top leaders should realize that fuel cell commercialization only requires government to make investments in basic research—nothing else. The fuel cell industry already has sufficient applied research, product development, and manufacturing capabilities, requiring no additional government assistance in those areas.

Any national leader must understand that such a large-scale project as the NFCDP would take an entire nation to implement, but success will hinge on whether the national leader demonstrates personal commitment through determined, visionary choices. The leader must be willing to commit ample national resources, ensure the project leader reports directly to him/her, give clear-cut direction, remove regulatory and legal barriers, and inspire the nation's scientific cadre to work on the project.

The following two examples show that top national leaders of the two largest world economies successfully accomplished goals such as these to promote fuel cell development:

- US President George W. Bush championed the launching of two ambitious PEMFC and hydrogen technology development programs. In January 2002, he implemented the Freedom CAR Initiative to develop fuel cell vehicles that would run on hydrogen fuel. In the following year, the President in his State-of the-Union address in January 2003 announced another major initiative—the Hydrogen Fuel Initiative—which was aimed to develop hydrogen technologies and infrastructure to produce, store, and distribute hydrogen to support fuel cell vehicles. The President was ambitious; he proposed allocating a total of $1.7 billion over 5 years to fund two programs to make commercial fuel cell vehicles possible by 2015.[49] These two initiatives increased global interest in fuel cells and inspired a fierce sense of competition in the world.

- Japanese Prime Minister Junichiro Koizumi, also a firm believer in fuel cells, might have been inspired by the US fuel cell initiatives to lend a hand to further accelerate the global fuel cell race. In February 2002, Prime Minister Koizumi implemented the PEMFC System Technology Development Project (2002–2004), known as the "Koizumi Initiative." He declared that "The fuel cell is the key to opening the doors to a hydrogen economy. We will aim to achieve its practical use as a power source for vehicles and households within three years."[50] He next launched the Large-Scale Stationary PEMFC Demonstration Project (2005–2009). In May 2009, Japan introduced a commercial model of a residential PEMFC system. Without the advocacy and patronage of a national leader such as Prime Minister Koizumi, Japan's fuel cell industry might not have achieved the leading position that it enjoys today.

It is also critical that the top national executive appoint the highest qualified individual to lead the NFCDP. The NFCDP leader will need all the powers that a top executive can give because the project is a truly difficult task. The project leader is responsible for running an expensive, complex, long-term program involving a large number of national laboratories, universities, corporations, and research cadre. He is also faced with overcoming daunting barriers between basic research and applied research and product development. To pull this off, he will need to have extraordinary qualifications:

- He (or she) must have a prominent and highly respected standing in the national and international fuel cell communities, encompassing all academic, scientific, industry, and government circles. He must be innovative and perceptive and be able to draw upon the accumulated knowledge of all the scientific disciplines to explain the mysteries of the fuel cell. In addition, he must be a quick and resolute decision maker, an energetic go-getter, and a strong and effective leader.

9.7.5. Three Possible National Options for the NFCDP Project Implementation

There are three countries in the world that could successfully implement a project on the scale of the NFCDP. They are Germany, Japan, and the United States. They each are global fuel cell leaders that have sufficient scientific, financial, and industrial resources for the project.

- Europe as a whole is too large, disparate, and unwieldy to implement such a difficult project quickly and effectively. Europe perhaps will be best served by delegating responsibility to Germany to represent European interests.

Although one possible option would be to organize the NFCDP as an international project, such an approach would be too arduous and time consuming to be successful. A project such as the NFCDP that involves complex regulatory and legal issues would take a lengthy period of time to reach the necessary international agreements. The following example shows how long an international project could take and how expensive it would be:

- In 1985, the then Soviet Union offered a proposal to the United States for an international nuclear fusion R&D project—initially €5 billion over 30 years.[51] More than 20 years later, in 2006, an international consortium of seven members—EU, India, Japan, China, Russia, South Korea, and the United States—signed a formal agreement to build the International Thermonuclear Experimental Reactor (ITER). The construction of facilities began in 2007. The first plasma is expected in 2019. ITER would be followed by "DEMO," a commercial demonstration fusion power plant, followed by a "PROTO" prototype power station, which perhaps is expected to be online in the post-2050 timeline. The rising price of raw materials and the changes to the initial design have already tripled the initial estimate to more than €16 billion.

Alternatively, if Germany, Japan, and the United States simultaneously implemented their own NFCDP, there would be immense and unnecessary redundancy and waste. It would be far better if one country were delegated to lead the other two. But it also would take considerable time to reach agreement among the three. Time is of the essence, and the country that takes the first initiative in launching the NFCDP might be the world's fuel cell market leader in the next decade.

9.7.5.1. German Option

Forschungzentrum Julich perhaps is the best choice to lead Europe's NFCDP. Julich has a prestigious reputation as Europe's SOFC leader and possesses exceptional qualifications, having led many EU FP fuel cell development projects. It achieved the world's SOFC longevity record. Julich, however, would need additional scientists and physicists from multiple disciplines, recruited not only from German and European research institutes, universities, and corporations but also from other regions of the world. It would also need financial resources not only from the German central and state governments but also from the EC and other European governments. The German Chancellor would need to champion and become directly involved in the NFCDP project.

- Europe has had few prominent champions that have backed fuel cell development. Perhaps that is one of the reasons why Europe as a whole has failed to establish a cohesive fuel cell policy and program.

9.7.5.2. Japanese Option

AIST perhaps is the best choice in Japan to lead the NFCDP. AIST was established in 1882 and is Japan's largest and most prestigious research organization with over 40 research centers, institutes, and laboratories, and about 2300 scientists and physicists.[52] AIST places high priority on the pursuit of what it calls "full research," a unique approach of interdisciplinary and cross-disciplinary basic research to discover rules, laws, and principles behind natural phenomena and use this integrated knowledge to conduct applied research and product development.[53] AIST has been the most important participant in many METI fuel cell R&D projects, and AIST's principal SOFC researcher served as the project leader of Japan's renowned 2008–2012 Development of System and Elemental Technology on Solid Oxide Fuel Cell Program, a basic research program to explore SOFC degradation. Nonetheless, AIST would still need to recruit additional basic researchers from other laboratories, universities, and fuel cell companies.

- In view of the immense difficulties that Japan has faced since the 2011 Tōhoku Earthquake and Tsunami, any Japanese Prime Minister would leap at the chance to champion the NFCDP. Basically, Japan has fewer energy choices than other countries. Nuclear fission is highly unpopular in Japan and could soon be phased out, and wind turbines and photovoltaic arrays are not well suited because of lack of land. Moreover, Japan needs a competitive edge to strengthen its economy.

9.7.5.3. *US Option*

The United States might consider assigning selected national laboratories and universities to engage in an US NFCDP. US national laboratories and universities are equipped with world-class research facilities, equipment, and research cadre. It may be difficult, however, to find the proper place for the headquarters of the NFCDP where most science cadre would work because national research laboratories and universities are scattered throughout the country. Because the project would be decentralized, the project leader would also have greater difficulty managing the project than his (or her) German and Japanese counterparts:

- Julich and AIST each have a large, concentrated pool of top-notch fuel cell research cadre that can serve as the headquarters for their NFCDP projects. They have often served as the core for many EU FP fuel cell projects and METI fuel cell projects, respectively.

The US President would need to select the best qualified project manager with the highest degree of stature and invest in him authority and power. The NFCDP project leader would need to have direct and unrestricted access to the President, akin to the President's science advisor, the director of the current OSTP. The NFCDP leader, however, might need a separate position from the science advisor. An OSTP director is often an academic or leading researcher in a scientific field who is chosen to manage large, general S&T portfolios. The NCFDP manager would need specialized in-depth knowledge in scientific and mathematical disciplines and in interdisciplinary research that is expressly related to fuel cells. In addition, the NCFDP manager needs to be appointed for as long as 5 years, longer than is typical for science advisors.

9.7.6. The Outlook—Japan Will Likely Emerge as First Global Fuel Cell Market Leader for the Next Decade—But the World Will Be the Ultimate Winner

The sooner the NFCDP is implemented, the better off the world will be. The NFCDP should be able to accomplish its mission in about 5 years. The fuel cell industry thus will be able to work toward commercialization soon after that.

If no country implements the NFCDP, however, Japan would likely be the global fuel cell market leader. Without the NFCDP, fuel cell commercialization will likely take much longer than a decade, and additional investment will be needed to build upon incremental technological improvements. Some fuel cell companies globally may not survive during this period because they would make little or no profit. Some German companies largely depend on foreign fuel cell technologies and find it difficult to maintain investments with little economic return. Most US fuel cell manufacturers are small and dependent on fuel cell suppliers and government subsidies, making their longer-term survival problematic. Most Japanese fuel cell companies are larger and have greater staying power, so at least some of them would be able to make limited incremental improvements.

On the other hand, if the NFCDP project were implemented, Japan would have a better chance of success than its competitors. Japan has strong fuel cell basic research capabilities, a Prime Minister likely to support the project, and a massive, highly competent fuel cell industry. As soon as the NFCDP begins testing and verification, the Japanese fuel cell industry would have opportunities to produce prototypes for use by the NFCDP, which would give them a competitive edge over their foreign partners.

- Nevertheless, once the knowledge of fuel cell is attained, it will become global property. Fuel cell technology can be licensed globally to benefit the fuel cell community as a whole. Indeed, the United States, with its ready accessibility to large natural gas resources that could provide hydrogen fuel for fuel cell vehicles, likely would be the greatest economic beneficiary of the NFCDP. Regardless of who becomes the market leader, the world will be the ultimate winner because mankind will have access to cheap, clean, reliable energy.

References

1. O'Sullivan JB. *Fuel cells: the last 45+ years*. Honolulu, HI: Fuel Cell Seminar; 14 November 2006.
2. Bagotsky VS. *Fuel cells: problems and solutions*. John Wiley & Sons, Inc.; 2009.
3. *Fuel cell origins: 1840–1890. Collecting the history of fuel cells*, Smithsonian Institution. <http://americanhistory.si.edu/fuelcells/origins/origins.htm>; [accessed August 2012].
4. Ertl G. *Modern surface chemistry – fuel cells, artificial fertilizers and clean exhaust*. Press release. The Royal Swedish academy of sciences; 10 October 2007.
5. *Unexpected activity of fuel cell catalysts revealed, science daily*. 14 December 2007.
6. Fuel cell, hydrogen technology development front line, vol. 7. Japan: FocusNedo, NEDO, 26, No. 2, Fall 2007.
7. *UMass-Amherst wins federal funds for fuel cell research*. Associated Press; 27 November 2007.
8. Fueling the Future Center for Chemical Innovation. News & Events, the Massachusetts Center for Renewable Energy Science and Technology, University of Massachusetts Amherst. <http://www.chem.umass.edu/masscrest/ntnews.html>; [accessed August 2012].
9. *(Energy innovation program) PEMFC practical use promotion technology development basic plan*. Japan: NEDO; 10 February 2010.
10. Heisei 22, PEMFC Practical Use Promotion Technology Development, Research Development, Item 1. Basic Technology Development, Contractors List. <https://app3.infoc.nedo.go.jp/informations/koubo/koubo/FA/nedokouboplace.2010-06-14.0567338773/nedokoubo.2010-07-12.8038973727/(52256>dfb1)59d48a174e885b9a51484e0089a7.pdf>; [accessed 14 July 2010].
11. List of Contractors Engaged in Applied Research related to PEMFC Practical Use Promotion Technology Development/Basic Technology Development/Degradation Mechanism Analysis and Basic Material Research for High Efficiency Cells Using Nanotechnology, NEDO, Japan. <https://app3.infoc.nedo.go.jp/informations/koubo/koubo/FF/nedokouboplace.2010-08-30.7315445547/nedokoubo.2010-10-08.0070839540/301052256dfb130115171540c5b9f65bd4e885b9a51484e0089a7.pdf>; 12 October 2010.
12. On the opening of fuel cell innovation research center. Media release, University of Electro-Communication, Tokyo, Japan. <http://www.uec.ac.jp/about/publicity/media_release/pdf/100827.pdf>; 27 August 2010.
13. Rosenberg R. The publishable executive summary of the 1st periodic activity report: M1–M12: LARGE-SOFC towards a large SOFC power plant, Project No. 019739. VTT Technical Research Centre of Finland; 8 May 2008.
14. Steinberger-Wilckens R. (Forschungszentrum Jülich). Executive summary: realizing reliable, durable, energy efficient and cost effective SOFC systems (Real-SOFC). Real-SOFC, <http://www.real-sofc.org/datapool/page/390/Real-SOFC_PublicExecSummary_M37-M48_coordinator_final.pdf>; 31 March 2008.
15. Hosoi K, Nakabaru M. Status of national project for SOFC development. 216th ECS Meeting. Vienna, Austria; 4–9 October 2009.
16. Feck T (Forschungszentrum Jülich). Real-SOFC project successfully terminated. News. Real SOFC. <http://www.real-sofc.org/news_extended/news-12-2008>; 4 December 2008.
17. Forschungszentrum Jülich erzielt erneut Weltrekord mit Brennstoffzellen. Lange Lebensdauer bei hoher Leistung nachgewiesen/Großer Schritt in Richtung kommerzieller Einsatz. Press release, Forschungszentrum Julich, Germany; 15 May 2009.
18. Hannover Messe: fuel cells with world record-life; Over 20 000 hours of operation for fuel cell Jülich. Numerous interactive exhibits at the Jülich Stand. Press release, Forschungszentrum Julich. <http://www.fz-juelich.de/portal/index.php?cmd=show&mid=766&index=163>; 19 April 2010.
19. Recent results in solid oxide fuel cell development at Forschungszentrum Julich. B-7—Solid Oxide Fuel Cells XII (SOFC-XII), 219th ECS Meeting, Montreal, Canada; 2 May 2011.
20. Sato Y. SOFC R&D trend. FY 2008 SOFC Demonstration Research Result Report Workshop, New Energy Foundation; 3 March 2009.

21. Fusion Power. Article. <http://en.wikipedia.org/wiki/Fusion_power> Wikipedia; Last modified on 9 August 2012; [accessed August 2012].
22. Budget Justifications and Supporting Documents. Office of Budget—CF-30, Department of Energy. <http://www.cfo.doe.gov/crOrg/cf30.htm>; [accessed August 2012].
23. Top 7 suppliers of oil to the US. Energy, GlobalPost. <http://www.globalpost.com/dispatch/100726/top-7-us-oil-importers>; 28 July 2010.
24. Summary of weekly petroleum data for the week ending June 1, 2012, weekly petroleum status report, US Energy Information Agency. <http://ir.eia.gov/wpsr/wpsrsummary.pdf>; 1 June 2012.
25. EIA lowers its crude oil price forecast in the June STEO. News, U.S. Energy Information Administration. <http://www.cattlenetwork.com/cattle-news/EIA-lowers-its-crude-oil-price-forecast-in-the-June-STEO-158933425.html?ref=425>; 13 June 2012.
26. Nuclear power. Article. <http://en.wikipedia.org/wiki/Nuclear_power>; Wikipedia, last modified on 13 August 2012, [accessed August 2012].
27. Chadwick J. Article. <http://en.wikipedia.org/wiki/James_Chadwick>; Wikipedia, last modified on 15 August 2012, [accessed August 2012].
28. Manhattan Project. Article. <http://en.wikipedia.org/wiki/Manhattan_Project>; Wikipedia, last modified on 13 August 2012, [accessed August 2012].
29. Office of Scientific Research and Development. Article. <http://en.wikipedia.org/wiki/Office_of_Scientific_Research_and_Development>; Wikipedia, last modified on 14 August 2012, [accessed August 2012].
30. Lawrence Berkeley National Laboratory. Article. <http://en.wikipedia.org/wiki/Lawrence_Berkeley_National_Laboratory>; Wikipedia, last modified on 20 June 2012, [accessed August 2012].
31. Pegram GB. Article. <http://en.wikipedia.org/wiki/G._B._Pegram>; Wikipedia, last modified on 21 June 2012, [accessed August 2012].
32. Preuss P, An Historical Perspective on the Lab's Legacy: A Year-Long Series in the View. Berkeley Lab, 75 years of World-Class Science, Lawrence Berkeley Lab. <http://www.lbl.gov/Publications/75th/files/04-lab-history-pt-2.html>; no date, [accessed August 2012].
33. Los Alamos National Laboratory. Article. <http://en.wikipedia.org/wiki/Los_Alamos_National_Laboratory>; Wikipedia, last modified on 3 August 2012, [accessed August 2012].
34. Oak Ridge National Laboratory. Article. <http://en.wikipedia.org/wiki/Oak_Ridge_National_Laboratory>; Wikipedia, last modified on 27 July 2012, [accessed August 2012].
35. Argonne National Laboratory. Article. <http://en.wikipedia.org/wiki/Argonne_National_Laboratory>; Wikipedia, last modified on 18 July 2012, [accessed August 2012].
36. Ames Laboratory. Article. <http://en.wikipedia.org/wiki/Ames_Laboratory>; Wikipedia, last modified on 10 April 2012, [accessed August 2012].
37. The Atomic Age. Page. <http://www.atomicheritage.org/mediawiki/index.php/The_Atomic_Age>; Atomic Heritage [Last modified on 27 September 2011, accessed August 2012].
38. Tour Brookhaven's History, Brookhaven History, Brookhaven National Laboratory. <http://www.bnl.gov/bnlweb/history/>; [accessed August 2012].
39. The Cosmotron. Brookhaven History, Brookhaven National Laboratory. <http://www.bnl.gov/bnlweb/history/cosmotron.asp>; [accessed August 2012].
40. Alternating Gradient Synchrotron. Brookhaven History, Brookhaven National Laboratory. <http://www.bnl.gov/bnlweb/history/AGS_history.asp>; [accessed August 2012].
41. Brookhaven National Laboratory. Article. <http://en.wikipedia.org/wiki/Brookhaven_National_Laboratory>; Wikipedia, last modified on 6 August 2012, [accessed August 2012].
42. About Brookhaven National Lab. About BNL. <http://www.bnl.gov/bnlweb/about_BNL.asp>; last modified 26 September 2011, [accessed August 2012].
43. Sandia National Laboratories. Article. <http://en.wikipedia.org/wiki/Sandia_National_Laboratories>; Wikipedia, last modified on 17 June 2012, [accessed August 2012].
44. Kirtland Air Force Base. Article. <http://en.wikipedia.org/wiki/Kirtland_Air_Force_Base#Manhattan_Project>; Wikipedia, last modified on 12 August 2012, [accessed August 2012].
45. FY2006 JARI Domestic Interview Report. Reference Material, Japan Automobile Research Institute, March 2007.
46. Review of the research program of the FreedomCAR and fuel partnership: third Report, ISBN: 0-309-15684-X, Committee on Review of the FreedomCAR and Fuel Research Program, Phase 3; National Research Council. <http://www.nap.edu/catalog/12939.html>, 2010.
47. One bad energy subsidy expires. Editorial, The New York Times. <http://www.nytimes.com/2012/01/07/opinion/one-bad-energy-subsidy-expires.html?_r=1&nl=todaysheadlines&emc=tha211>; 6 January 2012.
48. Large Scale Energy Conservation Technology R&D Project (Moonlight Project). Phosphoric Acid Fuel Cell Power Generation R&D, 1981–1990, Evaluation Report. Industrial Technology Council, METI; May 2000.

49. Vehicle Technologies Program Overview. NREL, EERE, DOE. <http://www.nrel.gov/programs/vehicles.html>; Last Updated, 22 September 2010, [accessed December 2010].
50. Takahara I (Director General, Agency for Natural Resources and Energy), Japan's approach to commercialization of fuel cell/hydrogen technology. IPHE Steering Committee, September 2005. <http://www.iphe.net/docs/Meetings/Japan_9-05/Japan_Statement.pdf>.
51. ITER. Article. <http://en.wikipedia.org/wiki/ITER>; Wikipedia, last modified on 3 August 2012, [accessed August 2012].
52. About AIST. National Institute of Advanced Industrial Science and Technology (AIST)<http://www.aist.go.jp/aist_e/about_aist/index.html>; [accessed August 2012].
53. National Institute of Advanced Industrial Science and Technology. Article. <http://en.wikipedia.org/wiki/National_Institute_of_Advanced_Industrial_Science_and_Technology>; Wikipedia, last modified on 14 May 2012, [accessed August 2012].

Abbreviations

A

AC alternating current
ADEME French Environment and Energy Management Agency
AES air electrode supported
AFC alkaline fuel cells
AFRL Air Force Research Laboratory (United States)
AGS Alternating Gradient Synchrotron, Brookhaven National Laboratory (United States)
AIST National Institute of Advanced Industrial Science and Technology (Japan)
ANL Argonne National Laboratory (United States)
ANR National Research Agency for promotion of basic and applied research, public–private sector partnerships, and technology transfers (France)
ANRE Agency for Natural Resources and Energy, Ministry of Economy, Trade, and Industry (Japan)
ANVAR National Agency for innovation in small and medium-sized enterprises (France)
APS Advanced Photon Source, Argonne National Laboratory (United States)
APU auxiliary power unit
ARGE DFC European Direct Fuel Cell Consortium
ARO Army Research Office (United States)
ARRA American Recovery and Reinvestment Act
ASC anode-supported cell
ata atmospheres absolute
ATLAS Argonne Tandem Linac Accelerator System, Argonne National Laboratory (United States)
atm atmospheric pressure
ATV all-terrain vehicle

B

BEV battery electric vehicle
BG biogas
BHEL Bharat Heavy Electricals Ltd.
BIS Department for Business, Innovation and Skills (United Kingdom)
BMBF German Federal Ministry of Education and Research
BMFT German Federal Ministry of Research and Technology
BMVBW German Federal Ministry of Building and Urban Affairs
BMW Bayerische Motoren Werke AG
BMWA German Federal Ministry of Economics and Labor
BMWi German Federal Ministry of Economics and Technology
BoP balance of plant
BSE Border States Electric Supply (United States)
B&W Babcock & Wilcox

C

CaFCP California Fuel Cell Partnership
CARB California Air Resource Board
CCD charge-coupled device
CCEF Connecticut Clean Energy Fund
CCGT combined cycle gas turbine
CEA Commissariat a l'Energie Atomique, Atomic Energy Commission (France)
CEP Clean Energy Partnership (Germany)
CERL US Army Construction Engineering Research Laboratory
CESI Centro Elettrotecnico Sperimentale, an Italian experimental electrotechnical center
CET Centro per gli Studi di Tecnica Navale, an Italian company specializing in maritime research in shipbuilding and marine propulsion
CFCL Ceramic Fuel Cells Limited
CFN Center for Functional Nanomaterials, Brookhaven National Laboratory (United States)
CGCRI Central Glass and Ceramic Research Institute (India)
CGO cerium gadolinium oxide
CHF clean hydrocarbon fuel
CHP combined heat and power
CISE Centre for Innovation and Economical Development (Italy)
CNR-ITAE National Council of Research-Institute of Advanced Energy Technologies (Italy)

CNRS Centre National de la Recherche Scientifique (France)
COE center of excellence
COS carbonyl sulfide
CPOX catalytic partial oxidation
CPUC California Public Utilities Commission
CRADA Cooperative Research and Development Agreement (United States)
CRIEPI Central Research Institute of the Electric Power Industry (Japan)
CSIRO Commonwealth Scientific and Industrial Research Organisation (Australia)
CTA Chicago Transit Authority
CTI Commission for Technology and Innovation (Switzerland)
CVD chemical vapor deposition

D

DARPA Defense Advanced Research Projects Agency (United States)
DC direct current
DFC direct carbonate fuel cell
DGA Délégation Générale pour l'Armement, French defense procurement agency
DICP Dalian Institute of Chemistry Physics (China)
DIR direct internal reforming
DLA Defense Logistics Agency (United States)
DLR German Aerospace Center
DMFC direct methanol fuel cell
DMS dimethyl sulfide
DND Canada's Department of National Defense
DNES Department of Non-Conventional Energy Sources (India)
DNV Det Norske Veritas (Norway), a world-leading consulting and certification company in the energy, power generation, transmission, and distribution sectors
DoD US Department of Defense
DOE US Department of Energy
DPUC Connecticut Department of Public Utility Control
DTI UK Department of Trade and Industry
DTU Technical University of Denmark

E

EBZ Entwicklungs- und Vertriebsgesellschaft Brennstoffzelle GmbH
EC European Commission
ECN Energy Research Center of the Netherlands
ECU European currency unit
EDB/ELSAM a consortium of Dutch and Danish utilities
EDF Electricity of France, a French electric utility company
EERE US DOE Office of Energy Efficiency and Renewable Energy
EIA US Energy Information Administration
EMC Electron Microscopy Center, Argonne National Laboratory (United States)
EMPA Swiss Federal Laboratories for Materials Testing and Research
EMWD California Eastern Municipal Water District
EnBW Energie Baden-Württemberg AG
ENEA Italian National Agency for New Technologies, Energy and Sustainable Economic Development
EPDC Electric Power Development Company, currently J-Power (Japan)
EPFL Swiss Federal Institute of Technology in Lausanne
EPRI Electric Power Research Institute (United States)
EPSRC Engineering and Physical Science Research Council (United Kingdom)
ERA European Research Area (Europe)
ERC Energy Research Corporation, currently FuelCell Energy
ERDA Energy Research and Development Administration (currently US Department of Energy)
ERDC US Army Engineer Research and Development Center
ESA European Space Agency
ESC electrolyte supported cell
ETL Electrotechnical Laboratory (Japan)
ETTA Energy and Telecommunications Training Australia
EU European Union
EV electric vehicle
EVD electrochemical vapor deposition

F

FC fuel cell
FC/GT fuel cell and gas turbine
FCB fuel cell bus
FCC Federal Communications Commission (United States)
FCCG Fuel Cell Commercialization Group (United States)
FCCJ Fuel Cell Commercialization Conference of Japan
FCE FuelCell Energy
FCEV fuel cell electric vehicle
FCG-1 Fuel Cell Generator 1 Project (United States)
FCH JTI Fuel Cells and Hydrogen Joint Technology Initiatives (Europe)
FCH JU Fuel Cells and Hydrogen Joint Undertaking (Europe)
FCHEV fuel cell hybrid electric vehicle
FCHV fuel cell hybrid vehicle
FCT Fuel Cell Technologies Limited
FCTec US DoD Fuel Cell Test and Evaluation Center
FCV fuel cell vehicle
FE US DOE Office of Fossil Energy

FIT fee-in-tariffs
FN Fabbricazioni Nucleari, an Italian company for manufacturing fuels for nuclear power plants
FOGA Swiss Gas Industry
FP EU Framework Program
FTA Federal Transit Administration (United States)

G

GDF Gaz de France
GE General Electric
GEF United Nations Global Environment Facility
GET United Nations Global Environment Fund
GM General Motors
GPS global positioning system
GRI/IGT/GTI Gas Research Institute, Institute of Gas Technology, currently Gas Technology Institute (United States)
GT gas turbine
GTI see GRI
GTL gas-to-liquids
GTT Gas Turbine Technologies (Italy)

H

HFC hydrofluorocarbon
HFIR High Flux Isotope Reactor, Oak Ridge National Laboratory (United States)
HFITP Hydrogen, Fuel Cells, and Infrastructure Technologies Program (United States)
HFP European Hydrogen and Fuel Cell Technology Platform
HHV higher heating value
HLG High Level Group (Europe)
HRSG heat recovery steam generator
HVAC heating, ventilation, and cooling

I

ICE internal combustion engine
IFC International Fuel Cells, a joint venture company for phosphoric acid fuel cell development established by United Technologies Corporation and Toshiba
IGCC integrated gasification combined cycle
IGFC integrated gasification fuel cell
IGT/GRI/GTI Institute of Gas Technology, Gas Research Institute, currently Gas Technology Institute (United States)
IIR indirect internal reforming
IKTS Fraunhofer Institute for Ceramic Technologies and Systems (Germany)
IMHEX Internally Manifolded Heat Exchanger
IGFC Integrated Gasification Fuel Cell
IPHE International Partnership for Hydrogen and Fuel Cells in the Economy
ISM integrated stack module
ITC Investment Tax Credit
ITER International Thermonuclear Experimental Reactor

J

JARI Japan Automobile Research Institute
JFCC Japan Fine Ceramics Center
JGU Japanese Gas Utility Consortium
JHFC Japan Hydrogen and Fuel Cell Demonstration Project
JPEC Japan Petroleum Energy Center
JUG Japanese Gas and Electric Utility Group

K

KEPCO Korean Electric Power Company
KEPRI Korean Electric Power Research Institute
KIER Korea Institute of Energy Research
KIST Korea Institute of Science & Technology
KOH potassium hydroxide
KTI Kinetic Technology International (The Netherlands)

L

LADWP Los Angeles Department of Water and Power
LANL Los Alamos National Laboratory (United States)
LBS Ludwig-Bölkow-Systemtechnik (Germany), an independent consultant in the fields of energy and environment
LCR lanthanum chromite
LHV lower heating value
LPG liquefied petroleum gas
LSC lanthanum strontium cobalt oxide
LSCF lanthanum strontium cobalt ferrite
LSM lanthanum strontium manganite

M

MBB Messerschmitt-Bölkow-Blohm, currently CFC Solutions
MCFC molten carbonate fuel cell
mCHP micro combined heat and power
MCP M-C Power
MEA membrane electrode assemblies
MELCO Mitsubishi Electric Corporation
MEMS MicroElectroMechanical Systems, Sandia National Laboratories (United States)
MES Mitsui Engineering and Shipbuilding
METI/MITI Japanese Ministry of Economy, Trade, and Industry, formerly Ministry of International Trade and Industry

MGT micro gas turbine
MGTCC micro gas turbine combined cycle system
MHI Mitsubishi Heavy Industries
MKE Korean Ministry of Knowledge Economy, formerly the Ministry of Commerce, Industry, and Energy (MOCIE)
MLIT Japanese Minister of Land, Infrastructure, and Transport
MMC Mitsubishi Materials Corporation
MNES Ministry of Non-Conventional Energy Sources (India)
MOCIE see MKE
MOLB Mono-Block Layer Built Planar SOFC
MOST China's Ministry of Science and Technology
MPa megapascal
MPGe miles per gallon equivalent
MRC TUBITAK-MRC Institute of Energy (Turkey)
MSRI Materials and Systems Research, Inc.
MTBF mean time between failure
MTTR mean time to recovery
MTU Motoren Und Turbinen Union Friedrichshafen, currently CFC Solutions

N

NASA US National Aeronautics and Space Administration
NEBUS New Electric Bus
NECAR New Electric Car
NEDO New Energy and Industrial Technology Development Organization (Japan)
NEF New Energy Foundation (Japan)
NFCDP National Fuel Cell Development Project, a proposed national fuel cell basic research program
NETL US DOE National Energy Technology Laboratory
NG natural gas
NFCBP National Fuel Cell Bus Program (United States)
NFCRC National Fuel Cell Research Center (United States)
NIMC National Institute of Materials and Chemical Research (Japan)
NIMTE Ningbo Institute of Material Technology and Engineering (China)
NIP National Innovation Program for Hydrogen and Fuel Cell Technology (Germany)
NIRE National Institute for Resources and Environment (Japan)
NOW National Organization for Hydrogen and Fuel Cell Technology (Germany)
NRC National Research Council (United States)
NRE new and renewable energy
NREL US DOE National Renewable Energy Laboratory
NSLS National Synchrotron Light Source, Brookhaven National Laboratory (United States)
NTT Nippon Telegraph and Telephone Corporation (Japan)
NUMMI New United Motor Manufacturing Inc., a joint venture between General Motors and Toyota
NYPA New York Power Authority
NYSERDA New York State Energy Research and Development Authority

O

OCV open circuit voltage
OEM original equipment manufacturer
OPAC Office Public d'Aménagement et de Construction, Public institutions of low-income housing (France)
ORNL Oak Ridge National Laboratory (United States)
OSRD Office of Scientific Research and Development, currently Office of Science and Technology Policy (United States)
Oxy Occidental Petroleum

P

PAFC phosphoric acid fuel cell
PATAC Pan Asia Technical Automotive Center (China), a joint venture between General Motors and SAIC of China
PBI polybenzimidazole
PDA personal digital assistant
PEFC proton exchange fuel cell, polymer electrolyte fuel cell
PEMFC polymer electrolyte membrane fuel cell, proton exchange membrane fuel cell
PEP Spanish Fuel Cell Program
PFSA perfluorosulfonic acid
PG&E Pacific Gas and Electric Company (United States)
PHEV plug-in hybrid electric vehicle
PNGVs Partnership for a New Generation of Vehicles (United States)
PREDIT Programme de Recherche sur le Développement des Transports Terrestres (France)
PSI Paul Scherrer Institute (Switzerland)
psi pounds per square inch
PV photovoltaics

R

RV recreational vehicle
R&D research and development
RHIC Relativistic Heavy Ion Collider, Brookhaven National Laboratory (United States)
RIST Korean Research Institute of Industrial Science & Technology
RIT Rochester Institute of Technology (United States)
RUPS rugged uninterruptible power supplies

S

SAIC Shanghai Automobile Industry Corporation
SASZ scandia-alumina stabilized zirconia
SCAQMD South Coast Air Quality Management District (United States)
SCE Southern California Edison
ScSZ of SSZ scandia stabilized zirconia
Sc-TZP scandia stabilized tetragonal zirconia polycrystals
SECA Solid Energy Conversion Alliance (United States)
SEM scanning electron microscope
SES Stuttgart Stuttgart city drainage company (Germany)
SFCV Shanghai Fuel Cell Vehicle Powertrain Company
SFOE Swiss Federal Office of Energy
SFPUC San Francisco Public Utilities Commission
SGIP California Self-Generation Incentive Program
SIC Shanghai Institute of Ceramics
SIMS secondary ion mass spectrometry microscope
SMUD Sacramento Municipal Utility District
SNS Spallation Neutron Source, Oak Ridge National Laboratory (United States)
SOFC solid oxide fuel cell
SUNY State University of New York
SUT Schumacher Umwelt- und Trenntechnik GmbH (Germany)
SUV sports utility vehicle
SWPC Siemens Westinghouse Power Corporation

T

TARDEC US Army Tank Automotive Research, Development and Engineering Center
TARGET Team to Advance Research for Gas Energy Transformation Project (United States)
TBM tetr-butyl mercaptan
TEAM Technology Early Action Measures Project (Canada)
TEPCO Tokyo Electric Power Company
TGI Tecnologia y Gestion de la Innovacion (Spain), a consulting company for Argentine companies in the field of advanced technology and resource management
TIWAG Tiroler Wasserkraftwerke Antiengesellschaft (Austria), a power utility services company
TMI Technology Management, Inc.
TNO Netherlands Organization for Applied Scientific Research
TVA Tennessee Valley Authority

U

UAS unmanned aerial system
UAV unmanned aerial vehicle
UCC Union Carbide Corporation
UEC University of Electro-Communications (Japan)
UITP L'Union internationale des transports publics
UNDP United Nations Development Program
UPS uninterruptible power supply
UTC United Technologies Corporation
UTM University of Toronto at Mississauga

V

VPE Program Véhicule automobile Propre et Econome en énergie, Clean Energy-Saving Automobile Vehicle Program (France)
VSE Verband Schweizerischer Elektrizitätsunternehmen, Swiss Associationof Electric Utilities
VTT Technical Research Centre of Finland
VW Volkswagen

W

WE Westinghouse Electric

Y

YSZ yttrium stabilized zirconia

Z

ZEBUS zero emission bus
ZeuS Project Zellen und Stacks Project (Germany)
ZEV zero emission vehicle
ZIP German Future Program for fuel cell demonstration and market preparation

Index

Note: Page numbers followed by "f" and "t" indicate figures and tables respectively

B

G

H

I

J

K

O

P

Q

R

S

U

V

W

Edwards Brothers Malloy
Ann Arbor MI. USA
November 27, 2013